International Conference on Materials Science and Manufacturing Technology (ICMSMT 2021)

IOP Conference Series: Materials Science and Engineering
Volume 1166

Coimbatore, India
8 - 9 April 2021

Editors:

Ramya Muthusamy
Amjad Anvari-Moghaddam

ISBN: 978-1-7138-4628-4
ISSN: 1757-8981

Printed from e-media with permission by:

Curran Associates, Inc.
57 Morehouse Lane
Red Hook, NY 12571

Some format issues inherent in the e-media version may also appear in this print version.

Institute of Physics
Dirac House, Temple Back
Bristol BS1 6BE UK

Phone: 44 1 17 929 7481
Fax: 44 1 17 920 0979

techtracking@iop.org

Additional copies of this publication are available from:

Curran Associates, Inc.
57 Morehouse Lane
Red Hook, NY 12571 USA
Phone: 845-758-0400
Fax: 845-758-2633
Email: curran@proceedings.com
Web: www.proceedings.com

TABLE OF CONTENTS

Preface

Materials science and manufacturing process is one of the significant symbols of modern day's high-tech revolution. The technological developments in the materials science paves the road for advancements in materials science and manufacturing technology across the world. Materials are the key factor for the emerging industries of all kind. In order for new material to get the actual application, exquisite preparation and manufacturing must be processed. No doubt manufacturing is critical in every walk of life. With rapid development computer technology, communications technology and network technology, traditional manufacturing process has evolved to intelligent manufacturing which is more intelligent, more precision and more efficiency. The Third International Conference on Materials Science and Manufacturing Technology 2021 (ICMSMT 2021) has provided a forum for researchers, scientists, academicians, scholars and practitioners around the world to present papers on recent developments in the broad fields of Materials Sciences and Manufacturing Technology. The ICMSMT series also provides a premier interdisciplinary platform for researchers, practitioners and educators to present and discuss the most recent innovations, trends, and concerns as well as practical challenges encountered and solutions adopted in the fields of Materials Science, Engineering and Manufacturing. The third edition of ICMSMT conference series (ICMSMT 2021) has been organized with partnered by Akshaya College of Engineering & Technology, Coimbatore and Diligentec Solutions, Coimbatore, India. The conference received potential number of submissions from across the world and the papers have been accepted by the editorial board of ICMSMT after the stringent screening and review process. ICMSMT is an annual technical event and has been aimed to be conducted in the second week of April every year. Due to the COVID-19, outbreak in all over the world followed by the national and international travel restrictions, the conference has been organized in online model. Each presenter were given ten minutes for presentation and three minutes for discussion on the queries. The entire meeting has been conducted using Google meet platform. The conference has been successfully completed with all the presentations and keynote sessions. Being the third edition, ICMSMT 2021 had received submissions from different regions of the world and scholars from 9 countries have registered and participated.

Dr. Ramya Muthusamy

Editor / Chair – TPC

ICMSMT 2021

ICMSMT 2021- Committeess

Executive Committee

Conference Chairs

Dr. Thangaprakash Sengodan, Diligentec Solutions, Coimbatore, India

Dr. Suguna N, Joint Director, Akshaya College of Engineering Technology, India

General Chairs

Dr. Baharudin Bin Ismail, Universiti Malaysia Perlis, Malaysia

Dr. Jaya J, Principal, Akshaya College of Engineering and Technology, India

Organizing Chairs

Dr. Senthilkumar S, Diligentec Solutions, Coimbatore, India

Dr. Kathiravan N, Dean(Academics), Akshaya College of Engineering and Technology, India

Editorial Committee

Editor

Dr. Ramya Muthusamy, Diligentec Solutions, Coimbatore, India

(ramsthangam@gmail.com)

Guest Editor

Dr. Amjad Anvari-Moghaddam, Aalborg University, Denmark

(aam@et.aau.dk)

Members

Dr. Sivakumar R, Akshaya College of Engineering and Technology, India

Dr. Sendhil Kumar S, Akshaya College of Engineering and Technology, India

Dr. Anithasree A, Akshaya College of Engineering and Technology, India

IOP Conf. Series: Materials Science and Engineering 1166 (2021) 011001 doi:10.1088/1757-899X/1166/1/011001

Review Committee

Chair

Dr. Markas Law Lee Lam, National University of Singapore, Singapore

Members

Dr. Rajkumar N, Akshaya College of Engineering and Technology, India

Dr. Viji C, Akshaya College of Engineering and Technology, India

Information Contact

Dr. Kapilan S, Akshaya College of Engineering and Technology, India

Publicity and Promotions

Mr. Sivakumar K, Diligentec Solutions, Coimbatore

Technical Program Committee Chairs

Dr. Joseph Xavier R, Advisor, Akshaya College of Engineering and Technology, India

Dr. Sabareesaan K J, Mechanical and Industrial Section, Nizwa College of Technology, Oman

Dr. Gopinth, S R Engineering College, Warangal, India

Dr. Rajamani R, PSG College of Technology, Coimbatore, India

Technical Program Committee

Dr. Jaroslaw Drelich, Michigan Technological University, USA
Dr. Chong Wen Tong, University of Malaya, Malaysia
Dr. Mohamed Arezki Mellal, M'Hamed Bougara University, Algeria
Dr. Khairel Rafezi Ahmad, University Malaysia Perlis, Malaysia
Dr. Mahanthi S D, Michigan State University, USA
Dr. Mohammed Henini, University of Nottingham, UK
Dr. Sharifah Shahnaz Syed Bakar, University Malaysia Perlis, Malaysia
Dr. Alexandru Bogdan Andreescu, Transilvania University of Brasov, Romania
Dr. Norinsan Kamil Othman, University Science Malaysia, Malaysia
Dr. Mitsuru Itoh, Tokiyo Institute of Technology, Japan
Dr. Sanjeev Kumar, Aalborg University, Denmark

Dr. Ramani Kannan, University Technology Petronas, Malaysia
Dr. Satyabrata Das, Indian Institute of Technology, Kanpur
Dr. Singh S P, Indian Institute of Technology, Delhi
Dr. Rajiv O. Dusane, Indian Institute of Technology, Bombay
Dr. Kamal K Kar, Indian Institute of Technology, Kanpur
Dr. Ashish Kumar Mishra, Indian Institute of Technology, Varanasi
Dr. Anand K, Indian Institute of Technology Madras, India
Dr. Sushil Mishra, Indian Institute of Technology, Bombay
Dr. Ajit Panigrahi, CSIR-Institute of Minerals and Materials Technology (IMMT), Bhubaneswar
Dr. Somnath Bhowmick, Indian Institute of Technology, Kanpur
Dr. Patil P, Indian Institute of Technology, Varanasi
Dr. Amrita Bhattacharya, Indian Institute of Technology, Bombay
Dr. Akhilesh Kumar Singh, Indian Institute of Technology, Varanasi
Dr. Krishnakant, Indian Institute of Technology, Delhi
Dr. Alankar Alankar, Indian Institute of Technology, Bombay
Dr. Vanitha C, National Institute of Technology, Warangal
Dr. Tanmoy Maiti, Indian Institute of Technology, Kanpur
Dr. Jitendra Kumar, Indian Institute of Technology, Varanasi
Dr. Rer. Nat. Ravikumar, Indian Institute of Technology, Madras
Dr. Bikash Kumar, CSIR-IMMT, Bhubaneswar
Dr. Ramkumar J, Indian Institute of Technology, Kanpur
Dr. Jyoti Prasad Borah, National Institute of Technology, Nagaland
Dr. Chandana Rath, Indian Institute of Technology, Varanasi
Dr. Mani V N, C-MET, DIETY, Government of India, India
Dr. Bimal Prasad Singh, CSIR-IMMT, Bhubaneswar
Dr. Vivek Verma, Indian Institute of Technology, Kanpur
Dr. Parag Tandaiya, Indian Institute of Technology, Bombay
Dr. Honey John, Cochin University of Science and Technology, Cochin
Dr. Manas K Dalai, CSIR-IMMT, Bhubaneswar
Dr. Nagaraja S, Akshaya College of Engineering Technology, India
Dr. Mathan Kumar N, Akshaya College of Engineering and Technology, India
Prof. Selvaraju M, Akshaya College of Engineering and Technology, India
Dr. Barkunan S R, Akshaya College of Engineering and Technology, India
Dr. Balamurugan, Amrita School of Engineering, Coimbatore, India
Dr. Josephine R L, National Institute of Technology Trichy (NITT), India
Dr. Kulkarni S M, National Institute of Technology (NITK), Surathkal, India
Dr. Gururaj Punekar, National Institute of Technology (NI

IOP Conf. Series: Materials Science and Engineering 1166 (2021) 011002 doi:10.1088/1757-899X/1166/1/011002

Peer review declaration

All papers published in this volume of *IOP Conference Series: Materials Science and Engineering* have been peer reviewed through processes administered by the Editors. Reviews were conducted by expert referees to the professional and scientific standards expected of a proceedings journal published by IOP Publishing.

- **Type of peer review:**
  Double-blind review

- **Conference submission management system:**
  icmsmt2021@gmail.com

- **Number of submissions received:**
  328

- **Number of submissions sent for review:**
  139

- **Number of submissions accepted:**
  86

- **Acceptance Rate (Number of Submissions Accepted / Number of Submissions Received X 100):**
  26.21

- **Average number of reviews per paper:**
  Two

- **Total number of reviewers involved:**
  36

- **Any additional info on review process:**
  Double blind review was adopted by the ICMSMT 2021. Two stages of review have been conducted for the submitted papers. The first round of review was screened to check for the conference scope, technical relevance, plagiarism and originality by the experts from the internal technical program committee. Manuscripts which passed the first round of review were sent to the field experts for peer review. Reviewers were assigned maximum of 10 papers each to review. The corresponding author was notified the outcome of the review process and invited to register for the conference if the manuscript had been recommended for acceptance. If reviewers recommended for further improvements in the manuscript, those manuscripts were sent back to the corresponding authors for revision. Final decision on the manuscript was taken by the Conference Chair and the same was communicated to the corresponding authors.

- **Contact person for queries (please include: name, affiliation, institutional email address)**
  Dr. Ramya Muthusamy
  Chair – TPC, ICMSMT 2021
  Email: ramsthangam@gmail.com , icmsmt2021@gmail.com

Design of a viscoelastic layer by incorporating graphite wafers in a zig-zag pattern for the augmented hybrid damping treatment of beam

Abhay Gupta*, Rajidi Shashidhar Reddy and Satyajit Panda

Department of Mechanical Engineering, Indian Institute of Technology Guwahati, Guwahati-781039, India
***E-mail:** abhay.gupta@iitg.ac.in

Abstract. In the present work, the ascendancy of actively constrained zig-zag 1-3 viscoelastic composite (VEC) layer on hybrid damping is studied. For this analysis, structural beam is taken that consists of substrate layer/beam, constrained zig-zag 1-3 VEC layer and constraining extension actuated piezoelectric layer. The zig-zag 1-3 VEC is configured by incorporating graphite wafers in zig-zag pattern within conventional viscoelastic material (VEM). In order to investigate the hybrid damping characteristics, the finite element (FE) model for structural beam is evolved. In order to study the ascendancy of incorporating graphite wafers in zig-zag pattern within VEM layer, initially, bending and shear strain distributions are evaluated. Moreover, optimal geometrical configuration of present zig-zag 1-3 VEC layer is obtained based on maximal hybrid damping using exhaustive search optimization technique. The results reveal that the augmentation in energy dissipation or hybrid damping for the beam with zig-zag 1-3 VEC layer occurs due to rise in both bending and shear strain in VEM by incorporating graphite wafers in zig-zag pattern while for conventional VEM layer it occurs due to shear strain only. And this augmented energy dissipation ameliorates extension and shear counterparts as well as overall modal loss factor of beam with actively constrained zig-zag 1-3 VEC layer compared to conventional VEM layer.

Keywords. Structural vibration control, hybrid active-passive damping, viscoelastic composite, piezoelectric material, finite element method.

1. Introduction

Structural vibration is very harmful due to its negative consequences like fatigue, failure and damage. So, the control of structural vibration is one of the essential aspects in the design of structure. These vibrations can be reduced/controlled using different damping mechanisms in which unconstrained/free layer damping (UCLD) [1,2], passive constrained layer damping (PCLD) [3,4] and active constrained layer damping (ACLD) [5] are the commonly used damping mechanisms/treatments. Many researchers have been utilized these damping mechanisms/treatments to control the structural vibrations [6–10]. Although ACLD treatment/mechanism is advantageous and popular damping treatment due to its ameliorates damping capacity through passive viscoelastic material (VEM) as well as active piezoelectric material. These treatments also have excellent controlling capability over the wide range of frequency modes [11]. So, these (ACLD) mechanisms are also known as hybrid active-passive damping treatments. However, some other revamped hybrid treatments have been detailed in literature [12–15].

Besides, different passive and active damping materials such as viscoelastic material and piezoelectric material, correspondingly, are also obliging to ameliorate damping in structural vibration. In which VEMs consist of both viscous as well as elastic properties and having good energy absorption/dissipation capability under periodic loading. The commonly used VEMs for damping material are Polymer Blend, Butyl rubber, Viton-B, Styrene-butadiene rubber, etc. But these VEMs have very low stiffness as well as damping ability. Thus, viscoelastic composites (VECs) are used to alleviate damping as well as stiffness. Different VEC and its design configurations have been proposed for improved structural damping [16–21]. It may be noticed that in most cases damping occurs due to rise in shear strain in VEM only. However, very few VECs are available [16,17,21] where magnification in damping was expected to rise in bending and shear strain simultaneously in viscoelastic layer.

Thus, in the present work, the ascendancy on bending/shear strain in the magnification of hybrid damping by incorporating graphite wafers in zig-zag pattern within conventional VEM is investigated. Also damping performance of zig-zag 1-3 VEC is analyzed on ACLD treated beam structure. Initially, the finite element (FE) model of ACLD treated beam is evolved where constraining extension actuated piezoelectric layer is activated by providing external voltage correspondingly to velocity feedback control law. Then, to validate the present FE code in accounting for the geometrical location of incorporated graphite wafers within conventional VEM layer, the distributions of bending and shear strain of ACLD treated beam are computed. Further the ascendancy of incorporating graphite wafers in zig-zag pattern within conventional VEM layer on the hybrid damping (modal loss factor) is investigated. Moreover, an optimal geometrical configuration of zig-zag 1-3 VEC layer is obtained for the maximal hybrid damping using direct search optimization method. Both extension and shear counterparts of modal loss factor are computed for optimally configured zig-zag 1-3 VEC and compare the similar results for the conventional VEM layer. The frequency responses are also evaluated to study the beam's qualitative hybrid damping performance with actively constrained zig-zag 1-3 VEC layer over conventional VEM layer.

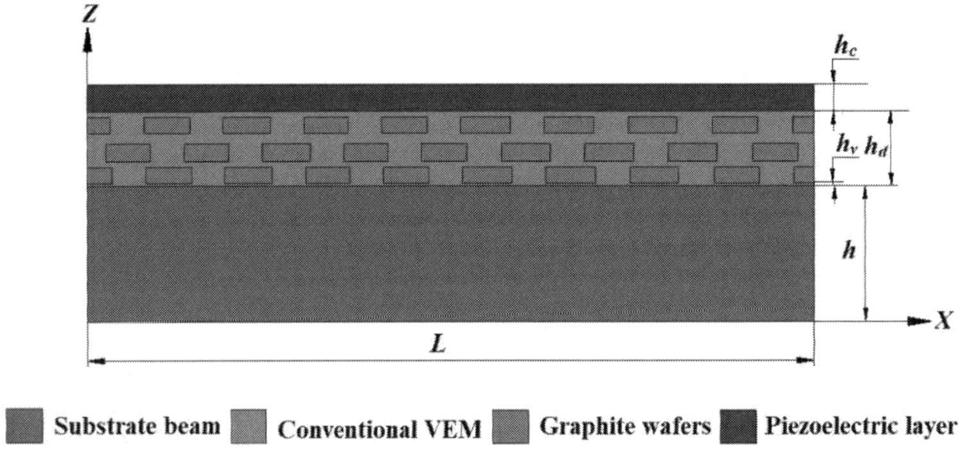

Figure 1. Schematic diagram of ACLD treated beam using zig-zag 1-3 VEC layer.

2. Finite element (FE) formulation of ACLD treated beam

The ACLD treated beam using zig-zag 1-3 VEC layer is represented diagrammatically in figure 1. The length/width and thickness of substrate beam/constrained layer/active constraining layer of ACLD treated beam are denoted by L/b and $h/h_d/h_c$, correspondingly. Geometrical configuration parameters such as number of graphite wafers, VEM layer thickness and the gap between two consequent graphite wafers within zig-zag 1-3 VEC are designated by n, h_v and Δ, correspondingly (figure 1). A transverse periodic point load (p_0) is considered at the middle of bottom surface of fixed beam. The ACLD treated beam is modelled by finite element (FE) method in two-dimensional (2-D)

xz-plane by considering the plane stress assumption [22] in y-axis (figure 1). So the state of strain/stress in terms of shear/bending counterparts (subscript s/b) within xz-plane can be expressed as,

$$\varepsilon_b = \{\varepsilon_{xx} \quad \varepsilon_{zz}\}^T,\ \boldsymbol{\varepsilon}_s = \gamma_{xz},\ \boldsymbol{\sigma}_b = \{\sigma_{xx} \quad \sigma_{zz}\}^T,\ \boldsymbol{\sigma}_s = \tau_{xz} \tag{1}$$

where, $\varepsilon_{xx}/\varepsilon_{zz}$ and σ_{xx}/σ_{zz} represents bending strain and stress in x/z axis, correspondingly; $\sigma_{xz}/\varepsilon_{xz}$ designate shear stress/strain in xz-plane. Since u/w denotes displacement components in x/z axis so according to linearized strain-displacement relations, the bending and shear strain ($\varepsilon_b, \boldsymbol{\varepsilon}_s$) is obtained in the form of operator matrix ($\boldsymbol{L}_b, \boldsymbol{L}_s$) and displacement vector ($\boldsymbol{d}$) in equation (2).

$$\varepsilon_b = \boldsymbol{L}_b\boldsymbol{d},\ \boldsymbol{\varepsilon}_s = \boldsymbol{L}_s\boldsymbol{d};\ \boldsymbol{d} = \{u \quad w\}^{\mathrm{T}},\ \boldsymbol{L}_b = \begin{bmatrix} \partial/\partial x & 0 \\ 0 & \partial/\partial z \end{bmatrix}^T,\ \boldsymbol{L}_s = \begin{Bmatrix} \partial/\partial z \\ \partial/\partial x \end{Bmatrix}^T \tag{2}$$

The constitutive relation for k^{th} material can be written as,

$$\boldsymbol{\sigma}_b^k = \boldsymbol{C}_b^k\boldsymbol{\varepsilon}_b^k;\ \boldsymbol{\sigma}_s^k = \boldsymbol{C}_s^k\boldsymbol{\varepsilon}_s^k;\ \text{where, } \boldsymbol{C}_b^k = \frac{E^k}{1-(\nu^k)^2}\begin{bmatrix} 1 & \nu^k \\ \nu^k & 1 \end{bmatrix},\ \boldsymbol{C}_s^k = E^k\big/\left(2(1+\nu^k)\right) \text{ for } k = 1, 2, 3 \tag{3}$$

$$\boldsymbol{\sigma}_b^k = \boldsymbol{C}_b^k\boldsymbol{\varepsilon}_b^k - \boldsymbol{e}_b^k E_z,\ \boldsymbol{\sigma}_s^k = \boldsymbol{C}_s^k\boldsymbol{\varepsilon}_s^k,\ D_z^k = (\boldsymbol{e}_b^k)^{\mathrm{T}}\boldsymbol{\varepsilon}_b^k + \epsilon_{33}^k E_z;\ \text{for } k = 4$$

$$\text{where, } \boldsymbol{C}_b^k = \begin{bmatrix} \bar{C}_{11}^k & \bar{C}_{13}^k \\ \bar{C}_{13}^k & \bar{C}_{33}^k \end{bmatrix},\ \boldsymbol{C}_s^k = \bar{C}_{55}^k,\ \boldsymbol{e}_b^k = \begin{Bmatrix} \bar{e}_{31}^k \\ \bar{e}_{33}^k \end{Bmatrix},\ \boldsymbol{e}_s^k = 0$$

$$\bar{C}_{11}^k = C_{11}^k - (C_{12}^k)^2/C_{22}^k,\ \bar{C}_{13}^k = C_{13}^k - (C_{12}^k C_{23}^k)/C_{22}^k,\ \bar{C}_{33}^k = C_{33}^k - (C_{23}^k)^2/C_{22}^k,$$
$$\bar{C}_{55}^k = C_{55}^k,\ \bar{e}_{31}^k = e_{31}^k - (e_{32}^k C_{12}^k)/C_{22}^k,\ \bar{e}_{33}^k = e_{33}^k - (e_{32}^k C_{23}^k)/C_{22}^k \tag{4}$$

where, $k =$ 1, 2, 3, 4 are indicated for substrate beam, graphite wafers, conventional VEM and piezoelectric layer, correspondingly. Here, material properties of conventional VEM are frequency-dependent complex quantity so it is modelled using complex stiffness method and expressed as $\boldsymbol{C}_b^k(\omega)$, $\boldsymbol{C}_s^k(\omega)$ (for $k =$ 3). In equation (4), $E_z(=-V/h_c)$ and D_z^k are electric field and displacement component along z-axis, correspondingly; $\boldsymbol{e}_b^k/\boldsymbol{e}_s^k$ and ϵ_{33}^k designate bending/shear coefficients and electrical permittivity, correspondingly of piezoelectric material.

The FE model of ACLD treated beam is meshed using nine-node isoparametric elements in xz-plane (figure 1). Thus displacement vector ($\boldsymbol{d}$) is equal $\boldsymbol{N}\boldsymbol{d}^e$, where $\boldsymbol{N}$ designates shape function vector.

$$\int_{t_1}^{t_2}\left(\delta T_K - \delta T_P\right)dt = 0 \tag{5}$$

The elemental equation of motion can be obtained in equation (8), by implementing potential and kinetic energy variation (δT_P, δT_K) (equation (6) and (7)) in extended Hamilton's principle (equation (5)).

$$\delta T_P = \sum_{k=1}^{4}\left[\int_{A_k}\left\langle\left(\delta\boldsymbol{\varepsilon}_b^k\right)^{\mathrm{T}}\boldsymbol{\sigma}_b^k + \left(\delta\boldsymbol{\varepsilon}_s^k\right)^{\mathrm{T}}\boldsymbol{\sigma}_s^k\right\rangle dA_k - \int_{A_k}\left\langle(\delta E_z)^{\mathrm{T}} D_z^k\right\rangle_{k=4} dA_k\right] - \left\langle(\delta w)(p_0/b)\right\rangle_{x=L/2,z=0} \tag{6}$$

$$\delta T_K = \sum_{k=1}^{4}\left[\int_{A_k}\left(\{\delta\dot{u} \quad \delta\dot{w}\}\rho^k\{\dot{u} \quad \dot{w}\}^{\mathrm{T}}\right)dA_k\right] \tag{7}$$

$$\boldsymbol{M}^e\ddot{\boldsymbol{d}}^e + \left(\boldsymbol{K}_b^e(\omega) + \boldsymbol{K}_s^e(\omega)\right)\boldsymbol{d}^e = \boldsymbol{P}_E^e V + \boldsymbol{P}_0^e;\ \text{where, } \boldsymbol{M}^e = \int_{A_k^e}\left(\boldsymbol{N}^{\mathrm{T}}\rho^k\boldsymbol{N}\right)dA_k^e,$$

$$\boldsymbol{K}_b^e = \int_{A_k^e}\left(\boldsymbol{N}^{\mathrm{T}}\boldsymbol{L}_b^{\mathrm{T}}\boldsymbol{C}_b^k(\omega)\boldsymbol{L}_b\boldsymbol{N}\right)dA_k^e,\ \boldsymbol{K}_s^e = \int_{A_k^e}\left(\boldsymbol{N}^{\mathrm{T}}\boldsymbol{L}_s^{\mathrm{T}}\boldsymbol{C}_s^k(\omega)\boldsymbol{L}_s\boldsymbol{N}\right)dA_k^e,$$

IOP Conf. Series: Materials Science and Engineering 1166 (2021) 012001 doi:10.1088/1757-899X/1166/1/012001

$$P_0^e = \left(\{0 \quad 1\}^{\mathrm{T}} (p_0/b)\right)_{x=L/2, z=0}, \quad P_E^{e} = \int_{A_k^e} \left\langle N^{\mathrm{T}} L_b^{\mathrm{T}} e_b^k \left(-1/h_c\right)\right\rangle dA_k^e \Bigg|_{k=4} \tag{8}$$

In equation (8), M^e and P_E^e / P_0^e represents elemental mass matrix and electro-elastic coefficient vector/mechanical load vector, correspondingly; K_s^e / K_b^e denotes shear/bending counterparts of elemental stiffness matrix; A_k^e is elemental area made of kth material. After assembly, the global FE equation of motion of ACLD treated beam is obtained in equation (9).

$$M\ddot{d} + K(\omega)\, d = P_E V + P_0 \text{ ; where, } K(\omega) = K_b(\omega) + K_s(\omega) \tag{9}$$

In equation (9), control voltage V is equal to ($-k_d \dot{w}_s$), where $\dot{w}_s = N_T \dot{d}$. So the final FE governing equation of motion can be obtained in equation (10).

$$M\ddot{d} + C\dot{d} + K(\omega)\, d = P_0 \text{ ; where, } C = P_E k_d N_T \tag{10}$$

In equation (10), k_d and N_T designate velocity-feedback control gain and transformation row matrix, correspondingly.

Through free vibration analysis ($p_0 = 0$), fundamental frequency (ω_n) and modal loss factor (η) are estimated using equation (11) . The solution procedure to estimate damping (modal loss factor) and natural frequency using equation (11) is detailed in [18,20].

$$\left[\left\langle K_b^R(\omega^{prev}) + K_s^R(\omega^{prev})\right\rangle + j\left\langle K_b^I(\omega^{prev}) + K_s^I(\omega^{prev})\right\rangle + \omega^{prev}\, \mathbf{C}\right]\psi = \left(\omega^{curr}\right)^2 M\psi \tag{11}$$

3. Results and discussion

The effect on hybrid damping of ACLD treated beam by incorporating graphite wafers in zig-zag pattern within conventional VEM is investigated (figure 1). The hybrid damping characteristics are estimated by modal loss factor and frequency response of ACLD treated beam structure. The dimensions of ACLD treated beam (figure 1) are considered as, $L =$ 0.4 m, $b = 4$ mm, $h = 5$ mm, $h_d =$ 2.5 mm and $h_c = 0.5$ mm. However, transverse periodic point load (p_0) of amplitude 0.5 N is considered at the middle of bottom surface of fixed beam. The active constraining layer and substrate beam are constructed with PZT-5H and Aluminium [20] while material properties of graphite wafers and Butyl rubber conventional VEM are taken from [18,23]. Initially, the FE formulation/code for CLD treated beam is verified for estimating damping (modal loss factor) and fundamental frequency. And, noticed that present results from FE code are nearby reference results [24] as shown in Table 1. Further the validation of FE code for piezoelectric electro-elastic coupling model is also done by considering a three-layered cantilever smart beam. The computed results are nearby reference results [25] as shown in Table 2.

Table 1. Verification of FE code for estimating damping (modal loss factor) and fundamental frequency of CLD treated beam.

Parameter	Present FE results	Ref. [24]
Natural frequency	740.48	740.5
Modal loss factor	0.00448	0.00448

Table 2. Validation of FE code for piezoelectric electro-elastic coupling model.

Normalized displacement	Present	Ref. [25]
$(u/h\,(10^{-4}))$	8.247	8.208
$(w/h\,(10^{-2}))$	-2.088	-2.040

In order to study the ascendancy of incorporating graphite wafers in zig-zag pattern within conventional VEM as well as to verify the present FE code in accounting the geometrical location of incorporation of graphite wafers, the distribution of bending/shear strain (ε_{xx}, γ_{xz}) of ACLD treated beam in xz-plane are computed. In strain analysis, the geometrical configuration of incorporated graphite wafers within conventional VEM layer are taken as, $n=3$, $h_v =$ 0.2 mm and $\Delta=40$ mm. While the material properties of conventional VEM is considered at constant operating frequency 1200 rad/s and room temperature $32^{\circ}C$. However, the extension actuated piezoelcteric layer is activated by applying the voltage $V =$ 25 volt at top surface of piezoelectric actuator layer. While magnitude of static transverse point load (p_0) is taken as 100 N. The results in figure 2(b) and 2(d) revel that bending/shear strain (ε_{xx}, ε_{xz}) appears according to these specified location of graphite incorporation and verify the realty of FE code. Moreover, similar results for ACLD treated beam using conventional VEM layer are also illustrated in figure 2(a) and 2(c). Figure 2 reveals that both bending and shear strain (ε_{xx}, ε_{xz}) ameliorate in zig-zag 1-3 VEC layer compared to conventional VEM layer. This magnification may be due to the incorporation of graphite wafers in zig-zag pattern within conventional VEM layer.

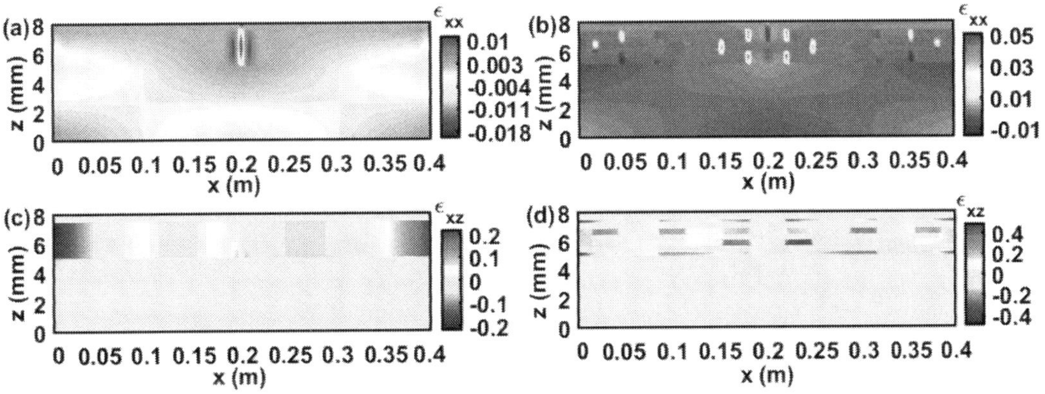

Figure 2. Strain distribution of (a)-(b) bending (ε_{xx}) and (c)-(d) shear (ε_{xz}) in xz-plane ((a),(c) conventional VEM layer and, (b),(d) zig-zag 1-3 VEC layer).

Figure 2 infers that the magnification in strains are dependent on geometrical configuration of zig-zag 1-3 VEC. Thus to investigate the ascendancy of incorporating graphite wafers within conventional VEM layer on hybrid damping, a geometrical configuration of zig-zag 1-3 VEC is decided based on maximal hybrid damping (modal loss factor). Based on manufacturing point of view, the feasible bounds of geometrical configuration parameters of zig-zag 1-3 VEC are taken as, $1 \leq n \leq 12$, 50 μm$\leq h_v \leq$200 μm, 50 μm$\leq \Delta \leq$350 μm. With these bounds of n, h_v, Δ, a 3-D domain with the values of modal loss factor (η) respective to these configuration parameters at control-gain $k_d =$ 100 are obtained and illustrated in figure 3. The maximum advantageous value of configuration parameters $n=$ 5, $h_v =$ 50 μm and $\Delta=$ 50 μm, are acquired with respect to maximal modal loss factor (η). At this optimal geometrical configuration the extension/shear (η_e, η_s) counterparts of modal loss factor of ACLD treated beam using zig-zag 1-3 VEC layer are evaluated and presented in Table 3. Similar results for conventional VEM are also evaluated and presented in same Table 3. The resuts in Table 3 reveal that hybrid damping of ACLD treated beam ameliorated using zig-zag 1-3 VEC layer compared to conventional VEM layer. And this magnification in hybrid damping is due to both extension/shear (η_e, η_s) counterparts of modal loss factor. It may be due to fact that extension/shear (η_e, η_s) counterparts of modal loss factor rises by the ameliorate in energy dissipation through bending/shear strain, correspondingly within conventional VEM by incorporating graphite wafers in zig-zag pattern (figure 2).

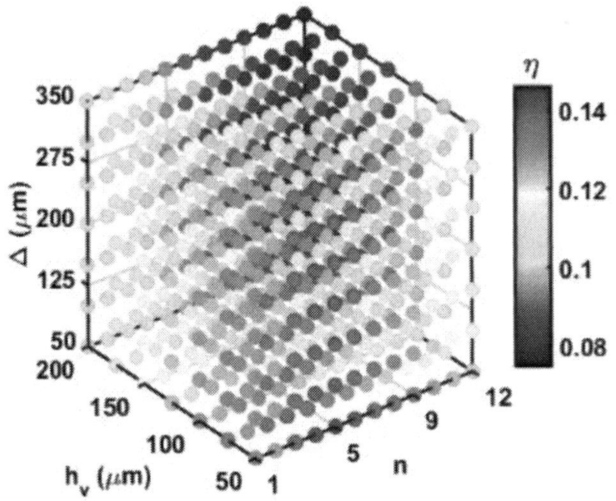

Figure 3. The values of modal loss factor (η) of ACLD treated beam using zig-zag 1-3 VEC layer at different geometrical configuration parameters (n, h_v, Δ) within 3-D contour (k_d = 100).

Table 3. Hybrid damping of ACLD treated beam using different viscoelastic layers.

Viscoelastic layer	η_e	η_s	η
Conventional VEM	0.0165	0.0515	0.0517
Zig-zag 1-3 VEC	0.0305	0.1269	0.140219

To study the hybrid damping performance of ACLD treated beam using zig-zag 1-3 VEC layer, the frequency response is evaluated by applying transverse periodic point load ($p_0 = 0.5$ N) at the center of bottom surface of the beam. Maximal transverse displacement amplitude (W_{max}/h) and respective required control-voltage (V_{max}) of ACLD treated beam with zig-zag 1-3 VEC layer is obtained around the resonant frequency domain (figure 4(a)-4(b)). Similar results for ACLD treated beam with conventional VEM layer are also presented in the same figure (figure 4(a)-4(b)). The results shown in figure (figure 4) reveal that attenuation ameliorates and respective control-voltage (V_{max}) reduces for zig-zag 1-3 VEC compared to conventional VEM. Thus, zig-zag 1-3 VEC layer may be beneficial for the magnification of hybrid damping treatment of beam.

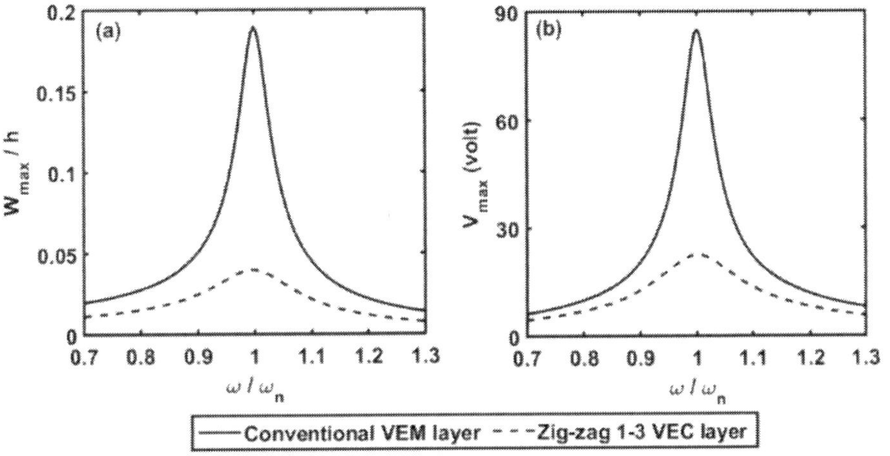

Figure 4. Variation of (a) maximal transverse displacement-amplitude (W_{max}/h) and respective required (b) control-voltage (V_{max}) of ACLD treated beam around the resonant frequency domain ($p_0 = 0.5$ N).

4. Conclusions

In this work, hybrid damping characteristics of ACLD treated beam using zig-zag 1-3 VEC layer are studied. The zig-zag 1-3 VEC layer consists by incorporating graphite wafers in zig-zag pattern within conventional VEM layer. For the analysis, the FE model of ACLD treated beam is formed. Initially, strain analysis is carried out to study the ascendancy of incorporating graphite wafers in zig-zag pattern within conventional VEM layer. The results reveal that both bending and shear strain within conventional VEM layer ameliorates by incorporating graphite wafers in zig-zag pattern compared to conventional VEM layer. Subsequently, optimal geometrical configuration of zig-zag 1-3 VEC layer is obtained for maximal hybrid damping of ACLD treated beam. It is found from the results that at an optimal geometrical configuration, the hybrid damping of ACLD treated beam is significantly ameliorated for the present zig-zag 1-3 VEC layer compared to conventional VEM layer. Moreover, the magnification in hybrid damping is due to both extension and shear counterparts of modal loss factor. Further qualitative hybrid damping performance of ACLD treated beam using zig-zag 1-3 VEC layer is also investigated. The result reveals that there is significantly ameliorated attenuation of resonant displacement-amplitude with lesser control voltage for optimal geometrical configuration of zig-zag 1-3 VEC layer compared to conventional VEM. These observations imply that present zig-zag 1-3 VEC is preferable for ameliorated hybrid active-passive damping treatment of beam.

5. References

[1] Oberst H and Frankenfeld K 1952 Damping of the bending vibrations of thin laminated metal beams connected through adherent layer *Acustica* **2** 181–94.

[2] Ungar E E and Kerwin Jr E M 1964 Plate damping due to thickness deformations in attached viscoelastic layers *J. Acoust. Soc. Am.* **36** 386–92.

[3] Swallow W 1939 An improved method of damping panel vibrations *Br. Pat. Specif.* **513** 171.

[4] Kerwin Jr E M 1959 Damping of flexural waves by a constrained viscoelastic layer *J. Acoust. Soc. Am.* **31** 952–62.

[5] Baz A and Ro J 1995 Performance characteristics of active constrained layer damping *Shock Vib.* **2** 33–42.

[6] Roy P K and Ganesan N 1996 Dynamic studies on beams with unconstrained layer damping treatment *J. Sound Vib.* **195** 417–27.

[7] Moita J S, Araújo A L, Soares C M M and Soares C A M 2018 Vibration analysis of functionally graded material sandwich structures with passive damping *Compos. Struct.* **183** 407–15.

[8] Kliem M, Høgsberg J, Vanwalleghem J, Filippatos A, Hoschützky S, Fotsing E-R and Berggreen C 2019 Damping analysis of cylindrical composite structures with enhanced viscoelastic properties *Appl. Compos. Mater.* **26** 85–113.

[9] Plattenburg J, Dreyer J T and Singh R 2015 Active and passive damping patches on a thin rectangular plate: a refined analytical model with experimental validation *J. Sound Vib.* **353** 75–95.

[10] Mevada J R and Prajapati J M 2018 Active vibration control of smart beam under parametric variations *J. Brazilian Soc. Mech. Sci. Eng.* **40** 394.

[11] Azvine B, Tomlinson G R and Wynne R J 1995 Use of active constrained-layer damping for controlling resonant vibration *Smart Mater. Struct.* **4** 1.

[12] Liu Y and Wang K W 2000 Active-passive hybrid constrained layer for structural damping augmentation *J. Vib. Acoust.* **122** 254–62.

[13] Liao W H and Wang K W 1996 A new active constrained layer configuration with enhanced boundary actions *Smart Mater. Struct.* **5** 638.

[14] Yellin J M and Shen I Y 1996 A self-sensing active constrained layer damping treatment for a Euler-Bernoulli beam *Smart Mater. Struct.* **5** 628.

[15] Kumar R 2012 Active vibration control of beams by combining precompressed layer damping and ACLD treatment: Performance comparison of various robust control techniques *J. Vib. Acoust.* **134** 21015.

[16] Kumar A, Panda S, Narsaria V and Kumar A 2018 Augmented constrained layer damping in plates through the optimal design of a 0-3 viscoelastic composite layer *J. Vib. Control.* **24** 5514-5524.

[17] Gupta A, Reddy R S, Panda S and Kumar N 2020 Damping treatment of beam with unconstrained/constrained 1-3 smart viscoelastic composite layer *Mater. Today Proc.* **26** 956-962.

[18] Gupta A, Panda S and Reddy R S 2020 Improved damping in sandwich beams through the inclusion of dispersed graphite particles within the viscoelastic core *Compos. Struct.* **247** 112424.

[19] Huang C-Y and Tsai J-L 2015 Characterizing vibration damping response of composite laminates containing silica nanoparticles and rubber particles *J. Compos. Mater.* **49** 545–57.

[20] Gupta A, Panda S and Reddy R S 2020 An actively constrained viscoelastic layer with the inclusion of dispersed graphite particles for control of plate vibration *J. Vib. Control* 1077546320956533.

[21] Gupta A, Rajidi S R and Panda S 2020 Design of a 1-3 smart viscoelastic composite layer for augmented constrained layer damping treatment of plates *IOP Conf Ser: Mater Sci Eng.* **872** 12067.

[22] Popov E P 1990 *Engineering mechanics of solids* (Prentice Hall).

[23] Jones D I G 2001 *Handbook of viscoelastic vibration damping* (John Wiley & Sons).

[24] Lall A K, Asnani N T and Nakra B C 1988 Damping analysis of partially covered sandwich beams *J. Sound Vib.* **123** 247–59.

[25] Vasques C M A and Rodrigues J D 2005 Coupled three- layered analysis of smart piezoelectric beams with different electric boundary conditions *Int. J. Numer. Methods Eng.* **62** 1488–518.

Review on solid-liquid separation using various conditioning methods

H R Rashmi and C P Devatha

Department of Civil Engineering, NITK, Surathkal, India.

E-mail: *rashusmile14@gmail.com, revacp@gmail.com*

Abstract. The increase in industrialization and population in the recent days the sludge production is also increased which has become difficult to handle it. Sludge used to be incinerated, compost and landfilling before disposal, but as it is not economic and environmentally friendly most of researchers are concentrating on the sludge dewatering process which are sustainable technologies so that sludge can be converted to semi solid which will be easy to handle and transport. Some sludge is used to produce to biogas, organic matter and also source of energy further processing was not done as it contains toxic compounds in it. Hence studies on sludge dewatering can be useful in terms of environmental and economic point of view. The present review work is on study of different properties of sludge, their effect on sludge dewatering, fundamentals of sludge and assessment of sludge dewaterability.

Keywords: Sludge, Dewatering, Conditioning, capillary suction time, free water.

1. Introduction

The increasing rate of urbanization and industrialization as resulted in the production of sludge. Sludge consists of various matrix of inorganic particles like contaminants, dirt and microorganisms, fibrous matters, organic particles and extracellular polymers particles which are bounded with various lipids, proteins, organic acids and polysaccharides. Improper disposal of sludge results in human health and environmental effects such as odour, pathogenic microorganisms, heavy metals leaching, groundwater contamination etc. hence it is very much necessary to dispose of the sludge in the safe, effective, and cost-efficient way. Composting, incineration, anaerobic digestion, sludge dewatering, landfilling, gasification are the various methods used for sludge processing and handling. The sludge quantity varies according to the population and sewerage expansions etc., the sludge content generation is increasing every year which leads to the increase in transportation and handling cost and also impacts on environment [1].

Various types of sludge's are available like primary sludge, biological sludge, mixed sludge, digested sludge, mineral sludge and physical-chemical sludge. Also, there are different forms of water in the sludge, shown in the Figure 1, which are as follows,

- Free water: - Water that can be easily isolated by gravitational settling. Free water is not chemically or mechanically bound to the sludge. Around 75% of the water in the sludge is free water.
- Interstitial water: - The water which is surrounded to the sludge particle as a floc and which can be removed by mechanical dewatering means. About 20% water in the sludge is of interstitial water.
- Vicinal water: - It is also very difficult to extract the water that is physically attached to the sludge particles.
- Water of hydration: - Water that attached to the sludge particles chemically. Around 2.5% water in sludge is of Water of hydration.

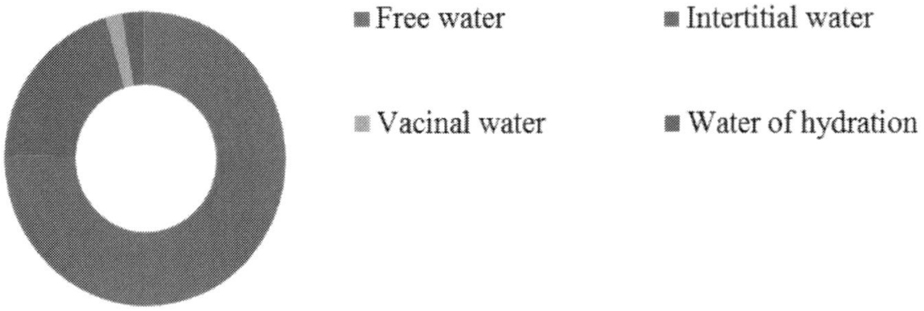

Figure 1. Different water forms in the sludge [2]

The sewage sludge characteristics and the quantities are depended on its sludge type and also level of treatment. Sludge's are classified into different forms

- Primary sludge: - The sludge collected from the settling chamber.
- Biological sludge: - Sludge obtained during the biological treatment.
- Digested sludge: - This are the sludge produced from the biological stabilization.
- Mixed sludge: - It is the mixture of primary and biological sludge.
- Physico-chemical sludge: - It is generated by the physico-chemical treatment of wastewater.
- Mineral sludge: - it is produced from the mineral industries, quarry etc.,

2. Characteristics of the sludge

Sludge has a very high convoluted nature due to which there is a vast change in their characteristics like physical, chemical and biological. Understanding the behaviour of sludge and its characteristics is utmost difficult issues in the treatment of sludge [3]. However, it is important to study its properties and understand the concept and also its effect on the treatment system.

In the early study it was shown that as the sludge quantity increases, phosphorous content decreases. So it was stated that sludge characteristic changes with respect to the source [4]. **Table 1** shows the various sludge

Table 1. Various characteristics of sludge

Physical Characteristics	Chemical Characteristics	Biological Characteristics
Odour	pH	Microbial community
Colour	Alkalinity	Surface polymers
Drainability	Solid concentration	Sludge stability
Specific gravity	Surface charge	
Settleability	Fertilizer and nutrient value	
Particle size and shape	Heavy metal and toxic organics	

Water distribution	Digestibility
Dewaterability	Fat content
Filterability	
Rheology	
Floc porosity	
Floc structure	
Floc density	
Thermal conductivity	
thermal content	
Compressibility	

After sludge conditioning, thickening and digestion, sludge is dewatered before processing it to incineration, landfilling and composting, which leads to reduction in sludge content and sludge transportation cost [5]. In this study, it is aimed to address on sludge dewatering options. Figure 2 shows the different stages of sludge treatment processes.

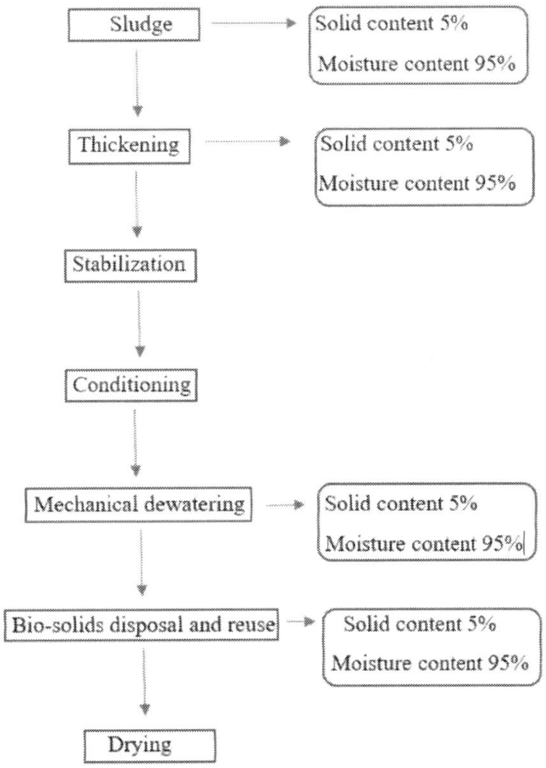

Figure 2 shows the main stages of sludge treatment processes

Sludge dewaterability is also sometimes referred as sludge filterability. Final solid content achievable by reducing the sludge volume is the main goals of dewatering. There are different ways to improve dewaterability of bio sludge. [6] reported that the filterability of sludge is used to check initially the output of dewatering. Also, he has analysed and concluded that sludge treatment, transportation of sludge and disposal of sludge requires 50 to 60% cost of wastewater treatment in the treatment plant.

3. Sludge conditioning

Solid liquid separation is necessary and can be achieved by mechanical thickening and sludge dewatering by conditioning. Physical processes, biological and chemical processes are the methods used for the sludge conditioning and these methods changes the characteristics of the sludge by gaining the dewatering performance by enhancing the settleability, filterability and also by releasing the trapped water.

3.1 Chemical conditioning

When certain chemicals are applied to either a raw or digested sludge or with other mixture, they develop a relatively insoluble flock that agglomerates suspended and colloidal particles, conditioning is carried out using organic or inorganic additives. Chemical conditioning is done in small scale researchers when compare to large scale treatment. The main intension of sludge conditioning is to create a rigid lattice structure so that the liquid content easily flows between the sludge particles without resistance [7].

The efficiency of Chemical conditioning is affected by various factors like properties of the sludge, type of sludge, type of conditioners used, pH, mixing rate, dewatering method etc., The conditioners for the sludge conditioning are selected based on the sludge type to be dewatered. The two main mechanisms which takes place in chemical conditioning are neutralization and particle bridging for the formation of floc structure. Usually, the wastewater sludge possesses an anion charge hence for the neutralization process a cation charge additive are used so that individual particles are flocculated. Following are the various chemical conditioning methods.

- Organic additives: Organic additives in terms of polyelectrolyte are used since many years so as to enhance efficiency of mechanical dewaterability during the conditioning of sludge. Two different mechanisms may clarify sludge conditioning by polyelectrolytes: charging neutralization and interparticle bridging [8].
- inorganic additives: inorganic additives are also considered quite expense that the organic counterparts, and can be widely used in various treatment practices, such as sludge dewatering, wastewater beverage industry. Since the coagulant is mainly based on aluminium or iron, it is highly useful in reducing turbidity and dewatering of the sludge [9].

3.2 Physical conditioning

From an economic and environmental point of view, physical conditioning techniques are usually more beneficial among the methods used to increase the dewater capacity of bio sludges [10].
A broad variety of systems achieve through the physical conditioning. Physical conditioners, including minerals and carbonaceous materials, which helps in developing voids so that a firm lattice structure is obtained. physical conditioners are are called as skeleton builders.

Table 2. Various physical conditioners

Form of filter aid	Skeleton materials
Inorganic	Alum sludge
	Fly ash
	Gypsum
	Lime
	Cement Kiln dust

Carbonaceous	Coal fines Rice husk biochar Corn Wheat dregs Rice shells Saw dust char

Non-chemical additives are also called physical conditioners. Various physical conditioners are listed in Table 2. Physical conditioners are the highly porous inert materials that minimize high compressibility, enhance solid permeability and enhance sludge mechanical power. Carbonaceous materials have a high advantage in the process of dewatering sludge when compared to mineral materials, since it contains high heating value, high porosity and low ash content[11]. Following are the various physical conditioning methods.

- Non-Chemical Additives: Non-Chemical Additives are the conditioners used in sludge dewatering which may be organic or carbonaceous. Non-Chemical Additives helps in mechanical dewatering as the load is applied to the sludge these additives helps in providing porous structure with more voids so that the liquid can be easily penetrated as well as the compressibility of the sludge is reduced. Hence these additives are also called as skeleton materials or filter aid[12].
- Cavitation pre-treatment: Cavitation is a mechanism through which the local pressure in the liquid phase is decreases below the equilibrium vapor pressure and some microcavities or microbubbles are created after reaching an unstable diameter, gradually rising and violently collapsing. This method generates a shock wave that locally causes high temperature[13].
- Thermal pre-treatment: Heat allows water to escape from the sludge when the organic sludge is heated. Thermal treatment systems release water that is bound within the sludge's cell structure, thus enhancing the sludge's dewatering and thickening functionality. In thermal pre-treatment the sludge is treated at 70-190^{0}C this range, thermal pre-treatment will split the cell wall and the biologically degrading proteins release [14].
- Freeze/Thaw treatment: Freeze / thaw treatment is one more physically effectual way to alter the arrangement of the floc and to decrease the content of bound water in the sludge. Meanwhile in the sludge is first frozen, about 15 C, and kept at this state for some time, and then thawed at room temperature. Water and solid particles separate during the formation of ice crystals [15].

3.3 Biological conditioning

By disrupting the gel structure of flocks by hydrolysis of EPS contained in the sludge, the biological or enzyme pre-treatment of sludge increases the dewaterability of sludge. As all enzymes are proteins and are often assumed to be environment conscious, these techniques could be very attractive, specifically when it substitutes neurotoxic acrylamide based synthetic polymers [16].

The influence, individually or in combination, of industrial microbial cellulase, protease and lipase enzymes on the decrease in solid disposal amount of anaerobically digested wastewater sludge. It has been found that a combination of these enzymes reduces total suspended solids by thirty to fifty percent by weight in equal proportion and increases solid settlement. To extract EPS from sewage sludge, enzymes and enzymes combined with sodium tripolyphosphate (STPP) were introduced. They reported that enzymes in conjunction with STTP released greater quantities of total EPS compared to enzymes on their own [17].

4. Sludge dewatering

Dewatering will be taken up before the incineration or thermal drying because it reduces the fuel consumption and also reduction in the leachate as the sludge disposal to landfill is less. Sludge dewatering helps to remove as much water from the sludge as possible, leading into the reduction of biosolids concentration and, subsequently, the transport costs. Dewatering methods widely used include mechanical methods and thermal processes Sludge often needs a conditioning procedure before mechanical dewatering to improve the efficiency of water content removal [18].

Based on the volume as well as characteristics of the sludge, different process are identified. Mechanical dewatering processes are compared to thermal ones when considering economy. There are two key criteria used as dewatering output representatives which are the concentration of solid cake and the consistency of centrate. After all, the vital role of the dewatering process is to maximize the solid sludge amount so as to decrease the volume of sludge, dry solid content has to be specified as a indicator that represents the process's efficacy. Although supernatant quality represents the efficiency of solid capture during the process of conditioning and dewatering, in particular, when the centrifuge is used as a dewatering process. During conditioning and dewatering, it is important to remove as many solids as achievable for the efficient operation of sewage treatment [19]

It minimizes solids disposal to the plant inlet when the liquid fraction is returned to the head – of – works. Meanwhile all the wastewater treatment plants, the aim of dewatering sludge is to establish a clean center while obtaining a cake high in solids.

5. Assessment of sludge dewaterability

Several tests used to measure the efficiency of various conditioners often provide useful information that can be used to determine sludge dewaterability. Assessment of dewaterability is an important factor in all treatment facilities and also the goal of the treatment facilities is to optimize dewatering process. Nonetheless, due to the volatile and mysterious nature of the sludge forms along with differences in dewatering methods, this work is fairly challenging. Various measurements for the dewatering process are already studied and tested out over years in combination with the dewatering technology before the dewatering assessments were created.

However, despite these developments, there is no formal dewatering index yet that can completely reflect the capacity to dewater sludge. An accurate dewatering index is believed to not only approximate the actual water separation whereas to estimate the overall achievable solid concentration of the sludge cake. One or both of these tend to lack traditional dewatering indices, and therefore they hardly properly convey the efficiency of dewatering. Following are the various indices for the assessment of sludge dewaterability.

- Specific resistance to filtration (SRF): Specific resistance Specific resistance is a general used to test dewaterability of sludge. The frequency at which water drains off from the sludge through an imposed vacuum is assessed and described in form of dewatering resistance of the sludge. Increase in specific resistance value indicates that it is less easy to dewater the sludge.
  SRF method has many significant benefits, as independent of the concentration of solids. It can be used particularly to estimate the concentration of the final cake after dewatering. Most of the techniques developed to determine the dewatering efficiency of filtration devices. These tests are time-consuming and the procedures are very complicated and involve a high degree of skill. Such

examinations are time-consuming and the techniques are very complex and require a high degree of expertise [20].

- Capillary suction time (CST): CST is simple to run, it provided a fast sign of filterability, and the findings found strong reproducibility. In CST test a small test tube open at both the ends where the one end is kept at the filter paper and the other samples are dropped. Paper's capillary suction extracts centrate from the sludge, dampen up surface. Filtrate time needed to flow 1 cm is reported as capillary suction time or CST.CST measuring is simpler and uncomplicated to test, and requires less expertise than Measuring, making it more common recently. CST was originally designed to calculate the dewatering rate as a replacement for Freshwater. Given that, in reality, CST is still not considered to be fundamental methods of measuring the dewaterability of sludge. Unlike SRF, CST is influenced by solid concentration and can't figure out the final cake solids that can be achieved by dewatering devices [21].
- Time to filter (TTF): The TTF test is an experiment where in vacuum application in a Buchner funnel filled and attached with filter paper, and measuring the time needed for the collection of samples in the graduated cylinder of a fixed volume of supernatant (usually 50 percent of the sample volume). The amount of time it takes to obtain the specified filtrate volume is the TTF. The minimum TTF value is set to imply optimum sludge conditioning. The Time-to-Filter (TTF) test is an attempt to simplify the complicated procedures involved in the more basic SRF testing. The TTF test involves measuring the time needed for filtration of a sludge or a conditioned sludge sample, rather than the actual resistances, using the same laboratory technique as the SRF test.
- Zeta potential: Zeta potential (ZP) is the electrical potential that occurs over a solid particle at the shear plane. The shear plane is a interface that behave like a obstacle bed of water that stays further from the particle that travels in alliance to the particle, with the solid and water. Due to its quantification of the electrical potential that induces interparticle repulsion, ZP measurements have been used to test particle stability and successful flocculation. A near zero ZP indicates minimal electrical repulsion, allowing attractive van der Waals forces to effect agglomeration of particles [22].
- Centrifugation: During centrifugation, the liquid and solid fractions are separated under the action of centrifugal force. The factors which have to be set during centrifuge are time, temperature and speed. Factors which are influenced in centrifuge is type of sludge, temperature and concentration of solids. Centrifuges are equipped with continuous and space-saving dewatering systems combined with high performance to isolate solids from sludge mixtures. A major use is the sludge dewatering in wastewater treatment plants. The spinning parts of the centrifuge, the drum and the scroll, are powered by two inverter-controlled electric motors and controlled by software running on a PLC [23].
- Rheology: Rheological characteristics are representative of floc intensity.In the management of sewage sludge, rheological parameters are very important, not only as the design of parameters in transport, landfilling ect., but also as the control of parameters in several treatments. While early research on the rheological behavior of sewage sludge has been carried out since the 1930s. Until recently, more studies have attempted to analyze the rheology of sludge dewaterability, The purpose of the conditioning and sludge dewatering is to anticipate, track and optimize [24].

Current dewatering indices are likely to lack and also dewatering efficiency was intended rarely. SRF is an unmanageable in connection with instrument and it is consuming the time, however after filtration, it only estimates the sludge content. Whereas the usage of CST is very much and it has become popular as it is very much easy in measuring the sludge dewaterability but only limitation is that concentration of solid cannot be predicted. CST and SRF are associated with free water, nearly 20 percent of total the water content will be taken part.

6. Conclusion

The characteristics of sludge cake should be thoroughly analysed with respect to water content, solid components, and floc characteristics in terms of understanding sludge. In order to achieve the optimum dewatering efficiency, a suitable porous and incompressible sludge cake structure can thus be developed and managed. Enhance the porosity and permeability of solid cake formation by using skeleton materials or filter aids such as slag or agricultural wastes. Carbon based physical additives are usually more effective in sludge dewatering than inorganic materials due to their increase porosity. Due to its low price and eco - friendliness of used materials, once it could give expected dewatering efficiency, this technique is encouraged. For sludge with strong compressibility, chemical conditioning itself will not always attain a sufficient improvement in dewatering quality. Physical conditioners that serve as skeleton constructors have been utilized to improve sludge dewaterability by providing more stable and incompressible structures to the sludge solids. For every sludge treatment system where enhancing the dewatering process is the goal, assessment of the dewaterability of sludge significantly necessary. As always, due to the erratic and elusive behaviour of all types of sludge, particularly bio-sludge, and also inconsistencies in solid-liquid separation methods, this study is relatively difficult. Although similar metrics for the very first dewatering indices are different for the method of dewatering has been studied and implemented over the years with the science of dewatering in tandem. There seems to be a lack of one or both of these traditional dewatering indices and, therefore, they mostly convey the performance of dewatering properly.

References

[1] Sanin, F. D., Clarkson, W. W. & Vesilind, P. A. (2011), *Sludge engineering: The treatment and disposal of wastewater sludges*, DEStech Publications, Inc.

[2] Mowla D, H.N. Tran, D.G. Allen(2013), *A review of the properties of biosludge and its relevance to enhanced dewatering processes*,Biomass Bioenergy, **58**, pp. 365-378.

[3] Colin, F., Leschber, R. & Mininni, G. 1988, *'Physical and chemical characterisation of sewage sludge', Sewage Sludge Treatment and Use: New Developments, Technological Aspects and Environmental Effects.* eds A. H. Dirkzwager & P. L'Hermite), Elsevier Applied Science.

[4] Xie, W., Wang, Q., Ma, H., Ohsumi, Y. & Ogawa, H. I. 2005, *'Study on phosphorus removal using a coagulation system', Process Biochemistry,* **vol. 40**, no. 8, pp. 2623-2627.

[5] Feng, X., Deng, J., Lei, H., Bai, T., Fan, Q. & Li, Z. 2009, *'Dewaterability of waste activated sludge with ultrasound conditioning'*, Bioresource echnology, **vol.100**, no. 3, pp. 1074-1081.

[6] Scholz, M. 2005, *'Review of recent trends in capillary suction time (CST) dewaterability testing research'*, Industrial and engineering chemistry research, **vol. 44**, no. 22, pp. 8157-8163.

[7] Wang, Hou-feng et al. (2019) *'Combined use of inorganic coagulants and cationic polyacrylamide for enhancing dewaterability of sewage sludge'*, Journal of Cleaner Production. Elsevier Ltd, **211**, pp. 387–395. doi: 10.1016/j.jclepro.2018.11.208.

[8] Thapa K, Qi Y, Hoadley A (2009b) *Interaction of polyelectrolyte with digested sewage sludge and lignite in sludge dewatering.* Colloid Surf. A-Physicochem. Eng. Asp **334 (1)**, 66–73.

[9] Lu MC, Lin CJ, Liao CH, Huang RY (2003) *Dewatering of activated sludge by Fenton's reagent.* Advances in Environmental Research **7 (3),** 667–670.

[10] Qi, Y. Thapa, K. B. & Hoadley, A. F. 2011, *'Benefit of lignite as a filter aid for dewatering of digested sewage sludge demonstrated in pilot scale trials'*, Chemical Engineering Journal, **vol. 166**, no. 2, pp. 504-510.

[11] Luo, H. *et al.* (2013) '*Effects of sawdust-CPAM on textile dyeing sludge dewaterability and filter cake properties*', *Bioresource Technology*. Elsevier Ltd, **139**, pp. 330–336. doi: 10.1016/j.biortech.2013.04.035.

[12] Can Gungoren, Yasin Baktarhan, Ilgin Kursun, Safak G. Ozkan, Orhan Ozdemir *Characterization of Flocs in Dewatering of Coal Plant Tailings*, Journal of Science and Engineering, 2018, **5 (3)** 215-218 ISSN NUMBER: 2148-4171 DOI: 10.17350/HJSE19030000097

[13] Meiqiang Cai, Jianqiang Hu, Guanghu Lian , Ruiyang Xiao, Quanyuan Wang, Dewen Luo, Zongsu Wei, 2018. *Synergetic pretreatment of waste activated sludge by hydrodynamic cavitation combined with Fenton reaction for enhanced dewatering.Ultrasonics Sonochemistry.* **Volume 42**, April 2018, Pages 609-618.

[14] Gokce Kor-Bicakci, Cigdem Eskicioglua. *Recent developments on thermal municipal sludge pretreatment technologies for enhanced anaerobic digestion.* Renewable and Sustainable Energy Reviews **110** (2019) 423–443.

[15] S. Reed, J. Bouzoun, W. Meddi *A rational method for sludge dewatering via freezing Journal* (Water Pollution Control Federation), **Vol. 58**, No. 9, Conference Issue (Sep., 1986), pp. 911-916.

[16] To, V. H. P., Nguyen, T. V., Vigneswaran, S., Nghiem, L., Murthy, S., Bustamante, H.& Higgins, M. J. 2014, '*Modified centrifugal index and sludge characteristics in assessing sludge dewatering', IWA 7th International Young Water Professional Conference,* Taipei, Taiwan.

[17] Vu Hien Phuong To, Tien Vinh Nguyena, Heriberto Bustamante, Saravanamuthu Vigneswaran a 2019 "*Deleterious effects of soluble extracellular polymeric substances on polyacrylamide demand for conditioning of anaerobically digested sludge* Journal of Environmental Chemical Engineering **7**, 102941

[18] Xiao, K. *et al.* (2019) '*Enhancemen"t of municipal sludge dewaterability by electrochemical pretreatment', Journal of Environmental Sciences (China)*. Elsevier B.V., **75**, pp. 98–104.

[19] Gao, N. et al. (2019) *A new method combining hydrothermal carbonization and mechanical compression in-situ for sewage sludge dewatering: Bench-scale verification'*, Journal of Analytical and Applied Pyrolysis. **Elsevier, 139**(February), pp. 187–195. doi: 10.1016/j.jaap.2019.02.003.

[20] Vesilind, P. 2000, '*A modified capillary suction time apparatus for measuring the filterability of super-flocculated sludges', Water Science and Technology*, **vol.42**, no. 9, pp. 135-139.

[21] Chen, G., Lin, W. & Lee, D. 2010, '*Capillary suction time (CST) as a measure of sludge dewaterability'*, Water Science and Technology**, vol.** 34, no. 3, pp. 443-448.

[22] Rashmi Hosurdoddi Ramachandra, Chella Purushothaman Devatha (2019) Experimental investigation on sludge dewatering using granulated blast furnace slag as skeleton material. Environ Sci Pollut Res, https://doi.org/10.1007/s11356-020-07614-w.

[23] To, V. H. P., Nguyen, T. V., Vigneswaran, S., Nghiem, L., Murthy, S., Bustamante, H & Higgins, M. J. 2014, 'Modified centrifugal index and sludge characteristics in assessing sludge dewatering', *IWA 7th International Young Water Professional Conference,* Taipei, Taiwan.

[24] Lidia Wolny, Pawed Wolski, Iwona Zawieja 2002 *Rheological parameters of dewatered sewage sludge after conditioning*. Desalination **222** (2008) 382–387

Validation and Similarity Evaluation of Leukocyte

P.Deepa and D.Deepa

Department of Electronics and Instrumentation Engineering, St. Joseph's College of Engineering, Chennai 600119, India
deepakumaran2002@gmail.com, deepaeie1997@gmail.com

Abstract. Generally the Examination of WBC to identify the foreign particles, abnormal conditions and it shows the status of theleukocyte. To avoid rick factors disease of brain such as aneurism, stroke and tumor at initial stage. It also show the types, region and severity. The individual examination of basophil(B), eosinophil(E), lymphocyte(L), monocyte(M) and neutrophil(N) predetermination of patient body condition are possible in clinical level examination to get a suggestion regarding the diseases to get treated and some of the techniques are used to reduce the complexity. Automated and semi-automated tool is useful for the process. The microscopic images of different classification of leukocyte is taken from the microscopic images. Each image is evaluated, validated, compared and verified with ground truth image.

***Keywords*:** leukocyte, predetermination, automated and semi-automated tool.

1. Introduction

Nowadays patient's body condition is monitored and bio-signals are recorded for further investigation. The recorded signal is viewed during the treatment periodby using the bio-images or bio-signal most of diseases are detected in initial stages.The way of classification and determination of severity can also be examine the immunity power,lower count indicates poor immune and higher count than the specified count chances of severe disease. The semi-automated technique work is proposed appraise image of Peripheral-Blood-Cell(PBC).It is analysed for testing purpose by using the Leukocyte-Image-For-Segmentation-Classification(LISC).The microscopic images are digitalized and recorded by comparing the given sample with ground truth. Ground truth are the doctor's feedback that is known value. Here the work incorporates on the social-group-optimization plus shannon's entropy (SGO+SE) based thresholding and chan-vese segmentation (CVS) to extract the images of foreign particles or stained portion from the PBC.

These kind of techniques will help the doctors to predicate the condition and can collect validated bio-image. Similarities of given sample with pre-processing, post-processing and then extracted image is compared with ground truth. Now the datasets are collectedand more number of images are taken for the examination. Each sets of images are threshold and segmented. The extracted stain portion from the given samples.

Traditionally blood detection of patient was a difficult thing to analysis the conditions.There are five different types of WBC on the LISC dataset are Basophil (B),Eosinophil (E),Lymphocyte

(L),Monocyte (M) and Neutrophil (N).Usually computer assisted diagnostic procedures are involved in up-coming procedures. PBC picture is examined to obtain a maximum accuracy of WBC. Peripheral-Blood-Cell (PBC) is normally stored with digital microscope. Further processes are integrated like Social-Group-Optimization, Shonnon's Entropy (SGO+SE), thresholding and Chan-Vese Segmentation (CVS) to differentiate each image for detection of WBC.

2. LITERATURE REVIEW

Nearly 250 images of Leukocyte Images for Segmentation and Classification (LISC) database were taken from theHematology-Oncology and BMT Research Center of Imam Khomeini Hospital in Tehran, Iran. Generally WBC has nucleus and cytoplasm were segmented by manual pixels of images contains 720x576 pixels and Gismo-Right technique is used to smeared and stained. This kind of simulation code used in vast areas like feature sets (Hydrodynamics, Black holes etc) where the images are acquired by a light microscope (Microscope-Axioskope 40).The digital analysis of blood cell(clinics in laboratory medicine the work of L.D Costa .et al(2015) used are Leukocyte nuclei segmentation using entropy function and chan-veseapproach. Semi-automated tool helps to optimal thresholding of image. Multi-thresholding(three level) processed(SGO+SE) sample image is now subjected to segmentation process with the help of CVS. Chan-vese segmentation method is used in multi-thresholded image. For a better performance hybridization of multi-thresholding and segmentation of given sample image of WBC. Leukocyte processed to mined nuclei to compare with ground truth image. The ground truth dataset image will helps for future evaluation.

Basically the ground truth prepared for 250 leukocytes to compare the both images. By comparing of extracted image of leukocyte and corresponding image of ground truth image as shown in **Figure 1** are used for optimal validation and similarities are Jaccard(JI),Dice(DC),Sensitivity(SE),Specificity(SP),Accuracy(AC),Precision(PR).

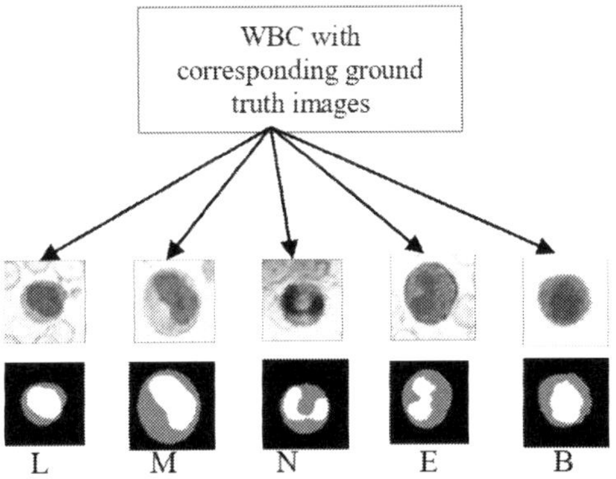

Figure 1 The sample leukocytes images

3. EXTRACTION TOOL

It helps to optimize the entire extraction procedure with Hybridization of SGO+SE and CVS techniques by using MatLab software for role of thresholding. In semi-automated tool development extract the strained portion by multi-thresholding level.Gathering of most suitable values to evaluate,in order to over-come a mathematical complexities,it acquires an enhancement of sample image. SGO attempts the various complex problems such as consistency in optimal solution, computation, optimal solution etc. the semi-automated tool. Shannon's entropy helps to obtain more random signals and organization of signal which is obtained from the Social-Group-Optimization. More numbers of data or signals aremeasured and the degree of freedom is reduced to get a solution. Chan-VeseSegmentation is used to mine the required portion and the extracted nuclei to compare the given sample with ground truth image. It shows the contour and object of both two dimensions and three dimensions.It has a property of enhancement of object and contour detection, goodcomparison of defective and non-defective images,calculating the intensity of image,area,perimeter,volume such information can be extracted to obtain measurement details.The hybridization of SGO+SE and CVS Segmentation to evaluate the Picture-Similarity-Mesurements(PSM) of JI,DC,SE,SP,AC,PR.

4. SOCIAL-GROUP-OPTIMIZATION

The Social-Group-Optimization is mainly used for the minimization of parameters are allowed. It is recently adopted in medical field for various applications. It reduces the complexity of many solutions in simple ways. Due to the reduced complexity of mathematics it is used widely in bio-medical applications.it also helps to find Global solution. The initial step of image with multi-thresholding

5. SHANNON'S ENTROPY

The Term Shannon's Entropy is storing a randomof uncertainly values in bit wise and it is also used to measure the randomness of random variable. Compression of data of image in bit wise. While compression process takes place there is a chance of lose in information.

6. CHAN-VESE-SEGMENTATION

In order to extract the stained portion of WBC after the thresholding process, minimal energy value is required to display the region which lies inside the image contour. Now the ground truth image and extracted image are compared to get computed values of PSM. The complete hybridization is done with the semi-automated tool. An Iteration process is carried until to get an identical pixels are obtained. Most of the information obtained regarding the similarities of leukocyte are evaluated.

7. BLOCK DIAGRAM

The methodology of hybrid image examination in Leukocyte image data sets are depicted in **Figure 2.**

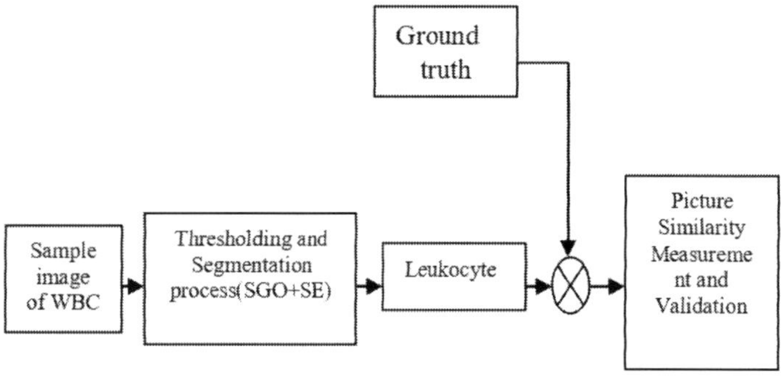

Figure 2. Block Diagram of hybrid image examination Methodology

8. RESULTS AND DISCUSSION

With the help of the Semi-Automated tool development extraction is compared with ground truth images obtained a computed values which are tabulated in **Table 1**. Each image is compared with the ground truth image to get a values of Jaccard,DC,SE,SP,AC,PR. With the hybridization of SGO+SE for the purpose of multi-thresholding Process and the CVS Segmentation process for the extraction of certain strained areas are achieved.

Table 1 Thresholding and Segmentation of leukocytes images

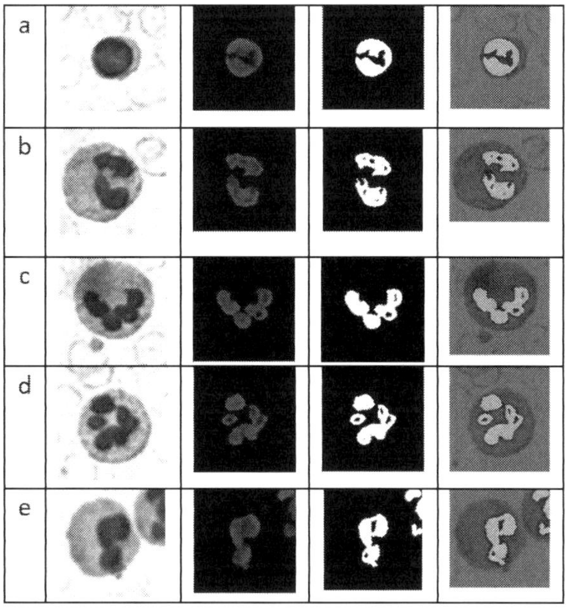

The leukocytes images are validated and the similarity are tabulated in **Table 2** for the further examination of disease, abnormality conditions occurs in the human body are clarified.

Table 2 The validation and similarities of leukocytes images

S.no	JI	DC	SE	SP	AC	PR
1.	0.043	0.082	0.049	0.001	0.043	0.258
2.	0.246	0.395	0.300	0.003	0.246	0.575
3.	0.191	0.320	0.250	0.005	0.192	0.445
4.	0.186	0.314	0.220	0.001	0.186	0.544
5.	0.282	0.441	0.336	0.318	0.332	0.640

9. CONCLUSION

The proposed method Semi-Automated performs the action to get a set of values which are obtained from the given sample images of leukocytes. An individual image was examined to validated the image and similarities are obtainedfor a given samples images.by which doctors can easily conclude the results. The above result executed well with the help of the tool MATLAB Software.This method useful in real time application. This technique can also be modified for other applications.

10. REFERENECS

[1] N. Dey, A.S. Ashour, F. Shi, S.J. Fong, R.S. Sherratt, Developing residential wireless sensor networks for ECG healthcare monitoring, *IEEE Transactions on Consumer Electronics*, **63(4)** (2017) 442-449.

[2] N. Dey, A.S. Ashour, S. Beagum, D.S. Pistola, M. Gospodinov, E.P. Gospodinova, J.M.R.S. Tavares, Parameter optimization for local polynomial approximation based intersection confidence interval filter using genetic algorithm: An application for brain MRI image de-noising, *Journal of Imaging*, **1(1)** (2015) 60-84.

[3] N. Dey, A.S. Ashour, F. Shi, R.S. Sherratt, Wireless capsule gastrointestinal endoscopy: direction-of-arrival estimation based localization survey, *IEEE reviews in biomedical engineering*, **10** (2017) 2-11.

[4] N. Dey, A.S. Ashour, F. Shi, S.J. Fong, R.S. Sherratt, Developing residential wireless sensor networks for ECG healthcare monitoring, *IEEE Transactions on Consumer Electronics*, **63(4)** (2017) 442-449.

[5] N. Dey, V. Rajinikanth, A.S. Ashour, J.M.R.S. Tavares, (2018). Social Group Optimization Supported Segmentation and Evaluation of Skin Melanoma Images. Symmetry, **10(2)** (2018) 51.

[6] V. Rajinikanth, N. Dey, S.C. Satapathy, A.S. Ashour, An approach to examine magnetic resonance angiography based on Tsallis entropy and deformable snake model, *Future Generation Computer Systems*, **85** (2018) 160-172. Doi: 10.1016/j.future.2018.03.025.

[7] L.B. Dorini, R. Minetto, N.J. Leite, Semiautomatic white blood cell segmentation based on multiscale analysis, *IEEE Journal of Biomedical and Health Informatics*, **17(1)** (2013) 250–256.

[8] S.S. Ahmed, N. Dey, A.S. Ashour, D. Sifaki-Pistolla, D. Bălas-Timar, V.E. Balas, J.M.R.S. Tavares, Effect of fuzzy partitioning in Crohn's disease classification: a neuro-fuzzy-based approach, *Medical & biological engineering & computing*, **55(1)** (2017) 101-115.

[9] N. Dey, A.S. Ashour, F. Shi, S.J. Fong, J.M.R.S. Tavares, Medical cyber-physical systems: A survey, *Journal of Medical Systems*, **42(4)** (2018)74.

[10] V. Rajinikanth, N.S.M. Raja, K. Kamalanand, Firefly algorithm assisted segmentation of tumor from brain MRI using Tsallis function and Markov random field, *Journal of Control Engineering and Applied Informatics*, **19(2)** (2017) 97-106.

[11] J. Prinyakupt, C. Pluempitiwiriyawej, Segmentation of white blood cells and comparison of cell morphology by linear and naïve Bayes classifiers, *BioMedical Engineering Online*, **14** (2015) 63. Doi: 10.1186/s12938-015-0037-1.

[12] S. Arslan, E. Ozyurek, C. Gunduz-Demir, A color and shape based algorithm for segmentation of white blood cells in peripheral blood and bone marrow images, *Cytometry A*, **85(6)** (2014) 480–490.

[13] L. Putzu, C.D. Ruberto, White blood cells identification and counting from microscopic blood image, *International Journal of Medical and Health Sciences*, **7(1)** (2013) 20-27.

[14] K. Manickavasagam, S. Sutha, K. Kamalanand, Development of systems for classification of different plasmodium species in thin blood smear microscopic images, *Journal of Advanced Microscopy Research*, **9(2)** (2014) 86–92.

[15] N.S. Balan, A.S. Kumar, N.S.M. Raja, V. Rajinikanth, Optimal multilevel image thresholding to improve the visibility of plasmodium sp. in blood smear images, *Advances in Intelligent Systems and Computing*, **397** (2016) 563-571.

IOP Conf. Series: Materials Science and Engineering 1166 (2021) 012004 doi:10.1088/1757-899X/1166/1/012004

Effect of annealing on structural and optical properties of SnS_2 thin films grown by thermal evaporation and post sulphur annealing technique

K Anlin Lazar [1,2], **V J Cicily Rigi** [3], **D Divya** [1], and **K J Saji** [3,4,5,6]

[1]*Government Victoria College (Affiliated to University of Calicut),Palakkad 678001, Kerala, India.*
[2]*Mercy College(Affiliated to University of Calicut),Palakkad 678006, Kerala, India.*
[3]*International School of Photonics, Cochin University of Science and Technology, Kochi 682 022, India.*
[4]*Centre of Excellence in Advanced Materials, Cochin University of Science and Technology, Kochi 682 022, India*
[5]*Inter University Centre for Nanomaterials and Devices, Cochin University of Science and Technology, Kochi 682 022, India.*
[6]Corresponding author: Email address: saji@cusat.ac.in (K.J. Saji), Telephone: +91 4842862721.

Abstract. Recently, two-dimensional layered semiconductors caught attention due to their tuneable thickness dependent properties and its optoelectronic device applications. This paper reports the effect of annealing temperature and annealing duration on structural, compositional and optical properties of tin disulphide thin films (SnS_2) fabricated using a bi- stage process. The process involves thermal evaporation of metallic tin precursor films and post sulphur annealing process in a non-vacuum furnace. The annealing temperatures ranged from 300 ^{0}C to 500 ^{0}C while annealing time varied from 0.5 to 3 hours. The annealing at 300^{0}C improved the crystallinity of SnS_2 thin films. Raman analysis showed a prominent peak at 207 cm^{-1}, while at higher temperatures nanocrystalline SnS_2 phase is observed. Optical bandgap got reduced as annealing temperature increased and at 300^{0}C, the bandgap was 2.7 eV. The above obtained properties of SnS_2 opens a pathway for using these films for potential applications.

1. Introduction

Recent studies in the field of flexible optoelectronic and display devices throws light to an innovative class of semiconductors, commonly called as two dimensional (2D) materials. Notable geometrical structure and mechanical properties exhibited by these materials makes them suitable for a wide range of applications in nanoelectronics industry. Since 2004, graphene, the monolayer of graphite, has been the most studied 2D material. The ability of graphene to form atomic sheets with attractive optical, electrical and mechanical properties caught the attention of scientists [1,2]. However, the absence of bandgap and low switching ratios made graphene less suitable for applications in optoelectronic devices. In this scenario, it became a necessity to find semiconductors which possess appreciable bandgap with magnificent electrical and mechanical properties. The demand for new class of 2D materials paved the path for development of other layered materials especially layered metal dichalcogenides (LMCs) [3]. LMC have been identified as a promising group of semiconductors due to possibility of processed in reduced dimension while maintaining appreciably better charge transport. It is observed that layered materials have better electrostatic control when the dimensionality of the material scales down to atomic thickness[4]. This property is utilized in bendable and wearable electronics which marks its superiority over other organic compounds [5]. Studies suggests that layered metal chalcogenides satisfies the criteria of both the tunable bandgaps as well as better electrical properties [6]. Their peculiarities adopt them in optoelectronic and flexible display device applications.

Two-dimensional materials generally possess strong intra layer covalent bond while the weak Van der Waals bond binds the inter layers enabling the isolation of bulk to single layers [4]. Even though MoS_2 takes the dominant position of being the first 2D semiconductor [7] which shows a transition from indirect bandgap in bulk to direct bandgap as it becomes monolayer [8], LMC also opens a novel field of semiconductors which has wider bandgap.

SnS_2 falls to this class of materials which is basically a non-transition dichalcogenides semiconductor. It is a representative of layered metal sulphides with a bandgap of 2.75 eV [9] , which is larger than most of the transition metal dichalcogenides (TMDCs). The wider bandgap has an added advantage on suppressing the short channel effects in field effect transistor (FETs). SnS_2 is isostructural with hexagonal cadmium iodide (CdI_2) crystal. S-Sn-S forms a stack of triple layer which are bonded together by strong covalent forces and the adjacent triple layers are coupled together by means of weak Van der Waals interactions [10]. This structure enables the peeling out of bulk crystal into different layers. SnS_2 poses high absorption coefficient ($>10^4$ cm^{-1}) in the visible region [11]. Importance of tin sulphide compounds is that basically they are nontoxic compounds and are abundantly available [12]. Novel fabrication methods for the preparation of SnS_2 with good quality and better properties caught the interest of many researchers. Tin disulfide thin films can be obtained by various methods like successive ionic layer adsorption and reaction (SILAR) [13], spray pyrolysis [14], thermal vacuum evaporation [15], plasma-enhanced chemical vapour deposition (PECVD) [16], dip coating [17] etc., each one with its advantages and disadvantages. Recently, SnS_2 has proven as an attractive material for its potential applications as a light absorber layer for dye sensitized solar cells, gas sensing and energy storage applications [18,19].

In this paper, we report a simple and cost-effective approach to prepare tin disulphide (SnS_2) thin films on glass substrates by thermal evaporation of tin powders as precursor followed by post sulphur annealing process. The influence of annealing temperature and annealing time on crystal structure, chemical composition and optical properties of the films has been examined.

2. Experimental Details

2.1 Growth of SnS_2 thin films

Glass substrates were cleaned by sequential ultrasonication for ten minutes each in acetone, isopropyl alcohol and distilled water. Metallic tin (Sn) precursor films were thermally evaporated using tin powders (99.995 % purity) in a vacuum system at a constant current of 40 A. The weight of tin precursor was varied from 0.08 g to 0.2 g. The coating time varied from 2 minutes to 8 minutes as the weight of tin precursor changed. The deposition was followed by annealing of the evaporated tin thin films in the presence of sulfur in a non-vacuum tubular furnace of 80 cm length. During annealing, the sulfur atmosphere was maintained by means of sulfur powder kept inside the tubular furnace. Both ends of the furnace were sealed with ceramic blocks and N_2 carrier gas was introduced into the tube using stainless steel pipes. The rate of N_2 gas flowing through the furnace was controlled by a needle valve. Sulfur powder (99.95% purity) was kept in alumina crucible at a distance of 10 cm to 20 cm from the center near the carrier gas inlet end of the furnace. Figure 1 represents the schematic diagram of annealing set up. The thermally evaporated tin thin films were kept at two different positions inside the furnace. The samples were loaded on top of alumina boat to elevate the sample to a certain height. The samples were placed at the centre of the furnace. When the furnace attained required temperature, it was held constant for required duration. The temperatures chosen ranged from 300 ^{0}C to 500 ^{0}C and duration of annealing varied from 0.5 to 5 hours. After the required annealing time was achieved, the samples were cooled slowly by leaving them overnight in the furnace. There was excess sulfur left in the boat after the annealing process is completed. Following the sulfurization, thin films were analyzed using different characterization techniques.

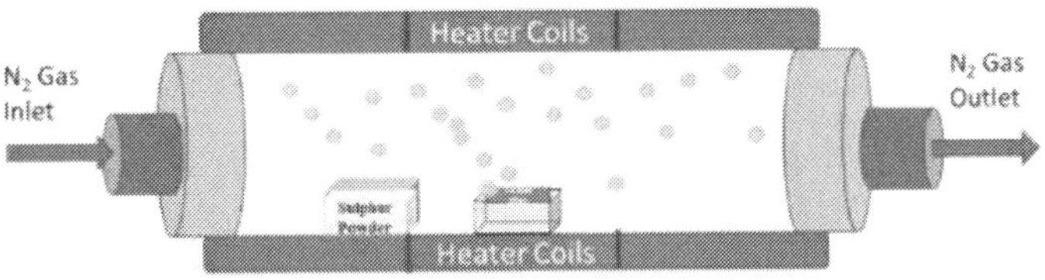

Figure 1. Schematic diagram of post sulfur annealing setup.

2.2 Characterization techniques

XRD data was measured using PANalytical (Aeris Research) system with Cu Kα radiation (λ= 1.540598 Å). The absorption was measured by UV-Vis spectrophotometer (Perkin Elmer Uv /Vis Lamba 365 double beam spectrophotometer) in the wavelength range of 190 - 1100 nm, and their optical bandgaps were estimated from the absorbance data. Raman analysis of the samples was done using Raman spectrometer (HORIBA JOBIN YVON Lab RAM HR) with Argon ion laser (514.5 nm) having a spectral resolution of 0.3 –1 cm^{-1}.

3. Results and Discussions

Processing temperature of 2D materials is crucial for different device applications. Low temperature processing is fairly recommended for the sensors and transistor applications, especially on flexible substrates. We deposited SnS_2 films at comparatively low temperature and post sulfur annealing treatment was done to improve the phase purity as well as crystallinity of SnS_2 thin films.

3.1 Structural Characterizations

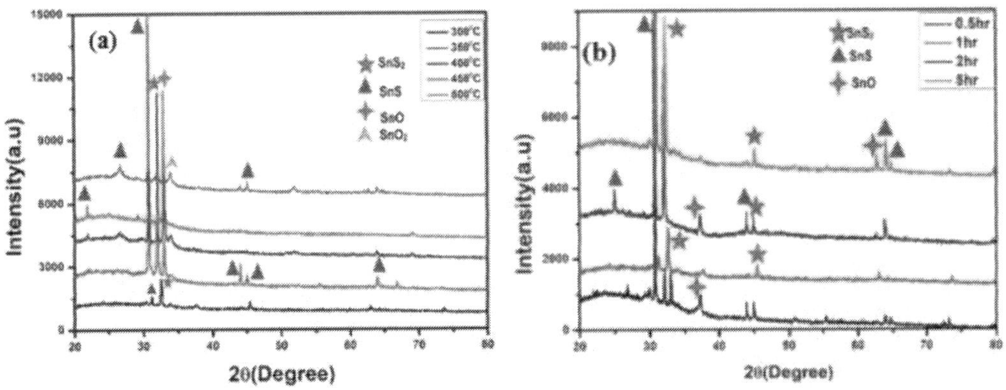

Figure 2. (a) Diffraction patterns of samples annealed at various temperatures for one hour at centre of the furnace. (b) Diffraction patterns of samples, kept at the centre of the furnace at 300^{0}C for different durations.

3.1.1 X-Ray Diffraction Analysis. Figure 2(a) shows the diffraction patterns of the films grown at different annealing temperatures. The tin disulfide films exhibited hexagonal structure similar to that of the mineral Berndtite. The diffraction patterns were indexed using ICDD file no 023-0677 [20]. The strongest peak at annealing temperature 300 ^{0}C indicates that the SnS_2 film is preferentially oriented along the (101) plane. Gedi et al [21] reported similar results and many other research grousps also supports this result [22]. The integral intensity ratio of SnS_2 to SnS obtained from XRD patterns is greater than unity for sample annealed at 300 ^{0}C for 1 hour. As the annealing temperature increased, mixed phases were formed along with SnS_2 phase which is in accordance with the results reported by Reddy et al [23]. Reports reveal that as the annealing temperature increases, Sn loss occurs in tin sulfide films. This is attributed mainly to the volatile nature of tin phase and hence they get decomposed according to the reaction mentioned below. [24]

$$2SnS_2 \rightarrow Sn_2S_3 + S \qquad (1)$$

$$Sn_2S_3 \rightarrow 2SnS + S \qquad (2)$$

Figure 2(b) shows the diffraction patterns of the films grown at 300^{0}C for different duration of annealing time. 30 minutes annealing resulted films with mixed composition which changes to dominant SnS_2 phase when the annealing time increased to one-hour duration. But as the annealing time was further increased to higher span, presence of the mixed phases again increased. The formation of SnS was noted as the annealing time increased and it is mainly due to re-evaporation of sulfur from the films. Since the furnace used for annealing was a non-vacuum system, the samples contained presence of SnO phase also. Thus, for samples which were annealed beyond 300 ^{0}C, the presence of SnS and SnO phases were detected. It has been reported that increase in sulfurization temperature favors the reaction of Sn and S phases to form SnS [25]. Even if excess sulfur is present in the furnace, annealing at temperatures higher than 350 ^{0}C could lead to high sulfur deficiency in the layers. The presence of mixed phases in the x ray diffraction patterns indicate that the post annealing process done in the furnace could not sulfurize the tin samples completely. Even if the ambient atmosphere of sulfur was provided inside the tubular furnace, the presence of oxygen added the oxide impurity along with the sulphide phases. Raman result also supports the formation of traces of other tin sulphide phases at higher annealing temperatures. In order to know more about the sub phases, Raman spectroscopy was also done.

3.1.2 W-H plot Analysis. Figure 3 shows the W-H plot analysis. Crystallinity variation in the film is manifested through the broadening of diffraction peaks. Peak width analysis helps to calculate the crystallite size and lattice strain. There are many reasons for strains in a crystal. Studies show that stress arises at the walls of grains and also during sintering or annealing treatments [26]. Lattice strain and crystallite size affects the peak width as well as the intensity and results in shifting of 2θ values. Crystallite size is usually calculated using Debye Scherrer formula, while the strain is calculated using Williamson formula. Hence by finding the FWHM of diffraction peaks from the x-ray diffraction, crystallite size and strain of the film can be evaluated. The Bragg peak breadth has a combined contribution from instrument and sample dependent effects.

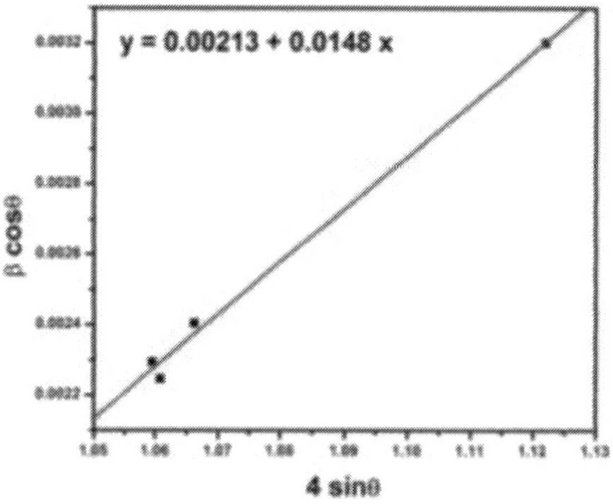

Figure 3. W-H plot for different annealing time.

In Williamson-Hall method, line broadening β_D of a Bragg reflection (h k l) originating from the small crystallite size follows Scherrer equation $\beta_D = \frac{0.94\lambda}{D\ cos\theta_{hkl}}$. Here λ is the X-ray wavelength, θ is the Bragg

angle and D is the effective crystallite size normal to the reflecting planes. Also, the strain induced broadening β_ε is given by the Williamson formula as
$\beta_\varepsilon = 4\varepsilon\ tan\theta_{hkl}$. Here ε is the root mean square value of the macrostrain. Assuming that the particle size and strain contributions to line broadening are independent of each other, the observed line breadth is calculated by adding the Scherrer equation and Williamson equation.

$$\beta_{hkl} = \beta_D + \beta_\varepsilon \qquad (3)$$

$$\beta_{hkl} = \frac{0.94\,\lambda}{D\cos\theta} + 4\varepsilon\tan\theta \qquad (4)$$

On re-arranging the equation (4),

$$\beta_{hkl}\cos\theta = \frac{k\lambda}{D} + 4\varepsilon\sin\theta \qquad (5)$$

$\beta \cos\theta$ was plotted with respect to $4\sin\theta$ for the peaks of SnS_2 annealed at different durations. Strain is obtained from slope and particle size is calculated from y-intercept of the fitted line. From the graph, strain obtained is 1.5×10^{-2} and studies show that positive slope is mainly due to tensile strain [27]. The average particle size calculated from the y-intercept of W-H plot is found to be around 68 nm.

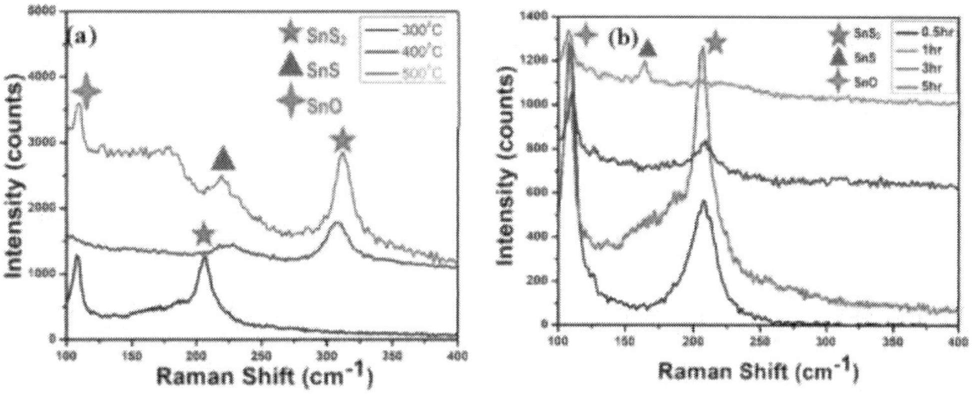

Figure 4: (a) Raman analysis of samples annealed at different temperatures for 1hour. (b) Raman analysis of samples annealed at 300 ^{0}C for different durations

3.1.3 Raman Analysis. Raman spectroscopy is a versatile experimental tool to analyse the 2D material structural properties which mainly includes identification of crystal polytype. Smith et.al reported the Raman spectra of different polytypes of SnS_2 [28]. Figure 4(a) is the Raman spectra of SnS_2 films annealed at different temperatures for a span of one hour. SnS_2 has two vibration modes named as E_g which is doubly degenerate and A_{1g} which is singly degenerate active mode. In A_{1g} active mode, the two sulfur atoms get displaced out of-phase parallel to c-axis with respect to each other keeping the Sn atom stationary while E_g mode arises due to the in-plane vibration mode of S-Sn-S of 2H SnS_2 [29]. Reports assign the peak at 311.5 cm^{-1} to the A_{1g} phonon mode of 2D SnS_2 crystals (30). According to the group theory analysis given by Lucovsky et al. [31], 311.5 cm^{-1} peak corresponds to the A_{1g} mode of nanocrystalline SnS_2 while E_g phonon mode at 207 cm^{-1} corresponds to bulk SnS_2 crystals. The dominance of E_g peak over the A_{1g} peak indicates that the bulk SnS_2 crystals are formed. It is reported that there is a disappearance of E_g peak due to the formation of nanocrystals of SnS_2 [29]. Here, as the annealing temperature increased, the E_g peak corresponding to 207 cm^{-1} disappeared while A_{1g} peak corresponding to 311.5 cm^{-1} peak appeared. Raman analysis also suggests the presence of traces of SnO and SnS which supports the results obtained from XRD. Figure 4(b) shows the Raman spectra of samples annealed at different durations at annealing temperature of 300^{0}C. The spectra show bands at 109 cm^{-1}, 163 cm^{-1} and 207cm^{-1}. Raman shift peaks at 109 cm^{-1}, 163 cm^{-1} and 207.2 cm^{-1} has been identified as SnO (B_{1g} mode), SnS (B_{2g} mode) and SnS_2 (E_{2g} mode) respectively(28). The peak at 207 cm^{-1} confirms the formation of SnS_2 compound at 300^{0}C for a duration of one hour [32].

IOP Conf. Series: Materials Science and Engineering 1166 (2021) 012004 doi:10.1088/1757-899X/1166/1/012004

3.2 Optical Studies.

The optical properties of the films were studied from the transmittance spectra of the films recorded in the wavelength range 190 -1100 nm. The absorption coefficient α and optical bandgap E_g are related by Tauc relation (33): $(\alpha h\nu)^m = A_n(h\nu - E_g)$ [6]

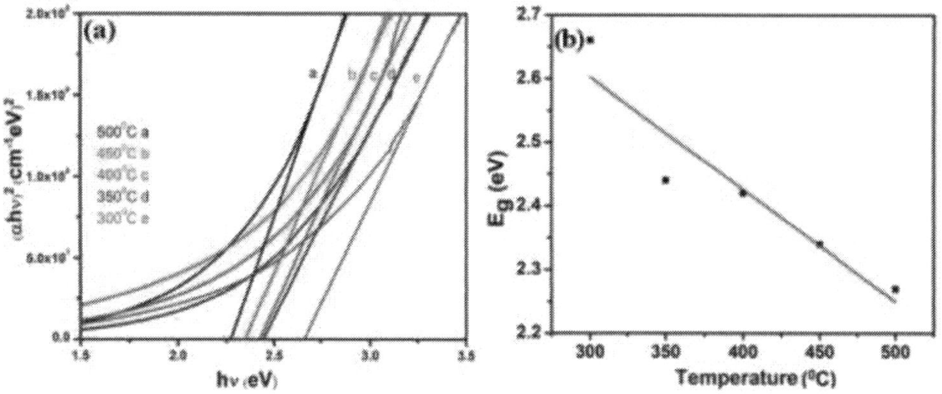

Figure 5: (a) Tauc plot of samples annealed samples at different temperatures for 1hour. (b) Variation of bandgap energy with annealing temperature.

where: A_n is an energy independent constant, the exponent *m* depends on the nature of the transition type. $m = 2$ for a direct allowed transition and $m = ½$ is for an indirect allowed transition. The direct bandgap E_g of the samples can be obtained by extrapolating the straight-line portion of the $(\alpha h\nu)^2$ vs hν plot to the energy axis which is shown in the figure 5a. The bandgap is found to be decreasing as the annealing temperature increased which is due to the change in the stichometry of the deposited films [34]. The variation of bandgap with annealing temperature is shown in figure 5b. The variation in the bandgap with the increase in the annealing temperature is showing a gradual decrease. This nature of optical bandgap is due to the change in the stoichiometry of thin films. XRD and Raman results confirmed the formation of mixed phases as the temperature increased. This trend is manifested in optical studies also through gradual lowering of bandgap.

4. Conclusion

Tin disulfide thin films have been prepared by physical evaporation of Sn powder and by post sulfur annealing at various temperatures from 300 °C up to 500 °C. XRD results confirmed the formation of SnS_2 thin films at 300 °C with hexagonal structure. As the annealing temperature increased, the formation of mixed phases was noted in XRD results. Raman studies of annealing time variation showed SnS_2 phase along with traces of SnS and SnO. But as the annealing temperature increased, the absence of E_g peak points to the formation of SnS_2 nanocrystals. The W-H plot was analysed and strain is obtained from the slope of the graph. Strain is found to be 1.5×10^{-2} and average particle size obtained from the W-H plot is 68 nm. The SnS_2 films showed direct bandgap which was found to vary in the range 2.7 eV to 2.3 eV gradually as the annealing temperature increased. This variation in the bandgap is due to the formation of mixed phases. Also, as the annealing temperature increased, there is a tendency for the formation of SnS_2 nanocrystals. From the experimental results, it can be concluded that SnS_2 can be opted for the fabrication of channel layer in nano electronic and optoelectronic devices.

Acknowledgements
Anlin Lazar K acknowledges DST FIST for providing the financial aid for procuring the vacuum thermal unit.

References:

[1]. Fiori G, Bonaccorso F, Iannaccone G, Palacios T, Neumaier D, Seabaugh A, et al. Electronics based on two-dimensional materials. *Nature Nanotechnology*. 2014;**9**(10):768–79.
[2]. Novoselov KS, Geim AK, Morozov S V., Jiang D, Zhang Y, Dubonos S V., et al. Electric Field Effect in Atomically Thin Carbon Films. *Science*. 2004;**306**(5696):666–9.
[3]. Ahn JH, Lee MJ, Heo H, Sung JH, Kim K, Hwang H, et al. Deterministic Two-Dimensional Polymorphism Growth of Hexagonal n-Type SnS_2 and Orthorhombic p-Type SnS Crystals. *Nano Letters*. 2015;**15**(6):3703–8.
[4]. Novoselov KS, Jiang D, Schedin F, Booth TJ, Khotkevich V V, Morozov S V, et al. Two-dimensional atomic crystals. *Proceedings of the National Academy of Sciences of the United States of America* [Internet]. 2005 Jul 26 [cited 2020 Mar 10];**102**(30):10451–3. Available from: http://www.ncbi.nlm.nih.gov/pubmed/16027370
[5]. Song HS, Li SL, Gao L, Xu Y, Ueno K, Tang J, et al. High-performance top-gated monolayer SnS2 field-effect transistors and their integrated logic circuits. *Nanoscale*. 2013;**5**(20):9666–70.
[6]. Ramakrishna Matte HSS, Gomathi A, Manna AK, Late DJ, Datta R, Pati SK, et al. MoS2 and WS2 analogues of graphene. *Angewandte Chemie* - International Edition. 2010;**49**(24):4059–62.
[7]. Radisavljevic B, Radenovic A, Brivio J, Giacometti V, Kis A. Single-layer MoS2 transistors. *Nature Nanotechnology*. 2011;**6**(3):147–50.
[8]. Qu B, Ma C, Ji G, Xu C, Xu J, Meng YS, et al. Layered SnS2-reduced graphene oxide composite - A high-capacity, high-rate, and long-cycle life sodium-ion battery anode material. *Advanced Materials*. 2014;**26**(23):3854–9.
[9]. Sharma D, Kamboj N, Agarwal K, Mehta BR. Structural, optical and photoelectrochemical properties of phase pure SnS and SnS_2 thin films prepared by vacuum evaporation method. *Journal of Alloys and Compounds*. 2020;**822**:153653.
[10].Mondal C, Ganguly M, Pal J, Roy A, Jana J, Pal T. Morphology controlled synthesis of SnS2 nanomaterial for promoting photocatalytic reduction of aqueous Cr(VI) under visible light. *Langmuir*. 2014;**30**(14):4157–64.
[11].Jing L, Xu Y, Chen Z, He M, Xie M, Liu J, et al. Different Morphologies of SnS2 Supported on 2D g-C3N4 for Excellent and Stable Visible Light Photocatalytic Hydrogen Generation. *ACS Sustainable Chemistry and Engineering*. 2018;**6**(4):5132–41.
[12].Burton LA, Whittles TJ, Hesp D, Linhart WM, Skelton JM, Hou B, et al. Electronic and optical properties of single crystal SnS2: An earth-abundant disulfide photocatalyst. *Journal of Materials Chemistry A*. 2016;**4**(4):1312–8.
[13].Deshpande NG, Sagade AA, Gudage YG, Lokhande CD, Sharma R. Growth and characterization of tin disulfide (SnS_2) thin film deposited by successive ionic layer adsorption and reaction (SILAR) technique. *Journal of Alloys and Compounds* [Internet]. 2007 Jun 14 [cited 2020 Mar 10];**436**(1–2):421–6. Available from: https://www.sciencedirect.com/science/article/abs/pii/S0925838806022444
[14].Amalraj L, Sanjeeviraja C, Jayachandran M. Spray pyrolysised tin disulphide thin film and characterisation. *Journal of Crystal Growth*. 2002;**234**(4):683–9.
[15].Shi C, Chen Z, Shi G, Sun R, Zhan X, Shen X. Influence of annealing on characteristics of tin

disulfide thin films by vacuum thermal evaporation. *Thin Soid Films*. 2012;**520**:4898–901.
[16].Wang S, Wang S, Chen J, Liu P, Chen M, Xiong H, et al. Influence of the deposition parameters on the properties of SnS_2films prepared by PECVD method combined with solid sources. *Journal of Nanoparticle Research*. 2014;**16**:2610.
[17].Panda SK, Antonakos A, Liarokapis E, Bhattacharya S, Chaudhuri S. Optical properties of nanocrystalline SnS2 thin films. *Materials Research Bulletin*. 2007;**42**(3):576–83.
[18].Bryushinin MA, Dubrovsky GB, Sokolov IA. Non-steady-state photocurrents in SnS2 crystals. *Applied Physics B: Lasers and Optics*. 1999;**68**(5):871–5.
[19].Zhou X, Zhang Q, Gan L, Li H, Zhai T. Large-Size Growth of Ultrathin SnS2 Nanosheets and High Performance for Phototransistors. *Advanced Functional Materials*. 2016;**26**(24):4405–13.
[20].Chao D, Liang P, Chen Z, Bai L, Shen H, Liu X, et al. Pseudocapacitive Na-Ion Storage Boosts High Rate and Areal Capacity of Self-Branched 2D Layered Metal Chalcogenide Nanoarrays. *ACS Nano*. 2016;**10**(11):10211–9.
[21].Gedi S, Minnam Reddy VR, Pejjai B, Park C, Jeon CW, Kotte TRR. Studies on chemical bath deposited SnS2 films for Cd-free thin film solar cells. *Ceramics International* [Internet]. 2017;**43**(4):3713–9. Available from: http://dx.doi.org/10.1016/j.ceramint.2016.11.219
[22].Khot K V., Ghanwat VB, Bagade CS, Mali SS, Bhosale RR, Bagali AS, et al. Synthesis of SnS2 thin film via non vacuum arrested precipitation technique for solar cell application. *Materials Letters* [Internet]. 2016;**180**:23–6. Available from: http://dx.doi.org/10.1016/j.matlet.2016.05.089
[23].Reddy M V., Babu P, Ramakrishna Reddy KT, Miles RW. X-ray photoelectron spectroscopy and X-ray diffraction studies on tin sulfide films grown by sulfurization process. *Journal of Renewable and Sustainable Energy*. 2013;**5**(3):031613.
[24].Weber A, Mainz R, Schock HW. On the Sn loss from thin films of the material system Cu-Zn-Sn-S in high vacuum. *Journal of Applied Physics*. 2010;**107**(1):013516.
[25].Ramakrishna Reddy KT, Purandhara Reddy P. Structural studies on SnS films grown by a two-stage process.. *Materials Letters* 2002;**56**(1–2):108–11.
[26].Zhang JM, Zhang Y, Xu KW, Ji V. General compliance transformation relation and applications for anisotropic hexagonal metals. *Solid State Communications*. 2006;**139**(3):87–91.
[27].Bharti B, Barman PB, Kumar R. XRD analysis of undoped and Fe doped TiO2 nanoparticles by Williamson Hall method XRD analysis of undoped and Fe doped TiO 2 nanoparticles by Williamson Hall Method. *AIP Conference Proceedings*. 2015;**1675**(030052):030025.
[28].Smith AJ, Meek PE, Liang WY. Raman scattering studies of SnS2 and SnSe2. *J Phys C: Solid State Phys*. 1977;**10**(8):1321–33.
[29].Abello L, Bochu B, Gaskov A, Koudryavtseva S, Lucazeau G, Roumyantseva M. Structural characterization of nanocrystalline SnO2 by X-ray and Raman spectroscopy. J*ournal of Solid State Chemistry*. 1998; (**1351**):78–85.
[30].Price LS, Parkin IP, Hardy AME, Clark RJH, Hibbert TG, Molloy KC. Atmospheric pressure chemical vapor deposition of tin sulfides (SnS, Sn2S3, and SnS2) on glass. *Chemistry of Materials*. 1999;**11**(7):1792–9.
[31].Lucovsky G, Mikkelsen JC, Liang WY, White RM, Martin RM. Optical phonon anisotropies in the layer crystals SnS_2 and $SnSe_2$. *Physical Review B*. 1976;**14**(4):1663–9.
[32].Huang E, Yao X, Wang W, Wu G, Guan N, Li L. SnS 2 Nanoplates with Specific Facets Exposed for Enhanced Visible-Light-Driven Photocatalysis . *ChemPhotoChem*. 2017;**1**(2):60–9.
[33].Tauc J, Grigorovici R, Vancu A. Optical Properties and Electronic Structure of Amorphous Germanium. *phys.stat.sol*. 1966;**15**(2):627–37.
[34].Hu W, Hien TT, Kim D, Chang HS. Enhancement in photoelectrochemical performance of optimized amorphous SNS2 thin film fabricated through atomic layer deposition. *Nanomaterials*. 2019;**9**(8):1083.

IOP Conf. Series: Materials Science and Engineering 1166 (2021) 012005 doi:10.1088/1757-899X/1166/1/012005

A study on magnetic and spin polarized electric transport properties of cation doped $LaCaMnO_3$perovskite ceramic system

Sam Rajan and Jayakumari Isac

Centre for Condensed Matter Physics, CMS College, Kottayam-686001, India

sam@cmscollege.ac.in

Abstract. Thepoly-crystalline$La_{0.5}Nd_{0.15}Ca_{0.35}MnO_3$ has been synthesized by a conventional solid reaction method through the cation doping in $LaCaMnO_3$. The analysis of the powder XRD of the composites revealed the formation of an orthorhombic structure in Pbnm space group. The cell parameters of the formed system has obtained and confirmed through Reitveld analysis and it is found to be a=5.4505,b=5.4457,c=7.6876.The lattice parameters and the spin ordering temperature T_C of this system are decreasing due to Nd substitution. The ac and dc magnetization studies and resistivity measurements in ZFC and FC mode gives good understanding to the magnetic and electric transport properties of the doped system. The magnetization studies show a paramagnetic to ferromagnetic transition at a temperature below the critical temperature T_C. The resistivity measurements reveals semiconducting to metallic transition below the temperature T_{MS}. The temperature dependent transport nature shown in this system is one type of the correlated polaron hopping mechanism and it can be modeled by an adiabatic small polaron hopping(SPH) mechanism.

1. Introduction

The compounds with perovskite structure and its usefulness in applied fields of science and technology has been intensively investigating since 1940 [1-3]. These applications includes ferro-electricity, magnetism, ionic conductivity and superconductivity, etc. The atomic ordering in perovskite structure plays an important role in structure formation, in order to achieve thermodynamic stability in these systems. There are evidence in literature about the modification of physical properties of systems due to this atomic ordering [4,5].The variation in the structure may vary degree of short and far range orders which may significantly affect magnetic, electronic, dielectric, ferroelectric, ionic and other useful properties of these systems[6].

The discovery of colossal magneto resistance in perovskite systems, initiated an organized effort in scientific community to investigate the interplay between crystal structure, magnetic and electric properties of these systems[7,8]. More over most of the complex oxides represent by these structures with ABX_3stochoimetry.The ideal type of this sort of structures are known as aristotype perovskite structures and it has $Pm\bar{3}m$ space group symmetry. The B cations form the octahedral framework with X anions , and these octahedra shares corners of a 3-dimensional network while A cations usually sits in 12 the coordinates of the cubo-octahedral cavities of this three dimensional framework . [9,10].Distortions from the aristotopic structures may produce significant changes in physical, magnetic, electric, and optical properties of these compounds. Moreover in 80-90% of all perovskites compounds, an octahedral tilting distortion present if A cation is too small in its octahedral cavities[11-15]. The distortions of similar nature also change the conduction bandwidth and strength of magnetic super exchange interactions, which intern helps to fine-tune the magnetic and electric properties for potential applications.[16-17].

In addition to the structural flexibility, these composites posses compositional flexibility[10].Chemical substitutions are possible in all the three sites of these perovskite structures. Oxynitrides and oxyflurides, etc. are examples for chemical substitution in anions. Cation substitution may lead to an ordered or disordered structural arrangement, which has profound effects on the properties of these compounds. The rare phenomenon of ferromagnetism in ordered peroviskite La_2MnMO_6(where M=Ni,Co) due to B cation ordering and the dependence ionic conductivity of $Li_{3x}La_{2/3-x}TiO_3$ on A cation ordering, etc. are examples for effect of cationic substitution in perovskite structures[17-19].The present study is interested to investigate the ordered or random cationic ordering in the structural framework to engineer new materials of useful properties. Cation ordering may be realized with either A or B site cations and there are significant differences in their ordering mechanism[10-16]. In comparison with B cation ordering, A cation ordering is less explored [10-20] and the noval properties of A cation ordering is the focus of present study.

The appearance of colossal magneto resistance in the compound LaCaMnO3[21,22] is the motivating criteria to study the magnetic and electric transport properties of this compound as the effect of cationic ordering . The other excitement in this investigation is that while going deep into this study , one may encounter different interrelated phenomena such as insulator-metal transition, orbital ordering, charge ordering, double exchange, lattice, and magnetic polaron and so on[23]. The problem of the interplay of magnetic, structural, and electric transport properties is an old one[24,25], but the discovery of high-temperature superconductivity brought it once again at the forefront. High Tc compounds usually have antiferromagnetic magnetic ordering while CMR compounds usually prefers ferromagnetic magnetic ordering[23-25].There are many reports about the doping of metal ions in manganites and its consequences of potential fluctuations of electrons in σ* bond, due to the larger difference in valance between $La^{3+}$and doped metal ions.[26,27]. Therefore the study of doped perovskite manganites may lead to the discovery of new compounds that demonstrates interesting magnetic and electric transport properties. In this paper, we study the electric and magnetic transport properties of the cation doped sample.

2. Experimental

2.1. Sample and its preparation

The poly crystalline powder sample $La_{0.5}Nd_{0.15}Ca_{0.35}MnO_3$ has been prepared using widely known ceramic technique called solid-state reaction method. The raw materials of high reagent graded (99.9%) purity chemicals wiz. lanthanum oxide (La_2O_3), calcium carbonate (Ca_2CO_3), neodymium oxide(Nd_2O_3), and manganese oxide(Mn_2O_3) were used as starting precursors. These high purity precursors were weighed in stoichiometric amounts and allowed for ball milling for a period of 48 hours. Thereafter the powder was mixed physically with an agate mortar and pestle in a medium of cyclohexane for a period of one week in order to achieve the required homogeneity. The pellet of this powder sample was then prepared under high pressure and it was again allowed for calcinations at 1100° C repeatedly with intermittent grindings to obtain sample of high crystalline in nature.

The room temperature powder X-ray diffraction pattern of the sample was collected with Bruker advanced diffractometer using Cu Kα radiation. The measurements were done in 10 to 100 range of 2θ with a step size of 0.02. The Reitveld refinement of this pattern was done with help of Fullprof software package. The AC and DC magnetic susceptibility of the sample was measured against temperature for a range of 10-300K using the SQUID Magnetometer. The resistivity of the sample were also taken with the help of superconducting magnets and four-probe arrangements.

3. RESULTS AND DISCUSSIONS

3.1 X-ray Analysis

The sharp and intense peak in the XRD pattern of the sample $La_{0.5}Nd_{0.15}Ca_{0.35}MnO_3$ is an indication of good crystallinity of the sample within the sensitivity limit of the diffractometer and it is corresponding to the perovskite structure. Fullprof software package was used to perform structural refinement. In the comparison of the observed pattern of the structure with the perovskite structure of existing manganites, it is possible to the conclude that this observed pattern is similar to the structure with ICDD card no.1521126 [27,28]. The refinement of the observed pattern given in figure: 1 indicate that it has an orthorhombic structure with Pbnm space group symmetry. The cell parameters thus obtained are a=5.4505, b=5.4457, c=7.6876. No secondary phase was observed in the sample, which strongly confirms the formation of a pure phase. The parameters used for the Rietveld refinement are a sixth-order Chebychev polynomial function for the background, Pseudo-vogit profile function, lattice parameters and atomic coordinates, etc. A good fit was obtained between the observed and calculated patterns. The fit parameters Rwp, Rp, and χ^2 represented a good fit.

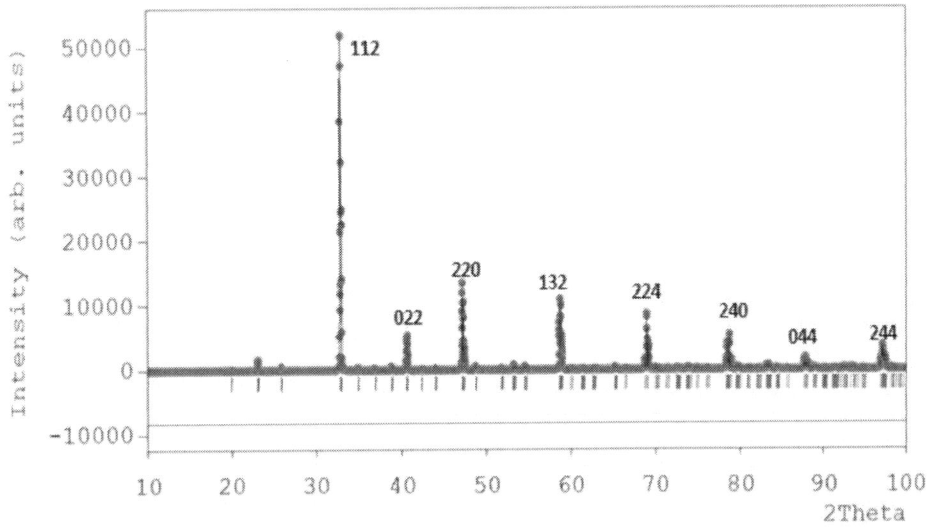

Figure 1. Rietveld refined powder XRD pattern of $La_{.5}Nd_{.15}Ca_{.35}MnO_3$

3.2 Magnetic properties

The AC and DC magnetization measurements have been conducted in Zero magnetic field(ZFC mode) and in the presence of magnetic field(FC mode). The variation of magnetic susceptibility with variation of temperature was measured in these modes. This variation of DC and AC magnetization with temperature are shown in figure 2 and 3.The AC magnetization measurement shown figure 3 indicate a Paramagnetic to Ferromagnetic transition with decrease of temperature as this increases during decreasing of temperature and this transition occurs below a temperature called Curie temperature T_C. The cation doping may introduce the mixed valency in Mn as result a sufficient change may happen the Mn^{3+}/ Mn^{4+} ratio. The mixed valency of Mn may result in double exchange which is believed to one of the reasons for ferromagnetism in manganites [29,30].The curie temperature can be obtained from the AC magnetization graph shown in figure 4 and it is found to be 206.3K. The observed Curie temperature T_C is below that of the parent compound and this result is similar that observed in the compound $La_{0.7}Sr_{0.3-x}Ag_xMnO_3$.[31].This is due to the super exchange interactions which become dominant over double exchange when Mn^{4+} concentration becomes more than 30% in total value Mn ions. The ferromagnetic material obeys Curie –Weiss law shown by equation(1), where the sample shows spontaneous magnetization below $T_C$and it behaves as a paramagnetic at temperature above T_C .

IOP Conf. Series: Materials Science and Engineering 1166 (2021) 012005 doi:10.1088/1757-899X/1166/1/012005

$$\chi_m = \frac{C}{T-\theta} \quad (1)$$

Figure .3 demonstrates the variation of $1/\chi_m$ versus temperature T. The high temperature region of graph above T_C is linear, which is clear indication of Curie- Weiss behavior of sample .The constant C in equation (1) can be obtained from the $1/\chi_m$ versus temperature T graph and it is found to be related to effective magnetic moment as

$$\mu_{eff} = 2.828\sqrt{C} (2)$$

The theoretical magnetic moments can be calculated from the following equation [40,41]

$$\mu_{cal} = \sqrt{.15\mu^2{}_{Nd^{2+}} + x\mu^2{}_{Mn^{3+}} + y\mu^2{}_{Mn^{4+}}} (3)$$

of where x and y are fractions of Mn^{3+} and Mn^{4+} . The fractions of Mn^{3+} and Mn^{4+} present in the sample is found from equation 2 and 3 and this value is an observable indication for preference of super exchange over double exchange .The Curie constant C can also express as $C= NP_f\frac{\mu_B^2}{3k_B}$ where P_f is the effective magneton number of the manganese ions of the sample. But the small polaron contributes the electronic and magnetic transport properties of the sample. At a temperature below T_C, the conductivity of the ferromagnetic phase is mainly due to the small polaron. The metal to insulator transition occurs at T_{MS}, which almost coincide with T_C is a sharp transition. At temperature below T_C, an ordering of magnetic moment occurs by Frank-Condon transition. This transition arises as the coexistence of parallel spin of the neighbouring sites. In polaron model the electrons move from Mn^{3+} ion to Mn^{4+} with the exchange of zero point phonons traveling in opposite direction for zero – momentum field transfer, which is the cause of phase coherence in magnetite [32] and it also requires a doubling of effective magneton number. The metallic behaviour of sample at a temperature below T_{MS} ,is also due to the mixed valence of Mn ion .

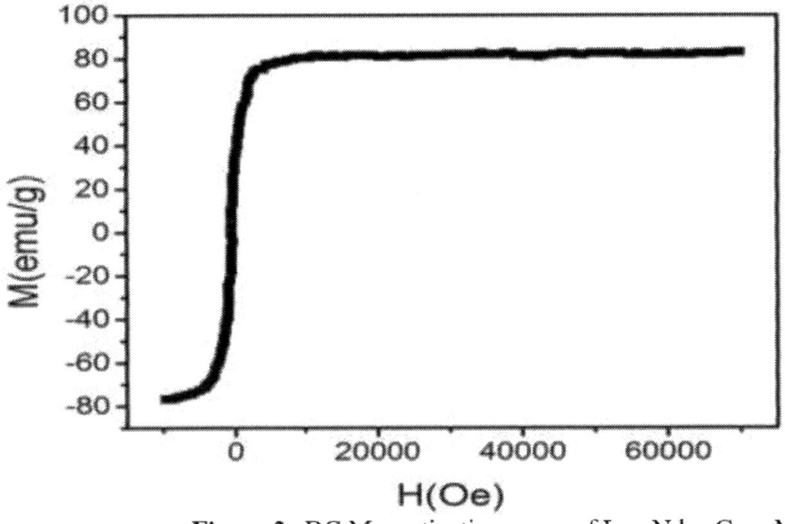

Figure 2. DC Magnetization curve of $La_{0.5}Nd_{0.15}Ca_{0.35}MnO_3$

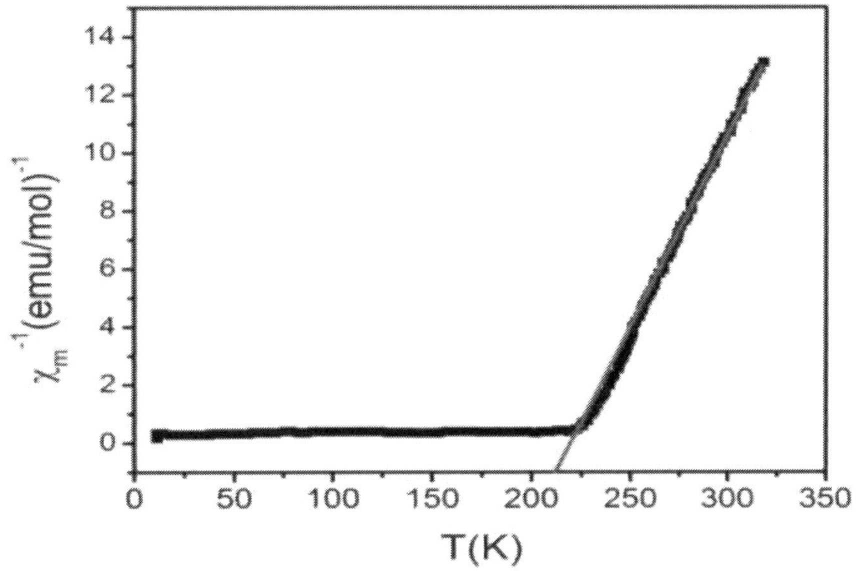

Figure 3. Plot of χ_m^{-1} with the temperature of $La_{0.5}Nd_{0.15}Ca_{0.35}MnO_3$ for T_C calculation

3.3 Electric Transport properties

Figure :4 represents the resistivity of the sample measured against temperature range (10K to 300K) in the presence of field (FC mode) and absence of field (ZFC mode) respectively . The resistivity measurements shown in the graph has good evidence of metal to semiconductor transition. The resistivity of the sample in the high temperature region of ρ-T graph shows an increase in resistivity with decrease in temperature, which is a clear indication of its semiconducting behaviour. Such a semiconducting behaviour has reported in many literature.[33,34]. But if we carefully observe resistivity of sample continuously during the decrease of temperature, the temperature coefficient becomes abruptly positive after a particular temperature and it is due to the metallic nature of the sample .This transition temperature is called metal to semiconductor temperature (T_{MS}). The magnitude of T_{MS} obtained from the graph is found to be close to T_C, which is a strong indication of mutual correlation of magnetic and electric transport properties of the prepared sample. This observations allow us to conclude that the sample is ferromagnetic- metallic at low temperature and paramagnetic-semiconductor at high temperature. The electric conductivity of the these sort of sample is mainly due to the hopping of electrons from the occupied Mn^{3+} site to unoccupied Mn^{4+}. The cation doping may change or tilt Mn-O-Mn bond angle and it may increase the hopping of electrons from the occupied Mn^{3+} site to unoccupied Mn^{4+}. This eventually might have decreased resistivity at low temperature.

There exist different models of conduction such as Arrhenius model, Small Polaron Hopping model and Mott's variable Range Hopping model to explain the electrical conductivity mechanism in perovskite structures .Figure 5 shows that the resistivity of prepared sample is very well fitted with Small Polaron Hopping model(SPH) and the equation of resistivity according to this model is $\rho = \rho_0 T exp^{\frac{E_P}{K_B T}}$, where E_P is the polaron hopping energy, T is the absolute temperature ,K_B is the Boltzmann constant and ρ_0 is the pre exponential part of resistivity. The polaron hopping energy can be calculated from the slope of the fitting curve between log (ρ/T) versus (1/T) and this values found to be good agreement with that of some of the previously reported cases.[34,35].

IOP Conf. Series: Materials Science and Engineering 1166 (2021) 012005 doi:10.1088/1757-899X/1166/1/012005

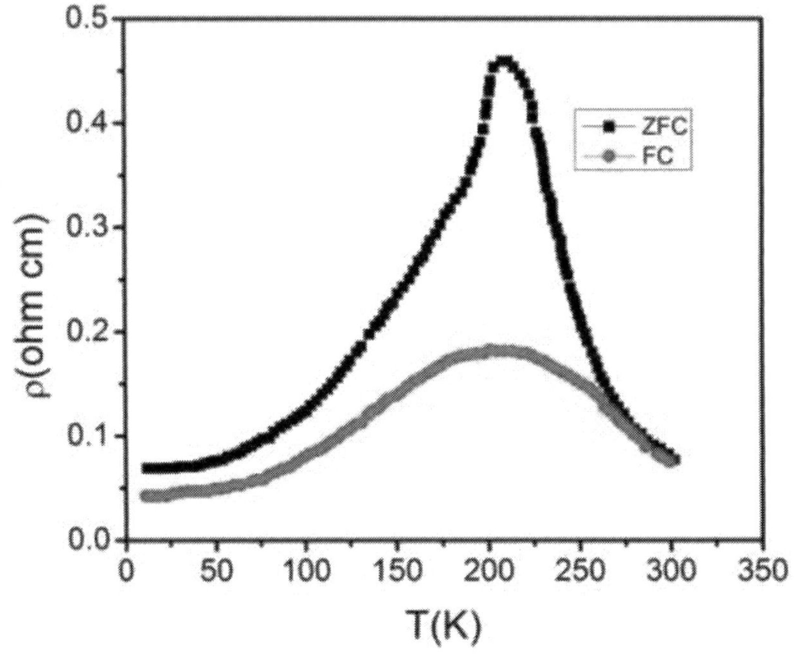

Figure 4. Resistivity versus temperature curve of $La_{0.5}Nd_{0.15}Ca_{0.35}MnO_3$

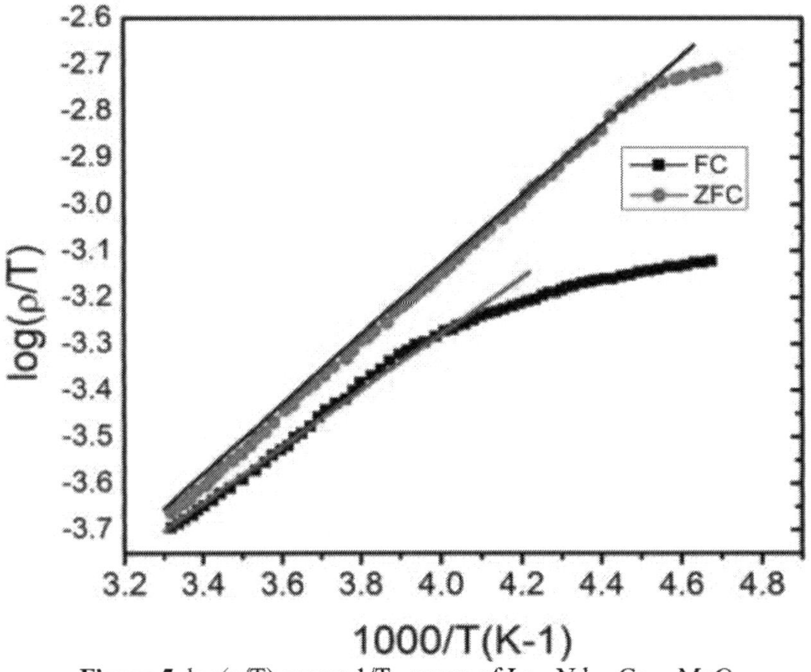

Figure 5. log(ρ/T) versus 1/T curve of $La_{0.5}Nd_{0.15}Ca_{0.35}MnO_3$

4. CONCLUSION

The conventional solid-state reaction method is used to prepare the cation doped sample $La_{0.5}Nd_{0.15}Ca_{0.35}MnO_3$. The structural, magnetic and electric transport properties of the sample has been investigated .The reitveld refinement of the XRD of sample revealed the formation orthorhombic perovskite structure in Pbnm space group symmetry. The correlated polaron model is used to explore the temperature dependence of the resistivity of the sample. The AC susceptibility and resistivity measurements revealed that the material has ferromagnetic-metallic behavior at low temperature and paramagnetic-semiconducting behavior at high temperature .The sharp decrease of magnetic susceptibility from its maximum value at a temperature T_C, is due to the Frank- Condon type of spin transfer of magnons of small order of energy. The charge transfer between Mn^{3+} and Mn^{4+}ions due to the exchange of phonon leads to the polaron type of electronic transport and magnetic order. Small Polaron Hopping model is found to be good to explain the electrical transport properties of the doped sample.

References

[1] Xue Q ZandZhang X 2003 *Phys. Lett. A* **313** 5-6
[2] Rabe KM, et al .2007*Physicsof Ferroelectrics: A Modern Perspective* (New York:Springer)
[3] Mitchel R H2002 *Perovskites, Modern and Ancient* (Canada: Almaz Press)
[4] King G and Woodward P M 2010 *J. Mater. Chem.***20** 5785-5796
[5] Akbas MA, Davies P K 1998 J. Amer. Ceramic Society**81** 670-76
[6] Talanov V M, Talanov M V, and Shirokov V B2014 *Crystallography Reports* ***59*** *650–661*
[7] Bingham NS, Lampen P, Phan M H, Hoang T D, Chinh H D, Zhang CL, CheongandS W, and SrikanthH. 2012 *Phys. Rev. B***86** 064420
[8] Nasri M, Trikia M, Dhahri E, Hlil E K and LachkarP 2013 *J. Alloys Compd.***576**404–408
[9] Xue Q Z, Zhang X, and Zhu D D2003 *Physica B***334** 216-220
[10] King G and Woodward P M2010 *J. Mater. Chem.***20** 5785-5796
[11] Howard C J and Stokes H T 1998 *ActaCrystallogr. Sect. B: Struct. Sci.* **54** 782-789
[12] Woodward P M1997 *ActaCrystallogr. Sect. B: Struct. Sci.* **53** 32-43
[13] LufasoM W and Woodward P M 2001 *ActaCrystallogr. Sect. B: Struct. Sci***57** 725-738
[14] Glazer A M 1972 *ActaCrystallogr.Sect. B: Struct. Crystallogr. Cryst.Chem.* **28** 3384-3392
[15] Eng HW, Barnes P W, Auer B M, and Woodward P M 2003 *J. Solid State Chem.***175** 94-109
[16] Zhou J S and Goodenough J B 2006 *Phys. Rev. Lett.* **96** 247201
[17] Kim B H and Min B I2009 *Phys. Rev. B: Condens. Matter Mater. Phys.***80** 064411-16
[18] TakataK, Azuma M, Shimakawa Y and Takano M 2005 *FuntaioyobiFunmatsu Yakin***52** 913–917
[19] Matar S F, Subramanian M A, Villesuzanne A, Eyert V and WhangboM H 2007 *J. Magn. Magn. Mater***308** 116–119
[20] Stramare S, Thangadurai V and Weppner W 2003 *Chem. Mater.***15** 3974–3990
[21] Von HelmholtR,et al. 1993 *Phys.Rev.Lett.***71** 2331-2333
[22] Jin S,et al. 1994 *Science***264** 413-415
[23] Khomskii D I and Sawatzky G A 1997 *Solid State Communications***102** 87-99
[24] JonkerG and van Santen 1950 *J. Physica***16** 337-349
[25] Kramers H A 1934 *J.Physica***1** 182-192
[26] Coey J M D, Viret M and Ranno L 1995 *Phys. Rev. Lett.* **75** 3910-3913
[27] Ramesha K, Smolyaninova V N, Gopalakrishnan J, and Greene R L 1998 *Chem. Mater* **10**1436-1439
[28] Volkov*a*M P and Stepanov*a*N N 2019 *Physics of the Solid State***61** 1416-1419
[29] Maatar S C, Nassri R.M, Koubaa W C, Koubaa M, and Cheikhrouhou A 2015 *J. Solid State Chem***225** 83-88
[30] Singh D and Mahajan A 2015 *J. Alloys Compd. 644*172-179
[31] Singh S, and Singh D J. 2017 *Alloys Compd.***702** 249–257
[32] Appel J 1968 *Solid State Phys.***21**, 31

[33] ShimuraT, Hayashi T, Inaguma Y, Itoh M 1996 *J. Solid State Chem.* **124** 250-263
[34] Lakshmi Y K, Venkataiah G, VithalMandReddy P V 2008 *Phys. B* **403** 3059–3066
[35] S. Bhattacharya, R.K. Mukherjee and B.K. Chaudhuri 2003 *Appl. Phys. Lett.* **82** 4101–4103

Impact of Ground Nut Shell Ash on Cobalt-Chromium metal matrix composites synthesized using Powder metallurgy process.

G R Raghav[1], Suraj R[2*], Sheeja Janardhanan[3], Vidya Chandran[4], K J Nagarajan[5] and Nikhil Asok[6]

[1,2,3,4,6]Department of Mechanical Engineering, SCMS School of Engineering and Technology, Ernakulam, Kerala, India.
[5] Department of Mechanical Engineering, KLN College of Engineering, Pottapalayam, Sivagangai Dt. Tamil Nadu, India 630612

**Corresponding Author email id:* *surajr@scmsgroup.org*

Abstract. Co-based composites are extensively utilized in the field of prosthesis and dental implants. Hybrid composites made using Powder metallurgy process, Co-10Cr-GNSA were studied. The surface morphology of the hybrid composites were studied using Scanning Electron Microscope. The elemental analysis was carried out using X-Ray Diffraction technique. The hybrid composites were analyzed for its various mechanical properties like microhardness, compressive strength, and density. Value of micro hardness of the composite materials showed slight improvement with addition of GNSA reinforcement. The value of density of the hybrid composites was found to be decreasing linearly with the addition of GNSA. Compressive strength of the materials showed a reasonable increment. Wear analysis to study the tribological characterization of the hybrid composites were done with the help of a pin on disc wear testing machine. The wear and COF studies show that with a rise in GNSA content, wear resistance increases because of the presence of oxides of GNSA particles. From the worn out surfaces of the hybrid composite it is concluded that the deformation of the composites takes places initially due to abrasive wear followed by plastic deformation. An electrochemical workstation was used to understand the corrosion characteristics of the hybrid composites in the presence of 3% NaClelectrolyticsolution.Co-5Cr-5GNSA hybrid composites exhibit better electrochemical corrosion resistance compared to other specimens.

Keywords: Powder metallurgy, Wear, Corrosion, GNSA

1. Introduction

Now a days more and more people suffer from osteoarthritis disorder, which makes them experience severe pain and discomfort. Recent survey suggests that there are nearly 50 million cases worldwide who are suffering from osteoarthritis disorder and in need of joint replacement surgery [1]. Co-Cr-Mo alloy is the extensively used artificial prosthetic material considering its higher value of wear, hardness and corrosion resistance. Even though Co-Cr-Mo alloys are excellent prosthetic material, still there are certain disadvantages such as wear of implants in the hip joints and problems related to bio compatibility since Mo is not a bio degradable material[2–5]. Therefore it is the need of the hour to produce a composite with much better wear and corrosion resistance which is also bio degradable and compatible to human body.

The ground Nut Shell Ash (GNSA) which is primarily a biological waste and is available in abundance all over the world. Moreover the GNSA particles have presence of MgSiO3 and AlSiO3 in high concentration .Hence it can be used to replace the hazardous Mo reinforcements[6].

There are many conventional methods to produce wear resistance artificial prosthetic implants such as plasma spraying, physical vapor deposition, electro deposition and chemical vapour deposition. Since these manufacturing processes includes more complex steps and requires costly equipments, the cost of the implants is high. The powder metallurgy technique has its own advantages which include uniform dispersion, low processing cost and ability to manufacture high melting point materials. Hence the powder metallurgy process possesses great potential for producing Co-Cr based hybrid composite materials with highly desirable mechanical properties along with wear and corrosion resistance[7–13].

This work aims to develop a Co-Cr-GNSA hybrid composite material with better wear, corrosion resistance and mechanical properties. In this study, four different compositions based on weight percent is formulated as follows Co-10Cr, Co-10Cr-2.5GNSA, Co-10Cr- 3.5 GNSA and Co-10Cr-5GNSA. The composite powders are mechanically milled and compacted and sintered in order to develop specimens of 8mm cylindrical pellets. The hybrid composites are then studied in order to explore their morphological properties using SEM. The mechanical behavior along with tribological and corrosion resistance behavior were studied and their mechanisms were reported.

2. Materials and Method

The materials CoCr (99.5% purity) which is used in this study were purchased from Mepco Ltd Tamil Nadu, India. The ground nut shell ash (GNSA) powder used in this work is prepared using heat treatment method which is discussed in our pervious paper [6]. Mechanical ball milling process was used for alloying the Co-Cr- GNSA hybrid composites. The process was carried out for two hours and was then compacted into 8 mm diameter pellet which is cylindrical in shape. The value of compaction pressure was set to 750 MPa consistently. After this, the soft green compacts were hardened by forcing them to sintering process at 1000oC for 2h.The morphology of the hybrid composites were studied using a Field Emission Scanning Electron Microscope (FE-SEM).ASTM: B962-13 standards were used to calculate the density of the Co-Cr- GNSA hybrid composites. The ASTM E384 standards were used to study the micro hardness of the hybrid composite pellets at a uniform load and dwell time of 1 kgf and 10 seconds respectively. Compressive strength of the hybrid composites were studied at a scan rate of 5 mm/min, with the help of a Universal Testing Machine (UTM). ASTM G99-05 standards were used to study the wear and friction behavior of the composites. EN 32 steel of hardness 65 HRC was used for the analysis. The specimens were cleaned using acetone solution before and after the wear test. The wear analysis of the composites was done at various sliding conditions such as the load, sliding distance and sliding speed. The electrochemical corrosion tests were simulated on a three electrode workstation using 3% NaCl solution as electrolyte[14–16].

3. Results and Discussion

3.1 Field Emission Scanning Electron Microscope Analysis

FE-SEM images of Co-10Cr- 3.5 GNSA & Co-10Cr-5GNSA hybrid Composites respectively are shown in Figure 1. There is a homogenous mixture of GNSA Particles with Co and Cr particles. The wettability of the GNSA particles was the major factor in achieving uniform amalgamation. It can be noted that due the milling operation the size of Cr particles have reduced to around 500 nm in size and are bonded strongly with Co matrix.

3.2 Microhardness

The microhardness test was done using a Vickers Micro Hardness Testing Machine with the test being conducted at five different points. Figure 2 shows the variation in the average value of microhardness of the composites at different configurations based on its composition, i.e. Co-10Cr, Co-10Cr-2.5GNSA, Co-10Cr- 3.5 GNSA and Co-10Cr-5GNSA. The microhardness of the composites varied from 320 HV to 340 HV. The hardness of Co-10Cr was found to be 320 HV and the introduction of GNSA resulted in an increase in the microhardness. The maximum microhardness was found to be in Co-10Cr-5GNSA composite with a value of 340 HV. The uniform amalgamation of GNSA particles was the major reason for this improvement in microhardness.

3.3 Compressive Strength and Density

With the addition of the GNSA reinforcement the density of the Co-10Cr –GNSA hybrid composites were found to be decreasing. The value of density for Co-10Cr composites was recognized as 8.1 g/cm3 whereas the density of the Co-10Cr-5GNSA hybrid composites were around 7.65 g/cm3 as shown in Figure.3. This reduction in density was attributed by the relatively soft nature of the GNSA particles. With the addition of GNSA particles, the compressive strength of the hybrid composite materials showed slight increase in its value. Figure.3 helps us understand the compressive strength of different combinations of Co-10Cr-GNSA hybrid composites. The compressive strength of Co-10Cr composite was established to be in the region of 380 MPa. The compressive strength has slightly increased to 401 MPa for the Co-10Cr- 5 GNSA hybrid composites which is due the presence of AlSiO3 particles in the GNSA ash content.

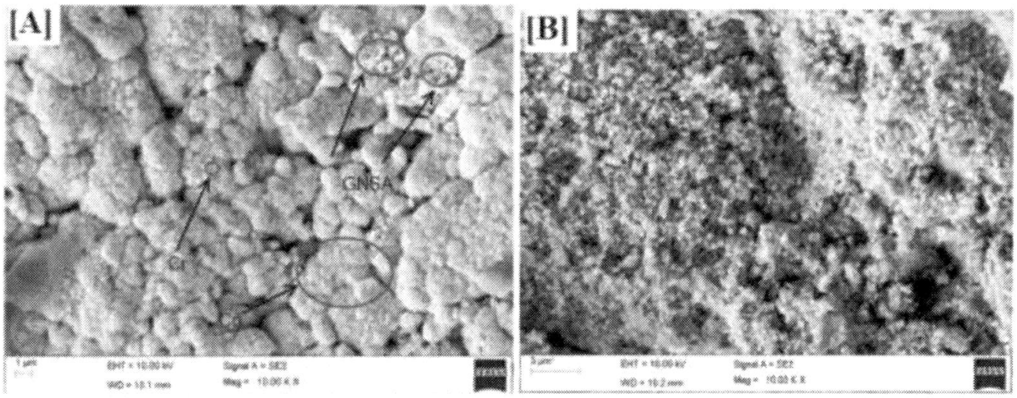

Figure 1.FESEM images of Co-10Cr- 3.5 GNSA & Co-10Cr-5GNSA hybrid Composite.

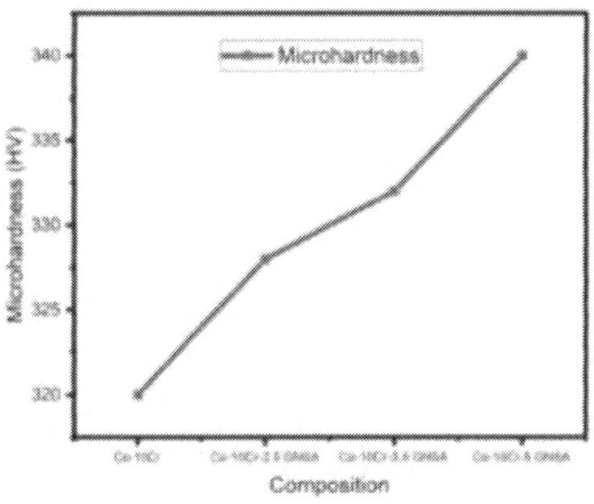

Figure 2. Graphical Representation of Co-10Cr-GNSA hybrid composites.

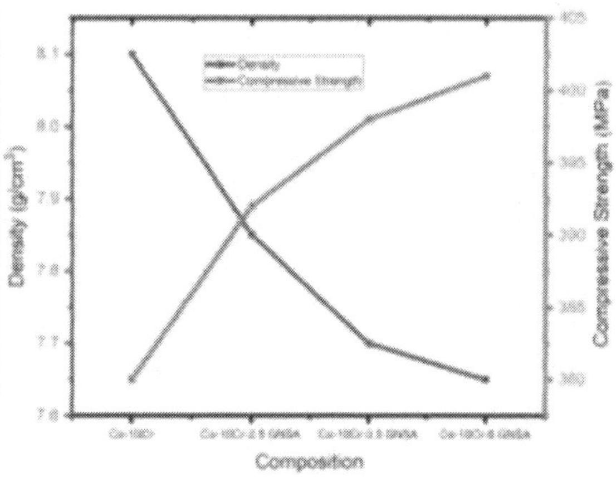

Figure 3. Comparison of Density and Compressive Strength of the Co-10Cr-GNSA hybrid composites.

3.4 Wear and COF Analysis

The loss of material due to wear of the Co-10Cr-GNSA hybrid composites is shown in Figure 4. The variation of wear loss of Co-10Cr-GNSA hybrid composites is depicted as graphical plots. The Figure 4 (A) indicates the wear analysis data of the Co-10Cr-GNSA hybrid composites at different loads (10N,15Nand 20N). The sliding speed (1.5 m/s) and sliding distance (1000 m) were kept constant. The Co-10Cr-5GNSA hybrid composites have witnessed very minimal wear loss at all loading conditions. The wear loss of Co-10Cr-GNSA hybrid composites at various sliding distance and speed is shown in Figure 4 (B&C) respectively. The wear loss has experienced similar trend.With the increase in GNSA concentration in the matrix there is definite resistance to wear and thereby the wear loss is very minimal for the Co-10Cr-5GNSA hybrid composites. The variation in coefficient of friction at different loads, sliding distance and sliding speed for Co-10Cr-GNSA hybrid composites is depicted in Figure5(A,B&C). It was observed that with an increase in load, the COF of the hybrid composites increased. Whereas, it reduced with an increase in sliding speed. Overall the Co-10Cr-5GNSA hybrid composites displayed better COF value. This improvement in Wear and friction characteristics is may be attributed to the presence of AlSiO3compounds in the composite material and also due to the tribo oxide surface layer formation on the surface of the composite specimen.The worn out surface analysis of the Co-10Cr-GNSA hybrid composites after wear analysis is represented in Figure 6. From the worn our surface analysis it can be concluded that there is plastic deformation experienced in hybrid composites which is preceded by abrasive wear.

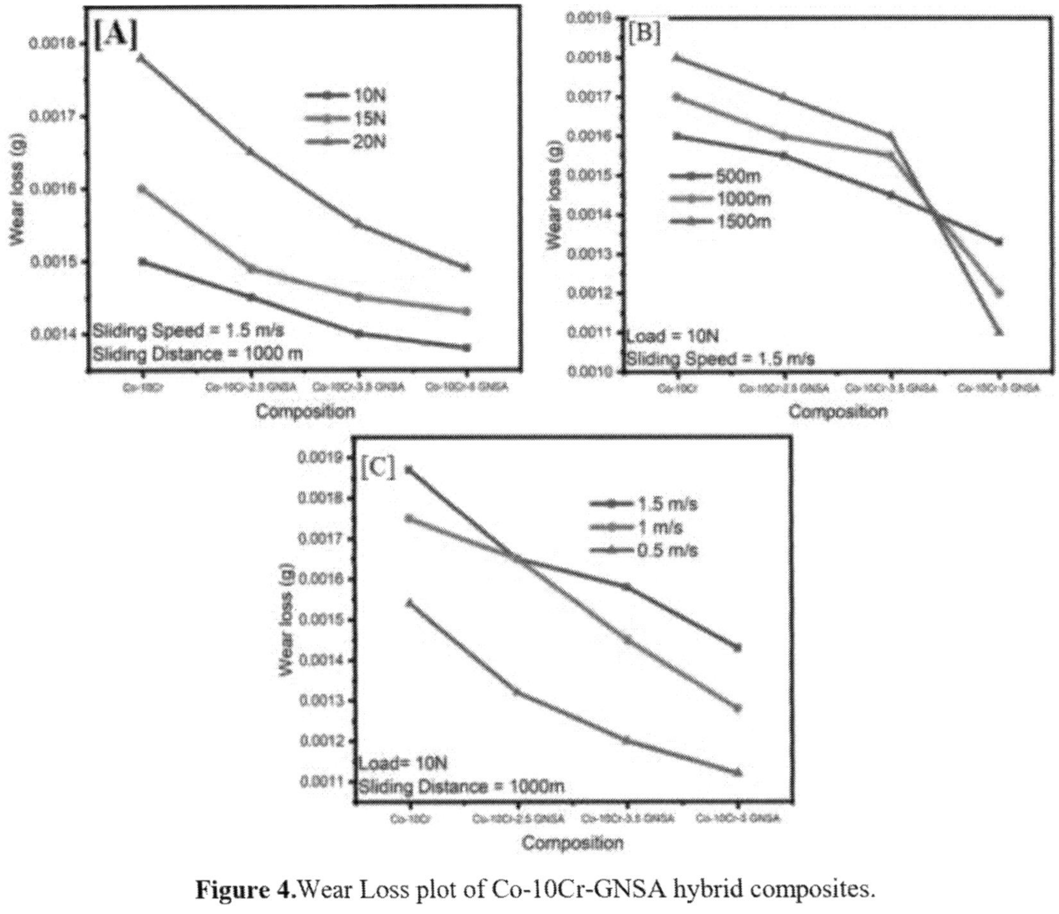

Figure 4. Wear Loss plot of Co-10Cr-GNSA hybrid composites.

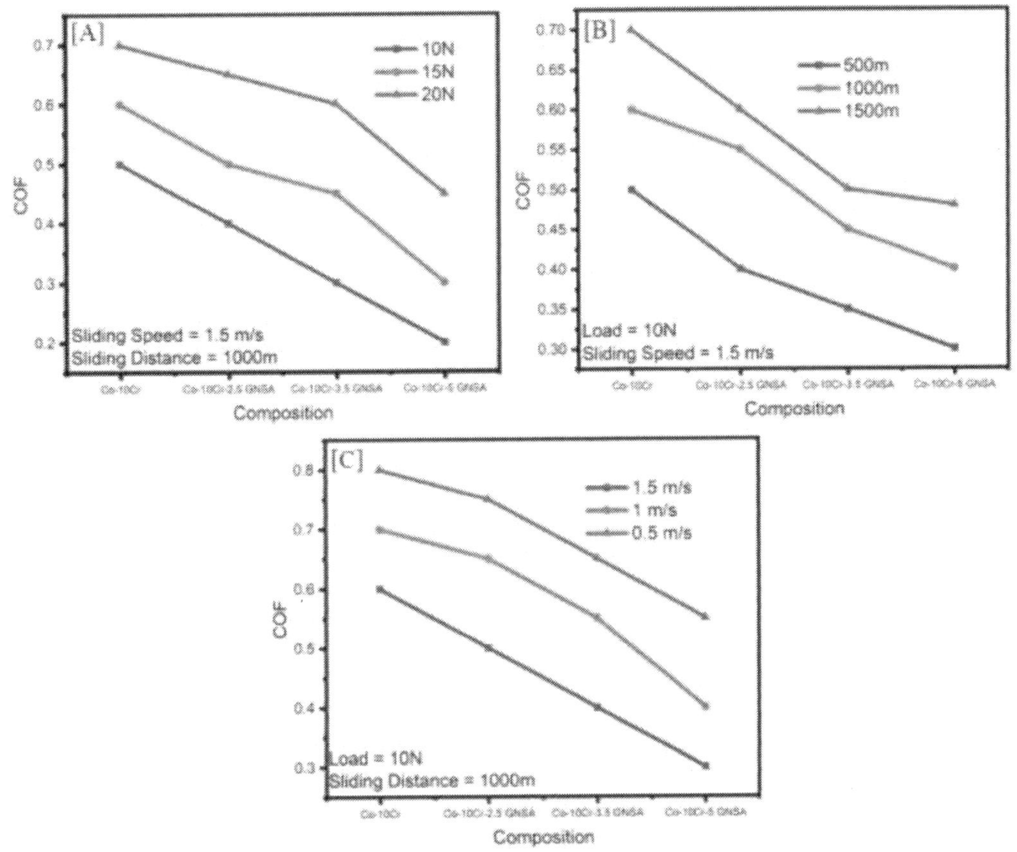

Figure 5.COF plot of Co-10Cr-GNSA hybrid composites.

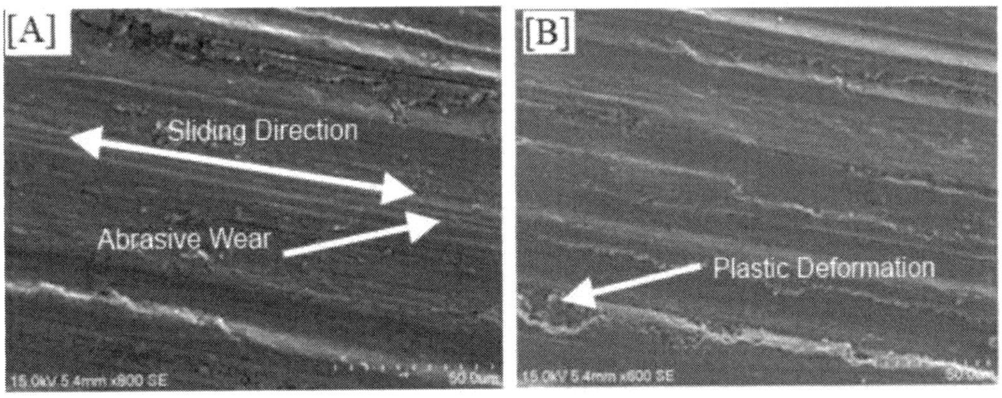

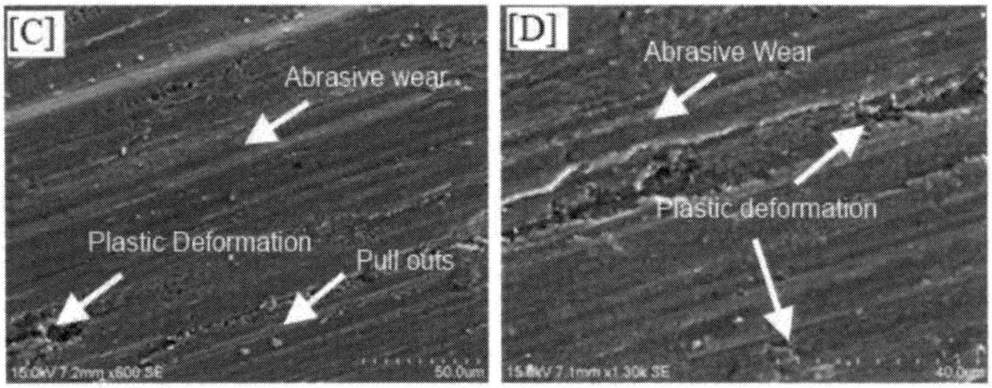

Figure 6. Worn out Surface analysis of Co-10Cr-GNSA hybrid composites.

3.5 Electrochemical Corrosion Analysis.

The corrosion analyses of the Co-10Cr-GNSA hybrid composites were done using an electrochemical work station with three electrodes. The electrolyte which was used in this study is 3% NaCl solution. The polarization curves are obtained by using tafel extrapolation methods as shown in Figure.7. The test results exhibit that the corrosion potential value, Ecorr and the corrosion current value, ICorr of Co-10Cr-5GNSA hybrid composites was found to be better compared to other combinations of hybrid composites. The Ecorr value of Co-10Cr-5GNSA hybrid composites was found to be -0.419 V and Icorr value was around -0.12 mA/cm^2. The corrosion performance of Co-10Cr-3.5 GNSA was also similar to that of Co-10Cr-5GNSA hybrid composites. The Co-10Cr composite shows lesser corrosion resistance than the hybrid composites as shown in Table.1.

Table 1. Tafel plot fallouts of Co-10Cr-GNSA hybrid composites.

S.No	Specimen	E_{corr} (*V*)	I_{corr} (*mA/cm^2*)
1	Co-10Cr	-0.442±0.051	0.5±0.020
2	Co-10Cr-2.5 GNSA	-0.437± 0.044	0.4±0.011
3	Co-10Cr-3.5GNSA	-0.420±0.021	-0.1±0.003
4	Co-10Cr-5GNSA	-0.419±0.0191	-0.1±0.002

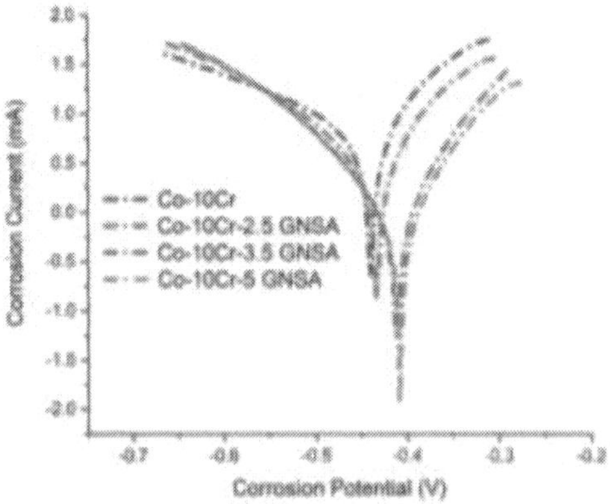

Figure 7. Potentiodynamic polarization plot of Co-10Cr-GNSA hybrid composites

4. Conclusions

The Co-10Cr-GNSA hybrid composites were studied and their r mechanical, Wear and corrosion mechanisms were reported.

- The addition of GNSA reinforcement resulted in an increment in the Microhardness of the Co-10Cr-5GNSA hybrid composites (340 HV) compared to Co-10Cr composites.
- The compression strength of the Co-10Cr-5GNSA hybrid composites (401 MPa) has improved considerably than the Co-10Cr composites.
- The value of density of the Co-10Cr-5GNSA hybrid composites showed a considerable decrement due to the addition of less dense GNSA reinforcement.
- The Co-10Cr-5GNSA hybrid composites exhibited a higher resistance to wear.
- Corrosion resistance of Co-10Cr-5GNSA hybrid composites was found to be better than the Co-10Cr composites from the electrochemical corrosion analysis.

References:

[1] Han Y, Liu F, Zhang K, Huang Q, Guo X, Wang C. *A study on tribological properties of textured Co-Cr-Mo alloy for artificial hip joints*. Int J Refract Met Hard Mater 2021;**95**:105463. https://doi.org/https://doi.org/10.1016/j.ijrmhm.2020.105463.

[2] Marques FP, Scandian C, Bozzi AC, Fukumasu NK, Tschiptschin AP. *Formation of a nanocrystalline recrystallized layer during microabrasive wear of a cobalt-chromium based alloy (Co-30Cr-19Fe)*. Tribol Int 2017;**116**:105–12. https://doi.org/10.1016/j.triboint.2017.07.006.

[3] Yamanaka K, Mori M, Torita Y, Chiba A. *Impact of minor alloying with C and Si on the precipitation behavior and mechanical properties of N-doped Co–Cr alloy dental castings*. Mater Sci Eng C 2018: **92**:112-120.. https://doi.org/10.1016/j.msec.2018.06.035.

[4] Zhou Y, Li N, Yan J, Zeng Q. *Comparative analysis of the microstructures and mechanical properties of Co-Cr dental alloys fabricated by different methods*. J Prosthet Dent 2018:1–7. https://doi.org/10.1016/j.prosdent.2017.11.015.

[5] Rodrigues WC, Broilo LR, Schaeffer L, Knörnschild G, Romel F, Espinoza M. *Powder metallurgical processing of Co – 28 % Cr – 6 % Mo for dental implants : Physical , mechanical and electrochemical properties*. Powder Technol 2011;**206**:233–8. https://doi.org/10.1016/j.powtec.2010.09.024.

[6] Raghav GR, Muthu Krishnan D, Sundar R, Ashokkumar R, Nagarajan KJ. *Investigation on mechanical, wear and corrosion properties of Fe-Co-Cr-W-GNSA hybrid composites synthesized using powder metallurgy process*. Eng Res Express 2020;**2**. https://doi.org/10.1088/2631-8695/ab9517.

[7] Gopinath S, Prince M, Raghav GR. *Enhancing the mechanical, wear and corrosion behaviour of stir casted aluminium 6061 hybrid composites through the incorporation of boron nitride and aluminium oxide particles*. Mater Res Express 2020;**7**:016582. https://doi.org/10.1088/2053-1591/ab6c1d.

[8] Prakash C, Singh S, Verma K, Sidhu SS, Singh S. *Synthesis and characterization of Mg-Zn-Mn-HA composite by spark plasma sintering process for orthopedic applications*. Vacuum 2018;**155**:578–84. https://doi.org/10.1016/j.vacuum.2018.06.063.

[9] Elkhoshkhany N, Hafnway A, Khaled A. *Electrodeposition and corrosion behavior of nano-structured Ni-WC and Ni-Co-WC composite coating*. J Alloys Compd 2017;**695**:1505–14. https://doi.org/10.1016/j.jallcom.2016.10.290.

[10] Stewart DA, Shipway PH, Mccartney DG. *Abrasive wear behaviour of conventional and nanocomposite HVOF-sprayed WC – Co coatings*,Wear 1999;**225-229**:789–798.

[11] Liu C, Su F, Liang J. *Nanocrystalline Co-Ni alloy coating produced with supercritical carbon dioxide assisted electrodeposition with excellent wear and corrosion resistance*. Surf Coat Technol 2016;**292**:37–43. https://doi.org/10.1016/j.surfcoat.2016.03.027.

[12] Bajat JB, Vasilic R. *Corrosion Stability of Oxide Coatings Formed by Plasma Electrolytic*

Oxidation of Aluminum : Optimization of Process Time 2013;**69**:693–702.

[13] Gadow R, Killinger a., Stiegler N. *Hydroxyapatite coatings for biomedical applications deposited by different thermal spray techniques*. Surf Coatings Technol 2010;**205**:1157–64. https://doi.org/10.1016/j.surfcoat.2010.03.059.

[14] Raghav GR, Balaji AN, Selvakumar N, Muthukrishnan D, Sajith E. *Effect of tungsten reinforcement on mechanical, tribological and corrosion behaviour of mechanically alloyed Co-25C Cermet nanocomposites*. Mater Res Express 2019;**6**. https://doi.org/10.1088/2053-1591/ab4f0a.

[15] Raghav GR, Balaji AN, Muthukrishnan D, Sruthi V, Sajith E. *An experimental investigation on wear and corrosion characteristics of Mg-Co nanocomposites*. Mater Res Express 2018;**5**:066523. https://doi.org/10.1088/2053-1591/aac862.

[16] Toptan F, Alves AC, Kerti I, Ariza E, Rocha LA. *Corrosion and tribocorrosion behaviour of Al-Si-Cu-Mg alloy and its composites reinforced with B4C particles in 0.05M NaCl solution*. Wear 2013;**306**:27–35. https://doi.org/10.1016/j.wear.2013.06.026.

Simple and cost-effective synthesis of activated carbon@few layers of graphene composite electrode for supercapacitor applications

Manoranjan Mandal[1], Subhasri Subudhi[1], Injamul Alam[1], BVRS Subramanyam[1], Santosini Patra[1] and Pitamber Mahanandia[1*]

[1]Department of Physics & Astronomy, National Institute of Technology, Rourkela, 769008, India

[*]***E-mail:*** *pitam@nitrkl.ac.in*

Abstract. We report a simple and low-cost technique to prepare a composite comprising of activated carbon and few layers of graphene sheets and fabricated as electrode material for supercapacitor (SC) applications. The composite is characterised by using Raman spectroscopy, powder X-ray diffraction (XRD), scanning electron microscope (SEM) and current-voltage (I-V) measurements. The synthesized composite is investigated by using galvanostatic charge-discharge (GCD) and cyclic voltammetry (CV) measurements in 3M KOH aqueous electrolyte to evaluate its electrochemical performance. The composite results high specific capacitance from 173 to 564 F/g at different scan rates (100-5 mV/s), analysed by CV and the capacitances resulted in GCD measurement were 196 to 587 F/g at various current densities (5-1 A/g). The cycle stability of the as-prepared AC/graphene composite is also tested and observed that 89% of capacitance is retained even after 5000 cycles which reveals the excellent long term charge-discharge stability of the prepared composite. Therefore, the composite performs very well in storage as well as life cycling. The obtained results suggest that AC/graphene composite can be a suitable electrode material for high-performance SC.

Keywords: Activated carbon, graphene, cyclic voltammetry, supercapacitor, scanning electron microscope

1. Introduction

The enormous population growth and development of portable electronic devices, a novel storage device with high energy and power efficient is the crucial demand. Supercapacitors (SCs) are the electric double-layer capacitor (EDLC), that bridges the gap between conventional batteries and traditional dielectric capacitors in terms of power/energy. It has high power density, fast charge/discharge rate, low equivalent series resistance, operation in extended temperature range and long cycle life (> 100000 cycles) that are essential for practical applications [1-5]. Therefore, SC has potential applications such as in industrial power supplies, hybrid electric vehicles, consumer electronics and memory back-up systems. However, the SC fall behind in many potential applications due to its significantly low energy density. Therefore, enhancement of energy short of dropping its power density for SC device is the most important challenge for the researcher now a day. People have given many efforts to design the SC devices taking various electrode materials to enhance the overall energy and power performance. SCs devices fabricated by carbon-based electroactive materials, metal oxide/hydroxide and electrochemical conducting polymers [6] are being investigated extensively. Although low cost, ease of fabrication, high capacitance and good conductivity have the advantages of electrochemical conducting polymers, however, they have very poor cycle life and relatively low mechanical stability. Similarly, metal oxide/hydroxide based single electrode pseudocapacitor show improved capacitance as well as energy performance, however, high materials fabrication cost, poor conductivity and degradation of the materials during redox chemical reaction are the serious concern. This limits the applications of oxide/hydroxide based pseudocapacitor in practical applications. On the other hand, carbon-based materials including carbon nanotubes [7, 8], activated carbon [9-11] and graphene [12-18] are being mostly used in EDLC because of their outstanding physical and chemical properties. Among various carbon materials, activated carbon (AC) is widely used as promising electrode material in SCs due to its large specific surface area around 500-3500 m^2/g, high packing density and cost-efficient. Nevertheless, the commercial application of AC limited to only small markets due to their poor electrical conductivity and low mesoporosity which produces a long diffusion pathway for ions which results in limited energy storage as well as very low rate capability. These drawbacks make AC unfavourable for the excellent cyclebility and process of rapid charge/discharge. So controlling narrow pore size distribution and increasing electrical conductivity without losing packing density and surface area are still challenging for AC.

Due to its excellent chemical stability, high surface area (2675 m^2/g), the high theoretical value of specific capacitance (up to ~ 550 F/g), good thermal and electrical properties, graphene has been given special attention towards SCs. However, experimentally the intrinsic capacitance (21 μF) of graphene was far away from the theoretical value [19]. Many efforts have been made to enhance the capacitance of SC based on modified graphene and graphene [20-22]. Unfortunately, based on graphene the experimentally obtained specific capacitance value far lower than the theoretical one due to the unavoidable tendency of graphene sheets to form restacking of layers during the preparation and consequent production processes that result in loss of their surface area and hence only the bottom and the top surface of graphene can accumulate electrolyte ions, which results into lower the capacitance as the area of the stacked surface of graphene could not be used [23]. The low packing density i.e. 0.005 g cm^{-3} is one more limitation of graphene that affect the volumetric capacitance of electrode materials. Also, due to its exceptional sheet-like morphology simultaneously achieving high surface area and volumetric density is very difficult for graphene. Therefore, enhancement of packing density without losing a surface area of graphene is highly necessary and imperious. However, the restacking tendency and low packing density of graphene could be overcome by integrating with AC and simultaneously the electrical conducting nature of AC could be enhanced and hence, better storage performance can be achieved in the composite electrode. Zheng et al. have prepared porous graphene/activated carbon composite where they found a maximum capacitance of 210 F/g and 103 F/g in aqueous electrolyte and organic electrolyte [24]. Chen et al. have prepared graphene–activated carbon composite by chemical activation and found maximum specific capacitance 122 F/g in an aqueous electrolyte [25]. Yu et al. have synthesized KOH activated carbon/graphene nanosheets composites and they have found maximum capacitance of 173 F/g and 205 F/g in organic and aqueous electrolyte [26].

Taking motivation of above reports, herein we have prepared AC/few layers of graphene (graphene) composite in a low cost as well as a simple method and achieved good cycle stability and the enriched storage performance for SCs electrode. Here graphene is being prepared using an electrochemical method where the few layers of graphene are achieved with mass scale and integrated with host AC, to be used in SC for commercial applications. By incorporating graphene over host AC, which not only increases the packing density but also avoids the restacking of graphene sheets by acting AC as a spacer between the sheets. Similarly, introducing graphene in host AC will especially increase the conductivity and reduces the long diffusion pathway of AC significantly. Moreover, the aqueous electrolyte has been used to get better cycle stability of AC/graphene composite electrode. The composite gives a specific capacitance of 564 F/g at 5 mV/s from CV and 536 F/g at 1 A/g from GCD in 3M KOH aqueous electrolyte. Therefore, the finding results claim that synthesized composite can be a suitable electrode material for SC devices.

2. Experimental

2.1 Used chemicals

The chemicals have been used without any further purifying. High-quality chemicals like nitric acid (HNO_3 with > 98% purity), oxalic acid (> 99 % purity) and graphite rods are procured from Sigma-Aldrich (India).

2.2 Synthesis of AC/graphene composite

Graphene has been prepared by the electrochemical method [27] in which the interaction between graphite rods and oxalic acid take place. The good quality of few layers of graphene sheets has been recovered after sonicating and filtering several times. Commercially available AC has been used after treating with HNO_3 for surface modification and to increase its porosity for better ions accumulation [28]. AC/graphene composite has been prepared by the sonication of AC and graphene individually for 3 hours to get the uniform dispersion. Then a mixture of AC and graphene (4:1 weight ratio) put into magnetic stirring for 12 hours at room temperature. Then using 100 mL of Teflon-line autoclave the mixture was closed and kept at 180^0C for 12 hours. The AC/graphene composite was obtained after drying in a muffle furnace at 110^0C for 24 hours.

2.3 Electrode fabrication

Three electrode set-up has been used where synthesized AC/graphene composite act as working electrode, the platinum wire is used as counter electrode and Ag/AgCl as a reference electrode in 3M KOH aqueous electrolyte for the electrochemical analysis of prepared composite. The working electrode was prepared as follows: at first AC/graphene was added with Nafion binder and carbon black in 8:1:1 weight ratio (to form a slurry) and the slurry (1mg) of prepared AC/graphene composite with carbon black and Nafion binder was pressed on Ni-foam of dimension 10mm ×10mm× 1mm and dried at 60^0C for 24 hours using the vacuum. Then the AC/graphene composite was investigated in an electrochemical workstation using both CV and GCD measurements.

3. Results and discussions

The microstructure of prepared AC/graphene composite, as well as AC and graphene, is characterised by using SEM shown in figure 1. AC shows a large number of interconnected pores and these pores are accumulating the electrolyte ions at the electrode surface (figure 1 (a)). Graphene shows that the layers are well exfoliated with orienting randomly and showing agglomeration of layers (figure 1(b)). Figure 1 (c) confirm that the graphene is well distributed over AC and some of the AC are found to be distributed in between the graphene layers.

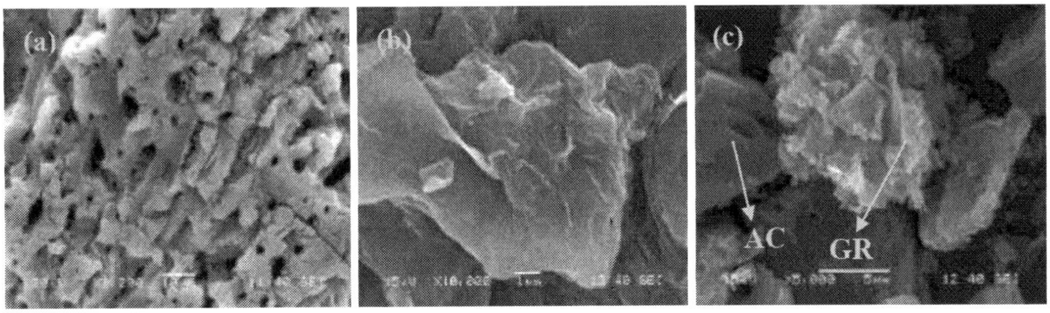

Figure 1. Surface morphology of (a) AC, (b) graphene and (c) AC/graphene composite.

The phase formation of synthesised materials such as AC, graphene and AC/graphene composite is characterised by using XRD shown in figure 2 (a). AC shows two broad peaks at $2\theta = 23.7^0$ and 43.5^0 with (002) and (101) plane which indicates that the graphite crystallite is present and also, confirm the commercial AC [29]. Graphene gives a very sharp peak at $2\theta = 26.78^0$ with (002) plane results in the presence of graphite crystallite with large crystallite size and also confirms the hexagonal structure. Two peaks at $2\theta = 26.34^0$ and 42.41^0 with (002) and (101) plane are found for AC/graphene composite where the peak at (002) plane is sharper i.e. less broadening than AC which reveals that the crystallite size increases in composite and further, reveals the presence of graphene in AC.

Raman spectroscopy study is very much essential to investigate the presence of structural defects and materials purity. Hence, the Raman spectra of AC/graphene composite, host AC and graphene have been shown in figure 2 (b). Two characteristics bands i.e. D (defect band) at ~ 1334 cm^{-1} and G (carbon band) at ~ 1583 cm^{-1} have been found for AC. Similarly, for graphene two bands i.e. D at ~ 1351 cm^{-1}, G at ~ 1587 cm^{-1} and the composite gives D at ~ 1353 cm^{-1}, G at ~ 1592 cm^{-1}. It has been found that by incorporating graphene on AC, both D and G peaks were slightly shifted towards the right probably due to the motion of carbon atoms in graphite layers. The ratio of D and G intensity (I_D/I_G) gives the materials purity and amount of structural defects. Lower the (I_D/I_G) ratio higher will be the purity of the material and less will be structural defects and vice-versa. The calculated (I_D/I_G) ratios for AC, graphene and AC/graphene composite are 1.06, 0.93 and 1.01. Therefore, by introducing graphene on AC the materials purity increases and structural defects decrease in the composite.

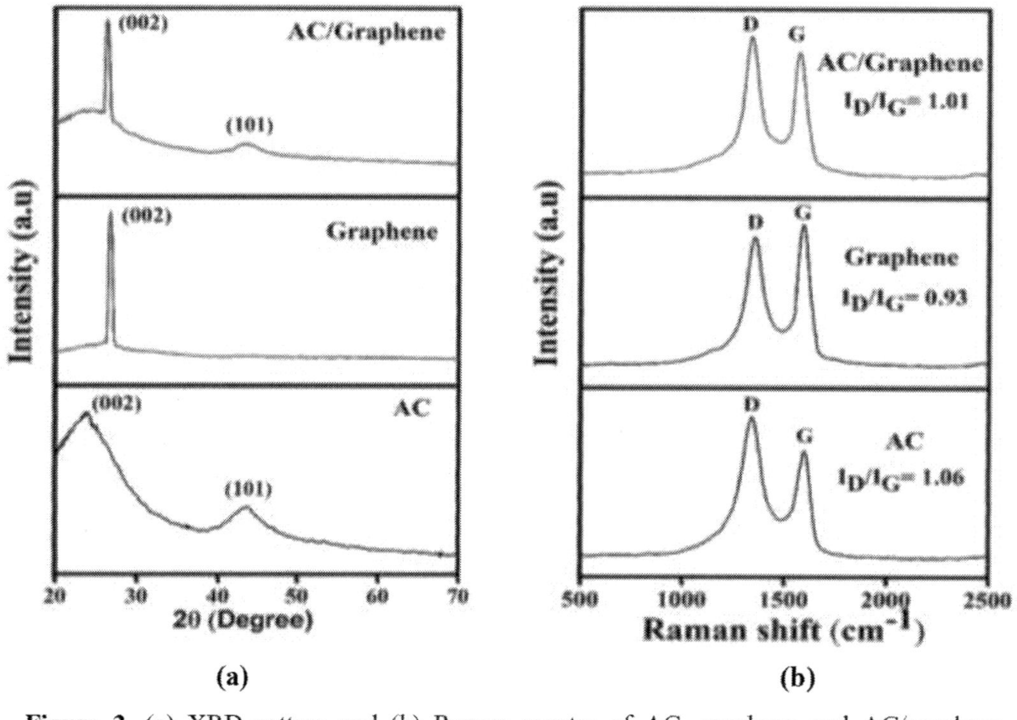

Figure 2. (a) XRD pattern and (b) Raman spectra of AC, graphene and AC/graphene

Figure 3 shows the change in current at various applied voltage of AC/graphene composite, AC and graphene with a maximum applied voltage of 7V. AC shows the linear change of current with applied voltage. It is observed that the I-V curve of AC looks like the current is not changing with applied voltage owing to a low current of AC i.e. 0.07mA at 7V compared to graphene and AC/graphene composite. Graphene shows Ohmic behaviour i.e. current changes linearly with voltage and the maximum current was 3.61mA at 7V. It has been found that AC/graphene composite gives a maximum current of 1.65mA at 7V. Therefore, by incorporating graphene on host AC the increase in current in AC is observed due to the highly conducting nature of graphene that results in an increase in the rate of ions transfer within the composite electrode.

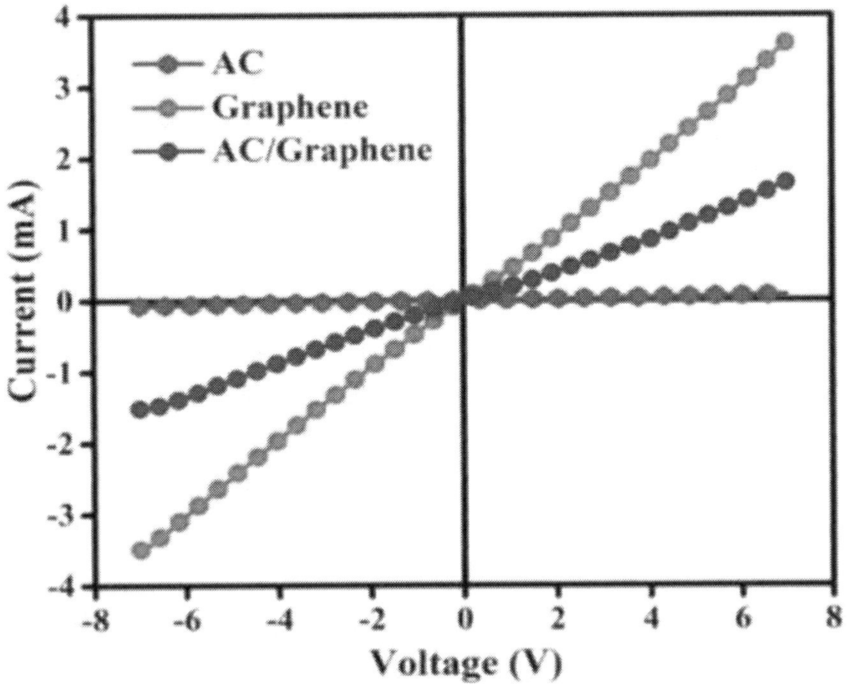

Figure 3. I-V curves of AC, graphene and AC/graphene composite at 7V.

The cell configuration in an electrochemical measurement acts a vital role to quantify the exact capacitance of electrode material for SCs application. Two electrode set-up gives more accurate results compared to three electrodes for asymmetric SC devices [30, 31]. However, we have studied in three-electrode set-up by using both CV and GCD in 3M KOH aqueous solution. Both CV and GCD have been used to evaluate the specific capacitance of prepared composite using Eq. (1, 2).

$$C_{sp} = \frac{Ai}{2.\Delta V.\frac{dv}{dt}.m} \quad (1)$$

$$C_{sp} = \frac{I.\Delta t}{m.\Delta V} \quad (2)$$

where Ai is the total area under CV curve in one cycle (AV), dv/dt is the scan rate (mV/s), C_{sp} is the specific capacitance (F/g) obtained from CV and GCD, ΔV is the potential window (V), Δt is the discharge time (S), m is the loading mass of material (mg), and I is the current density (A/g)

The CV images of AC, graphene and prepared AC/graphene composite at 5 mV/s in 3M KOH solution is shown in figure 4 (a) and the composite gives the CV curve with a large integrated area than the precursors (AC and graphene) which indicates the storage capacity of synthesised AC/graphene composite is decent. The CV curves of AC/graphene composite at different scan rate from 5 to 100 mV/s with a potential range within -0.1 to 0.6V is shown in figure 4 (b) and as the scan rate increases, the integrated area under the curve found to be increased which indicates the outstanding charge storage capacity of the composite [32]. Also, it has been found that the shape of CV images of the composite at various scan rate does not change when increasing the scan rate which confirms the fast transfer of charge takes place inside the AC/graphene electrode [33]. The CV curves of prepared composite demonstrate tends to rectangular shape (as expected for EDLC behaviour in carbon materials) without any redox peak is another indication of good storage behaviour of AC/graphene composite.

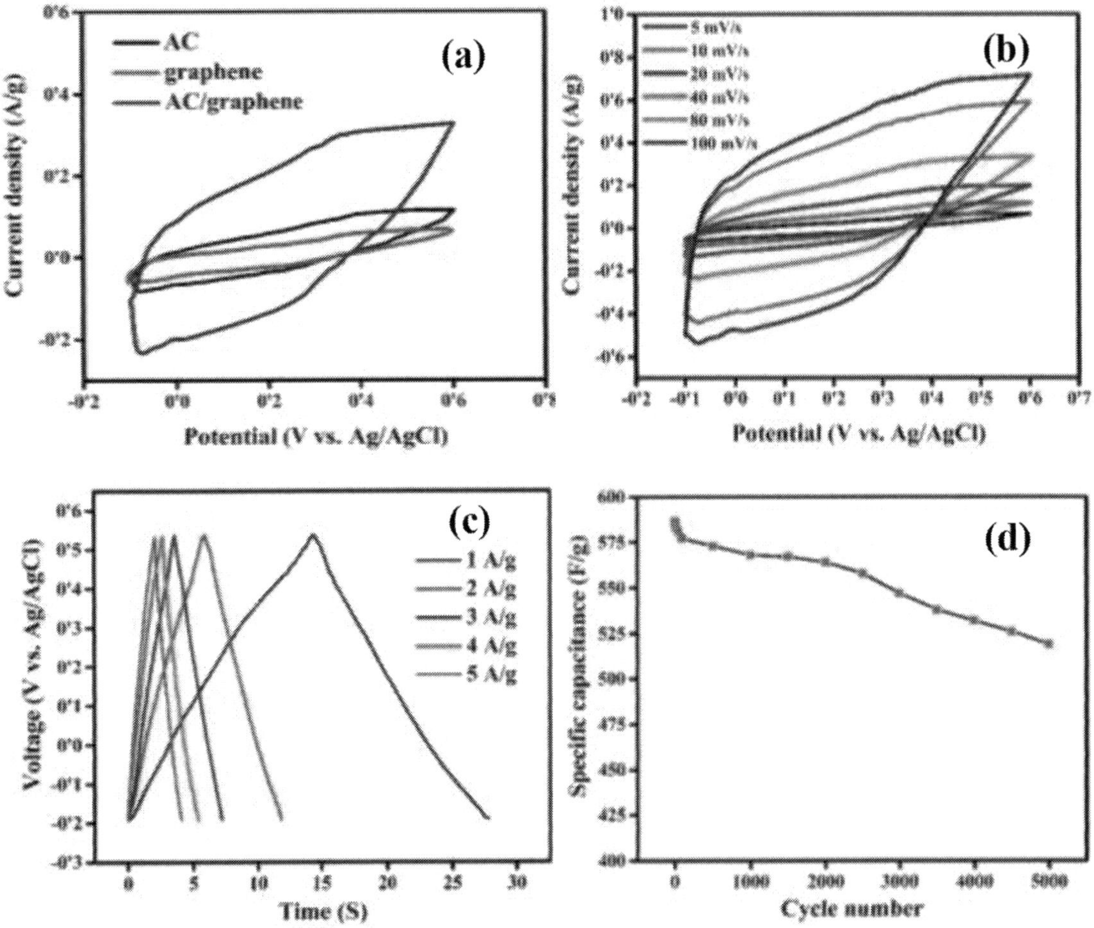

Figure 4. (a) CV images of AC, graphene and AC/graphene composite at scan rate of 5 mV/s, (b) CV images of AC/graphene composite at various scan rate (5-100 mV/s), (c) GCD images of AC/graphene composite at different current densities (1-5 A/g), and (d) Cycle stability of AC/graphene composite over 5000 cycles.

The composite has resulted in enhanced capacitances i.e. 564, 473, 377, 291, 215 and 173 F/g at 5, 10, 20, 40, 80 and 100 mV/s which were evaluated taking Eq. (1) from the CV. Hence, the maximum capacitance of 564 F/g at 5 mV/s is obtained for composite which is higher compare to AC (349 F/g) and graphene (296 F/g). The composite electrode showed enhanced capacitance compared to precursors owing to the incorporation of highly conducting graphene, which increases the rate of ions transfer and presence AC provides more packing density and surface area within the composite. GCD measurement has been performed to quantify the consistency of capacitance of AC/graphene composite obtained as of CV with the capacitance obtained from GCD. Figure 4 (c) shows the GCD curves of AC/graphene composite at a different current density from 1 to 5 A/g and the obtained curves are triangular shape as expected for EDLC behaviour. The capacitances obtained from GCD measurement were 587, 483, 379, 345 and 314 F/g at 1, 2, 3, 4 and 5 A/g, evaluated using Eq. (2). So the highest capacitance achieved as of the GCD is 587 F/g at 1 A/g and the obtained capacitances from GCD are consistency with CV which also reveals that the composite electrode performs excellently in terms of energy storage. The long-term cyclic performance of synthesised composite was demonstrated as shown in figure 4 (d) and 89% of capacitance is retained over 5000 cycles at 1 A/g that indicates the outstanding cycle stability of the AC/graphene electrode. Also, the composite performs very well in terms of storage and stability owing to the presence of aqueous electrolytes where more number of ions of aqueous electrolytes are able to access the electrode materials due to the compatibility of pore size of electrode material and ions size of aqueous electrolytes.

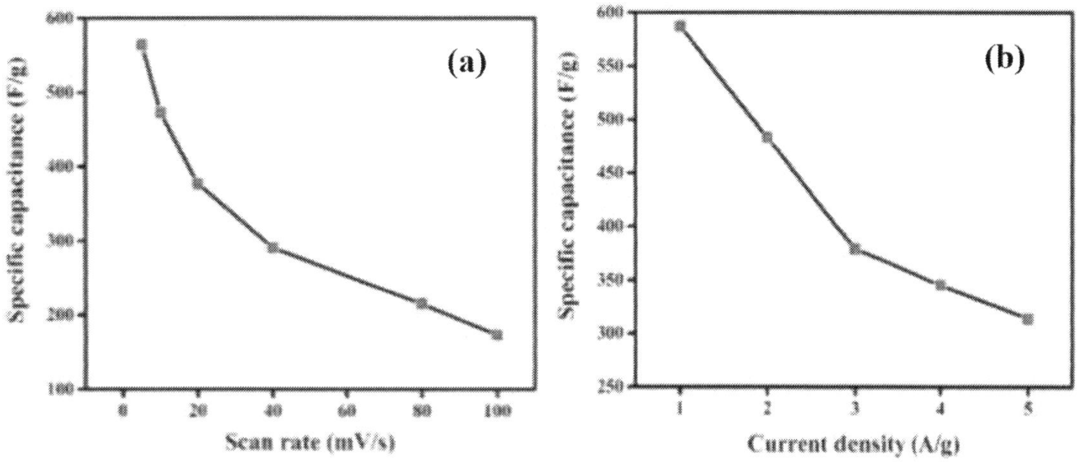

Figure 5. (a) Specific capacitance of AC/graphene composite at various scan rates and (b) Specific capacitance of AC/graphene composite at different current densities.

The change in specific capacitance at various scan rates of the AC/graphene composite shown in figure 5 (a) and observed that the capacitance gradually reduces as the scan rate increased probably owing to the deficiency of time to enter the small pores of electrode materials by the electrolyte ions and that results into lower specific capacitance at high scan rate [34]. Figure 5 (b) shows the deviation of capacitance at different current densities and found that capacitance decreases with increasing current density because at high current density very few electrolyte ions able to pass the electrode material.

4. Conclusions

AC/graphene composite is being prepared successfully in a low-cost as well as a simple method and the synthesised composite was investigated by using CV and GCD in 3M KOH solution in 3 electrode set-up. The composite showed the enhanced capacitance of 564 F/g at 5 mV/s and 173 F/g at 100 mV/s form CV. Also, using GCD the highest capacitance achieved by the composite was 587 F/g at 1 A/g and the obtained capacitances from GCD were reliability with the CV results which reveals the excellent charge storage capacity of AC/graphene composite. The large capacitance has been found in the composite electrode due to the presence of highly conducting graphene provide fast ions transfer and the addition of AC offers large packing density within the AC/graphene composite electrode. Also, the long terms cycling performance of synthesised composite was demonstrated and 89% of capacitance is retained over 5000 cycles at 1 A/g. Hence, the AC/graphene composite not only gives good cycle stability but also provide enhanced storage performance and can be a suitable electrode material for SC application.

Acknowledgements

I am thankful to the Minister of human resource development (MHRD) for the financial support of this work and also, grateful to IIT Bhubaneswar for giving me the facility of GCD and CV measurements.

References:

[1] Winter M. and Brodd R J 2004 *Chem. Rev.* **104** 4245
[2] Burke A 2000 *J Power Sources* **91** 37
[3] Miller J. R and Simon P 2008 *Science* **321** 651
[4] Zhang L. L and Zhao X. S 2009 *Chem. Soc. Rev.* **38** 2520
[5] Miller J. R and Burke A. F 2008 *Electrochem. Soc. Interface Spring* **17** 53
[6] Peng C, Zhang S, Jewell D and Chen G. Z 2008 *Natural Science* **18** 777
[7] Jurcakova D. H, Puziy A. M, Poddubnaya O. I, Garcia F. S, Tascon J. M. D and Lu G. Q 2009 *J. Am. Chem. Soc.* **131** 5026
[8] Kaempgen M, Chan C. K, Ma J, Cui Y and Gruner G 2009 *Nano Lett.* **9** 1872
[9] Hu L. B, Choi J. W, Yang Y, Jeong S, Mantia F. L, Cui L. F and Cui Y 2009 *Proc. Natl. Acad. Sci.* **106** 21490
[10] Frackowiak E and Beguin F 2001 *Carbon* **39** 937
[11] Frackowiak E and Beguin F 2002 *Carbon* **40** 1775
[12] Frackowiak E 2007 *Phys. Chem. Chem. Phys.* **9** 1774
[13] Novoselov K. S, Geim A.K, Morozov S.V, Jiang D, Zhang Y, Dubonos S.V, Grigorieva I.V and Firsov A.A 2004 *Science* **306** 666
[14] Zhu Y.W, Murali S, Stoller M.D, Ganesh K.J, Cai W.W, Ferreira P.J, Pirkle A, Wallace R.M, Cychosz K.A, Thommes M, Su D, Stach E.A and Ruoff R.S 2011 *Science* **332** 1537
[15] Liu C.G, Yu Z.N, Neff D, Zhamu A and Jang B.Z 2010 *Nano Lett.* **10** 4863
[16] Wang Y, Shi Z.Q, Huang Y, Ma Y. F, Wang C. Y, Chen M. M and Chen Y. S 2009 *J. Phys. Chem. C* **113** 13103
[17] Wang D. W, Li F, Zhao J. P, Ren W. C, Chen Z. G, Tan J, Wu Z. S, Gentle I, Lu G. Q and Cheng H. M 2009 *ACS Nano.* **3** 1745
[18] Sun Y. Q, Wu Q. O and Shi G. Q 2011 *Energy Environ. Sci.* **4** 1113
[19] Xia J. L, Chen F, Li J.H and Tao N.J 2009 Nat. *Nanotechnol.* **4** 505
[20] Huang Y, Liang J.J and Chen Y.S 2012 *Small* **8** 1805
[21] Ning G.Q, Fan Z.J, Wang G, Gao J.S, Qian W.Z and Wei F 2011 *Chem. Commun.* **47** 5976
[22] Zhang L.L, Zhao X, Stoller M.D, Zhu Y.W, Ji H.X, Murali S, Wu Y.P, Perales S, Clevenger B and Ruoff R.S 2012 *Nano Lett.* **4** 1806
[23] Stoller M. D, Park S. J, Zhu Y. W, An J. H and Ruoff R. S 2008 *Nano Lett.* **8** 3498
[24] Zheng C, Zhou X, Cao H, Wang G and Liu Z 2014 *Journal of Power Sources* **258** 290
[25] Chen Y, Zhang X, Zhang H, Sun X, Zhang D and Ma Y 2012 *RSC Advances* **2** 7747
[26] Yu S, Li Y and Pan N 2014 *RSC Adv.* **4** 48758.
[27] Mahanandia P, Simon F, Heinrich G and Nanda K. K 2014 *RSC Chem. Commun.* **50** 4613
[28] Satapathy D and Natarajan G.S 2006 *Indian journal of chemistry* **45A**. 2011
[29] Xie Z, Guan W, Ji F, Song Z and Zhao Y 2014 *Journal of Chemistry* **2014** 9
[30] Yao S. H, Zhang J. J, Shen D. K, Xiao R, Gu S, Zhao M and Liang J. Y 2016 *J. Colloid and Interface Sci.* **463** 118
[31] Frackowiak E, Khomenko V, Jurewicz K, Lota K and Beguin F 2006 *J. Power Sources* **153** 413
[32] Khomenko V, Frackowiak E and Beguin F 2005 *Electrochem. Acta* **50** 2499
[33] Dai C. S, Chien P. Y, Lin J. Y, Chou S. W, Wu W. K, Li P. H, Wu K. Y and Lin T. W 2013 *ACS Appl. Mater. Interfaces* **5** 12168
[34] Cheng Q, Tang J, Ma J, Zhang H, Shinya N and Qin L. C 2011 *Phys. Chem. Chem. Phys.* **13** 17615.

Influence of H_2S on the growth of sputter deposited Cu_2ZnSnS_4 thin film

P Hajara[1], T Priya Rose[1,2], M K Jayaraj[3] and K J Saji[1, 2, 3,4]

[1]International School of Photonics, Cochin university of Science and Technology, Kochi-22, India.
[2]Inter University Centre for Nano materials and Devices, Cochin University of Science and Technology, Kochi-22, India.
[3]Centre of Excellence in Advanced Materials, Cochin university of Science and Technology, Kochi-22, India.

[4]Corresponding Author: International School of Photonics, Cochin University of Science and Technology, Cochin- 682022, Kerala, India
E-mail address: saji@cusat.ac.in

Abstract. Cu_2ZnSnS_4 (CZTS) is considered as a suitable absorber layer for thin film solar cells. In the present study, the effect of flow rates of H_2S on the fabrication of CZTS thin films are studied in detail. RF magnetron sputtering in the presence of Ar and H_2S gas was adopted to deposit sulphides of Cu, Sn and Zn, followed by annealing at higher temperature in H_2S ambience in the same chamber to complete the crystallization process. Raman spectra showed peaks corresponding to kesterite CZTS and new impurity peaks observed while the H_2S flow rate increases. XRD result helped to identify various secondary phases present along with CZTS phase. It was revealed by FESEM images that the samples prepared at different H_2S flow rates exhibiting different nano shapes. Compositional study indicated that the Zn/Sn ratio decreases with H_2S flow rate. Surface roughness was calculated by AFM technique and big sulphur aggregates were present on the surface of the sample prepared at high H_2S flow rate. Optical bandgap value of the CZTS films was varied from 1.28 to 1.74 eV and electrical parameters were calculated for samples prepared at lower H_2S flow rate. This study presents the importance of choosing appropriate H_2S flow rates for the fabrication of CZTS thin films.

1. Introduction

The kesterite structured Cu_2ZnSnS_4 (CZTS), a quaternary semiconductor, has been emerged as a promising absorber layer for thin film solar cell due to its desirable optoelectronic properties. CZTS is an earth abundant and non-toxic p-type semiconductor with direct bandgap of around 1.5 eV, absorption coefficient more than 10^{-4} cm^{-1} [1]. Even though the predicted photovoltaic conversion efficiency of CZTS solar cell is 34%, the highest efficiency attained so far is 12.6 % only, which makes this material to be studied further [2]. One of the major challenges in the fabrication of CZTS is the formation of secondary phases such as Cu_2S, ZnS, SnS, SnS_2 and Cu_2SnS_3 [3]. CZTS thin films have been synthesized by several methods such as electron beam evaporation [4], thermal evaporation [5], DC and RF sputtering [6,7], pulsed laser deposition [8], successive ionic layer adsorption and reaction (SILAR) [9], sequential electro deposition [10] and sol-gel spin coating [11]. Among these, sputtering method is a widely preferred technique for large scale deposition due to its high deposition rate, uniformity, reproducibility and flexibility in tuning compositions [12,13].

IOP Conf. Series: Materials Science and Engineering 1166 (2021) 012008 doi:10.1088/1757-899X/1166/1/012008

Numerous studies have been conducted to improve the quality of CZTS thin films through the optimization of different growth parameters such as composition of the precursors [14], stacking order and thickness of precursors [15,16], pre annealing and ramping rate of annealing [17,18], sulfurization temperature and time [19,20] etc. Optimizing the sulfurization condition is one of the major approaches to achieve good quality CZTS films and researchers adopted different sulfurization methods. Generally, sulfurization processes of the films are carried out either with elemental sulphur powder or in the presence of H_2S gas. Annealing with H_2S gas gives an advantage of fabricating CZTS film under the controlled sulphur partial pressure. Moreover, there are reports of crystalline quality improvement with H_2S gas compared to elemental sulphur [21]. Awadallah et al fabricated CZTS films through sol-gel sulfurization with H_2S gas [22]. Rujun and co-workers prepared CZTS films by sulfurizing the stacked precursors in H_2S + N_2 gas and the phase formation was investigated by varying the temperature [23]. In a recent work, Idris and co-workers studied the effect of flow rate of H_2S on the properties of CZTS thin film by sulfurizing ZnS/Sn/Cu sequence by a two-step process. The surface of samples which are subjected to sulfurization at higher flow rate of H_2S was found to be more homogeneous and free of cavities [24]. However, the fabrication of CZTS films in H_2S ambience is not much explored and the formation mechanisms are not clearly investigated.

In the present study, we report the influence of the H_2S gas in the formation of CZTS thin films which are obtained by the sequential sputtering of Cu, Sn and Zn in Ar + H_2S ambience followed by sulfurization with different flow rate of H_2S in the same sputtering chamber. The Sn layer was selected as a middle layer to eliminate the loss of volatile Sn. Substrate temperatures during sputtering and post annealing were fixed to 200°C and 500°C respectively [25].

2. Experimental

Copper (99.95%), Zinc (99.9%) and Tin (99.9%) metallic targets of 2 inch diameter (Testbourn Ltd. UK) was sputtered sequentially onto a glass substrate in a custom designed RF magnetron sputtering system in the presence of H_2S and Argon gas. The substrate was ultrasonically cleaned in acetone, IPA and DI water respectively, for 15 min each and then dried in N_2 gas. The base and working pressure of the sputtering chamber were 2×10^{-6} and 8×10^{-3} mbar respectively. The substrate holder was rotated at 5 rpm and the substrate was kept at a distance of 10 cm from the target. Ar and H_2S gases were injected into the chamber in different ratios and the deposition was carried out at RF powers of 50 W for all the targets. The substrate temperature was kept at 200 °C during the sputtering and immediately after the sputtering, the substrate temperature was increased to 500 °C at a ramp rate of 30 °C per min. The annealing was done for 1 hour in H_2S ambience at $\sim2\times10^{-2}$ mbar pressure followed by natural cooling.

The stacking order Cu/Sn/Zn was selected considering the volatile nature of Sn, as it may evaporate quickly if it is the upper layer and CuS may form easily in Cu capped stacking order. The sputtering time for the deposition of Cu, Sn and Zn were fixed at 32, 17 and 6 min respectively to get a total thickness of around 600 nm. To understand the H_2S flow rate dependence, Ar:H_2S gas ratios were varied, keeping Ar + H_2S = 20 sccm. Four samples were prepared with H_2S flow rate of 2, 3, 5 and 8 sccm (Samples named as HS2, HS3, HS5 and HS8 respectively). The total thickness of each stack was around 500 nm.

Deposited CZTS thin films were subjected to different structural, morphological and optical studies. Raman spectra were recorded using Horiba LABRAM HR Raman spectrometer excited with the 514 nm line of an Ar^+ laser and XRD measurements were carried out using Bruker D8 Advance X-ray diffractometer. The film morphology and elemental analysis were studied using Carl Zeiss field emission scanning electron microscopy (FESEM) equipped with an EDS analyser and AFM images were obtained using Keysight 5500 Scanning Probe Microscope in non-contact mode. Transmission spectra were recorded with ParkinElmer Lambda 365 UV/Vis spectrophotometer. The electrical studies were performed using Hall effect measurements at room temperature.

3. Result and Discussion

The reaction mechanism of Cu, Zn and Sn metals in H_2S ambience can be written as,

$Sn + 2H_2S \rightarrow SnS_2 + 2H_2$

$Zn + H_2S \rightarrow ZnS + H_2$

$2Cu + H_2S \rightarrow Cu_2S + H_2$

$SnS_2 + ZnS + Cu_2S \rightarrow Cu_2ZnSnS_4$

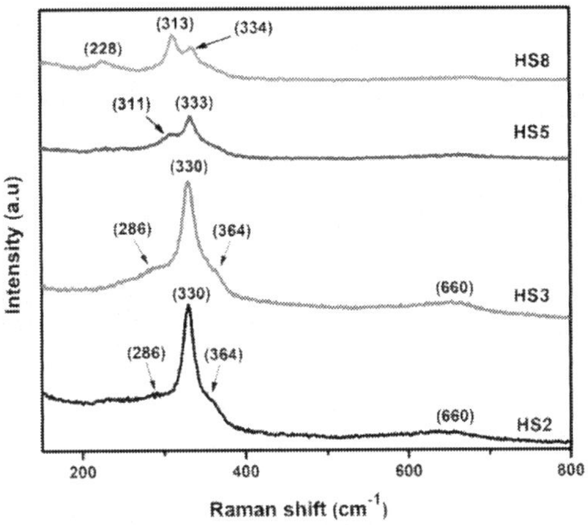

Figure 1. Raman spectra of CZTS (samples HS2, HS3, HS5 and HS8) thin films.

Raman spectroscopy is a powerful tool for detecting impurity phases in CZTS, because in the XRD pattern there is an overlap between the peaks of CZTS and some of the binary/ternary phases. Figure 1 shows the Raman spectra of CZTS thin films deposited at different H_2S flow rates. For the samples HS2 and HS3, spectra show major peak at 330 cm^{-1} with weaker peaks at 286 cm^{-1}, 364 cm^{-1} and 660 cm^{-1}. The peak at 364 cm^{-1} is attributed to the E(LO) mode and all other peaks corresponded to the vibrational A symmetry modes of kesterite CZTS [26]. The main A mode of Kesterite, stems from the vibration of S atom, has been reported between 330-338 cm^{-1} depending the synthesis method and 337-338 cm^{-1} is the most commonly reported value for the good quality single phase Kesterite CZTS films [27]. There is a shift for this major CZTS peak to 333 cm^{-1} in sample HS5 with a shoulder peak at 311 cm^{-1} which is ascribed to the SnS_2 phase [28]. The sample HS8 consists of three distinct peaks at 334 cm^{-1},313 cm^{-1} and 228 cm^{-1} which are assigned to kesterite CZTS, SnS_2 and SnS respectively [29]. As H_2S flow rate increases, there is a clear shift for major CZTS peaks from 330 cm^{-1} to 334 cm^{-1} indicating the improvement in the crystalline quality of CZTS [30]. However, the intensity of this peak gets weaker owing to the reduced amount of CZTS due to the change in stoichiometry as H_2S flow rate increases. Moreover new peaks of SnS_2 and SnS are emerging with H_2S flow rate which indicates the presence of high tin content in the sample. It was observed that while the H_2S flow increases to 5 sccm, adhesion of the film was poor resulting peel off from the substrate.

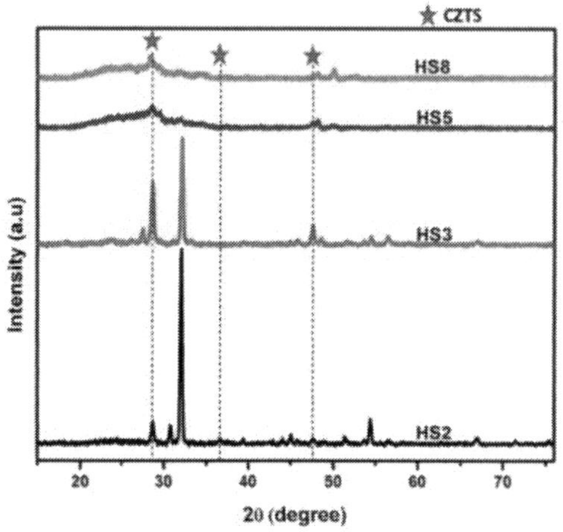

Figure 2. XRD patterns of CZTS thin films prepared under different H_2S flow rate

It is challenging to fabricate CZTS without secondary phases. Therefore XRD is employed for further phase identification of CZTS thin films. XRD patterns are given in Figure 2 and phases corresponding to each peak are listed in Table 1. It can be seen that all of the CZTS films consisted of a diffraction peak at $2\theta = 28.6°$ which is attributed to the (112) plane of CZTS phase (ICCD 00-026-0575) and the intensity of this peak is very low in samples prepared at high H_2S flow rate. The other characteristic peak of CZTS at $2\theta = 47.6°$ is present in all the samples except HS8. The diffraction peak at $2\theta = 36.6°$ present in sample HS2 is also assigned to the CZTS phase. In addition to the CZTS phase, all samples exhibited binary and ternary secondary phases. Both HS2 and HS3 samples revealed the existence of secondary phases like Cu_2S (ICCD 00-053-0522), CuS_2 (ICCD 00-018-0381), Cu_3SnS_4 (ICCD 00-036-0217) and Cu_2SnS_3 (ICCD 00-027-0198) and the samples HS5 and HS8 exhibited peaks of CuS_2, Cu_2S, Cu_4SnS_4 (ICCD 00-027-0196), SnS (ICCD 00-039-0354) and SnS_2 (ICDD 00-023-0677). Cu-S based secondary phases obtained in XRD could not be detected in Raman spectra due to the low penetration depth of laser. The distribution of Cu to the surface of the film may be low since Cu layer was at the bottom of our stacking order. In comparison, samples prepared at lower H_2S flow rates exhibit diffraction peaks of Cu-S related secondary phases along with CZTS phase. Whereas sample HS5 and HS8 have Sn related secondary phases in addition to CZTS phase. The presence of SnS and SnS_2 in these samples was revealed in the Raman analysis also.

Table 1. Materials correspond to diffraction angle of XRD.

Sample	2θ(°)	Material	Sample	2θ(°)	Material
HS2	28.69	CZTS	HS3	18.48	Cu_3SnS_4
	30.79	Cu_4SnS_6		27.56	Cu_3SnS_4
	32.09	Cu_2S		28.68	CZTS
	36.61	CZTS		32.17	Cu_2S
	39.45	Cu_2SnS_3		33.20	Cu_2S
	44.07	Cu_4SnS_4		39.45	Cu_2SnS_3
	45.06	Cu_2SnS_3		45.86	Cu_2S
	45.84	Cu_2S		47.58	CZTS
	47.68	CZTS		48.55	Cu_4SnS_4
	51.42	Cu_4SnS_6		53.66	Cu_2S
	53.68	Cu_2S		54.44	Cu_2S
	54.36	Cu_2S		56.43	Cu_3SnS_4
	56.56	Cu_3SnS_4		66.94	CuS_2

	66.87	CuS_2		75.33	CuS_2
	71.45	Cu_2SnS_3	HS8	28.46	CZTS
	75.37	CuS_2		29.46	Cu_4SnS_4
HS5	28.64	CZTS		31.96	SnS
	29.50	Cu_4SnS_4		48.15	Cu_2SnS_3
	30.76	Cu_4SnS_6			
	31.96	SnS			
	47.63	CZTS			
	48.13	Cu_2SnS_3			
	50.00	SnS_2			

Field emission scanning electron microscope with EDS was used to study the surface properties and to find out the elemental composition of CZTS thin films. The atomic percentage and atomic ratios of CZTS thin film samples fabricated at different H_2S flow rate are given in Table 2. It has been reported that the CZTS absorber layer with Copper-poor ($Cu/(Zn + Sn) < 1$) and Zinc-rich ($Zn/Sn > 1$) compositions are ideal for photovoltaic applications [31]. The composition $Cu/(Zn + Sn) \approx 0.8$ and $Zn/Sn \approx 1.2$ is suggested as the most suitable for high performance solar cell [32]. It can be seen in Table 2, with an increase of the H_2S flow rate, the atomic percentage of zinc is decreasing drastically, which indicates that the sputtering rate of zinc is very poor in the H_2S ambience while comparing with that of other metals copper and tin. Further study is needed to find out the reason behind the low sputtering rate of zinc. All of the CZTS films are found highly Sn-rich contrary to our expectation as Sn loss is a commonly observed process in the fabrication of CZTS. These can be explained by the fact that Sn is depositing as SnS_2 by the sputtering of Sn at 200°C in the presence of H_2S which is not volatile as metallic tin and this SnS_2 contributes to the further formation of CZTS. Whereas Sn loss is happening in two step process during temperature ramping due to the volatile nature of Sn. As a result of Sn-rich and Zn-poor composition, all our CTZS films are away from stoichiometry. Elemental mapping of the samples was examined and only HS8, prepared at higher H_2S ambience, consists of sulphur aggregates on the surface as seen in the Figure 3. Since the melting point of sulphur is very low it should have evaporated at higher temperature of 500 °C. The existence of sulphur can be explained through the formation process of SnS which is present in the sample prepared with high H_2S flow rate. Generally the formation of SnS is observed at a higher temperature, much above 500 °C, by the thermal decomposition of SnS_2 into SnS and Sulfur. But this is not the case here since the temperature never exceeded 500 °C and also the temperature is given in H_2S ambience. Zhang and co-workers reported that formation of SnS is favourable even in low temperature if the H_2S concentration is very high [33]. The H_2 released from H_2S acts as a strong reducing agent and decomposes SnS_2 in to SnS and Sulfur. This sulfur resides on the surface of the film as aggregates.

Table 2. Elemental compositions of CZTS films

Sample	Cu	Sn	Zn	S	Cu/(Zn + Sn)	Zn/Sn
HS2	17.51	28.14	11.01	43.49	0.45	0.39
HS3	19.19	22.50	6.17	51.90	0.67	0.27
HS5	19.62	28.68	2.48	49.04	0.63	0.09
HS8	18.18	27.69	2	52.2	0.61	0.08

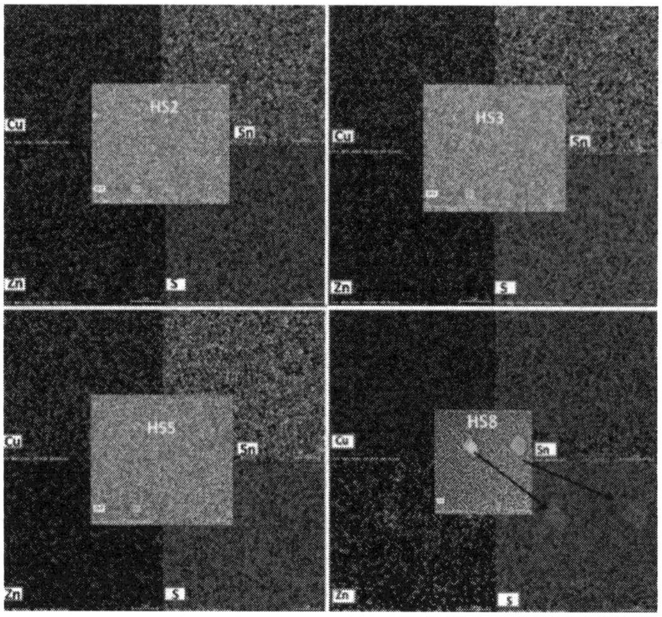

Figure 3. EDS mapping of CZTS samples HS2, HS3, HS5 and HS8.

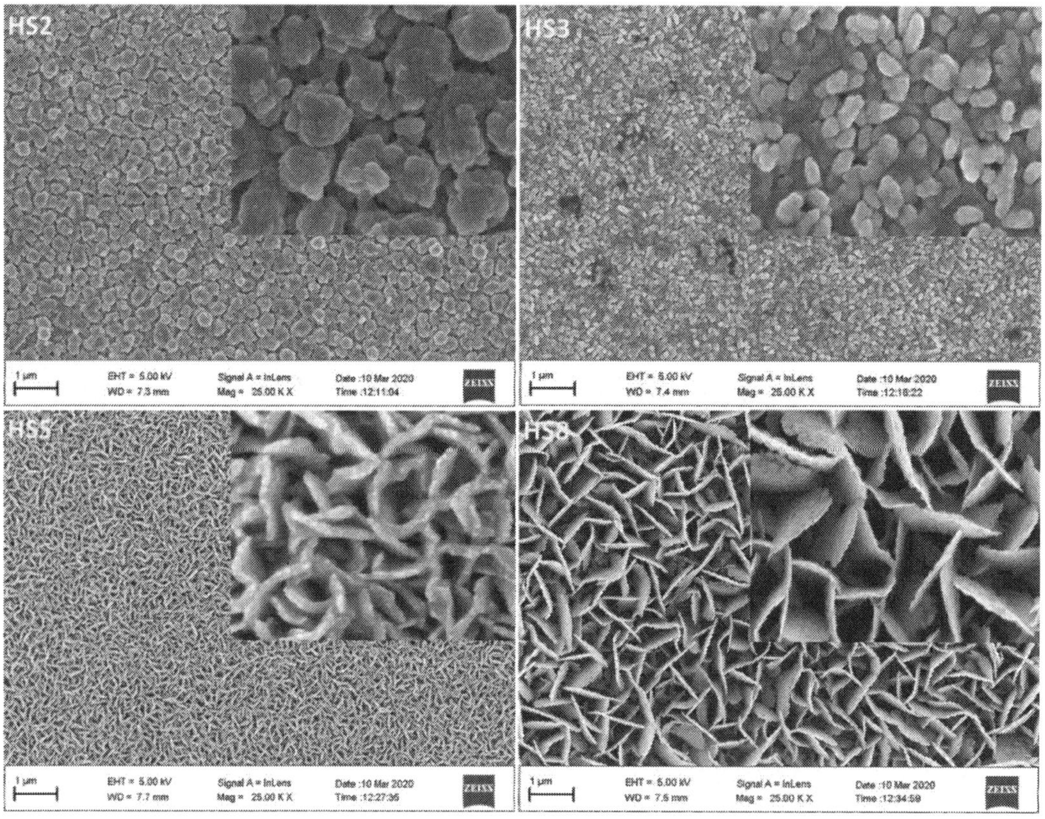

Figure 4. FESEM images of samples HS2, HS3, HS5 and HS8

Figure 4 presents the FESEM images of CZTS thin film samples and the magnified views of structures are given as an inset. As shown in the image, the surface morphologies of the films are not similar and consist of different nano structures with homogeneous surface as the flow rate of H_2S vary from 2 sccm to 8 sccm. The sample HS2, prepared at a lower H_2S flow rate consist of a flower like structure and morphology changed to nano flakes with sharp edges at higher flow rate. It has been reported that Sn has a tendency to form nano structures in the presence of sulphur [34]. Since EDS results confirmed the high atomic ratio of Sn in deposited films, we can conclude that the excess amount of Sn present in our sample is responsible for the formation of different nano structures. This result confirms that variation in H_2S atmosphere plays a crucial role in the formation of CZTS films.

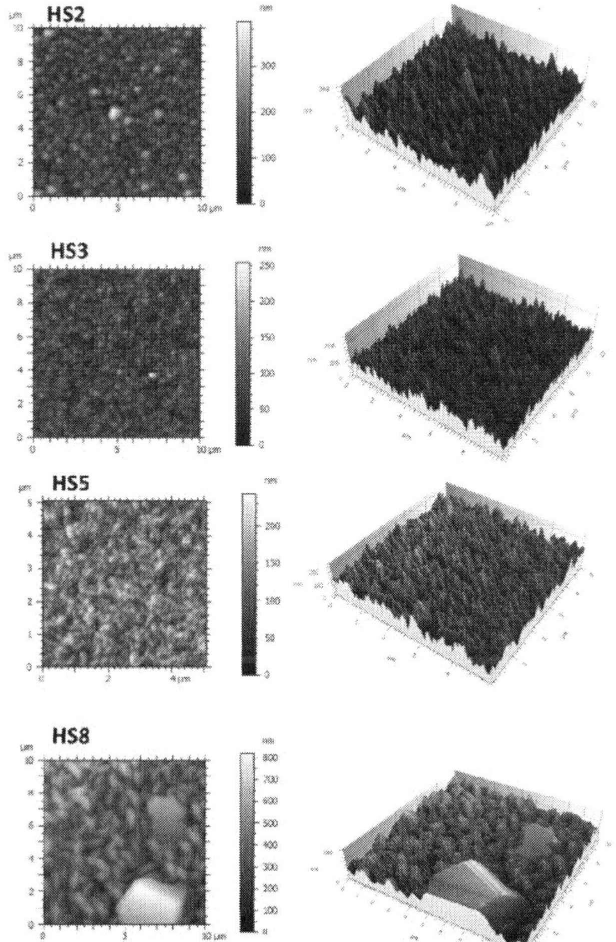

Figure 5. AFM images of samples HS2, HS3, HS5 and HS8

Surface topography of CZTS thin films is studied by non-contact atomic force microscopy. The 2D and 3D AFM images of the films deposited at various gas flows are shown in Figure 5. While these images are examined, it is seen that the grains are homogeneous and uniform for the samples with a lower H_2S flow, whereas samples deposited at a higher H_2S flow have big aggregates on surfaces. Surface roughness value is an important parameter for solar cell application and below 50 nm is the desirable value. The root mean square (RMS) and average roughness values of the samples are given in Table 3. The rms roughness values are found varying from 44.4 nm to 26.1 nm as the H_2S gas flow increases from 2 to 5 sccm whereas sample HS8 showed increased rms roughness value of 131 nm due to the formation of big sulphur cluster.

Table 3. Surface roughness values of CZTS films

Sample	RMS (nm)	Mean (nm)
HS2	44.4	34.2
HS3	26.1	20.8
HS5	26.1	20.6
HS8	131	102

Transmittance spectra of CZTS thin films for the wavelength ranging from 300 to 1100 nm is shown in Figure 6(a). The optical bandgap is calculated by extrapolating the straight line portion in the plot of $(\alpha h\nu)^2$ versus photon energy (hν) and values are found to be 1.45, 1.62, 1.74 and 1.28 eV for samples HS2, HS3, HS5 and HS8 respectively as seen in Figure 6(b). The lower bandgap for HS8 is due to the comparatively high concentration of SnS secondary phase which is having very low bandgap energy of 1.07 eV. The sample HS5 has a little higher bandgap value of 1.74 eV due to the presence of high band gap (2.3 eV) SnS_2 phase.

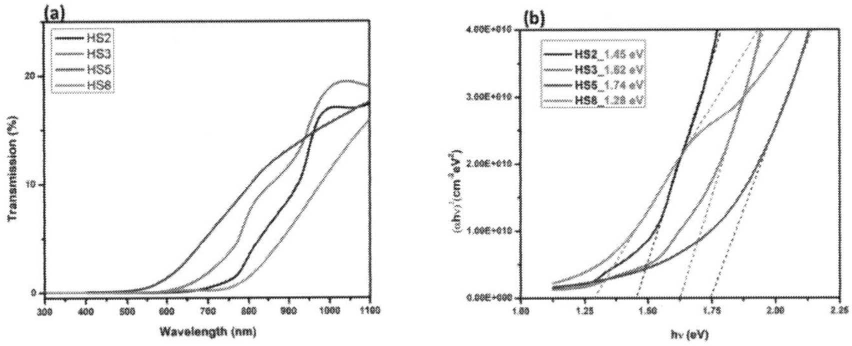

Figure 6. (a) Transmission spectra of CZTS thin film samples HS2, HS3, HS5 and HS8 (b) Tauc plot of CZTS film samples HS2 and HS3.

Electrical properties of the samples HS2 and HS3 were studied using Hall effect set-up. Films were p-type in nature and Table 4 shows the parameters sheet resistance, carrier density and mobility. HS3 exhibit higher mobility and sheet resistance with a carrier concentration of $4.9\times10^{18}cm^{-3}$. Hall measurements of other two samples were not done due to its peal off nature.

Table 4. Electrical parameters of CZTS films HS2 and HS3.

Sample name	Sheet resistance (Ω)	Carrier density(cm^{-3})	Mobility (cm^2/Vs)	Carrier type
HS2	647	$1.72 \pm 0.4 \times 10^{20}$	1.2 ± 0.26	P
HS3	2289	$4.94 \pm 1.01 \times 10^{18}$	11.14 ± 2.29	P

4. Conclusion

To summarize, CZTS thin films were grown by RF magnetron sputtering by sequentially depositing copper, tin and zinc in the presence of H_2S + Ar gas at 200 °C, followed by post annealing at 500 °C in H_2S ambience. Studies were carried out by changing the flow rate of H_2S to identify the impact of gas flow in the growth process. Raman spectra revealed the formation of CZTS phase in all the samples and confirmed the emergence of secondary phases of SnS_2 and SnS peaks while the H_2S flow rate increases. Various secondary phases were identified from XRD analysis in addition to CZTS phase and noted that Cu-S related phases are dominant in the films prepared in lower H_2S flow rate whereas Sn related phases are formed while the H_2S flow rate increases. These secondary phases occur because of non-stoichiometric compound caused by the Zn-poor and Sn-rich composition as evident from EDS analysis. It is observed that there is a significant loss of Zn while H_2S flow rate increases and Sn loss was not happening since the sputtering was carried out in H_2S ambience. It was revealed by FESEM images that the CZTS films had different morphologies depending on different flow rate of H_2S. Films fabricated at higher H_2S flow rate contain sulphur aggregates due to the decomposition of SnS_2 into SnS and Sulfur. Since non-stoichiometric compound are not favourable for photovoltaic applications a further optimization is needed to fabricate stoichiometric CZTS thin films by eliminating the formation of secondary phases. However, this study demonstrates the importance of choosing H_2S flow rate as it is a crucial factor in the fabrication process of CZTS thin films. The understanding of critical role of H_2S atmosphere may help researchers in improving the quality of CZTS thin fims.

References

[1] Kumar M, Dubey A, Adhikari N, Venkatesan S and Qiao Q, Strategic review of secondary phases, defects and defect-complexes in kesterire CZTS-Se solar cells 2015 *Energy Environ. sci.* **8** 3134-3159.

[2] Wang W, Winkler M T, Gunawan O, Gokmen T, Todorov T K, Zhu Y and Mitzi D B, Device characteristics of CZTSSe thin-film solar cells with 12.6 % efficiency 2014 *Adv. Energy Mater.* **4** 1301465.

[3] Jung H R, Shin S W, Gurav K, Gang M G, Lee J Y, Moon J H and Kim J H, Evolution of detrimental secondary phases in unstable Cu_2ZnSnS_4 films during annealing 2016 *Electron. Mater. Lett.* **12** 139-146.

[4] Peksu E, Terlemezoglu M, Parlak M and Karaagac H, Characterization of one-step deposited Cu_2ZnSnS_4 thin films derived from a single crystalline powder 2019 *Renew. Energy* **143** 1133-1142.

[5] Shi C W, Shi G Y, Chen Z, Yang P F, Yao M, Deposition of Cu_2ZnSnS_4 thin films by vacuum thermal evaporation from single quaternary compound source 2012 *Mater.Lett.* **73** 89-91.

[6] Li. J, Zhang Y, Wang H and et al. , On the growth process of $Cu_2ZnSn(SSe)4$ absorber layer formed by selenizing Cu-ZnS-SnS precursors and its photovoltaic performance 2015 *Sol. Energy Mat. Sol. Cells* **132** 363-371.

[7] Wang J, Li S, Cai J, Shen B, Ren Y and Qin G, Cu_2ZnSnS_4 thin films: facile and cost effective preparation by RF-magnetron sputtering and texture control 2013 *J. Alloy.Compd.* **552** 418-422.

[8] Vanalakar S A, Agawane G L, Shin S W et al , A review on pulsed laser deposited CZTS thin films for solar cell applications 2015 *J. Alloys Compd.* **619** 109-121.

[9] Henry J, Mohanraj K and Sivakumar G, Electrical and optical properties of CZTS thin films prepared by SILAR method 2018 *J. Asian Ceram. Soc.* **4** 81-84.

[10] Tao J, Chen L, Cao H, et al., Co-electrodeposited Cu_2ZnSnS_4 thin-film solar cells with over 7 % efficiency fabricated via fine tuning of the Zn content in absorber layers 2016 *J. Mater. Chem. A* **4** 3798- 3805.

[11] Haass S G, Diethelm M, Werner M, Bissig B, Romanyuk Y E, Tiwari A N, 11.2 % efficient solution processed kesterite solar cell with a low voltage deficit 2015 *Adv. Energy Mater.* **5** 1-7.

[12] Dalapati G K, Masudy-Panah S, Chua S T, Sharma M, Wong T I, Tan H R and Chi D, Color tunable low cost transparent heat reflector using copper and titanium oxide for energy saving applications 2016 *Sci. Rep.* **6** 20182.

[13] Selvakumar N, Biswas A, Rajaguru K, Gouda G M, Barshilia H C, Nanometer thick tunable AlHfN coating for solar thermal applications: transition from absorber to antireflection coating 2015 *Sol. Energy Mater. Sol. Cells* **137** 219-226.

[14] Zhuk S, Kushwaha A, Wong T K S, Panah S M, Smirnov A and Dalapati G K, Critical review on sputter-deposited Cu_2ZnSnS_4 based thin film photovoltaic technology focusing on device architecture and absorber quality on the solar cell performance 2017 *Sol. Energy Mater. Sol. Cells* **171** 239-252.

[15] Thota N, Gurubhaskar M, Anantha Sunil P, Prathap P, Subbaiah Y P V and Tiwari A, Effect of metal layer stacking order on the growth of Cu_2ZnSnS_4 thin films *2017Applied Surface Science* **396** 644-651.

[16] Dalapati G K et al. Impact of molybdenum out diffusion and interface quality on the performance of sputter grown CZTS based solar cells 2017 *Sci. rep* **7** 1350.

[17] Jiang J, Zhang L, Wang W, Huang X and Hong R, The role of pre-annealing in the sulfurization of Cu-Zn-Sn stacked metal layer prepared by magnetron sputtering 2018 *Material Science in semiconductor processing* **83** 125-132.

[18] Ge J, Yu Y H, Zhang C J, Zuo S H, Jiang J C, Ma J H, Yang P X and Chu J H, Comparative study of the influence of two distinct sulfurization ramping rates on the properties of Cu2ZnSnS4 thin films 2012 *Appl Surf Sci.* **258** 7250-7254.

[19] Zhao Q, Hao R, Liu S, Yang M, Liu X, Chang F and Lu Y, Fabrication and characterization of Cu_2ZnSnS_4 thin films by sputtering a single target at different temperature 2017 **Physica B** *523* 62-66.

[20] Jiang J, Zhang L, Wang W, Huang X and Hong R, The role of pre-annealing in the sulfurization of Cu-Zn-Sn stacked metal layer prepared by magnetron sputtering 2018 *Materials science in semiconductor processing* **83** 125-132.

[21] Scragg J j, Dale P J and Peter L M, Synthesis and characterization of Cu_2ZnSnS_4 absorber layers by an electrodeposition-annealing route 2009 *Thin solid films* **517** 2481-2484.

[22] Awadallah O and Cheng Z, Formation of sol-gel based Cu_2ZnSnS_4 thin films using ppm-level hydrogen sulphide 2017 *Thin solid films* **625** 122-130.

[23] Sun R, Zhuang D, Zhao M, Zhang N, Xie M, Wei Y, Ren G, Wu Y, Gong Q and Wei J, Phase formation of Cu2ZnSnS4 thin films by sulfurizing stacked precursors by sputtering from Cu-Zn and Cu-Sn targets 2019 *Thin Solid Films* **690** 137561.

[24] Akyuz I, Atay F, Aydin R and Kose S, Production and characterization of CZTS films: On the role of H_2S flow rate 2019 *Solar energy* **194** 709-715.

[25] Zhao Q, Hao R, Liu S, Yang M, Liu X, Chang F, Lu Y and Wang S, Fabrication and characterization of Cu2ZnSnS4 thin films by sputtering a single target at different temperature 2017 *Physica B* **523** 62-66.

[26] Fernandes P A, Salome P M P and da Cunha A F, Study of polycrystalline CuZnSnS4 films by Raman scattering 2011 *J. Alloy. Comp.***509** 7600-7606.

[27] Parkin I P, Price L S, Hibbert T G and Molloy K C, The first single source deposition of tin sulphide coatings on glass:aerosol-assisted chemical vapour deposition using $[Sn(SCH_2CH_2S)_2]$ 2001 *J. Matter. Chem,* **11** 1486-1490.

[28] Ahn J H, Lee M, Heo H, Sung J H, Kim K, Hwang Hand Jo M, Deterministic Two-Dimensional Polymorphism Growth of Hexagonal n-Type SnS_2 and Orthorhombic p-Type SnS Crystals 2015 *Nano Lett.* **15** 3703–3708.

[29] Sinsermsuksakul P, Heo J, Noh W, Hock A S and Gordon R G, Atomic layer deposition of tin monosulfide thin films 2011 *Adv. Energy Mater.* **1** 1116–1125.

[30] Zhao Q, Hao R, Liu S, Yang m, Liu X, Chang F and Lu Y, Fabrication and characterization of Cu_2ZnSnS_4 thin films by sputtereing a single target at different temperature 2017 *Physica B* **523** 62-66.

[31] Katagiri H, Jimbo K, Maw W S, Oishi K, Yamazaki M, Araki H and Takeuchi A, Development of CZTS-based thin film solar cells 2009 *Thin Solid Films* **7** 2455-2460.

[32] Chen S, Walsh A, Gong X G and Wei S H, Classification of lattice defects in the kesterite Cu_2ZnSnS_4 and $Cu_2ZnSnSe_4$ earth-abundant solar cell absorbers 2013 *Adv. Mater.* **11** 1522-1539.

[33] Zhang H, Balaji Y, Mehta A N, Heyns M, Caymax M, Radu I, Vandervorst W and Delabie A, Formation mechanism of 2D SnS_2 and SnS by chemical vapor deposition using $SnCl_4$ and H_2S 2018 *J. Mater. Chem. C* **6** 6172.

[34] Singh O P, Parmar R, Gour K S, Dalai M K, Tawale J, Singh S P, Singh V N, Synthesis and characterisation of petal type CZTS by stacked layer reactive sputtering *Superlattice. Microst.* **88** 281-286.

Acknowledgments
The authors would like to acknowledge DST-PURSE, Govt of India for the financial support for the purchase of sputtering equipment. One of the authors P. Hajara would like to thank MNRE for providing NREF Fellowship.

IOP Conf. Series: Materials Science and Engineering 1166 (2021) 012009 doi:10.1088/1757-899X/1166/1/012009

Minimizing surface reflectance of crystalline silicon: A simulation based study

Tiasa Das, Sandip Kumar, Anamika Upadhyay, Arjyajyoti Goswami

Dept. of Mechanical Engineering, National Institute of Technology, Durgapur, West Bengal, India

E-mail: tiasadas2013@hotmail.com

Abstract. Silicon has been a popular choice for optoelectronic devices and solar cells since its discovery mainly due to its abundance in nature. However, methods to improve light trapping and hence increase the efficiency of these devices has always been a subject of great interest. The objective of this study has been to minimize the surface reflectance of thin film silicon through use of anti-reflection coating and surface patterning. In this study, the most effective anti-reflectance coating (ARC) is determined amongst three well known materials namely, Silicon Nitride(Si_3N_4), Titanium Dioxide(TiO_2) and Magnesium Fluoride (MgF_2) by simulation through S4 software by implementing Rigorous Coupled Wave Analysis Algorithm (RCWA). The best possible combination of the substrate and ARC is determined for a working wavelength range of 300 nm to 900 nm by varying the thickness of ARC. The best possible combination of substrate and ARC has been subjected to further investigation by imparting patterns and minimum average reflectance of 0.135986 was obtained experimentally.

1. Introduction

Since the mid twentieth century, the use of silicon has taken a big leap. Nowadays, Silicon is practically used everywhere from wristwatches to international space stations. Silicon has played a major role in converting solar energy to electrical energy.

The very first solar cell was developed in1954 at Bells Labs, which in theory could give a 23% efficiency provided the physical parameters were ideal [1]. Although, the energy conversion efficiency obtained was about 4% which at the least was a proof of a new concept. Apart from overcoming the band gap, there were mainly two hurdles faced by the leading physicists- one that it was very difficult to make a proper electrical contact with the silicon due to the poor adhesion with metals and secondly the high reflectance by the silicon wafers itself.

Since, bare Crystalline Silicon shows a very low efficiency in the absorption of light, i.e most of the energy of the incident light is lost due to reflection, depending on various physical factors such as surface texture, refractive indices of the mediums to the Silicon, thermodynamic parameters, etc. In order to decrease the reflectance usually mainly two widely used approaches are taken in combination

- Introduction of patterns or surface textures.
- Application of an antireflection coating(ARCs) on the top bare Silicon surface.

In the former approach regular patterns are introduced using techniques such as wet etching, dry etching, photolithography, etc[2]. These patterns help in entrapment of the incident light predominantly by increasing the net internal reflection [3].

In the later approach a coating of suitable material is applied on the top layer of the bare Silicon which can transmit light as much as possible and reduce the reflectance of the incident light. The entrapped light waves scattering in the bulk of the ARC then undergo coupling and quantum interference effect helps in absorption in the bulk [4]. Ideally, an ARC should be able to minimize the reflectance for a range of wavelengths [5].

By using ARCs the overall efficiency can be increased by upto 40% [6]. Since reflectance of an ARC depends on its thickness and its refractive index (with respect to Silicon). So, it is in our best interest to study variation of reflectance with respect to the variation of the thickness and the material. Optimization of the above said parameters can reduce the loss in overall reflectance and in turn increase the efficiency of the solar cell.

In addition to adding an ARC, patterns can be introduced on the topmost layer to decrease the reflectance further, as per Fresnel Theory. Although, introducing patterns or textures on the ARC itself opens the possibility of many configurations that can occur in practice which may not necessarily increase the overall efficiency of the cell. So this makes it crucial to study the variation of reflectance in respect to these configurations as well [7][8].

This paper presents the investigation of various anti reflective coatings for minimization of the reflectance with polychromatic sources in respect to two physical factors – surface textures (patterns) and thickness of the antireflection coatings .Also, it presents the study of the introduction of patterns on ARCs with different configurations.

2. Methodology

A significant and well proven method to promote light absorption and reduce surface reflectance has been the use of effective anti- reflection coatings and creation of patterns on layer. In this study, a comparative analysis is performed to determine the most suitable configuration of substrate and coating which produces the least mean surface reflectance and hence provides the most efficient combination for use in devices. The analytic work carried out in this study is performed on the data obtained by a series of simulations carried out in the S4 software, The detailed procedure is enlisted below in sections for easier understanding.

2.1. Comparative Analysis of Anti Reflection Coatings(ARCs)

A comparative study is performed to determine the most effective anti-reflection coating on the bare silicon amongst three well known materials for the role, namely, Magnesium Fluoride (MgF_2), Titanium Dioxide (TiO_2) and Silicon Nitride (Si_3N_4). A 1000nm Silicon substrate along with a single layer of one of the three ARCs is considered (with an initial thickness of 50nm). Surface reflectance values for a wavelength range of 300nm to 900 nm with steps of 3nm for this configuration is obtained. After this, keeping the thickness of the Si substrate constant, reflectance values using thickness of the ARC layer as 60nm,70nm,90nm,110nm,130nm and 150 nm was obtained to identify the best working thickness for the given substrate thickness.

After this, reflectance values for each of the above mentioned ARC layer thickness is calculated by changing Si substrate thickness to 10,000nm, 25,000nm and 50,000nm. This same process is repeated for each of the materials as the ARC and the average reflectance value for all the combinations of each material is calculated. The material, substrate and ARC thickness combination which provides the least average reflectance is selected to further investigate the effect of patterns on the surface reflectance.

2.2. Comparative analysis of different configurations for the selected Substrate ARC combination

The substrate and ARC thickness combination that provides the least mean surface reflectance is chosen for further investigation. Five different configurations for the selected pair is simulated and the mean reflectance values are compared to obtain the best possible configuration for reduced reflectance. These five configurations are explained in details below along with their cross sectional views.

2.2.1. Case 1

This configuration(fig 1) is simulated first. It is simply the combination of the substrate and ARC that produces the minimum mean reflectance without any patterns.

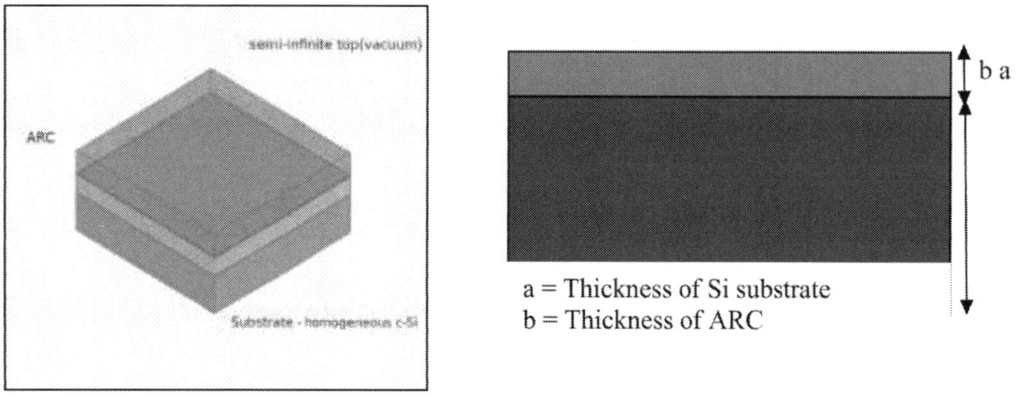

FIGURE 1: Case 1 Configuration

2.2.2. Case 2

In this configuration (fig 2), circular holes are made on the structure with a diameter of 100nm, pitch of 350 nm and thickness being the same as the ARC thickness. This is accomplished by creating a circular pattern in the centre of the 350 nm*350 nm space lattice in the layer 1 which is the ARC layer and filling it with vacuum.

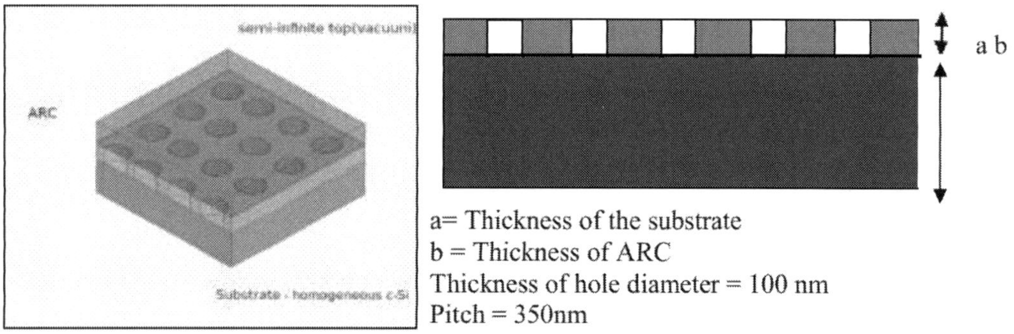

FIGURE 2. Case 2 Configuration

2.2.3. Case 3

In this configuration(fig 3), circular holes are made in the structure with the same diameter of 100 nm and pitch 350 nm but with a thickness of an additional 200 nm into the substrate apart from the entire ARC thickness. The results are obtained by simulating a three layered structure with a circular pattern in the center of 350 nm*350 nm space lattice with the material as vacuum running through layers 1 and 2 which are the ARC layer and 200 nm of substrate respectively. The remaining thickness of the substrate forms the third layer. This case is a representation of an ideal configuration which is not possible to produce in real life.

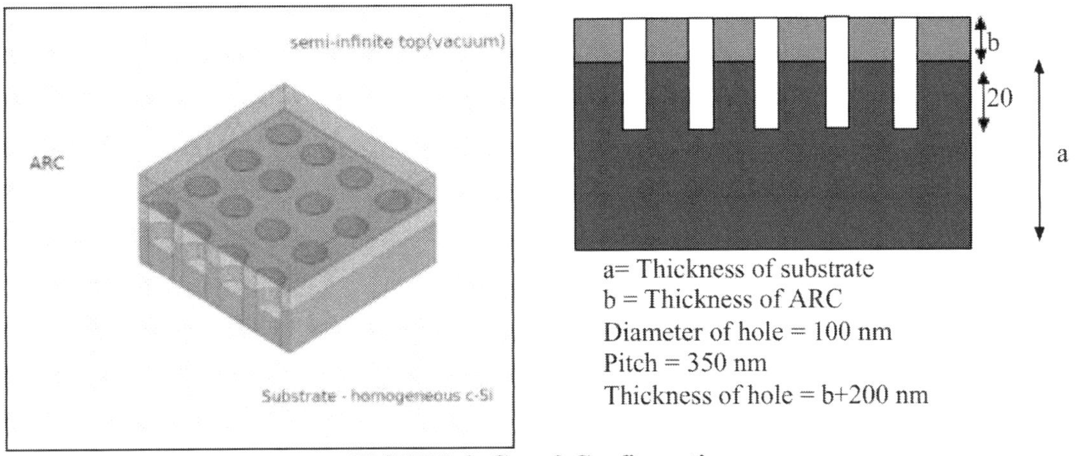

FIGURE 3. Case 3 Configuration

2.2.4. Case 4

This configuration is feasible in reality and hence is referred to as the practical case. In this configuration (fig 4), circular holes are made in the structure with the same diameter of 100 nm and pitch 350 nm. A thickness of an additional 200 nm is made into the substrate (apart from the entire ARC thickness) and the removed ARC thickness settles within the created void in the substrate. The results are obtained by simulating a four layered structure with a circular pattern in the center of 350 nm*350 nm space lattice with the material as vacuum running through layers 1 and 2 which are the ARC layer (let, thickness= b nm) and (200- b) nm thickness of substrate respectively. The third layer consists of substrate with thickness same as the ARC thickness (=b nm) with a circular pattern in the center of the lattice filled with material as the ARC. The remaining portion of the substrate constitutes the fourth layer.

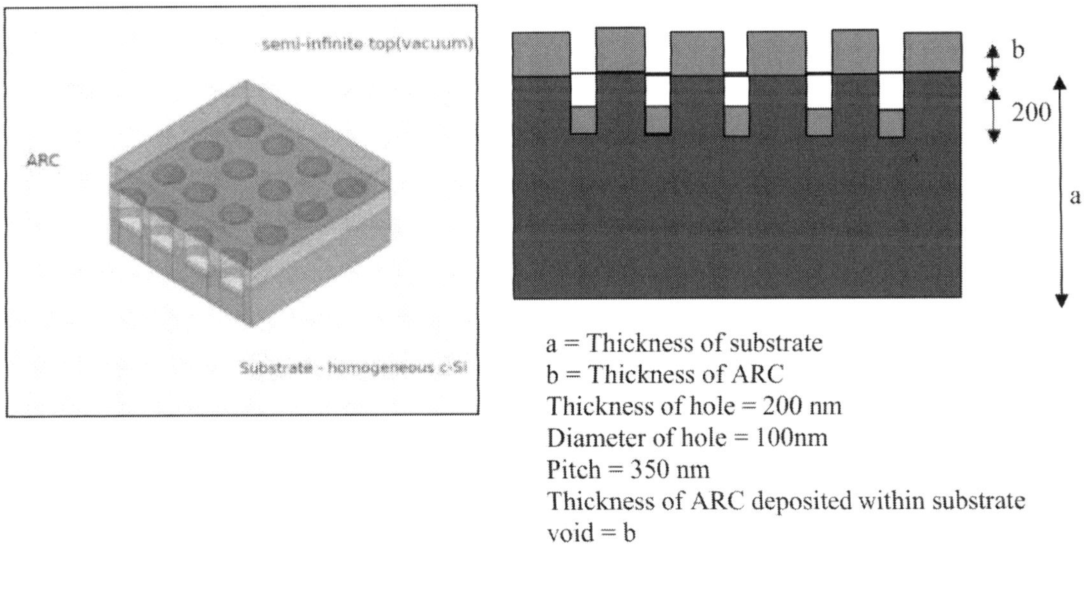

FIGURE 4 Case 4 Configuration

IOP Conf. Series: Materials Science and Engineering 1166 (2021) 012009 doi:10.1088/1757-899X/1166/1/012009

2.2.5. Case 5

This configuration is the extension of the prior configuration wherein the top layer of residual ARC after creating the holes is removed, known as 'top cleaning'. This configuration (fig 5) can be simulated by disabling the first layer only and defining the other parameters same as the preceding case.

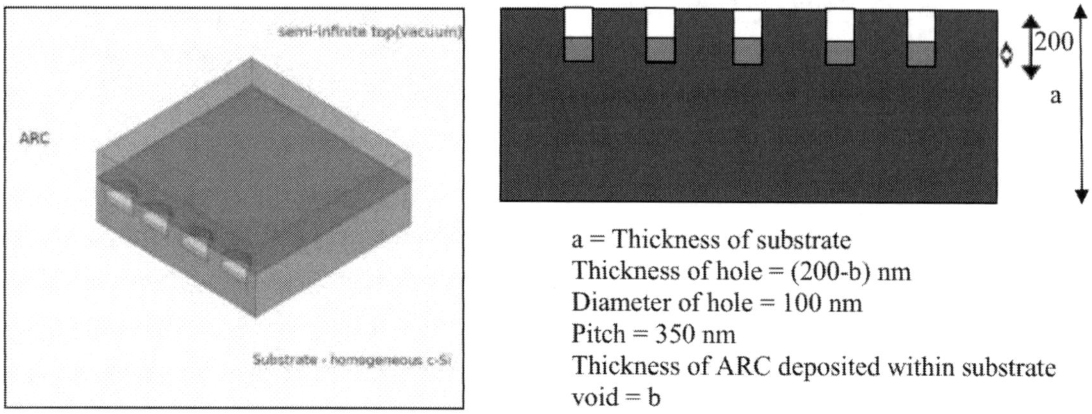

FIGURE 5. Case 5 Configuration

On obtaining the data for all these cases with the specifications as mentioned for a wavelength range of 300 nm to 900 nm with a step of 3 nm, compute the mean surface reflectance values for all the cases and compare them to find the best suitable configuration for reduced reflectance of the combination of the silicon substrate and best performing anti- reflection coating.

3. Simulation of optical properties

Minimization of surface reflectance has always been a critical challenge to resolve the issue of optical losses in solar cells where crystalline silicon is used mainly. In this method, we have attempted to minimize the surface reflectance for silicon substrate by running simulations using an open source software available on Nanohub, 'Stanford Stratified **Structure Solver'** which is more commonly known as **S4**.

4. Results and Discussions

As mentioned earlier, three materials are taken as Anti- reflection Coating(ARC) – Silicon Nitride(SiN_4)), Magnesium Fluoride(MgF_2) and Titanium Dioxide(TiO_2). Simulations were run for bare ARC with no pattern, taking the substrate as c-Si. A comparative study was done to check the reflectance capability for all the above three ARCs individually at different thicknesses.

An average reflectance is calculated at every thickness of the ARC for a constant substrate thickness. This is done to find the material which has the least reflectance value. The following table (Table 1) depicts the best combination of Substrate ARC thickness for achieving minimum surface reflectance value.

IOP Conf. Series: Materials Science and Engineering 1166 (2021) 012009 doi:10.1088/1757-899X/1166/1/012009

Table 1. Minimum surface reflectance combinations for all 3 ARCs.

ARC USED	SUBSTRATE THICKNESS (LAYER 2) (nm)	ANTI-REFLECTION COATING(ARC) THICKNESS (LAYER 1) (nm)	AVERAGE REFLECTANCE VALUE
Silicon Nitride (Si_3N_4)	50000	60	0.13997
Magnesium Fluoride (MgF_2)	50000	90	0.222356
Titanium Dioxide (TiO_2)	25000	50	0.235154697

Comparing all the mean reflectance values for different materials at different thicknesses and different substrate thicknesses we have found out the least values for all the three materials. The ideal results for all the above materials are mentioned in bold in the table itself. This is represented graphically in Figure 6.

From the plot, we can see that Silicon Nitride (Si_3N_4) gives the least reflectance. The cost of patterning is very high. So for the subsequent cases with patterns, Silicon Nitride is selected as the most appropriate material as an anti-reflection coating.

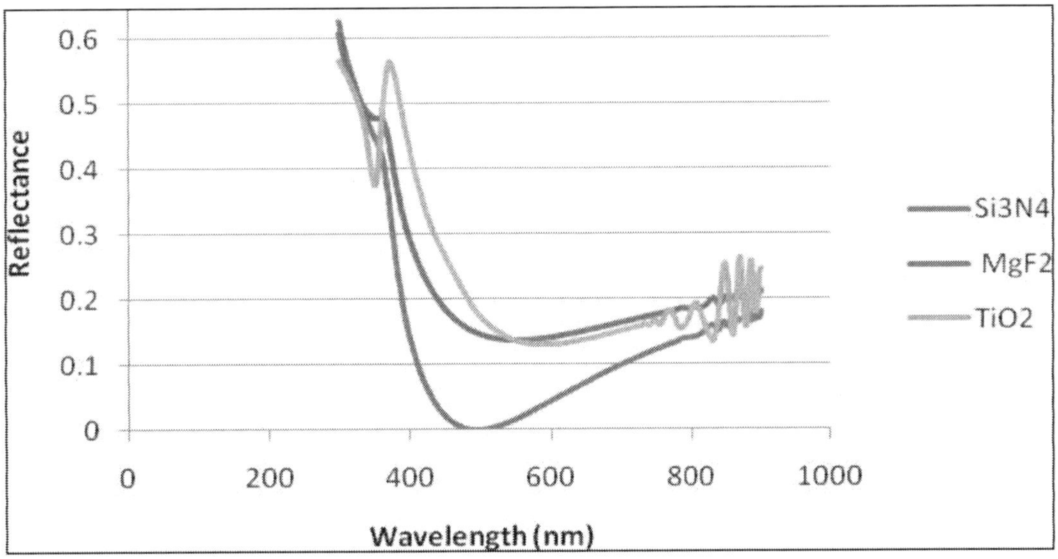

FIGURE 6. Average surface reflectance plot for MgF_2, TiO_2 and Si_3N_4

After selecting the best anti-reflection coating as Silicon Nitride, we move on to the several cases that have been discussed in Section 2.2. These configurations are considered for study to find out the influence of patterns in the surface reflectance result. The first case is the most basic configuration and hence its surface reflectance value is taken as the comparison yardstick for the other cases involving patterns. In keeping with our objective, we perform a comparative analysis of all the configurations.

This is done to select the best possible patterned configuration for Silicon. Simulations are run in the S4 tool and the results which are obtained are depicted below in the result table 2.

Table 2: Compact result table for all the cases.

CASE	FRONT VIEW	MEAN SURFACE REFLECTANCE VALUE	COMPARATIVE ANALYSIS
1	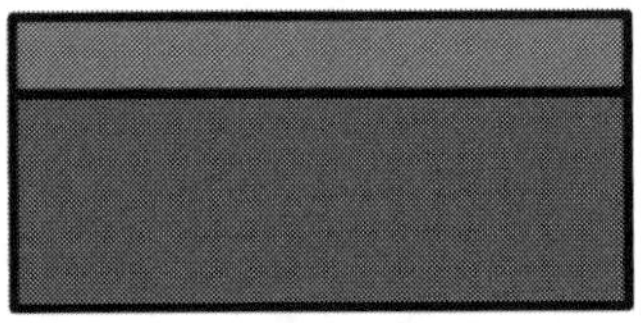	0.13997	Taken as the initial value and is compared with other cases
2	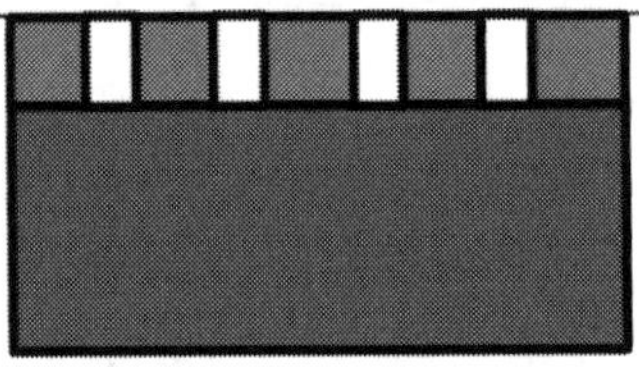	0.1405032	Increases by 0.38%
3	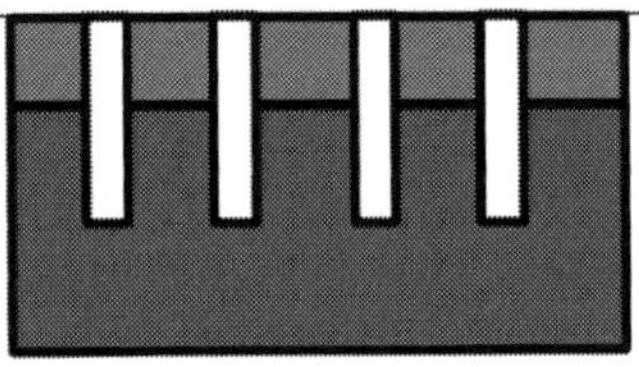	0.136691	Decreases by 2.34%
4	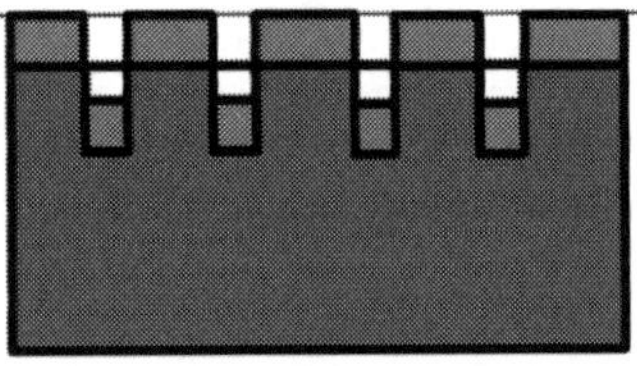	0.1359801	Decreases by 2.85%
5	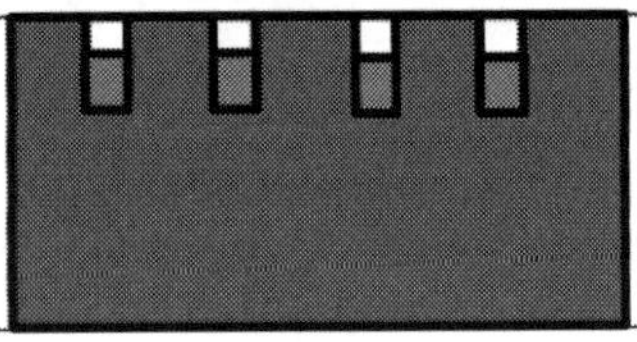	0.3828156	Increases by 173.5%

From Table 2, we can see that the mean reflectance value decreases in Cases 3 and 4. Case 2 shows a slight increase in surface reflectance. In Case 5, reflectance increases rapidly. This is due to the fact that the top part i.e, the Silicon nitride layer is top cleaned. As there is no definite anti-reflection coating layer, so the light doesn't get trapped. So the amount of light reflected back increases swiftly. Comparing the surface reflectance values for all the cases we can say that Case 4 exhibits the best result amongst others with the least surface reflectance. Here the light gets trapped inside the hole causing a very small portion of it to be reflected back. It is the closest depiction of the real case. In actual practice during etching, small amounts of the ARC gets trapped in the cylindrical structures formed on Silicon.

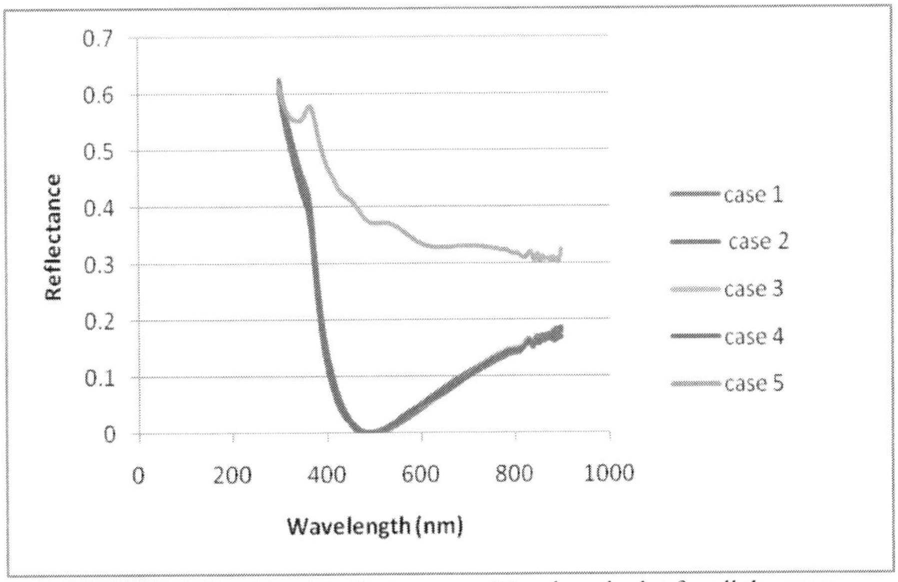

FIGURE 7. Combined Reflectance vs. Wavelength plot for all the cases

A combined plot for all the different cases is shown in Figure 7. From the plot, we notice that almost all the cases (except for Case 5) show the same nature, while Case 5 shows a completely different curve nature as there is rapid increase in reflectance at all wavelength values due to the absence of a definite ARC layer as compared to the other cases.

The surface reflectance is almost zero in the wavelength range of ~470-520nm which basically means that the patterned and the coated substrate will absorb all the incident light in this range. In addition to photovoltaic, such a substrate can have applications for absorbing incident green laser light, which can create problems in aviation industry [9].

5. Conclusion

This study was done as an attempt to reduce the surface reflectance of crystalline silicon through use of anti-reflective coatings (ARCs) and hence determining the most effective configuration. The method was successful as the average surface reflectance was reduced by determining the best performing ARC and the suitable configuration. The fourth configuration where the ARC gets trapped inside the hole gives the best result with a total 2.85% reduction in surface reflectance.

6. References

[1] D. M. Chapin; C. S. Fuller & G. L. Pearson (May 1954). "A New Silicon p- n Junction Photocell for Converting Solar Radiation into Electrical Power". *Journal of Applied Physics*.**25 (5)**: 676–677.

[2] Mehul C. Raval and Sukumar Madugula Reddy (May 15th 2019). *Industrial Silicon Solar Cells, Solar Cells, Majid Nayeripour, Mahdi Mansouri and Eberhard Waffenschmidt, IntechOpen,* DOI: 10.5772/intechopen.84817.

[3] Moona, Girija & Kapruwan, Pankaj & Sharma, Rina. (2017). Silicon Wafer Surface Reflectance Investigations by Using Different Surface Texturing Parameters. *Proceedings of the National Academy of Sciences, India Section A: Physical Sciences.* **88.** 10.1007/s40010- 017-0384-3.

[4] Raut, Hemant & Venkatesan, Anand Ganesh & Nair, Sreekumaran & Ramakrishna, Seeram. (2011). Anti-Reflective Coatings: A Critical, In-Depth Review. *Energy & Environmental Science.* **4.** 3779-3804. 10.1039/C1EE01297E.

[5] Hashmi, Galib & Rashid, Mohammad & Mahmood, Zahid & Hoq, Mahbubul & Rahman, Md. (2018). Investigation of the impact of different ARC layers using PC1Dsimulation:

ftapplication to crystalline silicon solar cells. *Journal of Theoretical and Applied Physics*. **12**. 10.1007/s40094-018-0313-0.

[6] "From 40.7 to 42.8 % Solar Cell Efficiency". *University of Delaware*. July 30, 2007. Retrieved 2008-01-16.

[7] S. C. Baker-Finch, K. R. McIntosh and M. L. Terry, "Isotextured Silicon Solar Cell Analysis and Modeling 1: Optics," in *IEEE Journal of Photovoltaics*, **vol. 2**, no. 4, pp. 457-464, Oct. 2012, doi: 10.1109/JPHOTOV.2012.2206569.

[8] A. Goswami, S. Aravindan and P.V.Rao (December 2016) "Optimization of nanohole array parameters for improving the ultimate efficiency of nanohole structured c-Si solar cells" in *Proceedings of the Institution of Mechanical Engineers, Part N: Journal of Nanomaterials, Nanoengineering and Nanosystems,* **230,** 4, 231-240.

[9] "*https://www.boeing.com/commercial/aeromagazine/articles/qtr_01_10/3/*" retrieved on 9 February 2021.

An integrated approach forinteroperability of standards, condition monitoring methods, and research models used in the power generation sector

Hanumant P. Jagtap[1], Anand K. Bewoor[2], Rajkumar B. Patil[3] Ajinkya C. Dawale[1], Dnyaneshwar P. Kamble[1], Suhas B. Sawant[1]

[1]Zeal College of Engineering & Research,NarhePune, Maharashtra, India
[2]MKSSS's Cummins C.O.E. for women, Pune, Maharashtra, India
[3]AnnasahebDange C.O.E. and Technology, Ashta, Maharashtra

E-mail:*jagtaphp@gmail.com*

Abstract. With the rapid growth of modern technology, maintenance contributes significantly vitalfunction in many industries. The objectives of maintenance-related decision-making process will lead to select and use various condition monitoring techniques for different industrial applications. It is evident from the published literature that, effective use of ISO standards in isolation is very well documented, but after knowing the importance of the use of integrated condition-based maintenance (ICBM) strategy, it will be more important to know how interoperability of international standards, company standards, standard guidelines, and recent research models/recommendations will help to implement ICBM strategy more effectively.These paper overview relevant international standards related to condition monitoring &diagnosis of selected machines used in power generation and highlights case of how concept of interoperability of standards defined at different levels helpedtoselect suitable maintenance strategytoimprove the availability of the system.

1. Introduction

Condition monitoring is the process of monitoring the parameters of a state of the machine in order to detect major changes, indicating a growing fault [1]. Parameters related to machine performance such as vibration, acoustic emission, thermography, wear, temperature, etc. They are useful indicators of the state of the machine [2] .The successful implementation of the condition monitoring system allows the machine to operate without fail [3]. The state-of-the-art monitoring of energy production mainly consists of three distinct elements (i) monitoring of mechanical properties i.e. pins, bearing, and gear, pump, etc. of the program, (ii) application areas such as rotating machinery, wind turbines, mills, electrical equipment, etc., (iii) application of monitoring measures status for decision-making maintenance plan [4]. In recent years, there have been the developments of various effective surveillance and diagnostic techniques. Vibration Analysis, Acoustic Emission Monitoring, Wear Debris, Temperature Analysis, Ultrasonic Monitoring, Thermography, Non-Destructive Testing, Visual Inspection, Vehicle Monitoring, and Current Signature Analysis [5] various should be developed and published by the International Organization for Standardization (ISO) for rolling element bearings, shaft, pumps, fans, steam turbines, gears, centrifugal compressors, induction motors, screw compressors, large generator, steam turbine generator set, etc.

The published standards viz. ISO/7919 series and ISO/10816 series describe acceptable vibration limits for Evaluation of Vibration Severity explains for Mechanical vibration and evaluation criteria of non-reciprocating machines and non-rotating parts respectively [6]. As ISO 13373-1:2001 provides general guidelines for the measurement of machinery vibration for condition monitoring and diagnostics of machines. The recommendations are provided for the measurement methods, parameters, selection, location, and attachment of transducer, data collection, machine operating conditions, vibration monitoring systems, signal conditioning systems, interfaces with processing systems, continuous and periodic monitoring[7]. Also, ISO 17359:2003 provides general guidelines for the overall procedures to be considered when fixing a condition monitor program for Condition monitoring and diagnostics of machines. The choice of a maintenance policy features a great influence on the system's operational cost and its operational availability in thermal power plants (TPP) [8]. The upkeep task is assessed into three groups which are corrective, preventive, and predictive (or condition-based) maintenance [9]. There's a growing impact of standards on global commerce and widely utilized in industries. Therefore this paper reviews the utilization of condition monitoring techniques (CMT) and their application for fault detection, fault identification, and fault classification of machinery through a survey of the literature. This paper reviews and classifies CMT, with reference to the subsequent areas viz. maintenance scheduling, reliability improvement, availability improvement, fault diagnosis alongside their applications in several research and practical domains. This paper overview relevant international standard related to condition monitoring and diagnostics of selected machines used in power generation and represent a case to know how the concept of interoperability of standards defined at different levels that helped toselect suitable maintenance strategy to enhance the availability of the system.

2. Review of literature

Relevant publications to date have been reviewed for the purpose of studying CMT strategies used in plant components, to study international standards, company standards, and research models. The CMT practice in industrial use is aimed at the purposes of a) Maintenance Planning b) Reliability improvement c) Availability improvement. Condition monitoring techniques have been used for vibration analysis, wear debris analysis, ultrasonic analysis, and thermography. The practice of monitoring measures used in industrial applications is set out in Table 1.

Table 1. The trend of CMT

Objective	Vibration Analysis	Acoustic Emission Analysis	Wear Debris Analysis	Ultrasonic Analysis	Thermography
Maintenance Scheduling	[10][11][12][13][14][15][16]	[17][18]	[16]	[19][20]	[21][22]
Reliability Improvement	[23][24][25][26][27][28][29][30][13]	[29][30]	[27][13]		[31][32]
Availability Improvement	[33][34][35][3][36][37][38]	[34][35][36][38]	[35][37][38]	[38]	[38]

Maintenance, reliability, and availability are important tools for decision-making regarding the estimation and planning of maintenance strategies concerning manufacturing and production systems. the utilization of condition monitoring techniques makes it easier for decision-making. These are a number of the popularly used various monitoring tools allowing the upkeep to be supported the condition instead of equivalent operating hours, thereby extending the periods between overhauls. The reliability also can be increased by employing a mixture of monitoring tools altering the plant operators before faults are fully developed.

2.1 Standards utilized in Condition Monitoring

Standards are documented agreements containing technical specifications or other precise criteria to be used consistently as rules, guidelines, or definitions of characteristics, to make sure that materials, products, processes, and services are fit their purpose. within the field of Condition monitoring and diagnostics of machines, a spread of relevant standards are developed and published by ISO (International Organization for Standardization) for Vibration Analysis, Acoustic Analysis, Wear Debris Analysis, Ultrasonic Analysis, and Thermography, which is tabulated in Table No.2 as follows.

Table 2.ISO standards used in condition monitoring technique of machines

Objective	Vibration Analysis	Acoustic Analysis	Wear Debris Analysis	Ultrasonic Analysis	Thermography
International Standards worldwide	ISO 7919 Series, ISO 10816 Series, ISO 13373-1:2001, ISO 13374-1:2003 ISO 13379:2002 ISO 13381-1:2004 ISO 17359:2003	ISO 1996, ISO 1999, ISO 2533, ISO 3744, ISO 9614, ISO 12001, ISO 10843, ISO 22096 ISO 16148:2006	ISO 4406-1999 ISO 4407:2002 ISO 81400 ISO 3722:1976	ISO12715: 2014	ISO 9712, ISO 18434, ISO 18436
Standards followed by country wise	DIN 4150 JIS D 1601:1995 AS ISO 13381, AS 1359 IS 14817	DIN 45681 JIS Z 8732 AS 4748-2001 EN 15856:2010 IS 4758-1968	DIN EN ISO 3104 AS 4002:2001 IS 11083	AS 1065-1998 JIS Z 2345:2000 IS 11626	AS 1929-2009 IS 12782
Guidelines	a) Mechanical vibration of non-reciprocating machines and machine vibration-measurements on non-rotating parts b) To predict maintenance on a wide range of rotating machinery c) Vibration occurs in the 10Hz-10KHz band is considered the best parameters: monitoring structural problems of imbalance, looseness,	a) Perform route cause failure analysis. Find the source of the problem. Fix it once and fix it right b) Periodic Inspection of gas cylinder	a) Design and specification of gearboxes b) Coding the level of contamination c) Qualifying & controlling cleaning methods	a) Investigate the long term service degradation b) Test methods and quality classification	Requirements for training and certification of personnel
Component / Equipment	Steam Turbine, Generator, Bearing, Shaft, Gear, Motor, structure	Cylinder, shaft, bearing, boiler furnace water-cooling wall tube, Induction Motors	Wind turbines, steam turbine-generator, Gearboxes, Hydraulic fluid power systems	Boiler furnaces, Bearing, testing of carbon and low alloy steelplate	Monitoring of bearing, gearboxes, high tension motors.
Models	Maintenance scheduling model, Availability model (Markov model),	Maintenance and asset management model,	Life cycle cost management model, Fault	Maintenance Prediction Model	Fault detection model

	Reliability prediction model Fault prognosis model	Fault prognosis model	prognosis model

The published standards used for vibration analysis defining acceptable vibration limits are the ISO 7919 series (5 parts) and the ISO 10816 series (6 parts). The 7919 series is used for "non-reversing machine vibration i.e. rotating shaft measurements and the testing process" and the ISO / 10816 series is used for "Mechanical vibration testing machines in moderate parts". , in condition monitoring and diagnostic equipment, ISO13373-1: 2001 is used to monitor Vibration status. Provides general guidelines for vibration measurement systems. Also, ISO 17359: 2003 provides general guidelines for standard procedures to be considered when establishing a state-of-the-art monitoring and evaluation system.

In the case of acoustic analysis the ISO standards used are ISO 1996, ISO 1999, ISO 2533, ISO 3744, ISO 9614, ISO 12001, and ISO 10843. ISO 1996 describes the basic values that should be used to describe the environment of a noisy society and describes the basics of the testing process [39]. It also clarifies methods for assessing environmental noise and provides a guide to predicting the potential public response to long-term exposure from a variety of environmental sounds. In the case of dress waste analysis, ISO 4406-1999 has been used as basic guidelines for particle size in liquids [40].

The International Organization for Standardization has developed a 4406: 1999 purification code for measuring pollution levels of one milliliter of liquid in three sizes: 4μ [c], 6μ [c], and 14μ [c]. This ISO code is displayed with the numbers 3: 19/17/14. Each number represents the defining level code of the corresponding particle size. The code includes all particles of specified and larger size. It is important to note that each time the code increases the number of particles doubled.

For Ultrasonic Analysis ISO 12715:2014 is used asreference blocks and test procedure for characterisation of contact probe sound beams [41]. Also, ISO 9712 is used as personnel certification in non-destructive testing, ISO 18434 gives general procedures of thermography and ISO 18436 is used for requirements for training and certification of personnel[8].

After review of various standards as a reference for condition monitoring techniques, the next section reviews the importance of the term "Interoperability" for the maintenance scheduling and condition monitoring program for industrial applications.

It has been found that interoperability may be a must for the mixing of commercial automation systems with operation and maintenance systems so as to enhance the collaboration and interoperability for systems along the products/assets life cycle. Therefore, interoperability will guide for implementation of a condition monitoring program for systems with reference to a selection of standards, selection of condition monitoring techniques, and finally with the objective to improve availability, reliability, and selection of maintenance strategy.

3. Interoperability of standards for condition monitoring program and maintenance strategy

In the field of mechanical monitoring and analysis, various types of standards have been identified and published by ISO (International Organization for Standardization). The ISO is a global coalition of national standards frameworks from 145 countries and considers itself a bridge between the public and private sectors [42]. ISO standards for monitoring and vibration equipment are regulated by the TC108 technical committee whose SCs (subcommittees) Mechanical vibration and shock. Sub-committees include; SC1 measurement, including measuring instruments, SC2 measurement and vibration and shock testing as used in equipment, vehicles, and components, SC3 measurement and measurement tools, SC4 Personal Identification, SC5 Measurement and diagnostic equipment and - SC6 for Vibration and Shock Generating Systems. The stated scope of TC108 is Standardization in the field of mechanical vibration and shock as well as monitoring and diagnostic equipment. In view of all the basic international standards related to monitoring and diagnostic equipment, compliance standards are studied to apply the monitoring system to systems. The interdependence of standards will help in selecting appropriate remediation strategies, improving accessibility, and evaluating system reliability. The principle of cooperation is regarded as a law, regulation, test vessel, standard, or requirement.

Collaboration is the ability of two or more systems or components to exchange and use data. The principle of interaction can be applied to ISO standards for monitoring and diagnostic equipment. The published standards used for Vibration Analysis defining acceptable vibration limits are the ISO 7919 series (5 parts) and the ISO 10816 series (6 parts). Series 7919 is used for "Mechanical vibration of non-returnable machines i.e. Measurement of rotating shafts and testing process" and series ISO / 10816 is used for "Mechanical vibration by measuring machine vibration by measurement in non-rotating parts". ISO 7919-1 is a basic document that describes the general requirements for measuring and testing vibrator equipment using shaft measurements. Similarly, ISO 10816-1 is a basic document that describes the general requirements for inspecting vibrations using case and / or base measurements. Subsequent sections of each series of documents apply to various categories and types of equipment and include specific testing methods used to assess vibrational complexity.

- ISO / TC108 / SC4 define the standards of Condition Monitoring and Diagnostics of Machines in the various categories discussed below.
- ISO 13372 This standard provides Terminology for Fields of Condition Monitoring and Diagnostics of Machines as a basic guide.
- ISO 13373 This standard provides general guidelines for the measurement of vibration monitoring devices. In conjunction with ISO 13373, recommendations are provided for monitoring the status and diagnostic of equipment through ISO 13373-1: Part 1: as standard procedures and ISO 13373-2: Part 2 Analysis, analysis, and diagnostic of the machine.
- ISO 13374 This standard is used for monitoring and diagnostic equipment for data processing, communication and presentation. It takes a reference to the types of programs, ratings, converters, data presentations, advanced and analytical Annex A - types and areas of ratings, Annex B - material information, Annex C - entertainment, ratings of Annex D.
- ISO 13379: Condition monitoring and diagnostic equipment - General guidelines for data interpretation and diagnostic techniques are 1) condition setting and diagnostic setup, 2) failure mode indicators, 3) diagnostic equipment, 4) diagnostic methods.
- ISO 13380: This standard provides a variety of documents, including indicator standards, suitable for Performance Monitoring and Diagnostics of equipment 1) parameters observed, 2) measurement process, 3) error detection, 4)) error chart monitoring.
- µ ISO 17359: The standard level introduces adopted guidelines to be used in the development of a monitoring system. It is used as a standard checklist to make sure there are no big things to miss. General guidelines for setting up a situation monitoring system at various stages consider 1) critical analysis, 2) measurement methods, 3) measurement areas, 4) data collection and analysis, 5) maintenance action
- ISO 18436: This standard defines the certification process for vibration screening personnel who perform monitoring and diagnostic equipment. This standard introduces Guidelines for Training and Certification Staff Requirements to meet the required learning, training, and experience requirements, and successfully completes the qualification assessment and equipment diagnostics described in three sections as 1) Part I - verification requirements, 2) Part II - vibration analysis, 3) Part III - training body requirements.

After a critical review of ISO standards, it was found that ISO 17359 describes the methods used to create monitoring systems and the selection of equipment to repair equipment. The methods used to monitor the condition of the equipment are i) vibration analysis ii) acoustic analysis iii) wear waste analysis iv) Thermography. Defined definitions of ISO standards for monitoring systems are already described in section 2.1. The importance of the interaction of standards in choosing a remediation strategy is discussed as follows.

As "Interoperability" has great importance for implementing condition monitoring programs and selecting maintenance strategy for systems. The upkeep activities on critical equipment and systems have traditionally employed one among two maintenance philosophies which are preventative maintenance or corrective maintenance but, condition-based maintenance (CBM) or predictive maintenance represents a replacement maintenance philosophy, whereby maintenance activities are only performed when there's objective evidence of an impending fault or failure condition. The CBM approach wont to optimize the supply of high-value critical assets, whilst also reducing overall maintenance and logistics costs. Condition-based maintenance entails continuous monitoring of system data to supply an accurate assessment of the health, or status, of a component/system and performing maintenance supported its observed health. It involves using real-time system monitoring and processing. There are three levels of the CBM process viz. data acquisition, processing, and maintenance decision making.

In order to achieve improved maintenance decision-making, it is required to use condition monitoring for the execution of condition-based maintenance strategy. Recommendation of published literature and fields experience advocates the use of more than two CBM methods for correct diagnosis and condition predictions. But, it demands the use of different standards to monitor the condition of the same component/system and in such situations interoperability of such standards could be a challenging issue to tackle with. One such case of coordinated condition monitoring approach implementation is discussed in brief with a case study.

3.1 Case study

The growing requirement for electricity has indicated that it is crucial to keep the power generation resources in India with a higher preference. With several sources, TPP is principal source of power production. It's crucial to retain TPP in running condition. Providentially, this is not the case, because breakdown is inevitable,although it can be reduced through execution of accurate maintenance policy. The earlier researchers have conducted the study for the improvement of the reliability and availability analysis for subsystems of thermal power such as boiler-furnace subsystem [52], turbine-generator subsystem [53], condenser subsystem [54], coal-supply subsystem [55], and water-circulation subsystem [56].

Jagtap, et al. [38] presented a case of induced draft fan of TPP. In their study, a coordinated condition monitoring approach is proposed by the mixing of 4 techniques, namely, vibration analysis, noise analysis, wear analysis, and ultrasound analysis techniques for diagnosis and validated through a case study of induced draft fan utilized in a TPP. At the initial stage, the methodology focused on selection criteria of critical equipment of TPP using multi-criteria decision making techniques viz. analytic hierarchy process [56]. State of the ID fan is monitored by measuring various parameters measured such as vibration level, noise level, and size of wear particle. Then, ISO standards were followed for finding the deviation from a reference level of these parameters for vibration level [ISO 10816], noise level [ISO 1996], ultrasound level [ISO12715:2014], and particulates in oil sample [ISO 4407:2002]. At the initial stage of fault detection, the trends of vibration parameters are analyzed, and a comparison is formed between measured vibration parameters level with ISO-10816 standard. The measured vibration characteristics of the ID fan have provided quick and accurate information that hydraulic coupling is during a deteriorated state. At the final stage of methodology, through detailed study, the fault of the bearing of the hydraulic coupling (HC) of the induced draft fan is detected and it was recommended to remove the old HC bearing and suggested to install a new pack of HC bearing well in advance to avoid the failure of ID fan of TPP. An ISO 9712:2012 recommends and defines conditions for principles for the qualification and certification of employees who could perform industrial non-destructive testing at the plant to confirm the effectiveness of solution implementation. The result obtained from a case study helped the plant engineer to optimizethe maintenance plan and resource allocation at the nearest available opportunity without generation loss.

4. Summary

This study presents an overview of relevant published literature which deals with interoperability of international standards and therefore the models for condition monitoring-based maintenance strategy.The Paper also highlights the challenging area for implementing simultaneous use of ISO standards and company standards for condition monitoring, fault diagnosis, reliability improvement, availability improvement, maintenance modelling of the system used in various process industries such as TPPs. A case study reported in this paper presented guidelines for use of ISO standards, the recommendation for maintenance planning, and general guidelines for the skills required for the personnel at end-user. For precise fault finding and diagnosis and diagnosis, it is recommended that field operators/engineers must understand relevant ISO standards, company standards/guidelines, research models, and best practices followed in the field to define optimum maintenance strategy for subsystems/components of system/subsystem/component under consideration.

References

[1] Y. Han and Y. H. Song, "Condition Monitoring Techniques for Electrical Equipment—A Literature Survey," *IEEE Trans. Power Deliv.*, vol. 18, no. 1, pp. 4–13, 2003.
[2] R. S. Gurjar, A. Yadav, and P. Jayashwal, "Failure analysis of belt conveyor system in a thermal power plant," *Int. J. Adv. Technol. Eng. Res.*, vol. 2, no. 3, pp. 204–210, 2012.
[3] B. R. Kumar, K. V. Ramana, and K. M. Rao, "Condition monitoring and fault diagnosis of a boiler feed pump unit," *J. Sci. Ind. Res. (India).*, vol. 68, no. 9, pp. 789–793, 2009.
[4] S. Bagavathiappan, B. B. Lahiri, T. Saravanan, J. Philip, and T. Jayakumar, "Infrared thermography for condition monitoring - A review," *Infrared Phys. Technol.*, vol. 60, pp. 35–55, 2013.
[5] Z. K. Piyushsinh, "Application of monitoring approaches on the steam turbine of the thermal power plant," *Int. J. Mech. Eng.*, vol. 3, no. 2, pp. 63–67, 2014.
[6] I. Standard, "Mechanical vibration — evaluation of machine vibration by measurements on non-rotating parts," 2004.
[7] J. M. Robichaud and P. Eng, "Reference Standards for Vibration Monitoring and Analysis," *(Bretech Eng. Ltd)*, 2009.
[8] A. Standard, "Australian Standard ® Condition monitoring and diagnostics of machines — Thermography Part 1 : General procedures," 2014.
[9] Ramnarayan Sahu, Dr. A.C.Tiwari, and Anupama Yadav, "Implementation of Preventive Maintenance Program in Thermal Power Station: a Review and Case Study," *Int. J. Mech. Eng. Technol. (IJMET)*, pp. 685–692, 2012.
[10] Nikam Amol, Garud Amol, Simraan Khan, Sadakale Aniket, "Vibration-Based Condition Monitoring of FD-FAN of Boiler at Power Plant," *Int. J. Eng. Technol. Sci. Res.*, vol. 5, no. 3, pp. 1657–1667, 2018.
[11] X. Jin, H. Yang, and H. Wang, "Fault diagnosis for civil aviation engine test based on the vibration signal," *Procedia Eng.*, vol. 29, pp. 3002–3006, 2012.
[12] Z. Guan, P. Chen, X. Zhang, X. Zhou, and K. Li, "Vibration analysis of shaft misalignment and diagnosis method for rotating machinery," *Int. J. Performability Eng.*, vol.13, 4, pp. 337–347, 2017.
[13] S. Ebersbach, Z. Peng, and N. J. Kessissoglou, "The investigation of the condition and faults of a spur gearbox using vibration and wear debris analysis techniques," *Wear*, vol. 260, no. 1–2, pp. 16–24, 2006.
[14] N. Dileep, K. Anusha, C. Satyaprathik, B. Kartheek, and K. R. A. Professor, "Condition Monitoring of FD-FAN Using Vibration Analysis," *Int. J. Emerg. Technol. Adv. Eng.*, vol. 3, no. 1, pp. 170–186, 2013.
[15] R. Dutta, J. P. Dwivedi, V. P. Singh, and A. Ghosh, "Using Vibration Analysis to Identify & Correct an Induced Draft Fan ' s Foundation Problem of a Pollution Control Device - A Case Study," *Int. J. Appl. Eng. Res.*, vol. 13, no. 8, pp. 5831–5840, 2018.
[16] Z. Peng, N. J. Kessissoglou, and M. Cox, "A study of the effect of contaminant particles in lubricants using wear debris & vibration condition monitoring techniques," *Wear*, vol. 258, 11–12, pp. 1651–1662, 2005.
[17] C. K. Tan and D. Mba, "Identification of the acoustic emission source during a comparative study on the diagnosis of a spur gearbox," *Tribol. Int.*, vol. 38, no. 5, pp. 469–480, 2005.
[18] D. P. Jena, S. N. Panigrahi, and R. Kumar, "Multiple-teeth defect localization in geared systems using filtered acoustic spectrogram," *Appl. Acoust.*, vol. 74, no. 6, pp. 823–833, 2013.
[19] A. Khaira, S. Shrivastava, and A. Suhane, "Analysis of co-relation between ultrasonic testing and microstructural examination for condition-based maintenance system," *Int. J. Eng. Sci. Res.*, vol. 04, no. 02, pp. 820–823, 2013.

[20] T. Heo and S. H. Cho, "Thin-Plate-Type Embedded Ultrasonic Transducer Based on Magnetostriction for the Thickness Monitoring of the Secondary Piping System of a Nuclear Power Plant," *Nucl. Eng. Technol.*, vol. 48, no. 6, pp. 1404–1411, 2016.
[21] S. Bagavathiappan, T. Saravanan, N. P. George, J. Philip, T. Jayakumar, and B. Raj, "Condition monitoring of exhaust system blowers using infrared thermography," *Insight Non-Destructive Test. Cond. Monit.*, vol. 50, no. 9, pp. 512–515, 2008.
[22] A. S. N. Huda and S. Taib, "Application of infrared thermography for predictive/preventive maintenance of thermal defect in electrical equipment," *Appl. Therm. Eng.*, vol. 61, no. 2, pp. 220–227, 2013.
[23] Z. Peng and N. Kessissoglou, "An integrated approach to fault diagnosis of machinery using wear debris and vibration analysis," *Wear*, vol. 255, no. 7–12, pp. 1221–1232, 2003.
[24] V. N. A. N. V. M. S. Hussain, "Reliability Modeling for Rotor Systems with Imbalance Based on Vibration Analysis," *Proc. 2016 Int. Conf. Ind. Eng. Oper. Manag. Detroit, Michigan, USA, Sept. 23-25, 2016*, pp. 341–351, 2016.
[25] N. Mehala, "Condition Monitoring and Fault Diagnosis of Induction Motor Using Motor Current Signature Analysis Electrical Engineering Department National Institute of Technology Diagnosis of Induction Motor Using," *Electr. Eng.*, no. 2, pp. 1–194, 2010.
[26] C. Liu, D. Jiang, J. Chen, and J. Chen, "Torsional vibration and fatigue evaluation in repairing the worn shafting of the steam turbine," *Eng. Fail. Anal.*, vol. 26, pp. 1–11, 2012.
[27] C. Yibo, X. Xiaopeng, L. Yan, and D. Tianhuai, "A Study of the Correlation between Gear Wear and Vibration," *Proc. Cist*, no. 1, pp. 273–277, 2008.
[28] L. Deng and R. Zhao, "A vibration analysis method based on hybrid techniques and its application to rotating machinery," *Meas. J. Int. Meas. Confed.*, vol. 46, no. 9, pp. 3671–3682, 2013.
[29] Y. Qu, D. He, J. Yoon, B. Van Hecke, E. Bechhoefer, and J. Zhu, "Gearbox tooth cut fault diagnostics using acoustic emission & vibration sensors - A comparative," *Sensors*, vol. 14, 1, pp. 1372–1393, 2014.
[30] A. M. Al-Ghamd and D. Mba, "A comparative experimental study on the use of acoustic emission and vibration analysis for bearing defect identification and estimation of defect size," *Mech. Syst. Signal Process.*, vol. 20, no. 7, pp. 1537–1571, 2006.
[31] M. M. Ahmed, A. S. N. Huda, and N. A. Mat Isa, "Recursive construction of output-context fuzzy systems for the condition monitoring of electrical hotspots based on infrared thermography," *Eng. Appl. Artif. Intell.*, vol. 39, pp. 120–131, 2015.
[32] V. Leemans, M.-F. Destain, B. Kilundu, and P. Dehombreux, "Evaluation of the Performance of Infrared Thermography for On-Line Condition Monitoring of Rotating Machines," *Engineering*, vol. 03, no. 10, pp. 1030–1039, 2011.
[33] A. A. Mohamed, R. Neilson, P. MacConnell, N. C. Renton, and W. Deans, "Monitoring of fatigue crack stages in a high carbon steel rotating shaft using vibration," *Procedia Eng.*, vol. 10, pp. 130–135, 2011.
[34] J. Arroyo, M. Muñoz, F. Moreno, N. Bernal, and C. Monné, "Diagnostic method based on the analysis of the vibration and acoustic emission energy for emergency diesel generators in nuclear plants," *Appl. Acoust.*, vol. 74, no. 4, pp. 502–508, 2013.
[35] C. K. Tan, P. Irving, and D. Mba, "A comparative experimental study on the diagnostic and prognostic capabilities of Acoustics Emission, Vibration and Spectrometric Oil Analysis for spur gears," *Mech. Syst. Signal Process.*, vol. 21, no. 1, pp. 208–233, 2007.
[36] R. Christian, S. H. Song, and H. G. Kang, "Study of core support barrel vibration monitoring using ex-core neutron noise analysis and fuzzy logic algorithm," *Nucl. Eng. Technol.*, vol. 47, no. 2, pp. 165–175, 2015.
[37] S. Ebersbach, Z. Peng, and N. Kessissoglou, "Smart Condition Monitoring By Integration of Vibration, Oil and Wear Particle Analysis," *14th Int. Congr. Sound Vib.*, 2007.
[38] H. P. Jagtap, A. K. Bewoor, and R. Kumar, "Failure analysis of induced draft fan used in a thermal power plant using coordinated condition monitoring approach : A case study," *Eng. Fail. Anal.*, vol. 111, no. April 2019, p. 104442, 2020.

[39] measurement and assessment of environmental noise Acoustics-Description, "International Standards ISO 1996-1." 2003.
[40] "ISO 4406 : 99 Cleanliness Level Standards ISO 4406 : 99 Cleanliness Level Standards Oil Cleanliness levels may be reported using either a two or three-digit rating .," pp. 10–11.
[41] I. S. ISO and 12715, "Non-destructive testing-Ultrasonic testing," vol. 2014, 2014.
[42] L. W. Akerblom, "International Standards pertaining to Thermography Practices, Training and Certification 2 . The need for International Standards," *IV Conf. Panam. END Buenos Aires – Oct. 2007*, pp. 1–9, 2007.
[43] M. O. Kobra Heidarbeigi, Hojat Ahmadi, "Adaptive Vibration Condition Monitoring Techniques for Local Tooth Damage in Gearbox," *Mod. Appl. Sci.*, vol. 4, no. 7, pp. 104–110, 2010.
[44] C. Crabtree, Y. Feng, and P. Tavner, "Detecting incipient wind turbine gearbox failure: a signal analysis method for on-line condition monitoring," *Eur. Wind Energy Conf. (EWEC 2010)*, no. Ewec, p. 6, 2010.
[45] B. E. Brandon Van Hecke, Yongzhi Qu, David He, "Vibration-based wind turbine planetary gearbox fault diagnosis using spectral averaging A New Spectral Average-Based Bearing Fault Diagnostic," *J. Fail. Anal. Prev.*, no. January 2015.
[46] H. P. Jagtap, A. Bewoor, and R. Kumar, "Instrumentation Mesure Métrologie Thermal Power Plant Condenser Fault Diagnosis Using Coordinated Condition Monitoring Approach," vol. 18, no. 3, pp. 223–235, 2019.
[47] D. Mba and R. B. K. N. Rao, "Development of acoustic emission technology for condition monitoring and diagnosis of rotating machines: Bearings, pumps, gearboxes, engines, and rotating structures," *Shock Vib. Dig.*, vol. 38, no. 2, pp. 3–16, 2006.
[48] A. Joshuva and V. Sugumaran, "Classification of various wind turbine blade faults through vibration signals using hyperpipes and voting feature intervals algorithm," *Int. J. Performability Eng.*, vol. 13, no. 3, pp. 247–258, 2017.
[49] [56] O. Bennouna, N. Heraud, and Z. Leonowicz, "Condition monitoring & fault diagnosis system for Offshore Wind Turbines," 2012 11th Int. Conf. Environ. Electr. Eng., pp. 13–17, 2012.
[50] . Antoniadou, G. Manson, W. J. Staszewski, T. Barszcz, and K. Worden, "A time-frequency analysis approach for condition monitoring of a wind turbine gearbox under varying load conditions," Mech. Syst. Signal Process., vol. 64–65, pp. 188–216, 2015.
[51] International standards ISO 13372, Condition monitoring and diagnostics of machines.
[52] Hanumant Jagtap, Anand Bewoor, Ravinder Kumar, Mohammad Hossein Ahmadi, Giulio Lorenzini, "Markov-based performance evaluation and availability optimization of the boiler–furnace system in coal-fired thermal power plant using PSO", Energy Reports, 6,pp. 1124–1134, 2020.
[53] H.P. Jagtap, A.K. Bewoor, R. Kumar, M.H. Ahmadi, L. Chen, "Performance analysis and availability optimization to improve maintenance schedule for the turbo-generator subsystem of a thermal power plant using particle swarm optimization", Reliability Engineering & System Safety, 2020.
[54] Hanumant Jagtap, Anand Bewoor, Ravinder Kumar, Thermal Power Plant Condenser Fault Diagnosis Using Coordinated Condition Monitoring Approach, Instrumentation Mesure Métrologie 18 (3), 223-235.
[55] H.P. Jagtap, A.K. Bewoor, "Markov Probabilistic Approach-Based Availability Simulation Modeling and Performance Evaluation of Coal Supply System of Thermal Power Plant", Reliability, Safety and Hazard Assessment for Risk-Based Technologies, pp-813-824, 2020.
[56] Hanumant P. Jagtap, Anand K. Bewoor, Ravinder Kumar, Mohammad Hossein Ahmadi, Mamdouh El Haj Assad, Mohsen Sharifpur, RAM analysis and availability optimization of thermal power plant water circulation system using PSO, Energy Reports,2020.
[57] H.P. Jagtap, A. K. Bewoor, "Use of analytic hierarchy process methodology for criticality analysis of thermal power plant equipments", Materials Today: Proceedings, 4 pp. 1927–1936, 2017.

Structural and dc electrical transport study of Cu-Zn ferrialuminates prepared by chemical route.

U B Sankpal*[1], A M Kulkarni [1], S. S. Waghadare [1], K V Sukhatankar*[2], P P Kulkarni [1] and D J Sathe [3]

[1]Department of Chemistry, Gogate - Jogalekar College, Ratnagiri, Maharashtra, India.
[2]Department of Physics, Gogate - Jogalekar College, Ratnagiri, Maharashtra, India.
[2] Minor Research Project, University of Mumbai, Mumbai, Maharashtra, India
[3]Department of BSH, KIT'S College of Engg. (Autonomous), Kolhapur, Maharashtra, India.

dr.ubsankpal@gmail.com & ksukhatankar@gmail.com

Abstract. Ferrimagnetic ionic crystalline materials, with general formula $AB_2O_4$are termed as spinel phase ferrites. Electro – magnetic properties, concentration of different ions and surface morphology of mixed–metal oxides shows significant applications in the various fields. In above work, Cu doped Zn ferrialuminate were prepared by sol-gel auto-combustion route. Purity of the synthesized materials were confirmed by XRD analysis which reveals the formation spinal phase structure and by applying the Debye-Scherer's formula. Nanocrystalline nature of the samples were studied by transmission electron microscopy (TEM). The particle was found in the nanorange. The temperature dependent DC electrical conductivity was studied by two probe method. Semiconductor behavior with the rise in temperature were found in all the samples.

Keywords: Ferrimagnetic; XRD; Nano-sized particle; semiconductor;

1. Introduction

Nanostructured spinel ferrites, nominally AFe_2O_4 where A is a bivalent metal ion and often a transition metal ion, play a significant role in technologies, viz. high frequency transformers, antennae, various microwave and radar systems, high speed digital tape recording devices, catalysis and magnetic equipments [1-6]. The different physico-chemical properties of the ferrospinels having resourceful structure as well as proportional composition play a vibrant role to mark them applicable in several ways. Its spreading over tetrahedral (A) and octahedral (B) sites is key factor to make them tailor made material. These compounds have good ability to tune their electrical and magnetic properties which mainly depend upon the selection of metal ions, its distribution and their processing conditions [7-11].

Various magnetic properties of spinels are due to the strong interaction of metal ions, distributed over tetrahedral (A) and octahedral (B) sites [12-20]. They have strong potential to absorb high frequency radiations and hence are applicable in radio frequency devices [21-25]. Ferrsospinels are high resistivity materials and are suitable for microwave appliances also [26-30]

Our literature review expose that, numerous research reports on ternary system containing diamagnetic transition metal ion distributed over different sites but very limited reports on various composition system having four number of ions in single composition with single phase. With an assessment to study the four different element containing series of Zn^{+2} ion substituted Cu - ferrialuminates were formulated by soft chemical route using citric acid and investigated by various characterization tools.

2 Experimental Methodology

Cu-Zn ferraluminate [$Cu_{1-x}Zn_xFeAlO_4$ ($0.0 \leq x \leq 1.0$)] system with various compositions having nanoscale particle size were synthesized by one-step citrate sol–gel method using citric acid as oxidant as well as fuel [22]. Stoichiometric amount of $Cu(NO_3)_2.5H_2O$, $Zn(NO_3)_2.6H_2O$, $Fe(NO_3)_3.9H_2O$, $Al(NO_3)_3.9H_2O$, liquor ammonia and Citric acid ($C_6H_8O_7$) were uses as starting material. All the metal salts were dissolved in deionized double distilled water. The molar concentration of metal salts and fuel-oxidant (citric acid) were maintained as 1:2. Aqueous solutions of metal nitrates were added to the citric-acid solution and stirred for 2 Hrs to get homogenous mass. With the help of ammonia solution, the pH of the solution was maintained about 8. The above homogenous mass was heated slowly on hotplate at about 80°C to convert solution into gel with continuous stirring. Finally colored gel was further heated to convert into floopy powder by auto- combustion. Finely grinded powder form samples were analyzed by diffractometer by X-rays (XRD) and electron beams of transmission electron microscope (TEM) to focus on sample to produce images and confirm nanostructure techniques. The temperature dependent dc resistivity study was followed using two probe method and investigated in detail.

3 Characterization

3.1 Phase formation by XRD

The as prepared middle composition $Cu_{0.5}Zn_{0.5}FeAlO_4$ of the mentioned series was sintered at different temperature (500 – 900°C). Their X-ray pattern was scanned on Philips PW 1710 diffractometer with $CuK\alpha_1$ radiation (1.5406Å). Figure 1 shows that 700°C is desired temperature to get single phase spinel compound. Hence all the composition of the series was sintered at same temperature (Figure 2). All the samples show the presence of face centered cubic spinel structure. The peaks were indexed to plane (111), (220), (311), (222), (400), (411), (511), and (440) of pure single phase nanocrystalline structure. Lattice constant 'a' was seen to vary from 8.41 Å to 8.48 Å with increase in Zn^{2+} ion concentration (x = 0.0 to 1.0). It's because of large ionic radii of Zn^{2+} ion (0.82 Å) [31] compared to that of Cu^{2+} ion (0.72 Å). Well known Debye-Scherer's equation was followed to determine crystallite size and it's in the range of 56-76 nm.

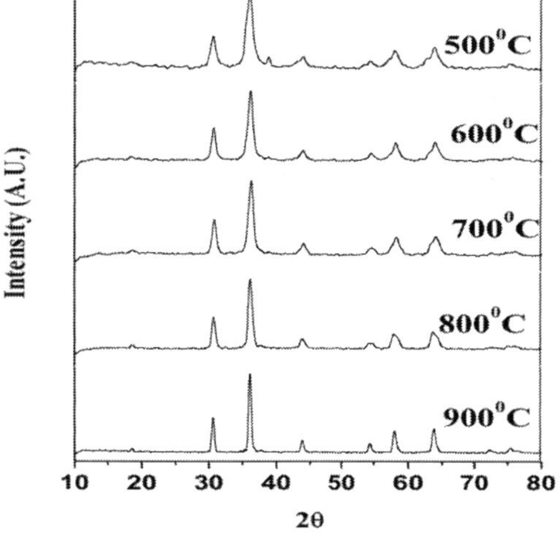

Figure 1. XRD Diffraction pattern of $Cu_{0.5}Zn_{0.5}FeAlO_4$ at various sintering temperature

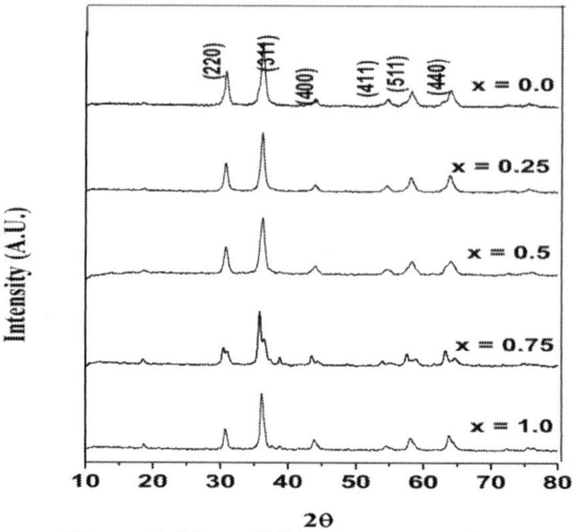

Figure 2. X-ray diffraction pattern of $Cu_{1-x}Zn_xFeAlO_4$ $(0.0 \leq x \leq 1.0)$.

3.2 Electrical resistivity

The liner plots without any disruption of log ρ vs 1000/T (Figure 3) reveals the semiconducting nature of all the samples. The dc electrical resistivity of all the samples varied from 10^7 - 10^{10}Ωcm while activation energy values differs from 0.46 to 081eV.

The nonlinear trend in electrical resistivity with respect to increase in concentration of diamagnetic Zn^{+2} ions is due to distribution of same or different ions over two different sites (tetrahedral and octahedral). The site preference energy of various ions is also accountable for its anomalous behavior. These plots of Figure 3, shows that, dc electrical resistivity increases gradually as Zn^{+2} ion concentration increases up to x = 0.5 ($Cu_{0.5}Zn_{0.5}FeAlO_4$) and then drops to lowest value at x = 0.75 ($Cu_{0.25}Zn_{0.75}FeAlO_4$). The last composition of the system $ZnFeAlO_4$ (x=1.0) shows most dc electrical resistivity. The hopping mechanism in the oxides helps to explain the conduction in mixed-metal oxides which takes place through charge carries over different sites (tetrahedral and octahedral) as well as probable variable oxidation states of same / different ions. In the initial compositions (up to x = 0.5) there will be absence of Fe^{+2} ions [32] to show continuous increase in resistivity while in further compositions, $Cu_{0.25}Zn_{0.75}FeAlO_4$ & $ZnFeAlO_4$. Zn^{+2} ions strongly desire tetrahedral (A) sites & Cu^{+2} ions favor the octahedral (B) sites while $Fe^{+2/+3}$ ions moderately locate mutually A and B sites [2, 33-40]. Gradual increase in Zn^{+2} ion concentration (at A site) results decrease in the Cu^{+2} ions concentration at B site which is responsible for relocation of some $Fe^{+2/+3}$ ions from tetrahedral (A) site to octahedral (B) site with reduction in content of Cu^{2+} ions from B sites. As a consequence of it, amount of ferrous and ferric ions at B site increases, which are responsible for conduction in ferrite by hopping mechanism.

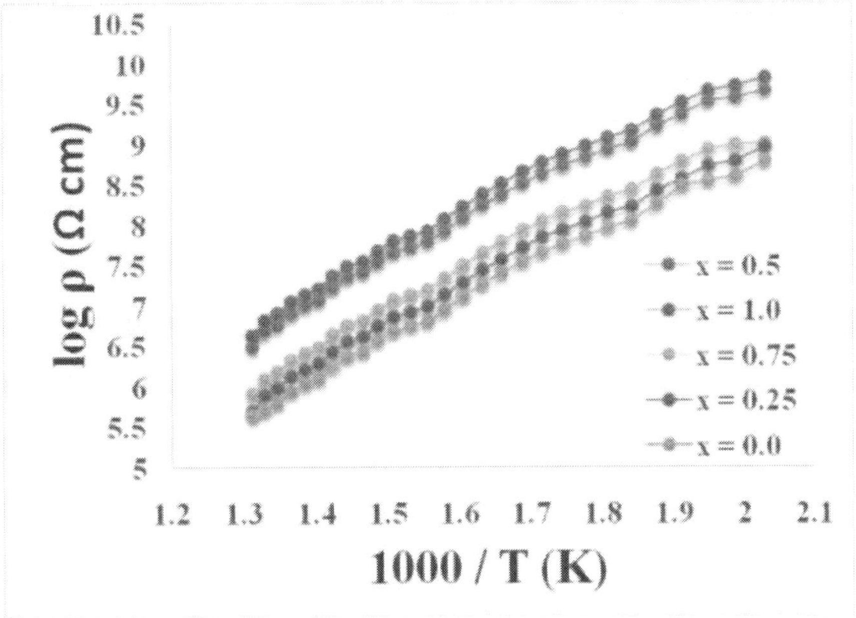

Figure 3. Resistivity plot of $Cu_{1-x}Zn_xFeAlO_4$ ($0.0 \leq x \leq 1.0$)

3.3 Surface Morphology

TEM images of $CuFeAlO_4$ and $ZnFeAlO_4$ are shown in Figure 4 (a) and (b) respectively. Particle dimensions of both the samples are in nanorange and found to be of 50 - 70 nm. The content of zinc ions and sintering temperature aids for interaction amongst grain to develop them appropriately. The higher energy barrier of $CuFeAlO_4$ than that of $ZnFeAlO_4$ results the large particles of Cu-ferrialuminates than Zn- ferrialuminates.

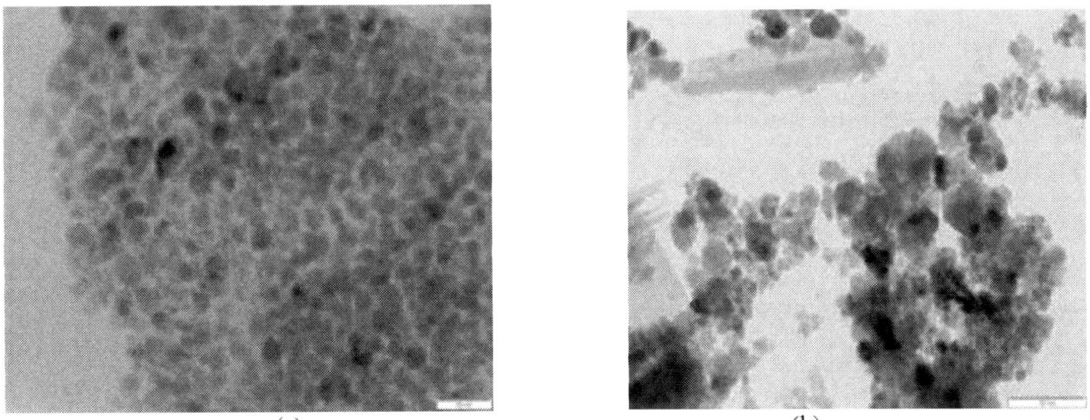

(a) (b)

Figure 4. Transmission electron microscopic images of samples: (a) $CuFeAlO_4$, (b) $ZnFeAlO_4$.

4. Conclusion

Soft and low temperature chemical route (sol-gel auto-combustion) is employed for synthesis of Cu-Zn-Fe-Al quaternary oxides with good stoichiometry. We are successful to get single-phase FCC spinel structure. DC resistivity is linearly increases with increase in concentration of zinc reveals the semiconducting nature of all the compositions. The hopping mechanism of electrons between the Fe^{+2} and Fe^{+3} ions explicates the deviation of dc conductivity with temperature. The investigated material has desirable electrical resistivity for use in microwave appliances.

References

[1] G. Blasse, Philips 1964 *Res. Rep. Suppl.*3 96.
[2] J. B. Goodenough, 1971 *Prog. Solid State Chem.*5 145.
[3] J. Smith and H. P. J. Wijn, 1959*Ferrites Physical properties of Ferrimagnetic Oxides in Relation to their Technical Applications* (John Wiley and Sons, New York).
[4] K. Jalaiah, K. Chandra Mouli, K.Vijaya Babu , R.V.Krishnaiah, 2019 *J. Science: Advanced Materials and Devices***4** (2) 310.
[5] R. J. Rennard and W. L. Kha, 1971 *J. Catal.***21** 182.
[6] S. P. Ghorapade, V. S. Darshane, S. G. Dixit, 1998 *J. Appl.Catal. A: General* **166** 135.
[7] B. A. Mulla and V. S. Darshane, 1983 *Proc. Ind. Acad. Sci. (Chem. Sci.)***92** (2) 139.
[8] Santosh Kalunge, Ashok V. Humbe, Mangesh V. Khedkar, S. D. More, A. P. Keche and A. A. Pandit, 2020 NANOMAT-2020 *Journal of Physics: Conference Series IOP Publishing* **1644** 012017.
[9] S.Dasgupta, K. B.Kim, J. Ellrich, J. Manna 2006 *J. Alloys Compd.* **429** 13.
[10] Alex Goldmann, 1999*Handbook of Modern Ferromagnetic Material*, (Kulwar Academic Publisher, Boston).
[11] S Z Zhang and G C Messing, 1990 *J. Am. Ceram. Soc.***73** 61.
[12] M Sugimoto 1999 *J. Am. Ceram. Soc.***82** 269.
[13] C Kittel Introduction to Solid State Physics,7th Ed. John Wiley and Sons 1996.
[14] A C F M Costa E. Tortella, M.R. Moreui, R. Kiniman 2003 *J. Magn. Magn, Mater***256** 174.
[15] B M Berskoy, V F Medvedev, M S Kralov, *Magnetic Fliuid Engineering Application* (Oxford University Press) 19393.
[16] M S Hasan, M I Arshad, A Ali, K Mahmood, N Amin, S S Ali, M I Khan, G Mustafa, M J Khan and M Saleem, 2018 *Materials Research ExpressIOP Publishing*, **6**(1) 016302.
[17] PP Hankare, RP Patil, UB Sankpal, K M Garadkar, R Sasikala, A K Tripathi,2010 *J. Magn. Magn. Mater.***322** 2629.
[18] R C Kamble, P A Shaikh, C H Bhosale, Y D Kolekar, 2009 *Smart Mater. Struct.***18** 115028.
[19] N Pompadian, N Narayanswamy, 2002*J. Appl. Phys.***95** (5) 2770.
[20] R S Devan, Y D Kolekar, B K Chougule, 2006 *J. Phys. Condens. Matter***18** 9821.
[21] B A Mulla, V S Darshane, 1983 *Proc. Ind. Acad. Sci. (Chem. Sci.)***92** (2) 139.
[22] A G Bhosale, B K Chougule, 2006 *Mater. Lett.* **66** 3912.
[23] D Ravinder, 2000 *Mater. Lett.* **43** 129.
[24] MB Shelar, P A Jadhav, S S Chougule, MM Mallapure, B K Chougule, 2009 *J. Alloys Compd.* **476** 760.
[25] A K M Akther Hossain, ST Mahmud, M Seki, T Kawai, H Tabata, 2007 *J. Magn. Magn. Mater.* **312** 210.
[26] N K Radhakrishnan and C R Murthy, 1981 *Bull. Mater. Res.* **3** 217.
[27] B H Devmunde, A V Raut, S D Birajdar, S J Shukla, D R Shengule, and K M Jadhav, 2016 *J. Nanoparticles,*2016 special issue Article ID 4709687, 8 2016.
[28] P P Hankare, R P Patil, U B Sankpal, S D Jadhav, PD Lokhande, K M Jadhav, R Sasikala, 2009 *J. Solid State Chem.***182** 3217.
[29] P P Hankare, R P Patil, U B Sankpal, S DJadhav, I S Mulla, K M Jadhav, B K Chougule, 2009 *J. Magn. Magn. Mater.***321** 3270.
[30] Preeti Thakur, Deepika Chahar, Shilpa Taneja, Nikhil Bhalla, and Atul Thakur, 2020 *Ceram. Int.***46**(10) 15740.

[31] R D Shanon and C T Prewitt, 1970 *Acta Crystallgr.* **1326** 1026.
[32] M U Islam, T Abbas, M Asharat Chaudhary, 2002 *Mater. Lett.* **53** 30.
[33] A Miller, 1959 *J. Appl. Phys.* **30** 245.
[34] K Jalaiah, K Chandra Mouli, R V Krishnaiah, K Vijaya Babu, P S V Subba Rao, 2019 *Heliyon* **5** (6) e01800.
[35] G Joshi, A Khot and S Sawant, 1998 *Solid State Comm.* **65** 1593.
[36] E J W Verway and J H de Boer, 1936 *Rec. Trans. Chem. Phys.* **55** 531.
[37] R J Begum, S M Srivastava, 1952 *Phys Rev.* **87** 240.
[38] M Raghasudha, D Ravinder, P Veerasomaiah, K M Jadhav, Mohd Hashim, Pramod Bhatt, Sher Singh Meena , 2017 *J. Alloys and Comp.* **694**, 366.
[39] K Rama Krishna, 2012 *Advances in Materials Physics and Chemistry* **2**(03)185.
[40] M A Ali, M N I Khan, M M Hossain, F U Z Chowdhury, M NHossain, RRashid, MA Hakim, S M Hoque and M M Uddin, 2020 *Materials Research Express* **7** 036101.

Immobilization of diastase on PVA- $CoFe_2O_4$ nanocomposite film for improving stability and recycling

A M Kulkarni[1]*, U B Sankpal[1]*, V A Ghadyale [2], N N Bhatkar [1], S S Waghdhare[1], S V Pathare [1]

[1]Department of Chemistry, Gogate - Jogalekar College, Ratnagiri, Maharashtra, India
[2]Department of Bio-Chemistry, Gogate - Jogalekar College, Ratnagiri, Maharashtra, India

**draparna.gjc@gmail.com*

Abstract. Enzyme immobilization on a suitable support is one of the strategies used to improve enzyme stability and recovery. In present communication, PVA-CoF_2O_4 nanocomposite film has been used as a support to develop highly active, stable and magnetically separable immobilized diastase. For preparing magnetically separable support, first Cobalt ferrite nanoparticles have been prepared by using co-precipitation method. Prepared nanoparticles have been dispersed in polyvinyl alcohol (PVA) solution to get a PVA-$CoFe_2O_4$ film. The detail characterization of cobalt ferrite nanoparticles and PVA-$CoFe_2O_4$ film has been carried out. Enzyme diastase has been immobilized on the PVA-$CoFe_2O_4$ film and the catalyst activity of immobilized enzyme has been tested in hydrolysis of starch to maltose. The stability of immobilized diastase has found to be higher than free enzyme. Reusability study of the immobilized diastase has been carried out. Improved stability and reusability of immobilised diastase offer the promising applicability as magnetically separable biocatalyst for industry.

Keywords: PVA - $CoFe_2O_4$ Nano composite, Immobilized enzyme, Magnetically separable biocatalyst

1. Introduction

Designing catalysts that will meet the demands of green and sustainable chemistry is a key task for development of greener and environmentally sound synthetic protocols **[1]**. Biocatalysts particularly enzymes satisfy most of the requirements of the sustainable and green catalyst and therefore are expedient substitutes for traditional catalysts. These biocatalysts offer significant advantages like high specificity **[2]**, mild reaction conditions **[3]** and environmental friendliness **[4]**. Though enzymes have ability to perform most complex chemical processes under environmentally benign conditions, their use in chemical industries is still limited. The problems associated with use of enzyme as industrial biocatalysts are enzyme instability at high temperature and in toxic solvents, difficulties with recovery from reaction mixture and high cost of isolation **[5]**. The enzyme recovery and reuse are an important aspect for economically feasible processes that may compensate the high cost of enzyme. Enzyme immobilization is one of the strategies used to improve enzyme stability and recovery. Enzyme immobilization is a technique in which catalysts are attached to a solid support that is insoluble in the reaction mixture **[6]**. Binding of enzymes to solid supports not only improves the stability but also convert catalytic systems into heterogeneous form which enables easy separation of the biocatalytic system from the reaction mixture. Thus, main goals of immobilization of enzymes are improved enzyme stability, easy recovery and reuse. In this context, development of suitable immobilization technique for a particular enzyme is an interesting area of research. Designing a suitable immobilization technique is a key task in development of industrial biocatalysts which requires multidisciplinary approach. Various immobilization techniques viz. adsorption, covalent binding, encapsulation, cross-linking and entrapment have been developed **[7].** Range of materials like inorganic, organic, hybrid and composite materials have been used as stable and efficient supports for biocatalysts **[8]**. Recently, use of magnetically separable nanoparticle (MNP) as a support for enzyme immobilization has been studied extensively in view of an easy recovery of immobilized enzyme and possible reuse. MNPs provide large surface area and hydroxyl group on their surface allows easy modification of the surface and possibility of strong binding with the enzyme. MNPs are advantageous as the biocatalytic system immobilized on magnetic support can be easily separated using an external magnet **[9]**. Functionalized magnetic nanoparticles including graphene-Fe_3O_4**[10]**, silica- Fe_3O_4**[11]**, and polymer grafted- Fe_3O_4**[12, 13]** have been used as support. Variety of enzymes viz. Lipase **[14]**, glucose oxidase **[15]**, trypsin **[16]**, Amylase **[17]** etc. have been immobilized on functionalized magnetic nanoparticles.

In recent years there has been growing interest in the use of commercially available polymeric films due to their easily tunable properties, good porosity and a large surface area for attachment of variety of enzymes. The well-defined pore sizes and structure of polymeric films facilitate the immobilization of biomolecules not only on the surface of the support but also in its pores **[18]**. Poly (vinyl alcohol), poly(ethylene glycol), polyurethane and poly (vinylidene fluoride) has been used to prepare polymeric membrane for enzyme support **[19,20, 21]**.

In present communication we have developed a Poly vinyl alcohol (PVA) and $CoFe_2O_4$ (MNP) nanocomposite film to combine the advantages of both the types of supports. The PVA-MNP film has been used to immobilize diastase. The resulting biocatalytic system has been tested in hydrolysis of starch to glucose and the stability and activity of immobilized enzyme has been compared with free enzyme.

2 Material and Methods

2.1 General Remarks

Fourier transform-infrared spectra of the catalysts reported in this study were recorded on Shimadzu 8300 FTIR at ambient conditions. The spectra were recorded between 400 and 4000 cm^{-1}. The XRD patterns of the catalyst samples were taken using PHILIPS (PW3710) X-ray diffractometer with Cu $K\alpha_1$ radiation (λ = 1.5424 A^0). SEM micrographs of samples were taken on JEOL- JEM - 6360 microscope. The room temperature (300 K) magnetic properties of the prepared $CoFe_2O_4$ nanoparticles were investigated by the VSM technique in the range of approximately –15 to +15 kOe.

2.2 Reagent and materials

Ferric chloride anhydrous LR ($FeCl_3$) and Polyvinyl alcohol LR (PVA) were supplied by THOMAS BAKER (Mumbai, India). Cobalt (II) chloride extra pure (Hexahyrate) ($CoCl_2.6H_2O$), Sodium hydroxide (NaOH), Starch soluble (Ex potato) $(C_6H_{10}O_5)_n$ were supplied by MOLYCHEM (Mumbai India).

2.3 Preparation of cobalt ferrite nanocatalyst

$CoFe_2O_4$ was synthesized by the reported method **[22].** In a typical procedure, to a well stirred solution of stoichiometric amount of $FeCl_3{\cdot}7H_2O$ and $CoCl_2{\cdot}6H_2O$ in water (in 2:1 molar proportion) was added drop wise aqueous 4M solution of sodium hydroxide. The resultant precipitate was filtered and washed several times with de-ionized water to remove chloride ions. It was dried under vacuum at 60^0C for six hours. The resultant cobalt ferrite particles were sintered between 800- 1000 C for 2 hours.

IOP Conf. Series: Materials Science and Engineering 1166 (2021) 012012 doi:10.1088/1757-899X/1166/1/012012

2.4 Preparation of PVA – Cobalt ferrite film

$CoFe_2O_4$ nanoparticles prepared using co-precipitation method were dispersed into 20 ml of PVA (50 mg/ml) solution in distilled water. The hot solution was poured on petri plate to get uniform film and allowed it to dry for 2 days.

2.5 Immobilization of Diastase on PVA – Cobalt ferrite film

30 mg of diastase (Himedia) enzyme diluted in 3 ml saline was poured uniformly over PVA – Cobalt ferrite film. The film was allowed to stand for 24 hours at temperature 4 ^{0}C. After 24 hours, the film was washed with saline. Protein content in washing was checked by Lowery method **[23]**. Similar procedure had been used for 20 mg , 40 mg, 80 mg and 120 mg enzyme concentration.

2.6 Testing of Enzyme Activity

The resulting biocatalytic system was tested in hydrolysis of starch to glucose. After washing, 1ml of 1% starch and acetate buffer (pH 4.6, 0.2 M) were poured on film and solution was incubated at 37^{0}C for exactly 15 min. The solution was collected and digested reducing sugars were tested with DNSA method. Enzyme activity on same film was rechecked after 10 days.

2.7 Comparison with free enzyme

To compare the activity of immobilized enzyme, 30 mg free enzyme and 30 mg enzyme immobilized on PVA – Cobalt ferrite film were tested in hydrolysis of starch to maltose.

3. Result and discussion

The Fourier Transform Infrared Spectroscopy was performed to confirm the formation of spinel ferrite (Figure 1). The FTIR spectrograph of cobalt ferrite sample **[Fig. 1]** shows two characteristic absorptions at 583 cm^{-1} and 464 cm^{-1} for octahedral and tetrahedral metal – oxygen stretching in spinel ferrites.

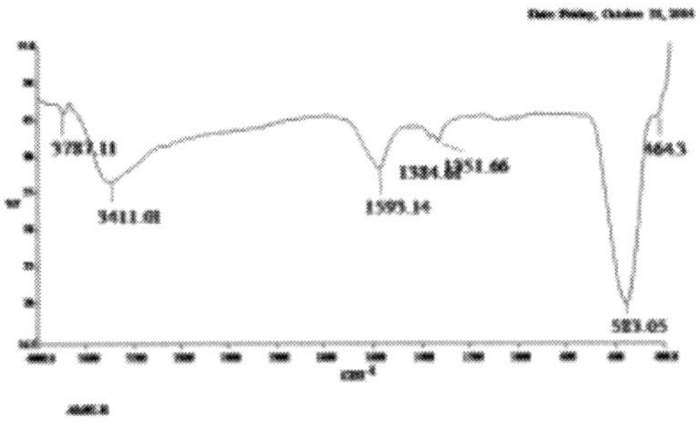

Figure 1. IR spectrograph of $CoFe_2O_4$

The phase purity of the prepared catalyst was analysed by studying powder X-ray diffraction pattern. The powder XRD patterns for Cobalt ferrite NPs and PVA - $CoFe_2O_4$ film are displayed in **Figure 2**. In case of $CoFe_2O_4$, **Fig. 2(a)** all the detectable peaks are indexed as $CoFe_2O_4$ with an inverse spinel structure and are in good agreement with the standard data (JCPDS NO. 22-1086). The crystallite size of the sample was found to be 40 nm.

IOP Conf. Series: Materials Science and Engineering 1166 (2021) 012012 doi:10.1088/1757-899X/1166/1/012012

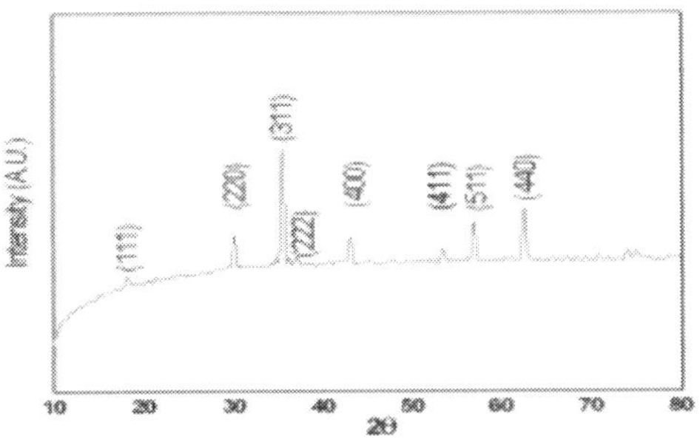

Figure 2. (a) XRD pattern of $CoFe_2O_4$–MNPs

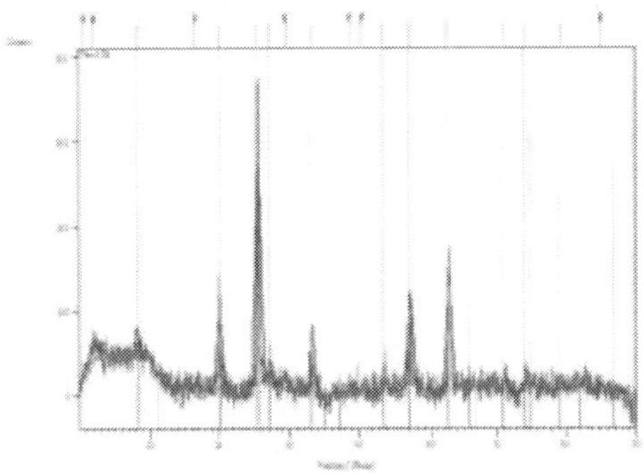

Figure 2. (b) XRD pattern of PVA - $CoFe_2O_4$ film

Figure 3 depicts the microstructure of the cobalt ferrite and PVA - $CoFe_2O_4$ film sample. The SEM topographs of $CoFe_2O_4$ catalyst shows that all the grains are in uniform size and distributed uniformly over the surface. The SEM image of PVA - $CoFe_2O_4$ film depicts rough and granular surface. From SEM images it can be concluded that PVA - $CoFe_2O_4$ film provides large surface area for binding of enzymes.

Figure 3. (a) SEM image of $CoFe_2O_4$ – MNPs

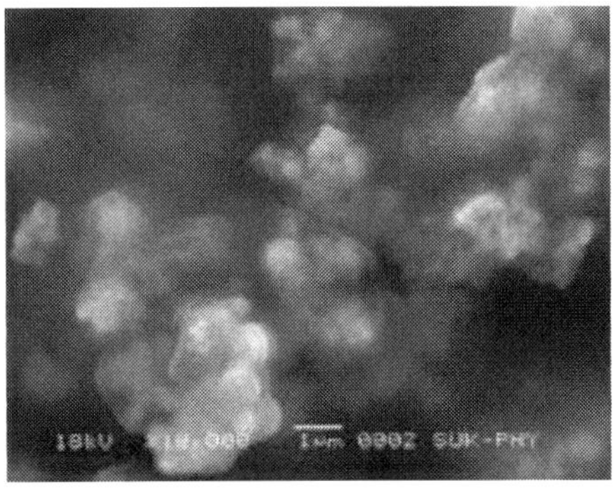

Fig.3 (b) SEM image of $CoFe_2O_4$ – MNPs film

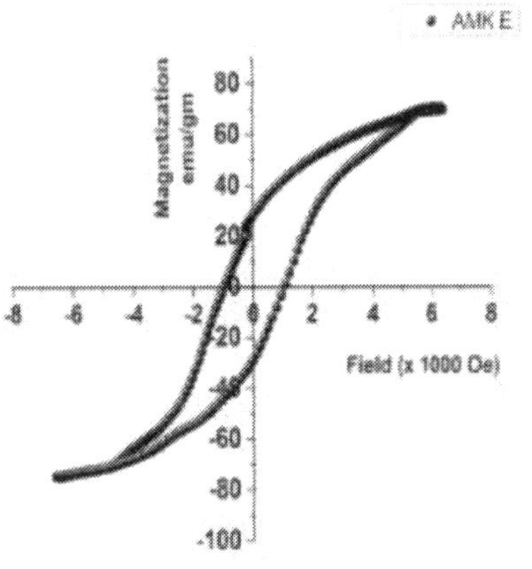

Figure 4. Hysteresis loop for $CoFe_2O_4$

Figure 4 shows hysteresis loops of the $CoFe_2O_4$ – MNPs and confirms the ferrimagnetic behaviour of the particles with saturation magnetization value (M_s) of 70.10 emu / gm which is an important property for easy separation of the catalyst using an external magnet for its reuse.

To check the maximum retention capacity of PVA – Cobalt ferrite film, increasing amount of enzyme viz. 20 mg, 30 mg, 40 mg, 80 mg, and 120 mg were immobilized on PVA – Cobalt ferrite film. Each film was tested for enzyme activity and washings were tested to check protein loss after immobilization. **Fig. 5** shows enzyme activities of different concentration of diastase immobilized on PVA-Cobalt ferrite film and enzyme activity of each film after 10 days.

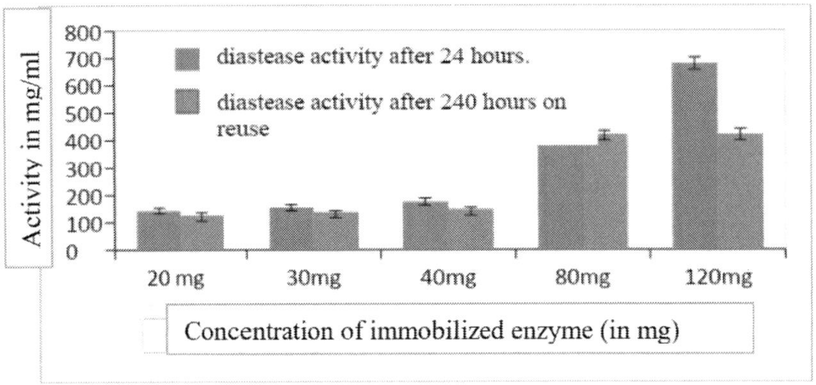

Figure 5. Activity of different concentration of immobilized diastase

Increasing enzyme concentration showed gradual increase in enzyme activity. At 80 mg enzyme concentration, almost 5fold increase in enzyme activity was observed as compared to 20 mg enzyme concentration. Though enzyme activity increased even after 80 mg to 120 mg, the washings of films showed presence of protein. At 120 mg 0.6818±0.07 mg protein loss was observed. It was concluded that up to 80 mg the enzyme gets trapped into pores and get retained on the support. However beyond 80 mgs the enzyme remains on the surface and hence resulted in protein losses. Up to 80 mg 100% immobilization was observed (protein content was nil) due to entrapment but above 80 mg protein losses was observed. Thus, maximum rate of adsorption 1.1658 mg/cm² was observed for diastase.

To compare activity of immobilized enzyme with free enzyme, 30 mg free enzyme and 30 mg enzyme immobilized on PVA – Cobalt ferrite film were tested in hydrolysis of starch to maltose. It was found that 30 mg free enzyme showed 150.11±12.3 units per mg activity while the immobilized enzyme showed 154.66±11.2 units per mg activity. Thus, activity of immobilized enzyme is comparable with free enzyme.

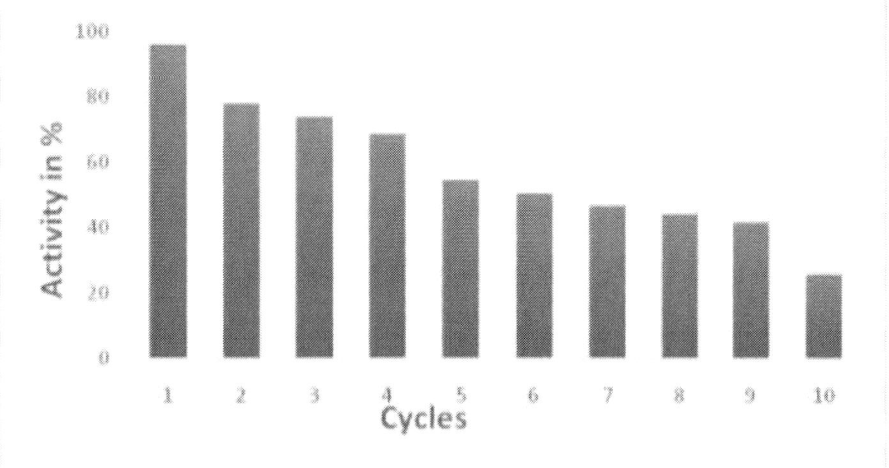

Figure 6. The recyclic performance of immobilized enzyme

Reusability Study of immobilized enzyme: Main aim behind the physical immobilization of enzyme is reusability. In this case plates were stored at 4°C and reused after 10 days. It was found that at lower concentrations almost 80 to 90% activity of enzyme has been retained while at higher concentration i.e. 80mg and 120 mg almost 40% loss of enzyme activity has been observed. Porosity of film might be protecting the enzyme from washing loss at lower concentrations. Reusability study suggested that up to 40 mg immobilized enzyme has appropriate activity, stability and reusability up to 5 runs **(Fig.6)**.

4. Conclusion

Highly active, stable and magnetically separable immobilized diastase has been developed using PVA-Nano CoF_2O_4 film as support. The method developed for immobilization of enzyme is effective and the stability of immobilized diastase has found to be higher than free enzyme. The activity of diastase immobilized on PVA – nano Cobalt ferrite film has found to be comparable with free enzyme. The immobilized diastase can be reused effectively over period of 10 days.

References

[1] Nair N U, Tang W L, Eriksen T D and Zhao H 2010 *Industrial Applications of Enzymes as Catalysts*, in: Baltz R H, Demain A L and Davies J E (Eds.) *Manual of Industrial 8 Microbiology and Biotechnology*, (ASM press, Washington), 480.

[2] Zhu Y T, Ren X Y, Liu Y M, Wei Y, Qing L S and Liao X 2014 *Mater. Sci. Eng. C Mater. Biol. Appl.***38** 278.

[3] Hong J, Xu D, Gong P, Sun H, Dong L and Yao S 2007 *J. Mol. Catal. B Enzym.***45** 84.

[4] Cowan D A and Fernandez-Lafuente R 2011 *Enzyme Microb. Technol.***49** 326.

[5] Hartmann M and Jung D 2010 *J. Mater. Chem.***20** 844.

[6] Jesionowski T, Zdarta J and Krajewska B 2014 *Adsorption* **20** 801.

[7] Sheldon RA 2007 *Adv.Synth. Catal.***491** 289.

[8] Zdarta J, Meyer A S, Jesionowski T and Pinelo M A 2018 *Catalysts* **8** 92.

[9] Cao M, Li Z, Wang J, Ge W, Yue T, Li R, Colvin V L and Yu W 2012 *Food Sci. Technol.* **27** 47.

[10] Yang D, Wang X, Shi J, Wang, X, Zhang S, Han P and Jiang Z. 2016 *Biochem. Eng. J.* **105** 273.

[11] Zhang Q, Han X and Tang B 2013 *RSC Adv.***3** 9924.

[12] Zhu W, Li Y, Zeng F, Yin H, Wang L and Zhu H 2015 *RSC Adv.***5** 23039.

[13] Wang J, Ji F, Xing J, Cui S, Bao Y and Hao W. 2014 *Chin. J. Chem. Eng.* **22**1333.

[14] Mehrasbi M R, Mohammadi J, Peyda M and Mohammadi M 2017 *Renew. Energy* **101** 593.
[15] Aber S, Mahmoudikia E, Karimi A and Mahdizadeh F 2016 *Water Air Soil Pollut.* **227** 93.
[16] Atacan K, Cakiroglu B and Ozacar M 2016 *Food Chem.* **212** 460.
[17] Dhavale R P, Parit S B, Sahoo S C, Kollu P, Patil P S, Patil P B and Chougale A D 2018 *Mater. Res. Express* **5** 075403
[18] Kuo C H, Chen G J, Kuo T Y, Liu Y C and Shieh C J 2012 *Ind. Eng. Chem. Res.***51** 5141.
[19] Vitola G, Mazzei R, Fontananova E, Porzio E, Manco G, Gaeta S N and Giorno L 2016 *J. Membr. Sci.* **516** 144.
[20] Konwarh R, Karak N, Rai S K and Mukherjee A. K 2009 *Nanotechnology* **20** 225107
[21] Donato L, Algieri C, Rizzi A, Giorno L 2014 *J. Membr. Sci.***454** 346.
[22] Kulkarni A M, Pandit K S, Chavan P V, Desai U V and Wadgaonkar P P 2015 *RSC Adv.***5** 70586.
[23] Lowry O H, Rosebrough N J, Farr A L, Randall R J. 1951 *J. Biol. Chem.* **193** 265.

Effect of polycarboxylate ether-based superplasticizer dosage on fresh and hardened properties of cement concrete

Ajmal Paktiawal[1*] and Mehtab Alam[2]

[1]Department of Civil Engineering, Faculty of Engineering and Technology, Jamia Millia Islamia, Jamia Nagar, New Delhi-110025, India.
[2]Department of Civil Engineering, Faculty of Engineering and Technology, Jamia Millia Islamia, Jamia Nagar, New Delhi-110025, India.

*Email: ajmal.paktiawal@gmail.com

Abstract. Ready-mix concrete nowadays is available almost everywhere in the country. RMC plants are indispensably using high-range water-reducing admixtures namely, sulfonate naphthalene formaldehyde condensate, and Polyether polycarboxylate, for making standard cement concrete. These admixtures are used to economize the cost of concrete by reducing the quantity of water. Commonly used high range water reducer admixture, the polycarboxylate ether-based (PCE) superplasticizer has been used to investigate its influence on workability and strength of concrete by measuring slump and ultrasonic pulse velocity as well as crushing strength. The cement of grade 43 has been used in this work. This superplasticizer was added to the mix to make concrete of grade M60 by weight of cement with 0.15%, 0.25%, 0.35%, 0.45%, and 0.55% without altering the w/c ratio. Experimental results reveal that the restricted dosage of the superplasticizer provided in Plain and reinforced concrete code of practice in Indian standard, IS456:2000, restricts to 1% is on the higher side. The slump, ultrasonic pulse velocity, and crushing strength of M60 grade standard concrete with water to binder ratio of 0.40 gives the optimum dose 0.45% by weight of cement, which is quite less than the restricted quantity of dosage.

1. Introduction

High-range water-reducing admixtures are commonly utilized in concrete where demand for pumpable high strength increment is required [1]. In self-consolidating concrete also high range water reducing superplasticizers are employed to produce require flowability and inherent compactability [2]. Mini slump and Marsh funnel tests with varying percentage of PCE SP dose were conducted to find out the saturation dosage, which is the maximum spread diameter beyond which there is no increase, on PPC and OPC+25% fly ash. Saturation dosages of PCE SP were found to be 0.4 and 0.6 in Mini slump and Marsh funnel tests respectively. The value of saturation dosage found out in Mini slump and Marsh funnel tests were reported 0.6 for OPC+25% fly ash [3]. In this experimental work, the influence of SP on plastic shrinkage of conventional and silica fume cement concrete with 7.5% of Type 1, Type 2, and Type 3 having the dosage of PCE SP 0.4, 0.7, and 0.66% respectively were investigated. It was reported that PCE SP showed the highest compatibility concerning to its massive retardation Type 1 and Type 3 silica fume cement concrete [4]. The incorporation of superplasticizer in fine recycled concrete aggregate was studied for mechanical properties. In this study, two types of superplasticizers called SP1, containing lignosulfonate and SP2, whose chemical composition is polycarboxylate was added with 1% by cement weight. Fine aggregate was substituted with fine recycled aggregate with 20, 40, 60, 80, and

100% by weight. It was reported that higher water reducing the power of the additive gave a worse relative performance of FRA in fine recycled concrete aggregate [5]. Polycarboxylate ether-based superplasticizer with 0.3, 0.5, 0.7, and 1% by weight of cement was used to investigate its influence on hydration, microstructure, and mechanical response of cement paste. It was reported that 1% of PCE-based SP in the cement paste produced a higher percentage of silicate at 2 days. Based on the microstructural analysis, it was inferred that the PCE-based superplasticizer slightly decreases the porosity of cement paste [6].

2. Experimental arrangement

2.1. Materials

Wonder brand OPC cement of grade 43, fresh lot complying with [7] was purchased from the authorized supplier. This OPC cement was examined for its physical characteristics conforming to BIS [8-12]. The test findings of cement are given in Table 1. Fine and coarse aggregate as natural river sand of zone II as per [13] and crushed stone of size 20mm and 10mm respectively were tested for their physical properties as shown in Table 2 had been found out [14, 15]. Natural river bed coarse sand was used as fine aggregate in this study. Other tested characteristics of the coarse aggregate as shown in Table 2 complying with [14, 15]. Polycarboxylate ether-based superplasticizer (HRWR) was procured from the authorized supplier (Kunal Chemical Company) used [16]. The characteristics of the superplasticizer used are presented in Table 3.

Table 1. Tested properties of OPC Cement

S. No.	Property	Result		Requirements as per IS Code
1	Normal consistency	31%		-
2	Initial setting time	116minutes		≥ 30minutes
3	Final setting time	229minutes		≤ 600minutes
4	Specific gravity	3.12		-
5	Fineness	98.0%		> 90%
6	Compressive strength	3days	24.0 MPa	23 MPa
		7days	32.7 MPa	33 MPa
		28days	47.3 MPa	43 MPa

Table 2. Physical characteristics of F. A. and C. A.

S. No	Property	Test result	
		F. A. (Zone II)	C. A.
1	Specific gravity	2.54	2.74
2	Water adsorption	1.94	0.65
3	Moisture content	0.95	0.25
4	Fineness Modulus	2.58	-

Table 3. Physical and mechanical properties of Admixture

S. No	Tests carried out	References	Results obtained
1	Appearance	BIS-9103	Light brown viscous liquid
2	Main chemical ingredients	BIS-9103	Polycarboxylate base
3	Dry material content (%)	BIS-9103	50.10
4	Relative density	BIS-9103	1.100
5	Chloride content (%)	BIS-6925	0.017
6	pH	BIS-9103	6.70

2.2. Concrete mix and its ingredients

Concrete for the grade of M60 with the desired workability of 100mm was mix design conforming to [17, 18]. Cube specimens of standard size with 100mmx100mmx100mm were casted using polycarboxylate ether-based (PCE) superplasticizer with 0.15%, 0.25%, 0.35%, 0.45%, and 0.55% by weight of cement and abbreviated as PCESP. In the present work, the concrete of grade M60 was prepared in the laboratory with varying percentages of superplasticizer to investigate its effect on the fresh and hardened state of the concrete. The binder to aggregate ratio is obtained as 1:1.73:2.48 for water to cement ratio of 0.35. The quantity of concrete ingredients required for 1m^3 of concrete mix of M60 grade is presented in Table 4.

Table 4. Mix proportioning for 1m^3 of concrete

Mix ID	Cement (kg/m^3)	F. A. (kg/m^3)	C. A. (kg/m^3)	w/c ratio	Admixture (%)
PCESP-0.00%	449	777	1113	0.35	0.00
PCESP-0.15%	449	777	1113	0.35	0.15
PCESP-0.25%	449	777	1113	0.35	0.25
PCESP-0.35%	449	777	1113	0.35	0.35
PCESP-0.45%	449	777	1113	0.35	0.45
PCESP-0.55%	449	777	1113	0.35	0.55

2.3. Allocation of test specimens and test conducted

Concrete moulds of a standard size of 100mmx100mmx100mm were prepared to cast concrete for compressive strength and ultrasonic pulse velocity conforming to [19]. To assess the effect of dosage of PCE based superplasticizer on the fresh and hardened state of concrete, the concrete in its fresh state was tested for slump as per [20], and the hardened concrete was tested for ultrasonic pulse velocity and crushing strength complying with [22, 21].

3. Findings and discussions

3.1. Workability

The slump test was conducted to scale the effect of the superplasticizer on workability. Polycarboxylate ether-based SP with 0, 0.15, 0.25, 0.35, 0.45, and 0.55% was added by weight of cement. Scaled slumps of the PCE-based SP concrete are depicted in Fig.1. The addition of a superplasticizer increased the slump. Concrete with 0.15% of SP has given 25mm slump, the degree of workability as per IS456:2000 clause 7.1 is low [18]. Concrete without SP having water to cement ratio of 0.4 was found very low and having a slump of about 15mm may be called zero slump concrete. The concretes with 0.25% and 0.35% of PCE-based SP having a slump of 50mm and 70mm respectively belong to the medium degree of workability. The concretes with 0.45% and 0.55% having a slump of 105mm and 130mm are of a high degree of workability. Fig.2 shows the desired slump (0.45% of SP) of the concrete.

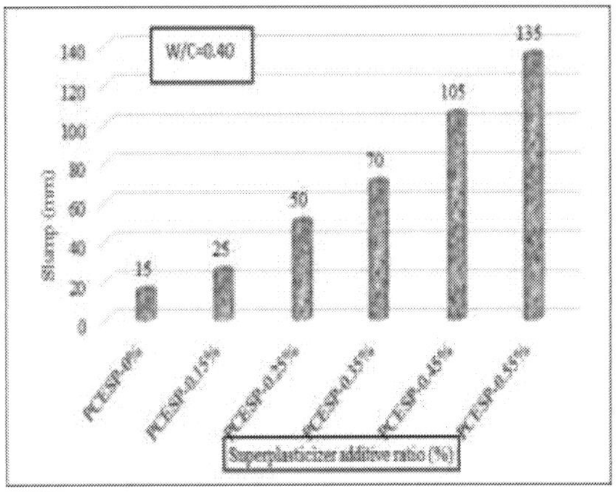

Figure 1. Shows slump value of concrete

Figure 2. Concrete of desired workability with 0.45% of SP

3.2. Workability

The density of concrete without superplasticizer was 2301kg/m^3. The concretes with additive superplasticizer of 0.15, 0.25, 0.35, 0.45, and 0.55% by weight of cement were found as 2331, 2420, 2434, 2443, and 2430kg/m^3. It can be mentioned that the density of concrete increases with increasing dosage of superplasticizer by 1.3, 5.17, 5.78, 6.17, and 5.60% as compared to concrete without superplasticizer.

3.3. Compressive strength

Control concrete (without SP) and concretes with varying dosages of SP were tested for crushing strength after 7 and 28 days of curing. The compressive strength findings are given in Table 5 and plotted in Fig. 3 and Fig. 4. The improvement of crushing strength of concrete with superplasticizer of 0.15, 0.25, 0.35, 0.45, and 0.55% after 28 days is found to be 5.68, 11.06, 18.32, 31.93 and 12.94% as compared to concrete without SP. The desired strength as per IS10262:2019 was given by 0.45% of SP content. The concrete with 0.45% of SP gives higher compressive strength. This improvement of compressive strength can be ascribed to decrease in air-entrained, the probable consolidation of the mix due to higher workability, and sweeping hydration of cement owing to dispersion ability of this optimum dosage of superplasticizer. The addition of SP beyond 0.45% dose caused a little bleeding in the mix, which decreased the strength.

Table 5. Compressive strength findings after 7 and 28 days

Mix ID	Age (Days)	Samples test Result (MPa)					Mean Strength (MPa)	Std Dev (Mpa)	C.O.V (%)	Strength Increase (%)
		S1	S2	S3	S4	S5				
PCESP-0.00%	7	30.0	29.8	30.5	29.6	31.1	30.20	0.60	1.98	0.00
PCESP-0.15%		33.0	33.4	29.2	32.6	32.4	32.12	1.68	5.23	0.25
PCESP-0.25%		36.3	32.1	35.4	31.8	34.8	34.08	2.02	5.93	5.84
PCESP-0.35%		34.6	37.4	35.5	36.3	33.4	35.44	1.54	4.35	10.10
PCESP-0.45%		39.3	38.6	40.4	40.3	39.6	39.64	0.74	1.87	23.11
PCESP-0.55%		36.3	38.8	37.3	37.1	36.8	37.26	0.94	2.52	15.71
	28									
PCESP-0.00%		47.6	48.5	47.5	46.9	49.1	47.92	0.87	1.82	0.00
PCESP-0.15%		49.2	53.2	49.9	52.6	48.3	50.64	2.15	4.25	5.68
PCESP-0.25%		56.8	54.4	51.5	50.9	52.5	53.22	2.40	4.51	11.06
PCESP-0.35%		54.2	56.5	59.6	58.4	54.7	56.70	2.32	4.09	18.32
PCESP-0.45%		61.7	65.1	64.4	63.3	61.6	63.22	1.57	2.48	31.93
PCESP-0.55%		57.3	51.1	52.9	57.6	51.7	54.12	3.11	5.75	12.94

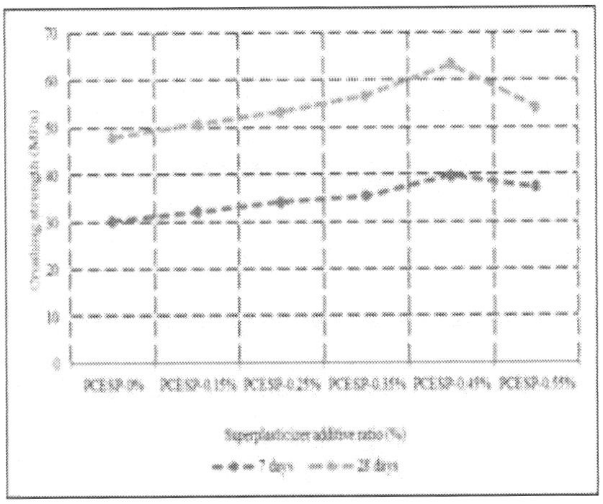

Figure 3. Comp. str. results after 7 and 28 days

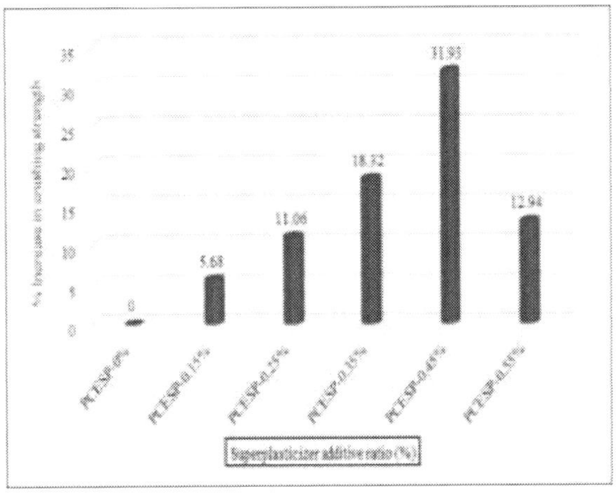

Figure 4. Percentage increase in comp. str.

3.4. Ultrasonic pulse velocity

The UPV test is conducted to investigate the quality of hardened concrete for homogeneity, cracks, and voids complying with [22]. The concrete samples were taken out of the curing tank after 28 days of normal curing and placed in the concrete laboratory room for 24 hours under an average temperature of 26°C. The UPV test readings are taken on four faces of cube specimens of standard size of 100mmx100mmx100mm with one reading on each face. To ascertain the quality grading of concrete with respect to the pulse velocity, Table 6 gives the recommendations [22]. Concrete with and without superplasticizer was tested for ultrasonic pulse velocity after 28 days. The findings for UPV are tabulated in Table 7, and Fig. 5. Results show that concrete with 0 and 0.15% of SP are of good quality. The concrete with 0.25, 0.35, 0.45, and 0.55% are found to be of excellent quality. It is pertinent to note here that the concrete with 0.15, 0.25, 0.35, 0.45, and 0.55% of SP after 28 days, the UPV test results improved by 3, 4.62, 5.31, 9, and 4.85% as compared to concrete without superplasticizer as plotted in Fig. 6.

Table 6. Velocity measurements for concrete grading as per BIS

S. No.	UPV (m/s)	Concrete grading (Quality)
1	Above 4500	Excellent
2	3500 to 4500	Good
3	3000 to 3500	Medium
4	Bellow 3000	Doubtful

Table 7. Ultrasonic pulse velocity findings after 28 days

Mix ID	Age (Days)	Samples test Result (MPa)					Avg. UPV (km/s)	Std. Dev (km/s)	C.O.V (%)	UPV Increase (%)
		S1	S2	S3	S4	S5				
PCESP-0.00%	28	4.32	4.28	4.35	4.42	4.29	4.33	0.06	1.39	0.00
PCESP-0.15%		4.59	4.53	4.24	4.48	4.48	4.46	0.13	2.91	3.00
PCESP-0.25%		4.55	4.48	4.39	4.65	4.59	4.53	0.10	2.21	4.62
PCESP-0.35%		4.48	4.65	4.53	4.50	4.65	4.56	0.08	1.75	5.31
PCESP-0.45%		4.98	4.65	4.72	4.59	4.65	4.72	0.15	3.18	9.00
PCESP-0.55%		4.59	4.48	4.53	4.53	4.59	4.54	0.05	1.10	4.85

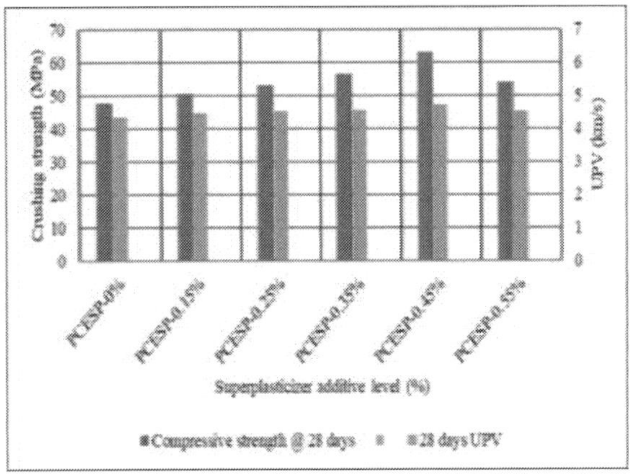

Figure 5. Relationship between comp. str. and UPV

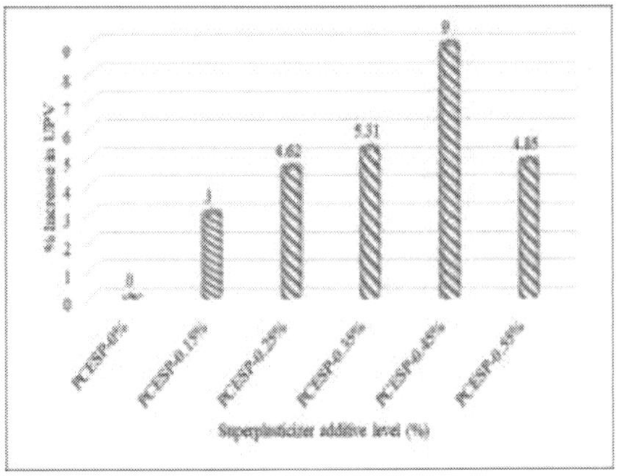

Figure 6. % increase in UPV results due to SP dose

3.5. Moudulus of elasticity

The 28 days elastic modulus of concrete has been obtained from a cube specimen of size 100mmx100mmx100mm under the compression testing machine at the recommended rate of loading 2.3kN/s. The deformation was measured by the deflectometer attached to the steel member welded to the lower platen of the CTM. The load and deformation were captured in the video by an 8-megapixel camera installed. The test arrangement is presented in Fig. 7. The load-deformation data had been used to plot stress-strain curves. The plotted curves are shown in Fig. 8. The moduli of elasticity of concrete with varying percentages of PCE-based superplasticizer have been calculated considering the linear component of the curves. The findings are given in Table 7. The improvement of elastic modulus value of concrete with superplasticizer of 0.15, 0.25, 0.35, 0.45, and 0.55% after 28 days is found to be 37.5, 52.2, 54.8, 54.7, and 28.2% as compared to concrete without SP.

Table 8. Modulus of elasticity results

Mix ID	Modulus of elasticity (MPa)
PCESP-0.00%	$1.88x10^4$
PCESP-0.15%	$2.58x10^4$
PCESP-0.25%	$2.86x10^4$
PCESP-0.35%	$2.91x10^4$
PCESP-0.45%	$2.91x10^4$
PCESP-0.55%	$2.41x10^4$

Figure 7. Test set up for stress-strain curve measuring

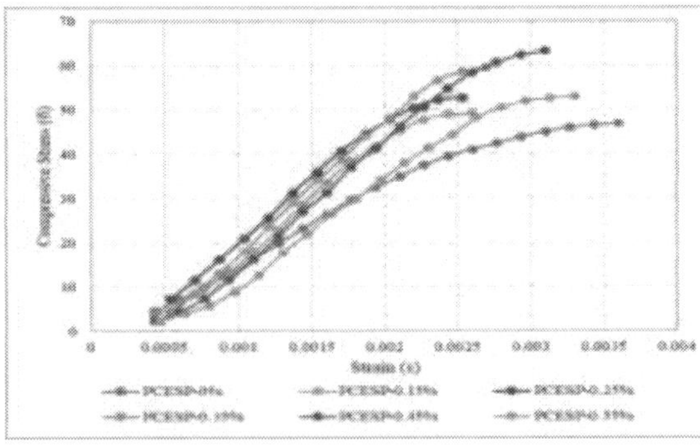

Figure 8. Stress-strain curve with various SP level

4. Conclusion

This study experimented the role and influence of PCE based superplasticizer dosage on the properties of fresh and hardened standard M60 concrete with water to binder ratio of 0.40. On the bases of 28 days experimental findings the following conclusions can be summarized.

- The superplasticizer dose of 0.45% in this concrete supported the desired workability with a slump of 105mm. Concrete with a w/c ratio of 0.4 without SP was found giving a slump of 15mm and called very low workable concrete. Concrete with 0.15% of SP dose gave a mix of low workability having a slump of 25mm. Concrete with 0.25%, and 0.35% of SP dose gave a mix of medium workability having a slump of 50mm and 70mm respectively. Concrete with 0.55% of SP dosage caused a slight bleeding and related to the class of high workability.
- The concrete with a dose of SP less than 0.45% entrapped the air giving a lower density of the concrete. Ultrasonic pulse velocity and crushing strength results of concrete having a dose of SP less than 0.45% follow the trend of density. The improvement of crushing strength of concrete containing superplasticizer with 0.15, 0.25, 0.35, 0.45, and 0.55% after 28 days is found to be 5.68, 11.06, 18.32, 31.93, and 12.94% as compared to concrete without SP.
- Concrete with 0 and 0.15% of SP are of good quality. Concrete with 0.25, 0.35, 0.45, and 0.55% are found to be of excellent quality, however ultrasonic pulse velocity of concrete with 0.45% of SP is the maximum. The increase of ultrasonic pulse velocity of concrete with 0.15, 0.25, 0.35, 0.45, and 0.55% of SP after 28 days is 3, 4.62, 5.31, 9, and 4.85% as compared to concrete without superplasticizer.
- Code of practice IS456:2000 in its amendment no. 4 restricts the dosage of superplasticizer to 2% and polycarboxylate-based admixture to 1%. However, experimental results reveal that the maximum percentage of polycarboxylate-based admixture is on the higher side. The experimental results suggest that separate recommendations to restrict the dosages of superplasticizer and polycarboxylate-based admixture be provided for compacting concrete and self-compacting concrete.
- The slump, UPV, and crushing strength of concrete of grade M60 with water to cement ration of 0.40 gives the optimum dose 0.45% by weight of cement, which is quite less than the restricted quantity of the dosage.

Acknowledgments
The author is very grateful to the Civil Engineering Department for their support in providing laboratory equipment and allocating a quiet library for research scholars.

References

[1] ACI 212.4R-93 (1998). ACI Committee, Report on Guide for the Use of High-Range Water-Reducing Admixtures (Superplasticizers) in Concrete, *American concrete institute.*

[2] Jurg Schlumpf, (2013). "Sika Concretge Handbook." Edition 03/ 2013. *Sika Services AG Corporate Marketing Service.*

[3] Dhanya Sathyan, Kalpathy Balakrishnan Anand, Influence of superplasticizer family on the durability characteristics of fly ash incorporated cement concrete, *Construction and Building Materials.* **204** (2019) 864-874.

[4] Omar S. Baghabra Al-Amoudi, Taofiq O. Abiola, Mohammed Maslehuddin, Effect of superplasticizer on plastic shrinkage of plain and silica fume cement concretes, *Construction and Building Materials.* **20** (2006) 642-647.

[5] P. Pereira, L. Evangelista, J. de Brito, the effect of superplasticizers on the mechanical performance of concrete made with fine recycled concrete aggregates, *Cement & Concrete Composites.* **34** (2012) 1044-1052.

[6] F. Puertas. H. Santos, M. Palacios, S. Martı´nez-Ramı´rez, Polycarboxylate superplasticiser admixtures: effect on hydration, microstructure and rheological behaviour in cement pastes, *Advances in Cement Research.* **17** (2005) 77-89.
[7] IS8112:2013, Ordinary Portland Cement, 43 Grade Specification, *Bureau of Indian Standards.*
[8] IS4031(Part 4):1988, Method of physical tests for hydraulic cement, Determination of consistency of standard cement paste, *Bureau of Indian Standards.*
[9] IS4031(Part 5):1988, Method of physical tests for hydraulic cement, Determination of initial and final setting times, *Bureau of Indian Standards.*
[10] IS4031(Part 15):1991, Method of physical tests for hydraulic cement, Determination of fineness by wet sieving, *Bureau of Indian Standards.*
[11] IS4031(Part 6):1988, Method of physical tests for hydraulic cement, Determination of compressive strength of hydraulic cement other than masonry cement, *Bureau of Indian Standards.*
[12] IS650:1991, Standard sand for testing cement specification, *Bureau of Indian Standards.*
[13] IS383:1970, Specification for coarse and fine aggregate from natural sources for concrete, *Bureau of Indian Standards.*
[14] IS2386(Part 1):1963, Method of teste for aggregates for concrete, particle size and shape, *Bureau of Indian Standards.*
[15] IS2386(Part 3):1963, Method of test for aggregate for concrete, specific gravity, density, voids, absorption and bulking, *Bureau of Indian Standards.*
[16] IS9103:1999, Concrete admixtures-specification, *Bureau of Indian Standards.*
[17] IS10262:2019, Concrete mix proportioning-Guidelines, *Bureau of Indian Standards.*
[18] IS456:2000, Plain and reinforced concrete- code of practice, *Bureau of Indian Standards.*
[19] IS10086:1982, Specification for moulds for use in tests of cement and concrete, *Bureau of Indian Standards.*
[20] IS1199:1959, Methods of sampling and analysis of concrete, *Bureau of Indian Standards.*
[21] IS516:1959, Methods of tests for strength of concrete, *Bureau of Indian Standards.*
[22] IS13311(Part 1):1992, Non-destructive testing of concrete-Methods of test, Ultrasonic pulse velocity, *Bureau of Indian Standards.*

Some new applications of β-Laplace integral transform

Amit Gaur[1,*] and Garima Agarwal[2]
[1]Department of Mathematics and Statistics, Manipal University Jaipur, Jaipur, India.
[2]Department of Mathematics and Statistics, Manipal University Jaipur, Jaipur, India.

*E-mail: *gaur84amit@gmail.com*

Abstract. The β-Laplace integral transform(β-LIT) has been introduced recently by Gaur et. al. which is a generalization of Laplace transform. This new β-LIT shows some special properties over Laplace integral transform. In the current paper, we investigate the applicability of the β-LIT technique to solve partial differential equation such as initial value problem of first order and second order PDE, wave equation, Diffusion equation, Potential and current equation, Heat conduction equation, Axisymmetric Heat conduction equation, Inhomogeneous partial differential equation, inhomogeneous wave equation, Stroke and Rayleigh problem in fluid dynamics and illustrate by some examples.

1. INTRODUCTION

It is obvious that many of physical science and engineering science processes are best described by Partial differential equation. Partial differential equation has many applications in the various fields such as electrical engineering, chemical science, physical science, life science, mechanics, control theory, economics, image processing, signal processing vibration and control, statistics, kinetic model, Riesz potential, geophysics, fluid dynamics, bio-medical engineering, chaos theory and many more [1,2,5,14,15,16,17].

Laplace integral transform has been effectively used for almost 150 years in mathematics, physics, and engineering science for solving numerous problems [7,8,9,18,20].

Recently, Gaur et al. introduced a generalization of Laplace transform (LT) and named β-Laplace integral transform(β-LIT) which shows some mathematical advantage over Laplace transform. This new β-LIT is very interesting transform and shows many special properties [11,12,13].
For any exponentially bounded and sectionally continuous function $\psi: [0, \infty) \to \mathcal{R}$, the β-LIT is defined by [12]

$$\mathcal{L}_\beta\{\psi\}_{(s)} = \int_0^\infty \beta^{-s\xi}\psi(\xi)d\xi\,, \;\; \beta > 1, \Re(s) > \frac{\alpha}{log\beta}. \qquad (1)$$

Where α is exponential order of the function $\psi(x)$, and $s \in \mathbb{C}$.

In the current paper, we investigate the applicability of the β-Laplace integral transform technique to solve partial differential equation such as initial value problem of first order and second order PDE, wave equation, Diffusion equation, Potential and current equation, Heat conduction equation,

Axisymmetric Heat conduction equation, Inhomogeneous partial differential equation, inhomogeneous wave equation, Stroke and Rayleigh problem in fluid dynamics and illustrate by some example [3,4,6,10,19].

In our previous paper [11], we have seen that in case of ODE, the β-LIT gives rise to an algebraic equation for the transform which can be easily solved by the ordinary methods. In case of Partial Differential Equation (PDE), We shall show that the β-LIT gives rise to an ODE for the transform which can be easily solved by the ordinary methods.

2. MAIN RESULTS

Theorem 2.1. Let the function $u(x,t)$ be defined for $a \le x \le b$ and $t > 0$ and $\mathcal{L}_\beta\{u(x,t)\} = U_\beta(x,s)$, then

(i) $\mathcal{L}_\beta\left\{\frac{\partial u}{\partial t}\right\}_{(s)} = slog\beta U_\beta - u(x,0).$

(ii) $\mathcal{L}_\beta\left\{\frac{\partial^2 u}{\partial t^2}\right\}_{(s)} = (slog\beta)^2 U_\beta - slog\beta u(x,0) - \left\{\frac{\partial u}{\partial t}\right\}_{t=0}.$

(iii) $\mathcal{L}_\beta\left\{\frac{\partial u}{\partial x}\right\}_{(s)} = \frac{dU_\beta}{dx}.$

(iv) $\mathcal{L}_\beta\left\{\frac{\partial^2 u}{\partial x^2}\right\}_{(s)} = \frac{d^2 U_\beta}{dx^2}$

Proof.

(i) By definition

$$\mathcal{L}_\beta\left\{\frac{\partial u}{\partial t}\right\}_{(s)} = \int_0^\infty \beta^{-st}\frac{\partial u}{\partial t}dt,$$

$$\mathcal{L}_\beta\left\{\frac{\partial u}{\partial t}\right\}_{(s)} = \lim_{r\to\infty}\left\{[\beta^{-st}u(x,t)]_0^r + slog\beta\int_0^r \beta^{-st}u(x,t)dt\right\},$$

$$\mathcal{L}_\beta\left\{\frac{\partial u}{\partial t}\right\}_{(s)} = slog\beta U_\beta - u(x,0). \tag{2}$$

(ii) Let $\frac{du}{dt} = v$, then

$$\mathcal{L}_\beta\left\{\frac{\partial^2 u}{\partial t^2}\right\}_{(s)} = \mathcal{L}_\beta\left\{\frac{\partial v}{\partial t}\right\}_{(s)} = slog\beta\mathcal{L}_\beta\{v\}_{(s)} - v(x,0),$$

$$\mathcal{L}_\beta\left\{\frac{\partial^2 u}{\partial t^2}\right\}_{(s)} = slog\beta\left[slog\beta\mathcal{L}_\beta\{u\}_{(s)} - u(x,0)\right] - \left(\frac{\partial u}{\partial t}\right)_{t=0},$$

$$\mathcal{L}_\beta\left\{\frac{\partial^2 u}{\partial t^2}\right\}_{(s)} = (slog\beta)^2 U_\beta - slog\beta u(x,0) - \left(\frac{\partial u}{\partial t}\right)_{t=0}. \tag{3}$$

(iii) By definition

$$\mathcal{L}_\beta\left\{\frac{\partial u}{\partial x}\right\}_{(s)} = \int_0^\infty \beta^{-st}\frac{\partial u}{\partial x}dt,$$

Applying Leibnitz rule [16]

$$\mathcal{L}_\beta\left\{\frac{\partial u}{\partial x}\right\}_{(s)} = \frac{d}{dx}\int_0^\infty \beta^{-st}u\,dt = \frac{dU_\beta}{dx}, \tag{4}$$

(iv) Let $\frac{\partial u}{\partial x} = v$, then

$$\mathcal{L}_\beta\left\{\frac{\partial^2 u}{\partial x^2}\right\}_{(s)} = \mathcal{L}_\beta\left\{\frac{\partial v}{\partial x}\right\}_{(s)} = \frac{d}{dx}\left[\mathcal{L}_\beta\{v\}_{(s)}\right] = \frac{d}{dx}\mathcal{L}_\beta\left\{\frac{\partial u}{\partial x}\right\}_{(s)} = \frac{d}{dx}\left(\frac{dU_\beta}{dx}\right) = \frac{d^2U_\beta}{dx^2}, \quad (5)$$

3. ILLUSTRATIVE EXAMPLE

3.1 Initial and Boundary Value Problem of First Order

Example 3.1.1. $\frac{\partial u}{\partial x} = 2\frac{\partial u}{\partial t} + u.$

Given $u(x,0) = 6e^{-3x},\ x > 0, t > 0.$

Solution. Taking β-Laplace Integral transform both the sides of the given PDE,

$$\mathcal{L}_\beta\left\{\frac{\partial u}{\partial x}\right\}_{(s)} = 2\mathcal{L}_\beta\left\{\frac{\partial u}{\partial t}\right\}_{(s)} + \mathcal{L}_\beta\{u\}_{(s)},$$

Using theorem (2.1),

$$\frac{dU_\beta}{dx} = 2\left[slog\beta U_\beta - u(x,0)\right] + U_\beta,$$

By the given boundary conditions

$$\frac{dU_\beta}{dx} - (2slog\beta + 1)U_\beta = -12e^{-3x},$$

Which is an ordinary linear differential equation, after solving, we obtain

$$U_\beta = ce^{(2slog\beta+1)x} + \frac{6}{slog\beta+2}e^{-3x},$$

Where c is the constant of integration.

Since, when $x \to \infty,\ u(x,t)$ must be bounded,

Therefore, when $x \to \infty, U_\beta(x,s)$ should also be bounded,

Hence

$$c = 0.$$

From above equation, we get

$$U_\beta(x,s) = \frac{6}{slog\beta+2}e^{-3x},$$

Now taking inverse β-LIT of both sides

$$u(x,t) = 6e^{-2t-3x}.$$

Example 3.1.2. Consider the equation

$$x\frac{\partial u(x,t)}{\partial x} + \frac{\partial u(x,t)}{\partial t} = x, \qquad t > 0,\ x > 0,$$

With the conditions

$$u(0,t) = 0 \quad \text{for } t > 0.$$
$$u(x,0) = 0 \quad \text{for } x > 0,$$

Solution. Apply the β-LIT

$$slog\beta U_\beta(x,s) + x\frac{dU_\beta}{dx} = \frac{x}{slog\beta}, \qquad U_\beta(0,s) = 0.$$

after solving,

$$U_\beta(x,s) = Cx^{-s} + \frac{x}{slog\beta(slog\beta+1)},$$

Where C is integral constant, Since $U_\beta(0,s) = 0, \Longrightarrow C = 0$ for a bounded solution.

Therefore,

$$U_\beta(x,s) = \frac{x}{slog\beta(slog\beta+1)} = x\left(\frac{1}{slog\beta} - \frac{1}{slog\beta+1}\right).$$

taking Inverse β-LIT

$$u(x,t) = x(1 - e^{-t}).$$

Example 3.1.3. Consider the equation

$$\frac{\partial u(x,t)}{\partial x} + x\frac{\partial u(x,t)}{\partial t} = x, \qquad t > 0, \ x > 0,$$

With the conditions

$$u(0,t) = 0 \quad \text{for } t > 0.$$

$$u(x,0) = 0 \quad \text{for } x > 0,$$

Solution. Apply the β-LIT

$$xslog\beta U_\beta(x,s) + \frac{dU_\beta}{dx} = \frac{x}{slog\beta}, \qquad U_\beta(0,s) = 0.$$

after solving, obtain

$$U_\beta(x,s) = \frac{1}{(slog\beta)^2} + C\beta^{-\frac{1}{2}sx^2},$$

Where, C is an integral constant, since solution is bounded. Therefore,

$$U_\beta(0,s) = 0, \Longrightarrow C = -\frac{1}{(slog\beta)^2}.$$

After putting the value of C,

$$U_\beta(x,s) = \frac{1}{(slog\beta)^2}\left[1 - \beta^{-\frac{1}{2}sx^2}\right].$$

taking Inverse β-LIT

$$u(x,t) = t - \left(t - \frac{1}{2}x^2\right)H\left(t - \frac{x^2}{2}\right).$$

Or, equivalently

$$u(x,t) = \begin{cases} \frac{x^2}{2}, & if \quad t > \frac{x^2}{2} \\ t, & if \quad t < \frac{x^2}{2}. \end{cases}$$

3.2 Initial-Boundary Value Problem of Second Order

Example 3.2.1. Let consider equation,

$$\frac{\partial^2 u}{\partial x^2} - \frac{\partial^2 u}{\partial t^2} = xt,$$

Given $$u(x,0) = 0, \left(\frac{\partial u}{\partial t}\right)_{t=0} = 0.$$

Solution. Taking β-LIT both the sides of the given PDE

$$\mathcal{L}_\beta\left\{\frac{\partial^2 u}{\partial x^2}\right\} - \mathcal{L}_\beta\left\{\frac{\partial^2 u}{\partial t^2}\right\} = \mathcal{L}_\beta\{xt\},$$

And we get

$$\frac{d^2 U_\beta}{dx^2} - \left[(slog\beta)^2 U_\beta - slog\beta u(x,0) - \left(\frac{\partial u}{\partial t}\right)_{t=0}\right] = \frac{x}{(slog\beta)^2},$$

Applying the boundary conditions,

$$\frac{d^2 U_\beta}{dx^2} - (slog\beta)^2 U_\beta = \frac{x}{(slog\beta)^2},$$

after solving,

$$U_\beta = c_1 \beta^{sx} + c_2 \beta^{-sx} - \frac{x}{(slog\beta)^4},$$

From the boundary conditions, $x \to \infty, U_\beta = 0 \Longrightarrow c_1 = 0,$
Again when $x = 0, U_\beta = 0 \Longrightarrow c_2 = 0,$
After we get

$$U_\beta = -\frac{x}{(slog\beta)^4},$$

Now taking inverse β-LIT of both sides

$$u(x,t) = -\frac{xt^3}{6}.$$

3.3 *The Heat Equation*

Definition 3.3.1. Heat Equation
Heat equation (one dimensional) is given by the PDE

$$\frac{\partial u}{\partial t} = k\frac{\partial^2 u}{\partial x^2}.$$

Where $u(x,t)$ denote the temperature of the body.
k is Diffusivity and $k = \frac{K}{c\rho}$, K is Thermal Conductivity, c is specific heat, ρ is Density.All these have been assumed constants.

Example 3.3.2. The temperature $u(x,t)$ of the body is given by the BVP

$$\frac{\partial u}{\partial t} = k\frac{\partial^2 u}{\partial x^2}, \quad x,t > 0.$$

Boundary conditions:

$$u(x,0) = 0, \quad u(0,t) = U_0.$$

Solution. Since the temperature of the body is finite.
therefore, $\exists K(> 0), \forall x,t;$ such that

$$|u(x,t)| < K.$$

Taking β-LIT of both sides

$$\mathcal{L}_\beta\left\{\frac{\partial u}{\partial t}\right\}_{(s)} = k\,\mathcal{L}_\beta\left\{\frac{\partial^2 u}{\partial x^2}\right\}_{(s)},$$

$$slog\beta U_\beta(x,p) - u(x,0) = k\frac{d^2 U_\beta}{dx^2}, \quad \text{where } \mathcal{L}_\beta\{u(x,t)\} = U_\beta(x,s)$$

$$\frac{d^2 U_\beta}{dx^2} - \frac{slog\beta}{k}U_\beta = 0$$

After solving, we get

$$U_\beta(x,s) = c_1 e^{\sqrt{\frac{slog\beta}{k}}x} + c_2 e^{-\sqrt{\frac{slog\beta}{k}}x},$$

Where c_1 and c_2 are the constants of integration.
To determine c_1 and c_2:

Since $u(x,t)$ is finite as $x \to \infty$
Then $U_\beta(x,s)$ is finite when $x \to \infty$ otherwise the temperature will be infinity

$$\therefore\ c_1 = 0$$

Applying the β-LIT and the condition $u(0,t) = U_0$,

$$\mathcal{L}_\beta\{u(0,t)\} = U_0 \mathcal{L}_\beta\{1\},$$

$$U_\beta(0,s) = \frac{U_0}{slog\beta},$$

Taking $x = 0$, gives $c_2 = \frac{U_0}{slog\beta}$.
Now using equations

$$U_\beta(x,s) = \frac{U_0}{slog\beta} e^{-\sqrt{\frac{slog\beta}{k}}x},$$

Now taking inverse β-LIT

$$u(x,t) = U_0\left\{1 - erf\left(\frac{x}{2\sqrt{kt}}\right)\right\},$$

3.4 Diffusion Equation in Finite Medium

Example 3.4.1. Let consider the diffusion equation

$$\frac{\partial u}{\partial t} = \kappa \frac{\partial^2 u}{\partial x^2},\quad 0 < x < a,\ t > 0,$$

With conditions,

$$u(x,0) = 0,\ 0 < x < a,$$
$$u(0,t) = K,\ t > 0,$$
$$\left(\frac{\partial u(x,t)}{\partial x}\right)_{x=a} = 0,\ t > 0,$$

Where, K is a constant.

Solution. Applying the β-LIT with respect to t and obtain

$$\frac{d^2 U_\beta}{dx^2} - \frac{slog\beta}{\kappa} U_\beta = 0, \qquad 0 < x < a,$$

$$U_\beta(0,s) = \frac{M}{slog\beta},\quad \left(\frac{d}{dx}(U_\beta(x,s))\right)_{x=a} = 0.$$

The general solution of the differential equation is

$$U_\beta(x,s) = Acosh\left(x\sqrt{\frac{slog\beta}{\kappa}}\right) + Bsinh\left(x\sqrt{\frac{slog\beta}{\kappa}}\right),$$

After applying the given condition, we get the value of constants A and B.
Now we have

$$U_\beta(x,s) = \frac{K}{slog\beta}\cdot\frac{cosh\left[(a-x)\sqrt{\frac{slog\beta}{\kappa}}\right]}{cosh\left(a\sqrt{\frac{slog\beta}{\kappa}}\right)}.$$

Finally, taking inverse β-LIT

$$u(x,t) = K\mathcal{L}_\beta^{-1}\left\{\frac{1}{slog\beta}.\frac{cosh\left[(a-x)\sqrt{\frac{slog\beta}{\kappa}}\right]}{cosh\left(a\sqrt{\frac{slog\beta}{\kappa}}\right)}\right\}_{(t)}.$$

The inverse β-LIT can be found by applying Cauchy Residue Theorem.
Then we have

$$u(x,t) = M\left[1-\frac{4}{\pi}\sum_{n=1}^{\infty}\frac{1}{(2n-1)}sin\left\{\left(\frac{2n-1}{2a}\right)\pi x\right\}e^{\left\{-(2n-1)^2\left(\frac{\pi}{2a}\right)^2\kappa t\right\}}\right].$$

3.5 Wave Equation

Definition 3.5.1. Wave Equation (One dimensional equation)
Wave equation (one dimensional)

$$\frac{\partial^2 u}{\partial t^2} = c^2\frac{\partial^2 u}{\partial x^2}$$

Where, $u(x,t)$ is displacement which is function of x and t and c is a constant.

Example 3.5.2. Let consider a wave equation

$$\frac{\partial^2 u}{\partial t^2} = c^2\frac{\partial^2 u}{\partial x^2},\ x \geq 0, t > 0$$

With the conditions

$$u(0,t) = Af(t),\ \ t \geq 0,$$
$$u(x,t) \to 0 \text{ as } x \to \infty,\ \ t \geq 0,$$
$$u(x,t) = 0 = \frac{\partial u(x,t)}{\partial t},\ \text{ at } t = 0 \text{ for } 0 < x < \infty.$$

Solution. Taking β-LIT

$$\mathcal{L}_\beta\left\{\frac{\partial^2 u}{\partial t^2}\right\}_{(s)} = c^2\mathcal{L}_\beta\left\{\frac{\partial^2 u}{\partial x^2}\right\}_{(s)},$$
$$(slog\beta)^2 U_\beta(x,s) - slog\beta u(x,0) - \left(\frac{\partial u(x,t)}{\partial t}\right)_{t=0} = c^2\frac{d^2 U_\beta}{dx^2},$$

Where,

$$\mathcal{L}_\beta\{u(x,t)\}_{(s)} = U_\beta(x,s).$$

By the conditions,

$$\frac{d^2 U_\beta}{dx^2} - \frac{(slog\beta)^2}{c^2}U_\beta = 0, \text{ for } 0 \leq x < \infty,$$
$$U_\beta(x,s) = A\mathcal{L}_\beta\{f(t)\}_{(s)} \text{ at } x = 0,$$
$$U_\beta(x,s) \to 0 \text{ as } x \to \infty.$$

After, solving the differential equation

$$U_\beta(x,s) = A\mathcal{L}_\beta\{f(t)\}_{(s)}\beta^{-\left(\frac{xs}{c}\right)}.$$

Taking inverse β-LIT

$$u(x,t) = Af\left(t-\frac{x}{c}\right)H\left(t-\frac{x}{c}\right)$$

Solution can be re-written as

$$u(x,t) = \begin{cases} 0, & t < \frac{x}{c} \\ Af\left(t - \frac{x}{c}\right), & t > \frac{x}{c}. \end{cases}$$

3.6 Equation of Current and Potential in a Transmission Line

Example 3.6.1. Let consider potential $V(x,t)$ and current $I(x,t)$ satisfy the following system

$$L\frac{\partial I}{\partial t} + RI = -\frac{\partial V}{\partial x},$$
$$C\frac{\partial V}{\partial t} + GV = -\frac{\partial I}{\partial x}.$$

In a transmission line. Where, L, R, C, G denote the inductance, resistance, capacitance, leakage conductance of the transmission line.

Solution. After eliminating V or I, we get the same equation which is in the following form

$$\frac{1}{c^2}u_{tt} - u_{xx} + au_t + bu = 0.$$

Where $c^2 = (LC)^{-1}$, $a = LG + RC$, and $b = RG$.
Above equation is known as Telegraph Equation.
To solve the equation, there may be some cases:

(a) Transmission line without loss ($R = 0$ and $G = 0$)
If $R = 0$ and $G = 0$ then classical wave equation

$$u_{tt} = c^2 u_{xx}.$$

The solution of this equation can be carried out from example(?) with the following conditions in the potential $V(x,t)$:

$$V(x,t) = V_0 f(t) \text{ at } x = 0, \ t > 0.$$

With conditions

$$x = 0 \text{ for } t > 0,$$

and

$$V(x,t) \to 0 \text{ as } x \to \infty \text{ for } t > 0.$$

Then the solution is given by

$$V(x,t) = V_0 f\left(t - \frac{x}{c}\right) H\left(t - \frac{x}{c}\right).$$

(b) Ideal submarine cable (The kelvin ideal cable)
For kelvin ideal cable, $L = 0$, and $G = 0$ equation reduces to the classical diffusion equation

$$u_t = \kappa u_{xx},$$

Where $\kappa = a^{-1} = (RC)^{-1}$.

Solve as same as Example. After using the given conditions, we get potential $V(x,t)$

$$V(x,t) = V_0 erfc\left(\frac{x}{2\sqrt{\kappa t}}\right)$$

(c) A Non-inductive leady cable

For a non-inductive leady cable, $L = 0$ and $G \neq 0$. We get

$$V_{xx} - aV_t - bV = 0,$$

With conditions

$$V(0,t) = H(t) \text{ and } V(x,t) \to 0 \text{ as } x \to \infty.$$

The β-Laplace integral transform of the problem is

$$\frac{d^2V_\beta}{dx^2} = (aslog\beta + b)V_\beta,$$
$$V_\beta(0,s) = \frac{1}{slog\beta}, \qquad V_\beta(x,s) \to 0 \text{ as } x \to \infty.$$

Then the solution is

$$V_\beta(x,s) = \frac{1}{slog\beta} e^{\left[-x(slog\beta a+b)^{\frac{1}{2}}\right]}.$$

Taking inverse β-Laplace integral transform

$$V(x,t) = \frac{1}{2}e^{x\sqrt{b}} erfc\left(\frac{x}{2}\sqrt{\frac{a}{t}} + \sqrt{\frac{bt}{a}}\right) + \frac{1}{2}e^{-x\sqrt{b}} erfc\left(\frac{x}{2}\sqrt{\frac{a}{t}} - \frac{bt}{a}\right).$$

(d) Heaviside distortion less cable

For Heaviside distortion less cable, $\frac{R}{L} = \frac{G}{C} = \kappa = constant$, the potential $V(x,t)$ and the current $I(x,t)$ satisfy the same equation

$$u_{tt} + 2\kappa u_t + \kappa^2 u = c^2 u_{xx}, \quad 0 \leq x < \infty, \quad t > 0.$$

Apply the β-Laplace Transform and we get

$$\frac{d^2V_\beta}{dx^2} = \left(\frac{slog\beta+\kappa}{c}\right)^2 V_\beta \,,$$

Then the solution for $V_\beta(x,s)$ with the boundary conditions is

$$V_\beta(x,s) = V_0 \mathcal{L}_\beta\{f(t)\}_{(s)} e^{\left[-\left(\frac{slog\beta+\kappa}{c}\right)x\right]}$$

Taking inverse β-Laplace transform, and obtain

$$V(x,t) = V_0 e^{-\frac{\kappa x}{c}} f\left(t - \frac{x}{c}\right) H\left(t - \frac{x}{c}\right).$$

Where, wave propagating at a velocity $c = (LC)^{-\frac{1}{2}}$.

3.7 Axisymmetric Heat Conduction Equation

Example 3.7.1. Let consider axisymmetric heat conduction equation

$$u_t = \kappa\left(u_{rr} + \frac{1}{r}u_r\right),\ 0 \le r < a,\quad t > 0,$$

With conditions

$$u_t(r,0) = 0 \quad \text{for } 0 < r < a,$$
$$u(r,t) = f(t) \text{ at } r = a \text{ for } t > 0,$$

Where κ and T_0 are constants.

Solution. Apply the β-Laplace integral transform to equation gives

$$\frac{d^2U_\beta}{dr^2} + \frac{1}{r}\frac{dU_\beta}{dr} - \frac{slog\beta}{\kappa}U_\beta = 0.$$

Or,

$$r^2\frac{d^2U_\beta}{dr^2} + r\frac{dU_\beta}{dr} - r^2\left(\frac{slog\beta}{\kappa}\right)U_\beta = 0.$$

Above equation is known as Standard Bessel Equation, and the solution is given by

$$U_\beta(r,s) = AI_0\left(r\sqrt{\frac{slog\beta}{\kappa}}\right) + BK_0\left(r\sqrt{\frac{slog\beta}{\kappa}}\right)$$

Where $I_0(x)$ and $K_0(x)$ are modified Bessel function of zero order and A and B are integral constants. Since $K_0\left(r\sqrt{\frac{slog\beta}{\kappa}}\right)$ is unbounded at $r = 0$, then we get $B \equiv 0$. After putting the value

$$U_\beta\,(r,s) = AI_0\left(r\sqrt{\frac{slog\beta}{\kappa}}\right)$$

Using transformed boundary condition $U_\beta(a,s) = \mathcal{L}_\beta\{f(t)\}_{(s)}$, we obtain

$$U_\beta(r,s) = \frac{\mathcal{L}_\beta\{f(t)\}_{(s)} I_0\left(r\sqrt{\frac{slog\beta}{\kappa}}\right)}{I_0\left(a\sqrt{\frac{slog\beta}{\kappa}}\right)} = \mathcal{L}_\beta\{f(t)\}_{(s)}\mathcal{L}_\beta\{g(t)\}_{(s)}$$

Where,

$$\mathcal{L}_\beta\{g(t)\}_{(s)} = \frac{I_0\left(r\sqrt{\frac{slog\beta}{\kappa}}\right)}{I_0\left(a\sqrt{\frac{slog\beta}{\kappa}}\right)}.$$

By Convolution Theorem, the solution takes the form

$$u(r,t) = \int_0^t f(t-\tau)g(\tau)d\tau,$$

Where,

$$g(t) = \frac{log\beta}{2\pi i}\int_{c-i\infty}^{c+i\infty} \beta^{st}\frac{I_0\left(r\sqrt{\frac{slog\beta}{\kappa}}\right)}{I_0\left(a\sqrt{\frac{slog\beta}{\kappa}}\right)}ds.$$

This integral can be carried out by applying the theory of residue of complex analysis where the poles of the integrand are at the points $s = s_n = -\kappa\alpha_n^2, n = 1,2,3,...$ and α_n are the roots of $J_0(a\alpha) = 0$. The residue at pole $s = s_n$ is

$$\left(\frac{2i\kappa\alpha_n}{a}\right)\frac{I_0(ir\alpha_n)}{I_0'(ia\alpha_n)}e^{-\kappa t\alpha_n^2} = \left(\frac{2\kappa\alpha_n}{a}\right)\frac{J_0(r\alpha_n)}{J_0(a\alpha_n)}e^{-\kappa t\alpha_n^2},$$

So that

$$g(t) = \left(\frac{2\kappa}{a}\right)\sum_{n=1}^{\infty}\frac{\alpha J_0(r\alpha_n)}{J_1(a\alpha_n)}e^{(-\kappa t\alpha_n^2)}$$

The solution becomes

$$u(x,t) = \left(\frac{2\kappa}{a}\right)\sum_{n=1}^{\infty}\frac{\alpha J_0(r\alpha_n)}{J_1(a\alpha_n)}\int_0^t f(t-\tau)\, e^{(-\kappa t\alpha_n^2)}d\tau.$$

3.8 Inhomogeneous Partial Differential Equation

Example 3.8.1. Let an inhomogeneous problem

$$u_{xt} = -\omega sin\omega t, \quad t > 0$$

With conditions

$$u(x,0) = x, \quad u(0,t) = 0.$$

Solution. Applying the β-Laplace integral transform which gives

$$\frac{dU_\beta}{dx} = \frac{slog\beta}{(slog\beta)^2+\omega^2},$$

After, solving we get

$$U_\beta(x,s) = \frac{slog\beta x}{(slog\beta)^2+\omega^2} + A$$

Where A is integral constant. Since $U_\beta(0,s) = 0, \Longrightarrow A = 0$, and taking β-Laplace integral transform

$$u(x,t) = xcos\omega t.$$

3.9 Inhomogeneous Wave Equation

Example 3.9.1. Consider an inhomogeneous wave equation

$$\frac{1}{c^2}u_{tt} - u_{xx} = ksin\left(\frac{\pi x}{a}\right), \quad 0 < x < a, \ t > 0,$$
$$u(x,0) = 0 = u_t(x,0), \quad 0 < x < a,$$

$$u(0,t) = 0 = u(a,t), \quad t > 0,$$

Where a, c and k are constants.

Solution. Applying the β-LIT which gives

$$\frac{d^2 U_\beta}{dx^2} - \frac{(slog\beta)^2}{c^2} U_\beta = -\frac{k}{s} sin\left(\frac{\pi x}{a}\right),$$

$$U_\beta(0,s) = 0 = U_\beta(a,s).$$

After solving above differential equation, we get

$$U_\beta(x,s) = A\beta^{\left(\frac{sx}{c}\right)} + B\beta^{\left(-\frac{sx}{c}\right)} + \frac{ksin\left(\frac{\pi x}{a}\right)}{a^2 slog\beta\left((slog\beta)^2 + \frac{\pi^2 c^2}{a^2}\right)}.$$

In view of equation $A = B = 0$, and hence, the solution becomes

$$U_\beta(x,s) = \frac{ksin\left(\frac{\pi x}{a}\right)}{a^2 slog\beta\left((slog\beta)^2 + \frac{\pi^2 c^2}{a^2}\right)}.$$

Taking inverse β-LIT, then solution is,

$$u(x,t) = \frac{k}{(\pi c)^2}\left[1 - cos\left(\frac{\pi ct}{a}\right)\right] \sin\left(\frac{\pi x}{a}\right).$$

3.10 *Rayleigh and the Stokes Problem in Fluid Dynamics*

Example 3.10.1. Consider the Stokes problem,

$$u_t = v u_{zz}, \quad t > 0, \ z > 0,$$

With some following conditions

$$u(z,t) = U_0 e^{i\omega t} \text{ on } z = 0, \ t > 0,$$
$$u(z,t) \to 0 \text{ as } z \to \infty, t > 0,$$
$$u(z,t) \to 0 \text{ at } \forall z > 0, \ t \le 0. \ ,$$

Where, U_0 is a constant, and $u(z,t)$ represent the fluid velocity.

Solution. Applying the β-Laplace transform and given boundary conditions. Then we get

$$U_\beta(z,s) = \frac{U_0}{slog\beta - i\omega} e^{-z\sqrt{\frac{slog\beta}{v}}}$$

Using the inverse β-Laplace integral transform, we obtain the solution

$$u(z,t) = \frac{U_0}{2} e^{i\omega t}\left[e^{-z\sqrt{\frac{i\omega}{v}}} erfc\left(\frac{z}{2\sqrt{vt}} - \sqrt{i\omega t}\right) + e^{z\sqrt{\frac{i\omega}{v}}} erfc\left(\frac{z}{2\sqrt{vt}} + \sqrt{i\omega t}\right)\right].$$

4. References

[1] Asmar N 2005 *Partial Differential Equations*, (New Jersey: Prentice Hall)
[2] Bateman H 1944 *Partial Differential Equations of Mathematical Physics* (New York: Dover Publications)
[3] Blanchard P et al 2005 *Differential Equations* (Brooks Cole) 3rd ed. pp 559-595
[4] Churchill R 1958 *Operational Mathematics* (McGraw-Hill) 2nd ed.
[5] Coleman M 2013 *An Introduction to Partial Differential Equations with MATLAB* (CRC Press)
[6] Copson E T 1935 *An Introduction to the Theory of Functions of a Complex Variable* (Oxford University Press)
[7] Daubechies I 1990 *The wavelet transform, time-frequency localization and signal analysis* (IEEE Trans. Infor. Theory) vol. **36** pp 961–1005.
[8] Debnath L and Bhatta D 2015 *Integral transforms and Their Applications* 3rd edition. (Boca Raton: CRC Press, Chapman & Hall)
[9] Doetsch G 1970 *Introduction to the Theory and Applications of the Laplace Transformation* (New York: Springer-Verlag)
[10] Gakhov F D 1966 *Boundary Value Problems* (London: Pergamon Press)
[11] Gaur A and Agarwal G 2020 *Application of β-Laplace integral transform* (Talent Development and Excellence), vol. **12** no. 3s pp 1058–1068
[12] Gaur A and Agarwal G 2020 *On β-Laplace integral transforms and their properties* (International journal of advanced science and technology) vol. **29** pp 1481-1491
[13] Gaur A and Agarwal G 2020 *Relation of β-Laplace integral transforms with other integral transforms* (Test Engineering and Management) vol. **83** pp. 8653–8659
[14] Heaviside O 1899 *Electromagnetic Theory* 1950 reprint, (New York: Dover Publications)
[15] James G 1999 *Advanced modern Engineering Mathematics* (New York: Pearson education limited) 3rd ed.
[16] Kreyszig E 1998 *Advance Engineering Mathematics* (John Wiley and sons) 8th ed.
[17] Lawrence C. Evans 1998 *Partial Differential Equations* (AMS)
[18] Sneddon I N 1972 *The Use of Integral transforms* (New York: McGraw-Hill Book Company)
[19] Sourav K and Subrata K *Engineering Mathematics* (Tata McGraw Hill)
[20] Spiegel M R 1965 *Schaum's outline of theory and problems of Laplace Transforms* (Mc Graw Hill)

Design and fabrication of multi utility unmanned robot

Ashik Faizal[1], Aswin S Nair[1], Gokuldas V R[1], Jerin P Raju[1], Vishnu Sankar[2] and Joel George[3]
[1] Department of Mechanical Engineering, Rajagiri School of Engineering and Technology, Kochi, Kerala 682039, India
[2] Department of Mechanical Engineering, Rajagiri School of Engineering and Technology, Kochi, Kerala 682039 India
[3] Department of Mechanical Engineering, Rajagiri School of Engineering and Technology, Kochi, Kerala 682039, India

E-mail: vishnusankar9369@gmail.com

Abstract. Humans have dreamed of mechanical devices to ease their chores and assist them in performing repetitive tasks for thousands of years. This project can be identified as a key tool for reducing the vulnerability of human lives when we deal with emergencies. The human presence is very much important in dealing with all emergencies. Even if it is a fire accident, disinfection activities or surveillance in places where the accessibility is very much limited. Each case the first responders' lives are at risk. Their direct contact in the impact zones can put them in danger. This is where unmanned robots can be utilized. In this paper a multi utility unmanned robot equipped with a long range control system, nozzle system, camera for surveillance, and an all-terrain track to conquer different working environments was designed and fabricated. Thus the threat for first responder's and health workers can be reduced.

1. Introduction

The human presence is very much important in dealing with all emergencies. Even if it is a fire accident, disinfection activities or surveillance in places where the accessibility is very much limited. The risk for first responders is very high during these incidents. Many of these accidents occur at hazardous and risky areas where humans cannot even engage directly for rescue operations. This causes huge damage to both human life and property. Due to the increased number of high raised buildings, accessing the situation in a fire outbreak or any other incident is critical. Currently, there are a lot of limitations faced by humans who are engaged in such situations due to their direct exposure to the fire. MUUR (Multi Utility Unmanned Robot) focuses on limiting these potentially life-threatening situations that can occur to a first responder unpredictably. Hence MUUR avoids the direct involvement of a human in dealing with these situations. Complete automation in this field can be really expensive, complicated and human decisions taken during these situations are of equal importance in saving lives. So we suggest a semi-autonomous MUUR to engage the situation so that human intervention can also be made possible.

2. Design of chassis and all terrain mechanism

MUUR is identified to be the priority for infrastructure support for any surveillance warning systems prevention and rescue in view of a pandemic situation as well as disasters like flood, fire incidents and other natural and unnatural calamities. Considering a pandemic situation, one of the topmost problems affecting is the process of cleaning and disinfecting the public places that could increase the risk of getting affected by the exposed sites. MUUR with a long range control system will help humans to tackle this situation from a safe distance into a wide area. Similarly in any other situation, human safety will not be compromised. Figure.1 depicts the initial design of MUUR.

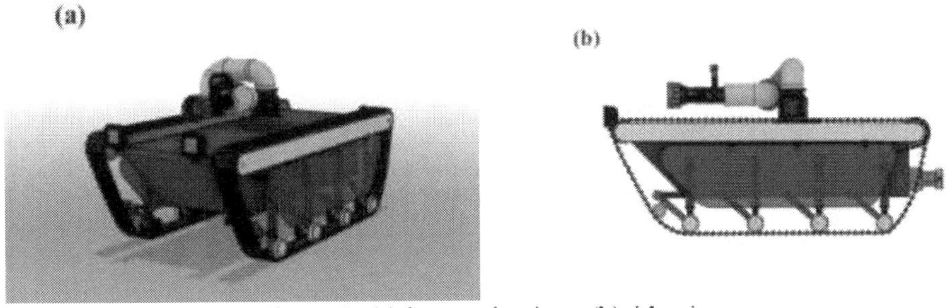

Figure 1. (a) isometric view; (b)side view.

2.1. MUUR frame

The MUUR frame is the most fundamental part of the MUUR. A rectangular frame is the most commonly used frame due to its simple structure. An inclination of 45 degrees is also provided in the front of the frame for an advantage in terrain interactions. A pair of motors on each side opposite to each other turn in the same directions. This configuration is selected based on the symmetry of the frame. The frame is designed in Autodesk Fusion 360. The frame size depends on the total mass including the payload that the MUUR has to carry. As the frame size increases, the inertia also increases [1]. For the ease of manufacture and material availability mild steel was selected as frame members. They also provide good material property. The payload will be the medical supplies or goods that have to be transported across an affected region. In total, 200kg of kerb weight is calculated for the MUUR.

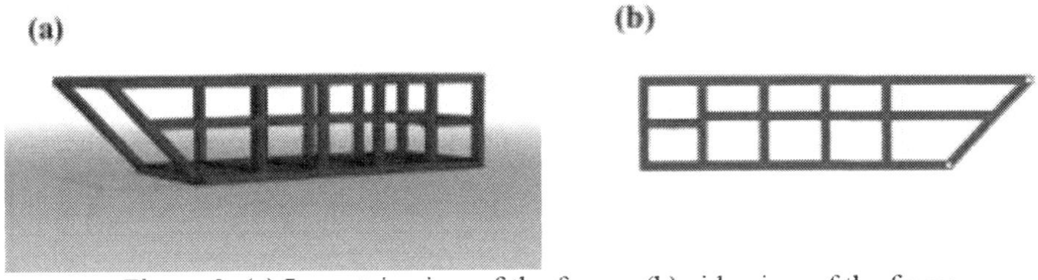

Figure 2. (a) Isometric view of the frame; (b) side view of the frame.

2.1.1. Analysis of frame members. MUUR frame has a hollow square pipe of the dimension 1 inch and 1.6mm thickness. Base member of the frame shown in figure 2, takes the maximum load in the chassis. So the maximum load that the chassis can bear has to be determined [2]. Considering the for fixed hollow beam, the total load capacity can be estimated using the equation (1) :

$$M = \frac{\sigma_{per} I}{Y} \tag{1}$$

The term M denotes the bending moment, σ_{per} is the maximum permissible stress for the material selected, I denote the area moment of inertia of the hollow beam and Y is the distance from the centre of the hollow pipe to the outer edge. The values for I can be calculated by equation (2).

$$I = \frac{BD^3}{12} - \frac{bd^3}{12} \tag{2}$$

Terms B and b gives the external and internal breadth of the section of the pipe and D and d gives the section height of the pipe. The maximum weight that can be loaded on the pipe for the length 360mm is calculated to be 238.32kg. Static stress analysis on the frame is shown in figure 3.

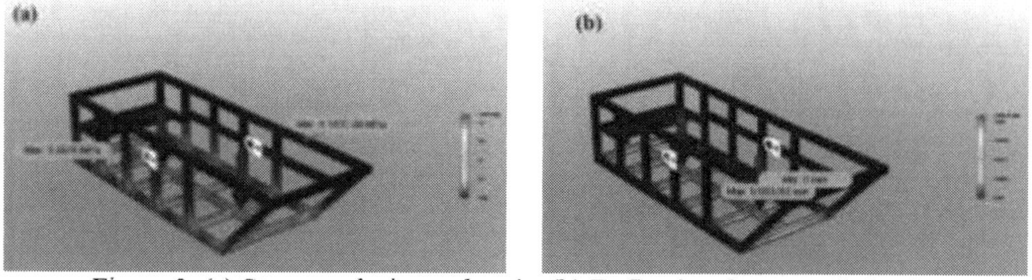

Figure 3. (a) Stress analysis on chassis; (b) Deflection due to static load

2.1.2. Analysis on the shaft of the road wheels. The strength of the shaft connecting the road wheels is a very important factor. The connecting rod arm for the rollers that are attached to the chassis is connected via a bearing to obtain a swing action for the suspension system. So the entire possible weight of the robot will be acting on the shaft connecting the rollers [3]. The minimum thickness of the required shaft can be calculated by equation (3):

$$M = \sigma_{per} Z \tag{3}$$

The term M denotes the bending moment of the shaft and σ_{per} is the permissible stress for the selected material for the shaft. The section modulus Z can be calculated for a solid cylindrical shaft by the equation (4) in which r is the radius of the cylinder.

$$Z = \frac{\pi r^3}{4} \tag{4}$$

2.2. Selection of chain and sprocket for drive. The MUUR uses two brushless dc motors for drive. Each motor is coupled to a chain drive system which will drive the all-terrain tracks. So calculating the required number of the chain links is essential to set the number of sprocket teeth required at each end. Considering a gear reduction ratio of 1:2 the sprocket that drives the tracks will have twice the number of teeth in comparison to the sprocket coupled with the motor. Setting the minimum distance between the sprockets is also important in deciding the chain length, pitch and the no. of chain links to avoid slippage [4].

The pitch of the chain link required can be calculated from the equation (5):

$$P \leq 10\left(\frac{60.67}{n_1}\right)^{\frac{2}{3}} \tag{5}$$

Term n_1 is the rotations per second at the motor end. This value is obtained when the rpm of the drive motor is calculated. Using the above equation the pitch selected for chain drive is 12.70mm. Now selecting the sprockets for the chain can be done using the equation (6):

$$D = P\left(\sin\left(\frac{180}{Z}\right)\right)^{-1} \tag{6}$$

The term D is the pitch diameter of the sprocket, P is the pitch of the chain and Z is the number of teeth on the sprocket. Considering the running life of chain and sprocket, it is advised to have an even number of chain links if the number of sprocket teeth is odd or vice versa. For a drive without any slippage, the centre to centre distance of the sprockets has to be precise [5]. The total number of chain links required for a 150mm centre to centre distance of sprockets can be calculated by equation (7):

$$L_n = 2\left(\frac{a}{P}\right) + \left(\frac{Z_1 + Z_2}{2}\right) + \left(\frac{Z_2 - Z_1}{2\pi}\right)^2 \times \left(\frac{P}{a}\right) \quad (7)$$

a is the assumed centre to centre distance and P is the pitch of the chain. The total number of chain links required is calculated to be 48. Now the centre to centre distance has to be corrected to avoid slip during the drive. This correction can be done using equation (8):

$$a = \frac{P}{4}\left\{\left[L_n\left(\frac{Z_1 + Z_2}{2}\right)\right] + \left\{\left[Ln - \left(\frac{Z_1 + Z_2}{2}\right)^2\right] - 8\left[\frac{Z_2 - Z_1}{2\pi}\right]^2\right\}^{\frac{1}{2}}\right\} \quad (8)$$

By following the equation (9) the centre to centre distance was corrected to 155.3mm. Chain and sprocket based on calculations sketched and shown in figure 4. For the required load that has to be driven with his chain, a single chain is only required based on the studies from different catalogues of different brands.

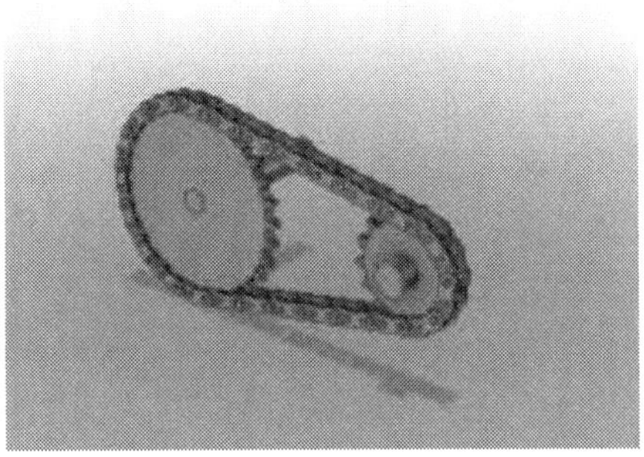

Figure 4. Chain and sprocket.

2.3. *Selection of bearings*

MUUR is designed with 4 road wheels on each side to ensure an even distribution of pressure to the ground and also to help with tackling rough terrains by improving traction. Thus the load bearing capacity for the bearings is very important. A compromise in the working of rollers affects the terrain interactions of MUUR. In the above section (2.1.2), the diameter of the shaft for rollers was calculated to be 20mm. The centre bore diameter of the required bearing will remain the same. Both static load and Dynamic loads have to be considered for the selection.

On static conditions, the load will be resting on the road wheel so in the long term this could damage the hull of the bearing if this load is not considered. The dynamic load will be maximum when the MUUR will be taking a tight corner turn. The entire load of the MUUR will get transferred to the outer road wheels of the turn. This situation could trouble the bearing if it is not selected accordingly. Radial loaders are the loads that act on the bearing under static condition. This load can be easily calculated by considering the weight distribution of MUUR. Here it is considered a situation where the entire weight of the MUUR will be acting on a single road wheel for static load. For dynamic load, the centripetal force acting on a road wheel is also considered during a tight turn. The equivalent dynamic load on the bearings can be calculated by the equation (9):

$$P_e = XF_r + YF_a \quad (9)$$

P_e is the equivalent dynamic load, F_r and F_a are the radial and axial forces on the bearing and X and Y are the radial and thrust factors. By doing trial and error method for finding values of X and Y from the design catalogues of bearing the calculated value for P_e is 5317.47N. So the dynamic load capacity can be calculated from the equation (10):

$$C = P_e(L_{10})^{\frac{1}{3}} \quad (10)$$

In the above equation L_{10} is the life of the selected bearing from the catalogue. For the design, it is decided to have the outer race of the bearing in rotation. Hence a race rotation factor 1.2 has to be included in the equation (10) for the radial force. From the above equations, the calculated dynamic load capacity of the bearing is 13829.96N.

2.4. *Selection of spring for suspension*

For the riding comfort through different terrain, MUUR requires a quality suspension that should not get compromised during the operation. The suspension cannot be very stiff nor very soft. Making a stiff suspension will make MUUR have sudden jerks when interacting with a sudden obstacle which could lose control of the system. Similarly, a very soft suspension system is also not adequate because of the load that has to be carried and the minimum ground clearance has to be maintained [6]. While designing the spring suspension system of MUUR, the width of the system was a major constraint. Since MUUR has to be driven through narrow paths the width has to be set minimum. This will help to steer in such narrow paths. So the length of the spring suspension was increased to reduce the width.

The load acting on the springs shall be considered as a fluctuating in nature. Therefore oil-hardened and tempered steel wire of SW grade is selected for this application. The spring index is assumed as (C) 6. The wohl factor K can be calculated from the equation (11):

$$K = \frac{4C-1}{4C-4} + \frac{0.615}{C} \tag{11}$$

The induced shear stress that will be acting on the spring upon fluctuated loading can be calculated from the equation (12):

$$\tau = K\left(\frac{8P_1C}{\pi d^2}\right) \tag{12}$$

The term P_1 is the maximum load that can be acted upon the spring and d is the wire diameter of the spring. The wire diameter is calculated by trial and error method by taking into account the maximum permissible shear stress (τ_d) that can be acted on in the spring. Induced shear has to remain lesser than the permissible shear stress ($\tau < \tau_d$).

The permissible shear stress can be calculated from the equation (13):

$$\tau_d = 0.3S_{ut} \tag{13}$$

S_{ut} is the ultimate tensile strength of the material selected for spring. Hence after doing many iterations, the calculated value for wire diameter (d) is 6mm. The total deflection (δ) of the spring upon static condition and the maximum compression is fixed as 40mm. The spring constant can be calculated for the fluctuating load with the equation (14):

$$k = \frac{P_{max} - P_{min}}{\delta} \tag{14}$$

Once the spring constant is calculated, then the number of coils required for the spring to function with the above mentioned constraints can be calculated by the equation (15):

$$N = \frac{Gd^4}{8D^3k} \tag{15}$$

The term G denotes the modulus of rigidity, D is the outer diameter of the coil. For a spring with square and grounded ends, an additional 2 coils have to be added to the above calculated value. Hence the number of coils required is calculated and now total deflection under the full load condition can be calculated with the equation (16):

$$\delta_1 = \frac{8P_{max}D^3N}{Gd^4} \tag{16}$$

The solid length of the spring can be calculated by taking simply the product of the number of coils and the wire diameter. Fixing a gap of 0.5mm between each coil the total free length of the coil spring for the above calculated constraints can be calculated [7] by the equation (17). Figure 5, figure 6 and figure 7, shows the designed shock absorber system for the MUUR.

$$\text{Total free length} = \text{Solid length} + \text{Total deflection} + \text{Total axial gap} \quad (17)$$

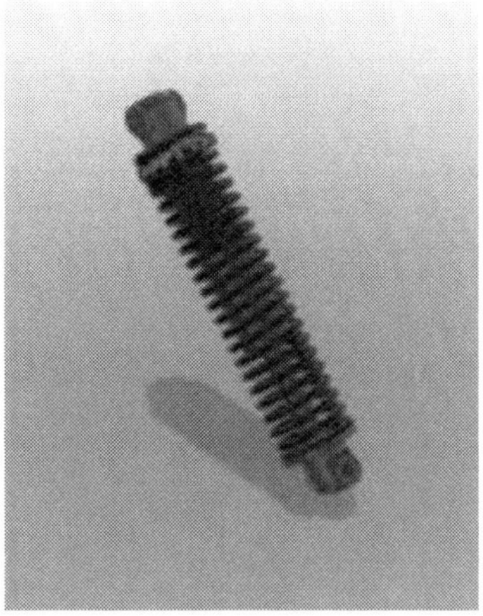

Figure 5. Shock absorber.

Figure 6. Road wheels attached to shock absorbers.

3. Analysis of MUUR power demand

Power requirement for the tracked robot is calculated by fixing primary constraints like the total load that has to be carried along the ride, maximum inclination that the robot should interact, acceleration and maximum velocity that it should travel with.

3.1. *Power demand under general condition.* The robot is driven by two brushless dc motors, the power requirement (p) in general can be calculated [8] by the following equation (18):

$$p = \left(mgf\cos(\alpha) + mg\sin(\alpha) + \frac{C_D A}{21.15}u^2 + \delta m \frac{du}{3.6dt} \right) \frac{u}{3600\eta_T} \quad (18)$$

In equation (18), m denotes mass, f is the coefficient resistance of rolling resistance, g is the acceleration due to gravity, $\boldsymbol{\alpha}$ is road gradient, C_d is the coefficient of air resistance, A is the frontal area for the robot, u is the maximum velocity, $\boldsymbol{\delta}$ is the rotating mass conversion factor, $\boldsymbol{\eta_T}$ is the mechanical transmission efficiency. Since the robot is targeted to achieve a very low maximum speed the term considering the air resistance can be removed [9].

3.2. *Power demand under a climbing condition.* The power requirement will keep varying according to the slopes. The maximum slope that the robot is designed to climb is 30°. With this constraint the power requirement to climb the slope with the max velocity can be calculated by the equation (19):

$$p = \left(mgf\cos(\alpha) + mg\sin(\alpha) + \delta m \frac{du}{3.6dt} \right) \frac{u}{3600\eta_T} \quad (19)$$

3.3. *Power demand for maximum velocity.* When the robot is travelling on a normal flat surface the power requirement is less as compared to when the robot is climbing a slope. This power can be calculated by altering the general equation.

Eliminating the slope factors, the general equation can be rewritten as (20):

$$p = \left(mgf + \frac{C_D A u_{max}^2}{21.15} \right) \frac{u_{max}}{3600 \eta_T} \tag{20}$$

3.4. *Calculating torque requirement.* The maximum requirement of torque will occur when the robot is climbing a slope. MUUR is designed to travel a slope of 30°. So the torque required at the wheels plays an important part. This torque can be calculated by equation (21):

$$T = \frac{F_t r}{i_g i_0 \eta_T} \tag{21}$$

In equation (21), r is the radius of the drive wheel, i_g and i_0 are the transmission gear ratios of the powertrain. For powertrain two brushless dc motors can be used based on the calculations. Two to three rechargeable batteries can be selected with respect to the total running time requirement of the robot. Rechargeable batteries will help the operators to quickly change them when the charge appears to be low. This will help them to overcome any delay caused due to charging in the midst of any operations. Motor driver can is an essential component in the powertrain. This will help the operator to control the robot more handy. The drivers can be installed with respect to the operator's requirement that decides the total speed, torque and wireless control of the robot helps to face the challenges.

4. Cad drawing of MUUR

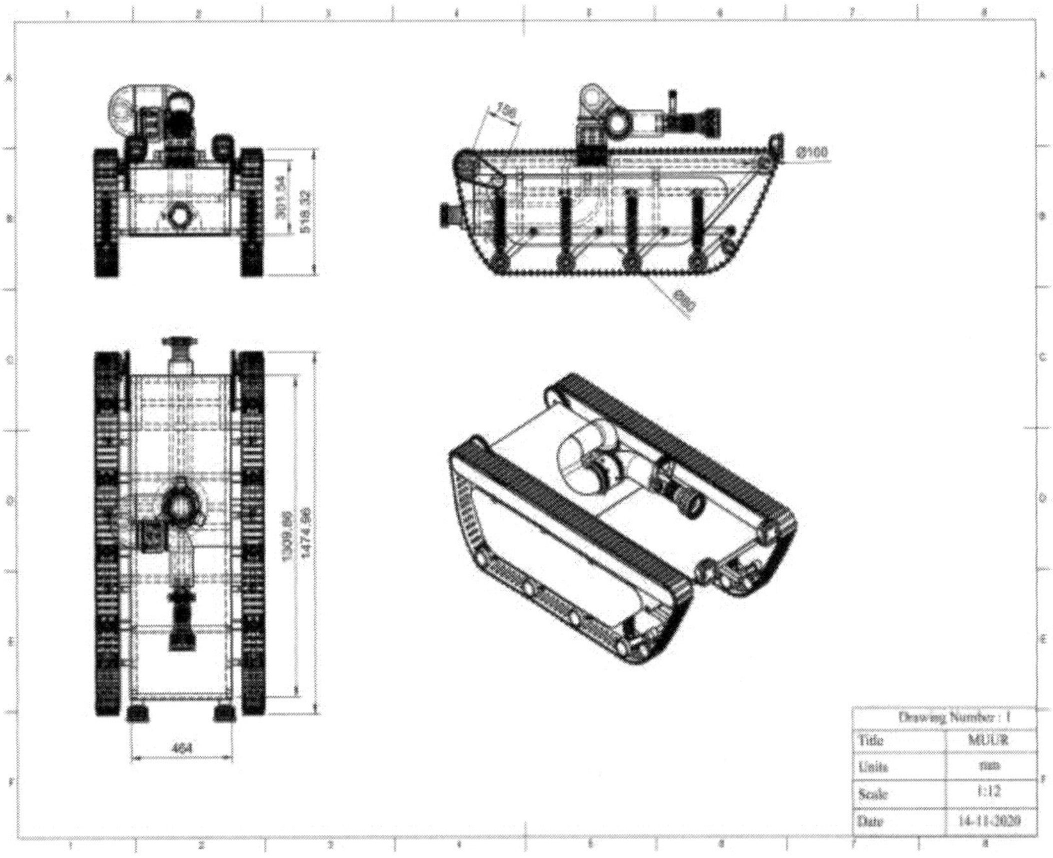

Figure 7. CAD model of MUUR

Figure 8. Rendered design of MUUR. (a)Isometric view from front; (b)Isometric view from rear.

Figure 7 shows the CAD model of MUUR and figure 8 shows the rendered design of MUUR. Thus design and fabrication of multi utility unmanned robot was completed.

5. Conclusion

Emergencies are always unpredicted and unexpected. Several casualties arise during one. The lives of the first responders and victims are equally important. MUUR with the suggested design the risk taken by the first responders can be minimized during such situations [10]. MUUR doesn't compromise the presence of humans for such an operation. Wireless control will enable the rescuers to visually inspect, analyze and act according to the situation strategically. Considering the case of fire or sanitizing an area, human intervention can be minimized, and hence the risk taken by the rescuers is eliminated to a certain limit. The All-terrain track of the MUUR helps to travel through a wide range of soft and rough terrains under different conditions. Incorporating a firefighting nozzle will help the firefighting application of MUUR convenient as the total liquid carrying capacity mentioned is limited. Such an update will help the MUUR.

6. References

[1] Vanmarcke E and Grigoriu M 1983 Stochastic finite element analysis of simple beams *J. Eng. Mech* **109** 1203-14.
[2] Wood R D and Zienkiewicz O C 1977 Geometrically nonlinear finite element analysis of beams, frames, arches and axisymmetric shells *Comp. Str* **7** 725-35.
[3] Lin Y H and Trethewey M W 1990 Finite element analysis of elastic beams subjected to moving dynamic loads *J. Sound. Vib* **136** 323-42.
[4] Conwell J C and Johnson G E 1996 *J. Mech. Mach. Thy* **31** 533-44.
[5] Conwell J C and Johnson G E 1996 Design, construction and instrumentation of a machine to measure tension and impact forces in roller chain drives *J. Mech. Mach. Thy* **31** 525-31.
[6] Gupta A, Jendrzejczyk J A, Mulcahy T M and Hull J R 2006 Design of electromagnetic shock absorbers *Int. J. Mech. Mat. Dgn* **3** 285-91.
[7] Wallaschek J 1990 Dynamics of non-linear automobile shock-absorbers *Int. J. Non-Li. Mech.* **25** 300-08.
[8] Li Jun, Zhao Ziling, Liu Minghui and Zheng Yihong 2011 *J. Jil. Uni.* **2** 298-302.
[9] Ning Guobao, Xiong Lu, Zhang Lijun and Y U Zhuoping 2013 Method of electric powertrain matching for battery-powered electric cars *Chi. J. Mech. Eng.* **26** 483-90.
[10] Choi, E. H., Ryoo, J. B., Cho, J. R., & Lim, O. K. (2010). Optimum suspension unit design for enhancing the mobility of wheeled armored vehicles *J. Mech. Sci. Tech* **24** 323-30.

Event Detection through Lexical Chain Based Semantic Similarity Algorithm

Swati Jain[1], Suraj Prakash Narayan[1], Nalini Meena[1], Rupesh Kumar Dewang[1], Utkarsh Bhartiya[1], and Varun Kumar[1], Arvind Mewada[1]*
[1]Department of Computer Science and Engineering, Motilal Nehru National Institute of Technology, Allahabad, Prayagraj-211004, India
*arvindmewada@mnnit.ac.in

Abstract. Twitter is a platform where millions of people tend to tweet about new events happening in their lives. Celebrities tweeting about the new product endorsement, Politicians tweeting about their views towards a policy or people, Natural calamities occurring are tweeted instantly. Studying this data can provide us with useful information. In this paper, we have proposed an Event detection using a lexical chain based semantic similarity algorithm, for detecting Events from Twitter streams. Lexical chains have been used to preserve the Lexical cohesion in a text. The Twitter data set was collected using "tweepy" API, then pre-processing was done, steps like tokenization, stop word removal, and stemming is carried out and stored the tweets in a text file. Then lexical chains were built using the tweets in the file. The formation of the key graph, with each node as a lexical chain, was carried out. Then the clustering algorithm 'SCAN' was used to cluster the lexical chains, the formed clusters represent the keywords of an event. The last summarization step was carried out for each cluster representing an event.

1. Introduction

Twitter is one of the most popular social networking sites currently, and over 500 million tweets are written by users every day and has more than 330 million monthly active users. All this data says a lot about twitter's popularity, and it generates a huge amount of data frequently. People on Twitter express their opinions, events surrounding them, social information, emotional reactions, etc. An event on Twitter is characterized by a huge number of tweets regarding the event within a time frame while engaging a considerable no of participants. Observations show that 'Twitter' is often the first medium to break events such as earthquakes, disease outbreaks, suicides, etc. This paper attempts to detect events using Twitter alone. People often want to know the events happening right now or recently in the past. The question that many people ask every day. People are more interested in those events that happen locally [3, 15]. Corporations are interested in selling their products to favourable customers [4, 14]. Event detection can answer these questions. Therefore, our project aims to do event detection using the tweets collected from 'tweepy' a python Twitter API. We also attempt to preserve the lexical cohesion among the words in a text, using the concept of lexical chains which are nothing but a list of words from a text connected by cohesion or relatedness.

2. RELATED WORK

Event detection has attracted significant research in information retrieval, machine learning, and social media modelling [6]. Along a similar concept of lexical chains work by Rupesh Kumar Dewang, Anil Kumar Singh [1]. International Journal of Engineering and Technology (IJET), It was related to spam review detection through lexical chain based semantic similarity algorithm for negative reviews. They preserved the semantic relation of the words in a text by using the concept of lexical chains and using them as features instead of keywords. The feature vector consisting of lexical chains and tweets fed into supervised learning algorithms gave an extremely high accuracy of 99.5% in SVM.

S. Sangeetha, R.S. Thakur, Michael Arock [2] used lexical chains to detect events from text documents. Their architecture followed three steps namely natural language pre-processing, lexical chain construction, and event detection. They also tried to preserve the lexical cohesion in the text.

Mario Jarmasz, Stan Szpakowicz [5] used Roget's Thesaurus to automate the construction of lexical chains. Paula Hatch, Nicola Stokes, Joe Carthy [7] work on topic detection using lexical chaining. They used lexical chaining to improve the overall performance by allowing them to exploit the context of a word and to disambiguate its senses.

3. Formatting the text

In this section, we give a detailed explanation of the methodology used.

A. Dataset
The Twitter data set was collected using 'tweepy' a python API for Twitter, live streaming of tweets was done for data collection.

B. Pre-processing
It was carried out using the python NLTK toolkit. Steps like tokenization, stop word removal, and stemming was carried out and the resultant tweets were stored in a text file.
  1. Stop words: They are commonly used words.
     Examples: 'a', 'an', 'the', 'or', 'but', 'how' etc.
  2. Stemming: Stemming is a process of reducing words to their base or stem.
     Example: Stemming operation on words 'cats', 'catlike', and 'catty' reduce them to 'cat'.
  3. Tokenization: means chopping into pieces. Tokenizing a sentence means separating words from it while in a word it means separating characters of the word.

C. Building lexical chains
Often, lexical cohesion occurs not simply between pairs of words but over a succession of several nearby related words that can span the entire text or a part of it. These sequences of related words are called lexical chains. Lexical chains may not end at sentence boundaries. They can connect a pair of adjacent words or range over an entire text. Consider an example highlighting the five parameters to determine lexical cohesion:
Examples [8]
  1. Reiteration with the identity of reference.
     (a) Lara bit into a mango.
     (b) Unfortunately, the mango was not ripe enough.
  2. Reiteration without the identity of reference.
     a. Lara ate some mangoes.
     b. She likes mangoes very much.
  3. Reiteration through superordinate.
     a. Lara ate a mango.
     b. She likes fruit.
  4. Systematic semantic relation
     a. Lara likes green bananas.
     b. She does not like yellow ones.
  5. Non-systematic semantic relation
     a. Lara spent three hours in the park today.
     b. She was plucking flowers.

The initial three relations are the simple repetition of the same word in the same sense (e.g., peach and peach), (e.g., peach and fruit). The last two relations include collocations i.e., semantic connections between words that regularly co-occur (e.g., green and red), (e.g., peach and fruit). We have used the ELKB (Electronic Lexical Chain Base) tool for the generation of lexical chains from the text file. The ELKB is based on Roget's Thesaurus. We have applied five types of thesaurus relations that suggest the inclusion of a candidate word in a lexical chain. The five thesaurus relations used are:

1. Inclusion in the same Head.
2. Inclusion in two different Heads linked by a Cross-reference.
3. Inclusion in References to the same Index Entry.

4. Inclusion in the same Head Group.
5. Inclusion in two different Heads linked to a Common third head by a Cross-reference.

The most important and widely used relations are relations 1 and 2. See Table 1.

A lexical chain is a sequence of related words in writing, spanning short (adjacent words or sentences) or long distances (entire text). A lexical chain is independent of the grammatical structure of the text and in effect, it is a list of words that captures a portion of the cohesive structure of the text. The main motive to use the lexical chain was to exploit the semantic relatedness of the words in any sensible text. Unlike the traditional methods like the bag-of-words model or the tf-idf model that work on keywords in a text disregarding the cohesiveness of words in it.

Algorithm-1 Construction of lexical chains

Input: Keywords (TW)
Output: set S of lexical chains

1. *TW = {TW1, TW2, ..., TWn,}* and $S \leftarrow NULL$
2. **For** t in *TW*, **do**:
3. $K \leftarrow$ *keywords (TW)*
4. **For** k in K
5. **If** k has a repetition of the same word relation or inclusion of the same paragraph
6. **then** $LCi \leftarrow LCi + k$
7. **Else If** No chain (*LCi*) is constructed
8. **then** Create a new chain (*LC*n) and $LCn \leftarrow LCn + k$ and $S \leftarrow S + LCn$
9. **Else** go to step 5
10. **End If**
11. **End If**
12. **End For**
13. **End For**

D. Formation of the feature matrix

For constructing the feature matrix, we used TF-Itf (Term frequency-inverse tweet frequency). If we have 'M' lexical chains and 'N' tweets, the feature matrix will be of N×M dimensions. For LCi i.e., i^{th} lexical chain and j^{th} tweet, we need to calculate TF-Itf and put it in corresponding j, i cell of the matrix. After filling all the cells, we get the feature matrix. See figure 4.

1. The TF-Itf model for the formation of the feature matrix

Term frequency (TF) is calculated for tweets. Hence, defined the name of the algorithm is Term Frequency-Inverse Tweet Frequency (TF-ItF). The Classical tf-idf Concept is modified and is based on the lexical chain. The total number of tweets is 't' and the total number of lexical chains generated is 'c'. This generated feature vector matrix is of dimension t × c.

a.) Term frequency (TF)

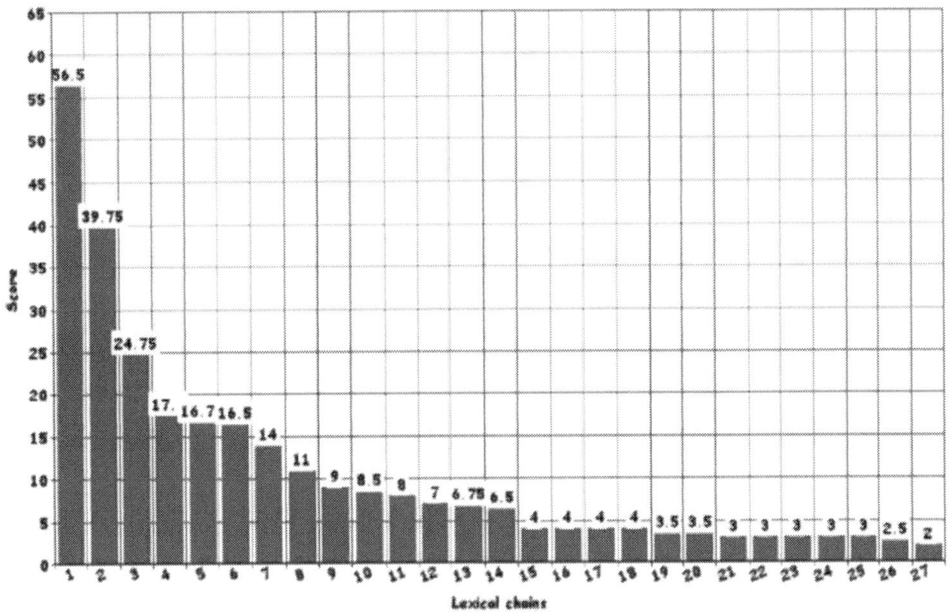

Figure1. The graph below represents a semantic score of lexical chains constructed using ELKB. Chains with lower scores were removed as they were weekly semantically related. score(chain) = length X homogeneity. Length = no. of all members in the chain. Homogeneity = 1 – (distinct members)/length.x

In TF-ItF, it measures how frequently the words of a lexical chain appears in a tweet. Let us consider a lexical chain LCj and a tweet Ti where 0<i< (number of tweets) and 0<j< (number of lexical chains) formed. The frequency for LCj and Ti is given by the following formula:

$$TF = \log(1 + sum(i, j)) \quad (1)$$

where sum (i, j) is the sum of the frequency of each word of LCi in a tweet Tj.

b.) Inverse Tweet Frequency (ItF)

The ItF is calculated by the following formula: $Itf = \frac{1}{\log(1+t)}$ (2)

Where 't' is the total number of tweets.

Example:

"Zuckerberg senior Facebook execute repeatedly congress parliament Facebook advertise want to happen Dr Dena Grayson Facebook Zuckerberg senior Facebook execute repeatedly congress parliament content clear UK parliament host intern parliament."

LC1 = [parliament, parliament, parliament, parliament, congress, congress, congress]
Lexical chain LC1 each word count in the tweet example:
('parliament',4), ('parliament',4), ('parliament',4), ('parliament',4), ('congress', 2), ('congress', 2), ('congress', 2)
Total no of count = 22
TF-tweet= $\log_{10}(1+count)$
TF-tweet = 1.36
tf (no. of tweets in which lexical chain appears) =1 Itf = $1/\log_{10}(1+ tf)$, ItF = $1/\log_{10}(1+1)$ = 3.32 TF-ItF = $1.36 \times 3.32 = 4.51$

Now say that in the feature matrix this Lexical chain appears on the jth column and this tweet appears in the ith row, then the value to be put in the i, j cell of the matrix = 2.78

E. Formation of Key graph

We calculated the semantic scores of each lexical chain and removed the ones that had scores below a threshold. The rest of the lexical chains are assumed to be the nodes of the graph. we created a graph variable which is a dictionary with key-list pair values. so, nodes LCi as the key will have nodes LCj in the list on value only if both lexical chains have TF-ItF scores in any one of the tweets is greater than 0.001. Now edge list is formed for the first graph where an edge is formed between two lexical chains only if that lexical chain pairs appear in the same tweet.

F. Formation of clusters

After forming the key graph, where each node is a lexical chain in the graph, we ran a clustering algorithm on the graph to cluster the same nodes together, to identify distinct events in the form of separate clusters. We used SCAN (Structural Clustering Algorithm for networks) for clustering the nodes. SCAN calculates the Structural similarity of surrounding nodes with the current node and uses a threshold 'ε' to either include it in the cluster on not. This is how the SCAN algorithm works on a high level, but there is more complexity to it which is out of the scope of this paper, for finding out the clusters of Lexical chains from the key graph based on direct connectivity of lexical chains. Therefore, at the end of this phase, we have clusters of lexical chains that consist of keywords that can together represent the keywords of an event.

G. Summarization of Events

Each cluster of Lexical chains may be considered as a synthetic document. Let us call it a key document for our purposes, now it can be used to produce a summarization of events. Documents in the original corpus (tweets) which are like this key document can be clustered, thus retrieving a cluster of topical documents. We used cosine similarity between two documents to discover document clusters for key documents. After clustering of original documents, we have applied a summarization algorithm in which each cluster of documents will be given as input and produces the summary of each cluster as output. Hence, the results will be displayed as a summary of each cluster of tweets where the summary represents the representative tweet that belonged to that cluster. See Figure 5.

The following key graph was obtained.

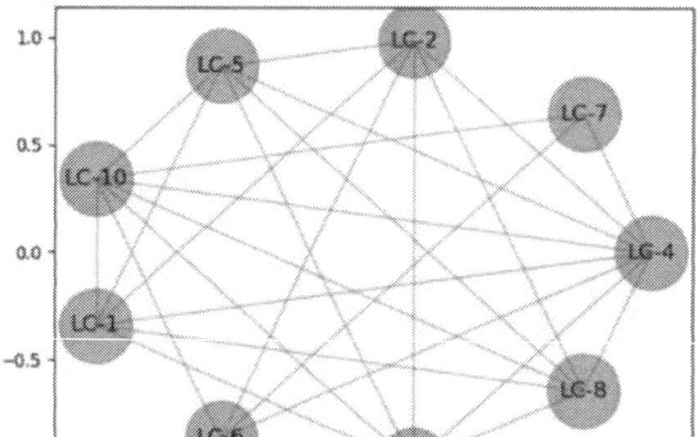

Figure2. This is a PCA plot and the axes shown here are two principle components. This is the plot of a key graph. Here LCi represents ith lexical chain. Two chains have an edge between them only if they both appear in the same tweet.

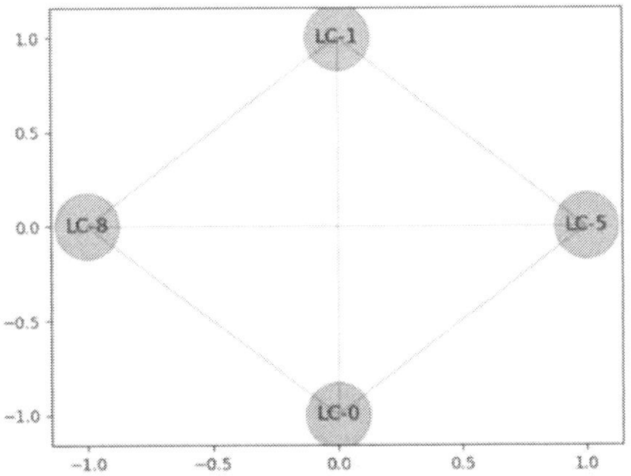

Figure3. This is a PCA plot and the axes shown here are two Principal Components. This plot is the result of the SCAN algorithm on the graph from figure 2. LC_i represents i[th] Cluster of lexical chains. These Clusters represent events.

	LC1	LC2	LC3	LC4	LC5	LC6	LC7	LC8	LC9	LC10
T1	0.0	0.0	0.21	0.0	0.0	0.0	0.0	0.0	0.0	0.0
T2	0.0	0.0	0.0	0.0	0.0	0.0	0.0	0.0	0.0	0.0
T3	0.0	0.0	0.0	0.0	0.0	0.0	0.0	0.0	0.0	0.0
T4	0.0	0.0	0.21	0.0	0.0	0.0	0.0	0.0	0.0	0.0
T5	0.68	0.517	0.307	0.0	0.42	0.42	0.0	0.42	0.0	0.42
T6	0.0	0.0	0.0	0.0	0.0	0.0	0.0	0.0	0.0	0.0
T7	0.0	0.0	0.0	0.0	0.0	0.0	0.0	0.0	0.0	0.0
T8	0.0	0.0	0.504	0.342	0.0	0.0	0.63	0.0	0.0	0.0
T9	0.685	0.517	0.307	0.0	0.42	0.42	0.0	0.0	0.42	0.0
T10	0.0	0.0	0.0	0.0	0.0	0.0	0.0	0.0	0.0	0.0

Figure4. The following matrix is the subset of the feature vector matrix formed from lexical chains (LC) and tweets (T). The i, j cell represents the TF-Itf score of ith tweet in j[th] lexical chain.

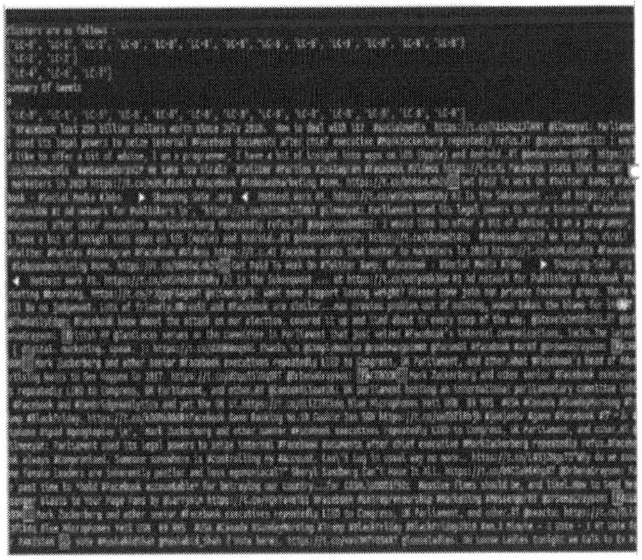

Figure5. Summarization was obtained on pre-processed tweets. In summarization basically, the most representative tweets were selected from each group of tweets that formed as a result of measuring cosine similarity between the tweets and the synthetic document. The synthetic document is nothing, but the clusters obtained, see (***Figure3***).

4. RESULTS

The following screenshot shows the summary obtained from a cluster. The clusters formed can be termed as a synthetic document, now each cluster formed is measured against each tweet for similarity, i.e., we find out cosine similarity of each synthetic document and tweet, we do it for every tweet and synthetic document and the tweets having the highest similarity with some synthetic documents are group into one and termed as an event, as all the tweets in that group represent an event. The summarization is obtained by selecting the most representative tweet from the groups formed.

Table 1. The below table shows the lexical chains Generated from the pre-processed tweet.

LC1	parliament	parliament	parliament	parliament	Congress	Congress	Congress
LC2	intern	intern	Intern	senior	senior	intern	rank
LC3	shop	shop	Break	trump	break	trump	shop
LC4	offer	offer	Pleas	Cover	Pleas	Cover	Pleas
LC5	Document	document	Support	Weight	Speak	Support	Weight
LC6	twitter	twitter	Twitter	strong	strong	strong	Twitter
LC7	Person	Person	Watch	Danger	Person	Person	Watch
LC8	Power	Chief	Power	Chief	Power	Biggest	Biggest
LC9	progress	progress	progress	progress	Profit	Upset	Profit
LC10	Insight	Insight	Insight	insight	insight	insight	insight
LC11	Matter	Matter	Import	Import	Import	Import	Import
LC12	Gentler	Fine	Gentler	Fine	Easiest	Easiest	Fine
LC13	Share	Share	Share	share	Leader	Leader	Parallel
LC14	Social	Social	Social	Social			
LC15	hottest	hottest	Hottest	hottest			
LC16	bigot	bigot	Bigot	bigot			
LC17	vector	vector	Vector	vector			
LC18	traffic	system	Traffic	traffic			
LC19	elect	elect	User	user			
LC20	attack	attack	Blast	blast			
LC21	host	host	Veteran	veteran			
LC22	club	start	Club	start			
LC23	annoy	annoy	Annoy				
LC24	album	album	Review				
LC25	Worth	worth					

The lexical cohesion can easily be seen among the formed chains:

For Example:

LC1 [parliament → parliament (Repetition)]; [congress → congress (Repetition)]
LC21 [attack → attack (Repetition)]; [attack → blast (Collocation)]

5. Conclusion

We used Lexical chains as a replacement to keywords used in previous works, as it takes care of semantic relations among words i.e., lexical cohesion among the words in a tweet. Also, it gives better features for a tweet than a bag of words model or tf-idf model (term frequency-inverse document frequency) which used tf-idf weights for keywords, we have modified it to suit our work and used tf-itf (term frequency-inverse tweet frequency) for weighing lexical chains. Here, we have worked upon a more optimal graph clustering algorithm for clustering of events when compared to the traditional event detection algorithms i.e., event detection using the co-occurrence of keywords.
Apart from the SCAN algorithm used for clustering, we used a lexical chain based semantic similarity algorithm for performing clustering in initial stages i.e., building lexical chains is also a kind of clustering based on semantic similarity of the words. In this algorithm, clustering is performed in such a way that all the keywords in all the tweets which are semantically related to each other fall into one chain. Here nodes of the graph are represented by lexical chains instead of keywords as in previous works done on event detection. Therefore, we can gain insight into events or topics that are extremely popular among their customers and hence, able to generate additional profits by doing event detection more efficiently by the cohesive structure of the text.

Hence, we conclude that by using lexical chains as features instead of keywords we were able to detect events more efficiently while taking care of the semantic relatedness of among the words in tweets. It was not possible with the keywords approach as it does not exploit the various forms of cohesion among the word in a meaningful text.

6. References

[1]. Rupesh Kumar Dewang et al. *International Journal of Engineering and Technology (IJET)*
[2]. S. S., Thakur R.S., Arock M. (2010) *Event Detection Using Lexical Chain*. In Loftsson H., Rögnvaldsson E., Helgadóttir S. (eds) Advances in Natural Language Processing. NLP 2010. Lecture Notes in Computer Science, vol **6233**. Springer, Berlin, Heidelberg
[3]. Q. Yuan, G. Cong, Z. Ma, A. Sun, and N. M. Thalmann, *Who, where, when and what: discover Spatio-temporal topics for Twitter users*, in Proceedings of the 19th ACM SIGKDD *international conference on Knowledge discovery and data mining*. ACM, 2013, pp. 605-613.
[4]. A. Farzindar, *Industrial perspectives on social networks*, in EACL 2012Workshop on *Semantic Analysis in Social Media*, vol. **2**, 2012.
[5]. Jarmasz M., Szpakowicz S. (2003*) Not as Easy as It Seems: Automating the Construction of Lexical Chains Using Roget's Thesaurus*. In: Xiang Y., Chaib-Draa B. (eds) Advances in Artificial Intelligence. AI 2003. Lecture Notes in Computer Science (Lecture Notes in Artificial Intelligence), vol **2671**. Springer, Berlin, Heidelberg.
[6]. *Event Detection and Tracking in Social Streams* by Hassan Sayyadi, Mathew Hurst, and Alexey Maykov (*Proceedings of the Third International ICWSM Conference 2009*).
[7]. Paula Hatch, Nicola Stokes, Joe Carthy. Nicola Stokes, Paula Hatch, Joe Carthy, *Topic Detection, a new application for lexical chaining?* In the Proceedings of *BCS IRSG Colloquium 2000, pp. 94-103, 2000.*
[8]. https://www.slideshare.net/uspansayuti/lexical-cohesion-28663231
[9]. Ellison, Nicole B. *Social network sites: Definition, history, and scholarship. Journal of Computer-Mediated Communication 13.1* (2007): pp.210-230.
[10]. Kwak, H., Lee, C., Park, H., Moon, S.: *What is Twitter, a social network, or a news media? Categories and subject descriptors*. Most 112(2), 591-600 (2010).
[11]. A. Java, X. Song, T. Finin, and B. Tseng, *why we twitter understanding microblogging usage and communities*, in Proceedings of the 9th WebKDD and 1st SNA-KDD 2007 workshop on Web mining and social network analysis. ACM, 2007, pp. 56-65.
[12]. X. Wang, M. S. Gerber, and D. E. Brown, *Automatic crime prediction using events extracted from Twitter posts, in Social Computing, Behavioural-Cultural Modelling, and Prediction*. Springer, 2012, pp. 231-238.

[13]. NALLAPATI, R., A. FENG, F. PENG, and J. ALLAN. 2004. *Event threading within news topics*. In Proceedings of the *Thirteenth ACM International Conference on Information and Knowledge Managem*ent, CIKM04, ACM, New York, NY, pp. 446-453.

[14]. Dewang, Rupesh Kumar, and Anil Kumar Singh. "*State-of-art approaches for review spammer detection: a survey*." Journal of *Intelligent Information Systems* 50.2 (2018): 231-264.

[15]. Sabeeh, Ahmed, and Rupesh Kumar Dewang. "*Comparison, classification and survey of aspect-based sentiment analysis.*" *International Conference on Advanced Informatics for Computing Research*. Springer, Singapore, 2018.

Improving the fatigue life of the tool-joint of drill pipes by optimizing the variable pitch of the box thread

V B Kopei, O R Onysko, A G Panchuk, A P Dzhus and V R Protsiuk

Ivano-Frankivsk National Technical University of Oil and Gas, Karpatska str., no. 15, 76019 Ivano-Frankivsk, Ukraine

E-mail: *vkopey@gmail.com*

Abstract. The reliability of oil drill-strings largely depends on the reliability of tool-joints. It is known that the load along the tool-joint thread is distributed unevenly and it is possible to equalize it by applying a variable thread pitch of a pin or a box. The article describes the optimization method of the pitch change function along the box thread by maximizing of the minimum value of fatigue life in the pin roots. Additionally, the value of the contact pressure on the shoulder, which determines the screw make-up force, is taken into account. To implement the method, we used the Abaqus/CAE and fe-safe software, the parametric axisymmetric finite element model of the tool-joint developed by the authors and the procedures for local (BFGS method) and global (Differential Evolution method) optimization from the SciPy package.

Keywords: tool-joint, fatigue, thread, variable pitch, optimization, finite element model

1. Introduction

It is known that the load in the threaded connections is mainly distributed unevenly – the first turns of the connection are loaded more than the last. This can lead to plastic deformation and fatigue failure in the area of the first turns. In [1] it is analytically proved that the load in the threaded connection of the bolt-nut is distributed according to the function of the hyperbolic cosine. The loads in the tapered thread tool-joint of drill pipes (API and double shoulder) are also uneven [2, 3, 4, 5]. The fatigue fractures (chipping and breakages) of threaded tool-joint are often appeared in the area of the first turns of the pin, and sometimes in the area of the last turns of the box [6]. They are caused by thread deviations (primarily operational wear) and insufficient make-up torque [7, 8].

2. Literature review

There are many design methods to increase the fatigue strength of thread shouldered connections: stress relief grooves [6, 9, 10], modification of the thread profile [11], chamfers on the first turns of the thread, plastic pre-deformation of the connection with a high screw torque, variable pitch diameter and thread pitch [1], double shoulder connections [3], make-up torque optimization [6], the use of coatings [6, 12, 13] and others.

In the work [14] the simulation of a double shoulder tool-joint is performed. It is shown that optimization of its parameters (including a constant pitch) allows to increase fatigue strength. In [15] the influence of the pitch on the stress distribution in the API casing connection is shown. But in these studies the influence of a variable thread pitch is not investigated. In [16] the contact stresses on the surfaces of the pin and box are determined depending on the size of the technological gap but not of the pitch. An analytical version of the theory of threaded joint, which is equipped with a spring-loaded collet, is developed in the paper [17] but not relatively tapered thread pitch. The study [18] discusses the problem of deformation simulation of thread joint parts and proposed easy measurement scheme but it does not includes fatigue strength. The increasing of the plasticity of the nut material causes the equalized loads in the joints with an increased nut pitch [1]. In [19] the parameters of the variable pitch thread, in particular the box thread pitch, are optimized using the three-dimensional finite element method. But such a thread has a variable profile and therefore it is less technological. In addition, the strength estimation is performed only by the distribution of equivalent stresses along the thread, and not by fatigue indicators.

The advantages of the method of variable pitch in comparison with other design methods of increase of fatigue strength are:

1. Any variable pitch thread is easily obtained on a machine with numerical program control by turning, i.e. it is technologically easy.
2. The developing of the new profile of the threading tool is not required [20].
3. Thread modification can be made only on the pin or only on the box [21].
4. Can be quite effective, as it directly addresses the problem of uneven loading along the thread [1, 19].
5. It is suitable for different types of threaded connections. For example, for threaded connections of sucker rods [21] and for casing joints [19].
6. Additionally it can improve the leak-tightness [22] and prevent self-loosening of the connection [23, 24]. Choosing a non-linear bolt pitch change function allows preventing the self-loosening of the connection more effectively [25].

The problem with this method is finding the optimal function for changing the pitch along the thread. For optimization it is necessary to develop a parametric finite-element model of the threaded connection with the possibility of automated calculation of fatigue life in unsafe areas [26].

3. Research methodology

The finite element method was used to study the stress-strain state. The three-dimensional FEM-model of the connection is more realistic, for example, it allows simulating asymmetric loads and geometry [19]. But the axisymmetric model requires fewer finite elements and allows you to create a fine mesh, which significantly increases the accuracy without large increasing the computational complexity. This is important for computationally complex optimization problems.

Abaqus/CAE 6.14, fe-safe 6.5 and Python-macro [27], developed by the authors, were used for modelling. The macro was used to automate the iterative optimization process. The parameters are changed, the model is built, the simulation is performed, and the results are obtained at the each iteration. Plastic deformations are not uncommon during normal operation of tool-joints [7], therefore, for adequate simulation, the models of the materials of the threaded joint must be elastic-plastic. Realistic models of the contact surfaces of the threaded connection are also needed. These features of the model can be implemented in Abaqus/CAE.

The parametric axisymmetric finite-element model of the drilling tool joint ZN-80 GOST 5286 with Z-66 GOST 28487 thread (2 3/8 REG API Spec. 7 equivalent) has been developed. The material of the parts is SAE-4140 steel with a Young's modulus E = 201 GPa, a Poisson's ratio ν = 0.33, a yield strength σ_y = 965 MPa, a ultimate tensile strength σ_t = 1076 MPa). Material plasticity and friction are simulated. The nonlinear section of the stress-strain diagram (σ-ε) is approximated by lines with points: (σ =965 MPa, ε = 0.0), (985, 0.002), (1005, 0.004), (1024, 0.007), (1044, 0.011), (1064, 0.016), (1084, 0.023), (1103, 0.032), (1123, 0.043), (1435, 0.511).

To simulate the make-up of an axisymmetric model of a threaded connection, the axial elongation ("bolt load") of the box shoulder [11] by the value Δ is performed. When changing the value of Δ in the model, the value of the contact pressure P_c on the shoulder should be monitored. Changing any connection parameters can also change the P_c value even for the same Δ value. For no-failure operation, you need $P_c >> 0$.

The Brown-Miller strain-life equation [28] is used to calculate the fatigue life N. It gives the most realistic life estimates for ductile metals, suitable for multiaxial, proportional and non-proportional loading. Here N is the number of cycles until the fatigue crack appears. The calculations are performed using the fe-safe 6.5 software. The fatigue characteristics of SAE-4140 steel (with a material endurance limit of 572 MPa) from the fe-safe material database are used. At the left end of the pin (Figure 1) the pressure L acts. It simulates the external tensile load. We consider the fatigue loading cycle $L_{min} = 0$ MPa, $L_{max} = 350$ MPa (roughly corresponds to a force of 1.6 MN).

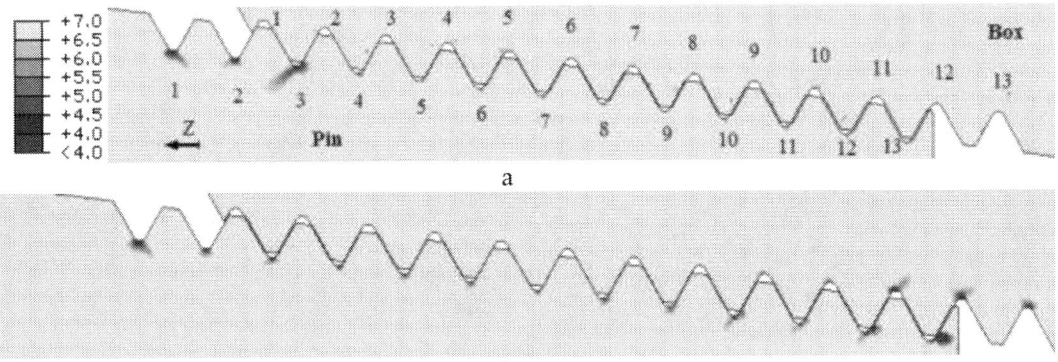

Figure 1. lg$N$ distribution: a – standard, Δ=0.25 mm; b – optimized on step 3, Δ=0.27 mm

Figure 1(a) shows a standard ZN80 tool-joint with numbered thread roots of the pin and box. The axial coordinate of the root with number i of the box thread can be determined as follows:

$$z(i) = -P \cdot i + x(i), \tag{1}$$

where i is the number of the thread root,

P is the thread pitch,

$x(i)$ is the function of deviation of the axial coordinate z of the root i. For the standard connections there is $x(i) = 0$.

Let's formulate the optimization problem:

$$f(x_1, x_2, \ldots, x_{13}) \rightarrow max, \tag{2}$$

where f is the objective function, the value of which y is the minima of the function lg$N(i)$ – the logarithm of the fatigue life N in the root i of the pin thread (Figure 1):

$$y = \min(\text{lg}N(i)), \tag{3}$$

the arguments x_i are the values of the function $x(i)$ for the turn i of the thread.

To optimize $x(i)$, we use the local (BFGS method) and global (Differential Evolution method) multivariate optimization procedures from the SciPy [29] package. Method BFGS uses the quasi-Newton method of Broyden, Fletcher, Goldfarb, and Shanno (BFGS) [30]. BFGS has proven good performance even for non-smooth optimizations [29]. But the objective function can have many minima, therefore, methods of global multivariate optimization should be applied additionally. For example, Differential Evolution is a stochastic population based method that is useful for global optimization problems. At each pass through the population the algorithm mutates each candidate solution by mixing with other candidate solutions to create a trial candidate [31].

Additionally, you should warn about the possible choice by the optimization algorithm of unacceptable values x_i (from the point of view of building a model in Abaqus). Differential Evolution has a *bounds* parameter to limit the values of variables. In the case of using BFGS, you can check the x_i values in f routine or apply the L-BFGS-B algorithm [29] for bound constrained optimization.

4. Results

Figure 2(a, b) shows the equivalent von Mises stresses σ in roots of the pin thread. An increase in Δ leads to an increase in stresses σ. However, for high make-up moments, the stresses increase insignificantly with increasing Δ. The highest stresses are observed in zones 1, 2, 3, 4. Dependences

σ(i) change their shape during the load cycle. Note the appearance of a minimum in the zone i = 9 for L = 350 MPa (Figure 2(b)). This change in shape creates the problem of choosing an optimality criterion. We suggest that it should be a min(lg$N$(i)) function (Figure 3(a)), not a max(σ(i, L=0)) or max(σ(i, L=350)).

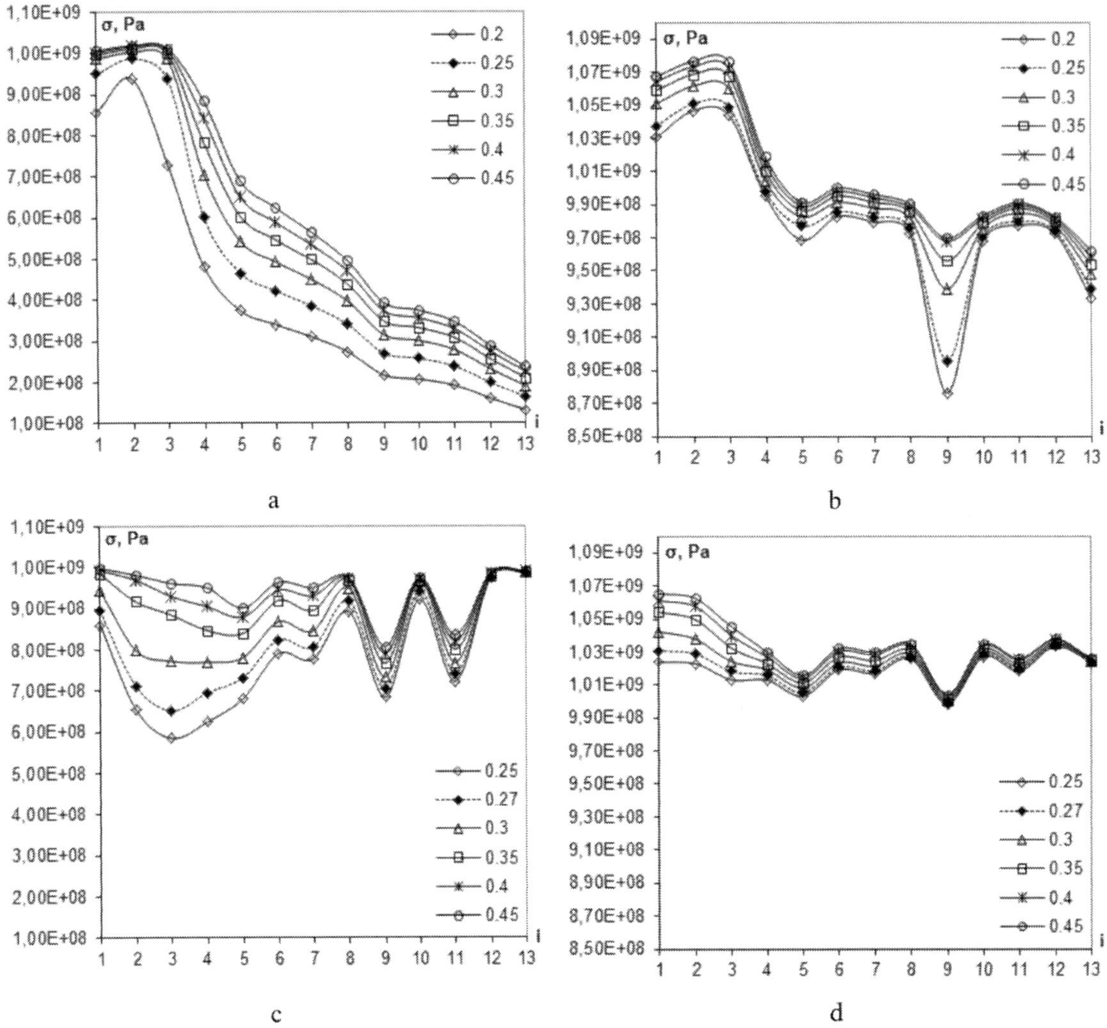

Figure 2. Values of von Mises stresses σ in roots i of pin thread for different values of Δ: standard (a, b), optimized on step 3 (c, d), L=0 MPa (a, c), L=350 MPa (b, d)

It is seen that the value of the decimal logarithm of the fatigue life lg$N$ is small in the roots 1, 2, 3 of the pin thread (Figure 1(a)). For Δ = 0.25 mm the zone with low values of lg$N$ (Figure 1(a)) has the largest area at the root 3 (lg$N$ = 3.64). The smallest values in the box (lg$N$ = 4.32) are in the zone of the root 12 (Figure 1(a), Figure 4(a)). The lg$N$ values in roots of the pin thread are sharply different (Figure 3(a)). The values are small in zones 1, 2, 3, 4, and in zones 5, 9, 13 the values are large. An increase in Δ usually decreases lg$N$ (Figure 4(a)). However, not all dependencies are monotonic. Note the minimum lg$N$ = 3.19 in the zone i = 2 for Δ = 0.3 mm. The value in the root 12 of the box thread remains are almost unchanged (lg$N$ = 4.3 ... 4.2). However, you should not choose the value Δ = 0.2 mm, since for it the contact pressure P_c on the box shoulder is equal to zero (Figure 5). Operation of the tool-joint with P_c = 0 is not permitted [6]. For example, let us choose the value Δ = 0.25 mm, which provides P_c = 81 MPa (Figure 5).

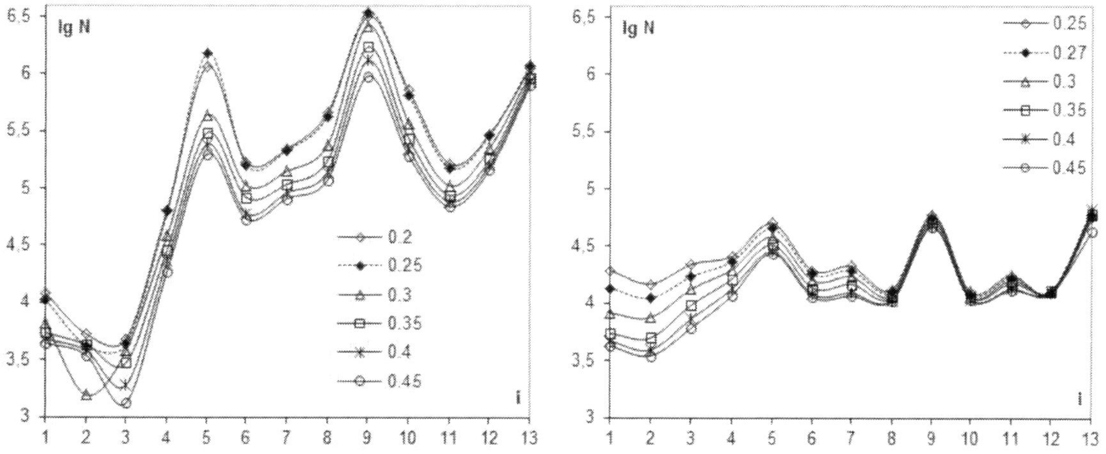

a b

Figure 3. Values of lg$N$ in roots i of pin thread for different values of Δ: standard (a), optimized on step 3 (b)

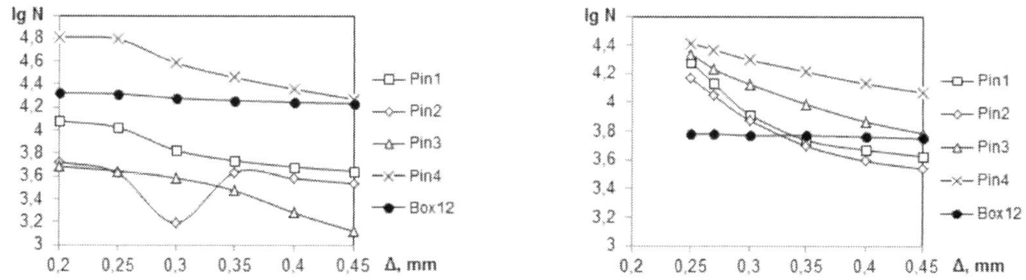

a b

Figure 4. Dependence of lg$N$ in thread roots of pin (roots 1-4) and box (root 12) on Δ: standard (a), optimized on step 3 (b)

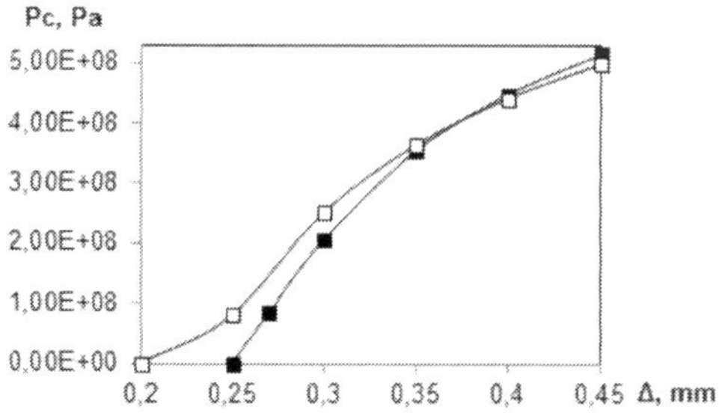

Figure 5. Dependence of P_c in the box shoulder on Δ: standard (□), optimized on step 3 (■)

In order to increase the load on the last turns, at the first step of optimization, an approximate (rough) function $x(i)$ was chosen, which is indicated in Figure 6(a) by number 1 (the standard thread is indicated by number 0). The results (Figure 6(b)) show a significant alignment of the lg$N$ values, but in roots 1, 2, 3, 4 there are almost no values have changed. In addition, a minimum appeared in the zone of pin root 11 (lg$N$ = 3.86).

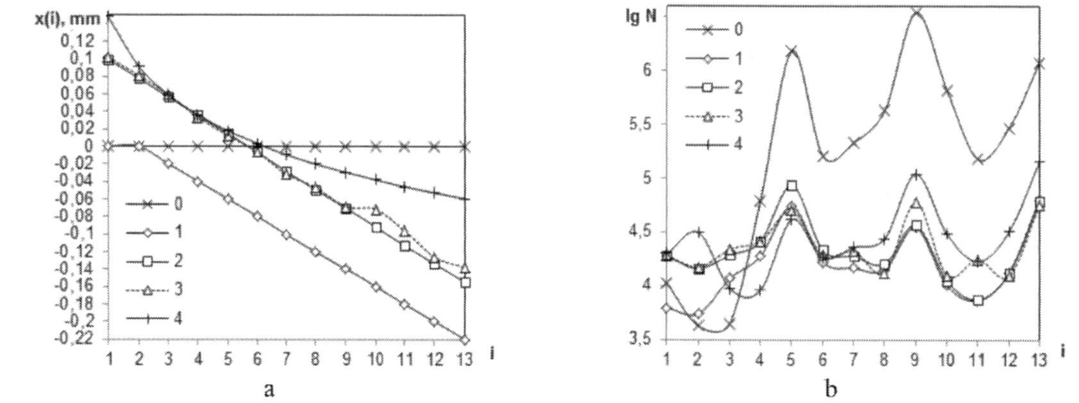

Figure 6. Values of $x(i)$ for the turns i of the box thread (a) and values of lg*N* in roots i of pin thread (b): standard (0) and optimization steps 1-4 (1-4)

At the second step, we approximate the first dependence by the line $x(i) = ai + b$ and search for the optimal values of *a*, and *b* using (2). We use the local optimization algorithm BFGS with initial values $a = -0.0193$ and $b = 0.0338$. As a result, the dependence $x(i) = -0.0212i + 0.1202$ is found and indicated in Figure 6(a) as number 2. It allows to increase the values in the first roots of the pin thread (lg*N* = 4.16), but the value in the root 11 remained almost unchanged (Figure 6(b)).

At the third step, we applied the Differential Evolution algorithm and directly searched for the optimal values of thirteen variables x_1-x_{13}. Algorithm parameters: the initial population is the points located near X_0 (X_0 equal to the values of the optimal function $x(i)$ found in the previous step), *bounds*=$X_0 \pm 0.1$, *popsize*=15 (population size), *polish*=False (do not use local optimization algorithm at the end). The result shows an increase in fatigue life in zone 11 (Figure 6(b)). In general, the results are satisfactory for the pin, but the lg*N* value in the zone 12 of the box decreased to 3.8 (Figure 4(b)). However, this is not so unsafe. In addition, the value $\Delta = 0.25$ mm in this case is insufficient, since the contact pressure P_c is equal to zero (Figure 5). The value of Δ should be increased to 0.27 mm and thus ensure $P_c = 83$ MPa (Figure 5). This will slightly decrease the lg*N* values. The distribution of values is shown in Figure 1(b). The danger zone near the pin root 3 has significantly reduced its area, however, dangerous zones with insignificant areas have appeared at the pin and box roots 11, 12, 13. An increase in Δ decreases the values of lg*N*, but not as significantly as in the original version (Figures 3(b), 4(b)). The stress values at the pin thread roots are shown in Figure 2(c, d). Significant stress equalization is noticeable.

The shape of dependence 3 (Figure 6(a)) is similar to the shape of the logarithmic dependence. Therefore, at the fourth optimization step, the last dependence (found at step 3) is approximated by the function $x(i) = a \cdot \ln(i) + b$ and the optimal values of the coefficients *a* and *b* are sought by the BFGS method (initial values: a = -0.097, b=0.1447). As a result, the dependence $x(i) = -0.081 \cdot \ln(i) + 0.1478$ is found (Figure 6(a)), which make it possible to increase the value of lg*N* in zone 12 of the box to 7 ($\Delta = 0.25$ mm), but to slightly reduce it in zones 3 and 4 of the pin (Figure 6(b)).

5. Conclusions

1. By choosing the optimal function of the coordinates deviations of the box (or pin) thread root *x*(*i*), it is possible to achieve a significant equalization of the loads along the thread of the tool-joint and to increase the value of the fatigue life *N* in unsafe zones. Under these conditions – from *N*=4200 to *N*=11000.

2. When optimizing this function *x*(*i*), you should also control the value of the contact pressure P_c on the shoulder, since changing the pitch can significantly reduce it, and also control the value of the fatigue life in the last roots of the box thread. It is recommended that the objective function takes these values into account.

3. Due to the complexity of the objective function, one should use combined optimization methods and be careful with the choice of the parameters of their algorithms. Due to the computational

complexity of the optimization problem, it is not possible to completely reduce the irregularity of lg$N$ values in the roots. The reader can try to refine the result using the above method.

4. In order to achieve the highest load equalization effect and increase the fatigue life, an attempt should be made to complement the variable pitch method with other methods of durability increasing. For example, to increase the value of the fatigue life in the area of free roots, you can try to apply a stress relief groove and increase its length.

5. The technological possibilities of reproducing the accuracy of the variable thread pitch require some further research.

References

[1] Birger I A and Iosilevich G B 1990 *Threaded and flange connections* (Moscow: "Mashinostroenie") p 368 (in Russian)

[2] Cheng J, Sun Y, Yu Y, Chen L and Ma X 2020 Nonlinear thermo-mechanical coupled analysis of high temperature effect on strength, contact stress and ultimate torque of tool joint *International Journal of Pressure Vessels and Piping* **188** 104221 DOI: 10.1016/j.ijpvp.2020.104221

[3] Liangliang Dong, Xiaohua Zhu and Desheng Yang 2019 Study on mechanical behavior of double shoulder drill pipe joint thread *Petroleum* **5** (1) 102–12 DOI: 10.1016/j.petlm.2018.01.004

[4] Shahani A R and Sharifi S M H 2009 Contact stress analysis and calculation of stress concentration factors at the tool joint of a drill pipe *Materials & Design* **30** (9) 3615-21 DOI: 10.1016/j.matdes.2009.02.022

[5] Onysko O, Kopei V, Medvid I, Pituley L and Lukan T 2020 Influence of the thread profile accuracy on contact pressure in oil and gas pipes connectors. In: Ivanov V., Trojanowska J., Pavlenko I., Zajac J., Peraković D. (eds) *Advances in Design, Simulation and Manufacturing III. DSMIE 2020* (Part of the Lecture Notes in Mechanical Engineering book series) (Cham: Springer) DOI: 10.1007/978-3-030-50794-7_42

[6] Severinchik N A and Kopej B V 1979 *Durability and reliability of exploration drill pipes* (Moscow: "Nedra") p 176 (in Russian)

[7] Lachinjan L A and Ugarov S A 1975 *Design, calculation and operation of geological exploration drill pipes and their connections* (Moscow: "Nedra") p 232 (in Russian)

[8] Sarojan A E 1990 *Theory and practice of drill string operation* (Moscow: "Nedra") p 263 ISBN 5-247-01598-3 (in Russian)

[9] Onyegiri I and Kashtalyan M 2018 Threaded connectors for sandwich pipes – Part 1: Parametric & comparative studies *International Journal of Pressure Vessels and Piping* **168** 117-24 DOI: 10.1016/j.ijpvp.2018.09.010

[10] Onyegiri I and Kashtalyan M 2018 Threaded connectors for sandwich pipes – Part 2: Optimisation of stress relief groove *International Journal of Pressure Vessels and Piping* **168** 125-31 DOI: 10.1016/j.ijpvp.2018.09.011

[11] Hoffman E L 1997 *Finite element analysis of sucker rod couplings with guidelines for improving fatigue life: Sandia report* Sandia National Laboratories p 66

[12] Shatskyi I P, Ropyak L Y, Makoviichuk M V 2016 Strength optimization of a two-layer coating for the particular local loading conditions *Strength of Materials* **48** (5) 726–30 DOI: 10.1007/s11223-016-9817-5

[13] Shatskyi I P, Perepichka V V and Ropyak L Y 2020 On the influence of facing on strength of solids with surface defects *Metallofiz. Noveishie Tekhnol* **42** (1) 69-76 (in Ukrainian) DOI: 10.15407/mfint.42.01.0069

[14] Xiaohua Zhu and Zhi Zhang 2017 Design of an ultra-high torque double shoulder drill-pipe tool joint for extended reach wells *Natural Gas Industry* B **4** (5) 374-81 DOI: 10.1016/j.ngib.2017.10.002

[15] Zhu Xiaohua, Dong Liangliang, Tong Hua, Gao Yuan and Pang Ran 2013 Mechanical behaviors of short-round thread of the API casing under combined load of stretching and

bending moment *Editorial office of ACTA PETROLEI SINICA* **34** (1) 157-63 DOI: 10.7623/syxb201301020

[16] Shats'kyi I P, Lyskanych O M and Kornuta V A 2016 Combined deformation conditions for fatigue damage indicator and well-drilling tool joint *Strength Mater* **48** 469–72 DOI: 10.1007/s11223-016-9786-8

[17] Shatskyi I, Ropyak L and Velychkovych A 2020 Model of contact interaction in threaded joint equipped with spring-loaded collet *Engineering Solid Mechanics* **8** (4) 301–12 DOI: 10.5267/j.esm.2020.4.002

[18] Pryhorovska T and Ropyak L 2019 Machining error influence on stress state of conical thread joint details *IEEE 8th International Conference on Advanced Optoelectronics and Lasers (CAOL)* Sozopol, Bulgaria pp 493-97 DOI: 10.1109/CAOL46282.2019.9019544

[19] Liangliang Dong, Jing Wang and Xiaohua Zhu 2019 Design and mechanical behavior study of ultrahigh-torque variable pitch casing joint *Advances in Mechanical Engineering* **11** (1) DOI: 10.1177%2F1687814018814084

[20] Kopei V Onysko O and Panchuk V 2020 The application of the uncorrected tool with a negative rake angle for tapered thread turning. In: Ivanov V. et al. (eds) *Advances in Design, Simulation and Manufacturing II. DSMIE 2019* (Part of the Lecture Notes in Mechanical Engineering book series) (Cham: Springer) DOI: 10.1007/978-3-030-22365-6_15

[21] Kopey V B Onysko O R and Panchuk V G 2019 Computerized system based on FreeCAD for geometric simulation of the oil and gas equipment thread turning *IOP Conf. Ser.: Mater. Sci. Eng.* **477** 012032 DOI: 10.1088/1757-899X/477/1/012032

[22] Yu Wang, Gang Chen, Haiyan Yang and Jun Jing 2021 A special thread design for failure prevention of expandable casing *Alexandria Engineering Journal* **60** (1) 213-25 DOI: 10.1016/j.aej.2020.07.001

[23] Gong H, Liu J and Feng H 2021 Review on anti-loosening methods for threaded fasteners *Chinese Journal of Aeronautics***14** (January) DOI: 10.1016/j.cja.2020.12.038

[24] Wang L 2012 *Analysis of anti-looseness properties and optimization design of variable thread nut* (dissertation) (Shenyang: Northeastern University)

[25] Ranjan B S C, Vikranth H N and Ghosal A 2013 A novel prevailing torque threaded fastener and its analysis ASME. *J. Mech. Des.* **135** (10) 101007 DOI: 10.1115/1.4024977

[26] Kopei V B, Onysko O R and Panchuk V G 2020 Principles of development of product lifecycle management system for threaded connections based on the Python programming language *J. Phys.: Conf. Ser.* **1426** 012033 DOI: 10.1088/1742-6596/1426/1/012033

[27] vkopey/ThreadsAbaqus: Abaqus/CAE python scripts for modelling threaded connections of oil and gas equipment URL: https://github.com/vkopey/ThreadsAbaqus

[28] Kandil F A, Brown M W and Miller K J 1982 Biaxial low cycle fatigue fracture of 316 stainless steel at elevated temperatures *Int. conf. on mechanical behaviour and nuclear applications of stainless steel at elevated temperatures* (London : The Metals Society) p 203-10

[29] SciPy — SciPy v1.2.0 Reference Guide URL: https://docs.scipy.org/doc/scipy-1.2.0/reference/

[30] Nocedal J and Wright S J 2006 *Numerical optimization* Second Edition (New York, NY: Springer)

[31] Storn R and Price K 1997 Differential evolution – a simple and efficient heuristic for global optimization over continuous spaces *Journal of Global Optimization* **11** 341–59 DOI: 10.1023/A:1008202821328

Identification and classification of faults in underground cables – A review

R Rengaraj[1], G R Venkatakrishnan[2], S Shalini[3], R Subitsha[4], S Suganthi[5] and S Sushmita Carolyn[6]

Department of Electrical Electronics Engineering,
Sri Sivasubramaniya Nadar College of Engineering, Kalavakkam - 603110
[6]*sushmita183001108@eee.ssn.edu.in*

Abstract. The need for power is constantly rising, with the current economic growth. In the power distribution system, transmission lines are the most important part as they play an indispensable role in transferring power from the power plant to the load centres. Transmission lines are done to faults which are largely caused by dampness in the paper insulation, weakness of the cable, or defects in splices or other accessories. The presence of these faults cases the power transmission efficiency to become low, giving rise to a huge annual power loss, Finding and designing new methods to identify the type of defect and its location, is the key to increasing the reliability and optimal performance of the system. This paper reviews about various approaches used for determining the location of underground cable faults and determining the faults using thermal imaging.

Keywords: Underground cables, classification, detection, locating fault.

1. Introduction :

The demand for power has grown exponentially over the years, due to the increase in industrial expansion and continuous growth in energy consumption. Power transmission is possible in both overhead as well as underground cables. Initially, transmission was carried out with overhead cables, but the performance of the cables was greatly affected by adverse weather conditions[1]. Due to their long service life and reliability, underground cables were found to be a replacement of overhead cables. However, if a fault occurs, it is very difficult to find the exact fault location as they are buried underground. Accurate detection of fault segment is required to enhance the dependability of the distribution system and to reduce the interruption time during the fault. Hence reliable fault detection techniques are needed to reduce outage time, minimize economic losses, significantly accelerate system improvement, and guarantee customer power quality [2].
Most of the transmission lines are laid by the overhead line method, but nowadays, an underground cable is widely used for the safety and small city development. Underground cable installations are expensive compared to overhead lines, but more are more reliable and also service life of underground cables is more as compared to overhead lines. Detecting faults in underground cable is difficult compared to overhead line cable. The underground cable is not affected by adverse conditions such as a rainstorm, snowfall and variations in temperature [3] - [6]. If the fault occurs in the underground cable, it becomes difficult to detect, so in this paper, the methods to determine different types of fault and its location is to be discussed.

This paperwork explains about the construction of underground cables, and the faults in underground cables and their types. It gives a brief description on why fault detection is important and describes the different methods to identify faults and techniques to locate them. Further, this paper describes how the faults can be detected using thermal imaging.

2. Overview of underground cables

An underground cable essentially consists of one or more conductors covered with suitable insulation and surrounded by a protecting cover which is shown in Fig. 1 [7]-[10].

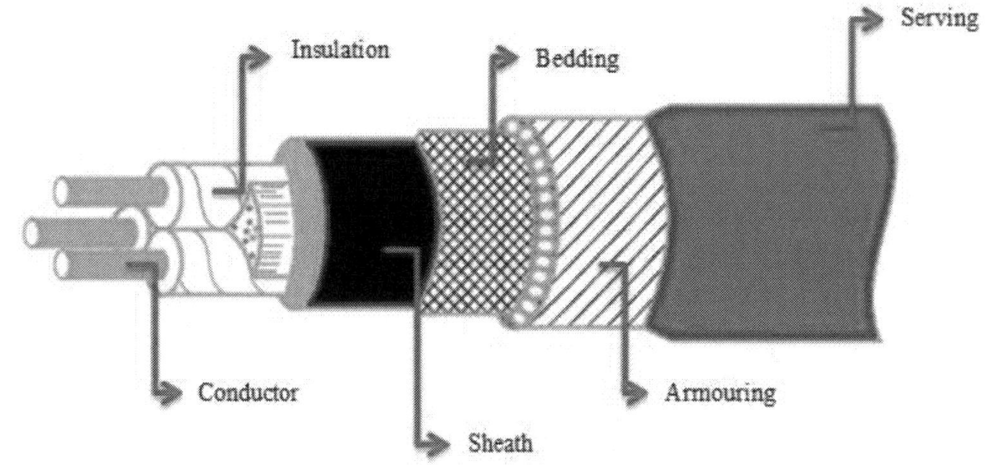

Figure 1. Construction of underground cable.

As shown in figure 1, the following are the parts of a three-core underground cable: (i) Core or Conductors (ii) Insulation (iii) Metallic sheath (iv) Bedding and (v) Armouring.

2.1. Core

The core or the central conductor, made up of aluminium or copper is used to carry the electric current. Depending on the application, a cable can have one or more cores. They are stranded in order to reduce proximity effect and skin effect, improve its capacity to carry current, and to keep the cable flexible.

2.2. Insulation

Each conductor has a provision of insulation of suitable thickness. The thickness of the layer of insulation depends upon the voltage rating of the cable.

2.3. Metallic Sheath

A metallic sheath made up of lead or aluminium surrounds the insulation. The sheath forms a radial barrier to prevent the penetration of moisture and other damaging liquids in the soil. The moisture and chemicals degenerate the insulation and give rise to certain chemical reactions, eventually causing the failure of the insulation.

2.4. Bedding

Bedding consists of a fibrous material like hessian tape or jute. It is used to protect the metallic sheath against corrosion and also from mechanical injuries caused due to armouring.

IOP Conf. Series: Materials Science and Engineering 1166 (2021) 012018 doi:10.1088/1757-899X/1166/1/012018

2.5. *Armouring*

Good mechanical strength and protection to the cable will be provided by the layers of high resistance galvanized steel wire called armouring. This layer will be placed over bedding.

The advantages of underground cable are: Lower transmission losses, Lower maintenance and cost, less damage during severe weather and to install it requires a small band of land. The disadvantages of underground cable are: Expensive installation, susceptible to damage due to underground water and moisture in the soil and it has less heat dissipation.

3. Faults in underground cables

Underground cables are vulnerable to different faults, which compromise the continuity of power supply, and hence, its reliability. When fault occurs, the fault must be cleared, before it causes disturbances in the system which further leads to extensive outages in the firmly interconnected system working within its limits. To raise the reliability indices, techniques have been developed to discover fault symptoms and to analyse the faults, so that the same fault is not allowed in the future. The different types of fault that can exist in underground cable are discussed in this section [11] - [15].

3.1. *Open circuit fault*

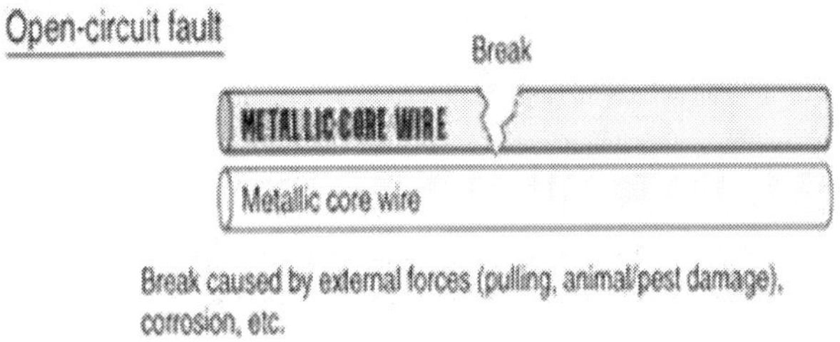

Figure 2. Open circuit fault

An open circuit fault occurs due to break in conductor as shown in figure 2. In that case, there will be zero current flows in the conductor. The Megger is used to check for open circuit faults. It can be checked using three conductors of the three-core wire at the end is shorted and grounded. Megger is used to measure resistance between conductor and ground. It is known to be not broken when megger indicates zero resistance in the conductor circuit. But when it indicates infinite resistance, it means the conductor it is broken and needs to be replaced.

3.2. *Short Circuit fault*

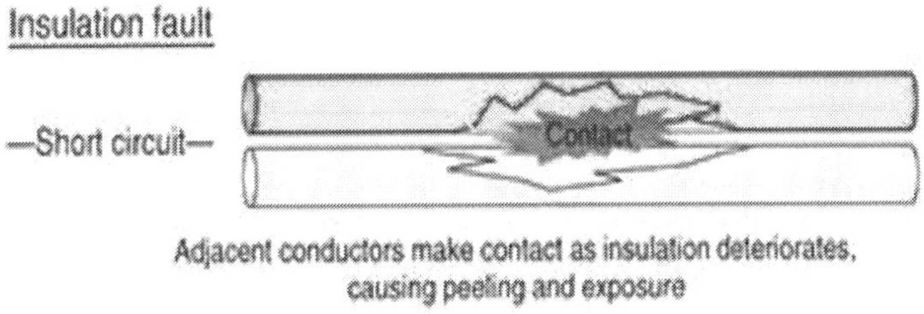

Figure 3. Short circuit fault

The two multi-core cable wires come into electrical contact with each other due to an insulation fault as shown in figure 3, then it is referred as a short circuit fault. For this circuit, the megger is also used to check the fault. The two terminals of megger are associated with any two conductors. In this method, the fault is detected when the megger gives a zero reading between the leads. It is further classified into 2 types as symmetric fault and asymmetric fault.

3.2.1. Symmetric fault. In this case all the three phases unit are short circuited.

3.2.2. Asymmetric fault. In this case the magnitude of the current is not equal and it does not move 120 degrees.

3.3. Earth fault

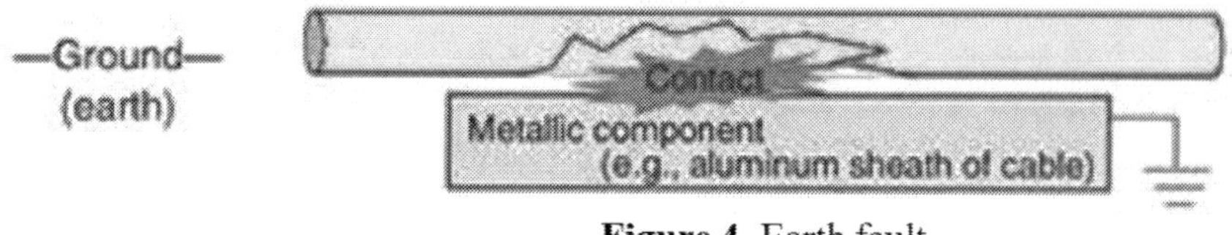

Figure 4. Earth fault

When those wire goes under contact for those ground, it will be known as ground fault. This failure occurs when the outer shell is damaged due to chemical reaction with the ground and mechanical vibration and crystallization as shown in figure 4. The fault is detected when two terminals of the megger are connected to conductor and ground. The ground fault is known when the megger indicates zero readings.

4. Importance of detection of faults

Cable faults are cable damage that affects cable resistance. If you continue, voltage changes may occur. Fault detection is necessary for the power system to eliminate the fault before it increases damage to the power system. Any failure in these systems can stop the supply of electricity, which is a big problem in today's life. Therefore, fault detection has become essential to provide an uninterrupted power supply.

5. Identification of fault detection methods

5.1. Online method

This method uses and processes the sampled voltages and current for determining the point of failure. The online method for underground cables is less than overhead lines [16].

5.2. Offline method

Offline method is divided into two methods, (e.g) the tracer method and the terminal method.

5.2.1. Tracer method. This method tests cable service in the field using a special instrument. The tracer method is a comprehensive way to locate a bad segment by going through the cable circles. A faulty segment can be determined from acoustic or electromagnetic signals and requires dispatch of crew members to the disruption area. There have been various techniques are largely used in the industries, including the tracking approach through acoustic, electromagnetic or current. Example:1) Tracking current method 2) sheath coil method [17].

5.2.2. Terminal method. It detects the fault location in a cable from either one end or both the ends without tracking. In this method, general areas of the fault are found, to speed up tracking in buried underground cables. Example: a) Murray loop method b) Impulse current method[9].

6. Locating Underground Cable Faults

Special techniques are required to find the exact location of the fault. The following are some of the methods used to locate faults in underground cables [18] - [21]:

a) Cable thumping
b) Time Domain Reflectometer (TDR) and
c) High voltage radar methods

The factors that affect these fault location methods include: remote source, cable length, effect of fault parameters, external faults, and load taps.

6.1. Cable thumping

Applying a high voltage supply to a faulty cable creates a high-current arc which makes a thumping sound at the fault location. The heat from this high current damages the insulation of the cable. These damages can be reduced by limiting the power supply through the cable. However, sustained cable testing can deteriorate the insulation to an intolerable level. The main advantage of cable thumping is its accuracy in detecting the location of the fault. Disadvantages of cable thumping are: i) It is infeasible for longer cables. ii) This method cannot detect short circuit faults.

6.2. A time-domain reflectometer (TDR)

A time-domain reflectometer (TDR) is used to measure the cable length and to determine the location of joints, low-resistive faults and intermittent faults. When a low voltage pulsed signal is sent by the TDR, if there is a characteristic change in the impedance, some part of the signal will reflect back to the source. Since the speed of propagation of the pulse is known, the time measured by the TDR can be converted to distance to the reflection. From this, the nature and location of the fault can be detected. Figure 5 shows a typical set up for a TDR measurement. One of the TDR leads is connected to one phase of the cable and the other TDR lead is connected to the cable sheath, which is grounded. The pulse is positively reflected at the end of the cable or at any interruption in the cable, negatively reflected at a short-circuit point, and indicated as a laid S when there is a change in impedance.

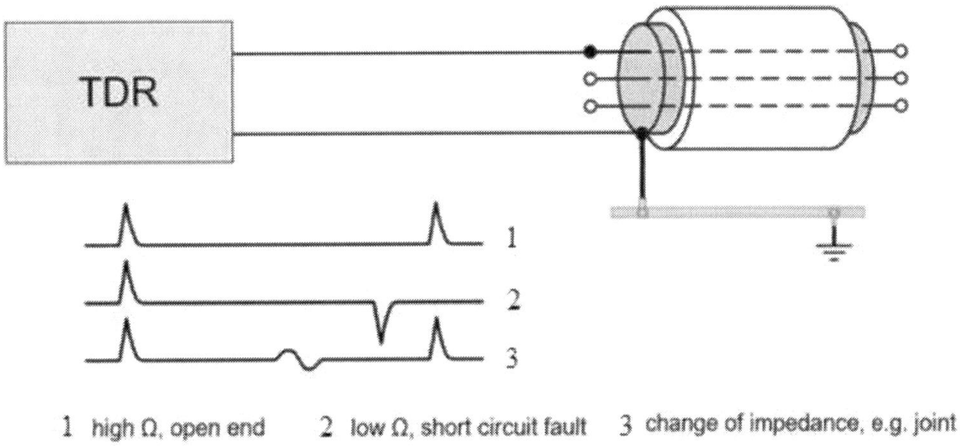

Figure 5. Typical set-up for a TDR measurement

The reflection coefficient is used to determine the magnitudes of the pulse reflections. Reflection coefficient, $r = (Z - Z_0) / (Z + Z_0)$, where Z is the impedance of a discontinuity and Z_0 is the characteristic impedance of the cable. For an open end $Z >> Z_0$ and r is a positive reflection and, for a short circuit $Z << Z_0$ and r is a negative reflection.

One of the major advantages of TDR is that, it does not degrade the insulation, as it sends a low energy signal. Disadvantages of TDR are: i) It cannot detect the exact location and gives only an approximate distance to the fault location. ii) TDR generally cannot detect high resistance ground faults. iii) TDR signal gets interfered by the surrounding electrical noise.

IOP Conf. Series: Materials Science and Engineering 1166 (2021) 012018 doi:10.1088/1757-899X/1166/1/012018

6.3. High voltage radar methods

To overcome the limitations of Time Domain Reflectometer, the following high voltage radar methods are used.

a) arc reflection method
b) surge pulse reflection method
c) voltage decay reflection method.

6.3.1. Arc reflection method. This method requires the use of a TDR along with a thumper or surge generator, and a filter. An arc is created across the shunt fault by the thumper. This arc results in a temporary short circuit, causing the TDR to show an effective downward deflection. The TDR is protected from the high voltage surge produced by the thumper, by the arc reflection filter. This filter also directs the low voltage signal through the cable. Arc reflection technique of locating the fault in underground cable is shown in figure 6.

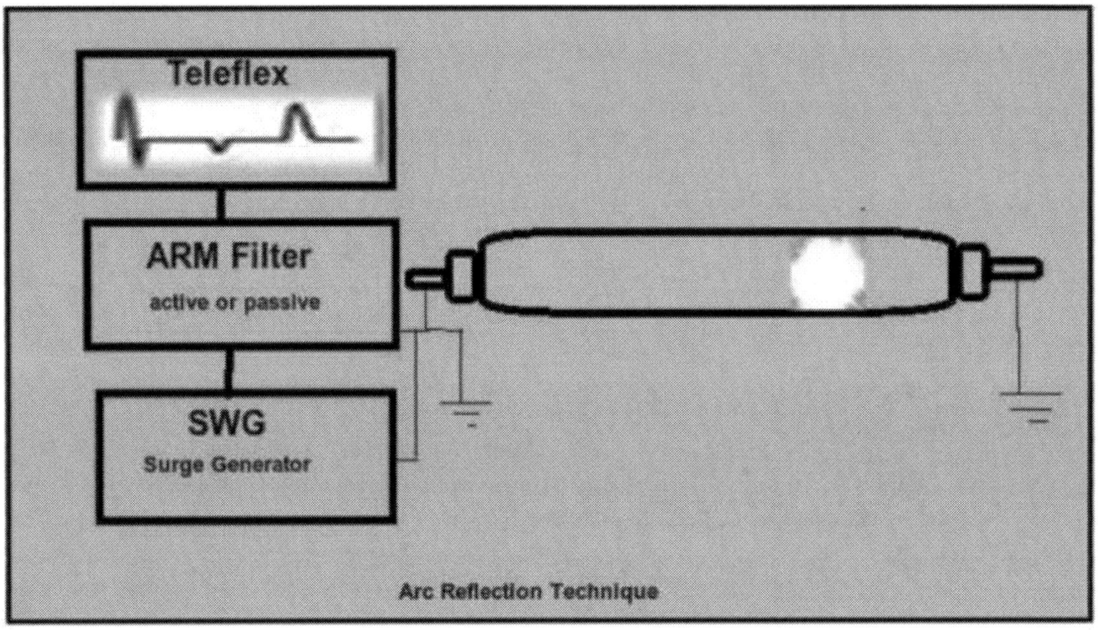

Figure.6 Arc reflection technique

6.3.2. Surge pulse reflection method. This method makes use of a thumper or surge generator, a current coupler, and an analyser or a storage oscilloscope. The thumper and the underground cable are directly connected without making use of a filter. A high-voltage pulse is transmitted down the cable, by the thumper, which creates an arc at the location of the fault. This causes energy to get reflected back to the surge generator. This reflection between the thumper and the fault continues back and forth till all the energy gets exhausted. The surge reflections are sensed by the current coupler and then captured by the storage oscilloscope and displayed on a screen.

6.3.3. Voltage decay reflection method. This method requires a voltage coupler, high-voltage dc test set and an analyser. Here, the flashover of dc voltage at the fault produces reflections which are sensed by the voltage coupler. The analyser acts as a storage oscilloscope to capture the reflections and display them.

IOP Conf. Series: Materials Science and Engineering 1166 (2021) 012018 doi:10.1088/1757-899X/1166/1/012018

7. Detection of faults using thermal imaging

Faults in underground cables can be detected using infrared camera images while the surrounding images of the faulty device are captured using the Charge Coupled Device (CCD) camera. Practically, the infrared camera image is easily affected by the temperature of the environment. Hence, the temperature range with minimum/maximum temperatures is set up for the thermal image, in order to have a diagnosis that is reliable and accurate. From this range of temperature, one pixel of palette is determined, which represents a particular temperature.

Figure 7. Images of a diagnostic program. [6]

7.1. Automatic Capturing of the Images

The infrared camera and the PC, which collects the infrared and CCD images as shown in figure.7, are connected together. The predetermined colour palette value of the thermal image between the maximum, tem_{high}, and the minimum, tem_{low}, is divided into 255 levels, which corresponds each RGB colour to a certain temperature. The temperature of each level of the palette is

$$tem_{level} = (tem_{high} - tem_{low})/256$$

After dividing the infrared image into 5600 pixels, the temperature of a pixel is determined by searching for the matched colour in the palette. The matched colour level is indicated as $data_{RGB}$. The temperature of each pixel is

$$tem = tem_{low} + (data_{RGB} \times tem_{level})$$

When this temperature, tem, goes above the error temperature, the CCD camera captures and saves the images, which can later be used for scheduling repairs, by the operator. [6] Figure 8 depicts the flowchart for the detection of the fault.

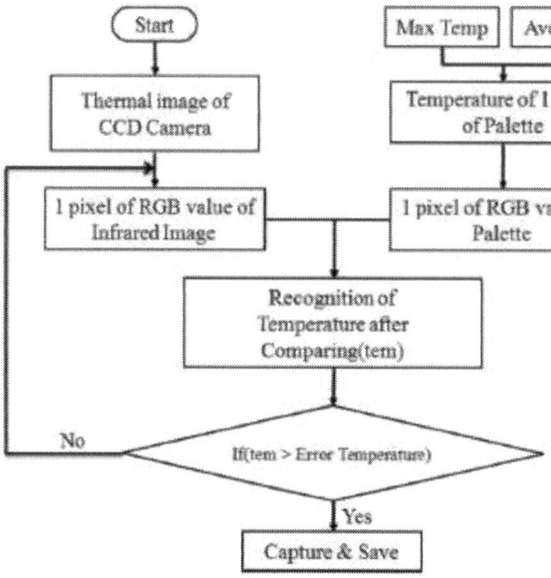

Figure 8. Flowchart for detection of faults [6]

7.2. To classify faulty and un-faulty conductors

A model developed using Convolutional Neural Networks (CNN) with LetNet architecture is utilized as an automated decision support system for the classification of faulty and un-faulty conductors. The performance metrics of the system can be calculated using the following equations

$$\text{Accuracy} = (TP + TN) / (TP + FP + TN + FN) \quad (1)$$

$$\text{Sensitivity} = TP / (TP + FN) \quad (2)$$

$$\text{Specificity} = TN / (TN + FP) \quad (3)$$

$$\text{Positive predictive value} = TP / (TP + FP) \quad (4)$$

$$\text{Negative predictive value} = TN / (TN + FN) \quad (5)$$

$$F - \text{Measure} = 2 * (\text{Precision} * \text{Recall}) / (\text{Precision} + \text{Recall}) \quad (6)$$

where, TP is True Positive, FP is False Positive, TN is True Negative, and FN is False Negative.

The adopted CNN classifier is trained using 200 thermal images from different instances and directions. In addition, 100 thermal images are made as the test image to analyse the performance of the CNN-based classification system. The 100 test images with un-faulted and faulted conductors are transferred to the CNN classifier.

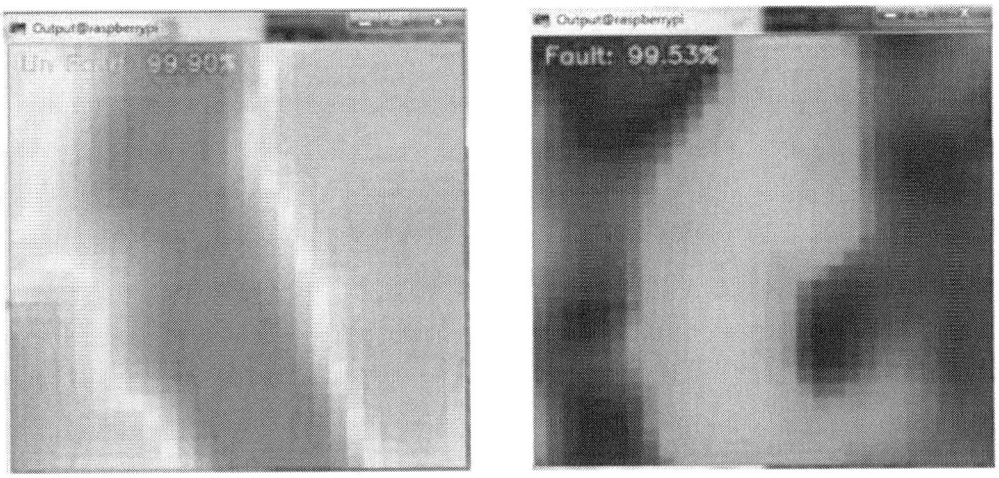

(a) Un-faulty condition (b) Faulty condition.

Figure 9. output of the CNN-based decision support system [5]

Figure. 9(a) and (b) shows the output of the CNN classifier, namely un-faulty condition and faulty condition respectively. When comparing the acquired thermal images of the faulted and un-faulted conductors, it is seen that there is a notable change, due to the generation of heat caused by the current flowing through the conductor. The quantity of current flow through the conductor varies with respect to its configuration. However, the change in thermal signatures cannot be seen or detected naturally; the CNN-based decision support system classifies faulty and un-faulty condition with a high degree of accuracy. The accuracy of the classifier can further be increased by increasing the number of test images. [5].

8. Conclusion

To enhance the power quality of the underground cables, and restore power supply in time, it is of great importance to locate and classify the fault rapidly and accurately. In this paper, various promising fault location and detection methods were discussed, which can be embedded in the existing relays. With the use of the Charge Coupled Device image, the infrared images are extensively used to detect and identify faults in underground cables. The faulty and un-faulty conductors can be classified using the output of the CNN classifier. Underground cables have been widely used in power distribution networks due to their limited environmental concerns and increased reliability. With accurate fault detection methods, the faults in underground cables can be minimized.

References

[1] P.Damodaran, R. Rengaraj, D.Rohit, G. R. Venkatakrishnan, G. Aadithya "A Simple Innovative Method To Reduce Extremely Low Frequency Magnetic Field By Conductor Splitting And Phase Mixing", *Journal of electrical Engineering*, Volume **18**, 2018.

[2] Damodaren P., Rengaraj R., Rohit D., Venkatakrishnan G.R., Santhoosh Aravind S., "Experimental Validation of Magnetic Field for Three-Phase Cables After Conductor Splitting and Phase Mixing", *Advances in Smart System Technologies, Advances in Intelligent Systems and Computing*, vol. **1163**, pp. 303 - 314, 2021.

[3] Furat Al-Obaidy, *"IC Testing Using Thermal Image Based On Intelligent Classification Methods"*, Thesis, Ryerson University.

[4] Dhivya Dharani.A, Sowmya.T, "Development of a Prototype Underground Cable Fault Detector", *International Journal of Electrical, Electronics and Computer Systems*, Vol. **2**, Issue. **7**, pp. 2347-2820, 2014.

[5] G G Rahul Ashok, Ramashankar Yadav, Pavan Kumar M L, Somashekar S N, P Sudir, "Under

Ground Cable Fault Detection Using IoT", *International Research Journal of Engineering and Technology*, Vol. **7**, Issue. **6**, pp. 4492-4496, 2020.

[6] A.A. Bhole, A.M. Paikrao, " An Overview on Different Methods of Fault Location in Underground Cable System", *International Journal of Electrical Engineering and Technology*, Vol. **5**, Issue. **3**, pp. 27-33, 2014.

[7] Paramasivam, A., Mohamed Shuaib, Y., Sonya, A., and Irum Fathima, "Identification of Electrical Faults in Underground Cables using Machine Learning Algorithm", In Proceedings of the *6th International Electronic Conference on Sensors and Applications*, Vol. 42, Issue. 1,2020

[8] Hyunuk Ha, Sunsin Han, and JangMyung Lee, "Fault Detection on Transmission Lines Using a Microphone Array and an Infrared Thermal Imaging Camera", in *IEEE Transactions on Instrumentation and Measurement,* Vol. **61**, Issue. **1**, pp. 267-275, 2012.

[9] Mohd Tahir Wasim, Lokesh M Devadig, Siddappa V Ramadurga, Iqbal Ahammed A K , "Under Ground Cable Fault Distance Locator", *International Research Journal of Engineering and Technology*, Vol. **6**, Issue. **5**, pp. 1992-1996, 2019.

[10] Barakat S, Eteiba MB and Wahba WI, "Fault Location in Underground Cables using ANFIS Nets and Discrete Wavelet Transform", *Journal of Electrical Systems and Information Technology*, Vol. **1**, Issue. **3**, pp. 198-211, 2014.

[11] Akshay Anil Sarawade, Nadir N. Charniya, "Analysis Of Electrical And Thermal Characteristics Of Load-Relay Circuit Using Thermal Imaging", *International Journal of Research in Engineering and Technology*, Vol. **7**, Issue. **12**, pp. 38-43, 2018.

[12] Ayush Sharma, Sneha Kumari, Indu Bhardwaj, "Analysis of Underground Cable Fault Location Identification Using Artificial Neural Network", *International Journal of Advanced Science and Technology*, Vol. **29**, Issue. **4**, pp. 5190-5201, 2020.

[13] O.A. Gashteroodkhania, M. Majidia, M. Etezadi-Amolia, A.F. Nematollahib, B. Vahidib, "A hybrid SVM-TT transform-based method for fault location in hybrid transmission lines with underground cables", *Electrical Power Systems Research*, Vol. **170**, pp. 205-214, 2019.

[14] Dr. D.G.Khairnar, Ruchika Rathod, Tina Patil, Slesha Madkaikar,"Design and Development of Underground Cable Fault Detection and Localization using IoT" , International Research Journal of Engineering and Technology, Vol. 6, Issue. 4, pp. 2238-2240, 2019.

[15] Saeid Khavari, Rahman Dashti, Hamid Reza Shaker, Athila Santos, "High Impedance Fault Detection and Location in Combined Overhead Line and Underground Cable Distribution Networks Equipped with Data Loggers", Energies, MDPI, Open Access Journal, Vol. **13**, Issue. **9**, pp. 1-15, 2020.

[16] Nikhil Kumar Sain, Rajesh Kajla, Mr.Vikas Kumar, "Underground Cable Fault Distance Conveyed Over GSM", *IOSR Journal of Electrical and Electronics Engineering*, Vol. **11**, Issue. **2** Ver. III, pp. 06-10, 2016.

[17] T.S.Sidhu and Zhihan Xu, "Detection and classification of incipient faults in underground cables in distribution systems," *2009 Canadian Conference on Electrical and Computer Engineering*, St. John's, NL, pp. 122-126, 2009.

[18] Durgesh Pathak, Mithil Rathod, Sandeep Vishwakarma, Harindar Maurya, Mahalaxmi Palinje, "Under Ground Cable Fault Detection Over IOT", International Journal of Engineering Research and Technology (IJERT) ICIATE, Vol. **5**, Issue. **1**, 2017.

[19] Roshani Shingrut, Dakshata Mokal , Shubham Shelar, Shekar Mhatre, Dr. Sharvari Sane, "Underground Cable Fault Detection", International Journal of Engineering Research and Technology, Vol. **9**, Issue. **2**, pp. 277-280, 2020.

[20] T. Nandhini, J. Shalini, Sai Sangeeetha, D.Gnanaprakasam, "Underground Cable Fault Detection using Arduino", International Journal of Engineering Science and Computing, Vol. **7**, Issue. **4**, pp. 6460-6463, 2017.

[21] Mr. Nilesh S. Wani, Dr. R.P. Singh, "Transmission Line Faults Detection- A Review" , International Journal of Electrical Engineering and Technology, Vol. **7**, Issue. **2**, pp. 50-58, 2016.

Effect of payload security in MQTT protocol over transport and application layer

Arunima Varma* and Srija UniKrishnan

Department of Electronics Engineering, Fr. Conceicao Rodrigues College of Engineering, Fr. Agnel Ashram, Bandstand, Bandra (W), Mumbai – 400050, Affiliated to University of Mumbai, India

E-mail: *arunimavarma20@gmail.com*

Abstract. Digital revolution has made our life dependent on devices that are can be connected to the internet. Hence, making Internet of things (Iot) an inevitable part for now and years to come. The transition from normal devices to those devices that can be connected to internet, has been exponential, in contrast, the technique used by the protocols in those devices to prevent cyber-attacks and keep user data intact has not been in the same pace as that of its usage. This concerns the end user as it risks the data, allowing it to be misused. In this paper we discuss, MQTT (message queue telemetry transport) a lightweight messaging protocol, wherein the payload is acting as the driver for the data to be carried. This payload can include all sort of information, private as well as public. The paper experiments with the vulnerability of the payload when exposed to a cyber-attack, here, Man in the middle (MITM) attack when the same payload is provided with 2 different types of security. Initially at the transport layer and later at the application layer. At first a channel-based security using Transport security layer (TLS) is provided to the payload and later, an object-based security using Advanced Encryption Scheme (AES) is provided to the same payload. Both of these payload encryption technique for MQTT protocol is discussed elaborately.

1. INTRODUCTION

India has gone from providing mobile in every hand, to wifi/ internet in every place. Clearly, this is also what the India post pandemic aims to offer. This concludes that not only mobile will facilitate more than what is expected but also internet will come in handy to devices that can be connected to the internet. Giving an open opportunity to Internet of Things and its potential. More devices can be connected to the internet and can be controlled with or without mobile phone. A plethora of available Iot protocols, can be utilized for the same. The application's purpose conjunct with the correct protocol within Iot, will not only provide data security but will also ensure smooth functioning of the application end to end. [1]

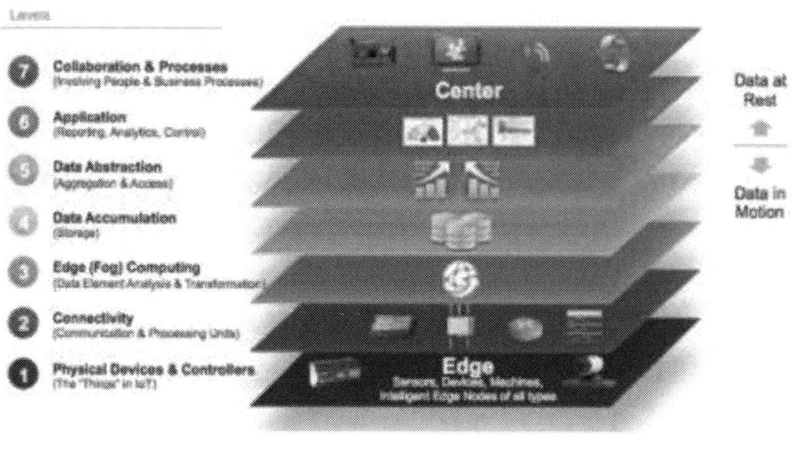

Figure 1: CISCO Seven-Level Reference Model (CISCO, 2014)

Sensitive information, always exists on network and as we can see, that we are now and henceforth, going to greatly rely on computer network to share information. It is no different for Internet of things as shown in figure 1. Assuming, each and every one will have connectivity to internet 24x7, it risks the Iot protocols, that are sending sensitive data to and fro. There is chance for the data being transferred from one Iot device to another Iot device to be at risk for any cyber-attack. For safeguarding sensitive data between two systems, that could potentially include personal information, we make use of secure socket layer (SSL). SSL is essentially a transport layer security of the OSI model, that makes use of encryption algorithms that will make hackers difficult to decrypt whether it is between two users or two websites or a user and a website. Transport layer security, updated version of SSL, also present at the transport layer, makes use of encryption techniques such as ECC (elliptic curve cryptography), RSA (Rivest–Shamir–Adleman) etc. These act as digital certificate to provide security between two systems at transport layer. [1] [2]

Protocol stack for Iot, visualization can be seen as an extended version of the TCP/IP protocol model and encompasses 6 layers namely physical layer, which is the primary one, followed by the link layer, which is the second layer, followed by the network layer, then later it has, transport layer. Last layer is an amalgamation of application layer and application services layer. One of the Iot application protocol is MQTT (message queue telemetry transport) a lightweight messaging protocol, that is responsible for M2M (machine to machine) communication. MQTT uses low power and bandwidth, based on subscriber-broker-publisher model and is extraordinarily light. This gives an extra edge for MQTT over other Iot protocols. [3] [4]

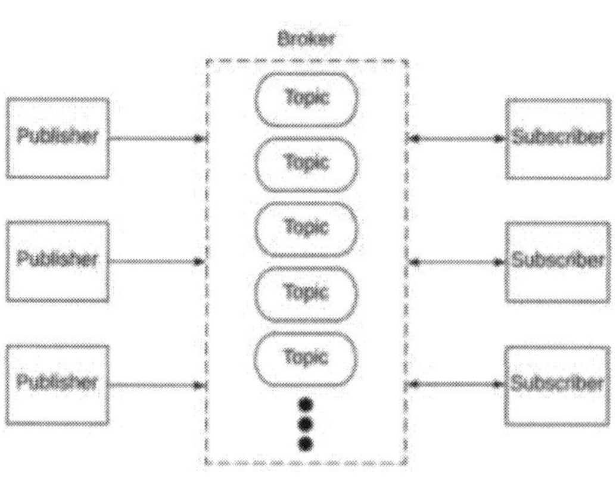

Figure 2. The architecture of MQTT

MQTT is an application layer protocol, message handover is promised by the elementary layer below it which is the transmission protocol layer (TCP/IP). The Quality of service (QoS) setting which is present for MQTT, has the functionality to deliver a message At Most Once, At Least Once or Exactly Once. However, conveying message handover does not ensure accuracy and consistency of data, seclusion or originality of the message. In order to ensure these considerations, a security layer must be implemented on top of the TCP/IP layer. To provide extra security features to the payload of MQTT is traditionally the responsibility of the implementer. [5]
Figure 2 represents the MQTT architecture. The following are the fundamental targets of this work:

- To analyse TLS security for the payload in MQTT against a cyber-attack such as MITM [7]
- To propose an AES encryption technique for MQTT protocol for wireless sensor networks [8]

We divide the paper into sections, which represent: Section 2 presents related work in IoT with Mqtt protocol w.r.t security. Section 3 talks through the overall methodology/ implementation of two scenarios MQTT payload when exposed to Man in the middle attack. Lastly, Section 4 talks about the results and conclusion.

2. RELATED WORK

Security protocols in Iot network such as MQTT have notable amount of research being done on them. MQTT falls in the category of resource-constrained protocol, TLS (transport layer security), is the suggested protocol over MQTT. Providing end-to-end security by the means of payload encryption are the other regions of research, that is being concentrated on. The publish/ subscribe is a typical example which is useful in Iot scenarios because of its incident-built nature, disassociated reporting, high yield and flexibility. [9] [10]
In order to keep data private these factors raise doubts. Both messages and subscriptions must be kept confidential, if the broker is disturbed then, it might act as a point of failure. Due to the disassociated nature of reporting, end-to end security cannot be guaranteed by using TLS. When using a secure channel for providing security, the channel from publisher to broker and broker to subscriber, if the data is stored in the broker, it is still at a potential risk. In contrast, payload encryption will ensure object-based security, wherein irrespective of the secure or unsecure channel, the MQTT payload is still confidential to the probable cyber-attacks. [11] [12]
Whether it is using symmetric, asymmetric or hybrid technique, several arenas of research that facilitate security to Iot's protocol – MQTT are in trend. Let us look at the first proposed technique. For using over MQTT an adaptation of the Augmented Password-Only Authentication and Key Exchange (AugPake) which is based on symmetric framework is proposed. The algorithm of AugPAKE is such that between the client and broker, it hands on, authenticated session keys. Hence, using the established session key, a secure symmetric-key encryption data is then sent and because the initial authentication of clients is made simpler compared to TLS and the final transmission overhead is decreased. However, because of the fact that the communication is only between client and broker, and not end-to-end this method is not full proof. Now, let us look at asymmetric key encryption method [2]. This makes use of lightweight attribute-based encryption (ABE) instead of elliptic curves. Only approved clients can decrypt the data, since ABE encrypts the data and works on access control policy. Hence, even though the broker has access privileges, it cannot decrypt it. Thus, assuring end-to-end encryption as only valid subscribers may decrypt the data. Here, one encryption process is required to convey the message to all subscribers, hence proving advantageous for its support in broadcast encryption. The suggested protocol is secure against most of the familiar cyber-attacks. Although, a significant processing overhead is incurred for implementing ABE. [13]
MQTT, as discussed is not efficient enough to provide any data encryption to the payload hence, puts a potential threat on payload. Whether the payload system uses any authentication mechanism explicitly, the attacker, if a man in the middle, per say, can intrude the data. To suit the Iot environment many areas of research look into tailoring TLS. TLS, does not exist on the application layer but in the transport layer hence it is possible for TLS and payload encryption can work hand in hand. [1] [14]

3. IMPLEMENTATION

The official document of MQTT v3.1.1 says "As a transport protocol, MQTT is concerned only with message transmission and it is the implementer's responsibility to provide appropriate security features. This is commonly achieved by using TLS." So, on one hand while the protocol document suggests us to make use of TLS, it also suggests that using TLS alone might not suffice. Hence, it asks us to make use of additional payload encryption techniques such as cipher encryption or use of public/private key encryption techniques. [2]

To conclude which payload encryption technique is appropriate, we deal with 2 different cases, each using same data and same security threat (man in the middle attack) but different encryption technique for the payload. Our first scenario includes usage of TLS certificate for payload and its ability to prevent a MITM attack and second scenario is where the encryption of payload will take place using a private key pair and its ability against MITM attack will be checked.

Pre-requisites for scenario 1 and 2 are:

Installing Paho client, CA certificate, Kali Linux and ARP spoofing.

Paho client will be the subscriber to a topic on the broker (broker can be the local machine or the ip address of another laptop/ computer) and kali linux will execute urlsnarf to sniff the data that is subscribed by the Paho client.

Create your own certificate using openssl, this will help creating CA certificate and server certificate.

The CA certificate installation steps are shown in figure 3, figure 4, figure 5, figure 6, figure 7, figure 8 and figure 9:

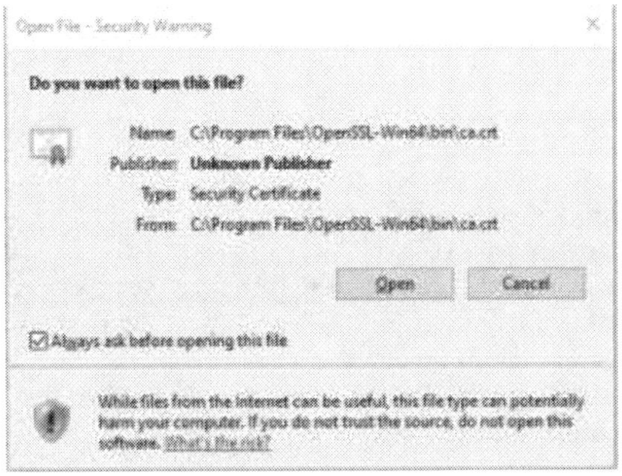

Figure 3. Step 1: Open CA certificate after installing OpenSSL

Figure 4. Step 2: Choose the location to store the certificate

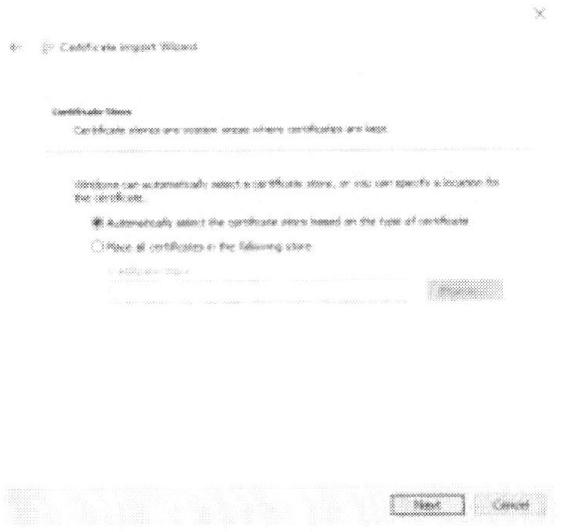

Figure 5. Step 3: Specify the location

Figure 6. Step 4: Click finish for the specified settings

Figure 7. Step 5: Click ok for a successful import

Figure 8. Check the expiry of the certificate

Summarising the CA certificate installation steps in a flow chart:

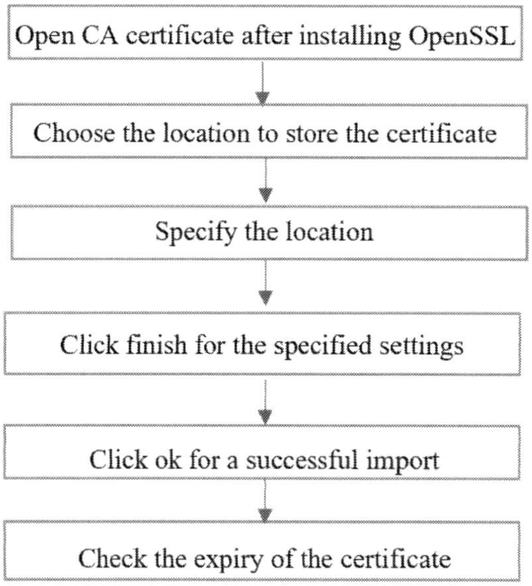

Scenario 1: Digital certificate is installed in chrome, IE as well as edge. A topic is subscribed by Paho client and simultaneously ARP spoof tried to urlsnarf. Since this is a TLS level encryption, a http url is easily attacked by Man in the Middle.

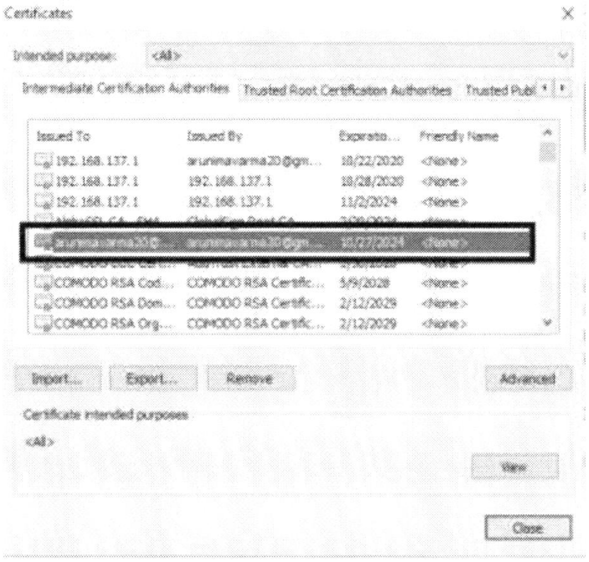

Figure 9. Check on the web browser if the certificate got imported

The GUI at the Python IDE (as shown in the figure 10) where we (PAHO client) would select the TOPIC to be PUBLISHED. The TOPICS will direct to the respective http url that will give information about that TOPIC. So here the dropdown has 2 topics to select from, each giving information of that particular topic to the subscriber by re-directing to the url that holds information.

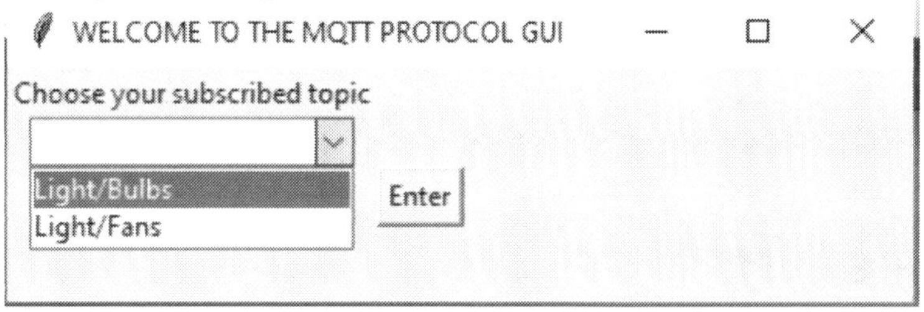

Figure 10: MQTT GUI with topic dropdown

Once the SUBSCRIBER has selected the TOPIC, the messages between the subscriber and the broker will be as follows as shown in figure 11:

- CONNECT (from the subscriber to the Broker)
- CONACK (from the broker to the subscriber), here the connection between subscriber and broker is achieved.
- SUBSCRIBE (from the broker to the publisher)
- PUBLISH (from the broker to the subscriber)
- SUBACK (from the broker to the publisher), here the broker has published the topic for the subscriber
- DISCONNECT (from the broker to the subscriber)

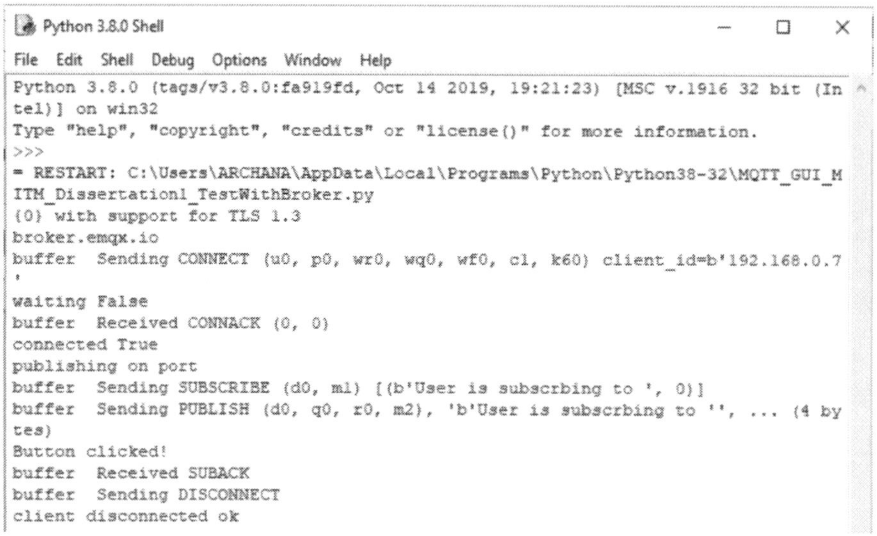

Figure 11: Paho client connecting to online broker and acknowledgement messages

After the topic is selected from the dropdown, the arp spoofing already set in KALI LINUX, which will try to spoof into the re-directed url i.e. the payload or the topic to be published to the subscriber.

Scenario 2: A topic subscribed by Paho client is encrypted using AES encryption. The Man in the middle attack fails

The pseudo code looks like:

1. Define the Tinter GUI
2. Define all components for the GUI
   app = tk.Tk()
   app.geometry('400x100')
   labelTop = tk.Label(app, text = "Choose your subscribed topic")
   labelTop.grid(column=0, row=0)
   comboExample.grid(column=0, row=1)
   bt = tk.Button(app, text = "Enter", command= onclick)
3. Define Paho client
   Eg: cname="192.168.56.1"
4. Define broker
   broker="sig.ddns.net"

5. Subscribe to topic
   client1.subscribe(topic)
6. Publish the topic
   client1.publish(topic,webbrowser.open('http:// /'))

4. RESULT

A simulation with 100 PUBLISH ‘http’ url TOPICS was done, one of which result is shown in here in figure 12 and 13. Each TOPIC was 4 bytes which was redirected as a url once the dropdown of the python GUI was clicked. A MITM cyber attack was launched for both cases, that is, case 1 or scenario 1, when a CA certificate was provided to the server as security to payload and the TOPIC was PUBLISHED and later in case 2 or scenario 2 when the AES encryption is done for the payload.
Since the payload of MQTT has the actual data from publisher to subscriber, it is more vulnerable of a cyber-attack. Here, since we are using MITM, the system is analysed under two lenses, the suitability

for the MQTT-based IoT environment when the payload is provided with TLS certificate on the transport layer and when the payload is provided with encryption using AES technique on the application layer.

The Paho client PUBLISHES the TOPIC to the SUBSCRIBER using client1.publish() as shown in the pseudo code. The topic could be any http url.

KALI Linux is already running in the background, it has been set up with the ARP spoofing for MITM attack. So, now we analyse, the success or failure of the spoofing using MITM for a PUBLISHED topic such as this one.

Below is the **Table 1**. that shows the topics (out of the 100 TOPICS attempted, 5 are mentioned below) that were given as dropdown menu on the python GUI where, the subscriber chooses (subscribes to the TOPIC):

Topics to Be Published	Encryption Applied	MITM attack
TOPIC 1 http://www.foxnews.com/	TLS certification AES Encryption	Failure Success
TOPIC 2 http://cg.nic.in/rajnandgaon	TLS certification AES Encryption	Failure Success
TOPIC 3 http://cesembedded.com	TLS certification AES Encryption	Failure Success
TOPIC 4 http://scratchpads.eu/explore/sites-list	TLS certification AES Encryption	Failure Success
TOPIC 5 http://icmsmt.com/	TLS certification AES Encryption	Failure Success

The following Table 1. Can be shown in a column bar graph as well, wherein, 1 stands for success and -1 is failure.

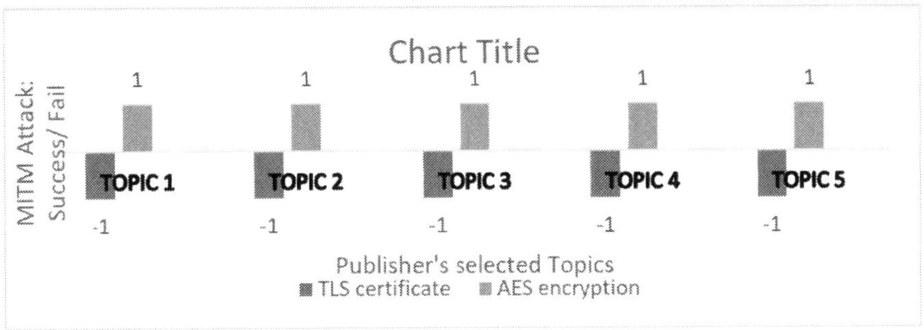

Table 1: Column graph depicting MITM attack success and failure.

One of the topic has been illustrated below:

- The scenario 1 experiment conducted on the payload of MQTT while publishing and using TLS protocol on the transport layer, which was the installation of the certificate on the browser, for both client and publisher proved to fail in case of a URL sniffer in terms of a Man in the Middle attack

Figure 12: MITM succeeds in sniffing the url

- While, the digital certificate existed on both ends the client and publisher itself, the ARP spoof was successfully conducted by the means of MITM attack.
- Hence, the security at the transport layer level protocol on this attack i.e. MITM failed to secure the MQTT payload
- On the other hand, scenario 2 experiment, wherein the payload (PUBLISHED TOPIC) was provided AES encryption was able to prevent a MITM attack. The PUBLISHED topic here, was using cipher key generation for AES encryption.

Figure 13: MITM fails in sniffing the url

5. CONCLUSION

The experiment conducted on the payload of MQTT while publishing the topic and using TLS protocol for both client and publisher proved to fail to provide security to the payload at transport layer. When the payload is secured using channel-based security that is, while, the digital certificate existed on both ends the client and publisher itself, the ARP spoof, was successfully conducted by the means of MITM attack. Although, when the same payload topic was secured using AES encryption, MITM sniffing failed. Hence, the security at the object level i.e. at the application layer should be considered for the payload rather than at the transport layer (TLS certificate) so as to make MQTT protocol more secured.

6. REFERENCES

[1] *A.Banks, R. G. (2014). Mqtt version 3.1.1. OASIS Standard.*

[2] *Monali S. Gaigole, Prof. M. A. Kalyankar, IJCSMC, Vol. 4, Issue. 5, May 2015, The Study of Network Security with Its Penetrating Attacks and Possible Security Mechanisms, A Monthly Journal of Computer Science and Information Technology*

[3] *OLASEHINDE Olayemi O., ALESE B. K., ADETUNMBI A.O., International Journal of Computer Science and Information Security (IJCSIS),* **Vol. 16, No. 12, December 2018, A Machine Learning Approach for Information System Security**

[4] *Rodrigo Emiliano, Mário Antunes, The 10th International Conference on Computer Science & Education (ICCSE 2015) July 22-24, 2015. Fitzwilliam College, Cambridge University, UK, Automatic network configuration in virtualized environment using GNS3*

[5] *Aadil Al-Mahrouqi, Sameh Abdalla, Tahar Kechadi, (IJACSA) International Journal of Advanced Computer Science and Applications,* **Vol. 6, No. 6, 2015, Cyberspace Forensics Readiness and Security Awareness Model**

[6] *R. Mohtasin , P.W.C. Prasad , Abeer Alsadoon , G. Zajko , A. Elchouemi , Ashutosh Kumar Singh, IEEE WiSPNET 2016 conference, Development of a Virtualized Networking Lab using GNS3 and VMware Workstation*

[7] *N. Angelescu, D.C. Puchianu, G. Predusca, L.D. Circiumarescu, G. Movila, ECAI 2017 - International Conference – 9th Edition Electronics, Computers and Artificial Intelligence 29 June -01 July, 2017, Targoviste, ROMÂNIA, DMVPN simulation in GNS3 network simulation software.*

[8] *Premkumar S and Saminadan V, International Journal of Applied Engineering Research ISSN 0973-4562* **Volume 12, Number 14 (2017) pp. 4443-4447 © Research India Publications. http://www.ripublication.com, Impact of Denial of Service (DoS) attack in Smart Distribution Grid Communication Network**

[9] *Jihyun Kim, Jaehyun Kim, Huong Le Thi Thu, and Howon Kim, 2016 IEEE, Long Short Term Memory Recurrent Neural Network Classifier for Intrusion Detection.*

[10] *Thomas Rausch, Stefan Nastic, Schahram Dustdar, 2018 IEEE International Conference on Cloud Engineering, EMMA: Distributed QoS-Aware MQTT Middleware for Edge Computing Applications*

[11] *Mahrouqi, P. Tobin, S. Abdalla, and T. Kechadi, International Journal of Computer Theory and Engineering,* **Vol. 8, No. 3, June 2016, Simulating SQL-Injection Cyber-Attacks Using GNS3**

[12] *Abdessamad MEKTOUBI, Hicham LALAOUI, Abdelouahed ZAKARI, 2016 IEEE, New approach for securing communication over MQTT protocol A comparaison between RSA and Elliptic Curve*

[13] *Joseph Jose Anthraper, Jaidip Kotak, International Conference on Sustainable Computing in Science, Technology & Management (SUSCOM-2019), Security, Privacy and Forensic Concern of MQTT Protocol*

[14] *Tetsuya Yokotani, Yuya Sasaki, The 2016 International Conference on Control, Electronics, Renewable Energy and Communications (ICCEREC), Comparison with HTTP and MQTT on Required Network Resources for IoT*

[15] *Stefan Profanter, Ayhun Tekat, Kirill Dorofeev, Markus Rickert, Alois Knoll, An-Institut Technische Universitat M¨unchen, Munich, Germany, OPC UA versus ROS, DDS, and MQTT: Performance Evaluation of Industry 4.0 Protocols*

Microwave Irradiation Assisted Synthesis of Gallium Aluminum Oxide Using Metal Acetylacetonate Precursors, Peculiar Observations and Safety Guidelines for Young Scientists

Pooja Punetha and Piyush Jaiswal

Department of Nanoscience and Technology, Centre for Advanced Studies, AKTU, Lucknow, 226031, India

E-mail: *punetha1@gmail.com*

Abstract. Microwave-assisted nano-particle synthesis has seen a boom in recent decade. Different variants are being tested all over the world, be it microwave-irradiation-assisted solvothermal synthesis, thin-film deposition using microwave plasma-assisted CVD, microwave-assisted combustion synthesis, or mist CVD. This technology induces self-catalysis, faster reaction time, less energy and is considered safer from other purely chemical variants. Gallium-Aluminum Oxide is an oxide of vital importance. Gallium Oxide in its most stable form, β-Ga_2O_3, has a band-gap of ~4.8 eV and cannot be used for deep ultra-violet (UV) applications (i.e., below 250 nm). Aluminum oxide is a wide band-gap material. The ionic radius of Aluminum ions is similar to that of Gallium ions. Therefore, according to Hume-Rothery rule, Aluminum ion can substitute Gallium in its lattice structure and vice-versa. This gives the possibility of band-gap engineering from 4.6 eV (band-gap, β-Gallium Oxide) to 8.8 eV (band-gap, α-Aluminum Oxide), extending the application of Gallium Oxide to deep UV applications. In present work, three different compositions of Gallium-Aluminum Oxide thin films (varied by mass) are deposited, using microwave-irradiation-assisted solvothermal thin-film deposition technique and metal acetylacetonate precursors. A stoichiometric ratio of ~5:1 is obtained for 90:100 variation. For 50:50 and 10:90, the stoichiometric ratio remained close to 5:5 and 1:9, respectively, showing good agreement with the Gallium to Aluminum ratio in precursor solution. The investigation is done by using energy dispersive X-Ray analysis. A particle variation from 5 nm to 500 nm was obtained as calculated from Field Emission Scanning Electron Microscopy (FESEM) results. In second part of the work, the corresponding author wants to strongly argue that "Microwave in kitchen is not same as microwave in lab" and report some of the observations from lab to pave a way towards a truly Green or safer technology.

1. Introduction

Microwave-assisted synthesis and deposition have drawn much curiosity among researchers worldwide. Many of the questions are still unanswered, and with these unanswered questions, the technology is slowly maturing. Early researchers reported cracking of vessels, generation of sparks, formation of hot-spots, super-heating phenomenon, faster reaction time, plasma generation inside a microwave, and attempted to explain these phenomena using different theories [25, 26]. Slowly microwave-assisted technology started to be used widely in laboratories, especially in calcination of ceramics, in organic chemistry, fusing glass rods, nano-particle synthesis and thin-film deposition.

Today microwave-assisted synthesis is a rapidly evolving technology for drug discovery, material processing and is being more extensively used in chemistry and nanotechnology. With more sophisticated instruments available, it is easier to tune/monitor the properties of materials (to be synthesized or deposited).

The advantages of microwave-assisted synthesis over other thin-film deposition methods are (a) Faster reaction time [22, 23, 29] (b) comparatively lower cost than CVD (Chemical Vapor Deposition), ALD (Atomic Layer Deposition) and other high-quality thin-film deposition methods (c) higher yield [25, 26] (d) deposition on dielectric or non-metallic surface [27, 28, 29] (e) high-quality thin-film deposition, specifically in case of metal oxides, metal sulfides [31, 32] or metal ion-substituted metal oxides (e.g., ZnO [30], Ga_2O_3 [22], ITO [27]).

Gallium-Aluminum Oxide is an important oxide for deep UV applications. With variation of composition, if right stoichiometric ratio can be obtained, there is a possibility of band-gap engineering, i.e., using Gallium-Aluminum Oxide for applications beyond 200 nm, which is currently not possible with Gallium-Oxide alone.

Al_2O_3 and Ga_2O_3 exist in multiple phases. Al_2O_3 has six known phases, α, γ, δ, κ, ε and θ (including other debatable phases). The phase, α-Al_2O_3 is thermodynamically stable phase and ε, κ are polar analogous to Ga_2O_3. β-Ga_2O_3 (thermodynamically stable), α-Ga_2O_3, γ-Ga_2O_3, and δ-Ga_2O_3, ε-Ga_2O_3, and κ-Ga_2O_3 phases (meta-stable phases) are the known phases of Ga_2O_3 [1-5]. The ε-Ga_2O_3 phase is the second most stable phase of Ga_2O_3 and is polar in nature.

Attempts to deposit Gallium-Aluminum Oxide have been done in past [6-9]. Some of the very recent attempts are shown in **table 1.**

Table 1. Recent attempts to deposit Gallium Aluminum Oxide thin-films

Product	Substrate	Method	Remarks	Year of Publication and References
γ-$(Al_xGa_{1-x})_2O_3$ thin-films	Defect spinel	Mist CVD	Band gap of film varied up to 6 eV on increasing Al content (lattice matching with misfit locations, reported).	2020 [10]
β -$Ga_{2-x}Al_xO_3$ and intermediate α, β polymorphs	No Substrate-Powder	Solvothermal Synthesis Method	Transformation of γ-$Ga_{2-x}Al_xO_3$ into monoclinic β-$Ga_{2-x}Al_xO_3$ for $0 \leq x \leq 1.3$ above 750 °C, intermediate α and β polymorphs for $1.3 < x < 1.8$ and α-$Ga_{2-x}Al_xO_3$ for x=1.8	2020 [11]
Aluminum-Gallium Oxide thin-films	c-plane Sapphire	Pulse Laser Deposition (PLD)	Average roughness of thin-film decreases to 0.69 nm, and 0.70 nm after annealing in Oxygen and Nitrogen atmosphere, respectively, at 900° C. Also, optical band-gap increases from 5.14 eV to 5.41-5.46 eV.	2020 [12]
ε-$(AlGa)_2O_3$	Sapphire (0001)	Pulse Laser Deposition (PLD)	Solar-blind photo-detector, fabricated by depositing a pure phase of ε-$(AlGa)_2O_3$ on sapphire substrate.	2021 [33]
β-$(GaAl)_2O_3$	c-Sapphire	Thermal Annealing	Al substitution up to 0.81, highly homogeneous ternary thin-films obtained at temperature ≤1200 °C as reported.	2021 [13]

Attempts to deposit Gallium-Aluminum Oxide films by microwave-assisted solvothermal method, has not been reported till the date of reporting this work, to best of author's knowledge. In this work, an attempt to deposit thin-films of $(Ga_xAl_{1-x})_2O_3$ (where, x ϵ {0.90, 0.50, 0.10}) and maintain stoichiometric ratio thereof is attempted, on Silicon p-type (400) substrate, to explore the possibility of this low cost and easily accessible deposition technology.

2. Method

The experiment is divided into two parts; the precursor preparation and the thin-film deposition. The primary aim is to understand thin-film deposition assisted using microwave-irradiation and possibility of composition variation.

2.1. Precursor Preparation

The precursors are prepared by adding 0.085 M, 0.047 M, 0.009M, of Gallium (III) Acetylacetonate (CAS-14405-43-7, Sigma Aldrich, 99%) to 0.011 M, 0.053 M, 0.096 M, of Aluminum (III) Acetylacetonate (CAS-13963-57-0, Sigma Aldrich, 99%) respectively. The solutions are prepared in 30% Ethanol and 70% of MilliQ ultra-pure DI (De-ionised) water. Acetic acid (99.9%), at stirring condition of 500 rpm, at 50°C, is added drop-wise to these solutions until a clear liquid is obtained. These solutions (now termed as precursors) are used for deposition and named as Ga90Al10, Ga50Al50 and Ga10Al90, for the purpose of documentation. The composition is varied by mass to ensure scalability and repeatability of experiments.

2.2. Substrate Preparation

Substrates used for deposition are cleaned by RCA-2, Piranha (3:1) solution, including cleaning by acetone, IPA (Isopropyl alcohol) solution and in-between with the jet of ultra-pure DI water. No additional pre-treatment is used on the substrate.

2.3. Thin-film Deposition Set-up

Fig.1. shows the thin-film deposition set-up used in this experiment. The microwave used for the experiment is single mode domestic microwave with magnetron frequency 2.45 GHz (IFB, Model No. 17PMMEC1) and cavity of 31.5 × 19.8 × 29.4 cm^3. Round bottom flask with 24/29 connector (Borosil®, 4380C21) is used as a reaction vessel. The vessel-condenser assembly is sealed with Teflon-tape to avoid any leakage inside/outside the chamber.

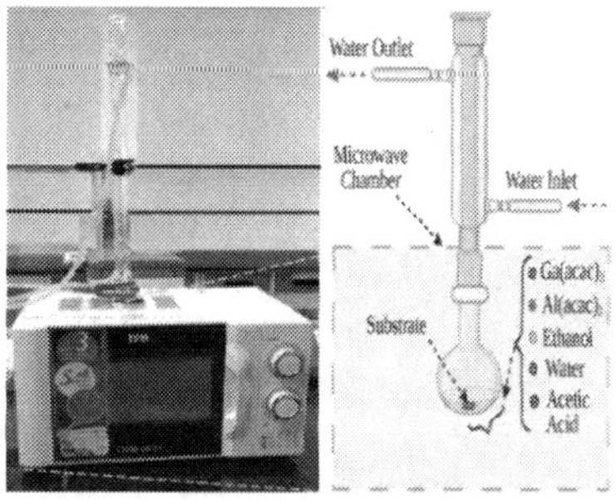

Figure 1. Experimental set-up for solvothermal based microwave-assisted deposition

2.4. Thin-film Deposition

The films are deposited by immersing the substrate in the corresponding solutions (Ga90Al10, Ga50Al50, and Ga10Al90). The cleaned substrate is immersed in the solution and irradiated by 2.45

GHz, 700W microwave power for 12 minutes, as shown in **Fig.1**, to ensure a film of sufficient thickness can be obtained and can be analyzed by accessible characterization tools.

3. Characterization

3.1. X-Ray Diffraction (XRD) Spectroscopy

The X-Ray Diffraction (XRD) spectroscopy is an important tool to understand crystal structure, phases and approximate particle size of thin-films deposited. XRD results of deposited films along with p-type Silicon substrate are shown in **fig.2.** The high-intensity broad doublet peak of p-type Silicon Substrate at 69.03° and 69.23° corresponds to (400) plane [14]. The peak at 32.85° and 32.94° corresponds to forbidden (200) plane reflections [15], the intensity of these peaks is such that they are nearly undetectable in presence of 69.03° and 69.23° doublet peaks.

The peaks at 61.58°, 65.77° and 65.88°, intensity 29.119 cps, 24.26 cps and 18.96 cps, respectively, are low intensity peaks with an additional peak of 68.74° with intensity 532 cps are assigned to substrate. These peaks are likely to originate from multiple reflections coming back from detector and diffracted by multiple planes of substrate due to very high intensity 69.03° and 69.23° doublet peak ($\sim10^4$) (**Fig. 4.**).

Since most of the peaks of blank substrate match with corresponding or shifted peaks of different phases of Ga_2O_3 and Al_2O_3 (JCPDS 00-046-1215, 01-086-1410, 01-074-1776), it was important to detect all the reflections from the blank substrate. Also the obtained data and results are reported 'as obtained' in **fig.2.**, due to following reasons (a) peak positions appear to be shifted after using smoothening function, (b) change in peak intensity, in different thin-films, was not detectable if normalization function is being used, (c) scaling up leads to peak broadening and scaling down to narrowing and (d) peaks of substrate and the peaks of deposited thin-films are very closely related and 2θ corrections may lead to misinterpretations.

With thin-film deposition, the intensity of XRD peaks increased, and the peaks appear to be shifted, as can be seen from **fig.2** and **fig.3**.

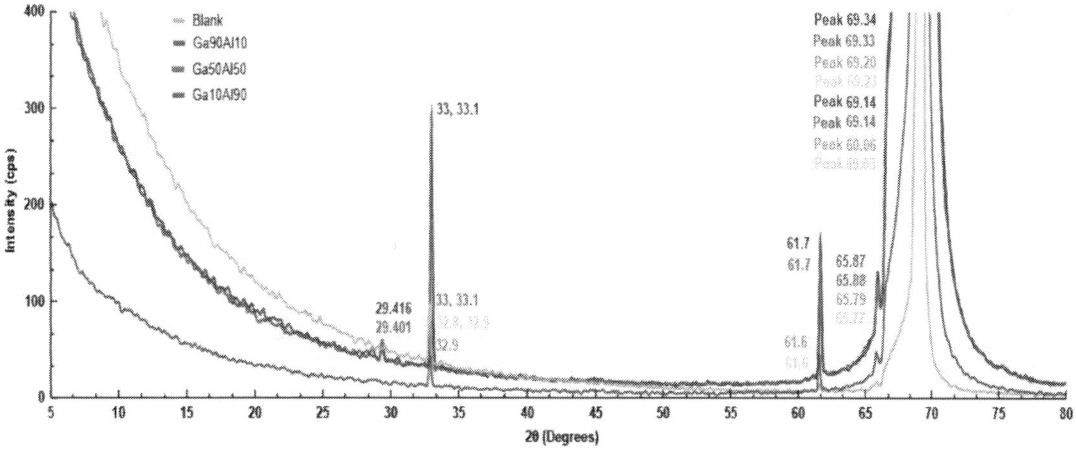

Figure 2. X-Ray Diffraction pattern of deposited thin-films on p-type Silicon substrate along with diffraction pattern of substrate (Yellow)

The shift in peaks can be attributed to increased compressive strain. Aluminum ions are smaller than Gallium ions, and if substituted in lattice, they most likely will cause compressive strain, i.e., reduced spacing between planes, thus increase in 2θ value. Also, due to d-block contraction, the atomic radius of Gallium atom is comparable to Aluminum atom (radii (Al) = 1.1667Å, radii (Ga)= 1.3514Å) [16]); ionic radii too follows the same trend, radii (Al^{3+})=0.468 Å and (Ga^{3+})=0.547 Å (Shanon's Radii) [17]. The reason XRD peaks of same phase of Al_2O_3 and Ga_2O_3 are close together can be derived from here, (during substitution the strain in lattice is lower, thus lesser change in lattice structure).

As can be seen from **fig.3.** the peaks are closely matching with $Al_{22}Ga_2O_{24}$ ((JCPDS 00-022-0615), a=b=5.60, c=22.71, α=β=90, c=120, space group P63/mmc). Using simulation results it was observed that the 2θ values of Ga10Al90 corresponds closely to cell parameters a=b=5.97, c=21.7, α=β=90, γ=120 (space group P63/mmc), the 2θ values were obtained by varying 'c' with respect to constant 'a' and 'b' *vice-versa*. No clear 2θ shift can be obtained in between $Al_{22}Ga_2O_{24}$ and simulated results. The shifted peaks of Aluminum Oxide are also present in Ga10Al90, which is equally likely as Aluminum to Gallium ratio is higher (can also be seen from EDX results **para 1, section 3.2.**).

In Ga50Al50, the peaks 33.05°, 68.78°, 69.13° corresponds to $Al_{22}Ga_2O_{24}$, where 69.13° is highest intensity peak of both i.e., thin-film and $Al_{22}Ga_2O_{24}$. In 2010 **Y. Wang *et al.*,** reported nano-nets [18] with matching peak positions of $Al_{22}Ga_2O_{24}$ (Ga_2O. $11Al_2O_3$), (according to FESEM results similar nano-nets with much smaller size (~7-21 nm) can be seen in cracked microstructures, in Ga50Al50 thin-films (**Fig. 6** (d)). The peaks are shifted from the original values, citing towards the fact that there is a possibility of change in formula or possibility of change in crystal structure. The peak 61.7° corresponds to β-Al_2O_3 (JCPDS 10-0414), the other corresponding peaks from **fig.2.** are assigned to $Al_{22}Ga_2O_{24}$ and $AlGaO_3$ (as shown in **fig.3.**).

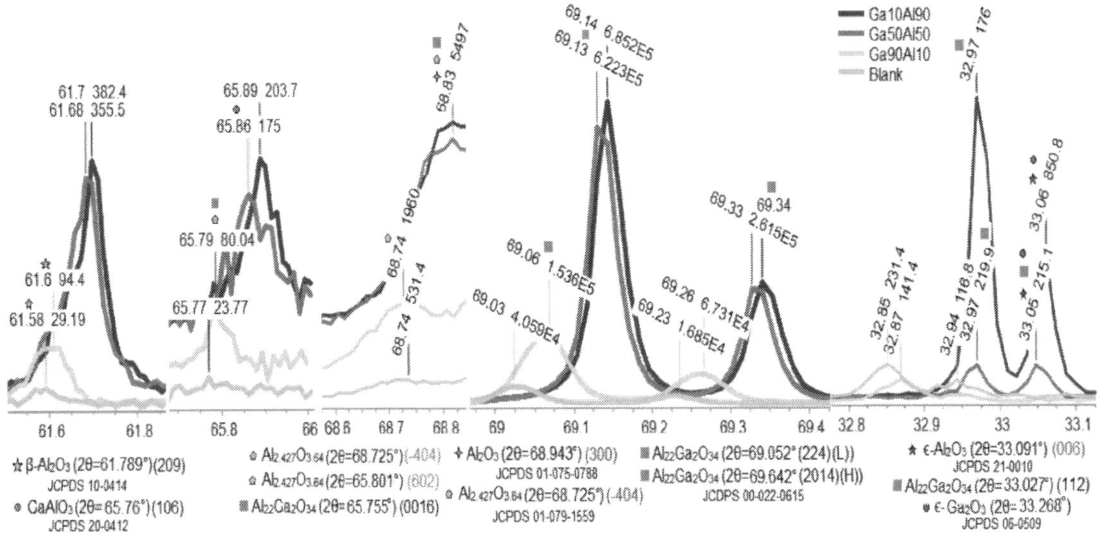

Figure 3. High Resolution X-Ray Diffraction spectra of deposited thin films with respect to silicon p-type substrate (blank)

Ga90Al10 peaks are lower in intensity than Ga10Al90 and Ga50Al50, the morphology of this thin film is also different (as shown in fig. 9.), the peaks are assigned to β-Al_2O_3 (JCPDS 10-0414) and $Al_{22}Ga_2O_{24}$ as shown in fig.3. The peak 32.87° cannot be assigned directly to any phase or JCPDS card number, but it can be assigned closest to ϵ-Ga_2O_3 (JCPDS 06-0509, 33.268) or β-Ga_2O_3 (JCPDS 11-0370, 33.525°). The peak 29.401° corresponds to ϵ-Ga_2O_3 (JCPDS 06-0509) which is not present in any other phases of Ga_2O_3 or Al_2O_3, and also absent in blank, a shifted peak of the same can be seen in Ga10Al90 (29.416°) and Ga50Al50 (29.401°), which indicates towards possibility of presence of localized substituted phases of ϵ-Ga_2O_3 along with substituted $Al_{22}Ga_2O_{24}$ and $AlGaO_3$ oxides.

3.2. Energy Dispersive X-Ray (EDX) Spectroscopy

The EDX spectra were obtained using Bruker SDD (Silicon Drift Detector)-EDX configured with Nova NanoSEM 450 FESEM. The results obtained for thin-films are tabulated in **table 2**.

Larger percentage of SiK was observed in thin-films with cracks (Ga50Al50), as larger Silicon area was exposed due to crack formation. In smaller area the percentage of Gallium, Aluminum and Oxygen remained smaller, however with larger area the percentage of Gallium, Aluminum and Oxygen was higher, this observation is due to presence of larger number of atoms in larger area and *vice-versa*.

Table 2. The EDX results obtained from thin-films (Ga10Al90, Ga50Al50, Ga90Al10) on p-type Silicon Substrate

Elements	**Ga10Al90**		**Ga50Al50**		**Ga90Ga10**	
	Weight %	Atomic %	Weight %	Atomic %	Weight %	Atomic %
OK	41.835	56.345	9.696	16.532	26.25	51.528
GaL	3.345	1.03	8.57	3.348	51.19	23.075
AlK	17.1	13.645	3.976	4.028	3.58	4.1675
SiK	37.725	28.975	77.756	76.088	18.98	21.233

Fig.4. the pie-chart obtained by plotting GaL, AlK values, from three composition variations.

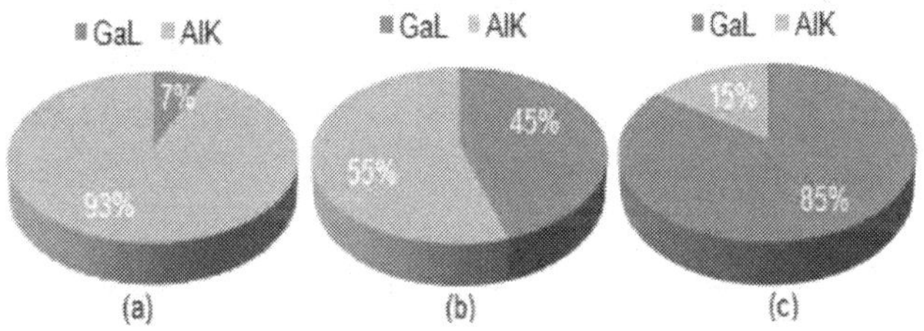

Figure 4. The percentage of Gallium to Aluminum in deposited thin-films (a) Ga10Al90 (b) Ga50Al50 and (c) Ga90Al10

However the results obtained may seem encouraging from **fig.4.** but the concept of percentage can be misleading. Thus, revisiting **table 2**, the ratio of Ga to Al atoms in different compositions can be re-written as ~ 1:13, ~1:1.20, and ~5.5:1 for Ga10Al90, Ga50Al50 and Ga90Al10, also presence of Hydrogen cannot be detected with EDX.

3.3. Field Emission Scanning Electron Microscopy (FESEM)

FESEM is a powerful technique to analyze surface morphology of thin-films; the images are captured using Nova NanoSEM 450 FESEM. The films deposited are not entirely uniform, but are reported as it, such that it can be understood what can be directly obtained from a domestic microwave oven, using present solvent system.

Fig.5. shows the FESEM image of Ga10Al90 thin-films, the average grain size distribution ranges from ~2.7 nm ~34 nm. The fine grain-like structures can be seen in the grown thin-films, however haphazardly grown structures 75 nm to 272 nm, with an average thickness of 141 nm are also obtained as can be seen from **fig.5** (left). To understand the morphology better 3D plots of the thin-film surface are obtained using ImageJ 1.53c [21], with the help of inverted structure it can be seen, that the growth

is likely to occur in net like structures. The grains follow a nano-hump like patterns and the spaces around are following inverted-hump (net-like) patterns (**fig.5** (right)).
As XRD results indicate towards possibility of presence of nano-nets, all the images obtained were thoroughly searched for net-like morphology, however Ga10Al90 showed nano-humps like morphology with size as small as 2.7 nm.

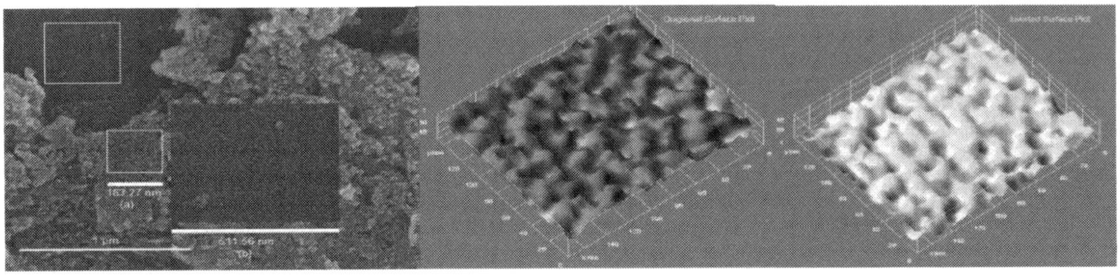

Figure 5. The FESEM image of Ga10Al90, 3D surface plot and corresponding inverted plot

Ga50Al50 thin-films are shown in **fig. 6.**, the films at lower magnification where appeared to be nearly uniform, on magnification, two kinds of cracks were observed, smaller cracks (**fig.6.**(b), ~400nm) and bigger cracks (**fig.6.** (c),~2 μm). It is important to understand which morphology can be obtained if the microstructures could be removed, to understand this in better context image (**fig. 6** (d) and **fig.6** (e)) were zoomed in, and the results are shown in **fig.7.**

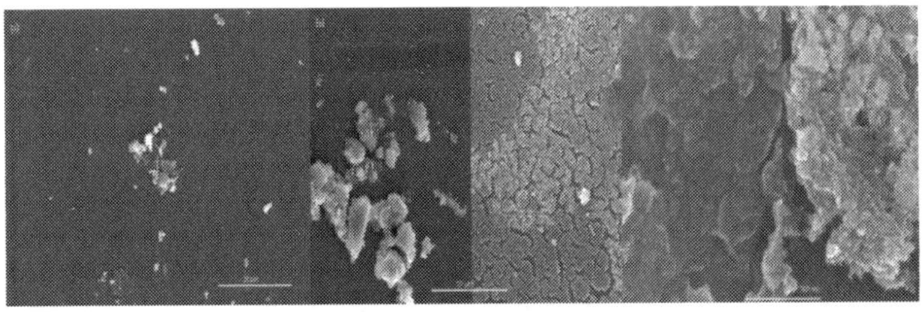

Figure 6. FESEM images of Ga50Al50 ((a) (5000 x); (b), (c) (10000 x); (d), (e) (200 000x))

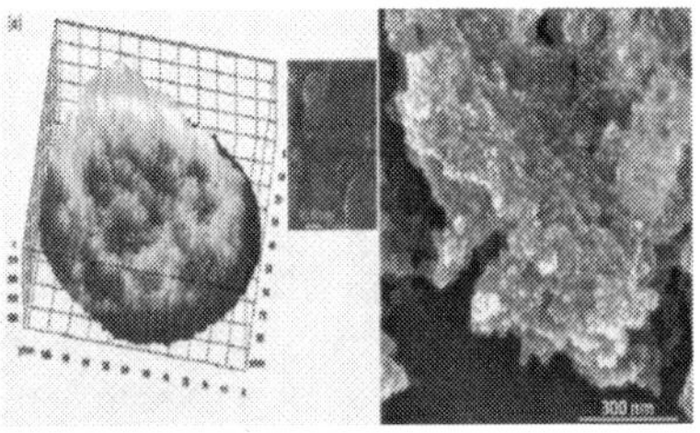

Figure 7. (a) Sectional zoom fig. 6 (d); **(b)** sectional zoom fig.6 (e)

Fig.7(a) shows net like structures with diameter varying from ~5 nm to 27 nm. The films deposited below the cracks are of similar grain-like morphology as that of Ga10Al90 (**fig. 5)** thin-films (**fig. 7(b)**). Ga90Al10 thin-films are shown in (**fig.8.**). Ga10Al90 and Ga50Al50 show similar kind of deposition in the form of debris/ microstructures, but rod and flower like microstructures (length varying from ~500 nm to ~4 μm) are observed in Ga90Al10.

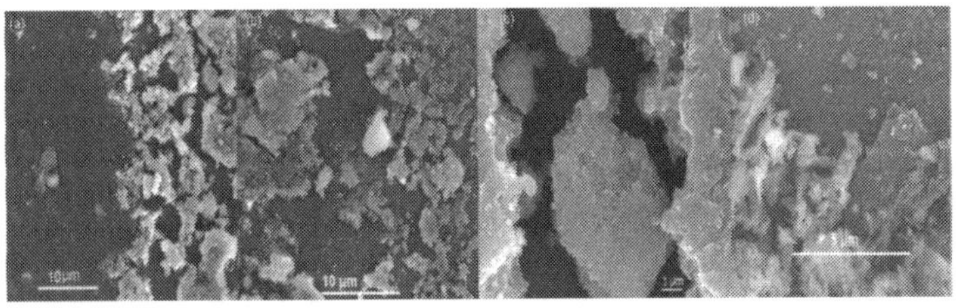

Figure 8. The FESEM images of Ga90Al10 in different magnification

Figure 9. Sectional zoom **fig. 8** (d) and **fig. 8** (b)

These flowers like structures can also be seen buried inside uniform looking part of thin-film (**fig.8** (a)), indicating towards a different morphology that other two counterparts. An important point to note here is that while the substrate is dipped in the solution and irradiated by microwaves the actual-film growth is completed within first few seconds/minutes of irradiation (depending upon the reaction mixture used, irradiation time, irradiation power, volume of solution and the substrate) and a saturation point is reached wherein the excess precursor not used in the film formation starts participating in deposition and falls and collects over the actual film surface in the form of debris. These debris have much looser compaction and are much thicker and bigger in size than actual thin-film. The thin-films grown show clear sign of this over-deposition.

3.4. Solvent System and UV-Visible Spectroscopy

The aim of study was to use minimum number of solvents during the study. Thus, solution for deposition was firstly prepared with ethanol and metal acetylacetonates only. The solubility of metal acetylacetonate was found to be poor with water and even with water-ethanol solution.

Since, solubility affects the quality of thin-film deposited therefore acetic acid was used as a solvent. High dielectric constant of Ethanol helped the solvent system to get increased heat, due to increased dielectric heating. Both are polar protic solvents and used on purpose to stabilize the carbo-cations. Water is used to ensure lab safety.

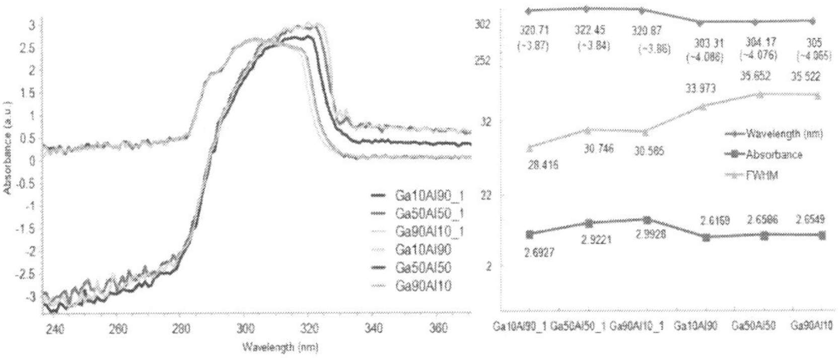

Figure 10. UV-Visible Spectra of thin-films with and without acetic acid (Left), change in FWHM (nm), Wavelength (nm), Absorbance (Right)

The percentage of ethanol (99.9%) should be below 50% of the solution, to avoid any burning, explosion or spilling of solution inside microwave, when operating at high power and high-temperature conditions, increased ratio of water helped with the same. The solutions prepared, by using metal acetylacetonates and 30% ethanol-water solution with 12 minute, 2.45 GHz, 700 W microwave-irradiation, are represented by Ga10Al90_1, Ga50Al50_1 and Ga90Al10_1 in **fig.10.**

The UV Visible spectroscopy gives very important details about electronic configuration of a material. Therefore, the solutions prepared before and after adding acetic acid with 12 minute microwave-irradiation, were tested and the results are shown in **fig. 10**; with addition of acetic acid the Full Width Half Maxima (FWHM) increased and the UV-Visible plots shifted by 17.18 nm on an average towards a lower wavelength i.e., towards higher energy.

4. Discussion, Conclusion and Future Scope

The study aimed to find the possibility of composition variation using microwave-assisted solvothermal synthesis method, the study resonates with the possibility. The XRD peaks of various phases of Ga_2O_3 and Al_2O_3 are very close to each other. Taking blank as reference the prominent peaks of Ga10Al90, Ga50Al50 and Ga90Al10, are shifted by 0.12, 010 and 0.03 2θ degrees (on average) as can be seen from **fig.3**. With decrease in Aluminum content the change in 2θ degrees is also decreasing, which indicates towards Aluminum substitution in Gallium lattice (**para. 3, section 3.1**). But, as the peak intensity decreases, the change in 2θ values of peaks become nearly indistinguishable.

During substitution it is difficult to tell exactly which phase and which compound has been formed using XRD i.e., *XRD alone is insufficient technique for thin-film characterization of Gallium-Aluminum Oxide,* due to overlapping of peaks in different phases of Ga_2O_3 and Al_2O_3.

Uniform films obtained below the microstructures (debris) as can be seen from **fig. 5, 7** and **8**, providing strong indication of uniformly grown thin-films up to a saturation point. The evidences of nano-humps beneath cracks (**fig. 6.** (e)) also provide a strong indication of a uniformly grown layer. Due to current limitations of detection a sufficiently thick layer was grown, however authors want to look forward the possibility of deposition of thin-films and observe crystal structures thus formed using same solvent system at lower deposition power and time of irradiation. Also, study the effect of post-processing including ultrasonication to remove sub-micron/micro structures.

The microwave-assisted solvothermal synthesis has potential to grow metal-substituted metal oxide thin-films with current metal acetylacetonates, acetic acid and ethanol-water based system. Future work may look forward in improving the thin-film quality and use this low cost microwave technology for device application, as demonstrated by **Piyush *et al's*** group in 2018 [22] with single solute-multiple-solvent system.

The change in UV-Vis spectra with addition of new solvent (**fig. 10**) indicate towards role of solvents in defining property of material inside microwave. It will be interesting to explore more about the effect of addition of new solvents and change in band-gap of deposited thin-films thus obtained.

5. Peculiar Observations and Lab Safety Guidelines

The experiments used in this research work and subsequent observation in working with microwave, multi-solvent systems and three different substrates over a period of more than 18 months has provided with following observations crucial for lab safety, and are reported in this work for the greater benefit of scientific community.

5.1. Cracking of Vessels and Effect of Contamination

a). Not only reaction mixture but the substrate also plays an important role in synthesis of nano-particle in microwave-assisted synthesis and variants i.e., everything inside microwave matters, even the substrate used for deposition, ignoring this fact may lead to accidents inside microwave/or in the lab. The vessel (chemical resistant) shown in **fig.11** (f) cracked as Silicon was used for deposition instead of Glass or cloth material. The crack took place within 30 seconds of irradiating substrate with 2.45GHz, 700W irradiation.

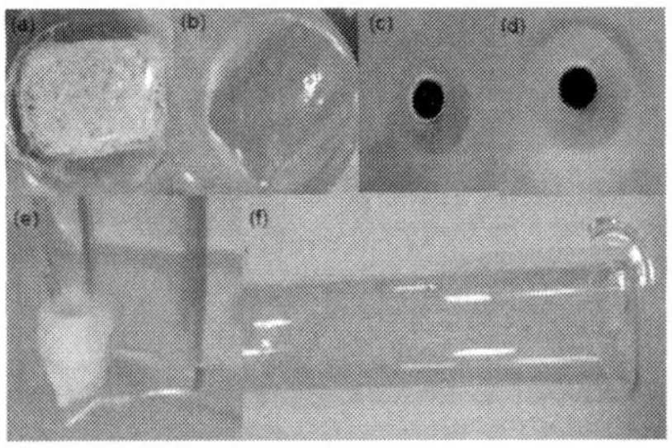

Figure 11. (a) Fused micro slide inside Quartz boat (top-view) (b) top-side-view (c) burned coating (before cleaning) (d) after cleaning (e) cracked vessel (top-view) (f) cracked vessel (side view)

b). Contamination may change a microwave transparent material to microwave absorbing material. Thus, working with microwaves one needs to be aware of any possible source of contamination. While accidently using a Quartz crucible (used previously for heating human hair (at 800°C in a heating furnace)), inside microwave for coating Glass substrate with a modified microwave-assisted thin-film deposition method, the Teflon tape wrapped outside the nozzle burned instantly after 30 seconds of 2.45 GHz, 700 W irradiation. The temperature inside microwave reached above 370°C (ignition temperature of Teflon) and the microwave coating burned (**Fig. 11** (c)).

c). After 15 minutes when the incident was reported to head-supervisor and microwave power as turned off instantly, the microwave door was opened as the door was opened the glass slide fused inside the Quartz boat (**Fig. 11** (a)) with a crackling sound, the temperature of Quartz boat was still above 50°C i.e., the temperature inside microwave would had gone much higher than 370°C, as when the microwave power was closed instantly after that incident the glass inside was red hot, as can be seen from microwave door inside, due to safety reasons the door was kept closed for a calculated time.

5.2. Commercial Microwave, Modification and Lab-safety

a). Commercial/Domestic microwaves come with lab/domestic safety guidelines and ensure maximum customer safety. However before attempting a major experiment, or while self-modifying system, make sure the coating inside is intact and test the equipment 'under expert supervision'.

b). If using vapors or liquid inside microwave, make sure the periodic structures in the doors of microwave system are free of contamination, when they are being used again; especially when microwave has been used to deposit different chemicals over a long period of time, or is been used too frequently (*if any part of microwave, especially door is heating frequently, there may be chances of microwave leakage and a need to discard the existing equipment*).

c). Vapors coming out of microwaves should be treated with same precautions as fumes coming out of a chemical reactor are being treated, especially when dealing with heavy metal compounds. Have a close look in changing health patterns while working with microwave, early detection can help one to take additional precautions.

d). Do not use domestic microwave equipments, or chemical resistant borosilicate petri-dishes for heating bare Silicon/Sapphire substrate inside microwave for chemical experiments, the petri-dishes may crack instantly or while attempting to open microwave door, this can be dangerous.

e). Do not directly heat the substrate/material, obtained after microwave irradiation, the material may explode or the vessel/dish in which it is to be heated may crack.

f). Always arrange and wear proper safety equipments, P-100 respirators or N-95 masks, proper chemical resistant safety goggles and necessary PPE kits while performing an experiment in microwave.

Corresponding author wants to strongly argue that lab safety is equally important in microwave-irradiation assisted synthesis, as everything inside microwave plays an important role and any carelessness or ignorance may lead to a major accident or affect the health of researcher severely over time. The corresponding author has skipped accidents nearly on modifying and experimenting with new materials inside microwave. Thus, as a moral responsibility these safety guidelines are included while reporting this research work.

Acknowledgement

Authors would like to thank home-institute, Centre for Advanced Studies, AKTU Lucknow for providing with necessary facilities to carry out this research work and anayticsir portal, to provide with an online platform to access testing/characterization facilities, available at CSIR laboratories (including CSIR-NBRI, Lucknow and CSIR-IITR Lucknow) as an external user on paid basis.

Corresponding Author is thankful to BSIP Lucknow and IIT, Kanpur, for providing with XRD and FESEM-EDX facilities as an external academic user on paid basis. This is to acknowledge that **fig.1** is created by using beta-testing program of biorender.com, corresponding author is thankful for their program. Corresponding author would like to extend a token of gratitude to *Dr. Friedrich Menges* for providing with academic license of Spectragryph v1.2.15 [24] that remained an important tool for data analysis.

References

[1] Yao, Yao, Serdal Okur, Luke AM Lyle, Gary S. Tompa, Tom Salagaj, Nick Sbrockey, Robert F. Davis, and Lisa M. Porter. "Growth and characterization of α-, β-, and ϵ-phases of Ga_2O_3 using MOCVD and HVPE techniques." *Materials Research Letters* 6, no. 5 (2018): 268-275.

[2] Cora, Ildikó, Francesco Mezzadri, Francesco Boschi, Matteo Bosi, Maria Čaplovičová, Gianluca Calestani, István Dódony, Béla Pécz, and Roberto Fornari. "The real structure of ε-Ga 2 O 3 and its relation to κ-phase." *CrystEngComm* 19, no. 11 (2017): 1509-1516.

[3] Ahmadi, Elaheh, and Yuichi Oshima. "Materials issues and devices of α-and β-Ga2O3." *Journal of Applied Physics* 126, no. 16 (2019): 160901.

[4] Chen, Zimin, Zeqi Li, Yi Zhuo, Weiqu Chen, Xuejin Ma, Yanli Pei, and Gang Wang. "Layer-by-layer growth of ε-Ga2O3 thin film by metal–organic chemical vapor deposition." *Applied Physics Express* 11, no. 10 (2018): 101101.

[5] Zinkevich, Matvei, Francisco Miguel Morales, Heiko Nitsche, Martin Ahrens, Manfred Rühle, and Fritz Aldinger. "Microstructural and thermodynamic study of γ-Ga2O3." *Zeitschrift für Metallkunde* 95, no. 9 (2004): 756-762.

[6] Wang, T., Li, W., Ni, C. and Janotti, A., 2018. Band-gap and Band Offset of Ga 2 O 3 and (Al x Ga 1− x) 2 O 3 Alloys. *Physical Review Applied*, *10*(1), p.011003.

[7] Tanen, Nicholas, Celesta Chang, Vladimir Protasenko, Jonathan McCandless, David Muller, Huili Xing, and Debdeep Jena. "Quantum Transport in Epitaxial Ultra Wide Bandgap Aluminum Gallium Oxide Tunnel Heterostructures." *Bulletin of the American Physical Society* 65 (2020).

[8] Shimizu, K.I., Satsuma, A. and Hattori, T., 1998. Selective catalytic reduction of NO by hydrocarbons on Ga2O3/Al2O3 catalysts. *Applied Catalysis B: Environmental*, *16*(4), pp.319-326.

[9] Haneda, M., Kintaichi, Y., Shimada, H. and Hamada, H., 2000. Selective reduction of NO with propene over Ga2O3–Al2O3: Effect of sol–gel method on the catalytic performance. *Journal of Catalysis*, *192*(1), pp.137-148.

[10] Horie, Ryuto, Hiroyuki Nishinaka, Daisuke Tahara, and Masahiro Yoshimoto. "Epitaxial growth of γ-(AlxGa1-x) 2O3 alloy thin films on spinel sub0strates via mist chemical vapor deposition." *Journal of Alloys and Compounds* 851: 156927.

[11] Cook, Daniel S., Joseph E. Hooper, Daniel M. Dawson, Janet M. Fisher, David Thompsett, Sharon E. Ashbrook, and Richard I. Walton. "Synthesis and Polymorphism of Mixed Aluminum–Gallium Oxides." *Inorganic chemistry* 59, no. 6 (2020): 3805-3816.

[12] Tien, Ching-Ho, Bo-Wen Hsiao, Chien-Ming Chen, Mu-I. Chen, Jung-Lung Chiang, and Dong-Sing Wuu. "Nitrogen and oxygen annealing effects on properties of aluminum-gallium oxide films grown by pulsed laser deposition." *Ceramics International* 46, no. 15 (2020): 24147-24154.

[13] Che-Hao Liao, Kuang-Hui Li, Carlos G. Torres-Castanedo, Guozheng Zhang, Xiaohang Li "Wide range tunable bandgap and composition β-phase (AlGa)2O3 thin film by thermal annealing" *Journal of Applied Physics Letters* 118, no.3. (2021) 032103.

[14] Okamura, Koshi, Babak Nasr, Richard A. Brand, and Horst Hahn. "Solution-processed oxide semiconductor SnO in p-channel thin-film transistors." *Journal of Materials Chemistry* 22, no. 11 (2012): 4607-4610 (Electronic Supplementary Material).

[15] Zaumseil, Peter. "High-resolution characterization of the forbidden Si 200 and Si 222 reflections." *Journal of applied crystallography* 48, no. 2 (2015): 528-532.

[16] Ghosh, Dulal C., and Raka Biswas. "Theoretical calculation of absolute radii of atoms and ions. Part 1. The atomic radii." *International Journal of Molecular Sciences* 3, no. 2 (2002): 87-113.
[17] Ghosh, Dulal C., and Raka Biswas. "Theoretical calculation of absolute radii of atoms and ions. Part 2. The ionic radii." *International Journal of Molecular Sciences* 4, no. 6 (2003): 379-407.
[18] Wang, Yu, Wen Wen, and Kai Wu. "A Ga 2 O· 11Al 2 O 3 nanonet prepared by interfacial reaction growth approach and its application in fabricating GaN nanowires." *Science China Chemistry* 53, no. 2 (2010): 438-444.
[19] WinPLOTR: a Windows tool for powder diffraction patterns analysis Materials Science Forum, Proceedings of the Seventh European Powder Diffraction Conference (EPDIC 7), 2000, p.118-123, Ed. R. Delhez and E.J. Mittenmeijer
[20] FullProf.98 and WinPLOTR: New Windows 95/NT Applications for Diffraction Commission For Powder Diffraction, International Union for Crystallography, Newletter N°20 (May-August) Summer 1998
[21] *Schindelin, J.; Arganda-Carreras, I. & Frise, E. et al. (2012), "Fiji: an open-source platform for biological-image analysis", Nature methods **9(7)**: 676-682, PMID 22743772*
[22] Jaiswal, Piyush, Usman Ul Muazzam, Anamika Singh Pratiyush, Nagaboopathy Mohan, Srinivasan Raghavan, R. Muralidharan, S. A. Shivashankar, and Digbijoy N. Nath. "Microwave irradiation-assisted deposition of Ga2O3 on III-nitrides for deep-UV opto-electronics." *Applied Physics Letters* 112, no. 2 (2018): 021105.
[23] Horikoshi, Satoshi, Tomoki Watanabe, Atsushi Narita, Yumiko Suzuki, and Nick Serpone. "The electromagnetic wave energy effect (s) in microwave–assisted organic syntheses (MAOS)." *Scientific reports* 8, no. 1 (2018): 1-10.
[24] F.Menges "Spectragryph- optical spectroscopy software", version 1.2.15, 2020, http://www.effemm2.de/spectragryph/
[25] Horikoshi, S., Osawa, A., Abe, M. and Serpone, N., 2011. On the generation of hot-spots by microwave electric and magnetic fields and their impact on a microwave-assisted heterogeneous reaction in the presence of metallic Pd nanoparticles on an activated carbon support. *The Journal of Physical Chemistry C*, *115*(46), pp.23030-23035.
[26] Gutmann, B., Schwan, A.M., Reichart, B., Gspan, C., Hofer, F. and Kappe, C.O., 2011. Activation and deactivation of a chemical transformation by an electromagnetic field: evidence for specific microwave effects in the formation of Grignard reagents. *Angewandte Chemie International Edition*, *50*(33), pp.7636-7640.
[27] Hammarberg, E., Prodi-Schwab, A. and Feldmann, C., 2008. Microwave-assisted synthesis of indium tin oxide nanocrystals in polyol media and transparent, conductive layers thereof. *Thin Solid Films*, *516*(21), pp.7437-7442.
[28] Huang, H., Zhang, S., Qi, L., Yu, X. and Chen, Y., 2006. Microwave-assisted deposition of uniform thin gold film on glass surface. *Surface and Coatings Technology*, *200*(14-15), pp.4389-4396.
[29] Leonelli, C. and Mason, T.J., 2010. Microwave and ultrasonic processing: now a realistic option for industry. *Chemical Engineering and Processing: Process Intensification*, *49*(9), pp.885-900.
[30] Sahoo, S., Barik, S.K., Gaur, A.P.S., Correa, M., Singh, G., Katiyar, R.K., Puli, V.S., Liriano, J. and Katiyar, R.S., 2012. Microwave assisted synthesis of ZnO nano-sheets and their application in uv-detector. *ECS Journal of Solid State Science and Technology*, *1*(6), pp.Q140-Q143.
[31] Xiong, W.W., Yang, G.H., Wu, X.C. and Zhu, J.J., 2013. Microwave-assisted synthesis of highly luminescent AgInS 2/ZnS nanocrystals for dynamic intracellular Cu (ii) detection. *Journal of Materials Chemistry B*, *1*(33), pp.4160-4165.
[32] Husham, M., Hassan, Z., Mahdi, M.A., Selman, A.M. and Ahmed, N.M., 2014. Fabrication and characterization of nanocrystalline CdS thin film-based optical sensor grown via microwave-assisted chemical bath deposition. *Superlattices and Microstructures*, *67*, pp.8-16.

[33] Gao, Yangyang, Qian Feng, Zhaoqing Feng, Yan Zuo, Yuncong Cai, Yachao Zhang, Jing Ning et al. "Epitaxial growth of ε-(AlGa) 2 O 3 films on sapphire substrate by PLD and the fabrication of photodetectors." *Optical Materials Express* 11, no. 2 (2021): 219-230.

Experimental analysis of effect of speed and time on aluminium nanoparticles fabricated by high energy ball milling

Vijay B. Sarode[1*] and Ravindra D. Patil[1]

[1]J.T.Mahajan College of Engineering Faizpur, Research Centre, Kavyitri Bahinabai Chaudhari North Maharashtra University, Jalgaon, INDIA 425 524

vijaysarode1975@gmail.com, rdp.ravindra@gmail.com

Abstract. Metal Matrix Composites are technologically advance material identified in recent era as a substitute for traditional engineering materials because of their excellent properties. Out of the available material, demand for Aluminium Matrix Composites are increasing in the sector of aerospace, automobile and defense applications. The Aluminium Matrix Composites are majorly accepted because of their excellent physical and mechanical properties. There are various manufacturing techniques developed for the fabrication of AMCs. High energy ball milling is a one of the most effective and versatile routes for fabrication of nano aluminium powder. The Al powder of 99.55% purity was taken for experimentation. The 60gm of Al powder was taken in stainless steel vial with BPR as 5:1. The stainless steel balls of diameter 4mm was used as a grinding media. The ball milling was performed at different speed and time in presence of 60ml toluene as process control agent. The characterization of powder particle was done with XRD and SEM. The analysis of peak were done on origin software. The optimum speed and milling time for fabrication of Al nanoparticles of size 36 nm were found to be at 250rpm and 3 to 6 hours.

Keywords: Aluminium, High Energy Ball Milling, X-Ray diffraction, Scanning Electron Microscope, Particle size

1. Introduction

The nano scale technology is very young and key technology and having potential application in the field of engineering, cosmetics, food product, textiles, biomedicine and domestic applications. Many research has adopted various methods of manufacturing out of that mechanical alloying is the best suitable method, which is also known as high energy ball milling. The main purpose of mechanical alloying is change the shape of particles, reduce particle size by continuous cold welding and fracturing. To achieve this it is require to use ductile material for MA process [2]. The aluminium (Al) and its alloys are frequently examined material because of its exceptional properties like corrosion resistance, low density and high strength to weight ratio. The ultrafine metallic powder have reported a wide applications in engineering because of large specific area and high reactivity and having particle size in nm [12]. A substantial improvement in mechanical properties are reported with particle size reaches to nanometer [4].

There are basically two terms associated with in high energy ball milling which are Mechanical Alloying and Mechanical Milling. Mechanical Alloying (MA) indicates mixture of one or two metal particles

(different metal) are milled together in ball milling machine to obtain the homogenous alloy. Mechanical Milling is associated with milling of pure metals, intermetallics or prealloyed powders where homogenization is not an important factor. Mechanical milling is used to reduce size of metal particles [3]. The main difference between mechanical alloying and milling are that the time required for milling is half than the alloying also intention of milling is to reduce the particle size or transformation of particle by mechanical means [1].
The various types of high energy ball millings (HEBMs) are available for manufacturing of nano crystalline and amorphous powder having different capacity, efficiency and controlled conditions with cooling arrangement to maintain the temperature of process. The widely used HEBMs for production of Al alloys are attritor, planetary, and Spex shaker mills. Attriators are used for manufacturing of larger quantities upto 100 Kg. The planetary mills are used for milling volume with variety of options. In planetary mill input energy can be changed by changing speed of mill upto 1000 rpm. Spex shakers ball mills are used for small quantities in a very short time. In this experimental work, researcher was used planetary ball mill. [11].The high energy Planetary ball milling consist of a container (Vial), lid, Supporting disc, grinding balls, material to be grind. The container is kept on rotating supporting disc, both are rotating opposite to each other as these are installed on planet like drive mechanism. Hence it is called as planetary ball mill. The centrifugal force developed by vial and supporting disc acts on the vial content. Due to rotation, grinding balls collide with each other and container wall. The small amount of powder trapped between grinding balls, wall and grinding balls. Due to this action, morphology of powder particle could change [11].
In this work, high energy ball milling of Al powder performed at different speed and time in presence of process control agent (PCA). When liquid is used during grinding, process is known as wet grinding [3]. The XRD and SEM analysis of ball milled Al powder has been performed to know the present of carbide and oxide phase and to reveal the change in morphology of Al powder after the ball milling [7] [14].
The characterization of nano materials is very important from application point of view by understanding its properties. The performance of composites changes as the size of particle changes. Different methods are available to reveal the characteristics of materials.

Characterization Techniques:
1) Chemical Characterization
2) Structural Characterization

Chemical characterization is performed to internal chemical structural details and to know shape, size, crystallinity and lattice constant structural characterization is used. In this work, for characterization of ball milled aluminium powder, structural characterization method is adopted. In structural characterization, X-Ray diffraction and Scanning Electron Microscopy (SEM) are used.

2. X-Ray Diffraction Technique

X-ray diffraction is a non-destructive technique which gives very important information about lattice structure of a crystalline structure like unit cell dimensions, bond angle, chemical compositions and crystallographic structure of natural and fabricated materials. The XRD machine works on well-known

Bragg's Law ($\lambda = 2dsin\theta$) -- (1)

where λ is the wavelength of incident X-ray, d= interplanar spacing and θ is the angle of incidence. In this case X-ray source is used for identification and characterization of crystalline solids. It is used to obtain the average size of the particles using Scherrer formula:

$a = 0.9\,\lambda/Bcos\theta$ -- (2)

where a= average particle size, λ is the wave length of incident X-ray, B= full width of half maxima in radians and θ is the angle of incidence [10] [5].

The XRD experimentation was conducted at UGC-DAE Consortium for Scientific Research, Indore on the D8 advanced XRD machine.

3. Scanning Electron Microscopy

In case of scanning electron microscope, develops the image by scanning the sample surface with high-energy beam of electron. The electron beam scanned over the surface of sample to measure the electrons that are scattered back. The high energy electron interact with atoms on the surface which producing signal that contains the surface topography of the sample [13]. The SE M analysis were conducted at UGC-DAE Consortium for Scientific Research, Indore on JEOL Scanning Electron Microscope.

4. Experimental Detail

In this study Aluminium is selected as matrix material as it is emerged as important material for industrial and commercial applications. Aluminium possesses excellent properties, light in weight, strength is equal to construction steel, easy to form into rolled sheet. Here aluminium powder is mixture of spherical and elongated shape particle having average diameter of 23.99 micron and average length of 31.38 micron. The Al powder is received from PASHWAMANI METALS, Mumbai having purity of 99.55% and particle size of 20 micron.The selection of process parameters namely ball milling speed (rpm), ball milling time (hrs) and CNT diameter (nm) were made on the basis of literature survey [6]. Three process parameters are hence selected to conduct this experiment namely ball milling speed (rpm), ball milling duration (hours) and CNT diameter (nm). Table 1 indicates the process parameters and the values for the same.

Table 1. Process parameters and their values.

Factors	LEVELS		
	LOW	MEDIUM	HIGH
Ball milling speed (rpm)	150	200	250
Ball milling time (hours)	3	6	12
CNT diameter (nm)	8	20	50

An L9 orthogonal array was chosen for 3 factors and 3 levels with the primary intention of optimizing the number of experiments. The Minitab 17 software was used to carry out the design of experiments. Table 2 illustrates the final design of experiments using the L9 orthogonal array.

Table 2. Design of experiments using L9 orthogonal array.

Expt. No.	Control Factors		
	S (rpm)	T (hours)	D (nm)
1	150	3	8
2	150	6	20
3	150	12	50
4	200	3	50
5	200	6	8
6	200	12	20
7	250	3	20
8	250	6	50
9	250	12	8

With above parameters, ball milling of Al powder was performed in FRITSCH Pulverisette 6 make planetary ball milling machine in 250ml stainless steel jar with stainless steel balls. The weight of each ball was 4g and ball to powder ration was taken as 5:1 during milling operation. The dwell time was takes as 15 minutes after 1hour and reversal of rotation after each hour so that powder is milled properly. After rotation of 1 hour, ball milling is given rest for 15 minutes in order to avoid rapid increase in temperature during milling operation. The toluene (C6H5CH3) was used as a process control agent to control the sudden rise in temperature inside the vial during milling process. The 60ml toluene quantity was taken for wet milling. The purpose of PCA are to control the cold welding of powder to ball and vial. The XRD diffraction pattern was recorded from 20° to 90° on D8 advance XRD system having 2.2Kw sealed tube of Cu as X-ray source.

5. Results and Discussions

5.1 *Characterization of Powder*:

The characterization of raw and ball milled Al powder were done by XRD and SEM to determine the changes in shape and size of Al particles. The aluminium powder were ball milled at 150, 200 and 250rpm for 3,6,12 hrs. The crystalline nature of milled aluminium powder was examined by using XRD technique. Fig. 3 shows XRD patterns of the raw and ball milled aluminium powder for different time and speed reveals that there was an appreciable change in particle size of aluminium powder. The average particle size for raw and ball milled aluminium powder were estimated by Scherrer formula equation(1).

The crystallite size for ball milled sample considered for study is given in table 3. The fig.1 shows the X-ray diffraction pattern of raw aluminium and ball milled aluminium powder prepared by high energy ball milling in planetary ball milling at different time and rpm. It was observed that raw aluminium powder has an average particle size of 44.35 nm estimated by Scherrer formula. The fig. 1 XRD pattern of aluminium powder before and after milling of aluminium powder at 150 rpm, 200rpm and 250rpm corresponding to 3,6 and12 hrs and enlarged section of first (Largest) peak of XRD. There are four sharp peaks detected corresponds to FCC structure of aluminium. The higher intensity of peak in Fig. 1(a) shows aluminium powder is more crystalline in nature and after ball milling, reduction in peak intensity followed by broadening of peak might be due to fine size of aluminium powder [9]. Compared to the pattern of raw aluminium powder, the milled powder shows broaden intensity peaks, indicates that powder size is effectively reduced [8]. The small shift observed in enlarged section of stacked largest peak is due to macrostrain in lattice. The change in crystallite size at particular rpm and hours is due to lattice strain and ductile nature of aluminium responsible for cold welding and deformation of aluminium particles. From fig. 2 it is found that with increase in RPM and milling time, there is a reduction in particle size. But at certain stages, there is increase in particle size due to the agglomeration, subsequent cold welding and fracture.

Table 3. Particle size of Aluminum nano-powder by Debye-Scherrer formula and particle size by Image J software from SEM images.

Sample	RPM	Hrs	Crystallite Size [nm]	Particle Size [μm] by ImageJ
Raw Aluminium			44.55	20
Ball Milled Aluminium Powder	150	3	52.64	78.24
		6	54.61	63.87
		12	51.01	71.20
	200	3	61.69	70.42
		6	43.82	165.06
		12	42.13	227.20
	250	3	36.84	128.32
		6	36.03	446.36
		12	43.45	333.90

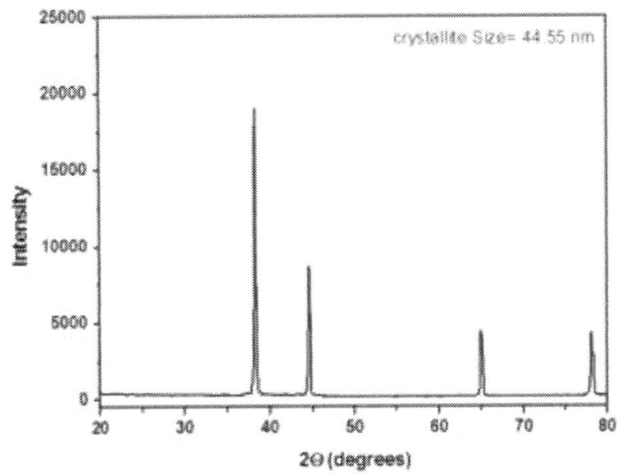

Figure (a)

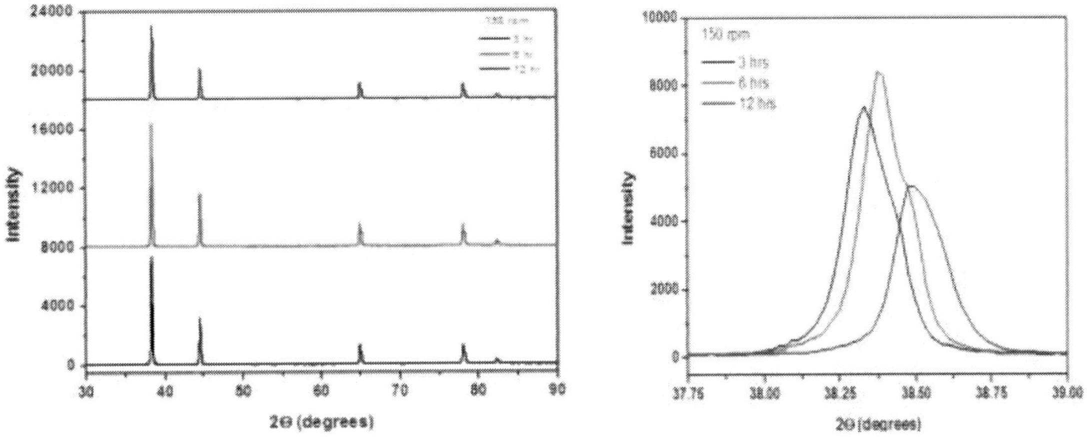

Figure(b)

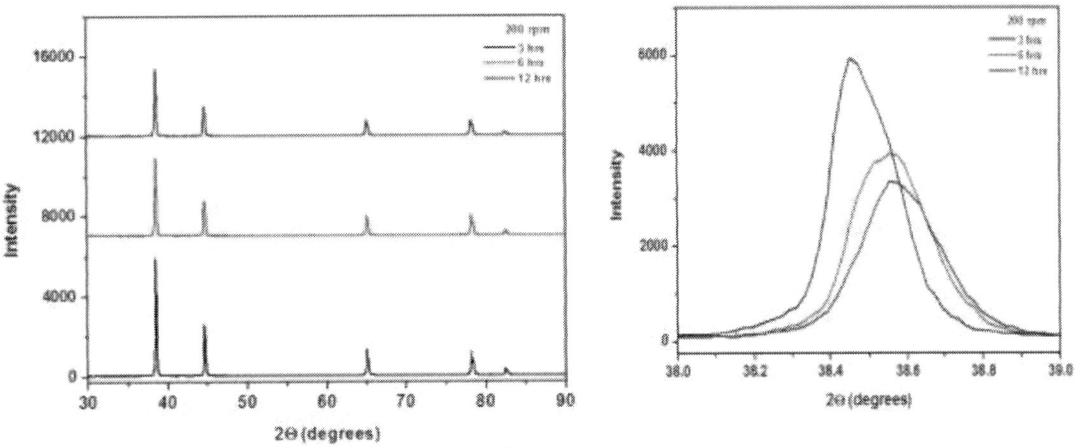

Figure(c)

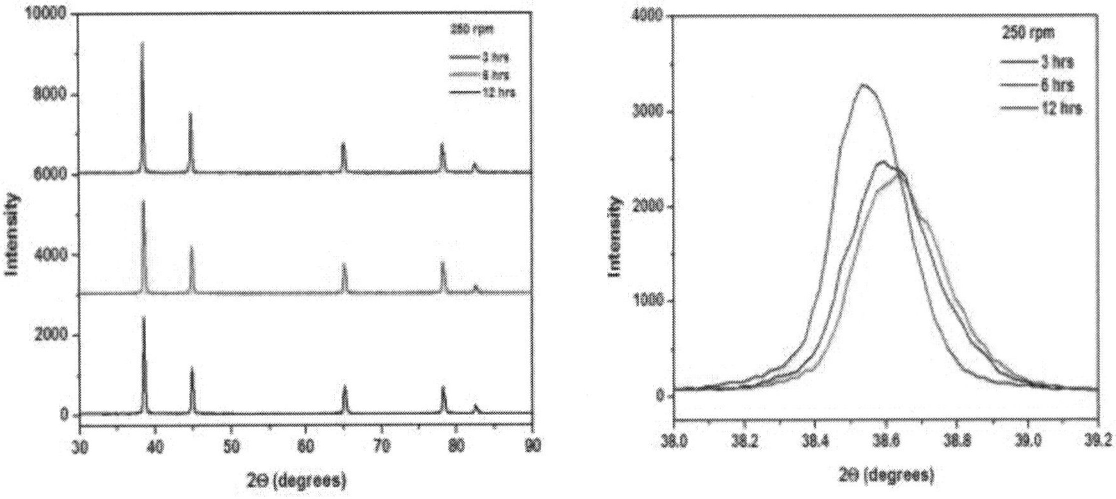

Figure.(d)

Figure 1. X-ray Diffraction pattern of Raw and Ball Milled Aluminium Powder by High Energy Ball Milling. (a) Raw Aluminium Powder (b) 150rpm at 3,6,12 hours (c) 200 rpm at 3,6,12 hours (c) 250rpm at 3,6,12 hours and enlarged section of stacked largest peak.

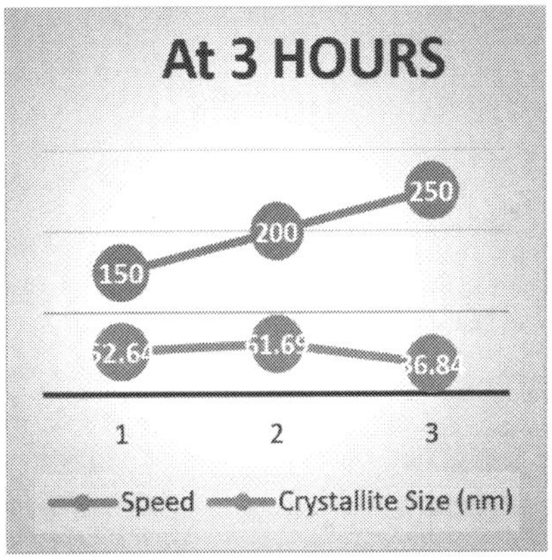

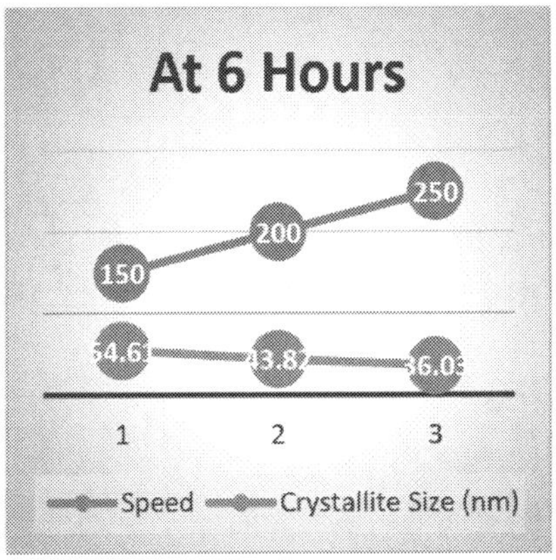

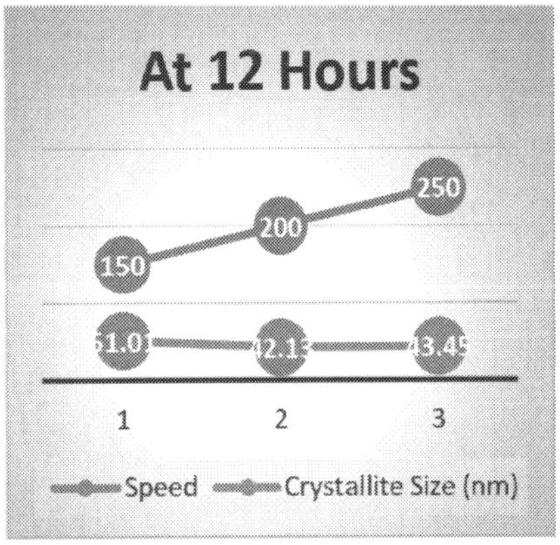

Figure 2. Variation of Crystallite size versus milling time (hours) for Ball Milled Aluminium Powder.

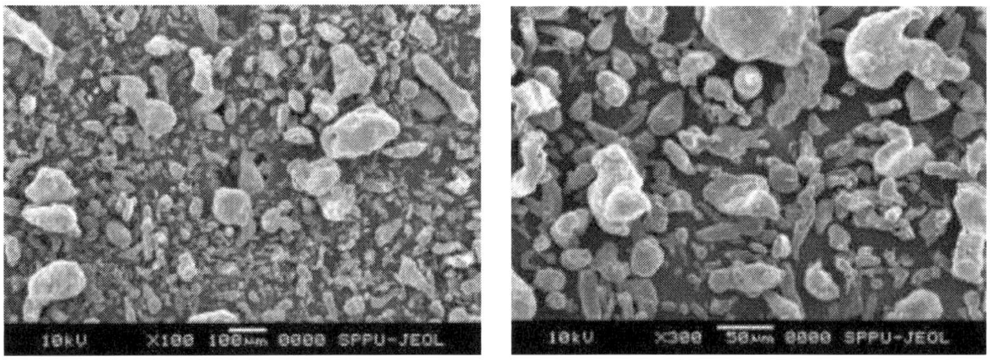

(a) SEM images of raw aluminium power

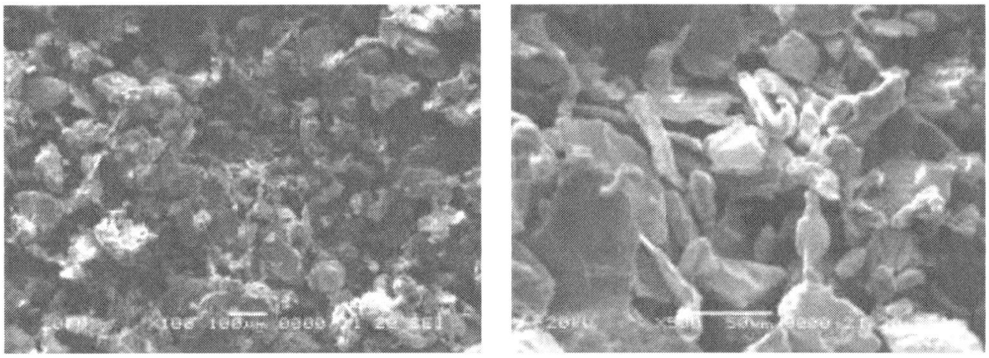

(b) At 150 RPM 3 Hours

(c) At 150 RPM 6 Hours

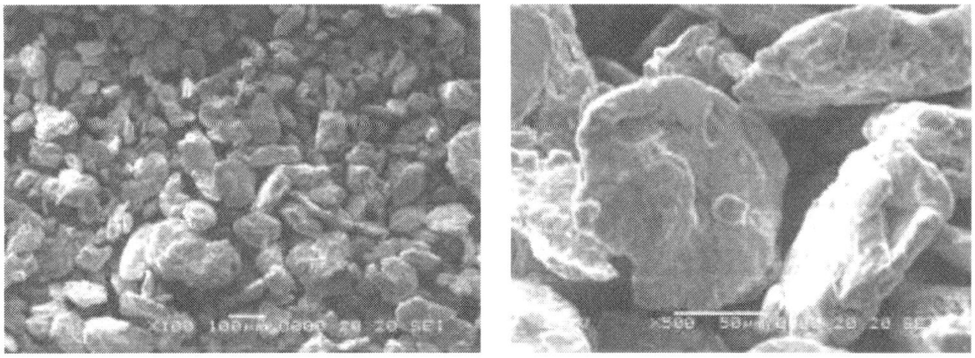

(d) At 150 RPM 12 Hours

(e) At 200 RPM 3 Hours

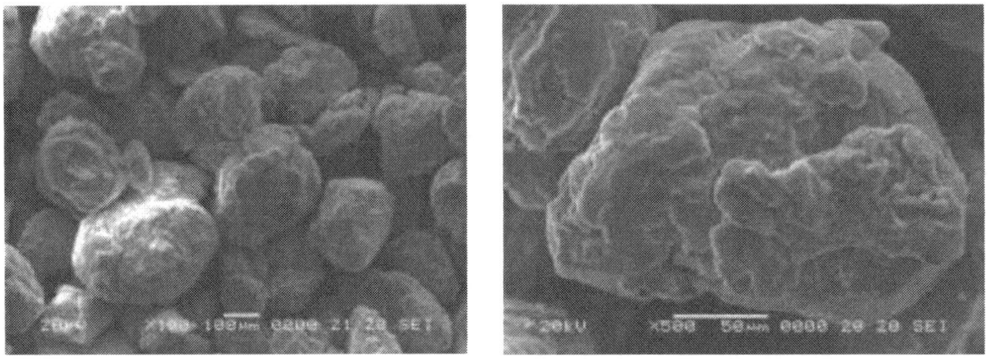

(f) At 200 RPM 6 Hours

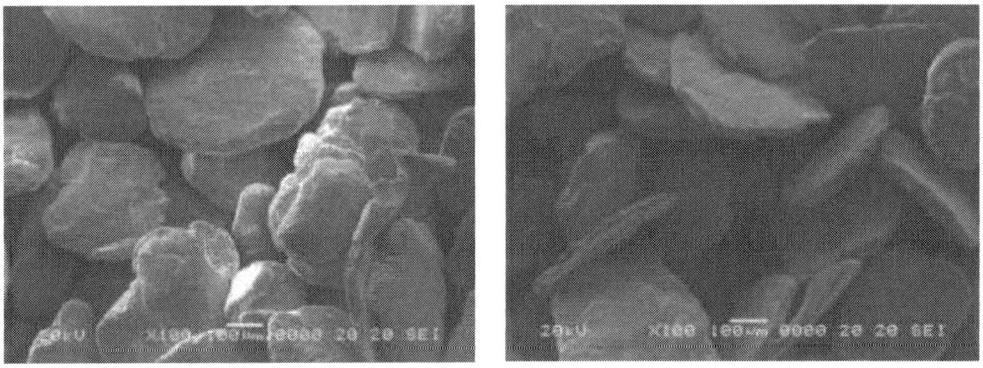

(g) At 200 RPM 12 Hours

(h) At 250 RPM 3 Hours

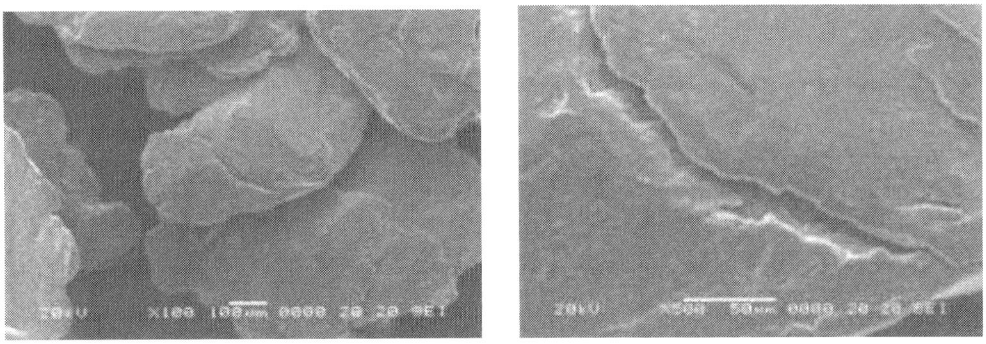

(i) At 250 RPM 6 Hours

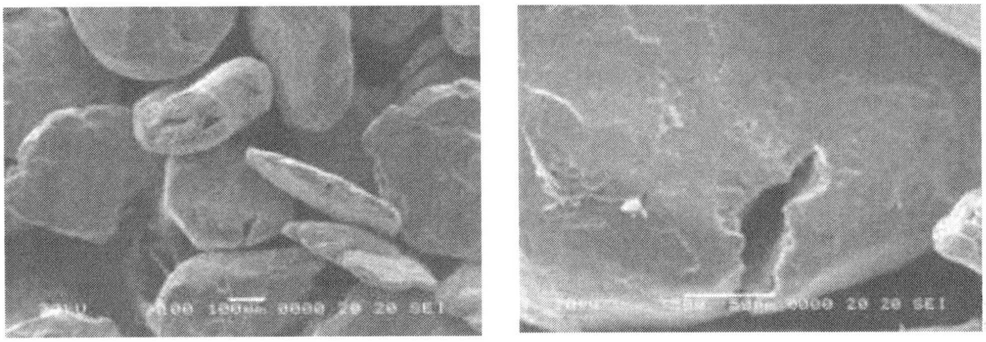

(j) At 250 RPM 12 Hours

Figure 3. SEM images of Aluminium powder (a) Raw Aluminium Powder. At 150 RPM (b) 3Hours (c) 6Hours (d) 12 Hours. At 200 RPM (e) 3 Hours (f) 6 Hours (g) 12 Hours. At 250 RPM (h) 3 Hours (i) 6 Hours (j) 12 Hours.

Fig. 3 shows SEM images of raw aluminium powder and ball milled aluminium powder at 150, 200 & 250 RPM corresponding to 3, 6, 12 hours of milling time. The SEM analysis of raw aluminium powder has been done to check the surface morphology and shape of particles. From Figure (a) it is observed that aluminium particles are spherical and elongated in shape has an average grain size is 20 micron. The SEM analysis of ball milled aluminium powder has been done to check the change in morphology and flatness of the powder.
The initial powder size was 20 micron and after the ball milling particle size increases to 52 micron at 150 rpm 3 hours. Again slightly increases at 6 hours and further decreases to 51 micron. The changes were occurred due to due to welding of aluminium powder and fracture mechanism were less than agglomeration. Because of this, at 200 rpm 6 hours and 250 rpm 6 hours, there were sudden increase in particle size (measured on Image J software) and same time aluminium powder particle surface became flaky and fracture mechanism initiated as shown in fig 3 (i) and (j).

6. Conclusion

Aluminium matrix composites have wide applications in engineering and high energy ball milling is a most popular method of production of aluminium nanoparticles of low crystallite size 36nm. The morphology of ball milled aluminium powder is studied by XRD technique and Scanning Electron Microscope (SEM).The particle size is calculated by X-ray diffraction technique and the change in morphology is analyzed through SEM. At 3hrs and 6hrs, with further increase in rpm, crystallite size decreases. At 12 hrs, with increase in rpm, crystallite size decreases at 200 rpm and further increases at 250 rpm. This is due to cold welding phenomenon. Also from SEM analysis it is observed that at 200 rpm 12 hrs and 250 rpm 6 hrs, particle became flaky as increase in surface area of particles. At 250 rpm

12hrs facture mechanism is initiated and with further increase in time there may be decrease in particle size.

References:

[1] A. W. Weeber, H. B. (1986). The Preparation of Amorphous Ni-Zr Powder by Grinding the Crystalline Alloy. *EUROPHYSICS LETTERS* , 445-448. doi:https://doi.org/10.1209/0295-5075/2/6/006

[2] Benjamin, J. (1992). Fndamentals of Mechanical Alloying. *Materials Science Forum* **Vols**, 88-90. doi:doi:10.4028/www.scientific.net/MSF.88-90.1

[3] C.Suryanarayana. (2001). Mechanical alloying and milling. *Progress in Materials Science, 46*, 1-184. doi:https://doi.org/10.1016/S0079-6425(99)00010-9

[4] Dinesh Kumar Koli, G. A. (2015). Advanced Aluminium Matrix Composites: The Critical Need of Automotive and Aerospace Engineering Fields. *Materials Today Proveedings*, 3032-3041.

[5] G.Cao. (2004). *Nanostructures and nanomaterials: Synthesis,properties and applications.* Imperial College Press.

[6] Jinzhi Liao, M.-J. T. (2011). Mixing of carbon nanotubes (CNTs) and aluminum powder for powder metallurgy use. *Powder Technology*, 42-48. doi:doi:10.1016/j.powtec.2010.12.001

[7] Lubna Rais, D. R. (2013). Synthesis and structural characterization of Al-CNT metal matrix composite using Physical Mixing. *IOSR Journal of Applied Physics*, 54-57.

[8] M. Ramezani, T. N. (2012). Mechanical milling of aluminum powder using planetary ball milling process. *Journal of Achievements in Materials and Manufacturing Engineering*, 790-798.

[9] Meysam Toozandehjani, K. A. (2017). Effect of Milling Time on the Microstructure,Physical and Mechanical Properties of Al-Al2O3 Nanocomposite Synthesized by Ball Milling and Powder Metallurgy. *Materials, 10*, 1-17. doi:doi:10.3390/ma10111232

[10] Pradeep, T. (2007). *Nano: The Essentials.* McGraw-Hill Education .

[11] Rajeev Kumar Gupta, B. M. (2017). *An overview of Ball Milled Nanocrystalline Aluminium Alloys.* Springer Briefs in Materials. doi:DOI 10.1007/978-3-319-57031-0

[12] S, R. M. (2018). EFFECT OF MILLING TIME ON PRODUCTION OF ALUMINIUM NANOPARTICLE BY HIGH ENERGY BALL MILLING. *International Journal of Mechanical Engineering and Technology (IJMET)*, 646-652.

[13] Satish Bodolla, S. T. (2018). A review on Characterization techniques of Nanomaterials. *International Journal of Engineering, Science and Mathematics*, 169-174.

[14] Seong-Hyeon Hong, D.-W. L.-K. (2000). Manufacturing of aluminum ¯ake powder from foil scrap by dry ball milling process. *Journal of Materials Processing Technology*, 105-109. doi:doi:10.1016/s0924-0136(99)00469-0

Acknowledgement

The author is very much thankful to J.T.M. College of Engineering, Faizpur, Research Centre KBC North Maharashtra University, Jalgaon for partial support and facilities.

A Review on the effects of PVD RF sputtering parameters on rare earth oxide thin films and their applications

Pramod Mandal[1], Udai P. Singh[2] and Sudesna Roy[1,*]

[1]School of Mechanical Engineering, KIIT University, Bhubaneswar 751024 India
[2]School of Electronics Engineering, KIIT University, Bhubaneswar 751024 India

* Corresponding Author: sudesna.royfme@kiit.ac.in

Abstract: This article provides a review on effect of processing parameters on the quality and performance of thin films of rare earth materials deposited through PVD RF sputtering technique. The applications of these thin films include optical, electronics, medical films, defence and renewable energy technologies. A systematic review of the literature revealed that the sputtering parameters such as plasma power and substrate temperature have the most impact on the grain size of the deposited thin films. Larger grain size was observed to be at higher sputtering power and higher substrate temperatures, while lower grain size was obtained when the sputtering power is decreased. The optical property, in particular the band energy gap (E_g) was found to improve with the film thickness. Thin films produced at higher doping concentration and subsequently annealed at higher temperature was found to increase the crystallinity of thin films.

Keywords: *Physical vapour deposition (PVD), RF sputtering, Rare earth materials.*

1. Introduction

In recent year, rare earth oxide (REO) materials have been emerges as one of the demanding materials worldwide for various applications because of their unique scintillation, phosphorescent, magnetic and catalytic properties. As per report almost 79% of rare earth in worldwide is produced by China. As recent increases demands of rare earth oxide over worldwide, China becomes the largest supplier of these materials. These oxide elements are a group of seventeen chemical elements of lanthanides in periodic table from atomic number 57 to 71. Their atomic masses subsequently vary from 138.9055 AMU for La to 174.967 AMU for Lu. They have been established as unique materials because of their electronic structure of the materials, where their generalized electronic structure may be represented as (Xe) $4f^n$ $6s^2$, where n varies from 1 to 14. Their unique optical and electronic properties are dictated by their valence electrons in the partially filled 4f orbitals. These include applications in optical, electronics, medical and renewable energy devices because of their excellent magnetic, scintillating, phosphorescent, and catalytic properties [1,2,3].

The applications of these rare earth oxides include computed topography (CT) scanner x-ray imagining system [4], ultrahigh-resolution imaging and light emitting diodes(LEDs) [5,6,7], TFT(thin-film transistor

) devices [8], high-efficiency silicon solar cell [9], Si-based optoelectronics devices [10], spintronics fields [high-speed magnetic filters, sensors, quantum transistors] [11], light emitting diode (LEDs) [12], active devices like non-volatile memory, pyro-electric detectors and microwave devices [13], ramp junctions & field effect devices [14], oxygen sensing instruments [15], dynamic random access memories (DRAMs), spintronics and humidity sensors [16,17], metal-oxide-semiconductor (CMOS) devices [18], switching mechanism for logic devices and memories [19] and high-efficiency silicon solar cells applications [20,21]. Table 1 shows the applications of some rare earth materials.

Table.1 Application based on research area [1]

Research Area	Applications
Opto-electronics and electronics	Television screen, LEDs (light emitting diode), CFLs (compact fluorescent lamps), baggage scanners, Lasers, optical glass, fiber optics, masers, radar detection devices.
Manufacturing	Metal alloys, High strength magnet, stress gauges, polishing powder .
Medical and Bio-medical	X-ray machines, nuclear medicine imaging, X-ray tubes, dental laser, cancer treatment applications, genetic screening tests,
Industrial	Ceramic pigments, colorants in glassware, chemical oxidizing agent, polishing powders,
Renewable energy technologies	Wind turbines, batteries, catalysts and electric cars

In this report, review was done on effects of process parameters and applications of rare earth oxide thin film deposition by Physical vapour deposition (PVD) rf sputtering. PVD rf sputtering is an effective method for depositing high quality thin films of oxides, carbides and nitrides of metals and semiconductors. The process is capable of obtaining high quality thin films with high deposition rates, good adhesion, uniformity in surface morphology, homogeneity, ease of operation and economical for depositing even large areas at room temperature without the need for melting as in thermal spraying. The main disadvantages that accompany this process are target material poisoning and possibility of arcing from the plasma during operation [22,23,24].

A typical flow chart of the deposition process has been explained in Figure 1. Here initial setup such as target material setting and substrate cleaning is conducted, then the chamber is evacuated and argon gas (at a predetermined flow rate) is introduced into the chamber. The plasma is initiated inside the chamber by striking with the power source. As a result energetic ions (argon ions) are formed in the plasma that bombards the target materials (cathode of discharge) atoms. As a result of the plasma bombardment, molecules of the target material are ejected/vaporized from its surface in a steady stream that is deposited on the substrate (anode).

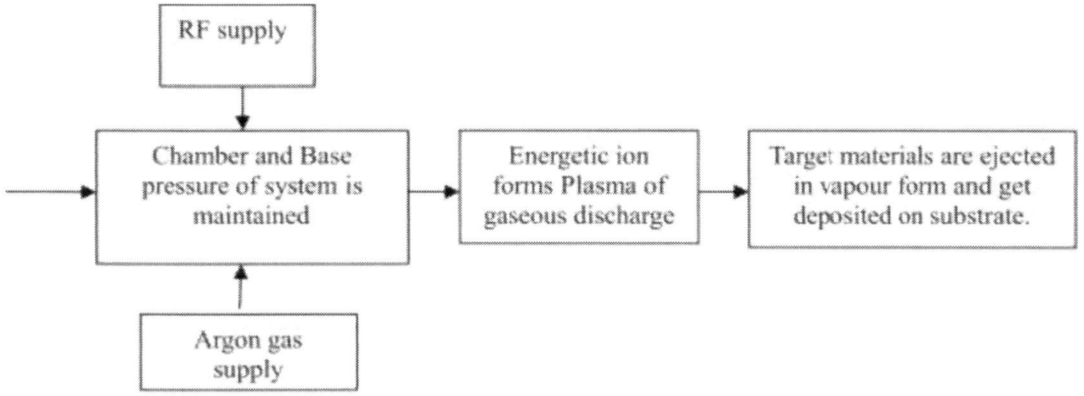

Figure 1. Flowchart of PVD rf sputtering working principal

Other than the PVD RF sputtering method there are other methods used for deposition namely pulsed electron beam, pulsed laser deposition (PLD)[25,26], spray pyrolysis[27], sol gel[28,29], DC sputtering[30], and metal organic vapour phase epitaxial[31]. Morea et al.[25] and Prakash et al.[26] grown rare earth materials by PLD method and thin films was good quality but some irregularities were observed when deposited in large area surface. Similarly Anand et al.[27] have grown using spray pyrolysis and observed that grain size of thin films produced was not interconnected to each other. Zhuang et al.[28] and Zargouni et al[29] grown rare earth by sol gel method and noticed that surface roughness of the thin films was in critical stage than required criteria. Thus compared to the other processes this is capable of growing smooth, uniform surfaces with excellent adhesive properties than the other methods [4-21]. This technique involves less time for fabrication and capable for mass production for large area depositions whenever required. This is advantageous since in electronic and opto-electronics industry high quality thin films are needed with smooth and good surface finish.

There are many review articles on deposition of rare earth materials deposited through PVD rf sputtering method, i.e., Goh et al. [32] reviewed rare earth Samarium oxide (Sm_2O_3), Alam et al.[33] reviewed rare earth materials doped with ferrite garnet-type thin films, Gasgnier et al. [34] reviewed rare earth neodymium and samarium for permanent magnet and superconducting materials. Winick et al. [35] reviewed rare earth erbium and neodymium by rf sputtering method. However they have not correlated their processing parameters to their quality and performance.

RF sputtering deposition process is mainly affected by its parameters such as plasma power, substrate temperature, deposition time, substrate orientation, argon pressure, substrate to target materials distance, substrate rotation, target tilt etc [5]. These parameters can affects coated parts individually or by coupling others parameters also. However doping of materials on base materials and flow of gas has also affected thin surface morphology such as optical and structural properties. This paper therefore correlated through the review of published articles the effects of processing parameters on the quality and performance of thin films of rare earth materials deposited through PVD rf sputtering technique.

IOP Conf. Series: Materials Science and Engineering 1166 (2021) 012022 doi:10.1088/1757-899X/1166/1/012022

2. Review and Discussion

2.1 Sputtering Power

RF sputtering power has vital role in achieving smooth, adhesive, uniform thin films. The deposition rate of the coating largely depends on its sputtering power. Topping et al [4] have grown lutetium oxide on graphite substrate and shown that the maximum working sputtering power for deposition of lutetium oxide could be 100 W. At powers higher than 100 W the oxide target material was shown to get irreparably damaged. With copper backing and water cooled oxide target materials can resist sputtering power up to 100 W. The sputtering power significantly affects the surface morphology of thin films. Thin films deposited at 50 W plasma intensity provided low deposition rate, while at 75W produced balanced deposition rates, but that at 100 W produced high deposition rate. The higher deposition rates produced poor crystallinity due to insufficient time for atomic ordering. Thin films with poor or partial crystallinity, however, improved its crystallinity and its performance after sufficient heat treatment. Application of these thin films include for x-ray imaging system of computed topography (CT) scanner [4].

Nagarkar et al. [5] deposited europium oxide doped lutetium oxide on transparent optical ceramic and revealed that morphology of thin films largely dependent on input sputtering plasma power. High x-ray absorption efficiency of the thin films was capable to provide full absorption of the incident irradiation with minimal thickness that enables its use in ultra high-resolution imaging application [5]. Kaya et al. [18] have similarly grown samarium oxide thin films on silicon substrate and shown that the crystallization of the films improved as the sputtering power increased and moreover, the grain size also increases with the increase in plasma power. The samarium oxide thin films were monoclinic B-type structure and with increased crystallinity their performance in complementary metal-oxide-semiconductor (CMOS) devices improved. Application includes for metal-oxide-semiconductor (CMOS) devices [18].

2.2 Deposition time

The deposition time, like the sputtering power also affects the thin films thickness of films are largely depends on deposition time. However, the deposition time is usually counted after the ignited plasma is stabilised so a buffer time, depending on the instrument and its condition is kept that is not included in the deposition time. Chen et al.[15] have grown cerium oxide thin films on both glass and silicon substrate and suggested that the deposition time has influenced the films smoothness as well its thickness. The lower the deposition time the more the average roughness of the thin films. While at higher deposition time the surface shows a smoother surface and increased adhesion to the substrate. With decreased deposition time and oxygen flow rate the crystal (grain) size decreased which improved its performance in oxygen sensing instruments. Application of these thin films is for oxygen sensing instruments.

2.3 Substrate orientation

The substrate orientation, in case of the single crystal substrate, is very important in deciding the properties of the thin film. The crystallinity and residual stress due to lattice mismatch largely depends on substrate orientation. The substrate orientation should be chosen such that the lattice mismatch between the coating and substrate is minimised. Roy et al.[7] have shown that for europium doped lutetium oxide thin films when grown on different orientations of (100), (110) and (111) single crystal yttria-stabilized zirconia (YSZ) substrates, the surface morphology, roughness and crystallinity changed enormously for the same experimental conditions. Their radioluminescence (scintillation in incident x-rays) also changed greatly with the orientation which make them suitable for application in high resolution x-ray detectors used in computed topography (CT) scanners [7].

2.4 Substrate to Target material Distance

The distance of the substrate to target materials, also referred to as standoff distance, also affects the deposition rate as well as the grain size distribution. Mandal et al. [6] have shown that the microstructure of europium doped lutetium-gallium oxide could be varied greatly by changing the standoff distance from 8 cm to 20 cm. By increasing the standoff distance by more than twice, the thickness, roughness and adhesiveness increased structural propertied drastically. This also improved its optical properties for use in wide band gap semiconductors. The standoff distance is related directly to the ions evaporated by the plasma and hence affects its deposition rate. Application of these films includes for opto-electronics devices such as LEDs.

2.5 Substrate temperature

The substrate temperature plays a vital role in achieving better crystallinity in the thin films. It is well known that the substrate temperature improves the nucleation and growth kinetics of the thin film on the substrate which affects its grain size. With increasing substrate temperature the grain size increases as the crystallinity of the grains increase from amorphous to columnar to equiaxed [36]. Sibaja et al. [13] have shown that for lanthanum oxide deposited on silicon substrates the crystal structure changes from hexagonal at 80°C to FCC above 200°C. Moreover, with change in crystal structure the optical properties also improved for its application in sensory devices like non-volatile memory, pyro-electric detectors and microwave devices. Tsai et al.[16] have shown that cerium oxide deposited on silicon substrates the substrate temperature significantly affected its structural properties. Here the orientation of the thin films changed as the substrate temperature was increased which improved its optical performance in dynamic random access memories (DRAMs). Application includes for DRAMs devices

2.6 Doping of materials

The purity and hence concentration of dopant, affects its crystallinity and grain size. Ren et al.[8] deposited erbium doped tin oxide on quartz substrates and revealed that as the erbium concentration increases from 0 to 4.6%, its electrical performance improves drastically which improved its potential applications in thin-film transistor (TFT) devices. Miura et al. [9] on the other hand have also doped erbium in tantalum oxide and suggested that addition of erbium although improves the uniformity and adhesiveness of the coating, however, significantly affects its photoluminescence (PL) properties. Applications includes for high-refractive-index and light-emitting material of a multilayered photonic crystal and high-efficiency silicon solar cell.

Cerqueira et al. [10] grown erbium doped nano crystal silicon thin films and shown that erbium concentration of 0.03 % gives the characteristic of Er^{3+} emission, while 0.55 % do not give the characteristics emission. This indicated that with increase in the erbium concentration the metallic erbium clusters in the films do not get activated. Hence this affects its performance for Si-based optoelectronics devices. Fang et al. [20] have grown terbium doped zinc oxide on glass substrate and suggested terbium acts as good dopant for formation of transparent and conductive oxide (TCO) ZnO films. The cystallinity of the thin films were found to improve as the dopant concentration was increased up to 4.1% which improved its performance in thin film solar cells.

2.7 Flow of gases (Argon / Oxygen/Nitrogen gas)

The gas flow during the sputtering process also plays a vital role in the deposition process. The gas mixture can vary both the ratio of the oxygen to argon or nitrogen and also the flow rate of the gases for achieving the desired crystallinity of the thin films. Shalaan et al.[11] have grown gadolinium nitride films on glass substrate by keeping Ar/N_2 mixed ratio (5/1) and it was observed that the guttering during pre-sputtering reduces the oxygen ratio in the chamber resulting in formation of amorphous thin films with high band gap energy which makes it suitable for applications in spintronics such as high-speed magnetic filters, sensors and quantum transistors[11]. Deng et al. [12] have shown that when gadolinium-doped yttrium aluminum garnet deposited on boron doped silicon substrate their cathodoluminiscence (CL) properties were substantially improved when the oxygen flow rate was low thus improving its application in light emitting diode (LEDs)[12].

Owens et al. [14] on the other hand have shown that the structure and surface finish of the cerium oxide thin films deposited on YSZ substrate improved with decreasing Ar/O_2 ratio. Moreover the resistivity of the thin films results decreased substantially improving its superconductivity, which can be applied for ramp junctions & field effect devices[14]. Atta et al. [19] have also deposited samarium oxide thin films on quartz substrate observed that the band energy gap of the thin films were found to improve by increasing the Ar/O_2 ratio which improves its optical properties for potential use in application for switching mechanism of logic devices and memories.

3. Discussion

PVD rf sputtering deposition method has produced rare earth oxide thin films with smooth, uniform, good adhesive and defects free surface in short time of periods. Processing parameters of rf sputtering has significantly influences the thin films of rare earth materials deposited. Based on processing parameters such as plasma power, substrate temperature, deposition time, substrate orientation, flow of gases, substrate to target materials distance, and doping of materials the review is carried out. The effects of processing parameters and application of deposited thin films have been discussed here briefly. Throughout it was observed that each parameter has its own effectiveness over surface morphological behaviour of rare earth oxide thin films. Optimal choice of parameters had showed better surface morphological behaviour, better phase, electrical and optical properties. Rare earth materials has capability for wide application for various field such as opto-electronics, electronics device, medical films, industrial field and renewable energy technologies because of their excellent magnetic, scintillating, phosphorescent, and catalytic properties.

4. Conclusion

This report reviews the effects of rf sputtering parameters on surface morphology, structural and optical properties of rare earth oxide materials thin films and their applications. Effects of several parameters such as plasma power, deposition time, substrate orientation, substrate to target material distance, substrate temperature, doping concentration of materials, Argon/Oxygen/Nitrogen gas mixing ratio on rare earth thin films was studied. Through observation it was concluded that, Sputtering power has more influenced surface characteristics of thin films, as sputtering power was increased. Grain size as well as crystallinity of thin films surface showed excellent improvements upon sputtering power increased. 75W sputtering power showed a balanced and controllable than other sputtering power. Deposition time play an vital role for obtaining smooth and defects free surface finish, through study observation was concluded that higher deposition time has showed smooth, adhesive as well as well good surface roughness than lower deposition time. Substrate orientation affects the crystallinity of surface structure hence preferred orientation of substrate must be used during deposition process. Substrate to target materials has also shown impacts on surface morphology, lower distance showed thick and more surface roughness. Substrate temperature plays an important role for obtaining good optical properties. Doping concentration affects the surface morphology of thin films as the doping concentration was increases optical as well as electrical properties showed excellent improvements. Excellent photoluminescence properties of thin films showed excellent in low argon /oxygen mixing ratio.

The applications of rare earth oxide thin films are available for optical, electronics, medical films, defence and renewable energy technologies.

Acknowledgements

The authors sincerely thank DST-SERB, Gov. of India and KIIT Deemed to be University, Bhubaneswar for partly funding the project that used as a basis for this review work.

References

[1] V. Balaram, Rare earth elements: A review of applications, occurrence, exploration, analysis, recycling, and environmental impact *Geoscience Frontiers* **10** (2019) 1285-1303.

[2] Nawshad Haque, Anthony Hughes, Seng Lim, and Chris Vernon Rare Earth Elements: Overview of Mining, Mineralogy, Uses, Sustainability and Environmental Impact *Resources* 2014, **3**, 614-635.

[3] Skirrow R.G.; Huston D.L.; Mernagh T.P.; Thorne J.P.; Dulfer H.; and Senior A.B., Critical Metals for a High-Tech World: Australia's Potential to Supply Global Demand *Geoscience* 2013; pp. 10–13.

[4] Stephen G Topping, C H Park, S K Rangan, and V K Sarin, Lutetium Oxide Coatings by PVD Mater. *Res. Soc. Symp. Proc.*Vol.**1038** (2008) Materials Research Society.

[5] Vivek V. Nagarkar, S. G. Topping, S. R. Miller, B. Singh, C. Brecher, and V. K. Sarin Characterization of vapor-deposited Lu_2O_3:Eu^{3+} scintillator for x-ray imaging applications *Proc. of SPIE*. Vol. **7450** 745003-1.

[6] Pramod Mandal, Udai P Singh, and Sudesna Roy, Co-sputtering of Lu_2O_3, Eu_2O_3 and Ga_2O_3 for optoelectronics applications *IOP Conf. Series: Materials Science and Engineering* **872**(2020), pp. 012062.

[7] Sudesna Roy, S.Topping, and V. Sarin, Growth and Characterization of Lu_2O_3:Eu^{3+} Thin Films on Single-Crystal Yttria-Doped Zirconia *JOM*. Vol. **65**, No. 4, 2013.

[8] Jin-hua Ren, Kai-wen Li, Jie Shen, Chu-ming Sheng , Yu-ting Huang , and Qun Zhang, Effects of rare-earth erbium doping on the electrical performance of tin-oxide thin film transistors *Journal of Alloys and Compounds* **791** (2019) pp. 11-18.

[9] Kenta Miura, Yuki Arai, Kazusa Kano, and Osamu Hanaizumi, Fabrication of Erbium and Ytterbium Co-Doped Tantalum-Oxide Thin Films Using Radio-Frequency Co-Sputtering. *Materials Sciences and Applications*. 2015 **6** pp. 343-347.

[10] M. F. Cerqueira, T. Monteiro, M. J. Soares, A. Kozanecki, P. Alpuim, and E. Alves, Erbium-doped nanocrystalline silicon thin films produced by RF sputtering annealing effect on the Er emission *Phys. Status Solidi*. C **7** No. 3–4 683–687 (2010).

[11] E.Shalaan, H.Schmitt, Structural and optical properties of sputtered gadolinium nitride films, *Optics Communications* **260** (2006) 588–594.

[12] Y. Deng, J.D. Fowlkes, P.D. Rack , and J.M. Fitz-Gerald, Thin film rf magnetron sputtering of gadolinium-doped yttrium aluminum garnet ultraviolet emitting materials, *Optical Materials* **29** (2006) 183–191.

[13] S.B. Brachetti-Sibaja , S.E. Rodil , M.A. Domínguez-Crespo, A.M. Torres-Huerta, E. Rodríguez, and A.B. López-Oyama, Optical properties of nanocrystalline La_2O_3 dielectric films deposited by radio frequency magnetron sputtering *Thin Solid Films* **636** (2017) 615–621.

[14] J.M.Owens, E. J.Tarte, P. Berghuis and R.E.Somek, The Sputter Deposition of Cerium Oxide Thin Films for Superconducting Electronics, *IEEE Transactions on applied superconductivity*, Vol. **5** (1995), 2.

[15] Pao-Chi Chen, and Xiao An Wang, Cerium Oxide Film Growth Using Radio-Frequency Sputtering Process, *Materials Sciences and Applications*, **11**, 305-318.

[16] Wen-Chou Tsai, and Tseung-Yuen Tseng, Structural and electrical properties of cerium dioxide films grown by RF magnetron sputtering, *Journal of Materials Science Materials In Electronics*, **8** (1997) 313- 320.

[17] R.Murugana, G.Vijayaprasatha, P.Sakthivel, T.Mahalingam, and G.Ravi, Structural, Morphological and Optical Properties Of CeO_2 Thin Films Deposited By RF Sputtering, *AIP Conf. Proc.*, **1731**, 080029-1–080029-3.

[18] S. Kaya, E. Yilmaz, H. Karacali, A.O. Cetinkaya, and A. Aktag, Samarium oxide thin films deposited by reactive sputtering: Effects of sputtering power and substrate temperature on microstructure, morphology and electrical properties, *Materials Science in Semiconductor Processing*, **33** (2015) 42–48.

[19] A A Atta, M M EL-Nahass, Khaled M Esaba WY, M M ABD EL-Raheem, A M Hassanien, A Alhuthalli, Ali Badawi and Amar Merazga Optical characteristics of transparent samarium oxide thin films deposited by the radio-frequency sputtering technique *Pramana – J. Phys.* (2016) **87**: 72.

[20] Z.B.Fang,Y.S. Tan, H.X. Gong, C.M. Zhen, Z.W. He, and Y.Y.Wang Transparent conductive Tb-doped ZnO films prepared by rf reactive magnetron sputtering *Materials Letters* **59** (2005) 2611 – 2614.

[21] K. Miura , T. Osawa, Y. Yokota, T. Suzuki, and O. Hanaizumi Fabrication of Tm-doped Ta_2O_5 thin films using a co-sputtering method *Results in Physics* **4** (2014) 148–149.

[22] Moustafa M. Zagho, Hana D. Dawoud, Nasr Bensalah, and T. M. Altahtamouni A brief overview of RF sputtering deposition of boron carbon nitride (BCN) thin films *Emergent Materials* (2019) **2** pp.79–93.

[23] Ashwin Kumar Saikumar, Shraddha Dhanraj Nehate, and Kalpathy B Sundaram Review—RF Sputtered Films of Ga_2O_3 *ECS Journal of Solid State Science and Technology* **8**(7) (2019) Q3064-3078.

[24] J. F. Muth, P. Gollakota, A. Dhawan, H. L. Porter, Y. N. Saripalli, and L.M. Lunardi Gallium Oxide as a Host for Rare Earth Elements *Mater. Res. Soc. Symp. Proc.* Vol. **866**, 2005 Materials Research Society.

[25] R. Morea, J. Fernandez, R. Bald, and J. Gonzalo Pulsed laser deposition of rare-earth doped glasses: a step toward lightwave circuits *Proc. of SPIE* Vol. **9744** 974402.

[26] Saurav Prakash, Siddhartha Ghosh , Abhijeet Patra , Meenakshi Annamalai, Mallikarjuna Rao Motapothula, Soumya Sarkar, Sherman J.R. Tan, Jia Zhunan, Kian Ping Loh, and T. Venkatesan Intrinsic Hydrophilic Nature of Epitaxial Thin-film of Rare-earth Oxide Grown by Pulsed Laser Deposition *Nanoscale*. 2017, **00**, 1-3.

[27] V. Anand, A. Sakthivelu, K. Deva Arun Kumar, S. Valanarasu, V. Ganesh, Mohd Shkir , and S. AlFaify Rare earth Eu^{3+} co-doped AZO thin films prepared by nebulizer spray pyrolysis technique for optoelectronics, *Journal of Sol-Gel Science and Technology* **86**(2), 293–304.

[28] Jiaqing Zhuang, Qi-Jun Sun, Ye Zhou, Su-Ting Han, Li Zhou, Yan Yan, Haiyan Peng,Shishir Venkatesh, Wei Wu, Robert K. Y. Li, and V. A. L. Roy,Solution-Processed Rare-Earth Oxide Thin Films for Alternative Gate Dielectric Application *ACS Appl. Mater. Interfaces* 2016, **8**, 31128−31135.

[29] S. Zargouni, S. El Whibi , E. Tessarolo , M. Rigon, A. Martucci, and H. Ezzaouia Structural properties and defect related luminescence of Yb-doped NiO sol-gel thin films *Super lattices and Microstructures* (2020) *106361*. doi:10.1016/j.spmi.2019.106361.

[30] Putut Marwoto, Sugianto Sugianto, and Edy Wibowo Growth of europium-doped gallium oxide (Ga_2O_3:Eu) thin films deposited by homemade DC magnetron sputtering *Journal of Theoretical and Applied Physics* 2012, 6-17.

[31] M. Peres, E. Nogales,B. Mendez,K. Lorenz, M. R. Correia, T. Monteiro, and N. Ben Sedrine, Eu Activation in β-Ga_2O_3 MOVPE Thin Films by Ion Implantation, *ECS Journal of Solid State Science and Technology* **8** (7) Q3097-3102 (2019).

[32] Kian Heng Goh, A.S.M.A. Haseeb, Yew Hoong Wong, Lanthanide rare earth oxide thin film as an alternative gate oxide, *Materials Science in Semiconductor Processing*, 68 (2017), pp. 302-315.

[33] Mohammad Nur-E-Alam, Mikhail Vasiliev, and Kamal Alameh Optical constants of rare-earth substituted amorphous oxide-mix-based layers deposited to enable synthesis of magneto-optic garnets, Optical Materials Volume **98**,109309(2019).

[34] Gasgnier.M., Rare-earth elements in permanent magnets and superconducting compounds and alloys (except new high T c ceramics) as thin films, thin crystals and thinned bulk materials, *Journal of Materials Science* **26**(8) (1991).

[35] Kim A. Winick, Rare-earth-doped waveguide lasers in glass and $LiNbO_3$: a review, *SPIE*, Vol., **3280**, 0277-786, doi:10.1117/12.305393.

[36] B. A. Movchan, and A. V. Demchishin, Study of structure and properties of nickel, titanium, tungsten, aluminum oxide and zirconium dioxide thick vacuum condensates, *Fiz. Met. Metalloved.* **28**, 653–660 (1969).

Modeling anisotropic fracture in a metal-fiber reinforced composite system

Dhaladhuli Pranavi and Amirtham Rajagopal

Indian Institute of Technology,Hyderabad,India

E-mail: pranavijaan1996@gmail.com

Abstract. Hybrid structures consisting of metal and composites can be applied to specific requirements of different applications. The computational modeling of composites is quite complex compared to homogeneous and isotropic materials like metals because of the heterogeneity introduced due to the presence of different phases such as matrix, fiber and matrix-fiber interface, and anisotropy due to the fiber alignment. The crack propagation in a composite material depends on a combination of various damage modes, namely, fiber pull-out, matrix cracking, delamination. The strength and stiffness of the composite depend on the mechanical and fracture properties of the individual phases, and the fiber inclination. The metal-composite interface is modeled using the cohesive zone approach. A nonlocal diffused approach is proposed to model the anisotropic failure in composites reinforced with unidirectional and woven fibers and the interaction of the crack with the interface. Parametric studies are conducted to understand the role of fiber orientation and interface fracture properties of the system. The proposed model is illustrated through numerical examples to understand various failure mechanisms in a metal-composite system.

1. Introduction

Hybrid structures are multi material structures formed by stacking different materials for a specified purpose. Hybrid metal composite structures consist of alternative layers of metal and fiber reinforced composites(FRC). These structures have the advantages of both metals and composites, namely, good strength, light weight, ductility, fatigue resistance and can be designed according to the need. The motivation for developing these structures is to improve the mechanical properties and produce low weight materials, especially in marine, aerospace industries (see [1] and [2]). The role of the metal is to protect from environmental degradation, provide higher impact resistance and strength to the structure and composite structure can resist the fatigue failure due to its high strength to weight and high stiffness to weight ratios (see [3] and [4]). Studies are conducted to analyze the mechanical properties by conducting experimental tests and to understand the optimisation of these hybrid structures (see [5], [6], [7] and [8]). There exists an interface between the two materials as shown in Fig.1 and the strength and failure mechanism in these structures mostly depend on the interface and its fracture properties. Based on the relative value of fracture toughness of the interface to that of the bulk, interfaces are classified as stiff and soft interfaces [9]. The interfaces are present between the metal and FRC (metal-composite interface) and also among the constituents of the composites (matrix-fiber interface). The properties at the macroscopic level are mostly affected by metal-composite interface.

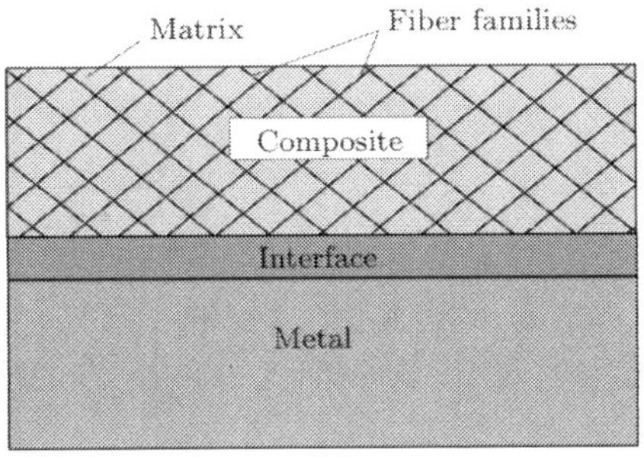

Fig.1: Fiber Metal laminates(FML) with composite reinforced with fiber families.

The failure modes of composite can be broadly classified as (a) Intralaminar (matrix/fiber failuer) (b) Interlaminar(delamination) (c)Translaminar(failure across different lamina). Several numerical methods have been developed to understand the complex fracture phenomena in composites such as continuum damage models [10], XFEM [11], cohesive zone models (see [12] and [13]), peridynamics [14], gradient damage models [15].

Smeared crack models like phase field method(PFM) have been widely applied in solving complex problems involving crack initiation, crack branching and multiple cracking phenomena [16]. It considers the regularization of the crack which is represented by a decaying exponential function. PFM have been adopted to model the brittle fracture (see [17] and [18]), ductile fracture [19], dynamic fracture [20] and cohesive fracture [21]. This approach has been applied to model anisotropic materials such as composites by considering anisotropic surface energy and structural tensors (see [22] and [23]). The intralaminar and translaminar fracture in the composites have been modeled using PFM [24]. It has also been applied to nanocomposites [25], composites with varied stiffness [26] and hyper elastic materials [27]. The interfaces can be modeled using a cohesive zone approach from the works of Dugdale [28] and Barenblatt [29]. The PFM together with the cohesive zone model have been developed to understand the interaction between crack and interface in brittle materials [30], microstructures [31] and to analyze the crack penetration-delflection events at the interface [32].

In the present study, both the crack and interface are regularized according to the phase field approach. Anisotropic crack surface density functional is introduced to model FRC. The metal-composite interface is modeled using an exponential cohesive zone law where the traction components (normal and tangential) depends on the displacement jump in both normal and tangential directions. The displacement jump is evaluated as a difference between the displacements across the interface based on the geometry of the element considered. A structural tensor depending on the orientation of fiber families is introduced to consider the direction dependent fracture. The main aim of the paper is to understand (a) the interaction of the crack with the metal-composite interface (b) the crack propagation in composites reinforced with one and two fiber families (c) the influence of the anisotropy parameters which indicates the extent of anisotropy considered on the mechanical behavior and crack propagation.

2. Methodology

A solid body Ω consisting of metal Ω_1 and fiber reinforced composite Ω_2 having two fiber families oriented at angles θ and $\bar{\theta}$ respectively is considered as shown in Fig.2(a). The solid body consists of sharp crack Γ and the metal-composite interface is represented by Γ_I.

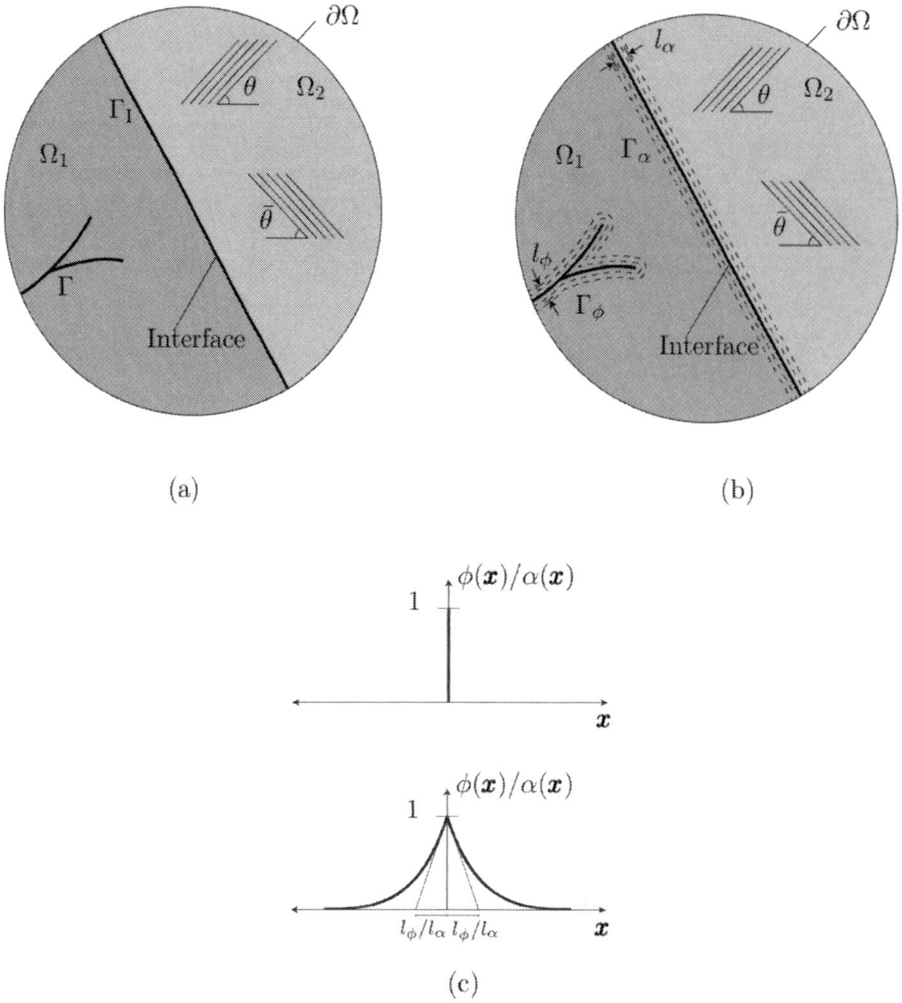

Fig.2: Schematic representation of the solid body Ω consisting of (a) sharp crack and interface, (b)diffused crack and interface and (c) Sharp and diffusive crack/interface topology.

The body forces and the surface traction are neglected for the analysis. The free energy functional can be written as

$$\mathsf{E} = \int_{\Omega} \Psi^e(\boldsymbol{\varepsilon})\, d\Omega + \int_{\Gamma} \mathcal{G}_c d\Gamma + \int_{\Gamma_{\mathrm{I}}} \Psi^{\mathrm{I}}(\llbracket \mathbf{u} \rrbracket, \boldsymbol{\kappa})\, d\Gamma_{\mathrm{I}} \tag{1}$$

The sharp crack and interface are diffused to Γ_ϕ and Γ_α as shown in Fig.2(b). The regularized crack can be written as (refer [33])

$$\Gamma_\phi = \int_{\Omega} \gamma(\phi, \nabla\phi)\, d\Omega, \quad \gamma(\phi, \nabla\phi) = \frac{1}{2l_\phi}\phi^2 + \frac{l_\phi}{2}(\nabla\phi \cdot \boldsymbol{\mathcal{A}}\nabla\phi) \tag{2}$$

The structural tensor $\boldsymbol{\mathcal{A}}$ can be written in terms of fiber orientation tensors $\mathbf{f}$ and $\bar{\mathbf{f}}$, anisotropy parameters β and $\bar{\beta}$ as [33]

$$\boldsymbol{\mathcal{A}} = \mathbf{I} + \beta\, \mathbf{f} \otimes \mathbf{f} + \bar{\beta}\, \bar{\mathbf{f}} \otimes \bar{\mathbf{f}}, \quad \mathbf{f} = [\cos\theta, \sin\theta, 0]^{\mathrm{T}}, \quad \bar{\mathbf{f}} = \left[\cos\bar{\theta}, \sin\bar{\theta}, 0\right]^{\mathrm{T}} \tag{3}$$

Both the crack phase field ϕ and interface phase field α are approximated as a decaying exponential function represented in Fig.2(c). The displacement jump $[\![\mathbf{u}]\!]$ in Eq.(1) is approximated as $\mathbf{d}(\boldsymbol{x})$ which has both normal and tangential components d_n and d_t respectively. The traction vector components corresponding to the interface can be obtained from the coupled exponential cohesive zone law (refer [34]) as

$$\mathbf{t}_n = \frac{\partial \Psi^{\mathrm{I}}}{\partial \mathrm{d}_n}, \quad \mathbf{t}_t = \frac{\partial \Psi^{\mathrm{I}}}{\partial \mathrm{d}_t} \tag{4}$$

3. Numerical problem

Consider a 0.5 mm × 1 mm specimen consisting of aluminium(metal) and fiber reinforced composite. The geometry and boundary conditions are depicted in Fig.3.

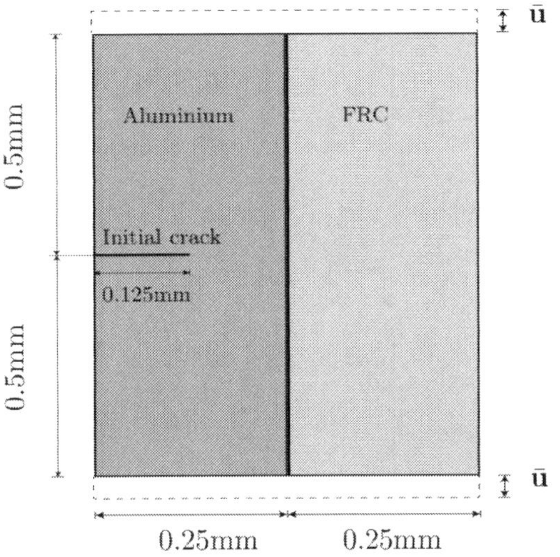

Fig.3: Geometry and boundary conditions of hybrid composite structure

The main aim of the study is to understand the effect of (a) fiber orientation (b) anisotropy parameter and (c) interface properties, on the mechanical response and crack propagation of the hybrid system. For this purpose, three cases are considered namely, CaseA: Unidirection fibers oriented at 30° with $\beta = 20$, CaseB: Unidirection fibers oriented at −30° with $\beta = 20$, CaseC: fiber families oriented at angles 30° and −30° with $\beta = 10$ and $\bar{\beta} = 30$ and CaseD: fiber families oriented at angles 30° and −30° with $\beta = 30$ and $\bar{\beta} = 10$. As a first study, in all the four cases, the interface is considered to be perfect (no displacement jump across the interface) and for the second study, the relative value of the fracture toughness of the interface is taken less which can be considered to be a soft interface. The material properties for the aluminim are taken as $E_m = 70$ GPa, poisson's ratio, $\nu = 0.33$ and fracture toughness $\mathcal{G}_c^m = 9.98$ MPa mm. For the woven fiber reinforced composite, the lamé's constants are taken as $\lambda = 5.2$ GPa, $\mu = 4.04$ GPa, $\mu_f = 64.6$ GPa, critical energy release rate attributed to fibers $\mathcal{G}_c^{\mathrm{aniso}} = 80$ MPa mm, critical energy release rate attributed to matrix $\mathcal{G}_{cI}^{\mathrm{iso}} = 5.5$ MPa mm [33]. The fracture toughness of the interface is taken as $\mathcal{G}_c^{\mathrm{I}} = 0.01\mathcal{G}_c^{\mathrm{aniso}}$ for the soft interface. Length scales are taken as 0.01 mm. Four noded bilinear elements are considered for the analysis.

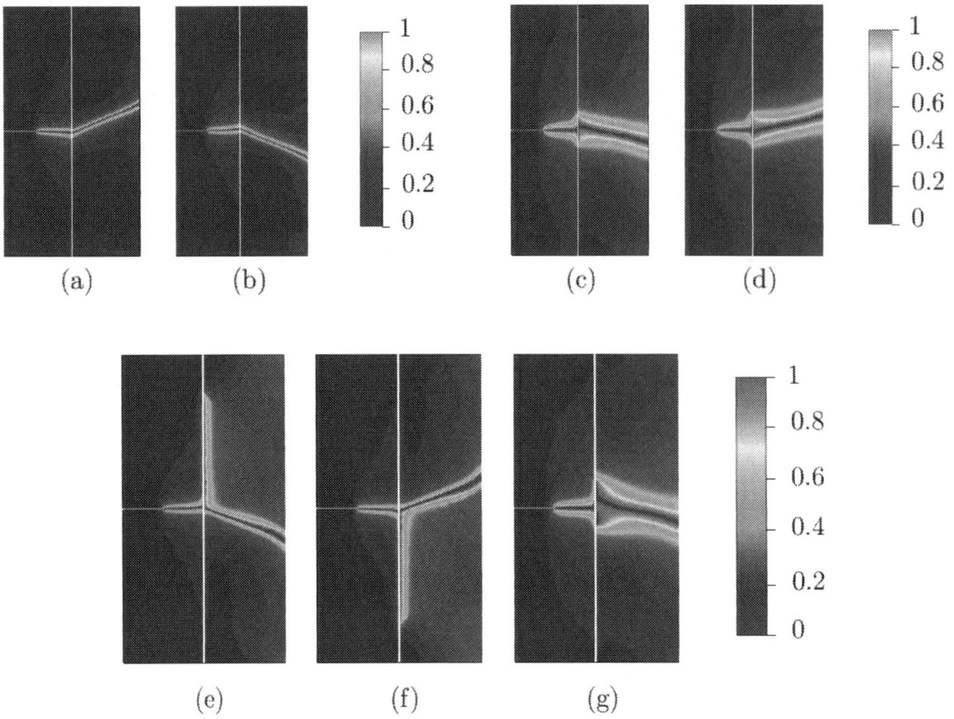

Fig.4: Evolution of the crack phase field for (i) perfect metal-composite interface (a) CaseA, (b) CaseB, (c) CaseC and (d) CaseD (ii) soft metal-composite interface (e) CaseA, (f) CaseB and (g) CaseC.

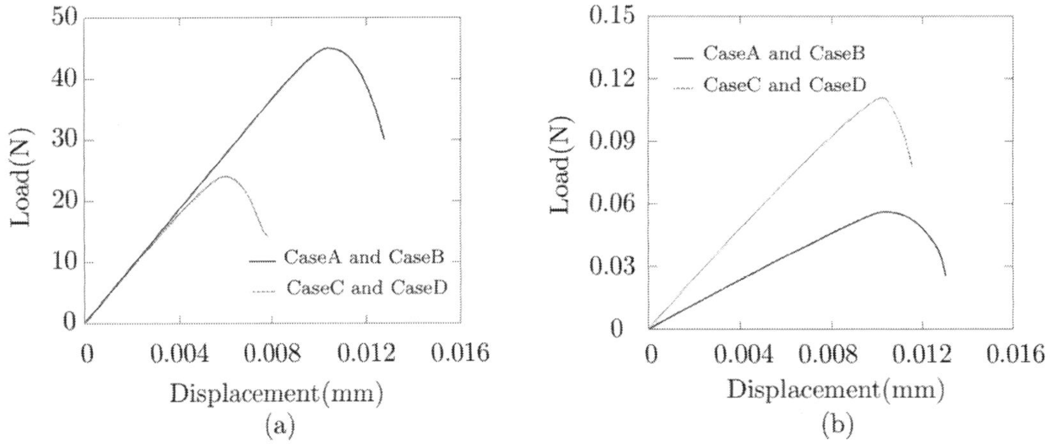

Fig.5: Load displacement curves for CaseA, CaseB, CaseC, CaseD for (a) perfect interface and (b) soft interface.

The crack propagation for unidirectional FRC (CaseA and CaseB) and woven FRC (CaseC and CaseD) for perfect interface and the crack propagation for CaseA, CaseB and CaseC for soft interface is shown in Fig.4. The load displacement curves for all the four cases for perfect interface and soft interface are plotted in Fig.5(a) and Fig.5(b) respectively.

4. Results and conclusions

From Fig.4(a)-(g), it is evident that the crack propagates at 0° in the metal phase and then penetrates into the composite in Fig.4(a)-(d), and deflects along the interface and then penetrates into composite in Fig.4(e)-(g). The crack propagates along the fiber direction in unidirectional FRC ((a) and (b)) while in woven FRC, the crack propagation is along the fiber family which has higher anisotropy parameter $\bar{\beta}$ in (c) and β in (d). The spread of the crack along the interface is less in woven FRC compared to unidirectional FRC ((e)-(g)). From Fig.5, it is clear that for the perfect interface, unidirectional FRC has higher strength and for soft interface, woven FRC has higher strength. It can also be observed that the load displacement plots for CaseA and CaseB, CaseC and CaseD are same. The failure loads for a soft interface are very less when compared to perfect metal-composite interface.

References

[1] Vlot A *Int. J. Impact Eng.* **18** 291
[2] Vlot A and Gunnink J 2001 *Fibre metal laminates* (Kluwer Academic Publishers)
[3] Bourlegat L L, Damato C, da Silva D, Botelho E and Pardini L *J. Reinf. Plast. Compos.* **29** 3392
[4] Lee D, Morillo C, Oller S, Bugeda G and Oñate E *Compos. Struct.* **99** 181
[5] Khalili S, Mittal R and Kalibar S *Mater. Sci. Eng. A* **412** 137
[6] Afaghi-Khatibi A, Lawcock G, Ye L and Mai Y *Comput. Meth. Appl. Mech. Eng.* **185** 173
[7] Nam H, Hwang W and Han K *J. Compos. Mater.* **35** 1654
[8] Peng W, Chen J, Wei J and Tu W *J. Compos. Mater.* **45** 237
[9] Nguyen T, Yvonnet J, Waldmann D and He Q 2019 *Eng. Fract. Mech.* **218** 106574
[10] Reinoso J, Catalanotti G, Blázquez A, Areias P, Camanho P and París F *Int. J. Solids Struct.* **126**
[11] Wang H, Zhou H, Ji H and Zhang X *Mater. Des.* **55** 191
[12] Parmigiani J and Thouless M *J. Mech. Phys. Solids* **54** 266
[13] Nian G, Li Q, Xu Q and Qu S *Compos. Sci. Technol.* **157** 195
[14] Ren H, Zhuang X and Rabczuk T *Comput. Methods Appl. Mech. Eng.* **318** 762
[15] Bazant Z and Pijaudier-Cabot G *J. Appl. Mech.* **55** 287
[16] Francfort G and Marigo J *J. Mech. Phys. Solids* **46** 1319
[17] Kasirajan P, Bhattacharya S, Rajagopal A and Reddy J 2020 *Comput. Methods Appl. Mech. Eng.* **366** 113019
[18] Rajagopal A, Fischer P, Kuhl E and Steinmann P *Comput. Mech.* **46** 471
[19] Ambati M, Gerasimov T and Lorenzis L D *Comput. Mech.* **55** 1017
[20] Schlüter A, Willenbücher A, Kuhn C and Müller R *Comput. Mech.* **54** 1141
[21] Verhoosel C and de Borst R *Int. J. Numer. Methods Eng.* **96** 43
[22] Li B and Maurini C *J. Mech. Phys. Solids* **125** 502
[23] Nguyen T, Réthoré J and Baietto M *J. Mech. Phys. Solids* **65** 279
[24] Quintanas-Corominas A, Reinoso J, Casoni E, Turon A and Mayugo J *Compos. Struct.* **220** 899
[25] Msekh M, Cuong N, Zi G, Areias P and Rabczuk X Z T *Eng. Fract. Mech.* **188** 287
[26] Natarajan S and Annabattula R *Compos. Struct.* **209** 424
[27] Mandal T, Nguyen V and Wu J 2020 *Int. J. Mech. Sci.* **188** 105941
[28] Dugdale D *J. Mech. Phys. Solids* **8** 100
[29] Barenblatt G *J. Appl. Math. Mech.* **23** 434
[30] Hansen-Dörr A, de Borst R, Hennig P and Kästner M *Comput. Methods Appl. Mech. Eng.* **346** 25
[31] Nguyen T, Yvonnet J, Zhu Q, Bornert M and Chateau C *Comput. Methods Appl. Mech. Eng.* **312** 567
[32] Paggi M and Reinoso J *Comput. Methods Appl. Mech. Eng.* **321** 145
[33] Denli F, Gültekin O, Holzapfel G and Dal H *Comput. Mech.* **65** 1149
[34] den Bosch M V, Schreurs P and Geers M *Eng. Fract. Mech.* **73** 1220

Kernel-based Attribute-aware Self adaptation and Multi thresholding for Rating Prediction

G Lavanya[1], T Sangeetha[2*], P Alaguvathana[3], G Prasanna[4], S Praveenkumar[5] and S Y Prabhu raam[3]

Sri Krishna College of Technology, Coimbatore, India
t.sangeetha@skct.edu.in

Abstract. In recommender systems, our main task is to predict the rating of a new product from the authorized user and then return the best rating for the particular item and this technique reduced the existing prediction error rate. Our proposed system is user-item rating matrix prediction based on Synergetic filtering techniques and this technique is more efficiency compared to other technique. Proposed System are providing personalized recommendation for help the users accord with information overburden problem .However, the techniques are the data insufficiency of the user-item rating matrix underlying for brand-new items and users are severely affected by Synergetic filtering technique. Since the character of common links and items between more accessible by the users in the Internet and this paper exploits the common links of users and the character of items to overcome the existing problems and to ease the rating insufficient effect. However they may need excessive computational moment, and they often accost the insufficient problem which negatively modify the ability of the system. Specifically, we initially propose a Kernel-based Attribute-aware Self adaptation and multi thresholding model to blend the character information of items into matrix factorization and then introduce self-adaptation and multi thresholding. KASM can find the indefinite interactions among characters, users, and items, which reduce the rating insufficient effect for brand-new items by nature. In this paper we suggest a quick recommendation algorithm based on self-adaptation and multi thresholding. Self-adaptation in its genuine meaning is a state-of-the-art method to alter the setting of control specification. It is called self-adaptive because the algorithm controls the setting of these specification itself sink them into a distinctive genome and emerging them. It is construct to deal with the specified drawbacks and enhance the prediction quality. Extended analysis on two real world data sets establish that our proposed method can attain necessarily improve performance than other state-of the-art-methods. In this method we get the accuracy rate of predicting user rating will be 95%

Index Terms—Rating prediction, matrix factorization, attribute aware, self-adaptation, multi thresholding.

1. Introduction

In present different applications including e-commerce websites, entry websites, social media and GPS based social networks have been entirely developed in the Internet, which considerably changes the method we live and think. Common business requirement and user attitude create overwhelming data, which contains precious information. However, the development of a big amount of data leads to information weigh down, the difficulty for a user to search a favorite items in an e-commerce website. To beat this problem, recommender systems which provide custom-fit items for users by sorting useful information in the data, have been suggested and widely studied in college. The aim of recommender systems is to as in regularly recognize user first choice within group of data, then use those option to make guidance that help with decisions. However, recommender systems be affected from data inadequate problem, which is specifically widespread in newly launched systems that have not still had adequate time to a mass required data. As a result, cross-domain recommender systems transmission know-how from a origin domain with comparably rich data to help recommendations in the destination domain.

In our proposed systems to anticipate the rating of a user is essential task to a brand-new item that the user has not been rating. To forward this task, synergetic filtering models utilize the synergetic power of the ratings provided by multiple users to make recommendation. Matrix factorization is one of the standard Synergetic filtering methods, which anticipate ratings over a user-item rating matrix. The major problem in matrix factorization is data inadequate that means the number of available ratings is relatively small and the rating matrix is extremely sparse. Therefore, it is not simple for matrix factorization to achieve adequate performance by simply using numeric rating data. When the predictions of multiple PMF models are precisely joined with the predictions of Restricted Boltzmann Machines models, we attain an error rate of 0.88 that is nearly 6.9% better than the result of Netflix's own system. The regression model is suggested by the kernel trick to definite model turn into a indefinite model and kernel trick act as a support vector machine[3]. In our analysis, we display that by combine geological neighborhood authority, lower prediction fault is achieved than the state-of-the-art models containing Biased MF, SVD++, and Social MF.

2. Related work

This related work reviews shortly the advances on predicting rating in our proposed systems. The proposed systems finding techniques implement to the ratings prediction of items on user or to make materialized suggestion services for the users. The techniques for predicting rating can be classified into three categories: Framework-aware technique, Synergetic filtering technique, Contented based technique. The Framework-aware techniques with the SF including context, context information and framework changed to attach event for rating [5]. However, they three major limitations endure generally including scope up to inefficiency of high measure of two-attribute interaction, attribute interaction and definite interaction between the attributes. The designs display rating prediction be a better performance for social networks based on location, but to predict the ratings is to be a place. Among these categories, Synergetic filtering technique is the famous approach for rating prediction is used, because it need only a rating matrix user-item as the input raw and the indefinite or definite ratings are available extensively in our proposed systems. Here to estimate the main task of correspondence of each pairs of items or user using the rating matrix user-item with correspondence of some determination such as Pearson correlation [3]. The hidden components of MF models on data large be through gained distributed multiplicative update or SGD. This is simple techniques based on the associate information. The user of a rating of an item in terms of anticipated the correspondence in between the item contents the user profile. This techniques are applied generally in suggestions of social media, in which the attributes can be product with absorbed from website pages, user rating reviews and item content.

3. Kernel-based attribute-aware self-adaptation and multi thresholding

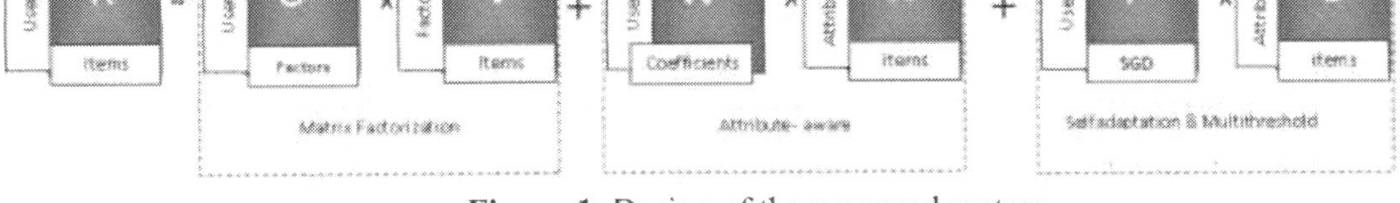

Figure 1. Design of the proposed system

The Figure 1 is briefly explained in the following section. In this section, we define the problem in Section 3.1, data collection and pre-processing in Section 3.2 and the matrix factorization in Section

3.3, Explain the attribute-aware, in Sections 3.4, we introduce the self-adaptation and in Section 3.5 describes the multi thresholding.

3.1 Problem statement

The primary task in a proposed System is rating of a user to predict a authorized rating of item and then return the top rating prediction of a new item to the user. In big data era, classical Matrix Factorization using only ratings suffer a serious drawback for not being able to exploit other accessible information such as the attributes of users/items/ratings [1]. Due to the data inadequate problem the user rating prediction efficiency would be decrease. To overcome this problem we introduce a self-adaptation and multi thresholding technique.

3.2 Data collection and pre-processing

The most initial component for any system is the input. Here the input is the data that is utilized in the user review details such as username, ratings, user-id, movie-id, genre. Then the data must be processed to table content discarding any false data. Dataset might contain noisy and missing values. Data Pre-processing techniques such as single imputation methods are used to change the missing and noisy data into a clean dataset. Single imputation methods are implemented using maximum occurrence attribute value substitution. This systems have become an integral part of e-commerce sites and other businesses like social networking, movie/music rendering sites. They have a big changes on the income earned by these businesses and also benefit users by decreasing the cognitive load of finding and sifting through an heavy load of data set collected from Amazon dataset.

3.3 Matrix factorization

Matrix factorization is one of most frequently used Synergetic filtering for producing recommendations to users. A main constraint of SF is that it entirely depends on noticed ratings and may abort if these noticed ratings are in definite amount called inadequate problem. Identifying this issues, cross-domain suggestion came into survival where transfer gaining mechanism is applied to ease inadequate problem and increase performance of the objective domain using other related origin domains. Matrix factorization (MF) has shown to be a superior ML idea for many problems such as dimensionality reduction, classification, latent topic modeling and manifold learning. In general, MF is a linear modeling method, so similar strategies, many of them based on kernel methods, have been recommended to extend it to non-linear modeling. Firstly, we lay on traditional matrix factorization (MF) method in origin domain to learn hidden factors of users and items through target function of MF[1]. After that, learned hidden factors of users are straightly transferred to destination domain. Modify target function of MF and learn users/items hidden factors of the objective domain. Finally, prediction on unobserved ratings in the objective domain is made using interior items of respective user and item hidden factors. Analysis results explained that our suggested method substantially work good from other without and with transmitting learning methods in terms of MAE and RMSE metrics.

3.4 Attribute awareness

Matrix factorization depend on the method of rating matrix user item with an amount of unknown user from undefined rating and generally data endure from the inadequate problem[2].A few users to check in restaurants while others choose to for different dishes for fun on the geological location to visit tourist attractions, indoors persons like visiting items in residential areas while choose travelling outdoors persons around to explore the world for new items. Therefore, the attribute-aware items on user complement the ratings of insufficient and more benefit bring for predicting rating[4]. In this technique, we using the attributes of products mainly focus because the attributes of users are more extensively, but our technique exploit the users of attribute in the identical way.

3.5 Self-adaptation

Self-adaptation in its natural meaning is a state-of-the-art method to adapt the setting of control framework. Self-adaptation is called self-adaptive because the algorithm controls the setting of these framework itself sink them into an exclusive genome and emerging them [5]. Self-adaptation technique analysis the user rating and predict the rating by SGD (Stochastic Gradient Descent) algorithm. Stochastic Gradient Descent algorithm is based on user id and movie id. Final output will be predicted by highest Stochastic Gradient Descent value based on the user input. In Self adaptation input will be acclimatization individually will help the result should be more fast and accuracy.

3.6 Multi thresholding

Multi-thresholding is a technique to predict the user rating with the help of training samples. This will helps us to reduce the execution time for predicting the user rating. This threshold is persistent by diminishing intra-class intensity variance, or equivalently, by expanding inter-class variance. Otsu's technique is a 1-Dimensional individual companion of Fisher's Discriminant Analysis, is associated to Jenks optimize technique, and is similar to a globally optimize k-means accomplished on the intensity histogram.

Algorithm 1 Incremental learning process for KASM

Input: Inadequate rating matrix R and characteristic matrix X

Output:

u_i - user
v_j- item
b_i- bias for user u_i
α_{ij}- Explicit coefficient associated user u_i and item v_j
n_{ui}- Number of training samples of each user u_i

1: The initializing phase
2: Set Dynamic arguments $\{u_i,v_j,b_i\}$ from (0,1) for each user u_i and item v_j
3: Initialize $\alpha_{ij} \leftarrow 0$ for each item v_j and user u_i
4: Initialize $n_i \leftarrow 0$ for the count of samples of user u_i
5: For each iteration repeat the following steps until convergence do
6: For each randomly selected known rating $r_{ij} \in R$ perform the following steps
7: SGD(u_i,v_j,r_{ij}) in Algorithm 2
8: Terminate the process if arrived at convergence
9: end for

Algorithm 2 Stochastic Gradient Descent: SGD(u_i,v_j,r_{ij})

Input: User u_i, item v_j, and rating r_{ij}
Output:

u_i - user
v_j- item
b_i- bias for user u_i
α_{ij}- Explicit coefficient associated user u_i and item v_j
n_{ui}- Number of training samples of user u_i

1: Predict rating $\hat{r}_{ij}$ based on Equation
$\hat{r}_{ij} = u^T_i v_j + \eta_3(1-\lambda_3\eta_3)^n \sum\_(J` = 1)^N \alpha_{ij}' \phi T(x_j')\phi(x_j) + b_i = u^T_i v_j + \eta_3(1-\lambda_3\eta_3)^n$
2: Compute error e_{ij} based on $e_{ij} = r_{ij} - \hat{r}_{ij}$
3: Update parameter u_i based on the equation
$u_i \leftarrow u_i + \eta_1(e_{ij} \cdot v_j - \lambda 1 \cdot u_i)$
4: Update parameter v_j based on the equation
$v_j \leftarrow v_j + \eta_2(e_{ij} \cdot u_i - \lambda_2 \cdot v_j)$
5: Update parameter bi based on equation
$b_i \leftarrow b_i + \eta_3 \cdot e_{ij}$

6: $n_{ui} \leftarrow n_{ui} + 1$
7: $\alpha_{ij} \leftarrow \alpha_{ij} + (1-\lambda_3\eta_3)-n_{ui}e_{ij}$

4. Simulation Results

The proposed algorithm is implemented by t a considering user data set as an input that contains some static and dynamic attributes. The static attributes are used as the base for calculation whereas the dynamic attributes changes with the training of the model in the incremental learning phase. The dynamic attributes plays a key role in this algorithm because they are used to induce the self-adaptation technique. When the dynamic values are changed by the self-adaptation methodology the Multi-threshold are varied, which is used to predict the values with high accuracy. Using Multiple thresholds in the experiment makes the system prone to more error rate as there are multiple attributes in place that is in need of tuning in the incremental learning phase. In the incremental phase the SGD algorithm performs the prediction and updates the value with attribute values in accordance with the self-adaptation technique. The SGD algorithm is comparatively faster than a normal Gradient Descent algorithm as it introduces a certain degree of randomness when it comes to picking up the points in the progress of the algorithm. After the prediction in the first iteration the error rate is computed, the dynamic parameters are updated for the next iteration to be performed. At the end, the algorithm generates the required user-item interaction matrix from which the output for recommendation is derived.

The data set shows the processed data based on the SGD value. The rating prediction error for our proposed system is 0.93

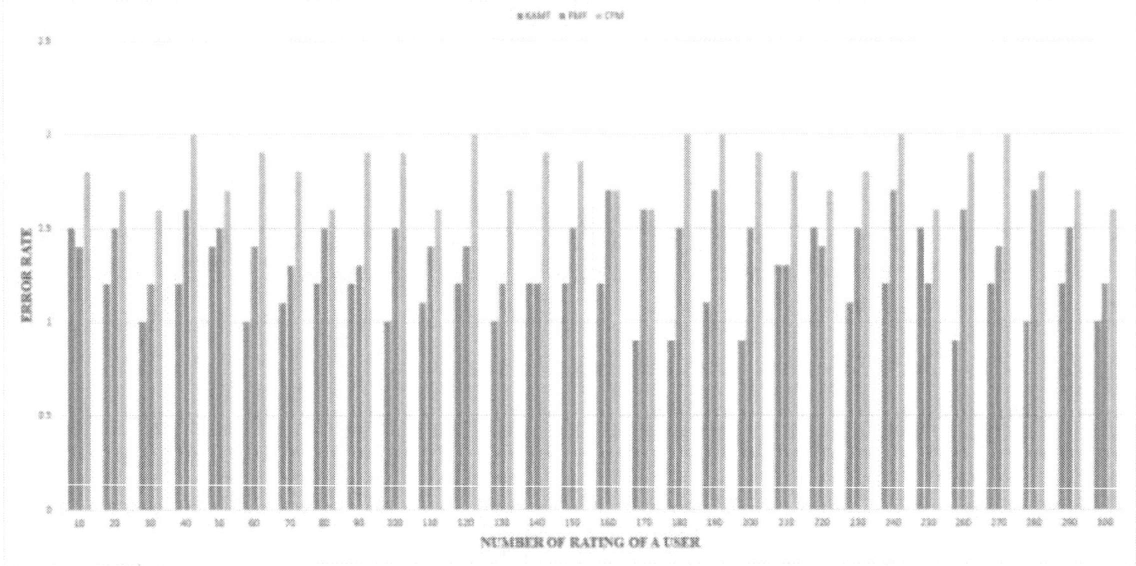

Figure 2. Existing System rating prediction error rate

Figure 2 shows that the prediction error rate using the existing methods, it is inferred from this graph that the prediction error rate of all the existing methods vary distinctively and the error rate is fluctuating to some extent starting from a minimal value of 0.8. It cannot be assured that the system always delivers an optimal performance. It can be seen that the KAMF is the most efficient among the already existing system.

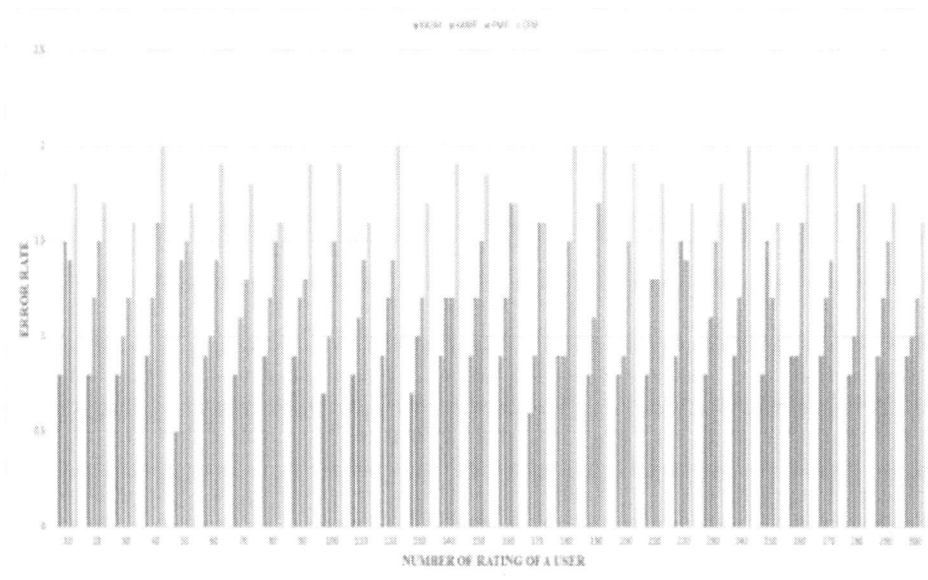

Figure 3. Proposed System of rating prediction error rate

Figure 3 also shows the comparison of the prediction error rate of different methods in addition with the proposed technique. It is evident from this graph that the proposed KASM technique is having the most minimalistic error rate of all the available methods.

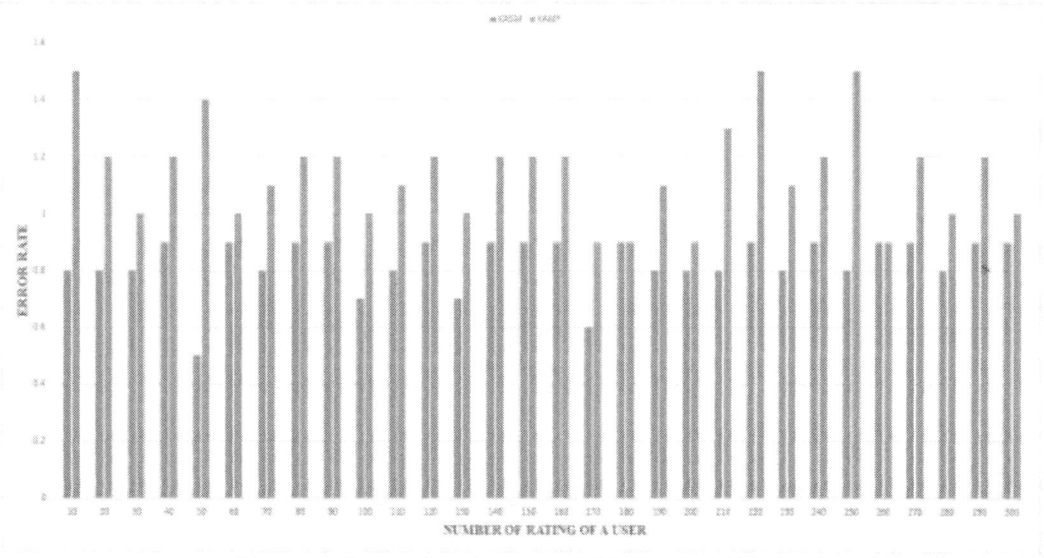

Figure 4. Comparison between KAMF vs KASM

Figure 4 provides a detailed comparison between the two methods KAMF (which is considered as the most efficient among the already existing methods) and KASM (the proposed method). The prediction error rates is considered as the metrics to measure the efficiency of the system. From this comparison it evident that the error rate of the proposed technique (KASM) is lesser than that of the already existing KAMF. The error is more stable than other techniques. It is even noted that the maximum error rate of KASM is lesser than that of the minimal error rate produced by the KAMF technique under the same circumstances.

This above graph shows the error rate of each technique based on the number of rating of a user. In Fig 2. show the error rate of existing system. In existing system KAMF produced minimum error rate.

In Fig 3. show the error rate of our proposed system KASM with the existing system. This will clearly show the error rate reduced in KASM and also produce a accurate rating for an new item. In Fig.4 show the comparison of existing system vs proposed system. In this graph error rate produced in KAMF minimum is 0.933 and our Proposed System of KASM produced error rate minimum is 0.5 and maximum is 0.9.

5. CONCLUSION

In this paper, Kernel-Based Attribute-aware Self-adaptation and Multi thresholding is a proposed model (KASM) to implement the items of attribute details into matrix factorization and self-adaptation. Kernel-based attribute-aware self-adaptation and multi thresholding to uncover the nonlinear interactions exploits the attributes of items among items, attribute, user, which the data sparsity effect of the rating matrix user rating on prediction rating. We used incremental algorithm to the model parameters estimate of KASM. KASM can find the indefinite interactions among characters, users, and items, which reduce the rating insufficient effect for brand-new items by nature. The major problem in matrix factorization is data inadequate was overcome using this self-adaptation and multi thresholding. In this technique we reduce the prediction error rate compared with matrix factorization. Our maximum error rate is 0.9% but existing system minimum error rate would be 1.0%. In this this technique we have reduced the data sparsity problem while predicting the user rating.

REFERENCE

[1] Y. Koren, "*Factorization meets the neighborhood: a multifaceted collaborative filtering model*," in ACM KDD, 2008.
[2] Y. Koren, R. Bell, and C. Volinsky, "*Matrix factorization techniques for recommender systems,*" Computer, **vol. 42**, no. 8, pp. 30–37, 2009.
[3] C. Cheng, F. Xia, T. Zhang, I. King, and M. R. Lyu, "*Gradient boosting factorization machines,*" in ACM RecSys, 2014.
[4] T. V. Nguyen, A. Karatzoglou, and L. Baltrunas, "*Gaussian process factorization machines for context- aware recommendations,*" in ACM SIGIR, 2014.
[5] S. Rendle, Z. Gantner, C. Freudenthaler, and L. Schmidt-Thieme, "*Fast context-aware recommendations with factorization machines,*" in ACM SIGIR, 2011.
[6] Ma, X., Lu, H., Gan, Z., & Zhao, Q. "*An exploration of improving prediction accuracy by constructing a multi-type clustering based recommendation framework*," in Neuro Computing 191,388-397(2016).
[7] Ait Hammou, B., & Ait Lahcen, "*A.A fast recommendation approach with improved prediction accuracy*" in Expert System with application, 87,90-97. (2017).
[8] Zuo, Y., Gong, M., Zeng, J., Ma, L., & Jiao, L. "Personalized recommendation based on evolutionary multi-objective optimization," in *IEE Computational Intelligence Magazine*, **10**, 52-62. (2015).
[9] Ortega, F., Hernando, A., Bobadilla, J., & Kang, J. H. "*Recommending items to group of users using matrix factorization based Synergetic filtering,*" in Information Sciences, **345**, 313-324.(2016).
[10] Zhang, J.-D., Chow, C.-Y., & Xu, J. "*Enabling kernel-based attribute aware matrix factorization for rating prediction,*" in *IEEE Transaction on Knowledge and data engineering*, **29**, 798-812.(2017).
[11] Ar, Y., & Bostanci, E. "*A genetic algorithm solution to the Synergetic filtering problem,*" in Expert System with application, **61**,120-128. (2016).
[12] Guo, G., Zhang, J., & Yorke-Smith, N." *Synergetic filtering with both the explicit and implicit influence of user trust and of item ratings*," in AAAI (2015).
[13] Sivaram Kumar V., Thansekhar M.R., Saravanan R., Miruna Joe Amali S." *Solving multi-objective vehicle routing problem with time windows by FAGA*" Procedia Engineering (2014)
[14] Jameer Basha A., Palanisamy V., Purusothaman T. "*Efficient multimodal biometric authentication using fast fingerprint verification and enhanced iris features*" Journal of Computer Science (2011) volume **7** issue 5 page no 698-706.

Preparation and structural, dielectric characterization of $CaBSi_2O_6$ (B = Mg, Co, Ni, Mn, Zn) ceramics

Tony Joseph[1,3], **K Anlin Lazar** [1,2], **D Divya**[1]

[1]*Government Victoria College (Affiliated to University of Calicut),Palakkad 678001, Kerala, India.*
[2]*Mercy College(Affiliated to University of Calicut), Palakkad 678006, Kerala, India.*

[3]Corresponding author: Email address: *tonyjrf@gmail.com*, Telephone: +91 9961482548.

Abstract. Silicates have received much attention because of their low relative permittivity and low dielectric loss properties for their applications in microwave and millimetre frequency region. This paper reports the preparation and structural, dielectric characterization of $CaBSi_2O_6$ ceramics. The ceramics are prepared by the conventional solid state ceramic route. X-ray diffraction studies were conducted to identify the phase formation of $CaBSi_2O_6$ ceramics. A scanning electron microscope was employed to observe the microstructure of the polished surfaces of the ceramics. The dielectric properties of the ceramics were measured at 1 MHz as well as at 8-12 GHz. The thermal expansion coefficient of the ceramics is also studied.

Keywords: Silicates, dielectric properties, SEM, XRD, solid state ceramic route

1. Introduction

Microwave dielectric ceramics have been widely used in various wireless communication systems such as resonators, filters, antennae, and substrates [1]. These applications particularly depend on the relative permittivity (ε_r) of the microwave dielectric ceramics. For example, materials having ε_r greater than 80 can aid miniaturization of mobile phone handsets, while those having ε_r in the range 25 to 50 and high-quality factor values ($Q_u \times f$) are suitable for increasing signal/noise ratio for the main application in mobile phone base stations. Dielectric ceramics with ε_r less than 20 and high $Q_u \times f$ (greater than 10^5 GHz) are desirable for high frequency applications like microwave wave communication. For such applications, low relative permittivity (ε_r) materials are essential to increase the signal propagation velocity, to reduce the transmission attenuation and also reduce the cross-coupling effect with a conductor. The resonant frequency decreases with increase in ε_r as well as with the increase in dimensions of the resonant structure. Hence for microwave communication applications ε_r should be in the range 6-20. This helps to achieve a resonator of a reasonable size for ease of fabrication as well as handling. Low dielectric loss (tan δ) is another critical requirement of materials useful for microwave communication systems. Usually, in the high frequency region, the dielectric losses increase. Thus, the development of microwave dielectric with very high-quality factor ($Q = 1/\tan \delta$) is necessary. Quality factor is a system property related to the efficiency of use of power supplied to the device. A high value of quality factor can also ensure excellent frequency selectivity. A near-zero temperature coefficient of resonant frequency (τ_f) is also important because it can offer the stability of the frequency against temperature. Usually, a value of τ_f in the range -10 to +10 ppm/°C is preferred to account for the subtle differences in the circuit design, construction and packaging of the similar components produced by different manufacturers. A low value of coefficient of thermal expansion is also desired. These stringent requirements limit the number of dielectric materials available for microwave applications.

In the case of low relative permittivity materials, silicates have received much attention because of the covalent nature of Si-O bond in the SiO_4 tetrahedron. In 1980 Linus Pauling reported that 50% of the Si-O bond in SiO_4 tetrahedron is of covalent nature. The report was based on the electronegativity difference between Si and O [2]. The covalent nature of Si-O bond ensures low values of relative permittivity as well as dielectric loss. Microwave dielectric properties of many of the silicates are reported earlier [1]. Tsunooka *et al.* reported the effects of TiO_2 on the sinterability and dielectric properties of high-Q forsterite ceramics [3]. High $Q_u \times f \sim 230000$ GHz and low $\varepsilon_r \sim 7$ were obtained for a composition containing forsterite and 1 wt% of TiO_2. The composition containing 24 wt% TiO_2 has $Q_u \times f = 82000$ GHz, $\varepsilon_r = 11$ and $\tau_f = 0$ ppm/°C [4]. The effect of calcination and sintering durations and the effect of Mn^{2+} substitution at Mg-site on the microwave dielectric properties of forsterite have also been reported [5,6]. The dielectric properties of several other silicates such as $MgSiO_3$, $CaSiO_3$, $CaMgSi_2O_6$, $CaAl_2Si_2O_8$, $NaAlSi_3O_8$, $Mg_2Al_4Si_5O_{18}$, $Ca_3(Zr_{1-x}Sn_x)Si_2O_9$, $Ba_2(Mg_{1-x}Zn_x)Si_2O_7$, $Sr_xBa_{1-x}Al_2Si_2O_8$, $Ca_xBa_{1-x}Al_2Si_2O_8$, $Sm_2Si_2O_7$, Li_2MgSiO_4 etc. were also investigated [7-15]

$CaBSi_2O_6$ belongs to a class of minerals called pyroxenes, having the general formula $ABSi_2O_6$. The pyroxene group is composed of minerals of the general formula $ABSi_2O_6$, where A and B are two divalent cations or a monovalent and a trivalent cation. If A and B are the same small-radius divalent cation like Fe or Mg the resultant structure is commonly orthorhombic, giving us the orthopyroxene group. At high temperature this structure inverts to C2/c high clinopyroxene which quenches to $P2_1/c$ low clinopyroxene. This gives the primitive clinopyroxene group. At the pure Mg composition, another orthorhombic structure with half the a-repeat of orthopyroxene is possible, which is called protoenstatitite. If A is a larger-radius cation like Ca or Na, the structure is monoclinic C2/c. If the B-site is too large, the resultant structures are triclinic giving the pyroxenoid group. The silicates, which are dealt with in this study, belong to the clinopyroxene group, the most important one being diopside or $CaMgSi_2O_6$.

There are no reports on the microwave dielectric properties of the clinopyroxenes, except for $CaMgSi_2O_6$. Sun *et al.* studied the dielectric properties of $(Ca_{1-x}Mg_x)SiO_3$ for values of x in the range 0.1 to 0.9 [9]. The sintering temperature of $CaMgSi_2O_6$-$CaTiO_3$ has been reduced from 1300 °C to 880 °C by using the additive Li_2CO_3-V_2O_5 [16]. Synthesis and microwave dielectric properties of $(Ca_{1-x}Mg_x)SiO_3$ ceramics prepared by the sol-gel process are also reported [17]. Numerous reports of glass-ceramics based on CaO-MgO-SiO_2 are also available [18-20]. In this work, we aim to study the microwave dielectric properties of $CaBSi_2O_6$ (B = Mg, Co, Ni, Mn, Zn) silicates prepared by the solid state ceramic route.

2. Experimental Details

The $CaBSi_2O_6$ (B = Mg, Co, Ni, Mn, Zn) ceramics were prepared by the conventional solid-state ceramic route. The chemicals used were $CaCO_3$, Co_3O_4, NiO, $MnCO_3$, ZnO, SiO_2, and $(MgCO_3)_4Mg(OH)_2.5H_2O$ (Aldrich Chemical Company, 99.9%). The raw materials were stoichiometrically weighed and were ball milled in polyethylene bottle using Yttria stabilized Zirconia balls in distilled water for 24 h. The slurry was dried at 100 °C in hot air oven and then calcined in the temperature range 1000-1150 °C/4h. The sintering was performed in the temperature range 1100-1325 °C/4h. In practice, the preparation conditions differ depending upon the composition, reactivity, grain growth etc. Hence it is necessary to optimize the processing conditions in order to get the final sintered compacts with desired properties.

The crystal structure and phase purity of the powdered samples were studied by X-ray diffraction technique and SEM micrographs recorded from the surface of the thermally etched sample using scanning electron microscope. Low frequency dielectric properties were measured using an LCR meter and the microwave dielectric properties (in the frequency range 7-12 GHz) using a scalar network analyzer (Aeroflex, IFR 6823 model). The relative permittivity and unloaded quality factor (Q_u) of the samples were measured by Hakki and Coleman and resonant cavity method respectively. The values of τ_f were measured in the temperature range 25-70 °C using the resonant cavity set up. The measured relative permittivity values were corrected for porosity using the equation derived by Penn *et al.*[21]. The coefficients of thermal expansion of the samples were measured using a dilatometer (NETZSCH DIL 402 PC) in the temperature range 30-500 °C.

3. Results and Discussions

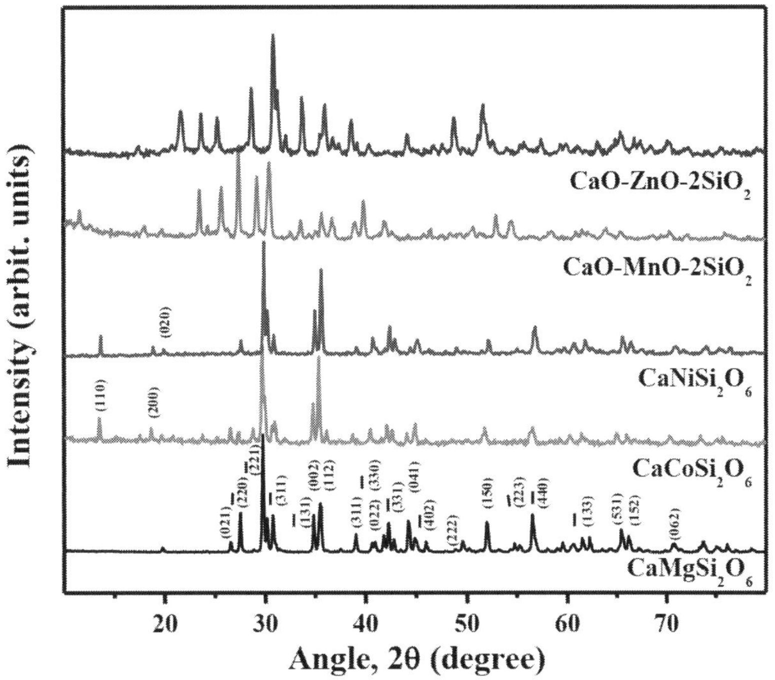

Figure 1. Powder X-ray diffraction patterns of $CaBSi_2O_6$ (B = Mg, Co, Ni, Mn, Zn) ceramics sintered at their respective sintering temperatures.

The powder X-ray diffraction patterns of $CaBSi_2O_6$ (B = Mg, Co, Ni, Mn, Zn) ceramics are shown in figure 1. The powder diffraction pattern of $CaMgSi_2O_6$ is indexed based on ICDD file card no. 19-0239. All recognizable reflection peaks could be readily indexed to the C2/c monoclinic structure with lattice parameters of a = 9.748 Å, b = 8.926 Å, c = 5.250 Å and β = 105.8°, which is consistent with the standard data file. The diffraction patterns of $CaCoSi_2O_6$ and $CaNiSi_2O_6$ can be matched with the ICDD file card no. 84-1288 and 84-1287 respectively. They also have the same structure and space group as that of $CaMgSi_2O_6$. The lattice parameters of $CaCoSi_2O_6$ are a = 9.806 Å, b = 8.950 Å, c = 5.243 Å and β = 105.45°, whereas that of $CaNiSi_2O_6$ are a = 9.734 Å, b = 8.891 Å, c = 5.228 Å and β = 105.87° according to the standard data files. Reflections from the planes (110), (200), (020) etc. are exhibited by $CaCoSi_2O_6$ and $CaNiSi_2O_6$ whereas they are absent for $CaMgSi_2O_6$. Variations in the intensity of certain reflections are also observed depending on the element at the B-site. Figure 1 also depicts the observed XRD patterns for B=Mn and B=Zn. The patterns are compared with the ICDD files 74-0336 for $CaMnSi_2O_6$ and 86-0748 for $CaZnSi_2O_6$. According to the ICDD files, $CaMnSi_2O_6$ and $CaZnSi_2O_6$ have monoclinic structure with C2/c space group similar to that of $CaBSi_2O_6$ (B = Mg, Co, Ni). However, the obtained diffraction patterns do not have any resemblance to the standard files. This indicates the absence of the formation of the phases $CaMnSi_2O_6$ and $CaZnSi_2O_6$.

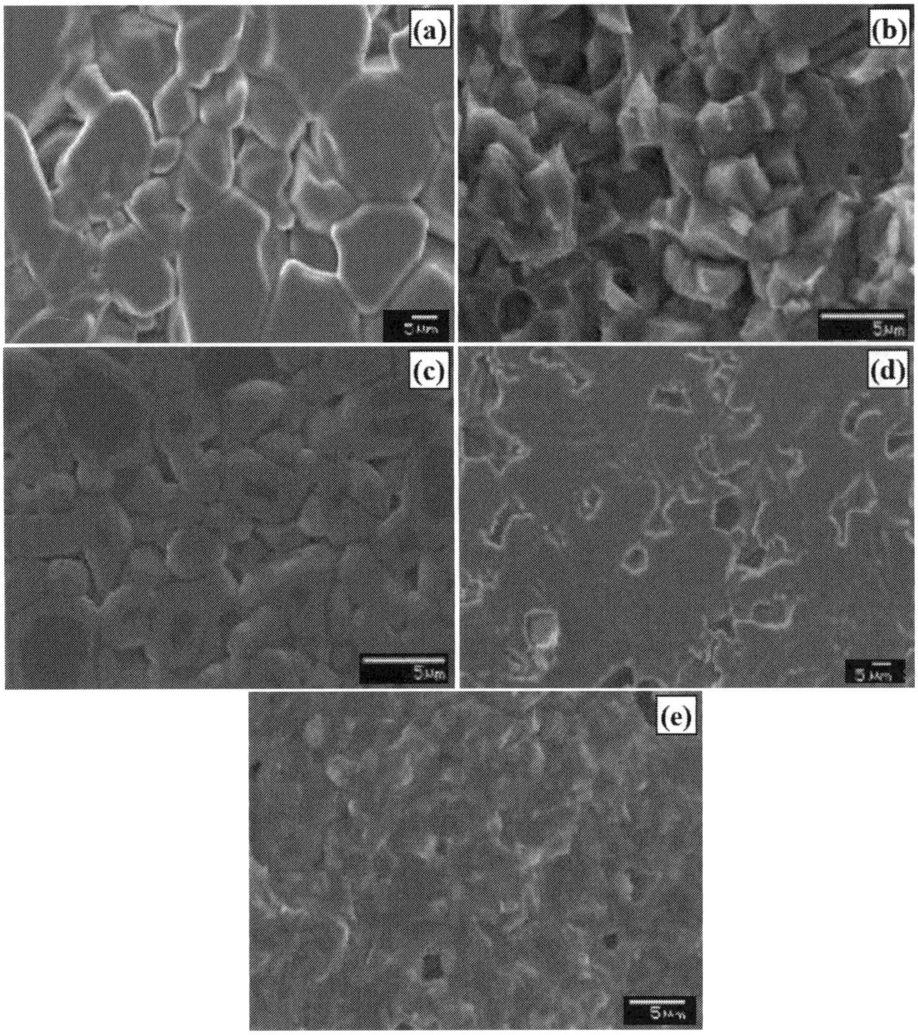

Figure 2. SEM images of (a) $CaMgSi_2O_6$ sintered at 1300 °C/2h (b) $CaCoSi_2O_6$ sintered at 1200 °C/2h (c) $CaNiSi_2O_6$ sintered at 1300 °C/2h (d) CaO-MnO-2SiO_2 sintered at 1200 °C/2h (e) CaO-ZnO-2SiO_2 sintered at 1150 °C/2h.

Figure 2 shows the SEM images of the $CaBSi_2O_6$ (B = Mg, Co, Ni, Mn, Zn) ceramics when heat treated at their respective sintering temperature. Figure 2(a) represents the polished and etched surface of $CaMgSi_2O_6$ ceramic, which exhibits grains of variable size starting from 5 μm to ~25 μm. Some porosity is also visible along the grain boundaries of the ceramic. Figure 2(b) shows the fractured surface of the $CaCoSi_2O_6$ ceramic. The fractured surface represents well packed grains of size less than 5 μm. Figure 2(c) depicts the surface of $CaNiSi_2O_6$ ceramic. Good microstructure is observed with grains of variable size. The smallest grain is about 1 μm in size while the largest grain is about 5 μm. Figure 2(d) shows the SEM image of CaO-MnO-2SiO_2, which exhibits grains of size starting from 5 μm to 20 μm for the largest grain. Some porosity is also visible in the photograph of CaO-MnO-2SiO_2. Figure 2(e) shows the fractured surface of the CaO-ZnO-2SiO_2 ceramic. The grain boundaries are not well visible and therefore represent a poor microstructure. Each of the images is of the same contrast and hence each composition $CaBSi_2O_6$ (B = Mg, Co, Ni, Mn, Zn) may represent a single-phase material. The formation of the phases $CaBSi_2O_6$ (B = Mg, Co, Ni) belonging to monoclinic structure with C2/c space group could be well identified from the XRD patterns, whereas those for B=Mn and Zn could not be recognized. More advanced studies will be required in this direction to identify the phases formed.

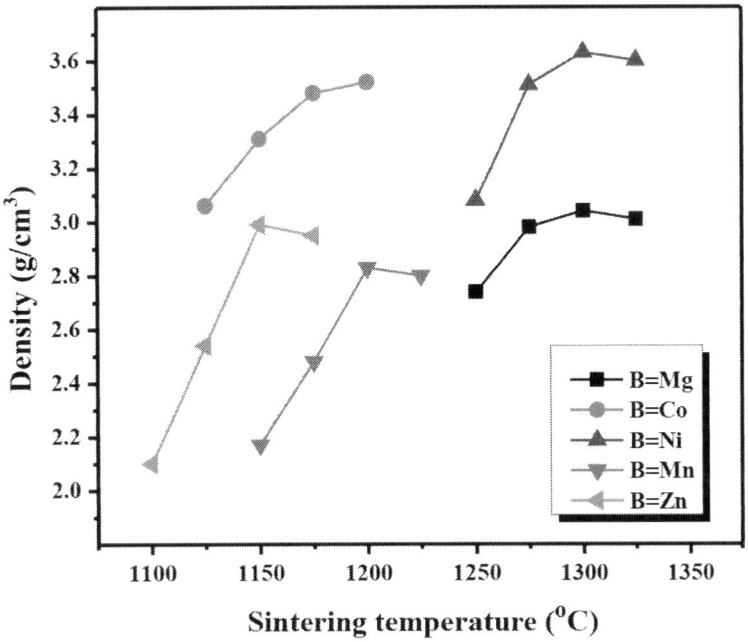

Figure 3. Optimization of sintering temperatures of $CaBSi_2O_6$ (B = Mg, Co, Ni, Mn, Zn) ceramics based on their bulk density.

Figure 3 represents the variation of density of $CaBSi_2O_6$ (B = Mg, Co, Ni, Mn, Zn) ceramics as a function of sintering temperature (ST). The density of all the silicates is found to be increased with sintering temperature. The increase in sintering temperature enhances grain growth, which results in the elimination of porosity and hence improvement in density. The highest density achieved for $CaMgSi_2O_6$ is 3.04 g/cm^3, when sintered at 1300 °C/2h. The bulk density obtained at 1300 °C/2h is 93.5 % of the theoretical value. Further increase in sintering temperature causes a decrease in density of the $CaMgSi_2O_6$ ceramics. In the case of $CaCoSi_2O_6$, the density increases up to a sintering temperature of 1200 °C/2h and with further increase in sintering temperature, the sample melts. Thus, the highest density achieved for $CaCoSi_2O_6$ is at 1200 °C/2h, which is 3.52 g/cm^3, 93.6 % of the theoretical value. Similarly, for $CaNiSi_2O_6$, the maximum density is achieved at 1300 °C/2h, which is 3.63 g/cm^3 (94.8%). The trend of variation of density with sintering temperature was also similar for $CaBSi_2O_6$ (B = Mn, Zn). However, the densities of the silicates achieved were much less than the theoretical values. For instance, $CaBSi_2O_6$ (B = Mn), when sintered at 1200 °C/2h, has a density of 2.83 g/cm^3, which is only 79.1% of the anticipated value of 3.58 g/cm^3 for $CaMnSi_2O_6$. Similarly, for B=Zn, the obtained density is 2.99 g/cm^3, which is only 77.7 % of the expected value of 3.85 g/cm^3 for $CaZnSi_2O_6$. The low values of relative density achieved are due to the absence of the formation of $CaMnSi_2O_6$ and $CaZnSi_2O_6$.

Table 1 gives the densification, dielectric properties (both at radio and microwave frequencies) and thermal expansion data of the $CaBSi_2O_6$ (B = Mg, Co, Ni, Mn, Zn) ceramics sintered at their optimized temperatures. As already mentioned, densities near to the theoretical values were achieved for B=Mg, Co and Ni, whereas for B=Mn and Zn, the expected phases were not formed. The relative permittivity of the $CaBSi_2O_6$ (B = Mg, Co, Ni) were ~ 9 and dielectric loss tangent of the order of ~ 10^{-3} at radio frequencies (1 MHz). The predominantly covalent nature of Si-O bonds might be the reason for the low values of relative permittivity and loss tangent. The microwave dielectric properties were measured in the range 8-12 GHz. As expected, the relative permittivity of $CaBSi_2O_6$ (B = Mg, Co, Ni) ceramics are less than those obtained at 1 MHz. This is due to the absence of polarization mechanisms which are active at radio frequencies [1]. Since the relative density obtained for all the silicates are less than 95%, the values of ε_r are corrected for porosity using the equation derived by Penn *et al.*, which are also given in table 1 [21]. The porosity corrected values of relative permittivity of $CaBSi_2O_6$ (B = Mg, Co, Ni) ceramics are 8.3, 8.5 and 8.3 respectively. A higher value of relative permittivity is obtained for $CaCoSi_2O_6$ compared to $CaMgSi_2O_6$ and $CaNiSi_2O_6$. This might be due to the fact that the ionic polarizability of Co^{2+} ion is higher than that of Mg^{2+} and Ni^{2+}. The ionic polarizability of Co^{2+}, Mg^{2+} and Ni^{2+} are 1.65, 1.32 and 1.23 $Å^3$ respectively [22]. Similar values of relative permittivity obtained for $CaMgSi_2O_6$ and $CaNiSi_2O_6$ can also be well understood based on approximately similar ionic polarizability of Mg^{2+} and Ni^{2+}. Since the relative permittivity obtained at microwave frequencies are less than 10 for all the silicates, high signal propagation speed can be achieved.

Table 1. Relative density, dielectric properties and thermal expansion data of $CaBSi_2O_6$ (B = Mg, Co, Ni, Mn, Zn) ceramics when sintered at optimum temperatures.

$ABSi_2O_6$	ST(°C)	Relative Density (%)	1 MHz		ε_r (8-12 GHz)		$Q_u \times f$ (GHz)	τ_f (ppm/°C)	α_l (ppm/°C)
			ε_r	tan δ	measured	porosity corrected			
B=Mg	1300/2h	93.5	9.11	0.006	7.5	8.3	53000	-45	+3.7
B=Co	1200/2h	93.6	8.94	0.001	7.7	8.5	28500	-55	0
B=Ni	1300/2h	94.8	9.05	0.005	7.7	8.3	23200	-41	+2.0
[a]B=Mn	1200/2h	--	7.96	0.009	7.2	--	18000	-31	--
[a]B=Zn	1150/2h	--	6.90	0.003	7.1	--	8000	-32	--

[a]The expected phases $CaMnSi_2O_6$ and $CaZnSi_2O_6$ are not formed.

Excellent $Q_u \times f$ values are also obtained for $CaBSi_2O_6$; 53000, 28500 and 23200 GHz respectively for B = Mg, Co and Ni. The high $Q_u \times f$ value of $CaMgSi_2O_6$ may be due to the low intrinsic loss of the $CaMgSi_2O_6$ ceramics compared to $CaCoSi_2O_6$ and $CaNiSi_2O_6$. The high $Q_u \times f$ values obtained for $CaBSi_2O_6$ (B = Mg, Co, Ni) ensures efficient use of power supplied to the device and also excellent frequency selectivity. The temperature coefficient of resonant frequency (τ_f) of the silicates are -45, -55 and -41 ppm/°C respectively for B = Mg, Co and Ni. Generally, silicates have a negative value of τ_f. The obtained values of τ_f are not suitable for immediate practical applications and therefore have to be tuned to near zero ppm/°C by suitable additives having a high positive τ_f such as TiO_2, $CaTiO_3$ or $SrTiO_3$ etc (Sebastian, 2008). The results of the measurement of coefficient of thermal expansion (α_l) of $CaBSi_2O_6$ (B = Mg, Co, Ni) ceramics are also shown in table 1. The measurement was conducted in the temperature range of 30-500 °C. The coefficients of thermal expansion of silicates are 3.7, 1.9 and 2.0 ppm/°C respectively for $CaMgSi_2O_6$, $CaCoSi_2O_6$and $CaNiSi_2O_6$.

4. Conclusions

$CaBSi_2O_6$ (B = Mg, Co, Ni, Mn, Zn) ceramics were prepared by the conventional solid state ceramic route. The phase formation of $CaMgSi_2O_6$, $CaCoSi_2O_6$ and $CaNiSi_2O_6$ were identified using the X-ray diffraction analysis. X-ray studies show the absence of the formation of $CaMnSi_2O_6$ and $CaZnSi_2O_6$. The microstructures of the ceramics contain grains of randomly varying size falling in the range 1-25 μm. The $CaMgSi_2O_6$ ceramics sintered at 1300 °C/2h has ε_r = 8.3, $Q_u \times f$ = 53000 GHz and τ_f = -45 ppm/°C in the frequency range 8-12 GHz. $CaCoSi_2O_6$ sintered at 1200 °C/2h has ε_r = 8.5, $Q_u \times f$ = 28500 GHz and τ_f = -55 ppm/°C in the same frequency range. Similarly, $CaNiSi_2O_6$ sintered at 1300 °C/2h has ε_r = 8.3, $Q_u \times f$ = 23200 GHz and τ_f = -41 ppm/°C. in the frequency range 8-12 GHz. The silicates $CaMgSi_2O_6$, $CaCoSi_2O_6$ and $CaNiSi_2O_6$ show small positive values of coefficient of thermal expansion 3.7, 1.9 and 2.0 ppm/°C respectively.

Acknowledgements

Dr. Tony Joseph acknowledges Dr. M. T. Sebastian, Former Scientist at National Institute for Interdisciplinary Science and Technology, Trivandrum, Kerala, India.

References:

[1] Sebastian M T 2008 *Dielectric materials for wireless communication* (Elsevier Publishers, Oxford, U. K)
[2] Pauling L 1980 *Am. Mineralogist.* **65** 321
[3] Tsunooka T, Androu M, Higashida Y, Sugiura H and Ohsato H 2003 *J. Eur. Ceram. Soc.* **23** 2573
[4] Ohsato H, Tsunooka T, Sugiyama T, Kakimoto K and Ogawa H 2006 *J. Electroceram.* **17** 445
[5] Sugiyama T, Tsunooka T, Kakimoto K I and Ohsato H 2006 *J. Eur. Ceram. Soc.* **26** 2097
[6] Ando M, Himura K, Tsunooka T, Kagomiya I and Ohsato H 2007 *Jpn. J. Appl. Phys.* **46** 7112
[7] Krzmanc M M, Valant M and Suvorov D 2005 *J. Eur. Ceram. Soc.* **25** 2835
[8] Krzmanc M M, Valant M and Suvorov D 2007 *J. Eur. Ceram. Soc.* **27** 1181
[9] Sun H, Zhang Q, Yang H and Zou J 2007 *Mater. Sci. Eng. B.* **138** 46
[10] Terada M, Kawamura K, Kagomiya I, Kakimoto K I and Ohsato H 2007 *J. Eur. Ceram. Soc.* **27** 3045
[11] Yokoi A, Ogawa H, Kan A and Nakamura Y 2007 *J. Eur. Ceram. Soc.* **27** 2989
[12] Song M E, Kim J S, Joung M R and Nahm S 2008 *J. Am. Ceram. Soc.* **91** 2747
[13] Wang H, Zhang Q, Yang H and Sun H 2008 *Ceram. Inter.* **34** 1405
[14] George S, Anjana P S, Deepu V N, Mohanan P and Sebastian M T 2009 *J. Am. Ceram. Soc.* **92** 1244
[15] Renjini S N, Thomas S, Sebastian M T, Kiran S R and Murthy V R K 2009 *Int. J. Appl. Ceram. Technol.* **6** 286
[16] Zhang Q L, Yang H and Sun H P 2008 *J. Eur. Ceram. Soc.* **28** 605
[17] Wang H P, Zhang Q L, Yang H and Sun H P 2007 *Acta Physico-Chimica Sinica* **23** 609
[18] Chang M W, Lyoo S H, Choo H S and Lee J M 2009 *Ceram. Inter.* **35** 2513
[19] Kim J, Hwang S, Sung W and Kim H 2009 *J. Electroceram.* **23** 209
[20] Marques V M F, Tulyaganov D U, Kothiyal G P and Ferreira J M F 2010 *J. Electroceram.* **25** 38
[21] Penn S J, Alford N M, Templeton A, Wang X R, Xu M S, Reece M and Schrapel K 1997 *J. Am. Ceram. Soc.* **80** 1885
[22] Shannon R D 1993 *J. Appl. Phys.* **73** 348

Development of a Simple Battery Management System for Cell Balancing

Calvin Immanuel S[1], **Paul Sathiyan S**[2]

[1]Department of Electrical and Electronics Engineering, Karunya Institute of Technology and Sciences, Coimbatore, India. *scalvinimmanuel.work@gmail.com*
[2]Department of Electrical and Electronics Engineering, Karunya Institute of Technology and Sciences, Coimbatore, India. *sathiyanme@yahoo.co.in*

Abstract. As the use of Lithium Ion (Li-Ion) battery is increasing, proper means of increasing the life cycle and reliability of the battery should be adopted. A discussion on the precautions to be carried out while integrating Li-Ion batteries are discussed. A Battery Management System (BMS) is developed to address the unbalance State of Charge (SoC) issue that arise in the battery pack. The proposed algorithm is simulated in MATLAB Simulink. The battery pack used in this simulation is inspired from Tesla. This paper can be used as a tutorial or as a reference for anyone getting started with Li-Ion batteries or battery management systems.

1. Introduction

With increase in Global warming and different types of pollutions because of increase in emission of carbon di oxide from factories, (ICE) Internal combustion engine vehicles, and also due to the rapid depletion of crude and fossil fuels a new source of energy must be sought for. This need for an alternate source of energy has led to the development of battery-based energy systems. These types of energy systems, are gaining much interest among researchers and thus paving way to the development of (EV) Electric Vehicles [1].

One of the most commonly preferred on board source of energy to power these applications is the Lithium-ion batteries (Li-ion). Over the due course of time, the Li-ion batteries have gained popularity in the global market because of their a) high energy density, b) well versed technology, c)desirable characteristics, d) no memory effect and d) least self-discharge rate [2]. Due to these characteristics, the Li-ion batteries are employed as either primary or secondary source in several electrical based energy applications [3], [4]. Though the Li-ion batteries have a lot of desirable features, they are quite intolerable when it comes to discrepancies in design and implantation such an a) over voltage, b) high temperature, c) over current or power, d) thermal runway etc., [5], [6]. Currently, many research works are being carried in the developing the chemical and physical characteristics of the Li-ion batteries both in terms of cost and performance in order to achieve maximum potential out of these batteries. Some of the most advanced and important ones among them are safely charging and discharging the batteries under safe State of Charge (SoC) values [7].

With Li-ion batteries becoming more and more popular for electric based applications, a need for a Battery Management System (BMS) arises to improve the performance, life span and safety of the battery to a much greater extent. A BMS is nothing but a protective circuit or a system or a module that controls the overall battery parameters, such as safe operating voltage, current, temperature, power etc., along with providing and monitoring real time data about fault diagnosis, amount of charge available, charge stabilization etc., [8], [9]. This paper discusses about the Li-Ion battery source and its wide application, issues arise due to integrating Li-Ion cells as battery pack, safety precautions while dealing with the battery pack, cell balancing strategies, BMS and its algorithm. Simulation of BMS system for the Li-Ion battery pack. A three cell and a five-cell battery pack model are tested for the proposed algorithm.

2. Lithium-Ion Batteries

The Li-ion batteries are an unmatchable choice when it comes to battery-based energy applications because of their high energy density and other such superior characteristics. They are being used in electric vehicles, portable tools, power electronics and so on. The Li-ion battery pack is nothing but a combination of individual lithium-ion cells connected in series formation in order to increase the voltage or in terms of parallel formation in order to increase the current or sometimes in combination of both to achieve higher ampere hour rating. Lots of cells can be included into a battery module, based on the voltage parameter required. For example, a Tesla car with an overall battery pack voltage of 85kWh has around 7104 cells. In basic terms a Li-ion cell individually consists of two electrodes namely a cathode and an anode, immersed in an electrolytic liquid or membrane in some cases, based on the type of cell and chemistry involved [10]. Right from the commercialization of the Li-ion batteries by Sony, the Li-ion batteries have gained a wide popularity in the global market [11], [12]. Now a question may arise what is so special about Lithium-ion batteries? The main features of the Li-ion batteries include less toxicity, longevity in terms of capacity and charge / discharge cycles, usability within a wide range of temperatures and environments etc., [13]. But, just like a coin has two sides, the Li-ion batteries also have two sides i.e. the positive and negative traits. The positive traits include the high energy and power density as already discussed above. But, the negative traits include mainly the cost and the protective systems that are needed for safe charging and discharging of the Li-ion Batteries [14]. While the Lithium and Nickel batteries were used in secondary systems of energy storage systems [15], learning about their wonderful utility, we started utilizing them as primary sources at present. There are a variety of materials used in the building of Li-ion batteries, often the cathode materials of these batteries include, amalgams of the oxides of Lithium metals, some of them consist, manganese (LMO), cobalt (LCO), combinations having a blend of aluminium such as $LiNiCoAlO_2$ [16]. Now we can make it clear that the High energy and power densities of the Lithium-ion batteries are because of the utilization of Lightweight materials and high potential electro chemicals in the construction of the batteries, thereby giving rise to voltages of 3.2 – 3.8V. In present days, extensive research works are being carried out on Li-ion batteries, to improve their performance in terms of power, energy and in terms of operating voltages too, as result of which, the operating voltages of up to 4.7 – 4.8 V have been achieved [17].

3. Safety and Precautions

Though Li-ion batteries have a lot of desirable characteristics, they are quite unforgiving when it comes to mishandling, error in designing and short circuits. The short circuit causes the current to flow in an unintended path thereby causing explosion, fires and even death. Hence, few safety measures need to be followed, they are as follows:

- At all times safety gear such as glasses, insulation gloves and aprons should be worn in the environment where works on Li-ion batteries are being carried out.
- Due concentration must be given while handling Li-ion batteries and chit chatting unnecessarily in the work place must be avoided.
- Never keep any metal tools just above the battery pack, because if the tool accidentally happens to fall on the battery and connects the battery terminals it could cause fatal short circuits.
- During designing proper calculations must be carried out in order to avoid mismatch of cells, over voltage / currents which could lead to thermal runaway.
- Battery packs should never be forced in to housings which are smaller in size comparing to the battery packs as these could result in cell deformations causing, cell venting, internal short circuits etc.,

- While measuring the Open Circuit Voltage (OCV), if the value comes to 0.0 V it could mean that wither the fuse is blown or the cell has been completely discharged [18].
- In the event of any signs of explosion such as flares, smokes, sparks etc., rush to the nearest exit.
- In case of unavoidable fires, never use water as the fire resulting will be an electrical and fire hence dry chemical extinguishers loaded with potassium bicarbonate or monoammonium phosphate must be used to put off the flames.
- Never ever mix Old and new cells together as they could have different levels of (SoCs) State of charges and could result in a mismatch hence a separate log should be maintained about the cells.
- In order to check the life of a cells, charge the cells and after one hour of resting the cell, measure the voltage of the cell. If the voltage shows out to be 4.2 V then the cells are good in condition to be used. Though this is a basic method, it can used in general and not to be relied up on all times.
- The cells and battery packs should be stored in a safe and closed environment under nominal temperature conditions.
- Suitable battery management systems with temperature, voltage and current sensors should be employed in any battery-based energy applications as this could not only boost the performance of the batteries but also can prevent fatalities due to poor handling of the batteries.

4. Cell Balancing

Cell mismatch or cell imbalance is one of the important phenomena that leads to the reduction in battery life of Li-ion batteries when using multiple cells together as described earlier for the meeting the power requirement of an application. It also leads to the reduction in battery performance and reduces its State of Health (SoH) [18]. Researches stated that in a string of series connected cells, the cell with the least SoH level, dominated the entire string of cells thereby causing safety issues and thermal runaway while the battery pack was discharged below its limit [19], [20]. In order to overcome such discrepancies that deter the overall performance of the cells and the battery pack as a whole, cell balancing techniques need to be used. The cell balancing techniques are of two types namely:

- Active cell balancing and
- Passive cell balancing

4.1. Active cell balancing

In general active cell balancing is a method in which, the cells are maintained in uniform charge levels, with the help of electronic storage elements such as capacitors, inductors etc., in which these elements distribute the charge from highly charged cells to the cells with lower charge, this operation takes place until the charge in all the cells become uniform [21]. The active cell balancing technique is of two types namely:

- Capacitor based active cell balancing
- Inductor and transformer based active cell balancing.

4.2. Passive cell balancing

Passive cell balancing technique employs the idea of using a venting or balancing resistor to vent out or discharge the extra charge in the highly charged cells and thereby balancing the entire string voltage by making the charge in the highest charged cell to the lowest cell voltage potential finally creating a balance in the entire string voltage. The passive cell balancing technique is of two types namely:

- Fixed shunt resistor method
- Switching shunt resistor method

4.2.1. Fixed shunt resistor method. In this method of passive cell balancing technique, a fixed resistor is placed in parallel across each series connected cells based on the current to balance in the cells. Since this method of passive cell balancing employs a simple circuit and very few components, the cost is very less. However, it has a disadvantage of heat dissipation through the cells during the cell balancing operation, where the excess energy is lost or vented out or discharged [22].

4.2.2. Switching shunt resistor method. employs resistors connected in parallel across each series connected cells, by means of a relay or semiconductor switch. In this method the imbalance in voltage can be sensed and based on that the cell balancing operation can be carried out accordingly This method is of two types namely the continuous mode in which the semiconductor switches are controlled simultaneously. Whereas, in the other mode known as the sensing mode, the cell balance operation is carried out once an imbalance in voltage is detected by the sensor module which decides the resistor to be shunted [22].

5. Battery Management System

A (BMS) Battery Management System is a very crucial part in any battery-based energy applications especially in (EV) Electric Vehicles. The main need for the BMS is to protect the battery pack from discrepancies in voltage/current/ power etc., along with collecting and providing real time log data on the battery conditions such as SoC, Temperature and so on. These parameters are very important when it comes to boosting the overall performance and efficiency of the battery packs. The main architecture of the Battery Management System can be split into two parts namely the Hardware and the Software parts. The Hardware part aims at Various types of sensors such as voltage, current, power, temperature sensors etc., these sensors need to have a high sampling rates in order to provide the values with precisely and with least errors. Whereas, the software part aims at the GUI through which the user can easily monitor battery parameters and the program codes which limit the hardware in terms of suitable values or parameters under within which the battery and the sensors are supposed to function [23]. Following are the features of a BMS:

- The BMS should prevent the voltage of the cells from exceeding the prescribe voltage value by limiting or blocking the charging current.
- It should also maintain the temperature of the batteries by ensuring proper cooling through cooling mechanisms such as current limiting during charging or using physical coolant liquids or gases to remove the heat from the battery pack.
- The BMS should ensure proper cell balancing by monitoring the SOH and SOC levels in the cells because failing to do so can affect the entire performance and the charging and discharging characteristics of the battery as a whole.
- The BMS should trigger alarms in case of any abnormalities with respect to the cells or the battery pack.
- Finally, the BMS should be able to notify the users about the cells which should be replaced, because a cell with low SOH value can Hinder the overall battery performance.

In order to design a Battery Management System, due calculations need to be done as for what applications, the BMS is needed or for what value of Battery voltage and so on. Though there are lots of methods in designing a Battery management System, the best method would be to use a simulation software to build the control blocks based on the battery parameters and to run a simulation on the model built. Finally, based on the test results readings and important details can be noted down as on how the practical Battery Management System can be implemented.

6. Design and Simulation of Passive Cell Balancing Circuit

In here, a circuit is built based on passive cell balancing principle and the output current, voltage and SoC values are analysed. The Battery employed here has parameters similar to that of Tesla's Li-ion cell. The cell characteristics is presented in Figure 1 and the cell specification is given in Table 1

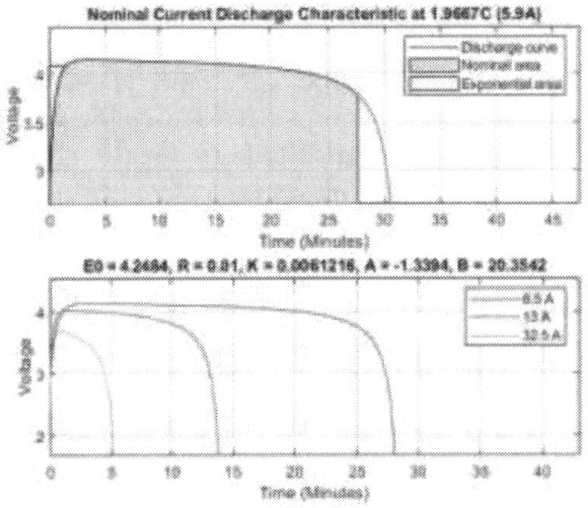

Figure 1. Cell Characteristics

Table 1. NCR 18650BE Cell parameters – Our simulation work employs a cell with similar characteristics [24]

Parameters	Specifications
Rated Capacity	3200mAh
Nominal Voltage	3.6V
Charging Voltage	4.2V
Weight (Max.)	48.5g
Diameter (D)	18.20mm
Height (H)	65.50mm

With reference to the concept of passive cell balancing, a three cell and five cell battery pack model with passive cell balancing technique have been developed using MATLAB Simulink as shown in Figure 2.

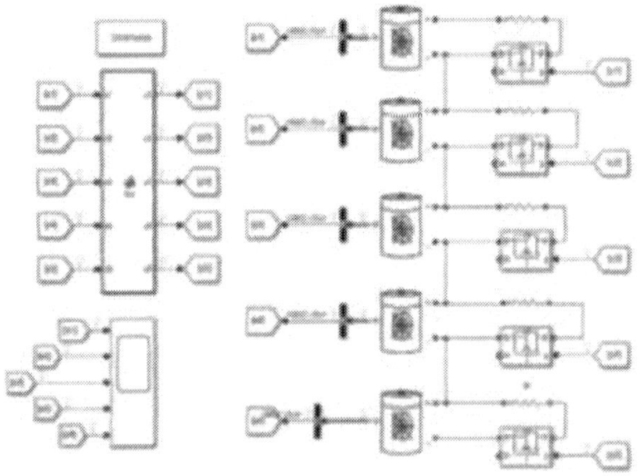

Figure 2. Five cell battery pack model with passive cell balancing based on [25].

The cell balancing algorithm is implemented using the MATLAB function black and for circuit simplicity and easy understandability, signal routing blocks are used. At the output terminals of the battery a series network having a power dissipation / balancing resistor and a solid-state switch is used. When the passive cell balancing algorithm is implemented, the output behaviour is studied by means of the graphs obtained from the scope. The passive cell balancing algorithm implemented in this work is presented in Figure. 3. The algorithm receives the SoC (Sx) of each cell in the battery as input and identifies the cell with the least SoC level and activates the passive cell balancing by turning on the MOSFET connected across the other cells having higher SoC level using the control signal (Yx). Where x indicates the number of cells being used in the battery pack.

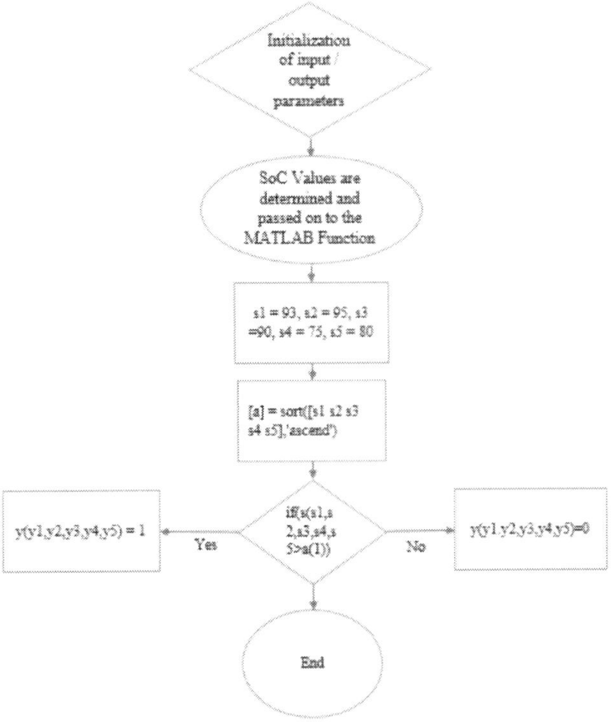

Figure 3. Flowchart.

Initially, the input (s1, s2, s3, s4, s5) and the output (y1, y2, y3, y4, y5) variables are initialized, following which we assign predefined values to the input 's' values and then with help of if else condition logic, we verify the values in each batteries, if the parameter in one battery is greater comparing to the other batteries then the parameter is vented out by the logic implemented by means of the bypass resistors in the circuit, this operation continues until the parameters are the same in all the five batteries.

The balancing takes place only when the battery power is not used for the application. The simulation is started considering that the battery pack has delivered power to the load and the SoCs of each battery pack is different as shown in Table 2. The simulation results for three cell balancing technique and five cell balancing techniques are presented in Figure 4 and 5. In both these cases, the SoC of every cell is discharged to the level of the cell having minimum SoC value.

Table 2. Initial values of SoC

Battery #	3 Cell Model	5 Cell Model
1	62	55
2	67	65
3	82	67
4	-	82
5	-	85
Time taken for cell balancing (s)	2650	3550

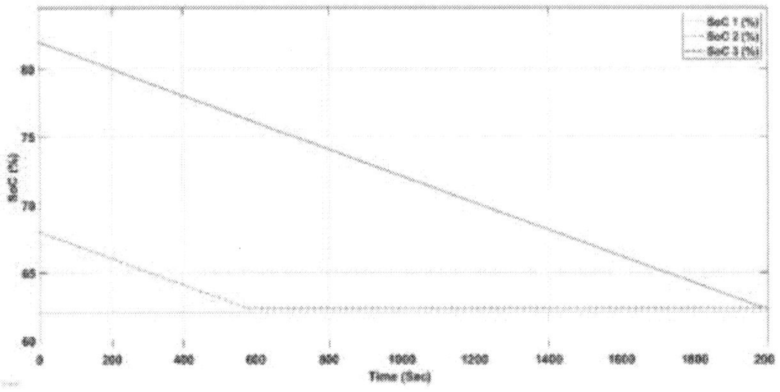

Figure 4. SoC stabilization for 3 cell battery pack.

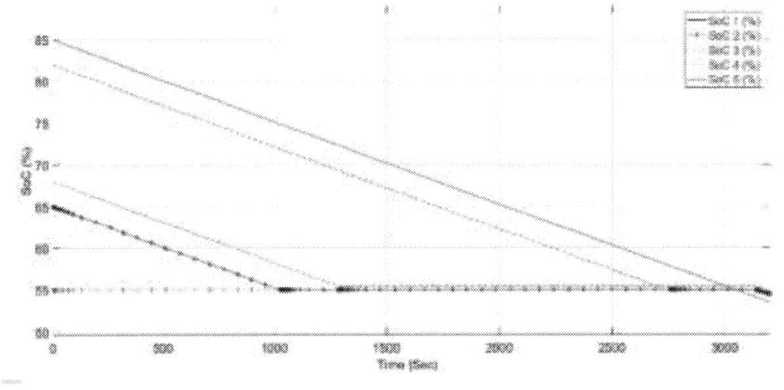

Figure 5. SoC stabilization for 5 cell battery pack.

From the output graphs, it is evident that how the battery parameters are stabilizing from their peak values to that of the values of the other batteries in the circuit.

7. Conclusion

The significance of Li-Ion batteries for electrical applications and the need for safe operation of these cells when used as battery packs are being discussed. In order to have longer life span for the battery pack, the need for cell balancing technique is also presented. A simple algorithm for performing cell balancing using passive balancing technique is presented and simulated using MATLAB Simulink This paper is aimed at serving as a tutorial paper that delivers complex information and concepts about batteries, battery management systems, cell balancing techniques etc., in a nutshell thereby eliminating the need for the reader to spend more time in learning about the working mechanism and concepts. In this paper we are focused more on the development and simulation of a passive cell balancing circuit in detail, while a glimpse of active cell balancing is given

References

[1] Hu X, Jiang J, Egardt B and Cao D 2015 Advanced Power-Source Integration in Hybrid Electric Vehicles: Multicriteria Optimization Approach *IEEE Trans. Ind. Electron.* **62** 7847–58
[2] Manzetti S and Mariasiu F 2015 Electric vehicle battery technologies: From present state to future systems *Renew. Sustain. Energy Rev.* **51** 1004–12
[3] A M, Fgw N, W W, M G and Du S 2016 Comprehensive study of the influence of aging on the hysteresis behavior of a lithium iron phosphate cathode-based lithium ion battery - An experimental investigation of the hysteresis *Appl. Energy* **171** 629–45
[4] Hoque M M, Hannan M A and Mohamed A 2017 Model Development of Charge Equalization Controller for Lithium-Ion Battery *Adv. Sci. Lett.* **23** 5255–9
[5] Shafiei N, Ordonez M, Craciun M, Botting C and Edington M 2016 Burst Mode Elimination in High-Power LLC Resonant Battery Charger for Electric Vehicles *IEEE Trans. Power Electron.*
[6] Xiong R, Sun F, Gong X and Gao C 2014 A data-driven based adaptive state of charge estimator of lithium-ion polymer battery used in electric vehicles *Appl. Energy* **113** 1421–33
[7] Hannan M A, Hoque M M, Hussain A, Yusof Y and Ker P J 2018 State-of-the-Art and Energy Management System of Lithium-Ion Batteries in Electric Vehicle Applications: Issues and Recommendations *IEEE Access* **6** 19362–78
[8] Anon (PDF) A review on the key issues for lithium-ion battery management in electric vehicles
[9] Xing Y, Ma E W M, Tsui K L and Pecht M 2011 Battery Management Systems in Electric and Hybrid Vehicles *Energies* **4** 1840–57
[10] Deng D 2015 Li-ion batteries: basics, progress, and challenges *Energy Sci. Eng.* **3**
[11] Aricò A S, Bruce P, Scrosati B, Tarascon J-M and van Schalkwijk W 2005 Nanostructured materials for advanced energy conversion and storage devices *Nat. Mater.* **4** 366–77
[12] Winter M, Besenhard J O, Spahr M E and Novak P 1998 Insertion electrode materials for rechargeable lithium batteries *Adv. Mater. Weinh.* **10**
[13] Kong D, Li X, Zhang Y, Hai X, Wang B, Qiu X, Song Q, Yang Q-H and Zhi L 2016 Encapsulating V2O5 into carbon nanotubes enables the synthesis of flexible high-performance lithium ion batteries *Energy Environ. Sci.* **9** 906–11
[14] Gallagher K G, Goebel S, Greszler T, Mathias M, Oelerich W, Eroglu D and Srinivasan V 2014 Quantifying the promise of lithium–air batteries for electric vehicles *Energy Environ. Sci.* **7** 1555–63
[15] Shimamura O, Abe T, Watanabe K, Ohsawa Y and Horie H 2007 Research and Development Work on Lithium-ion Batteries for Environmental Vehicles *World Electr. Veh. J.* **1** 251–7
[16] Anon Linden's Handbook of Batteries, Fourth Edition | McGraw-Hill Education - Access Engineering
[17] Su Y, Cui S, Zhuo Z, Yang W, Wang X and Pan F 2015 Enhancing the High-Voltage Cycling Performance of LiNi(0.5)Mn(0.3)Co(0.2)O2 by Retarding Its Interfacial Reaction with an Electrolyte by Atomic-Layer-Deposited Al2O3 *ACS Appl. Mater. Interfaces* **7** 25105–12
[18] Bentley W F 1997 Cell balancing considerations for lithium-ion battery systems *The Twelfth Annual Battery Conference on Applications and Advances* The Twelfth Annual Battery Conference on Applications and Advances pp 223–6
[19] Moo C, Ng K S and Hsieh Y 2008 Parallel Operation of Battery Power Modules *IEEE Trans. Energy Convers.* **23** 701–7
[20] Han W and Zhang L 2017 Battery Cell Reconfiguration to Expedite Charge Equalization in Series-Connected Battery Systems *IEEE Robot. Autom. Lett.* **PP** 1–1
[21] Anon (PDF) A new cell-to-cell fast balancing circuit for Lithium-Ion batteries in Electric Vehicles and Energy Storage System

[22] S H 2020 Overview of cell balancing methods for Li-ion battery technology *Energy Storage* e203
[23] Anon Battery Management Systems for Large Lithium-Ion Battery Packs
[24] Anon Comparing Panasonic 18650 BD vs. Panasonic 18650 BE *18650 Battery BATTERY BRO*
[25] Anon 2020 Passive Cell Balancing Simulation In Matlab Simulink *CS Electr. Electron.*

Effect of surfactant mediated inter-particle interactions on the magnetic properties of Manganese Zinc Ferrite ferrofluid

A. P. ReenaMary[1,5], D. Divya[1], M. G. Jestin[2], Imad Al Omari[3] and M R Anantharaman[4]

[1]*Department of Physics, Govt Victoria College Palakkad, (Affiliated to University of Calicut), Kerala, India 678001*
[2]*Department of Physics, Govt Engineering College Thrissur, (Affiliated to APJ Abdul Kalam technological University), Kerala, India 680 009*
[3]*Department of Physics, Sultan Qaboos University, Oman*
[4]*Department of Physics, cochin University of Science And Technology, Kerala, India -682022*
[5]Corresponding author: Email address: reenamaryap@gmail.com (Dr. Reena Mary A P), Telephone: +91 9495840114

Abstract. Ferrofluids of spinel ferrites presents a multitude of applications in engineering and biomedical fields. The nature of the divalent ions and their occupancy decides the magnetic properties of the spinel ferrites. The nanoparticles of these ferrites and their suspensions in various liquids are of great fundamental interest as well. This paper reports the synthesis and magnetic relaxation studies on fluid particles of Zinc substituted manganese ferrite. Nanoparticles of manganese zinc ferrite with a particle size around 5 -6 nm, are synthesized by chemical method and suspended in kerosene and water, with proper surfactants. The structural characterization is carried out by x-ray diffraction and electron microscopy. The magnetic properties are studied by employing a SQUID magnetometer. The temperature dependent static magnetic measurements and analysis reveal the inter particle interaction effects on the overall magnetic behavior of the constituent magnetic nanoparticles. The modification of the magnetic relaxation based on the surfactant is analysed.

Keywords: inter-particle interaction, surfactant, zero field cooled, brownian relaxation, superparamagnetism

1. Introduction

Manganese Zinc ferrite is a spinel soft ferrite with high saturation magnetisation and fast magnetic response. Nanoparticles and fluids of these ferrites have been studied for engineering applications such as heat transfer fluids, and bio applications like magnetic hyperthermia [1,2]. The saturation magnetisation and coercivity depends on the cation distribution in the lattice sites. The saturation magnetisation is found to increase upto 40% Zn substitution by chemical method and further the increase of Zn decreases the saturation magnetisation. As the size of the particles goes to nano regime, around 10 nm, they become superparamagnetic in nature. Nevertheless the surface effects and the inter-particle interactions decide the net magnetic response. Ferrofluids are magnetic nanoparticles suspended in suitable carrier liquid with proper surfactants. The surfactants separate the particles from one another and cause the particles to get suspended well to form stable ferrofluid. These fluids could also present different mechanisms including brownian relaxation, and the temperature dependent magnetization

IOP Conf. Series: Materials Science and Engineering 1166 (2021) 012027 doi:10.1088/1757-899X/1166/1/012027

reversal depends on the type of surfactants and carrier liquid that result in modified magnetic properties. The magnetic moment of the divalent anion (M) in the formula $(M_{1-x}Fe_x)[M_xFe_{2-x}]O_4$ and their preferred occupancy in tetrahedral or octahedral site (x being the degree of inversion) decides the overall magnetization[3,4] of the ferrite. They also exhibit surface anisotropy effects originating from the finite size effects and crystallinity[5] together with the inter-particle interactions and cation distribution. The synthesis routes decide the cationic occupancy in lattice sites and oxidation states of anions in Manganese Zinc ferrites[6,7,8].

Here we report the inter-particle interaction based on the magnetic studies of hydrocarbon based and water based manganese zinc ferrite fluid particles. The composition of $Mn_{0.6}Zn_{0.4}Fe_2O_4$, is which has been reported with maximum magnetization following the chosen synthesis route [4]. Nanoparticles of this particular composition are prepared and dispersed in both hydrocarbon base and aqueous base. Since in either case the surfactants are different and hence the inter particle separation and hence varied magnetic properties. Tailoring the interparticle interaction and understanding the magnetic relaxation is important in biomedical applications [2,3,9,10].

2. Experimental Details

The nanofluids under study are synthesized by controlled chemical co precipitation technique. Anhydrous ferric chloride ($FeCl_3$), Manganese chloride tetrahydrate ($MnCl_2.4H_2O$) and zinc chloride ($ZnCl_2$) in a stoichiometry of 2: 0.6: 0.4, is dissolved in water. This solution is made to react with 1 M boiling NaOH solution in water. This reaction mixture is stirred for five minutes. The precipitate is filtered from the supernatant solution to remove unreacted radicals and water soluble byproducts. 50 ml distilled water and 2 ml oleic acid are added to the mixture and stirred for another 90 minutes at 40°C. Then the resultant precipitate is allowed to cool while stirring and then a few drops of con. HCl are added so as to reduce the pH of the mixture below 7. The precipitate is coated with surfactant oleic acid and is washed several times with water and dried well with acetone and dispersed in kerosene by ultrasound agitation. The fluid obtained is named as MZFK.

The water based fluid is prepared from the precipitate obtained after supersaturating with NaOH and stirring for two minutes. The precipitate is magnetically decanted and to this 50ml distilled water, and 2g citric acid dissolved in 5ml water is added. The temperature is raised to 90°C and stirred for 60 minutes. The mixture is cooled and kept under a strong electromagnet to remove the supernatant solution. The mixture is washed with water to remove the excess of citric acid and at a pH of 7 the precipitate gets suspended in water.

The structural characterisation was carried out by X Ray diffraction(XRD) technique (Rgaku D Max at Cu Ka like), and Transmission Electron microscopy (JEM 2100F TEM) measurements. The magnetic measurements were carried out in in superconducting Quantum Interference Device (SQUID Quantum Design).

3. Results and Discussions

Figure 1 depicts the XRD pattern of the sample by drying off the carrier liquid. It's seen that the particles are crystallized in the fcc spinel structure. All the major diffraction peaks are identified (ICDD: 74-2401). The peak broadening evidences the nanocrystalline nature of the ferrite. The crystallite size calculated by employing the Debye Scherer's formula is 6 nm. The selected area electron diffraction (SAED) pattern of the MZFK is shown in figure 1b. The diffused rings indicate the nanocrystalline nature of the particles in the sample. The major crystallographic planes (311) (331) (220) (400) (422) and (440) of fcc spinel structure of $Mn_{0.6}Zn_{0.4}Fe_2O_4$ are identified. Transmission electron microscopy (TEM) conducted on the samples give the particle size and distribution of size. The samples for TEM are prepared by drying off the base liquids for both the samples.

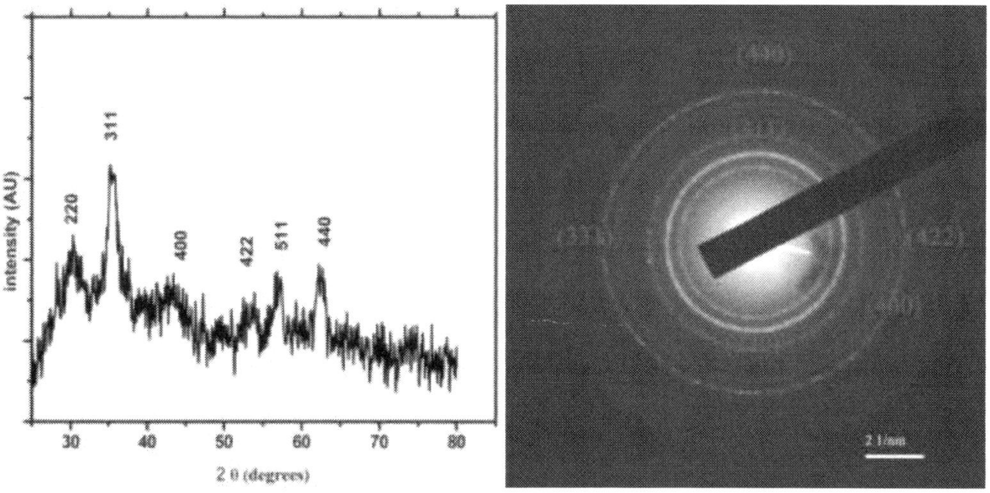

Figure 1(a) X ray diffraction pattern and **(b)** : SAED pattern of MZFW

Figures 2a depict the TEM images of MZFK and figure 2b is the particle size distribution histogram obtained from the images. The mean size is obtained as 5 nm with standard deviation of 0.7 nm. Figure 3a shows the TEM image of MZFW and figure 3b the size distribution with a mean size of 7 nm and standard deviation of 0.6nm. It could be seen that the particles are uniformly distributed in size in both cases. A slight increase in size for water dispersed fluid particles is attributed to the digestion conditions. The MZFW is heated at 90°C and allowed to crystallize after the citric acid is added. While in MZFK, the sample is heated at lower temperature and the growth is more inhibited.

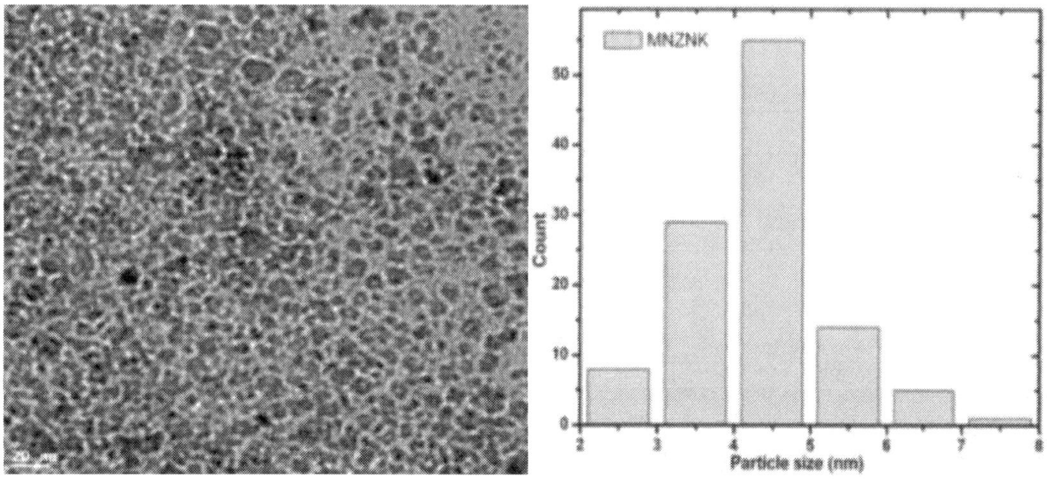

Figure 2(a): TEM image of MZFK **2(b)**: particle size distribution

A close examination of figure 2a reveals the formation of tiny clusters consisting of 6 to 8 particles and the clusters are well separated from one another. This is reported for nanoparticles passivated with steric surfactants[11]. Nevertheless each nanoparticle within the cluster is separately visible in TEM, since the particles are coated oleic acid which separates them.

IOP Conf. Series: Materials Science and Engineering 1166 (2021) 012027 doi:10.1088/1757-899X/1166/1/012027

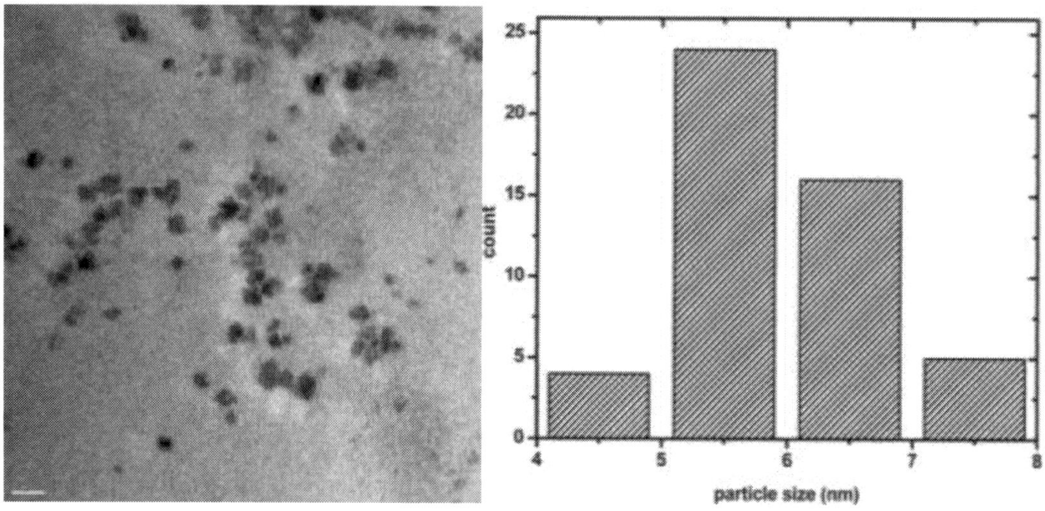

Figure 2(c): TEM images of MZFW **2(d):** particle size distribution

The TEM image of MZFW(fig 2c) shows well distributed particles and no pattern formation is seen due to clustering as in OA coated MZFK sample.

3.1 *Magnetic hysteresis measurements*

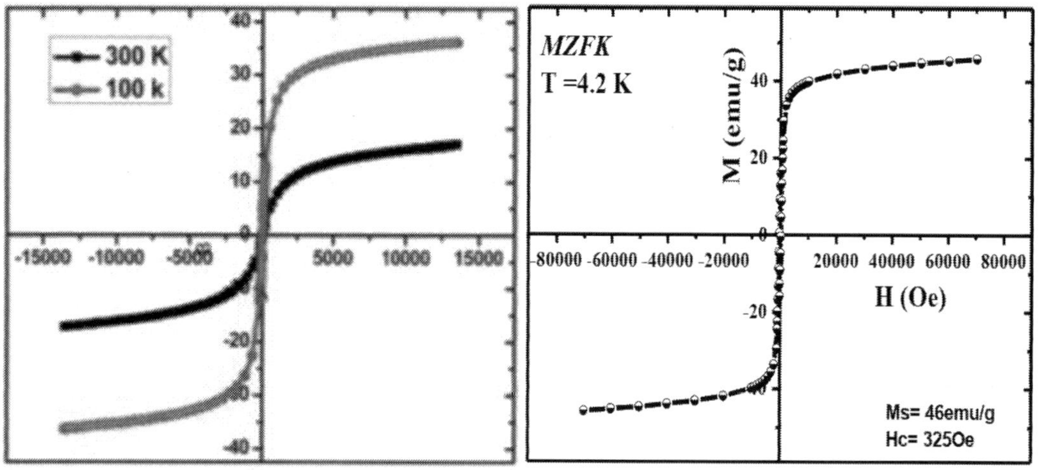

Figure 3(a): hysteresis loop (VSM) measurements on MZFK at room temperature and at 100K,
3 (b) : M H graphs(SQUID measurements) at 4.2K

Figure 3 depicts the magnetisation loops measured at different temperatures. The moment increases with decrease in temperature as expected. The moment is not saturated even at high applied magnetic fields, owing to the surface effects exhibited by the nanoferrites[12]. The magnetic order at the surface gets deformed from the bulk moment and when the surface becomes prominent these effects are reflected in the net magnetic behaviour.

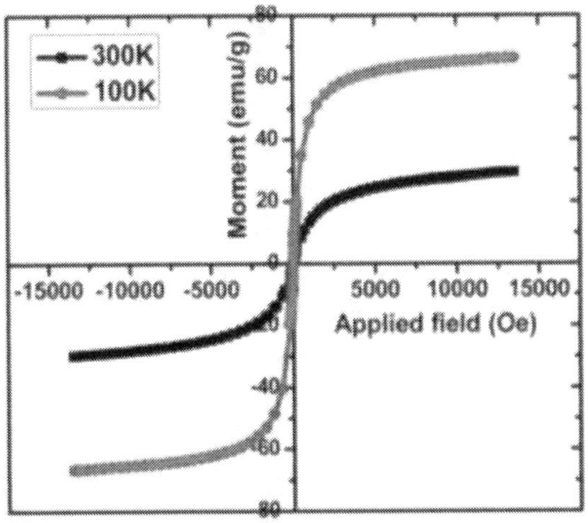

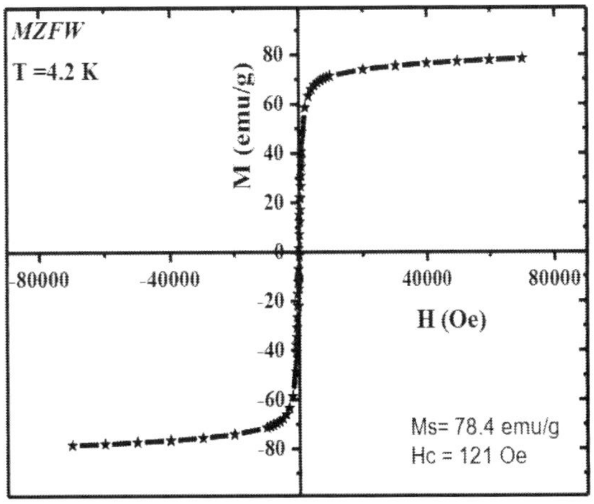

Figure 3(c): M-H loop (VSM) at room temperature and at 100K **Figure 3 (d) :** M (H) graphs (SQUID measurements) at 4.2K; on MZFW

Table 1 consolidates the saturation magnetisation of MZFK and MZFW at diffrent temperatures. The surafce spins in ferrites are reported to form a dead layer that decrease the net magnetisation of the material from the bulk. The saturation magnetisation for both the samples decreases with incearse of temperature, as expected by curie weiss law, while at all tempertures MZFK present a lower specific magnetisation than MZFW. This is because the oleic acid(with a molecular wt 282g) is a larger surfactant than citric acid (molecular wt 192 g) hence the weight fraction of nanoparticles is smaller in MZFK hence smaller value of technical saturation magnetisation for camparable particle sizes in both the samples. The magnetisation loop at room temperature presents negligible remenance and hysterisis showing the superparamagnetic nature of both samples as expected for a single domain particle with size of 5 - 6 nm

Table 1: comparison of magnetisation at varoius temperatures

Sample	300K	100K	4.2K
MZFK	20 emu/g	36 emu/g	46 emu/g
MZFW	30 emu/g	66 emu/g	78.4 emu/g

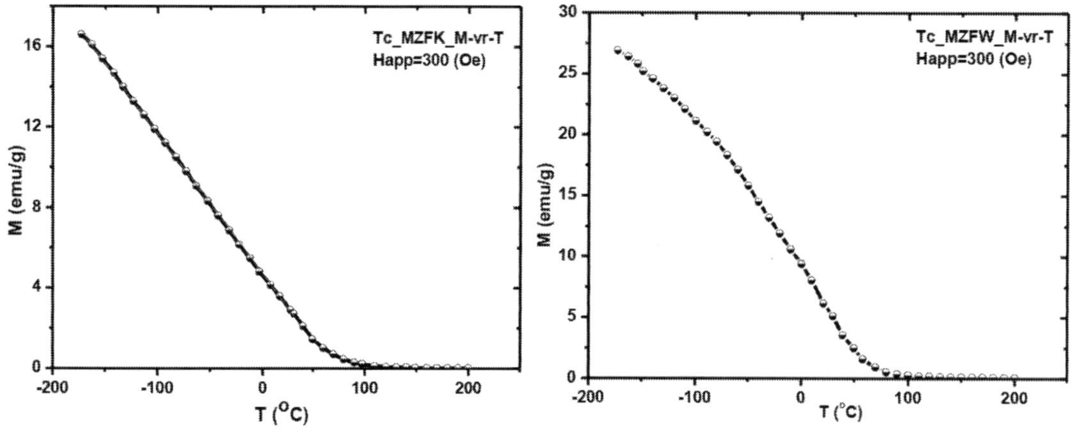

Figure 4: moment variation with temperature showing Tc(a): MZFK ,(b): MZFW

Figure 4 is the temperature dependent moment variation at an applied field of 300 Oe. The measurement has been recorded from liquid nitrogen temperature to 200°C, showing the curie temperature as the temperature at which the moment reduces to zero. The T_C values obtained is 120°C which is lowered from the bulk value [12,13].A detailed study regarding the curie temperature with respect to the zinc substitution shows that Tc increases for 30% zinc ions and then it decreases[13]. The variation in Tc could be originated from the cationic distribution in the sample. This requires further studies. The variation of graph 4(a) is typically paramagnetic in nature with a linear function of temperature as curie's law. This indicates the perfect langevin type beahaviour of non interacting superparamagnetic particles. The particles in MZFK are well separated from one another by the surfactant oleic acid which is covalently bound to the particle surface. Whereas the CA coated particles (figure 4b) shows slight deviation from the linear variation. This indicates the presence of interparticle interaction. since CA provides ionic/ electrostatic stabilization in water besides the nominal steric repulsion. So when the water is dried off, the electrostatic repulsion decreases and this causes the interaction among particles in MZFW.

3.2 *FC/ZFC measurements*

The field cooled (FC) and zero field cooled (ZFC) magnetic moment variation as a function of temperature are measured in a SQUID at two different applied fields. The results are presented in figures 5 and 6. In ZFC measurement, the sample is cooled down to 5K from room temperature with zero applied field. The particles get cooled with their moments in random directions. An external magnetic field corresponding to the initial magnetisation is applied and the moment of the sample is measured by increasing the temperature. In FC measurement, the sample is cooled in an applied field equal to the measurement field and the moment at different temperatures is recorded. Both ZFC –FC measurements are carried out for two applied fields for both the samples.

Figure 5(a) and 5(b) show the FC-ZFC measurements at two different applied fields on MZFK. Both T_B and T_{irr} get shifted to low temperature as the applied field increases. The blocking is affected by zeeman energy, reflected in the energy for the magnetisation reorientation. T_B is a function of the measuring field (inset of figure 5(b)). The ZFC moment increases, becomes a maximum at the blocking temperature and then monotonically decreases in all the cases. This shows the particles undergo magnetisation reversal more or less independently. The two temperatures, T_B and T_{irr} are closer with a separation of 23K in both fields. The interaction among the particles are less. This is evident from the TEM measurements that the particles are uniformly distributed and are well separated. The clustering effects could be a reason for the increased effective size of the particles, which is reflected in this

IOP Conf. Series: Materials Science and Engineering 1166 (2021) 012027 doi:10.1088/1757-899X/1166/1/012027

difference. The FC curve is measured from room temperature to 5K. The moment increases with decrease of temperature since the thermal fluctuation gets lowered, the moment variation deviates from ZFC at T_{irr} and continues to increase at and below T_B. moment continue to increase with a change of slope. This may be due to the cluster formation; in each cluster, the particles interacting through surfaces increase the anisotropy and hence are not free to rotate independently. This impedes otherwise free moments to align in the field and result in steady increase of magnetic moment with decrease of temperature. The narrow distribution of blocking temperatures (the T_B and T_{irr} are quite closer) indicate the uniform particle size and absence of strong interaction among the particles[14-17].

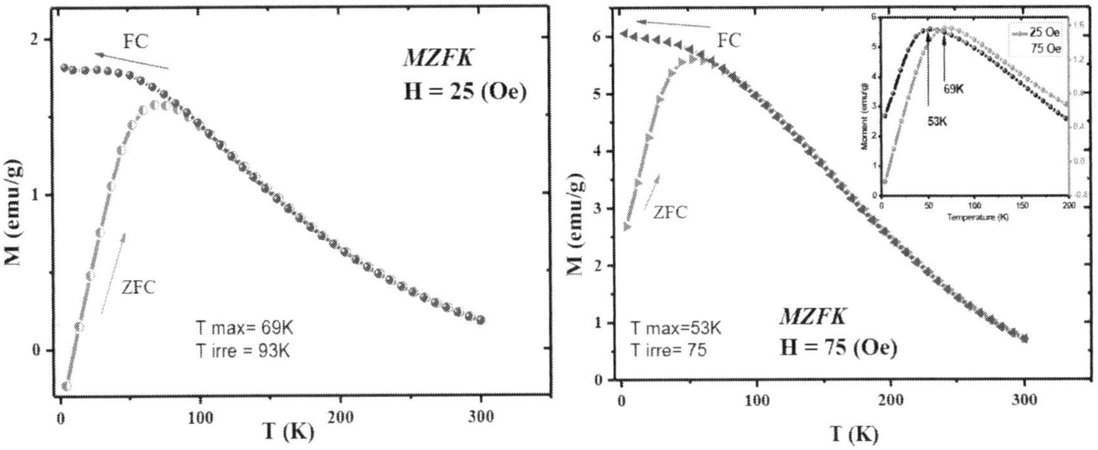

Figure 5: ZFC FC moment variation at applied fields **(a)** 25Oe, **(b):** 75Oe on MZFK

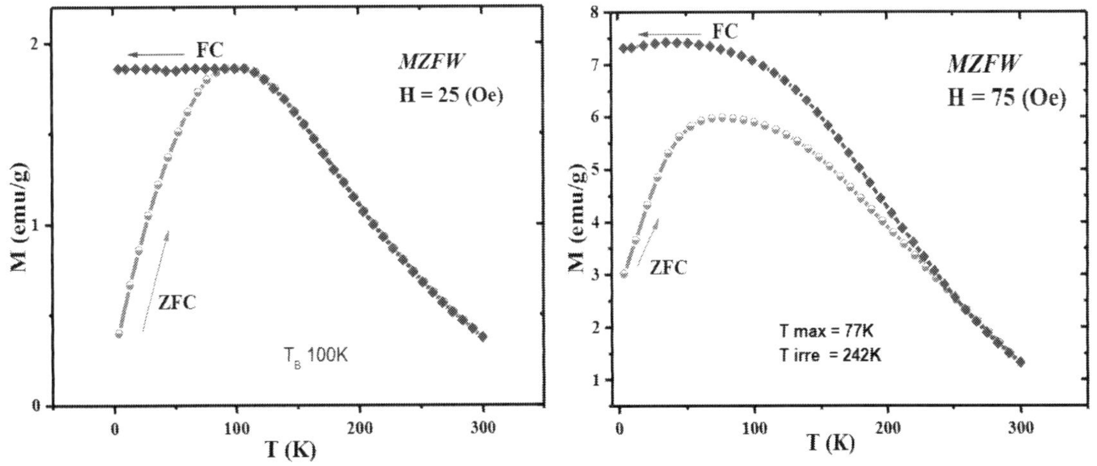

Figure 6: ZFC FC moment variation at applied fields **(a)** 25Oe, **(b):** 75Oe on MZFW

Figure 6 depicts the ZFC FC measurements on MZFW at two measuring fields. For low field, the blocking and irreversibility occurs at the same temperature which is spread around 100K at a measuring field of 25 Oe. The ZFC curve exhibits a broad maximum typically for superparamagnetic nanoparticles with uniform size distribution. The progressive deblocking of nanoparticles results in this phenomena. At $T > T_B$ both curves follow decrease of moment obeying a Curie -Weiss law. The FC moment for $T < T_B$ remains constant, which could be explained as a result of interparticle interaction which increases the anisotropy making the moments pinned to one another and the thermal energy is too small to overcome the energy barriers and cause a magnetisation reversal.

The curves obtained at 75 Oe measuring field shows lowering of blocking temperature which is quite expected as in the case of MZFK. However the bifurcation takes place at a far higher temperature from T_B. This shows the presence of larger particles in the sample. The strength of interaction depends on the moment of the particles and the increased measuring field causes stronger interaction resulting in an increase in larger effective volume and hence the higher T_{irr}. The ZFC curve shows a broader maximum and the magnetic relaxation for $T > T_B$ slightly deviates from exact Curie weiss law indicating the existence of dipole -dipole interaction among the particles.

4. Conclusions

Ferrofluids, both aqueous and hydrocarbon base fluids, of spinel manganese zinc ferrite particles of size around 5 nm, in the composition $Mn_{0.6}Zn_{0.4}Fe_2O_4$ is synthesized by supersaturating with sodium hydroxide at boiling temperature. It has been observed that the temperature, rate of addition and temperature during the whole reaction process are highly crucial in the formation of the magnetic phase. The fluids prepared are highly stable against sedimentation and possess good shelf life. Water based fluids are sufacted with citric acid which provides electrostatic stabilization which reduces when the carrier is dried off. The dipolar interaction is strong among the particles. The oleic acid coating of nanoparticles provides separation between particles and hence the dipolar exchange is found to be lowered. These fluids are expected to be good heat transfer nanofluids and can be good candidates for magnetic hyperthermia. The inter particle interaction among superparamagnetic particles are explained based on temperature dependent static magnetic measurements.

Acknowledgements
A P Reena Mary acknowledges T N Narayanan, TIFR Hyderabad for fruitful discussions and P M Ajayan for TEM measurements.

References:

[1] R. Arulmurugan, G. Vaidyanathana, S. Sendhilnathan and B. Jeyadevan , 2004 *J.Magn.Magn. Mater.* **298** 83
[2] Hassan Hejase, Saleh S.Hayek Shahnaz Qadric Yousef Haik 2012 *J. Magn. Mang. Mater.* **324** (22), 3620
[3] E. Veena Gopalan, I.A. Al-Omari, K.A. Malini, P.A. Joy, D. Sakthi Kumar, Yasuhiko Yoshida and M.R. Anantharaman 2009 *J. Magn. Magn. Mater* **321** 1092
[4] Shamima Nasrin, F.-U.Z.Chowdhury, S.M.Hoque, 2019 *Physica B: Conden. Matter.* **561** 54
[5] P. Poddar, H. Srikanth, S.A. Morrison, E.E. Carpenter, 2005 J. Magn. Magn. Mater **288** 443
[6] S. Calvin, E.E. Carpenter, B. Ravel, V.G. Harris, S.A. Morrison, 2002 *Phys. Rev. B* **66** 224405
[7] S.A. Morrison, C.L. Cahill, E.E. Carpenter, S. Calvin, V.G. Harris, 2007 *J. Appl. Phys.* **93** 7489
[8] D. J. Fatemi, V. G. Harris, V. M. Browning, and J. P. Kirkland, 1998 *J. Appl. Phys.* **83**, 6867
[9] Alison E.Deatsch and Benjamin A.Evans, 2014 *J. Magn.Magn.Mater.***354** 163
[10] A. P. Reena Mary, T. N. Narayanan, Vijutha Sunny, D. Sakthikumar ,Yasuhiko Yoshida, P. A. Joy and M. R. Anantharaman. 2010 *Nanoscale Res Lett* **5** 1706
[11] Y. Bao, A. B. Pakhomov, and Kannan M. Krishnan, 2006 *J. Appl.Phys.* **99**, 08H107
[12] G. F. Goya, T.S.Berquo, F.C.Fonseca, and M P Morales, 2003 *J. Appl. Phys.* **94** 3520
[13] Jing Feng, Li-Qin Guo, Xiaodong Xu, Shu-Yan Qi and Mi-Lin Zhang, 2007 *Physica B: Condensed Matter* **394** 100
[14] Jose. M. Vargas et al, 2005 *Nanotechnology* **16** S285

[15] G. Muscas, G. Concas, S. Laureti, A. M. Testa, R. Mathieu, J. A. De Toro, C. Cannas, A. Musinu, M. A. Novak, C. Sangregorio, Su Seong Leei and D. Peddis 2018 *Phys.Chem.Chem.Phys.* **20** 28634
[16] Gabriele Barrera, Paola Tiberto , Paolo Allia, Barbara Bonelli , Serena Esposito , Antonello Marocco , Michele Pansini and Yves Leterrie 2019 *Appl. Sci.* **9**, 212
[17] H. Mamiya, M. Ohnuma, I. Nakatani, and T. Furubayashim, 2005 *IEEE Trans. Magn.* **41** (10) 3394

Push system for Aluminium tube by using a stepper motor with the Arduino

P P Joshi[*]**, V S Jorapur**

Department of Mechanical Engineering , Fr. Conceicao Rodrigues College of Engineering, Fr. Agnel Ashram, Bandstand, Bandra (W), Mumbai–400050, India, Affiliated to Univ. of Mumbai.

E-mail: *pranavjoshi174@gmail.com*

Abstract: The research content of this paper is to design a push system for an Aluminium tube to be cut on the cutting machine. The system consists of one driving pulley, one idler pulley, and a timing belt. The stepper motor will drive the pulley and the stopper attached to a timing belt will push the Aluminium tube further. This paper describes the process of developing a push system using a stepper motor driver, membrane keypad, Nokia LCD, and a microcontroller board. Different types of stepping mode and the formula for the number of pulses to be sent to a stepper motor are derived and tested. The result of the paper shows that a stepper motor along with the electronic components improves accuracy for a push application and can be used.

1. INTRODUCTION

The stepper motor is a digital electromechanical system where a user sends an electrical pulse input which results in a rotation of the rotor by a specific angle called step angle[1]. The stepper motors are used in industries worldwide for robotics and automation due to their availability in different sizes and economical price. Applications of stepper motors include CNC machines, 3-D printers, flatbed scanners, security cameras, etc. These applications require precise positioning and accurate control over their position. The stepper motor works very precisely even without feedback by the position sensor. The open-loop operation of the stepper motor makes it cheaper.

A push system is used in this project for an extruded Aluminium tube of a rectangular cross-section to be pushed on the cutting station using a stepper motor. The motive is to push an Aluminium tube up to a required length using a stopper and after reaching its stop, the human operator will cut the tube which will be of a specific length entered using a keypad. The goal here is to assist the worker to push the tube further on the cutting station up to a required length that needs to be cut. This will save time as it eliminates the process of marking and then cutting.

2. LITERATURE REVIEW

The push system for the Aluminium tube needs very precise movements which comes in very handy by using a stepper motor. A stepper motor is used where accuracy is of utmost priority such as in robotic applications. A stepper motor is the most suitable motor for this system because the motor runs with respect to the number of pulses sent to a driver. The motor is very useful due to its properties such as high torque at low speeds, low cost, and reliability[1].

Stepper motors have some problems such as jerk produced by the motor while rotating the shaft by the bigger step angle. It creates dynamic instability in the motor and results in a very noisy operation. However, the step angle can be reduced by using microstepping mode. In microstepping mode, the full step angle is subdivided into multiple small-angle steps called microsteps. G. Baluta concludes that using microstepping mode, the noise was reduced substantially, position control became more precise and the high step resolution is achieved[2].

If the load torque is stable, operating at low speed, the stepper motor can be used in an open-loop as it requires a simple control algorithm[3]. The stopping position of the system is independent of load change and only depends upon the pulse frequency and the number of pulses sends to a stepper motor driver[4]. The stepper motor coils are always excited when the power is on which induces holding torque inside the stepper motor[5]. Due to a holding torque, a self-locking function is created by a stepper motor[4]. This self-locking function is very important as it will help to hold the position of the timing belt.

The stepper motor should move the stopper from one location to another location, generally through acceleration, constant speed, and deceleration. If the speed starts by activating the frequency exceeding the limit frequency of the stepper motor the step phenomenon occurs, which results in not starting correctly. If you suddenly stop to the end, due to inertia, the motor can overshoot a few steps which will reduce positional accuracy[6]. Therefore using steady acceleration and deceleration will provide better results.

3. CALCULATION OF NUMBER OF PULSES REQUIRED TO MOVE THE STOPPER

The stepper motor is also called a pulse motor. When a single pulse is sent to the motor, the motor shaft rotates by a certain degree of angle. Most of the stepper motors have a rotation of 1.8° per step, which means 200 steps per revolution.

In many applications, microstepping can increase system performance, and lower system complexity and cost, compared to full-step and half-step driving techniques. Microstepping can be used to solve noise and resonance problems, and to increase step accuracy and resolution. Microstepping modes are ranging from 1/2 to 1/32 of the full step. Higher the microstepping, lower the vibration and noise[7].

By using EasyDriver as a stepper motor driver, the microstepping up to $1/8^{th}$ of the full step can be achieved[8]. The EasyDriver will help to drive the stepper motor in $1/8^{th}$ microstepping mode. It is set as default settings.

The stepper motor runs on the pulses sent to it by a stepper motor driver. But the EasyDriver by default runs on microstepping mode. So either computer or PLC has to do the calculations for the required number of pulses.

3.1 Factors affecting the number of pulses

3.1.1 Stepping mode

Each pulse will result in a certain angle of rotation of the shaft. In this case, due to the use of an EasyDriver as the stepper motor driver, one full step will become 8 microsteps. The stepper motor takes 200 full steps to complete one revolution. It will need 1600 pulses or microsteps to complete one full revolution.

3.1.2 Pitch profile of the pulleys and the timing belt

The pulley and the timing belt will also be the factor in calculating the number of pulses. The pulley and the timing belt need to have the same tooth profile to interconnect with each other. The timing belt and pulleys have standard tooth profiles such as GT 2, GT 3, or GT 5 where the number indicates the pitch length. That means GT 2 pulley or timing belt will have a 2 mm pitch length as shown in figure

1.

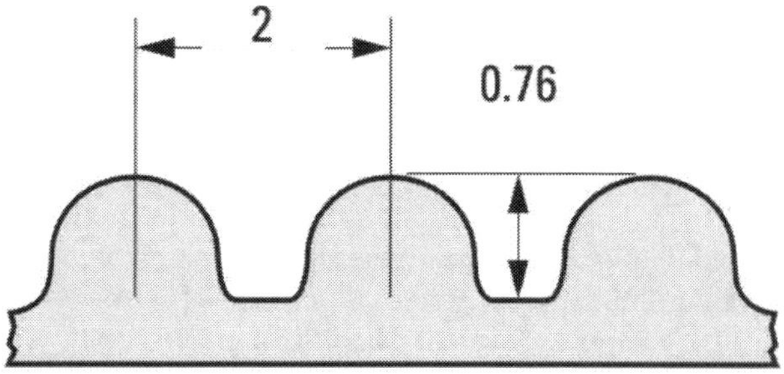

Figure 1. The tooth profile of the GT2 timing belt

3.1.3 Number of teeth on the pulleys

The number of teeth and the pitch diameter needs to be the same for driving pulley and idler pulley to push the Aluminium tube linearly on the cutting machine. The length of the timing belt will depend upon the diameter of the pulley and the distance between the driving pulley and the idler pulley.

Let,

The pitch of the timing belt and pulley = 2 mm i.e. GT 2 profile
Number of teeth on both pulleys = 20 teeth
Angle per full step taken by stepper motor = 1.8°
Total full steps to complete 1 revolution = 360°/1.8° = 200 full steps
Stepping mode = 1 / 8th step per pulse
Pulses need to move the rotor one full step = 8 pulses
Pulses need to move rotor one full revolution = 200 steps x 8 pulses = 1600 pulses

Now one revolution of the rotor is one revolution of the driving pulley as it will be interconnected
Hence,
One revolution of the driving pulley = Number of teeth x Pitch length
= 20 x 2 mm = 40 mm
Now the relation between length traveled by the stopper attached to the timing belt and the pulses is

1600 pulses = 40 mm Equation (1)

Equation (1) is completely dependent upon the three factors i.e. stepping mode, pitch profile of the pulleys, and the number of teeth on the pulleys.
By using Equation (1), the formula for the number of a required pulse will be

Required number of pulses for entered distance = ((Z x 1600) / 40),
Where, Z = Entered distance

4. CALCULATION OF TORQUE REQUIRED TO PUSH ALUMINIUM TUBE

A stepper motor should provide an output torque larger than load torque and be required to start and stop at a proper step rate against load inertia.

4.1 Types of torque

4.1.1 Detent torque

When a hybrid stepper motor is in a state of rest and not energized, a small torque must be applied to the rotor to break through the static equilibrium state which is called detent torque[5].

4.1.2 Holding torque

When the windings are powered and the rotor is at the balanced position, an external torque should be applied to a hybrid stepper motor to rotate the rotor one full step. This torque is called holding torque. Holding torque maintains the rotor position at equilibrium irrespective of load. And it depends upon the magnitude of the current[5].

So holding torque is a very important characteristic in the stepper motor. Also, the stepper motors are selected by calculating the holding torque required by the load.

In this case, the load is Aluminium tube, and the timing belt which is made of neoprene synthetic rubber.
The formula for torque is,

$T = F \times r \times \sin\theta$

Where T = Torque

r = Radius to apply the force

The stepper motor will be pushing the Aluminium tube parallel to the timing belt hence θ will become 90° and $\sin 90^{\circ} = 1$ as shown in figure 2.

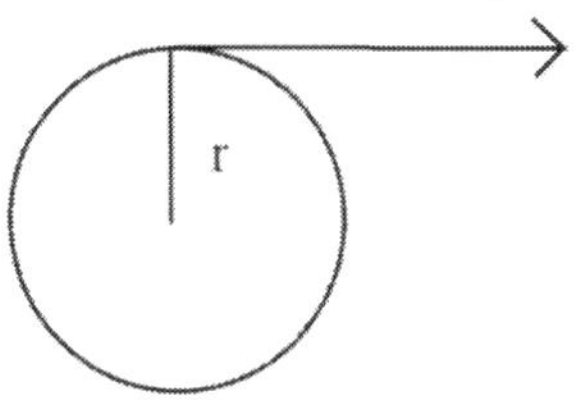

Figure 2. The direction of force acting on Aluminium tube

F = (Weight of Aluminium tube (Kg) + Weight of timing belt (Kg)) x 9.8 m/s^2
The weight of the Aluminium tube and timing belt is calculated in designing software, which approximately comes 2 Kg for Aluminium tube and 0.7 Kg for the timing belt of pitch 5 mm.
Weight of Aluminium tube = 2 Kg
Weight of timing belt = 0.7 Kg
Hence, F = (2 Kg+ 0.7 Kg) x 9.8 m/s^2

= 26.46 Kg.m/s^2 = 26.46 N

r = radius of pulley (m)
The diameter of the pulley depends upon the number of teeth and pitch.
Let us select a bigger pulley with a 5 mm pitch and 40 numbers teeth.
Diameter of pulley = Circumference / π

= 40 x 5/ 3.142 = 63.69 mm

Radius of pulley = 31.85 mm = 0.03185 m
Now multiplying Force with radius,

T = F x r

= 26.46 N x 0.03185 m

= 0.842751 N.m

4.2 Stepper motor selection based on torque ratings
Stepper motor torque will be needed to be more than the calculated torque as there are some factors which motor torque needs to overcome such as friction between pulley and timing belt, and friction between Aluminium tube and cutting machine. If the mechanical load increases above maximum electromagnetic torque, step loss will occur[9]. Also, the holding torque mentioned by the manufacturers is specified, when the full rated current is supplied.
Generally, the stepper motor torque is shown in Kg.cm for ease of understanding.
So, converting N.m toKg.cm

0.842751 N.m = 8.6 Kg.cm
Stepper motors are available in a very wide range of holding torque which is around from 1 Kg.cm to 500 Kg.cm. For the safety of a pushing system, selecting a Stepper motor twice the power will be suitable for this project i.e. 17.2 Kg.cm. NEMA 23 stepper motors have much higher holding torque than NEMA 17. Here, 23 indicate faceplate dimensions of stepper motor i.e. 2.3in x 2.3 in. Stepper motors are available in the standard faceplate dimensions such as NEMA 17, NEMA 23, NEMA 34, etc. The larger the faceplate, the larger the core will be, inducing higher torque.

5. SYSTEM DESIGN

The system design consists of software and hardware. The software implementation is done using Arduino IDE software which is based on the C programming language. The hardware consists of the components mentioned in table 1. The figure 3 shows the connection between various components

5.1 List of components

Table 1. List of electronic components used

Sr. No.	Components
1	Arduino MEGA 2560 R3
2	4 x 4 Membrane Keypad
3	Nokia 5110 Display (84 x 48 pixels)
4	EasyDriver Version 4.4

5.1.1 Arduino MEGA 2560 R3 Microcontroller Board

The Arduino is open-source hardware designed by the Arduino team in Italy. There are two types of Arduino which are mainly used viz. Arduino UNO, Arduino MEGA, etc. The only feature which apart MEGA from UNO is the processor. Arduino MEGA uses ATmega 2560 processor while UNO uses ATmega 328. ATmega 2560 is a 100 pin surface mount 8-bit microcontroller[10].

The processor can operate at 16 MHz that is 16 million transactions per second. The flash memory is 256K byte which is much more than the UNO. Flash memory is used for storing a program. Also, the number of ports on the MEGA is more than the UNO.

5.1.2 4 x 4 Membrane Keypad

The membrane keypad is a package with multiple switches. There can be multiple combinations of many switches on the keypad. Here 4 x 4 represents 4 rows and 4 columns. The keypad has flat cable lines for each row and each column. In this case, there are eight numbers of ports four for rows and four for columns.
When the key is depressed, the key's respective row and column get activated and the pulse is identified by the pins of Arduino. Each key has its row number and column number. The combination of a row and a column assertion can be decoded to determine which key is pressed [10].

5.1.3 Nokia 5110 Display

The Nokia 5110 display is a graphic LCD useful for a lot of applications. It was intended originally to be used as a screen for cell phones and was used in lots of mobile phones during the '90s era. It uses a low powered CMOS LCD controller/driver PCD8544, which drives the graphic display of size 84x48 pixels.[11] There are four backlight LED's fixed on the border of the display. The backlight intensity can be adjusted according to the need. The display uses a very low amount of current around 6 to 7 mA, so the use of resistors is suggested.[12]

5.1.4 EasyDriver Version 4.4 for stepper motor

EasyDriver Version 4.4 is the name of the driver for the stepper motor which is used in this project. EasyDriver can drive up to about 750mA per phase of a bipolar stepper motor. The driver by default use $1/8^{th}$ step microstepping mode. This setting can be easily overridden by tying the MS1 and/or MS2 pin to the ground to set the driver to use 1/8, 1/4, or 1/2 microstep mode. The driver only needs 2 types of signals from the Arduino i.e. number of steps to rotate the shaft and the direction in which the shaft is supposed to rotate. The driver can sustain voltage from 6 volts to 30 volts. The A3967 driver chip carefully controls the current to each coil to cause the rotation to move to positions in between the normal full step positions. This allows for smoother motion, reduced mid-band resonance, and higher positional accuracy and resolution.

The Easy Driver can operate in 1/8th, 1/4, half, and full step modes. These four modes are selected by the logic levels on the MS1 and MS2 input pins as shown in table 2. Normally, the pull-up resistors on the Easy Driver hold MS1 and MS2 high, which results in a default setting of 1/8th microstep mode. By making MS1 and/or MS2 ground, the above mentioned results can be achieved[8].

Table 2. Resolution setting of EasyDriver

MS1	MS2	RESOLUTION
Low	Low	Full Step
High	Low	Half Step (1/2)
Low	High	Quarter Step (1/4)
High	High	Eighth Step (1/8)

The selection of EasyDriver makes the default resolution of step $1/8^{th}$ of the full step. So, now 1 full step of the motor will be equal to 8 microsteps. And the rotation angle per step will become 1.8°/8. So by using the EasyDriver for the stepper motor, the motor will run with much less noise and vibration.

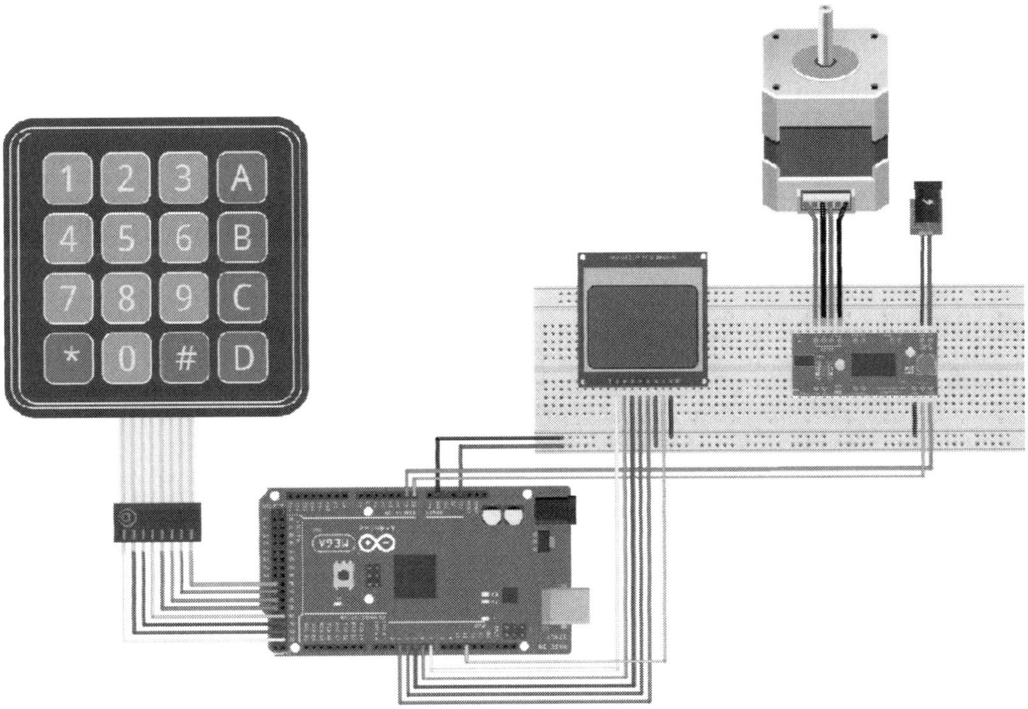

Figure 3. Connections between components

6. HARDWARE IMPLEMENTATION

The above mentioned components are going to be used with the stepper motor. There will be two pulleys at both ends and the timing belt will be attached in between them. The stepper motor shaft will be attached to the driving pulley. And at the other end, the idler pulley will be placed. Both the posts, the driving pulley post, and the idler pulley post will be bolted down to the table as shown in figure 4. In the figure, the stopper attached to the timing belt is shown red. The supporters are used as guiderails to guide the Aluminium tube to the cutting machine.

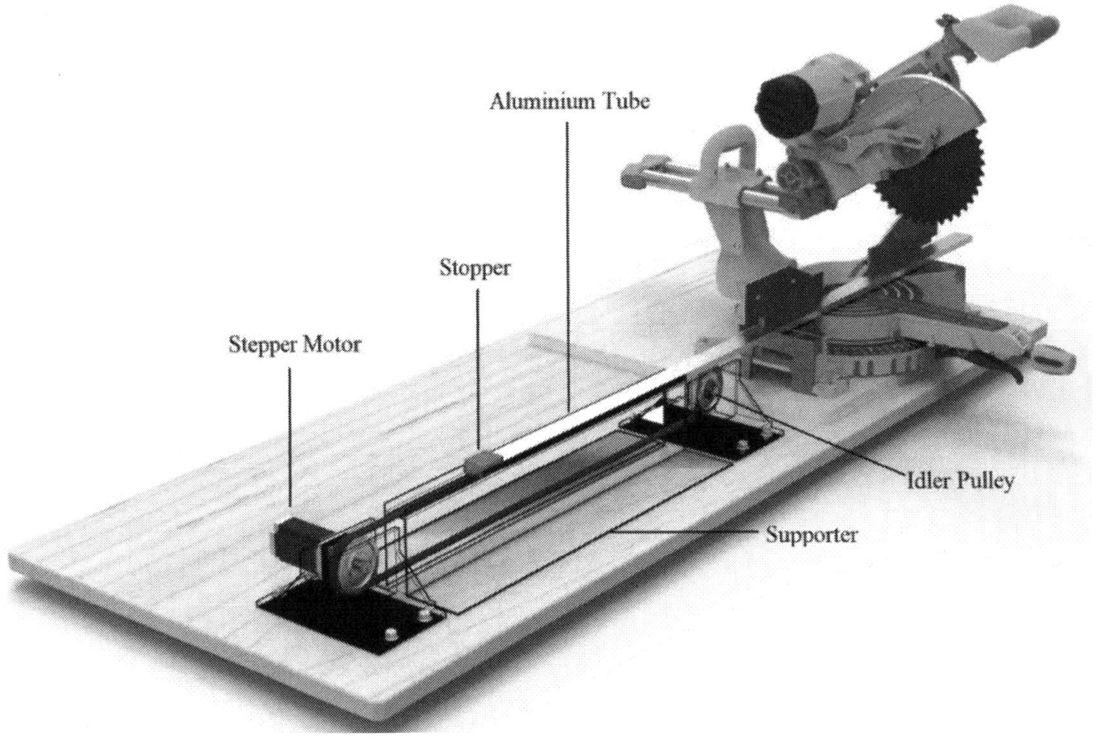

Figure 4. Rendered image of pushing system with Aluminium tube

The Aluminium tube will be kept on the supporter which will constrain its motion in only one direction. The Aluminium tube will slide on the supporter when the stopper will push the tube. Both posts and the supporter are shown transparent just for the good visibility of the pushing system.

7. MINI WORKING MODEL

The mini working model here is used for obtaining results by applying various hypotheses. The written program is compiled and ran using Arduino MEGA. The Nokia display is connected to Arduino MEGA without the use of level shifters as shown in figure 5.

The EasyDriver is used to drive the NEMA 17 stepper motor which needs a low amount of current than the NEMA 23 stepper motor. The temperature of the A3967 chip installed on the driver can go up to 139°C. However, the EasyDriver has a current limit pot set to limit a current flow which will regulate the current flowing to the stepper motor. The EasyDriver has a thermal shutdown temperature of 165°C[8] and if the heat sink is used then the driver can be used at higher currents also.

But a bigger stepper motor such as NEMA 23 will need a bigger driver. This EasyDriver has a limit of 750mA/Phase. So it is recommended to use a bigger version of the EasyDriver.

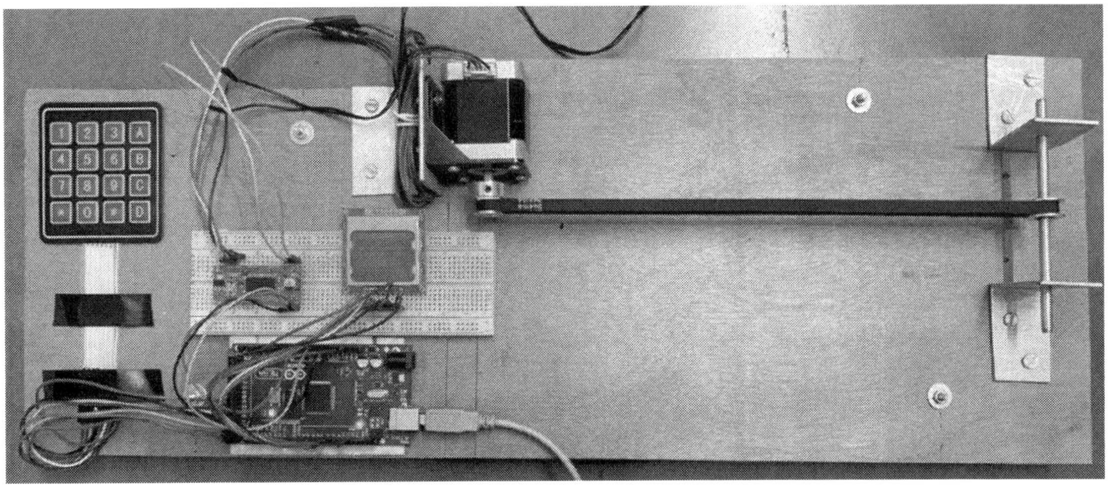

Figure 5. A miniature model with electronic components

The SparkFun website provides a bigger version of the EasyDriver which can withstand current up to 2A/phase. The Big Easy Driver, designed by Brian Schmalz, is a stepper motor driver board for bi-polar stepper motors up to a max 2A/phase. It is based on the Allegro A4988 stepper driver chip. Also, the advantage of using this driver is it runs by default on 16 steps microstepping mode which means more step resolution[13].

8. RESULTS

1) By using a stepper motor amongst all types of motors high holding torque and precision can be achieved.
2) The holding torque value depends upon the amount of current per phase.
3) Microstepping mode offers a low noise operation and high step resolution.
4) EasyDriver helps to drive the stepper motor in microstepping mode that is $1/8^{th}$ of a full step. And it is by default action taken by the EasyDriver
5) The gradual acceleration and deceleration can be achieved using the AccelStepper library in Arduino programming.
6) The calculated number of pulses rotates the stepper motor shaft precisely to its required location.

9. CONCLUSION

It is concluded from the above results, by using a stepper motor with the help of EasyDriver, effective control is achieved over stepper motor speed. Microstepping gives more step resolution. While the gradual acceleration and deceleration of the stepper motor speed help to remove jerk and noise. The calculated number of pulses provides the required precision to attain the required position by a stopper. The results are sufficient to conclude that this whole system can be used as a push system for an Aluminium tube.

ACKNOWLEDGMENT

I am grateful to Mr. Vishal Dakshini and Mr. Manthan Dakshini for always supporting and motivating me with their immense knowledge and for letting me do this project.

REFERENCES

[1] B. Aranjo, P. K. Soori, and P. Talukder, "Stepper motor drives for robotic applications," *2012 IEEE Int. Power Eng. Optim. Conf. PEOCO 2012 - Conf. Proc.*, no. June, pp. 361–366, 2012.
[2] G. Baluta, "Microstepping mode for stepper motor control," *ISSCS 2007 - Int. Symp. Signals, Circuits Syst. Proc.*, vol. 2, pp. 609–612, 2007.
[3] J. W. J. Ngoc Quy Le, "An Open-loop Stepper Motor Driver Based on FPGA," *Int. Conf.*

Control. Autom. Syst., no. 235, p. 245, 2007.
[4] X. Wang, "Research on Stepper Motor Control System based on Single Chip Microcomputer," vol. 50, no. Iceeecs, pp. 105–108, 2016.
[5] Y. Fan, "3D finite element solution of a hybrid stepper motor," *Powerconversion Intell. Motion*, vol. 16, no. 1, pp. 19–20, 22, 2019.
[6] X. Wei, "Acceleration and deceleration control design of step motor based on TMS320F240," *Procedia Eng.*, vol. 15, pp. 501–504, 2011.
[7] Japan Radio Corporations, "Microstepping," p. 11, 2015.
[8] "Easy Driver Stepper Motor Driver." [Online]. Available: https://www.schmalzhaus.com/EasyDriver/.
[9] J. De Viaene, S. Derammelaere, and K. Stockman, "Load angle estimation for dynamic stepping motor motion applications," *Mechatronics*, vol. 53, no. July 2018, pp. 229–240, 2018.
[10] S. Barrett, *Arduino Microcontroller Processing for Everyone!*, Third Edit. Laramie, Wyoming: Morgan & Claypool Publishers, 2013.
[11] D. T. Htwe, "ARDUINO BASED UV CARE SYSTEM WITH GSM MODULE."
[12] "Interface Nokia 5110 Graphic LCD Display with Arduino." [Online]. Available: https://lastminuteengineers.com/nokia-5110-lcd-arduino-tutorial/.
[13] Sparkfun, "Big Easy Driver." [Online]. Available: https://www.sparkfun.com/products/12859.

Building Machine Learning Model for Predicting Breast Cancer Using different Regression Techniques

K. Devasena[1] and J. Shana[2]

[1]Coimbatore Institute of Technology, Coimbatore, India
[2]Coimbatore Institute of Technology, Coimbatore, India

shana@cit.edu.in

Abstract. Breast cancer is one of the major life-threatening diseases in the world today. But early diagnosis would ensure timely treatment and even recovery. Apart from the medical procedures that help to identify cancer, techniques such as machine learning can be applied to predict the occurrence of the disease in patients by analyzing their cell data. In this work, an attempt is made to apply three different regression techniques Multiple, Ridge, and Polynomial regression techniques. The performance metrics reveal that Multiple regression has a better performance compared to the other models.

Keywords: Multiple Regression, Ridge regression, Machine Learning, Cancer data, Prediction

1. Introduction

Cancer has been portrayed as a heterogeneous sickness comprising a wide range of subtypes. The early analysis and anticipation of a malignant growth type have become a need in disease research, as it can encourage the resulting clinical administration of patients. The significance of ordering cancer patients into high or low has driven many exploration groups, from the biomedical and the bioinformatics field, to consider the utilization of AI (ML) techniques. In this manner, these methods have been used to show the movement and therapy of carcinogenic conditions. Furthermore, the capacity of machine learning algorithms to recognize key highlights from complex datasets uncovers their significance.

Researchers applied various techniques, for example, screening in the beginning phase, to discover kinds of malignant growth before they cause side effects. Also, they have grown new techniques for the early forecast of disease treatment results. With the appearance of innovations in the field of medication, a lot of malignant growth information has been gathered and is accessible to the clinical examination team.

Machine learning models have become so useful in every domain. It can deliver useful insights from datasets and allow effective decision-making. Linear regression and Logistic regression are the two most prominent techniques in machine learning. Regression analysis is a predictive modeling technique that analyzes the relation between the target or dependent variable and independent variable in a dataset. The different types of regression analysis techniques are used when the target and independent variables show a linear or non-linear relationship between each other. There are many types of regression analysis techniques, and the use of each method depends upon the number of factors.

1. Linear Regression
2. Logistic Regression
3. Ridge Regression
4. Lasso Regression
5. Polynomial Regression
6. Bayesian Linear Regression

The goal of the model is to predict the concavity of the cancer cell using different regression models. The larger the concavity, the larger the pressure on surrounding cells, since the cancer cell competes with adjacent normal cells for nutrients. The concavity of the cancer cell tells how much the cell can expand. The concavity of the cancer cell determines its tumor potential. It determines how fast this cancer cell divides. If it divides quickly, the tumor mass grows rapidly and thus it can invade other tissues and can also metastasize. If it divides slowly, mostly it will be benign and does not metastasize. Thus, by predicting the concavity of the cancer cells will help us in deciding how fast cancer would get severe in the future. Here various regression techniques have been applied and the performance of all these have been compared

2. Related Work

In [1] the authors have surveyed machine learning techniques applied for cancer data analysis. They include machine learning models such as Naive Bayes, Support Vector Machines, Neural Networks, etc. In [2] the authors have explained Multiple Linear Regression and the steps to do like finding out the missing values. In [3] the authors have explained Ridge regression regularization and its penalty term. The authors in [4] explore the different data mining approaches using classification which can be applied to Breast Cancer data to build deep predictions. Result analysis reveals that among all the classifiers Simple Logistic Regression yields the deep predictions. The authors in [5] have tried seven regression techniques and implemented them on the Breast Cancer dataset to know the best regression technique among those compared. The analysis found that Logistic Regression gives the best results to diagnose Breast Cancer with given attributes among seven regression techniques with least Relative Absolute Error (10.59%) and Least Root Relative Squared Error. (45.84%) using the Weka tool.

In [6] the authors have applied machine learning methods especially logistic regression on the datasets to determine and examine whether there are any signs or possibilities of cancer and if the person is examined as cancerous then the stage of cancer is also determined. And out of all the techniques logistic regression outperformed other techniques. The study in [7] used machine learning techniques to build models for detecting and visualizing significant prognostic indicators of breast cancer survival rate for a Malaysian hospital. Model evaluation using random forest algorithms yielded slightly better accuracy when compared to other algorithms. Nevertheless, accuracies displayed by all the algorithms appeared close.

Study in [8] compared the performance of artificial neural network and multivariable logistic regression models, in prediction of outcomes in head trauma and studied the reproducibility of the findings. Artificial neural network outperformed the other model in term of accuracy.

Authors in [9] investigate their applications in the Breast cancer dataset and compare the results through different algorithms using the Wisconsin breast cancer database (WBCD) which is the benchmark database for comparing the results through different algorithms. In [10], different machine learning and data mining techniques for the detection of breast cancer were proposed. Results obtained with the logistic regression model with all features included showed the highest classification accuracy (98.1%), and the proposed approach revealed the enhancement in accuracy performance. In [11] the authors have explained Polynomial regression, the relationship between the attributes and its evaluating metrics.

3. Regression Techniques

There are many regression techniques apart from linear regression as listed below. There exist linearity and nonlinearity between the variables so in order to model the exact variance of the data Multiple Linear Regression and Polynomial Linear regression both have been implemented and analyzed their performance in terms of variance the model explains (R^2 score) and accuracy of the predictions made by those metrics. In order to use Ridge Regression, VIF (Variance Inflation Factor) has been done to check the multicollinearity between the independent variables and it has multicollinearity.

3.1 Multiple Linear Regression

Multiple Linear regression (MLR) is a lot like simple linear regression in which it depends on one predictor variable to predict the criterion value whereas MLR depends on more than one predictor value to predict the criterion value. While doing MLR, the assumptions are as follows:

- Both predictor and criterion variables are in a linear relationship
- Predictor variable is not highly correlated with each other.
- Residuals should be the same at every explanatory level.
- Residuals should be normally distributed.

Suppose the model has 'n' independent variables, then Multiple linear regression Equation can be written as

$$y = b_0 + b_1x_1 + b_2x_2 + \cdots . b_nx_n + e \quad (1)$$

Where y = Criterion variable(Dependent Variable), b_0 = Y-Intercept(constant),
b_1,b_2 = Regression coefficients associated with x_1,x_2
x_1,x_2 = Explanatory variables
b_n = Slope and
e = Residual value(Error).

The advantage is it is an efficient prediction model with more than one independent variable as it will give a more precise output. The disadvantage is there is too much data and that may have false or incomplete data which is difficult to find.

3.2 Ridge Regression

Ridge Regression is a method that becomes useful when the dataset correlates with the independent variables in other terms when it has multicollinearity. It is an L2 regularization method that reduces the risk of Overfit and improves the model accuracy. As ridge regression is an L2 regularization method, it determines the parameters of the equation and minimizes the Sum of squared residuals with a penalty term which makes the ridge regression line way more efficient than the least square line.

$$RSS + \lambda \sum P_{i=1} \beta_i^2 \qquad (2)$$

In Ridge Regression, with a small amount of Bias due to a penalty, it reduces the variance of the model. The advantage is it reduces model complexity and provides a long-term solution. The drawback is it will shrink the lambda value close to 0 but never exactly 0. And it becomes computationally expensive.

3.3 Polynomial Linear Regression:

While modeling the data and finding it out as it is performing badly as it needed some curvature line for the data to fit, this is where Polynomial Linear regression comes in handy. It tells the relationship between the predictor and criterion variables as an nth degree polynomial. In Polynomial linear regression, it doesn't require the relationship between the predictor and criterion variables to be Linear.

$$y = b_0 + b_1 x_1^2 + b_2 x_2^2 + \cdots . b_n x_n^2 + e \qquad (3)$$

where n is the degree of the polynomial b is the set of coefficients.

The degree depends on the predictor and criterion variables. The polynomial linear regression with degree one, is just a simple linear regression. So, the degree for the polynomial linear regression should be more than one. The best degree for the model, in which it generates the lowest RMSE. In polynomial linear regression, as n takes the higher value it increases, its accuracy but also higher the degree, the higher the chance of overfitting. This regression fits a wide range of curvatures. But it is strongly sensitive to outliers.

4. METHODOLOGY

The process of building the regression model, in general, is depicted in Figure 1. The steps are elaborated in section 4.1. through 4.9.

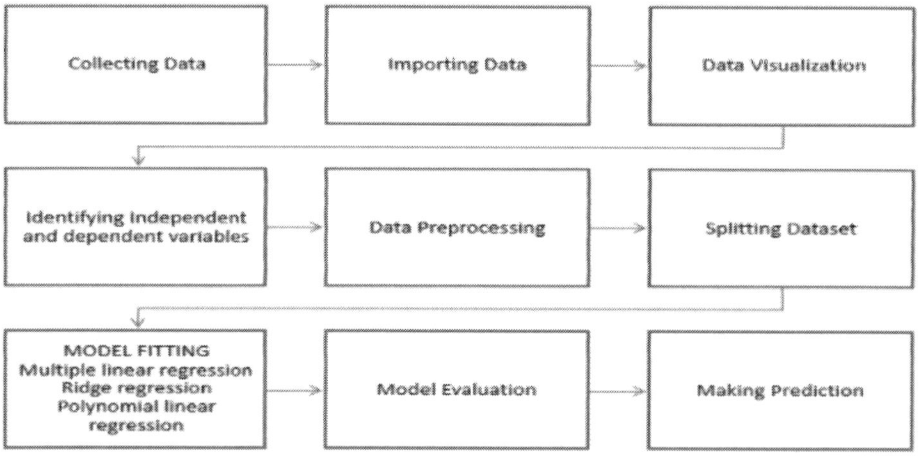

Figure 1: Methodology used in Machine Learning

4.1 Collecting Data:
Collecting data can be of Primary Source or Secondary Source. In primary sources, data is collected directly without any Third-party whereas, in Secondary sources, it takes the data from the primary source. The dataset for this study has been downloaded from the Kaggle website [12] and falls under secondary data. As it is a classification dataset, the dataset's features required for regression analysis have been separated from the original dataset. The attributes of the dataset are as follows:

- radius_mean - Mean of the distances from the center to the points on the perimeter of the cancer cell
- texture_mean - Standard deviation of greyscale values
- perimeter_mean
- area_mean
- smoothness_mean - Local variation in radius lengths
- compactness_mean - Perimeter^@2/area - 1.0
- concave points_mean - Number of concave portions of the contour
- fractal_dimension_mean
- concavity_mean - Severity of concave portions.

4.2 Importing Data:
Locate the CSV file from the system and import it using the pandas library with the pre-defined function named .read_csv("Filename.csv").

4.3 Data visualization:
The main goal of data visualization, is to make it easier to visualize the information and to identify patterns in the dataset. Charts and Graphs make the new data findings easier than the normal dataset.

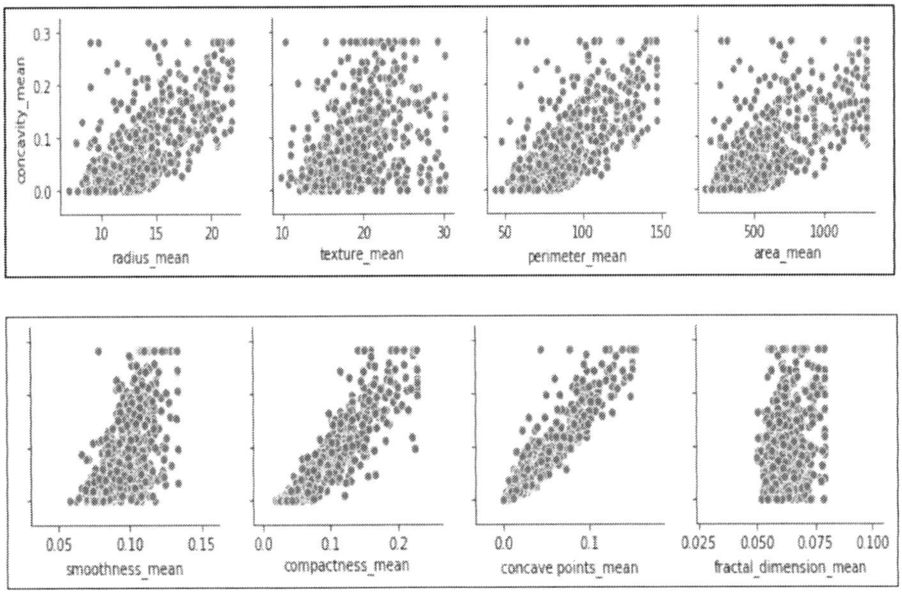

Figure 2: Plot between independent and dependent variables

Figure 2 shows the relationship between the concavity_mean and the radius_mean,area_mean, and some features.

4.4 Identifying Dependent and Independent Variables
In simple words, dependent variables are the cause, which means its value is dependent on the other and independent variables are the effect, which means its value changes the other's value. Here the independent variables are radius_mean, texture_mean, perimeter_mean, area_mean, smoothness_mean, compactness_mean, concave points_mean, fractal_dimension_mean, and the dependent variable is Concavity_mean

4.5 Data Preprocessing
Data preprocessing refers to the technique of preparing the raw data into data that is suitable to build the model. Without this step, the model's accuracy may reduce.

Missing values: Missing values can create problems. It is good if the missing value has been identified and replaced. These can be done with the mean/median method or by simply removing it but it may result in a dosage of data.

Feature Scaling: Feature Scaling is a technique used to normalize the range of features of the data. Feature Scaling brings all the values in the dataset to the same magnitude.
In this work, Min-max normalization is one of the techniques to perform feature scaling, it rescales a feature value with a distribution value from 0 to 1.

Removing Outliers: An outlier is an observation in a dataset that is different from other observations in the feature. It may occur due to input error or data corruption.

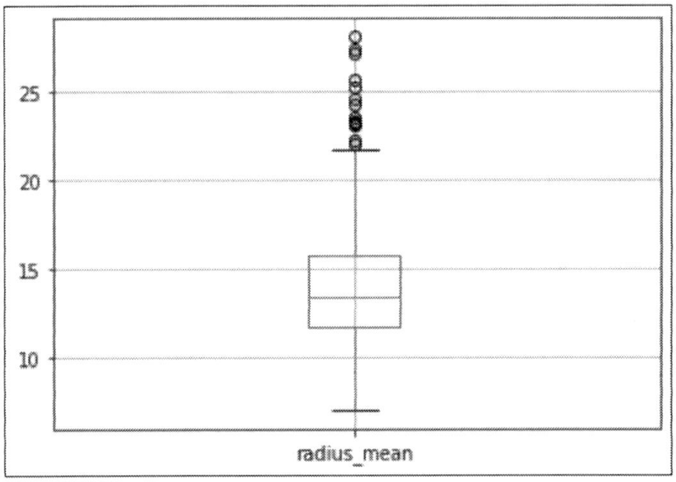

Figure 3: Before removing Outliers.

Figure 3 shows how the data will be visualized when it has outliers in the dataset

IOP Conf. Series: Materials Science and Engineering 1166 (2021) 012029 doi:10.1088/1757-899X/1166/1/012029

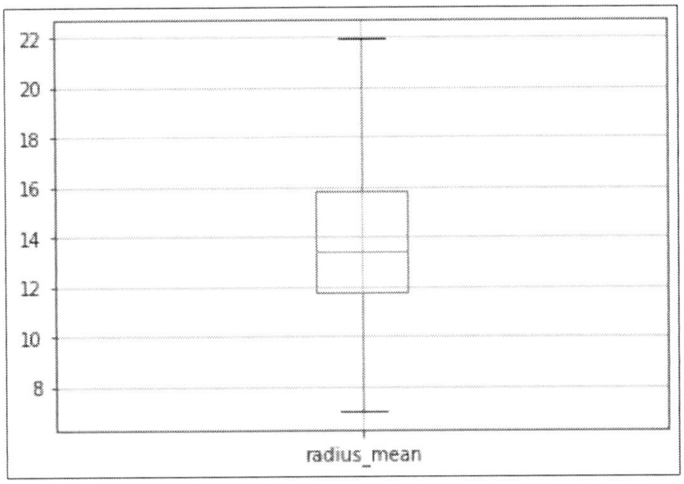

Figure 4: After removing Outlier

Figure 4 shows how the data will be visualized when it does not have outliers in the dataset. In this work, Outliers has been done to rectify the problem of input error.

4.6 Splitting Dataset

The dataset has been split into 2 separate sets namely the training set and test set. A training set is a subset that is used to train the model on the other hand Test set is the one that is used to test the predictions of the model. Usually, the dataset splitting happens in the ratio of 70:30 or 80:20 which means 70% or 80% of the data is for training the model and 30% or 20% of the data is for testing the model. The splitting can vary from dataset to dataset. In this work, the dataset has been divided into a ratio of 75% for the training dataset and 25% for the testing dataset.

4.7 Model Fitting

The model fitting is a measure of how well a machine learning model generalizes to similar data to that on which it was trained. A well-fitted model produces more accurate outcomes. A model that is overfitted matches the data too closely. A model that is under fitted doesn't match closely enough.

4.8 Model Evaluation

In Model Evaluation, the model is evaluated using different performance metrics such as Means Square Error (MSE) and Root Mean Square Error (RMSE). A model with good performance metrics is considered the best deployment model. Coefficient of determination (R^2) describes how the differences in the Independent variable can be affected by the differences in the Dependent variable. MSE reveals the average residual sum of squares. Mean Absolute Error specifies the average distance between the actual and predicted values. Root Mean Square Error (RMSE) - RMSE tells the Standard Deviation of the residuals. In this work, the model has used the above-mentioned metrics to evaluate the model performance.

4.9 Prediction

The model is used for predicting the disease. Given the new input data, the model will predict the concavity of the breast cancer cell.

5. Experimental Results

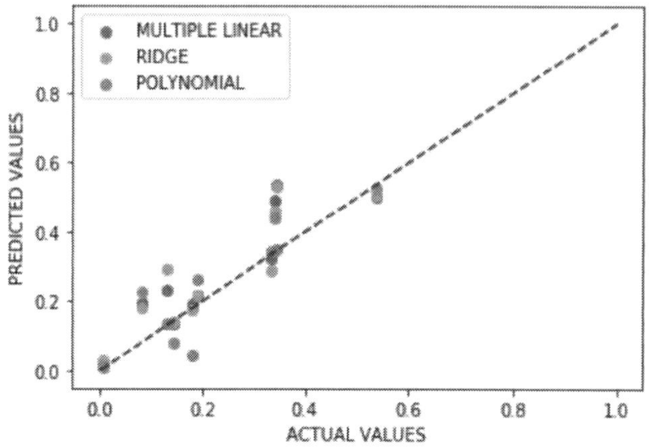

Figure 5: Actual values vs predicted values

Figure 5 represents the predictions made by the three different models. This graph shows that the actual and predicted values can differ from model to model. From the above observation, the Multiple Linear Regression model shows less difference in the actual and predicted values accurately than the Ridge Regression model or Polynomial Linear Regression Model.

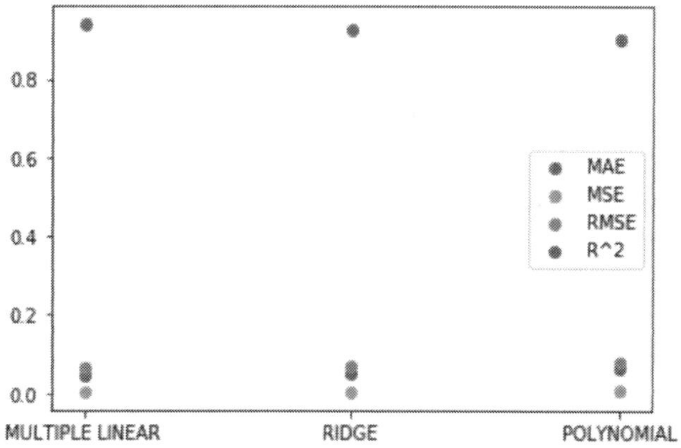

Figure 6: Prediction Metrics.

Figure 6, depicts the prediction metrics which are used to evaluate the models which are MAE, MSE, RMSE, R^2.

Table 1. Model Performance

Techniques	MSE	MAE	RMSE	Coefficient of determination
Multiple Linear Regression	0.00418	0.04794	0.06465	93.7%
Ridge Regression	0.00489	0.05147	0.06998	92.6%
Polynomial Linear Regression	0.00645	0.06253	0.08036	90.3%

Mean Squared value, tells variance of the Residuals (i.e how the data points are located either close or far away from the best fit line), the variance of the residuals is smaller if the data points are held closer to the best fit line of the model and the variance of the residuals is bigger if the data points are held far from the best fit line. Table 1 prediction of MSE tells, Lower the value higher the performance. The performance of Multiple linear regression > Ridge Regression > Polynomial Linear Regression.

Consider the case of Mean Absolute Error, which tells the Mean of the Residuals (i.e absolute distance from each point to the best fit line). In Table 1 prediction of MAE shows, lower the value higher the performance. The performance of Multiple linear regression > Ridge Regression > Polynomial Linear Regression.

Similarly, Root Mean Squared Error, which tells the standard deviation of errors (root of MSE). The prediction values of RMSE in Table 1 prove that the lower the value higher the performance. The performance of Multiple linear regression > Ridge Regression > Polynomial Linear Regression.

Take the case of the Coefficient of determination, which tells the proportion of variance explained by the model. Table 1, prediction of R^2 tells, higher the value higher the performance. The performance of Multiple linear regression > Ridge Regression > Polynomial Linear Regression.

6. Conclusion

In this work, the breast cancer dataset was analyzed using three regression techniques. Machine learning models can predict the target class with high accuracy. According to the performance metrics from the above observation and comparing with each other, Multiple Linear Regression with an accuracy of 93.7% has been performed better than Ridge Regression with an accuracy of 92.6% which performed better than Polynomial Linear Regression with an accuracy of 90.3%.

References

[1] Konstantina Kourou, Themis P.Exarchos Konstantinos P.Exarchos, Michalis V.Karamouzis, Dimitrios I.Fotiadis 2015 *Machine learning applications in cancer prognosis and prediction, Computational and Structural Biotechnology Journal*, 13.

[2] Gulden kaya Uyanik, Nese Guler 2013 *A study on Multiple Linear Regression Procedia - Social and Behavioral Sciences*, 106, 234 – 240.

[3] N S M Shariff, H M B Duzan 2018 *An application of proposed Ridge Regression Methods to real data problem,* International Journal of engineering and technology, 7,106-108.

[4] Sultana, J., 2018 *Predicting Breast Cancer using Logistic Regression and Multi-Class Classifiers,* International Journal of Engineering and Technology, 13,109-116.

[5] Ritu Tayal, Anshul Tickoo, 2014, *Performance Analysis of Regression Data Mining Techniques Implemented on Breast Cancer Dataset*, International Journal of Latest Trends in Engineering and Technology (IJLTET), 4(1), 188-195.

[6] Heena Nankani, Shruti Gupta, Shubham Singh, S. S. Subashka Ramesh, 2019 *Detection Analysis of Various Types of Cancer by Logistic Regression using Machine Learning*, International Journal of Engineering and Advanced Technology,9(1),45-53.

[7] Ganggayah, M.D., Taib, N.A., Har, Y.C. et al. 2019 *Predicting factors for survival of breast cancer patients using machine learning techniques*. BMC Med Inform Decis Mak 19, 48.

[8] Eftekhar B, Mohammad K, Ardebili HE, Ghodsi M, Ketabchi E. 2019 *Comparison of artificial neural network and logistic regression models for prediction of mortality in head trauma based on initial clinical data*. BMC Med Inform Decis Mak.19,48.

[9] Wenbin Yue, Zidong Wang, Hongwei Chen, and Annette Payne and Xiaohui Liu, 2018 *Machine Learning with Applications in Breast Cancer Diagnosis and Prognosis*, 2(13).

[10] Muhammet Fatih Ak, 2017 *A Comparative Analysis of Breast Cancer Detection and Diagnosis Using Data Visualization and Machine Learning Applications* Healthcare (Basel) 8(2).
[11] Eva Ostertagova 2012 *Modelling on Polynomial regression* Procedia Engineering, 48 500 – 506.
[12] https://www.kaggle.com/uciml/breast-cancer-wisconsin-data

Generation of Microtextured Characteristics using Maskless Electrochemical Micromachining

S. Kunar[1], G. Kibria[2], I. Shivakoti[3]

[1]Department of Mechanical Engineering,
Aditya Engineering College, Surampalem, India
[2]Department of Mechanical Engineering, Aliah University, Kolkata, India
[3]Department of Mechanical Engineering,
Sikkim Manipal Institute of Technology, Sikkim Manipal University, Sikkim, India

Corresponding author's e-mail: *sandip.sandip.kunar@gmail.com*

Abstract. To minimize and control friction and wear, the manufacture of microtextured characteristics on circular micropatterns is an innovative approach to extend the life of mechanical systems and components and enhance their reliability and performance. The form, size, and surface quality of circular micropatterns developed on stainless steel surfaces using maskless electrochemical micromachining directly affect the service life and function of textured surface. This article presents the experimental study on the generation by electrochemical micromachining of circular micropatterns on stainless steel. The effect of process variables such as inter-electrode gap (IEG), flow velocity and machining time on current density, current efficiency, material removal rate (MRR), surface roughness, and diametric taper angle are examined during the generation of circular micropatterns. The best parameters obtained from the experimental results are 100μm IEG, 4.4 m/s flow velocity and 3 minutes machining time for producing this micropattern. An attempt is also made to analyze the optical and 3D images of circular micropattern from the experimental results.

1. INTRODUCTION

Electrochemical micromachining based on anodic dissolution principle is very helpful to modify the shape of materials. The regulated modification of surface topography's form, size and surface quality produces the same micropattern, known as microsurface texture and used in many engineering fields, such as promoting benefits such as improving surface adhesion, improving tribological properties, increasing light absorption, reducing drag in aerodynamic applications, etc. [1]. Different microstructures having micro square and circular pattern are innovative loom for the dropping of rubbing in sliding mechanism [2, 3]. Micropatterns are acting as fluid reservoirs to hold the lubricants between two mating mechanical components for reducing friction. This texturing surface also reduces the increase in abnormal temperature during movement of automotive components. Two hybrid wire electro discharge machining processes and EMM are used to manufacture the collection of microelectrodes as cylindrical columns [4]. This process has some heat affected problems during generation of microelectrodes array and it takes more time for overall machining process. The micro

dimple array is produced successfully using a 275μm tool electrode by ECM process [5]. Jet-electrochemical machining method is used to make the dimple array on carbide metal alloys. This method is suitable for the controlled material removal due to restricted area of electric current by a confined area of jet [6]. Both processes produce one by one micro hole and take more time to produce the dimple collection. Through-mask EMM is used to make dimple arrays using PDMS mask [7]. This TMEMM method with laser machining is applied to construct dimple pattern on titanium alloys [8]. But the laser process scatters the uniformity of micropattern. The dry-film photoresist mask is employed to produce on a cylindrical inner microstructured surface using TMEMM. For masking of individual workpiece, photolithography process combines with number of stages [9]. This photolithography process is necessary for every machining. This technique is very time-consuming and costly for mass production. For machining of micro dimple array using TMEMM is very costlier than maskless EMM because masking is necessary for individual workpiece using UV photolithography, X-ray lithography, etc. Individual masking is time consuming matter. Various surface texturing patterns are successfully fabricated using sodium chloride electrolyte by maskless electrochemical texturing (MECT). For MECT tools, electrical discharge machining method is used on AISI 430 ferritic stainless-steel plate [10]. But this process has heat affected zone. Different elementary investigations have also been done by various surface structuring methods such as LBM [11], AJM [12], EMM [13, 14], etc. But electrochemical micromachining uses the anodic dissolution method for the removal of materials from the selective area of workpiece in electrolytic cell. So, electrochemical micromachining can be considered as a promising method for micro circular pattern generation due to high machining performance, material hardness and strength independence, etc.
The aim of this research paper is to create the experimental setup and developed flow system for producing microstructures. The effect of process parameters on current density, current efficiency, MRR, surface roughness and taper angle has been investigated i.e., IEG, machining time, and flow velocity. The study has been analyzed based on experimental analysis to obtain the best results.

2. EXPERIMENTAL PROCEDURE

The developed setup has EMM cell, electrical unit and electrolyte supply arrangement as shown in Figure 1. The cell has fixtured arrangements of electrodes, power unit and inlet and outlet ports. It is built by Perspex material. The vertical cross flow electrolyte system is devised for producing the better surface finish. The pulsed power is controlled by a constant voltage mode from the electrical unit (DOSF 20-60, Matsusada, Japan). The patterned tool containing micro circular pattern of 187μm has fabricated with SU-8 2150 mask using UV exposure system. The textured tool is employed with the precision micrometer to maintain inter electrode gap.
The correctness of patterned tool is more imperative for the circular patterned quality and accuracy. Stainless steel is one of the advanced materials in micro engineering applications and utilized as job material. The ranges of selected variables i.e., machining time, flow velocity and IEG are 3 to 6 minutes, 3.2 to 4.4 m/s and 100 to 250μm keeping other parameters fixed i.e., 10V voltage, 30% duty ratio, 20g/l mixed electrolyte concentration of NaCl and $NaNO_3$ and 190kHz frequency. Microfeatures of micropatterned surface are observed using profilometer and microscope. Taper Angle (°) is measured using the equation,

$$Taper\ angle(°) = tan^{-}\frac{R}{d} \quad (1)$$

Where, the radial overcut is R and d is the machined depth.

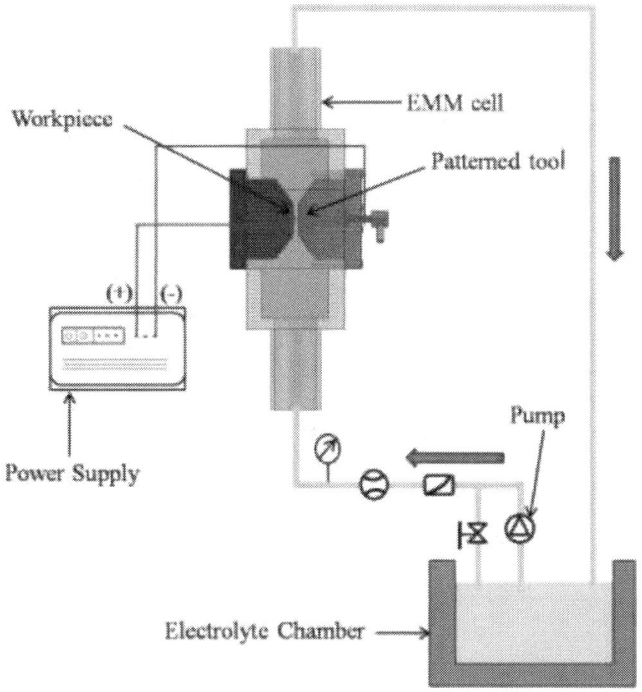

Figure 1. EMM setup

3. RESULTS AND DISCUSSION

The experimental study focuses on the observation in the form of graphs of the effect of IEG, flow velocity and machining time on the microtextured features of produced micro circular pattern.

Figure 2 depicts the effect of IEG on current density, MRR and current efficiency with fixed parameters such as flow velocity of 4.4m/s and machining time of 3 minutes. The current density, current efficiency and MRR reduce with rising IEG. The current density decreases due to higher ohmic resistance. With rising IEG, MRR decreases due to uneven etching for the distribution of uncontrolled current density. The current efficiency declines because of the non-uniform current flux distribution across the micromachining area. As a result, with the rise in IEG, anodic dissolution decreases progressively as the voltage drop increases through the electrolyte. Figure 3 depicts the influence of IEG on surface roughness and taper angle. The surface roughness and taper angle increase with increasing IEG. The surface roughness increases with increasing IEG due to non-uniform etching for unrestrained current flux distribution. The taper angle increases because the micropatterned periphery increases due to irregular current flux distribution with greater stray current effect and lower depth.

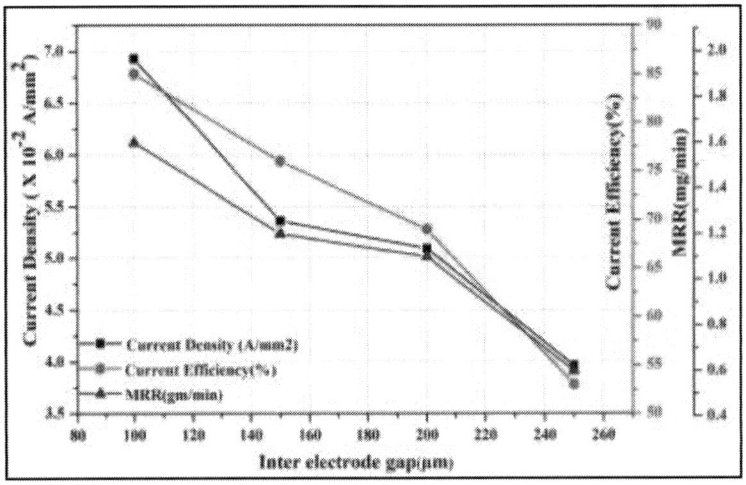

Figure 2. Effect of IEG on current density, current efficiency and MRR

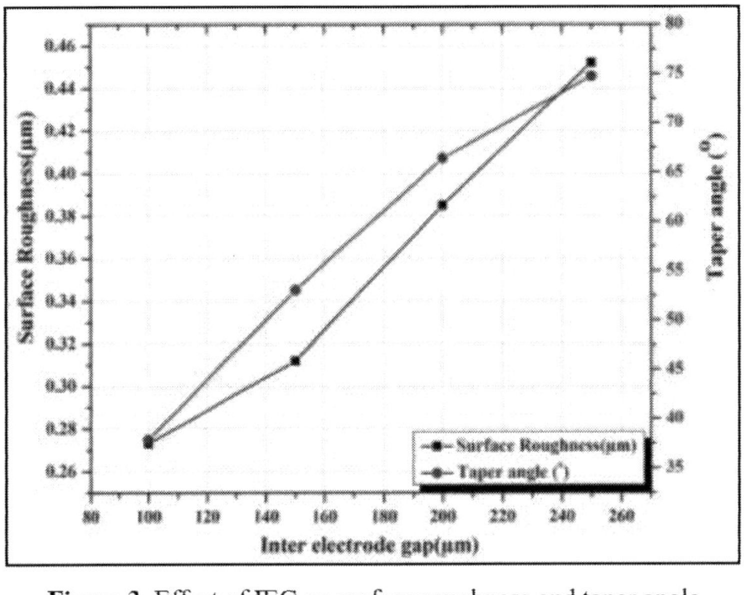

Figure 3. Effect of IEG on surface roughness and taper angle

Investigation is conducted to carry out the outcome of flow velocity on current density, MRR and current efficiency with fixed process variables i.e., 100μm IEG and 3 minutes machining time as shown in Figure 4. Due to higher anodic dissolution, the current density increases with rising flow velocity. The current efficiency rises with increasing flow velocity because of higher machining capability for the availability of higher current density. The MRR increases with augmenting flow velocity owing to controlled machining for the removal of sludges. Figure 5 shows the effect of flow velocity on the surface roughness and taper angle. The surface roughness increases with rising velocity because of uneven etching and then decreases due to uniform etching. The taper angle decreases with rising flow velocity owing to controlled machining across the micromachining zone and then increases due to non-uniform anodic dissolution.

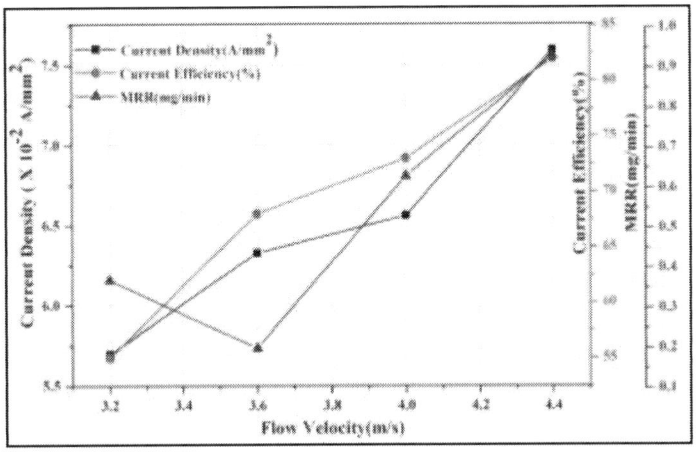

Figure 4. Effect of flow velocity on current density, current efficiency and MRR

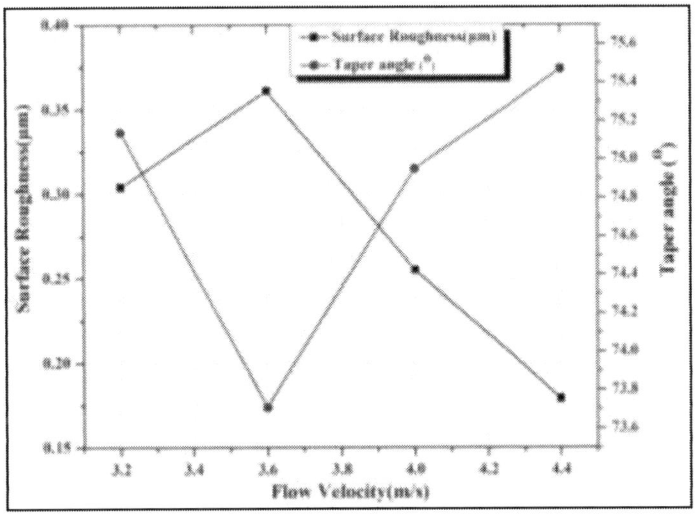

Figure 5. Effect of flow velocity on surface roughness and taper angle

The effect of texture time on current density, current efficiency and MRR with fixed parameters, i.e., 100μm IEG and 4.4 m/s flow velocity, is shown in Figure 6. The current density rises with increased machining time due to higher current flow. Because of higher anodic dissolution for extended machining time, the current efficiency increases. For regular machining, MRR increases and then decreases due to non-uniform machining. In Figure 7, the effect of machining time on surface roughness and taper angle is illustrated. The surface roughness decreases because of even etching and again increases for non-uniform machining. The taper angle increases owing to lower machining localization and then reduces for uniform machining and again increases for uncontrolled machining.

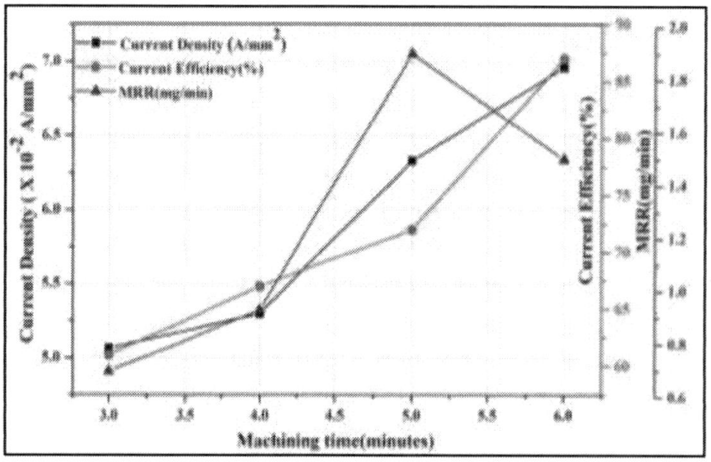

Figure 6. Effect of machining time on current density, current efficiency and MRR

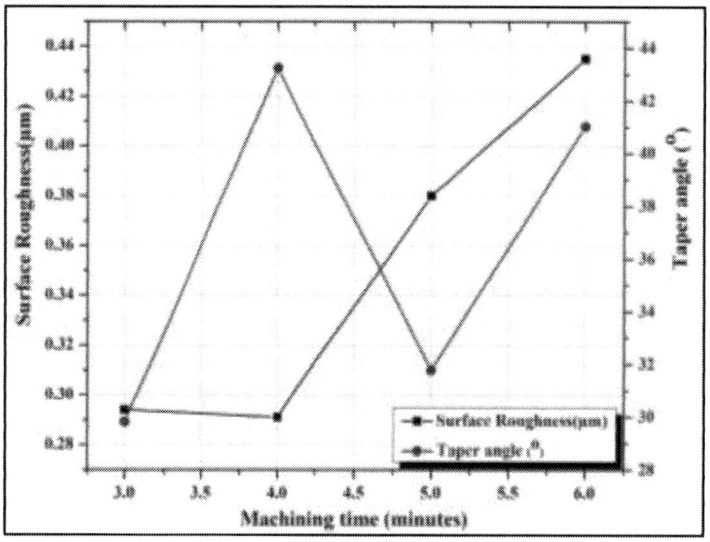

Figure 7. Effect of machining time on surface roughness and taper angle

4. ANALYSIS OF MICROGRAPHS

Investigation has been carried out with SU-8 2150 mask by EMM process. On stainless steel substrates, the micro circular pattern is shown before machining in Figure 8(a). During machining, this mask does not warp and reused nineteen times since this mask has a higher tolerance limit, greater sustainability, etc. as shown in Figure 8(b). These consequences imply that this mask has additional effectiveness for generating many micropatterns with higher shape accuracy and efficiency.

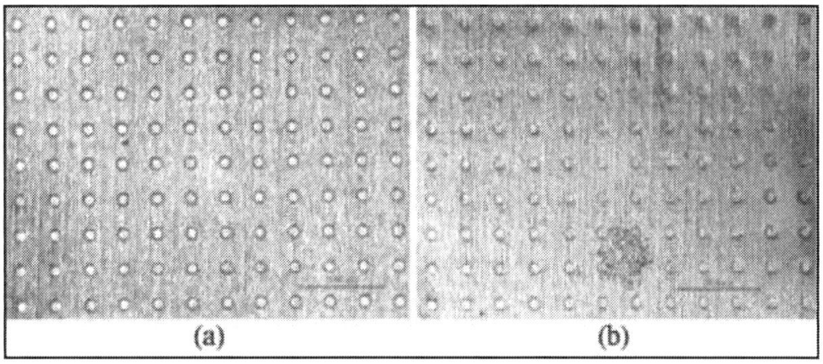

Figure 8. Masked tool (a) before machining (b) after machining

The microscopic image of the micro circular pattern is created at a particular parametric combination i.e., machining time of 3 minutes, IEG of 100μm, flow velocity of 4.4 m/s, as shown in Figure 9. Almost all micro impressions remain the appropriate and homogeneous patterned shape and size owing to regular anodic dissolution with better machining conditions. The machined depth produced is almost the same due to uniform etching for managed current density distribution. The surface quality is also consistent throughout the textures because of consistent etching along the pattern.

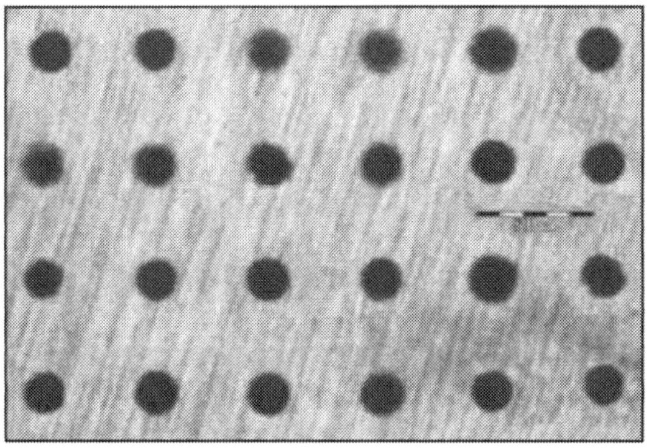

Figure 9. Regular micro circular pattern

Figure 10 displays the 3D image and 2D surface roughness profile of micro circular impression. The magnitude of surface roughness is 0.0713μm.

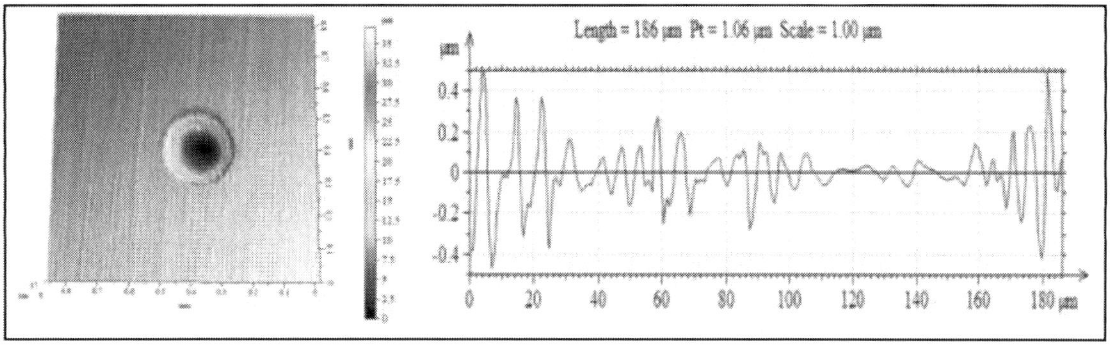

Figure 10. 3D micro circular profile with 2D surface roughness profile

5. CONCLUSIONS

The distinctive microtexturing process is used cheaply for the manufacture of high quality micropatterns using a single tool. The effect of process parameters is experimentally studied on machined features of micro circular patterns. It is promising to summarize the following conclusions:

- The masked tool is more prominent for generating nineteen machined samples using maskless EMM method.
- Using the established setup, higher flow velocity, lower machining time and lower IEG are suggested for producing high quality micropatterns.
- The best parametric setting of 50μm IEG, 3 minutes machining time and 4.4 m/s flow velocity can be obtained from the investigation.

However, to build different microtextured patterns in various advanced applications, it will further explore the development of intricate complex patterned tool and tool movement strategy.

REFERENCES

[1] Bruzzone AAG, Costa HL, Lonardo PM, Lucca DA. Advances in engineered surfaces for functional performance. *CIRP Annals–Manuf Technol.* 2008; **57**:750–769.
[2] Kunar S, Bhattacharyya B. Electrochemical micromachining of micro square pattern using reusable masked tool. *Mater Manuf Process.* 2018; **34**:487-493.
[3] Kunar S, Bhattacharyya B. Electrochemical microsurface texturing with reusable masked patterned tool. *Eng Sci Technol an Int.* 2018; 21: 1095-1103.
[4] Wang MH, Zhu D. Fabrication of multiple electrodes and their application for micro-holes array in ECM. *Int J Adv Manuf Technol.* 2009; **41**:42–47.
[5] Byun JW, Shin HS, Kwon MH, Kim BH, Chu CN. Surface texturing by micro-ECM for friction reduction. *Int J Precis Eng Manuf.* 2010; **11**:747–753.
[6] Hackert-Oschätzchen M, Martin A, Meichsner G, Zinecker M, Schubert A. Microstructuring of carbide metals applying jet electrochemical machining. *Precis Eng.* 2013; **37**:621–634.
[7] Chen X, Qu N, Li H, Xu Z. Electrochemical micromachining of micro-dimple arrays using a polydimethylsiloxane (PDMS) mask. *J Mater Process Technol.* 2016; **229**:102–110.
[8] Chauvy P-F, Hoffmann P, Landolt D. Electrochemical micromachining of titanium using laser oxide film lithography: excimer laser irradiation of anodic oxide. *Appl Surf Sci.* 2003; **211**:113–127.
[9] Qu NS, Chen XL, Li HS, Zeng YB, Electrochemical micromachining of micro-dimple arrays on cylindrical inner surfaces using a dry-film photoresist. *Chin J Aeronaut.* 2014; **27**:612–618.
[10] Parreira JG, Gallo CA, Costa HL. New advances on maskless electrochemical texturing (MECT) for tribological purposes. *Surf Coat Technol.* 2012; **212**:1–13.
[11] Wang X, Kato K, Adachi K, Aizawa K, The effect of laser texturing of SiC surface on the critical load for the transition of water lubrication mode from hydro-dynamic to mixed. *Tribol Int.* 2001; **34**:703–711.
[12] Wakuda M, Yamauchi Y, Kanzaki S. Material response to particle impact during abrasive jet machining of alumina ceramics. *J Mater Process Technol.* 2003; **132**:177–183.
[13] Kunar S, Bhattacharyya B. Influence of various flow methods during fabrication of micro ellipse pattern by maskless electrochemical micromachining. *J Manuf Process.* 2018; **35**:700-714.
[14] Kunar S, Bhattacharyya B. Investigation on surface structuring generated by electrochemical micromachining. *Adv Manuf.* 2017; **5**:217–230.

Photocatalytic Behavior of Ferroelectric Materials: Comparative Study of $BaTiO_3$ and Ag-loaded $BaTiO_3$ for Wastewater Treatment

Moin Ali Siddiqui[1] **and Piyush Jaiswal**[1]

[1]Nanotechnology Program, Centre for Advanced Studies, Abdul Kalam Technical University, Lucknow, 226031, Uttar Pradesh, India

E-mail: moinalisiddiqui@gmail.com (Moin Ali Siddiqui)

E-mail: piyushj@cas.res.in (Piyush Jaiswal)

Abstract. This work investigated the photocatalytic-induced degradation of organic dyes in the wastewater using barium titanate ($BaTiO_3$) supported silver (Ag) metal particles. $BaTiO_3$ ceramics were synthesized using a solid-state reaction route. Ag metal was loaded on $BaTiO_3$ particles using silver nitrate and polyethylene glycol reagents at room temperature without heat and light source. Ag-loaded $BaTiO_3$ powder was characterized using X-ray diffraction (XRD) and scanning electron microscopy (SEM) techniques. Ag-loaded $BaTiO_3$ powder showed more degradation of Rhodamine B (RB) dye as compared to $BaTiO_3$ powder. Around 95.05, 94.75, and 94.5 % of RB dye degradation were observed within 180 min under visible light during the first, second, and third photocatalytic cycles. The photon-induced strong surface plasmon resonance (SPR) effect of Ag metal has played a critical role in the induction of visible light photocatalysis in Ag-loaded $BaTiO_3$ powder.

Keywords: Photocatalysis, Rhodamine B, Ferroelectric, Silver., BaTiO3

1. Introduction

Water utilities worldwide are struggling with deteriorating water quality and quantity in the face of tightening regulation and heightened demand [1] . Water quality (pollution) is currently a major environmental issue in India [2] . One of India's primary reasons for water pollution is untreated sewage, the largest water pollution source [3]. The second primary reason most rivers, lakes, and surface water pollution is due to industries and agricultural run-off [4]. Researchers have found that untreated sewage discharge is the most critical source of surface and groundwater pollution [5]. There is a considerable gap between the generation and subsequent treatment of wastewater. Thus, worldwide wastewater generated is approximately 80% of water supplied by various- different sources. India accounts for 18% of the world population and 4% of world water resources. The country's only 30% water is recycled remaining water is taken from the surface or the ground [6, 7]. However, the population's dependence on water is not regularly circulated. For this, we must develop cost-effective and environment-friendly techniques [7, 8].

Dyes are produced and used in the textile, printing, paper industries, and coloring many other materials. It chemically bonds on the substrate surface [9]. They can be defined as substances that generate color when applied to these substrates. It also temporarily changes the crystal structure of elements. They stick to the well-matched surface by using a water solution. They create covalent bonds with salts or metals. Uses techniques are either physical adsorption or mechanical retention techniques [10]. Dyes are categorized based on their chemical structure, uses, and the groups of atoms (chromophores) present in the dye. The chromophores are diverse functional groups determining the color and defining the dyes' behavior, such as aril methane, azo, methane, nitro, and anthraquinone. Nowadays, many practical solutions are available to this severe problem, but photocatalysis is a very effective technique because this is non-toxic, economical, and renewable [11]. This catalyst has one more capability: the catalyst used can be reused just by cleaning it again. In this method, the visible rays of solar light energy are employed on the solution as an energy source that is green and clean, and this is also safe to use; these are abundant in the amount [12]. Now the big question that comes to mind that how we can exploit these solar energies. These are freely available energy, so the synthesis of natural and artificial photocatalysts comes into the picture. Because it attracted the world with tremendous and cost-effective activities, it has fascinated significant scientific interests in the world as a whole [13, 14]

Recently it has been observed that ferroelectric materials have shown photocatalytic performance due to the presence of an internal field, which restricts the recombination of electron-hole pair [15]. $BaTiO_3$ is one of the promising ferroelectric materials [16]. The photocatalytic performance of $BaTiO_3$ can be observed in the presence of UV light [17]. UV light has a drawback limiting its use in water cleaning [18, 19]. UV light can ensure cancer; thus, it can be avoided for water cleaning applications [20]. Therefore the photocatalytic activity in visible light can also be improved by morphology change deposition of noble metals, heterostructures construction, and doping of transition metal, composite, etc. [21–23] .

Photocatalysis is visible light obtained by using noble metal (Au, Ag, etc.) [24]. Due to the significant broadening of the light absorption range, Noble metals display SPR [25, 26] . This further load to photocatalysis and the deposition of noble metal restricts the electron/hole pair's recombination, enhancing the photocatalytic performance [27]. However, the low cost and ease of availability, and simple method to load on catalyst bring silver (Ag) one of the most satisfactory photocatalysis candidates [28, 29].

The various methods have been generally used to load Ag NPs, such as photoreduction [30, 31], conventional calcination, and impregnation chemical reduction methods, which act as external energy for heating. Luo et al. [32], proposed a novel way of loading Ag NPs on $BaTiO_3$ ceramic particles at room-temperature without imposing light or heat. The authors mentioned that $BaTiO_3$ particles exhibit a catalytic reduction of $Ag+$ ethylene glycol solution from this research point of view. We have adopted this method of loading $Ag+$ nanoparticles.

2. Methods

$BaTiO_3$ powder was synthesized through the solid-state reaction process [33]. In this process, the oxide powder of $BaCO_3$ and TiO_2 were weighed according to stoichiometric molar ratios. These powders were mixed and well-grounded in an agate mortar using acetone medium for 30 min. After this, the dried homogenous powder was calcined at 1100°C for four h in the Nabertherm furnace. Ag-loaded $BaTiO_3$ was synthesized using the method mentioned in reference [32]. In this method, 2g of each $BaTiO_3$ and $AgNO_3$ were added into 100 ml of polyvinyl glycol. The suspension was stirred using a magnetic stirrer for 12 h to obtain Ag loading on $BaTiO_3$. After this, Ag-loaded $BaTiO_3$ powder was filtered out using filter paper. The powder was thoroughly washed at least ten times and then dried in an oven at 150 C for two hours.

The phase of the sample was examined by the X-ray diffraction (XRD) technique using the Rigaku diffractometer (Rigaku Smart Lab, Japan) by which the pattern was collected for 2θ range from 20° to 80° with the scanning rate of 2°/min. The microstructure was observed with a scanning electron microscope (SEM; model JSM 7800F, JEOL, Tokyo, Japan). During the photocatalysis experiment, *Rhodamine B* (RB) dye solution was prepared in deionized water. RB dye was used as a model dye. The initial dye concentration was taken five mg/L. 10 ml dye solution was taken in a beaker, and a sample of 0.2g of sample was added to the dye solution. The beaker was irradiated with visible light during photocatalytic experiments under constant stirring at 500 rpm. Before performing the photocatalysis experiment, the adsorption equilibrium was achieved between dye solution and sample by placing the sample under dark for 24 h. During photocatalysis, the test samples were collected every 30 min. These samples were tested under the UV visible spectrophotometer to record the dye solution's corresponding absorbance value. The percentage of degradation was found by using the following formula [34].

$$Degradation\ \% = \left(1 - \frac{A}{A_0}\right) \times 100 = \left(1 - \frac{C}{C_0}\right) \times 100$$
$$Where\ A_0 = Initial\ Absorbance,\ A = Absorbance\ at\ time\ 't'$$
$$C_0 = Initial\ Concentration,\ C = Unknown\ Concentration$$

3. Results

XRD patterns of both samples ($BaTiO_3$ and Ag loaded $BaTiO_3$) are provided in Figure 1. XRD peak positions observed in both samples are precisely matched to the perovskite structure of $BaTiO_3$according to JCPDS File No. 01-087-0719. In Figure 1, no other addition peak in the XRD pattern revealed the non-existence of any impurity phases. Ag's peaks were also not observed in the XRD of Ag loaded $BaTiO_3$ powder, which may be due to Ag metal's low loading content on $BaTiO_3$ particles.

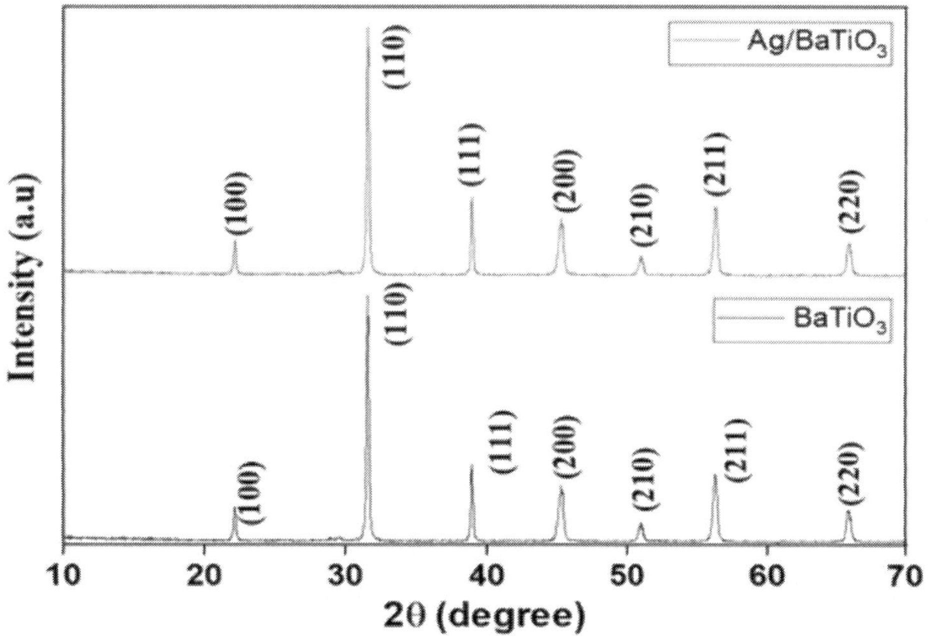

Figure 1. XRD of $BaTiO_3$ and Ag-loaded $BaTiO_3$

The morphology of $BaTiO_3$ powder and Ag-loaded $BaTiO_3$ powder is shown in Figure 2(a–b). The grain morphology of $BaTiO_3$ was random, as shown in Figure 2(a); however, the loaded Ag particles can be observed on $BaTiO_3$ particles as shown in Figure 2(b).

Figure 2. SEM image of $BaTiO_3$ and Ag-loaded $BaTiO_3$

The visible light was used for the photocatalysis experiment to photoexcite the $BaTiO_3$ and Ag-loaded $BaTiO_3$ powder samples. Figure 3(a) showed the decrease in absorbance peak intensity of RB dye with an increase in irradiation time using Ag loaded $BaTiO_3$ powder sample, which indicated the promising dye degradation capability of Ag loaded $BaTiO_3$ powder sample. For the same irradiation time, the dye degradation with $BaTiO_3$ powder and without any sample cases is shown in Figure 3(b).

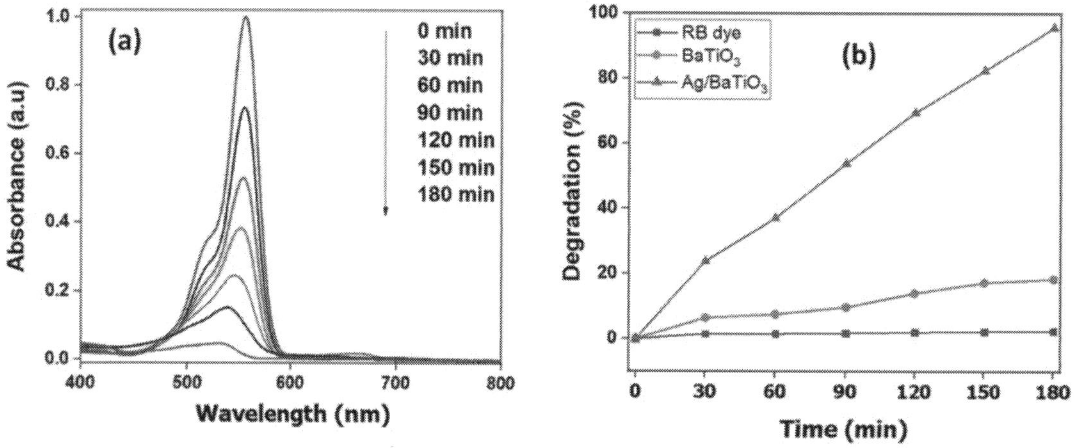

Figure 3. (a)(b) Photocatalysis performance of $BaTiO_3$ and $Ag/BaTiO_3$ ceramics

This indicated that Ag loaded $BaTiO_3$ sample possessed significantly higher dye degradation capability than two other control samples. Thus, it can be concluded that the loading of Ag metal particles has a critical role in increasing dye degradation capability. Maximum 95.5% dye degradation was recorded with Ag loaded $BaTiO_3$ sample under 180 min of visible light irradiation. A slight drop was recorded in the repeatability test of Ag loaded BaTiO3 powder sample (maximum of 2%) after five photocatalysis cycles, as shown in Figure 4. Even after the 5th cycle, Ag loaded $BaTiO_3$ sample powder gave 93% of dye degradation. This showed that the dye degradation performance of the Ag-loaded $BaTiO_3$ sample is repeatable.

IOP Conf. Series: Materials Science and Engineering 1166 (2021) 012031 doi:10.1088/1757-899X/1166/1/012031

4. Discussion

The transfer of an electron from the valence band to the conduction band of $BaTiO_3$ is not possible, as the bandgap of $BaTiO_3$ lies in the ultraviolet (UV) region, and here the visible is provided.

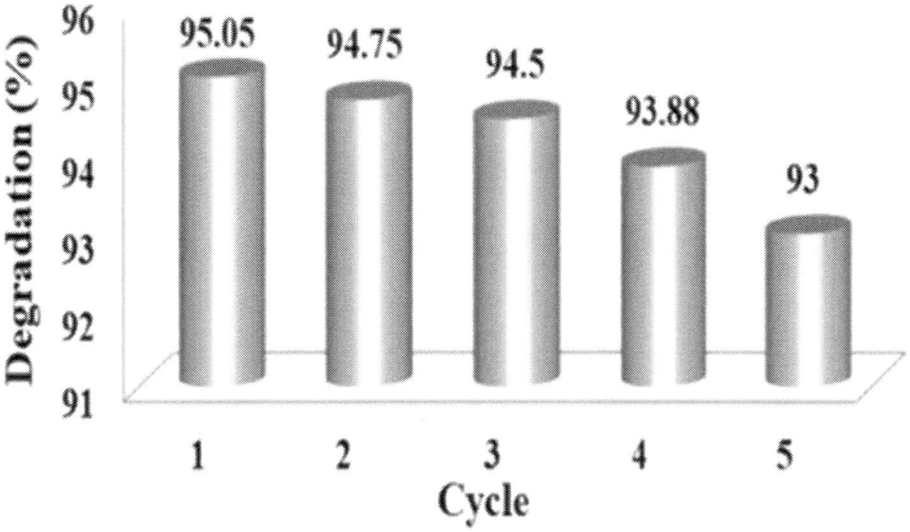

Figure 4. Repeatability graph of photocatalysis of Ag/$BaTiO_3$

The surface plasmon resonance (SPR) effect has been widely reported in the case of Ag metal, due to which hot electrons are produced from Ag metal. These electrons can directly transfer from Ag metal to the conduction band of $BaTiO_3$ ceramic and subsequently can reduce adsorbed oxygen (O_2) to superoxide ( ) radical. Similarly, the holes present in the valence band of BaTiO3 and/or Ag metal can oxidize adsorbed water (H_2O) to hydroxyl (• OH) radicals. These and • OH subsequently decomposed the RB dye into degradation products.

To further study, photocatalysis effects were explored by using the RB dye.

5. Conclusion

In this work, ferroelectric barium titanate ceramic was fabricated via a solid-state reaction route and loaded with Ag metal particles. Ag-loaded $BaTiO_3$ sample gave remarkable dye degradation performance under visible light irradiation, and the dye degradation performance of Ag-loaded $BaTiO_3$ was found to be repeatable.

Acknowledgements

This work is supported by Nanotechnology program at Centre for Advanced Studies, AKTU and School of Engineering, IIT Mandi.

References

[1] Elimelech M 2006 URL https://doi.org/10.2166/aqua.2005.064

[2] Paliwal R, Sharma P and Kansal A 2007 *Journal of Environmental Management* **83** 131–144 URL 10.1016/j.jenvman.2006.02.003;https://dx.doi.org/10.1016/j.jenvman.2006.02.003

[3] Goldar B and Banerjee N 2004 *Journal of Environmental Management* **73** 117–130 URL 10.1016/j.jenvman.2004.06.008;https://dx.doi.org/10.1016/j.jenvman.2004.06.008

[4] Kazi T G, Arain M B, Jamali M K, Jalbani N, Afridi H I, Sarfraz R A, Baig J A and Shah A Q 2009 *Ecotoxicology and Environmental Safety* **72** 301–309 URL 10.1016/j.ecoenv.2008.02.024;https://dx.doi.org/10.1016/j.ecoenv.2008.02.024

[5] Maria A, Version R and Technologies W 2003

[6] Rizwan U, Malik R N and Abdul Q 2009 *African Journal of Environmental Science and Technology* **3** 429–446

[7] Naidoo S and Olaniran A 2013 *International Journal of Environmental Research and Public Health* **11** 249–270 URL 10.3390/ijerph110100249;https://dx.doi.org/10.3390/ijerph110100249

[8] Chen G 2004 *Separation and Purification Technology* **38** 11–41

[9] Verma A K, Dash R R and Bhunia P 2012 *Journal of Environmental Management* **93** 154–168 URL 10.1016/j.jenvman.2011.09.012;https://dx.doi.org/10.1016/j.jenvman.2011.09.012

[10] Couto S R 2009 *Biotechnology Advances* **27** 227–235 URL 10.1016/j.biotechadv.2008.12.001;https://dx.doi.org/10.1016/j.biotechadv.2008.12.001

[11] Chaubey R, Sahu S, James O O and Maity S 2013 *Renewable and Sustainable Energy Reviews* **23** 443–462 URL 10.1016/j.rser.2013.02.019;https://dx.doi.org/10.1016/j.rser.2013.02.019

[12] Sibi M G, Verma D and Kim J 2020 *Catalysis Reviews* **62** 163–311 URL 10.1080/01614940.2019.1659555;https://dx.doi.org/10.1080/01614940.2019.1659555

[13] Şen Z 2004 *Progress in Energy and Combustion Science* **30** 367–416 URL 10.1016/j.pecs.2004.02.004;https://dx.doi.org/10.1016/j.pecs.2004.02.004

[14] Li R 2018 *Cuihua Xuebao/Chinese Journal of Catalysis* **39** 63104–63107

[15] Cui Y, Briscoe J, Wang Y, Tarakina N V and Dunn S 2017 *ACS Applied Materials & Interfaces* **9** 24518–24526 URL 10.1021/acsami.7b03523;https://dx.doi.org/10.1021/acsami.7b03523

[16] Ma N and Yang Y 2018 *Nano Energy* **50** 417–424

[17] Al-Assiri M S and El-Desoky M M 2012 *Journal of Non-Crystalline Solids* **358** 1605–1610 URL 10.1016/j.jnoncrysol.2012.04.023;https://dx.doi.org/10.1016/j.jnoncrysol.2012.04.023

[18] Astas S 2008 *Global NEST Printed in Greece. All rights reserved* **10** 376–385

[19] JAIN H 2004 *Ferroelectrics* **306** 111–127 URL 10.1080/00150190490458446;https://dx.doi.org/10.1080/00150190490458446

[20] Narayanan D L, Saladi R N and Fox J L 2010 *International Journal of Dermatology* **49** 978–986 URL 10.1111/j.1365-4632.2010.04474.x;https://dx.doi.org/10.1111/j.1365-4632.2010.04474.x

[21] Jiang D, Chen L, Zhu J, Chen M, Shi W and Xie J 2013 *Dalton Transactions* **42** 15726–15726 URL 10.1039/c3dt52008k;https://dx.doi.org/10.1039/c3dt52008k

[22] Paquin F, Rivnay J, Salleo A, Stingelin N and Silva C 2015 *J. Mater. Chem. C* **3** 10715–10722

[23] Meng X, Tian G, Chen Y, Zhai R, Zhou J, Shi Y, Cao X, Zhou W and Fu H 2013 *CrystEngComm* **15** 5144–5149

[24] Zhai Y, Zhang Y, Yin J and Fan X 2018 *Applied Surface Science* **484** 981–989

[25] Wang P, Huang B, Qin X, Zhang X, Dai Y, Wei J and Whangbo M H 2008 *Angewandte Chemie - International Edition* **47** 7931–7933

[26] Xiao F 2012 *Journal of Materials Chemistry* **22** 7819–7830

[27] Zhang X, Wang Y, Hou F, Li H, Yang Y, Zhang X, Yang Y and Wang Y 2017 *Applied Surface Science* **391** 476–483

[28] Shameli K, Ahmad M B, Jazayeri S D, Shabanzadeh P, Sangpour P, Jahangirian H and Gharayebi Y 2012 *Chemistry Central Journal* **6** 1–10 URL 10.1186/1752-153x-6-73;https://dx.doi.org/10.1186/1752-153x-6-73

[29] Zhao K, Feng Q and Chen G 1999 *Tsinghua Science and Technology* **4** 1570–1573

[30] Liang Y C, Wang C C, Kei C C, Hsueh Y C, Cho W H and Perng T P 2011 *Journal of Physical Chemistry C* **115** 9498–9502

[31] Malligavathy M, Iyyapushpam S, Nishanthi S T and Padiyan D P 2017 *Journal of Materials Science: Materials in Electronics* **28** 18307–18321 URL 10.1007/s10854-017-7777-6;https://dx.doi.org/10.1007/s10854-017-7777-6

[32] Luo S, Yu S, Fang F, Lai M, Sun R and Wong C P 2015 *RSC Advances* **5** 3377–3380 URL 10.1039/c4ra12665c;https://dx.doi.org/10.1039/c4ra12665c

[33] Simon-Seveyrat L, Hajjaji A, Emziane Y, Guiffard B and Guyomar D 2007 Re-investigation of synthesis of BaTiO3 by conventional solid-state reaction and oxalate coprecipitation route for piezoelectric applications URL 10.1016/j.ceramint.2005.07.019;https://dx.doi.org/10.1016/j.ceramint.2005.07.019

[34] Muruganandam G, Mala N, Pandiarajan S, Srinivasan N, Ramya R, Sindhuja E and Ravichandran K 2017 *Journal of Materials Science: Materials in Electronics* **28** 18228–18235 URL 10.1007/s10854-017-7771-z;https://dx.doi.org/10.1007/s10854-017-7771-z

A Review on Transition-metal Oxalate Based Electrode for Supercapacitors

Abhijeet Kumar Singh[1], **Piyush Jaiswal**[1] **and Preetam Singh**[2]

[1]Nanotechnology Program, Centre for Advanced Studies, Abdul Kalam Technical University, Lucknow, 226031, Uttar Pradesh, INDIA

E-mail: ak110singh@rediffmail.com (Abhijeet Kumar Singh)

E-mail: piyushj@cas.res.in (Piyush Jaiswal)
[2]Department of Ceramic Engineering, IITBHU, Banaras Hindu University, Varanasi, 221005, Uttar Pradesh, INDIA

E-mail: preetamsingh.cer@itbhu.ac.in (Preetam Singh)

Abstract. Transition-metal oxalates have been significantly showing the potential for charge storage in supercapacitor as an energy material. Transition-metal oxalate materials with framework structures, morphology and electrochemical properties offer an opportunity for sustainable energy storage material. Recently, transition metal oxalates, witnessed pseudocapacitive or battery type behaviour when it has been used as electrode for supercapacitor applications. In this review, we mainly address the recent study of transition metal oxalates as potential energy storage materials such as Co, Ni, Mn and Fe based oxalates and composites in the context of supercapacitors with electrochemical performances. Although, electrochemical performances of TMOx electrode can be enhanced through various methodology such as by preparing mixed TMOx structures or preparing nanocomposite with highly conductive secondary materials such as graphene and CNT. The generated idea after reading this paper will help the researchers to explore further usage of transition metal oxalates in the field of electrochemical energy storage devices.

Keywords: Transition Metal Oxalates, Open structure, Electrode, Pseudocapacitor

1. Introduction

Burning of fossil fuel leads to the emission of CO_2 which is one of the greenhouse gases that causes global warming. It has been observed that there is decrement in fossil fuel while an accompanying huge demand of energy increased in the society [1–3] . These issues can be minimized as the required society will utilize the sustainable and renewable resources. Electrochemical energy storage devices (EESD) as an alternative source which is more energy efficient and environmentally friendly [2, 4]. These EESD work on the principal of conversion of electrical energy into chemical energy and vice versa. Rechargeable batteries and supercapacitor are one of the peculiar EESD are currently being developed to resolve the issue of global demand of energy [5, 6].

Supercapacitors are EESD devices which has lower energy density as compare to batteries but it can deliver higher power [2, 3, 6].Supercapacitor can also overcome the limitation of batteries such as rate capability at higher current as well as cyclic stability [7, 8]. Supercapacitors, categories in double layer capacitors (EDLCs) and pseudocapacitors. EDLC performs by storing the ions on the surface of electrodes by electrostatic storage of charges. Mainly the carbon based electrode such as. porus carbon, activated carbon, graphene, Carbon Nanotube (CNT) has been used as an electrode for EDLC [3] [9] . Pseudocapacitor exhibits faradic redox reactions occurring at or near the surface of an appropriate electrode material lead to EDLC-like electrochemical features but the redox processes lead to much greater charge storage [10, 11].Transition-metal-oxides, conducting polymer, and nanocomposite materials have been also extensively used as an electrode for pseudocapacitor application [3, 12–15].

RuO_2 has been extensively studied as pseudocapacitive material for energy storage [12, 16]. Although RuO_2 exhibits having specific capacitance 600F/g with operating potential 1.2V in acidic medium [16]. Ruthenium is a costly element, therefore utilizing ruthenium in commercial supercapacitor is not feasible economically. Similarly, MnO_2 display in both amorphous and crystalline phase have been investigated for high-rate energy storage in aqueous electrolytes [12, 14]. MnO_2 have poor electrical conductivity which one of the reason having low specific capacitance [17]. In addition to that transition metal oxides suffer from irreversibility during charging and discharging due to the phase change [12, 18]. Layered material such as $Ni(OH)_2$ shows higher specific capacitance as compare to other oxides such as RuO_2 and MnO_2 but it has smaller potential window [19]. A part from that layered materials also have lower cyclic stability due to continuous phase change [20]. Transition Metal Dichalcogenide (TMDS) are usually two dimensional layered material which will offer both EDL as well as intercalation/de-intercalation of ions for faradic reaction [21]. Moreover, due to poor electronic conductivity, TMDs electrochemical performance decreases as electrolyte couldn't access more number of active sites which results poor capacitance. TMDs also suffer from rate capability and cyclic stability during charging/dis-charging cycle [22].Conducting polymer exhibit large pseudocapcitance in the range of ~ 300F/g which is more than few carbon material and metal oxide. Due to their high resistance by ion diffusion within the electrode slows down. As a result of that swelling and shrinking of polymers occurred during cyclic process [3, 13, 23].

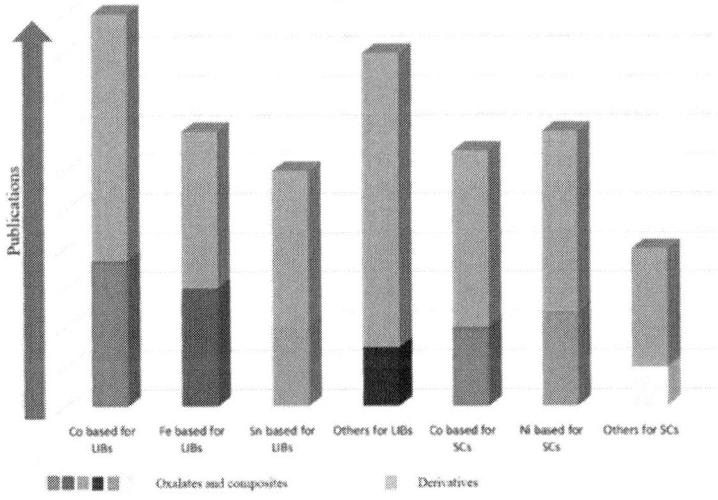

Figure 1. Bar chart for transitionmetal oxalates and their derivatives for LIBs and Supercapacitor.

In the recent years, the number of research has been conducted on transition metal oxalate (TMOx) materials and their electrochemical characteristics which defines the feasibility of TMOx as an energy storage material. Fig 1 clearly shows the recent work related to synthesis of nanostructured oxalate and oxalate-derived materials and their application in Lithium ion batteries (LIBs) and supercapacitors is a main focus [24].Tarascon et al. has mentioned the reversible Li^+ inseration capability in Fe(III)Oxalate tetrahydrate as an anode for Lithium ion batteries (LIBs) [25]. Reddy et al. reviewed that of oxalates and oxyanions based anodes for LIBs [26]. The electrochemical performance of TMOx has displayed equivalently capable, and offers higher specific capacities as compared to the identical oxide [27]. TMOx showed a great potential in the field of electrochemistry, mainly batteries and supercapacitors to acquire high energy and power density. Moreover, transition metal oxalates as an energy material offers an opportunity to explore electrodes that can possibly exhibits good capacity for charge storage. In account of that TMOx require more emphasised and in-depth study in the field of EESD more specifically supercapacitors. Thereby, we believe that presenting a concentrated review to summarize the significance of TMOx-based energy storage materials the further developments for pseudocapacitor applications. This review, consists comprehensively addressed the usage of oxalates and nanocomposites based on oxalate as an electrode for supercapacitors. In addition to that we have describe the type of electrode materials and structure of TMOx for charge storage mechanism.

2. A short briefing regarding the type of electrode materials

Electrochemical signatures including Cyclic voltammograms (CV) accompanied by galvanostatic charge/discharge (GCD) curves, the electrode materials can be categorised as EDLC, pseudocapacitive and battery-like as shown in fig2 which has been used for the fabrication of supercapacitors [28]. EDLC mainly show rectangular CV with scan rate response as well as linear charge discharge profile with respect to time. Pseudocapacitance electrode materials, such as MnO_2 and RuO_2 exhibits surface redox which produces approximately rectangular CV in aqueous media with almost-linear GCD with some inflation [12].Mostly Co/Ni based oxide, double layer hydroxide, and sulphide shows intercalation type also termed as battery-type electrode materials, with reversible redox peaks in their CV. In addition to battery type material show plateau in GCD curve.

3. Transition Metal Oxalates

Transition-metal oxalates compounds inculcates transition metal sites are coordinated with oxygen atoms of oxalate group $(C_2O_4)^{-2}$. Transition metal site exhibits redox coupling while oxalate group creates a framework structure that promote the diffusion of ions inside the structure during charge/discharge process. Moreover, different morphology of 2D TMOx nanomaterials also exhibits large number of redox active sites due to larger surface area. Hence, high surface area, framework structure, low price oxalates and their composites bring TMOx as sustainable candidate for electrode materials.

3.1. Cobalt based Oxalate

2D porous CoC_2O_4 thin sheet has been synthesised by Chen et al. by using simple hydrothermal method at 220^oC with different reaction time as shown in fig 3(a) which exhibits schematic diagram of synthesis of CoC_2O_4 thin sheet [29]. Sample S1.S2 and S3 of CoC_2O_4 2D thin sheet were prepared with reaction time of 5h, 10h and 20h respectively. Fig 3(b) display the FESEM image of S1(left),S2(right) and S3(bottom) of CoC_2O_4 which has a sheet like morphology. Electrochemical performance of CoC_2O_4 2D thin sheet has been measured using Ni foam as current collector in 6M KOH. From the fig 3(c) it has been S3 has higher discharge time as compare to S1 and S2 Fig 3(d) and (e) shows the CV and GCD of S3-CoC_2O_4 2D porous thin

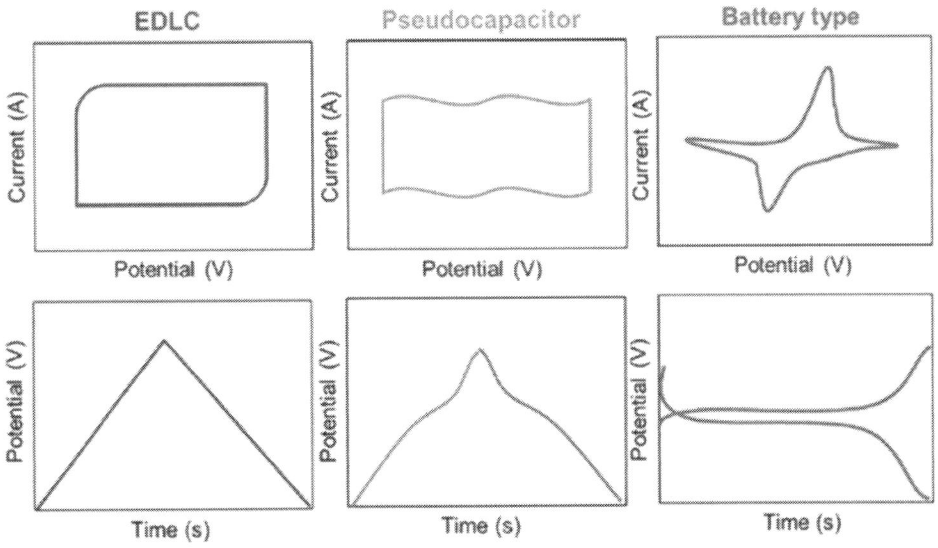

Figure 2. CV and GCD of EDLC, pseudocapacitive and battery typematerials. Reproduced (adapted) with permission [28]. Copyright 2018, RoyalSociety of Chemistry

sheet determines specific capacitance 1.631 F/cm^2 at the current 1.2 mA/cm^2 with excellent rate capability. Moreover, 2D porous thin sheet of S3-CoC_2O_4 was assembled with activated carbon to prepare complete cell as shown in fig 3(f) which exhibits maximum energy density and power density of 17.6Wh/Kg and 9000W/kg respectively. Cheng et al synthesised CoC_2O_4 by facile anodization strategy directly growing CoC_2O_4 on the cobalt foil [30]. From GCD fig 3 (g) calculated specific capacitance for CoC_2O_4 was 1269F/g at 6A/g of current density in 2M KOH. In addition, to that electrode displays good rate capability and cyclic stability i.e. 8.1% of decay after 100000 cycle as shown in fig 3(h).

3.2. Nickel based Oxalates

Nanostructure nickel oxalate can be prepared by dissolving Nickel foil in oxalic acid with various organic solvent and water [31]. Fig. 4 (a)-(e) exhibits change in the morphology of nanostructures nickel oxalate that are prepared in various solvents containing 1M oxalic acid.

The electrochemical test of nickel oxalate nanostructures conducted in 2M KOH medium for the pseudocapacitance application as shown in fig 4 (f). Subsequently, nickel oxalate annealed at higher temperature and converted into NiO which shows less intense peak as compare to nickel oxalate. Jung et al also investigated that the specific capacitance of the nickel oxalate structures is higher than that of the nickel oxide [31] . Zhao et al introduced 2D porous nickel oxalate (Ni–OA) thin sheets synthesized by a simple hydrothermal method [32]. NiC_2O_4 thin sheet were synthesised by hydrothermal method at constant temperature of 200 ^{o}C for 6, 12 and 24 h corresponding to sample was marked as S1, S2 and S3, respectively. Electrochemical test including CV and GCD were conducted in 6M KOH for S1, S2 and S3 at different scan rates and current densities respectively as shown in fig4 (g) and (h). Fig 4(i) CV curves reveals that as scan rates increases, peak current also increases for S1 Ni-OA. As a result of that it has been inferred that the 2D porous nickel oxalate based electrode exhibits the battery-type behavior. Furhter, fig4 (j) displays S1 Ni-OA electrode exhibits specific capacitance of 2835 F/g for 1 A/g of current density, and excellent cyclic stability which remains 92.5% at 10 A/g after 5000 charge–discharge cycles for supercapacitor as shown in fig 4(k).

IOP Conf. Series: Materials Science and Engineering 1166 (2021) 012032 doi:10.1088/1757-899X/1166/1/012032

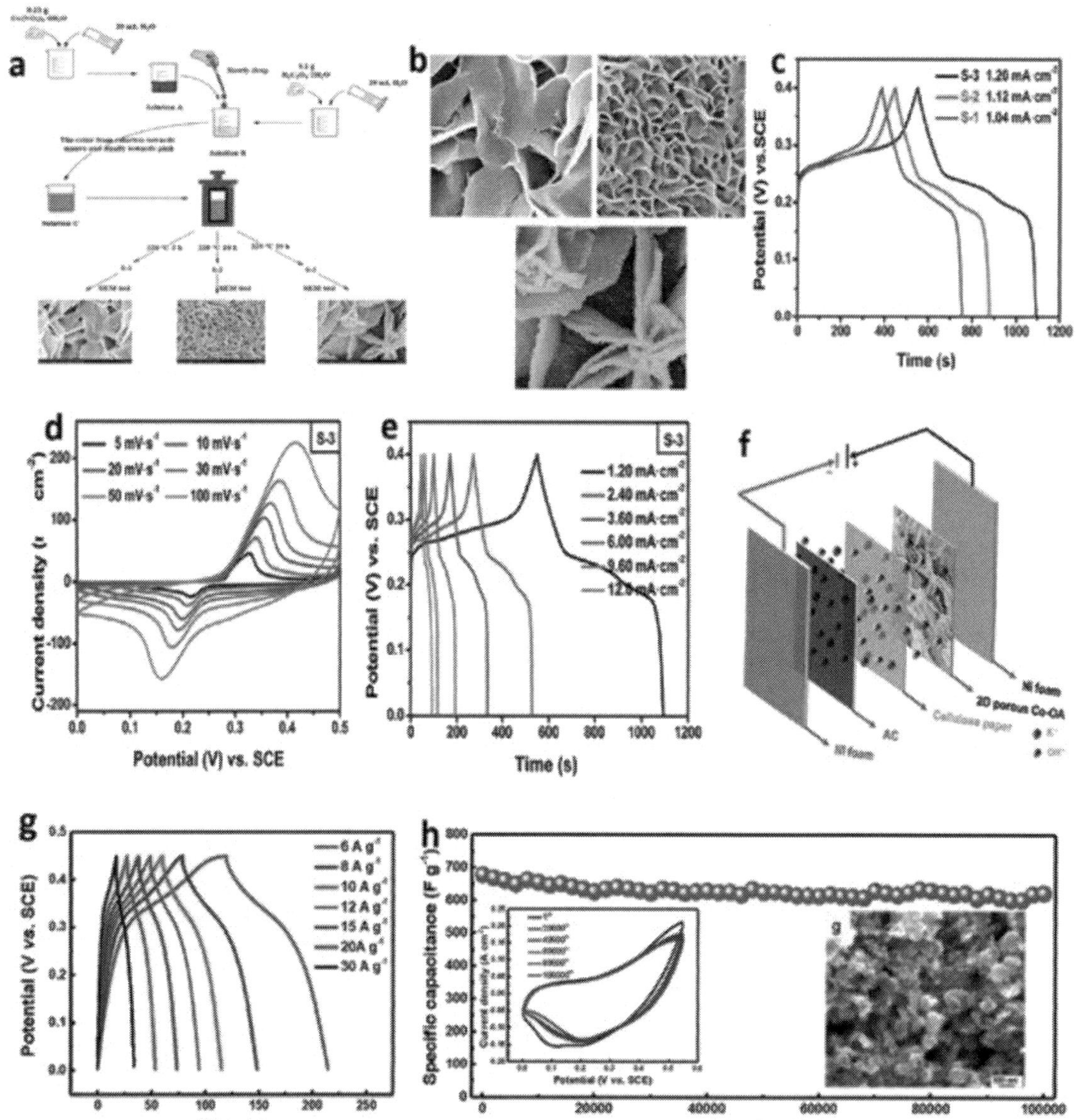

Figure 3. (a)Schematic diagram of Schematic diagram to illustrate the preparation of 2Dporous CoC_2O_4 thin sheets,(b) FESEM image of S1,S2 and S3of CoC_2O_4 thin sheets,(c) GCD of S1.S2 and S3.(d) CV ofS3 CoC_2O_4 thin sheets at different scan rate (e)GCD ofS3- CoC_2O_4 thin sheet at different areal current.Reproduced a, b, c, d, e and f with permission [29]. Copyright 2018, The Royal Society ofChemistry.(g) Galvanostatic charge/discharge curve at various currentdensities, (f) Nanostructured cobalt oxalate electrode after CV treatment of100,000 cycles along with SEM image showing the changed morphologies afternumber of cycle. Reprinted g and h with permission [30]. Copyright2016, Elsevier.

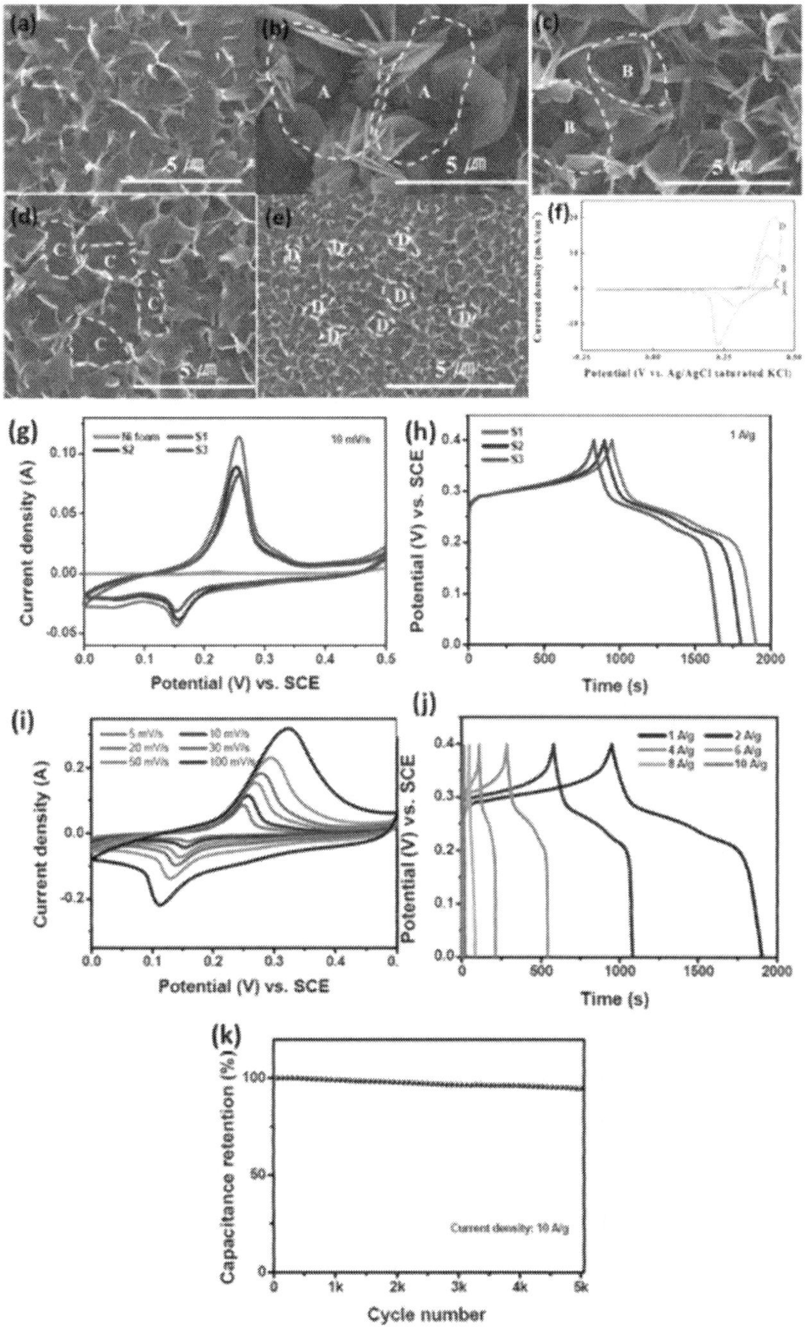

Figure 4. FE-SEMimages of nickel oxalate nanostructures with different morphology prepared in 1M oxalic acid dissolved in different solvents: (a) deionized water, (b) methanol,(c) ethanol, (d) n-propanol,(e) n-butanol. (f) CV of Nickle oxalate in 1M KOH.Reproduced a, b, c, d, e and f with permission [31]. Copyright 2010, The Royal Society ofChemistry. (g) CV of S1, S2and S3 sample of Nickel Oxalate thin sheet. (h) GCD curves of S1, S2 and S3 at 1.0 A/g,respectively, (i) CV curves of S1 at different scan rate (j) GCD curves of S1at different current densities. (k) Cyclic performance of S1 at a currentdensity of 10 A·g^{-1}. Reproduced g, h ,i , j and k withpermission [32]. Copyright 2020, The Royal Society ofChemistry.

3.3. Other Oxalates

Recently Liu and co-workers [33] reported of MnC_2O_4/GO materials as an electrodes in supercapacitor application. Hydrothermal decomposition of KMnO4 with ascorbic acid leads to the uniform distribution of the olive-like MnC_2O_4 particles on the surface of the GO sheets. The composite electrodes indicated a capacity of 122 F/g at 0.5 A/g of current density. Although from the CV as shown in Fig 5 (a) there was humps due to the presence of MnC_2O_4 which conclude the pseudocapacitance behaviour of MnC_2O_4/GO. Mixed oxalate the $Co_{0.5}Mn_{0.4}Ni_{0.1}C_2O_4{*}nH_2O$ micro polyhedrons prepared by single step co-precipitation method [34]. The specific capacitances attain by microployhedrons 990F/g at 0.6 A/g, in 3.0 M KOH solution. Subsequently $Co_{0.5}Mn_{0.4}Ni_{0.1}C_2O_4{*}nH_2O$ micro polyhedrons based electrode exhibited stable cycling performance maintaining specific capacitance of 968 F/g even after 6000 cycles.

Liu et al reported FeC_2O_4/rGO hydrogel composite as a negative electrode for supercapacitors [35]. The electrochemical performance of FeC_2O_4/rGO was conducted in aqueous Na_2SO_4 solution (1mol/L) using three electrode steup. The CV curves of rGO hydrogels were rectangular as rGO shows EDLC behaviour while redox peaks evidently appear for the FeC_2O_4/rGO hydrogel as shown in fig 5(b) indicating the pseudocapacitive of iron oxalate. The specific capacitance in Na_2SO_4 solution were 591F/g at 2A/g of current density with good rate capability.

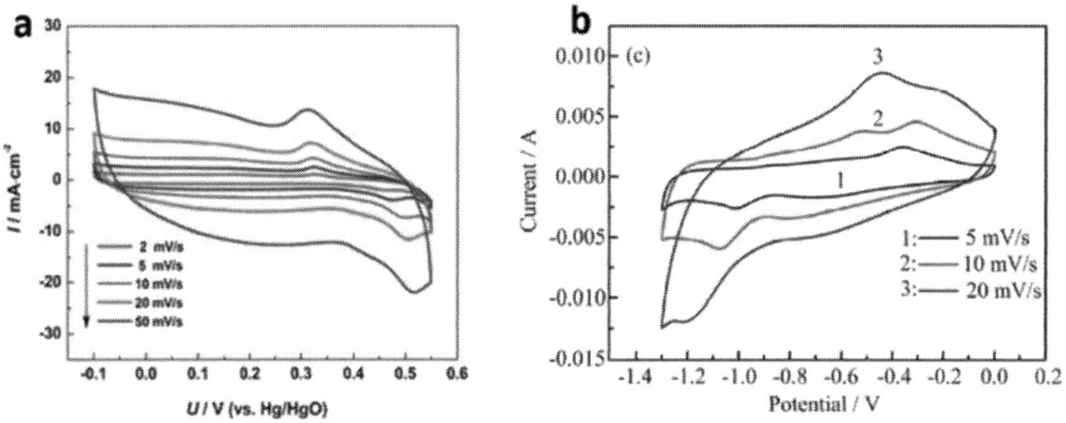

Figure 5. 5 (a) CV of MnC_2O_4/GO electrode in 6M KOH at different scanrate, (b) CV of FeC_2O_4/rGO in 1mol/L of Na_2SO_4solution at different scan rate. Reproduced (a) with permission [33]. Copyright 2013, Springer-Verlag Berlin Heidelberg; (b) Reproduced with permission [35]. Copyright 2018, Chinese Physical Society

Transition metal based oxalate has great potential of pseudo-capacitance applications as it showed the higher specific capacitance. Various type of electrode based on TMOx has been compared in Table 1.

4. Conclusion

In summary, the prime objective of this paper emphasised on the usage of TMOx based electrode for supercapacitors. In addition, TMOx synthesis process, morphology of nanostructures, as well as electrochemical performances has been conferred. Transition-metal oxalates emerges as a relevant energy material in the field of supercapacitor due to their cost effectiveness, environmental begin and open structure for ion transportation. Although, TMOx also have some

Table 1. Transition-metal oxalates based electrodes for Supercapacitor

Material	Morphology	Specific capacitance (F/g)	Elec-trolyte	Operating Potential(V)	Ref.
CoC_2O_4	Thin sheet	1269	6M KOH	0 to 0.5	[30]
NiC_2O_4	2D thin sheet	2835	6M KOH	0 to 0.4	[32]
MnC_2O_4/GO	Olive-like	122	6M KOH	-0.1 to 0.55	[33]
$Co_{0.5}Mn_{0.4}Ni_{0.1}C_2O_4*nH_2O$	Micro polyhedrons	990	3M KOH	0 to 0.4	[34]
FeC_2O_4/rGO	Porous rod	591	1M Na_2SO_4	-1.3 to 0	[35]

short coming such as less conductive and accumulation of particles degrades the electrochemical performance. In account of that, electrochemical performances of TMOx electrode can be enhanced through various methodology such as by preparing mixed TMOx structures or preparing nanocomposite with highly conductive secondary materials such as graphene and CNT. Hence, it is valuable to investigate more stable structures and highly electrochemically active oxalates for energy storage applications.

Acknowledgements

This work is supported by Nanotechnology program at Centre for Advanced Studies, AKTU and Department of Ceramic Engineering, IIT (BHU).

References

[1] Goodenough J B 2018 *Nat. Electron* **1** 204–204

[2] Winter M and Brodd R J 2004 *Chemical Reviews* **104** 4245–4270 URL 10.1021/cr020730k

[3] Simon P and Gogotsi Y 2010 *Mater. Sustain. Energy A Collect. Peer-Reviewed Res. Rev. Artic. from Nat. Publ. Gr* 138–147

[4] Li H B, Yu M H, Wang F X, Liu P, Liang Y, Xiao J, Wang C X, Tong Y X and Yang G W 2013 *Nature Communications* **4** 1–7 URL 10.1038/ncomms2932

[5] Cao X 2019

[6] Service R F 2006 *Science* **313** 902–902 URL 10.1126/science.313.5789.902

[7] Lee T H, Pham D T, Sahoo R, Seok J, Luu T H T and Lee Y H 2018 *Energy Storage Materials* **12** 223–231 URL https://doi.org/10.1016/j.ensm.2017.12.009

[8] Liu X, Shi C, Zhai C, Cheng M, Liu Q and Wang G 2016

[9] Liu X, Shi C, Zhai C, Cheng M, Liu Q and Wang G 2016 URL 10.1021/acsami.5b10781

[10] Chmiola J, Largeot C, Taberna P L, Simon P and Gogotsi Y 2008 *Angewandte Chemie International Edition* **47** 3392–3395 URL 10.1002/anie.200704894;https://dx.doi.org/10.1002/anie.200704894

[11] Costentin C, Porter T R and Savéant J M 2017 *ACS Applied Materials & Interfaces* **9** 8649–8658 URL 10.1021/acsami.6b14100;https://dx.doi.org/10.1021/acsami.6b14100

[12] Augustyn V, Simon P and Dunn B 2014 *Energy & Environmental Science* **7** 1597–1597 URL 10.1039/c3ee44164d;https://dx.doi.org/10.1039/c3ee44164d

[13] González A, Goikolea E, Andoni J and Mysyk R 2016 **58** 1189–1206 URL 10.1016/j.rser.2015.12.249

[14] Wang F, Xiao S, Hou Y, Hu C, Liu L and Wu Y 2013 *RSC Adv* **3** 13059–13084

[15] Chen G Z 2013 *Prog. Nat. Sci. Mater. Int* **23** 245–255

[16] Zheng J P, Cygan P J and Jow T R 1995 **142** 98–102

[17] Jabeen N, Xia Q, Savilov S V, Aldoshin S M, Yu Y and Xia H 2016 *ACS Applied Materials & Interfaces* **8** 33732–33740 URL 10.1021/acsami.6b12518;https://dx.doi.org/10.1021/acsami.6b12518

[18] Salavati-Niasari M, Mir N and Davar F 2009 *Polyhedron* **28** 1111–1114 URL 10.1016/j.poly.2009.01.026;https://dx.doi.org/10.1016/j.poly.2009.01.026

[19] Du H, Pan Y, Zhang X, Cao F, Wan T, Du H, Joshi R and Chu D 2019 *Nanoscale Advances* **1** 140–146 URL 10.1039/c8na00110c

[20] Tyagi A, Joshi M C, Shah A, Thakur V K and Gupta R K 2019 *ACS Omega* **4** 3257–3267 URL 10.1021/acsomega.8b03618;https://dx.doi.org/10.1021/acsomega.8b03618

[21] Pang H, Wei C, Li X, Li G, Li S, Chen J, Zhang J and Nis U 2014 1–8 URL 10.1038/srep03577

[22] Kumar K S, Choudhary N, Jung Y and Thomas J 2018 *ACS Energy Letters* **3** 482–495 URL 10.1021/acsenergylett.7b01169;https://dx.doi.org/10.1021/acsenergylett.7b01169

[23] Meng Q, Cai K, Chen Y and Chen L 2017 *Nano Energy* **36** 268–285

[24] Li N, Li Q, Guo X, Yuan M and Pang H 2019 *Chemical Engineering Journal* **372** 551–571 URL 10.1016/j.cej.2019.04.127;https://dx.doi.org/10.1016/j.cej.2019.04.127

[25] Ahouari H, Rousse G, Rodríguez-Carvajal J, Sougrati M T, Saubanère M, Courty M, Recham N and Tarascon J M 2015 *Chemistry of Materials* **27** 1631–1639 URL 10.1021/cm5043149;https://dx.doi.org/10.1021/cm5043149

[26] Reddy M V, Rao G V S and Chowdari B V R 2013

[27] Yeoh J S, Armer C F and Lowe A 2018 *Materials Today Energy* **9** 198–222 URL 10.1016/j.mtener.2018.05.010;https://dx.doi.org/10.1016/j.mtener.2018.05.010

[28] Bhoyate S, Kahol P K and Gupta R K 2020 120–149 URL 10.1039/9781788017053-00120

[29] Pu T 2018 *Dalt. Trans* **47** 9241–9249

[30] Cheng G 2016 *J. Power Sources* **312** 184–191

[31] Jung I, Choi J and Tak Y 2010 *Journal of Materials Chemistry* **20** 6164–6164 URL 10.1039/c0jm00279h;https://dx.doi.org/10.1039/c0jm00279h

[32] Zhao C, Jiang Y, Liang S, Gao F, Xie L and Chen L 2020 *CrystEngComm* **22** 2953–2963 URL 10.1039/d0ce00268b;https://dx.doi.org/10.1039/d0ce00268b

[33] Liu T, Shao G, Ji M and Ma Z 2014 *Ionics* **20** 145–149 URL 10.1007/s11581-013-1017-8;https://dx.doi.org/10.1007/s11581-013-1017-8

[34] Zhang Y Z 2014 *Sci. Rep* **5** 1–6

[35] Liu W, Song Y, Wang H, Wang H and Yan L F 2018 *Chinese J. Chem. Phys* **31** 707–716

Impregnation of Microwave treated exfoliated MoS_2 nanosheets into PMMA matrix by solution casting for reinforcing applications

Melbin Baby[1, 2*] and K. Rajeev Kumar[1,3]

[1] Department of Instrumentation, Cochin University of Science and Technology, Cochin 682022.
[2] Sophisticated Test and Instrumentation Centre, Cochin University of Science and Technology, Cochin 682022, Email: rajeev@cusat.ac.in.
[3]Centre of Excellence in Advanced Materials, Cochin University of Science and Technology, Cochin, India.
*Author for correspondence: *melbin_83@yahoo.co.in.*

Abstract. In the present work, poly methyl methacrylate (PMMA)-MoS_2 nanocomposite films were prepared by impregnation of exfoliated MoS_2 nanosheets into PMMA matrix by solution casting method. Synthesis of exfoliated MoS_2 nanosheets through microwave irradiation techniques was found to be simple, less time consuming, eco-friendly and cost effective. We have identified the optimization conditions by varying the parameters like microwave heating time, power and concentration of bulk MoS_2. The structural and optical properties of the exfoliated MoS_2 nanosheets were investigated. These optimized exfoliated MoS_2 nanosheets were introduced into the polymer matrix which was then subjected to different characterization techniques. The effect of exfoliated MoS_2 nanosheets on the mechanical properties of the polymer-MoS_2 composite was evaluated in detail.

Keywords: Exfoliation, MoS_2 nanosheet, Microwave heating, nanocomposite.

1. Introduction

Nowadays, polymer nanocomposite has been getting much attention in the areas of scientific and industrial fields. Polymer nanocomposite is a class of composite materials which contains a polymer matrix and an inorganic material that has at least one dimension in the nanorange [1]. The addition of inorganic nanomaterials into polymer matrix enhances their mechanical, thermal, barrier and fire-retardant properties [2]. This property enhancement is due to the reinforcement mechanism of the inorganic nanomaterials in polymer nanocomposite and strongly depends on the properties of the polymer matrix, nature and type of nanomaterials, concentration of polymer and nanomaterials, particle aspect ratio, particle size, particle orientation and particle distribution [2]. Polymer nanocomposite with layered reinforcing agents exhibit high storage modulus, increased tensile properties, heat distortion temperature and reduction in gas permeability. They also exhibit substantially improved self-extinguishing behavior, tunable biodegradability and enhanced optical and electrical properties when compared with pure polymer and conventional micro and macro composites [3, 4].

Compared to other polymers, poly methyl methacrylate (PMMA) is a one of the most widely used industrial plastics which is also known as acrylic glass [5]. PMMA is a transparent polymer which is easily soluble in different solvents like acetophenone, chloroform, toluene and dimethylformamide [6]. Due to its advantages such as high transparency, light weight, and shatter resistance, this polymer is predominantly used in optical and biomedical applications [5].

In recent years, exfoliated two dimensional nanomaterials such as graphene and transition Metal dichalcogenides (TMDCs) have been more focused due to their extraordinary optical and electrical properties [7]. In order to overcome the disadvantages of graphene due to its zero- bandgap, layered TMDC's which are analogues to graphene in several respects have gained considerable attention in recent time [8]. Among TMDCs family, MoS_2 is the most studied material because of its robustness [9]. The crystal structure of single layer MoS_2 consists of one layer of Molybdenum atoms sandwiched between two layers of Sulphur atoms. In contrast to graphene, bulk MoS_2 and monolayer MoS_2 have indirect bandgap of 1.2eV and direct bandgap of 1.9eV respectively [10]. There are two different strategies like 'top-down' and 'bottom-up' for synthesizing MoS_2 nanosheets. Top-down approach includes liquid phase exfoliation [11-12] micro-mechanical exfoliation [13-14], Taylor-Couette flow [15] etc. whereas bottom-up approaches are Physical Vapor Deposition [16], Solution Chemical Processes [17-18], Chemical Vapor Deposition [19], ALD [20] etc. Exfoliated MoS_2 material promise a wide variety of applications like sensing [21], photo catalyst [22], Li-ion battery etc. [23]. Their antibacterial activity [25] and bio marking [24] property make them suitable for several important biological applications.

The improvements in electrical, optical, thermal, mechanical, tribological, and flame-retardant properties of MoS_2-polymer nanocomposite strongly depend on the synergistic combination of MoS_2 and PMMA [26]. There are various methods such as physical blending and in-situ polymerization methods which have been widely used to fabricate PMMA–MoS_2nanocomposite [26]. The obtained MoS_2-PMMA nanocomposite has been explored for different applications from optical limiters [27], data storage [28] gas sensor [29] and photo detectors [30].

In this work, we synthesized the exfoliated MoS_2 nanosheets from bulk MoS_2 by Microwave irradiation techniques. It is simple, eco-friendly and less time-consuming method. A few papers have reported the exfoliation of MoS_2 nanosheets via microwave irradiation techniques. But none of them have reported on the structural and optical properties of exfoliated MoS_2 nanosheets. In our method, we used water as the solvent. Usually most of the papers related to microwave assisted synthesis methods, reports the use of ethylene glycol, DMF (Dimethylformamide) or other organic solvents as the solvent material due to their high susceptibility to microwave. But use of these organic solvents reduces the eco-friendly nature of this method. Hence, we chose water in our method for its ease of use and suitability in biomedical applications. For getting the optimized condition, bulk MoS_2 is subjected to different microwave power and heating time at constant quantity of raw materials. Structural and optical properties of the optimized MoS_2 nanosheets are characterized. In addition, mechanical properties of PMMA-MoS_2nanocomposite are also carried out.

2. Material and Methods

2.1 Materials

Molybdenum disulfide (MoS_2 powder, size<2 μm, molecular weight: 160.07 and 99% purity), PMMA (poly methyl methacrylate-$C_5O_2H_8$) and chloroform ($CHCl_3$) were purchased from Sigma Aldrich. In order to prepare aqueous solution of the sample, double distilled water was used.

2.2 Methods

2.2.1 Preparation of exfoliated MoS_2 Nanosheet

Domestic Microwave oven (model: Electrolux Solo, frequency 2450 MHz, output power 700W) was used for the synthesis of nanosheets.2.5μg bulk MoS_2 dispersed in 20ml distilled water(DW)was subjected to microwave irradiation for different power and time duration. The final product was washed with distilled water and centrifuged for 2 minutes at 5000 rpm and dried at 80^{0}C in an oven. The powder sample was used for further characterizations.

2.2.2 Preparation of PMMA and PMMA+ exfoliated MoS_2nanocomposite thin film

0.05g PMMA crystal was dissolved in 5ml chloroform and 0.013mg of exfoliated MoS_2 was added into the stirring PMMA solution. This was followed by drop casting onto glass substrate and was then kept in room temperature for 1hour for evaporation of the solvent.

Crystal structure of the PMMA-exfoliated MoS_2 nanocomposite was characterized using Powder X- ray diffraction (Bruker AXS D8 Advance) with CuKα radiation in the 2θ range of 3-80°, and scan speed of 3°/min. Surface Morphology of the samples was analyzed using Scanning electron microscopy (JEOL JSM 6390 LA, SEM) and transmission electron microscopy (JEOL JEM 2100, TEM). Raman Spectroscopy was used to characterize the vibrational modes of the samples (Horiba JY LabRAM HR 800 Micro Raman Spectrometer). Scanning probe microscopy was used to carried out the Nano mechanical properties of these samples (NanoGuru, N50). Elemental composition of the samples was identified using Energy dispersive analyzer (OXFORD XMX N, EDAX), UV-VIS spectrophotometer (Varian, Cary 5000) and photoluminescence spectroscopy (HORIBA FLUOROLOG FLUOROSCENCE SPECTROMETER) were used to investigate the optical properties of the samples.

3. Results and discussion

3.1 Optimization of exfoliated MoS_2 nanosheets

The control parameters viz microwave irradiation time and power was used to optimize the preparation conditions of exfoliated MoS_2 nanosheets. These parameters were optimized by the following method. A fixed concentration of 2.5mg bulk MoS_2 in 20ml distilled water was used for further experiments.

The microwave power and irradiation time were used as the optimizing parameters of exfoliation process, keeping the volume of the raw materials constant. In order to study the exfoliation details of the nanosheets, microwave treated six samples were marked as L2(low power-119W for 2minutes), M2(medium power-462W for 2 minutes), H2(high power-700W for 2 minutes), H4(high power -700W for 4 minutes), H6(high power-700W for 6 minutes) and H8(high power -700W for 8 minutes) respectively. The details of trials and parameters are tabulated in table1.

Table 1. Optimization of microwave power and time at constant volume of raw materials for synthesizing exfoliated nanosheets.

Volume of raw materials (0.0025g/20 ml DW)	Microwave power (W)	Time (minutes)	Raman spectra (cm^{-1})	Electron Microscopy	
				SEM	TEM
20 ml	L2 (119W)	2	26.6	Agglomerated	Agglomerated
20 ml	M2 (462W)	2	26.2	Agglomerated	Agglomerated
20 ml	H2 (700W)	2	25.9	Sheets are seen; but agglomerated	Agglomerated
20 ml	H4 (700W)	4	25.3	Sheets are seen; but agglomerated	Agglomerated
20 ml	H6 (700W)	6	24.8	Sheets are seen: but agglomerated	Agglomerated
20 ml	H8 (700W)	8	24.2	Sheets separated	Few layer sheets

The experiment was conducted by varying the power and time. Initially the samples were subjected to microwave irradiation at (low, medium, high) powers for 2 minutes. The results from above trials showed that compared to low and medium power, the sample irradiated at high power for 2 minutes resulted in more exfoliation of the sheets. Hence optimum power level was identified as 700W. Further trials were conducted for optimizing the time of irradiation. It was found that after 8 minutes of microwave irradiation at higher power (700 W), separate sheets were formed without agglomeration, and the solvent evaporated completely. Therefore, the optimized condition for the preparation of few layer MoS_2 nanosheets was selected to be H8 (high power-700W for 8 minutes). Electron microscopy (SEM and TEM) and Raman spectroscopy were characterization tools used for this optimization process. Henceforth, the optimized sample H8 is marked as MoS_2E.

Electron microscopy (SEM) Studies

Figure 1. SEM images of (a) Bulk (b) L2 (c) M2 (d) H2 (e) H4 (f) H6 and (g) H8.

Figure 1 shows the SEM images of MoS_2 samples of different trials. It is evident that, as explained in the optimization section, exfoliation is maximum at H8 condition. As we move from low power to high power, bulk MoS_2 is gradually exfoliated to nanosheets and the process is complete at H8 conditions.

Electron microscopy (TEM) Studies

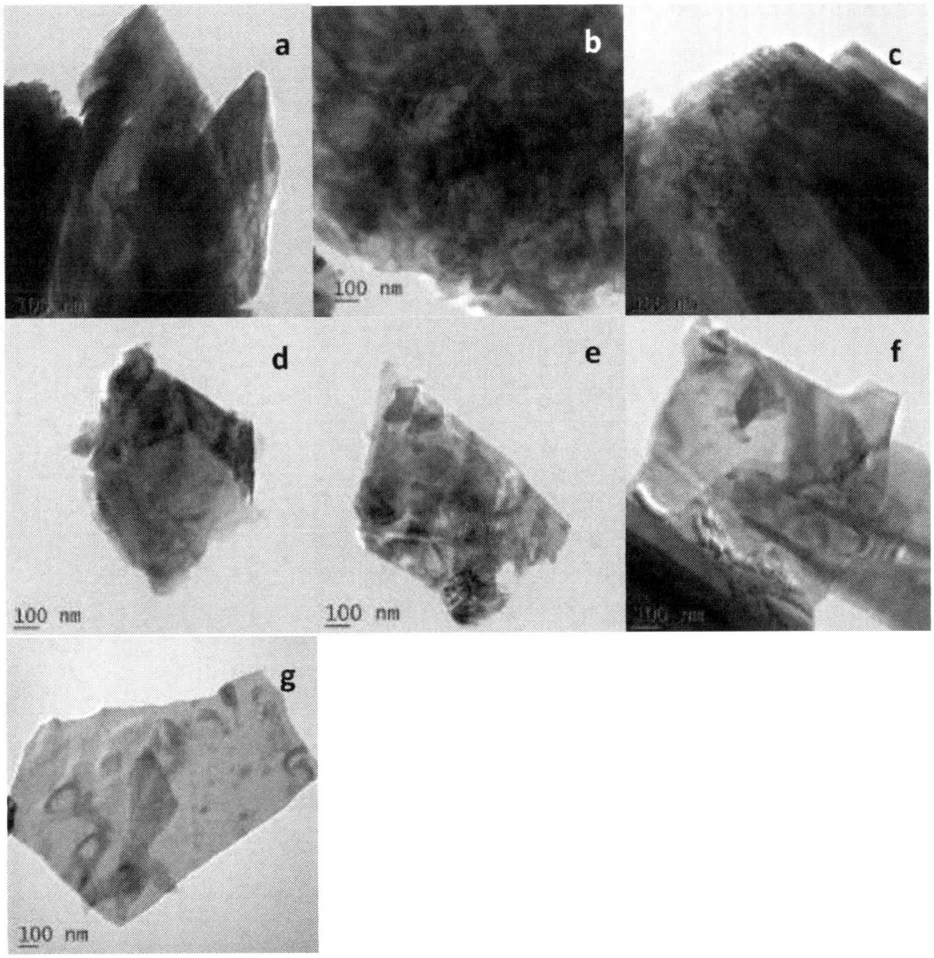

Figure 2. TEM images of (a) Bulk (b) L2 (c) M2 (d) H2 (e) H4 (f) H6 and (g) H8 (MoS_2E).

The SEM observations are confirmed by TEM images of Figure 2. As the microwave power and time of irradiation increase to condition H2 (ie. 700W, 2 minutes) the exfoliation starts. The process improves when the time of irradiation is further increased and neat separated sheets are obtained at eight minutes exposure time.

3.2 Raman studies

Raman spectrum of the layered MoS_2, which helps us to study the vibrational modes of MoS_2, exhibits thickness dependent behavior. Figure 3 shows Raman spectrum of the optimization trials (bulk to

H8) of the preparation of exfoliated MoS_2 nanosheets. In bulk MoS_2, the dominant vibrational modes are located at 377.34 $cm^{-1}(A_{1g})$ and 403.66 $cm^{-1}(E^1_{2g})$. As thickness of the MoS_2 decreases from bulk to exfoliated layers, the dominant vibrational modes of in-plane vibration and out-of-plane vibration modes are slightly shifted. The frequency difference between these two vibrational modes is found to be 24.01cm^{-1}(bulk to MoS_2E) when compared with bulk (26.32cm^{-1}). This value is close to the trilayer MoS_2 in reported literature [31].

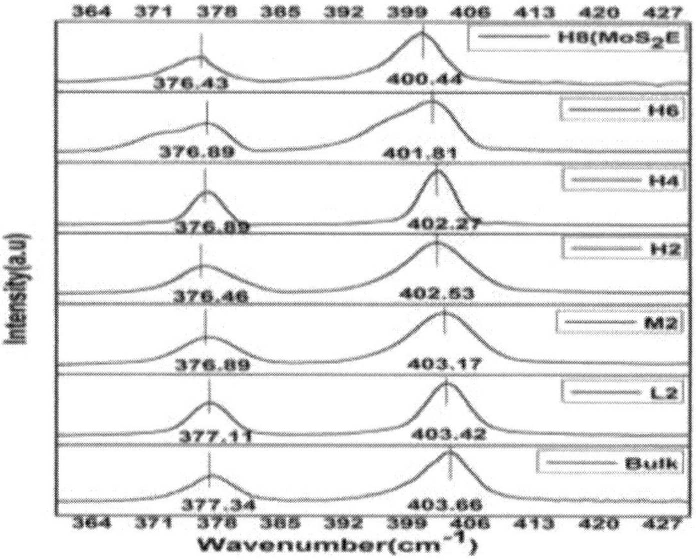

Figure 3. Raman spectra of Bulk, L2, M2, H2, H4, H6, and H8(MoS_2E) samples.

3.3 Structural properties

Figure.4 shows XRD patterns of both bulk and exfoliated MoS_2.The only difference between these two spectra is in Intensity of peaks. It is found that the reduction in XRD intensity of the (002) plane is due to the formation of exfoliated MoS_2 in the sample [32]. In addition to this, there is no other extra peak present in the samples.

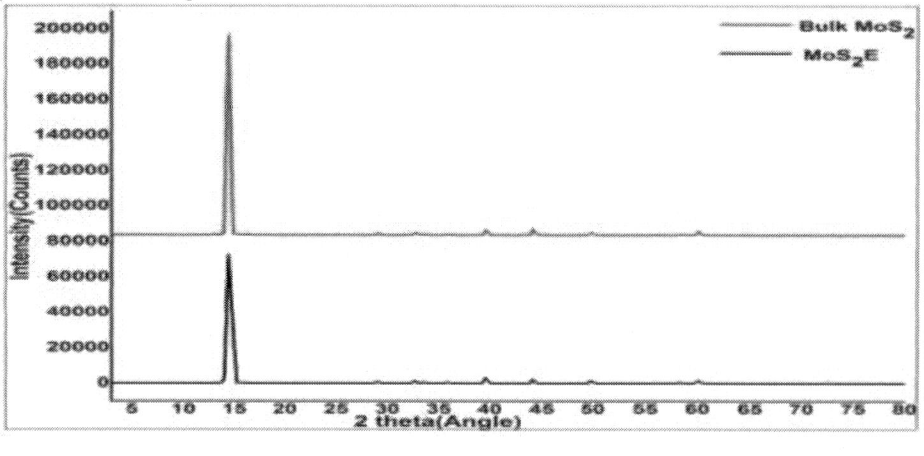

Figure 4. Shows XRD spectrum of bulk MoS_2 and MoS_2E.

Atomic Force Microscopy (AFM) technique was also used to further examine the morphology and thickness of the synthesized MoS_2 nanosheets. AFM topography images were captured with contact mode in air using cantilever tip and the images are shown in Figure.5. The theoretical thickness of MoS_2 monolayer is around 0.9-1.3nm [33]. From the AFM height profile of the areas marked in Figure. 5(b), the thickness of the MoS_2nanosheets is found to be below 5nm.This indicates existence of a few layers in the samples. During drying process, some particles show agglomeration.

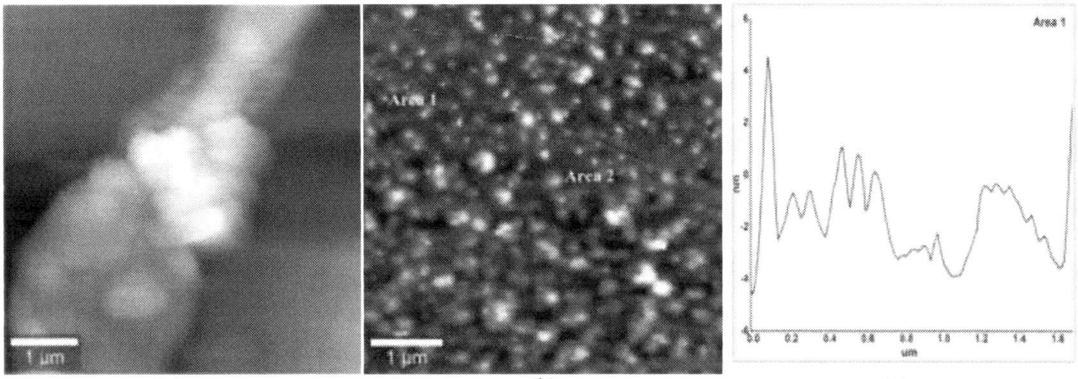

(a) (b) (c)

Figure 5. (a,b) AFM images of bulk MoS_2 and MoS_2nanosheets. (c) AFM height profile measured along the area 1 in (b).

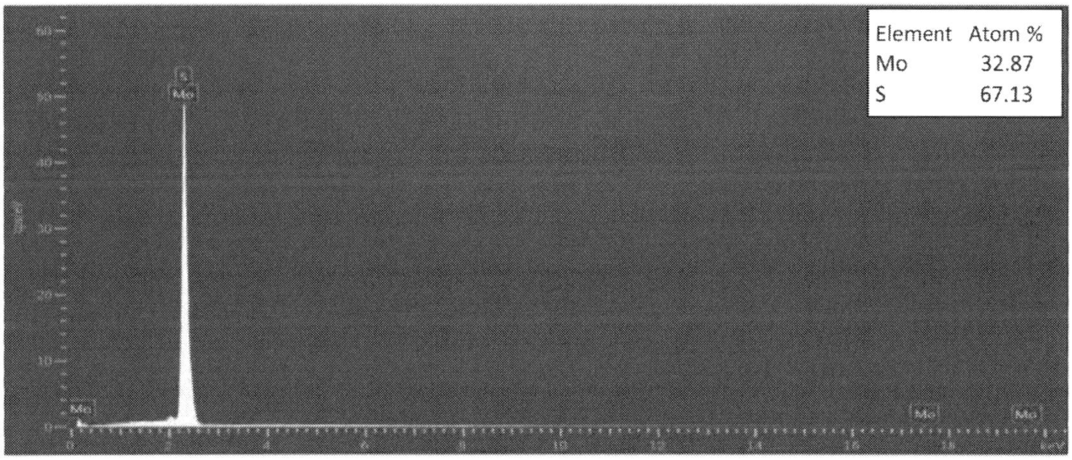

Figure 6. EDAX spectrum of MoS_2E.

It is confirmed from EDAX, that the sample MoS_2E (Microwave power 700W and irradiation time 8 minutes) has Mo to S stoichiometry of 1: 2.04 and is shown in figure 6.

3.4 Optical properties

Figure.7 shows the optical absorbance of the dispersed solution of bulk and exfoliated MoS_2 and Tauc's plot of exfoliated MoS_2.From the spectra, it is found that the exfoliated MoS_2 nanosheets exhibit higher absorbance than bulk. This is due to the fact that, being lighter the nanosheets have lower sedimentation rate than bulk MoS_2 and hence more surface area will be exposed to the UV-Vis radiation, which in turn increase the absorption in MoS_2E sample [34]. The absorption peaks at 631nm and 691 nm correspond to the exciton A and exciton B arising due to the direct band transitions [35]. The peak at 460 nm can be attributed to the optical transitions between the density of states peaks in the valance and conduction bands [36, 37].

According to Beer's law, the absorbance (A) = εlc. Where A is the absorbance of the material taken from UV-Vis spectra, ε is the molar extinction coefficient, l is the path length and c is the concentration of the solution [38].

ε=A/lc-- (1).

Using Tauc's plot, bandgap energy of the exfoliated $MoS_2$was calculated [39].

$(\alpha h\nu)=B(h\nu-E_g)^{\gamma}$ ....(2)

Where α is the absorption coefficient, which is a function of frequency ν. B is a constant that depends on the transition probability. The exponent$\gamma=1/2$ or 3/2 for direct transitions whereas, $\gamma=2$ or 3 for indirect transitions. E_g is bandgap energy. Here, molar extinction coefficient (ε) is molar absorption coefficient α.

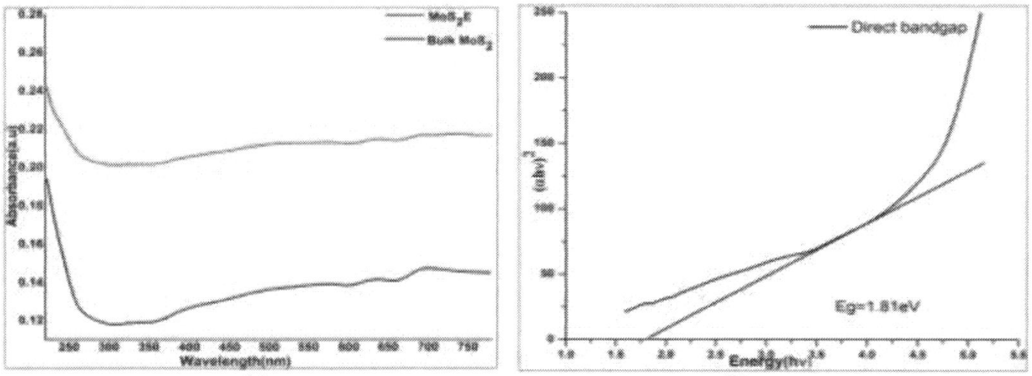

(a) (b)

Figure.7 (a). UV-VIS spectrum of bulk MoS_2 and MoS_2E. Fig.7 (b) Tauc plot curve of MoS_2E.

In bulk MoS_2, the bandgap energy is 1.2eV. For direct transition, the calculated optical bandgap energy of the exfoliated MoS_2 is 1.81eV which is close to the reported value [36]. Compared to bulk, the exfoliated MoS_2 sample has higher bandgap value due to quantum confinement [36].

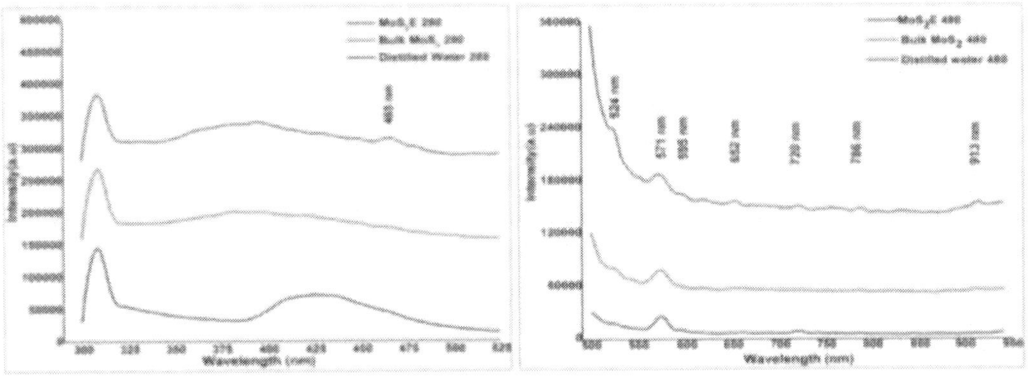

(a) (b)

Figure 8. PL emission spectrum of bulk MoS_2, exfoliated MoS_2 and distilled water, with excitation wavelengths (a) 280nm (b) 480nm.

The photoluminescence properties of dispersed solution of bulk MoS_2, and exfoliated MoS_2 nanosheets in distilled water were characterized and is shown in Figure 8. The samples were excited with two excitation wavelengths of 280nm and 480 nm and corresponding emission spectrum were recorded. At both wavelengths, the emission peaks of exfoliated MoS_2 nanosheets are observed to be at 465nm, 595nm,652nm,720nm,786nm and 913nm.The peaks observed at 595nm and 652nm are attributed to exciton A and exciton B respectively [40]. The peaks at 780nm (1.59eV) and beyond are ascribed to indirect bandgap transitions [41]. It is also observed from the PL spectra that the exfoliated MoS_2 samples show enhanced photoluminescence compared to bulk MoS_2.This enhancement is due to the lower relaxation rate of interband transition and it reflects significant changes in the electronic structure of the exfoliated MoS_2[42].

3.5 Mechanical properties

In order to investigate the effect of concentration of exfoliated MoS_2 in the PMMA matrix, the composite was analyzed using Nano indentation. In this technique, Berkovich diamond probe is used to indent and scan the surface of the samples. The topographical images of the samples are shown in Figure 9.

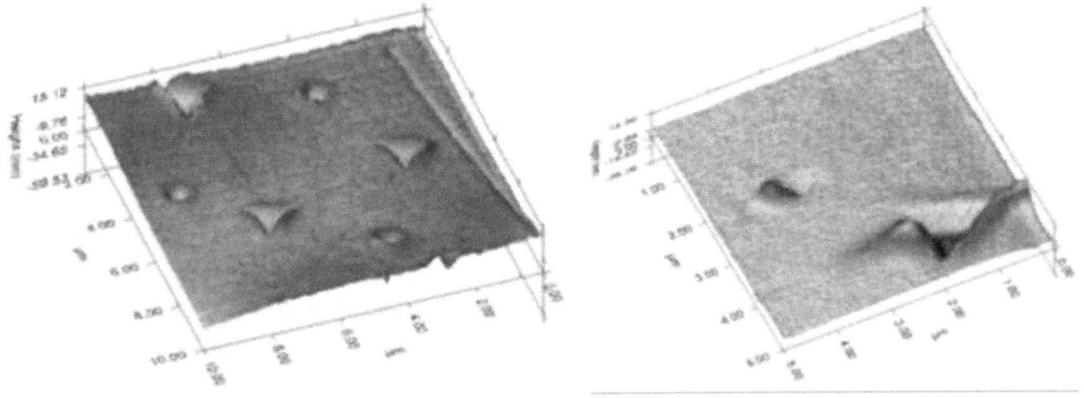

(a) **(b)**

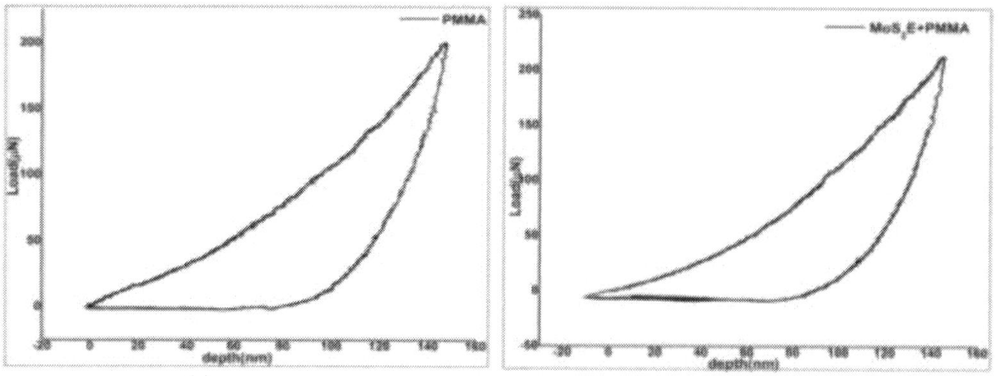

(c) (d)

Figure 9. SPM images of (a) MoS_2E-PMMA (b) PMMA. Force –depth curve of (c) PMMA and (d) MoS_2E-PMMA.

In this test, load- displacement of the indenter probe is measured continuously throughout the analysis. The data is used to extract hardness and elastic modulus of the materials. Hardness and Elastic modulus of the samples are calculated using the following equations.

Hardness (H)= Load maximum (μN)/contact area (nm^2) .... (3)

The probe area function of the ideal Berkovich indenter is given by

$A_c=24.6h_c^2$. Where h_c is the contact depth and A_c is the projected contact area at maximum load [43].

Elastic modulus (Er) $=\sqrt{\pi}/2\beta * S/\sqrt{A_c}$. ....(4)

Where β, is the Berkovich indenter geometry constant, is equal to 1.034 and S is the contact stiffness [43].

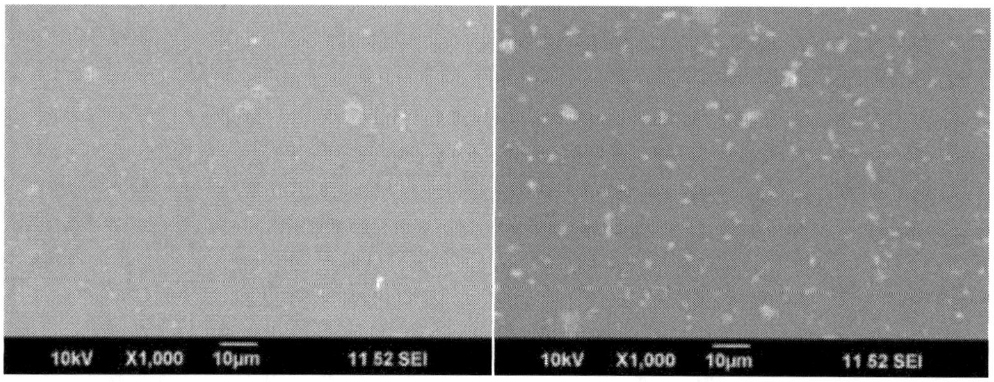

(a) (b)

Figure 10. SEM images of a) PMMA film. b) PMMA-MoS_2E film

The measured hardness of PMMA and nanocomposite samples is 0.2507GPa and 0.2755GPa respectively. Similarly, elastic modulus of PMMA and nanocomposite samples is found to be 6.7815GPa and 8.5547GPa respectively. Thus, it is clear that incorporation of exfoliated MoS_2 into PMMA matrix has caused significant improvement in the mechanical properties of the polymer. This improvement is due to the strong interaction between the polymer molecules and nanosheets. It is important to note here that no cracks and pores were visible in the film. This is confirmed by scanning electron microscopy images and given in Figure 10 shows SEM images of PMMA film and PMMA-MoS_2E film. Hence it is safe to conclude that efficient load transfer takes place between MoS_2 nanosheets and the PMMA matrix. This paves the way for using this nanocomposite material for load bearing applications [44] while retaining most of its transparency.

4. Conclusions

Exfoliated MoS_2 nanosheets are synthesized successfully by microwave irradiation method, which is a simple, eco-friendly, fast and cost-effective technique. The structural and optical properties of the exfoliated MoS_2nanosheets are characterized. The optimization of the exfoliation process of MoS_2 nanosheets is carried out with the help of Electron microscopy and Raman spectroscopy. Raman spectroscopy studies reveal that there is a small shift in the vibrational frequencies of E^1_{2g} and A_{1g} peaks of exfoliated samples. Using electron microscopy, few layer nanosheets are visualized. The stoichiometric ratio of exfoliated MoS_2 nanosheets (1:2) is confirmed using Energy Dispersive X-ray analyzer. Optical properties of MoS_2 nanosheets are investigated using UV-VIS Spectrophotometer and Photoluminescence Spectroscopy. PMMA-MoS_2 nanocomposite is prepared by incorporation of optimized MoS_2 nanosheets into the polymer matrix by solution casting. Compared to PMMA, the mechanical properties like hardness and elastic modulus of the nanocomposite are enhanced. From these observed results, one can safely conclude that this PMMA- MoS_2E nanocomposite is a suitable material for potential reinforcing applications.

Acknowledgement

The authors acknowledge 'Sophisticated Test and Instrumentation Centre (STIC)' and 'Sophisticated Analytical Instrumentation Facility (SAIF), Cochin', for providing analytical facilities and for valuable discussions. We are thankful to Dr. Chintan Bhat, Mr. Kiran Raphael for giving SPM facility (NanoGuru, N50) and also grateful to Dr. E.I. Anila and Miss Ebitha (UC College, Aluva) and Prof. M. K. Jayaraj and Miss. Anju (Department of Physics, CUSAT) for providing PL facility and Raman spectroscopy for this work.

Conflicts of Interest

All authors declare that they have no conflict of interest.

Financial funding

This research received no external financial funding and support.

References

[1] Passador, F. R. &Pessan, L. A. *Matrices and Lamellar Clays. Nanostructures* (Elsevier Inc., 2017). doi:10.1016/B978-0-323-49782-4/00007-3.

[2] Dantas de Oliveira, A. & Augusto Gonçalves Beatrice, C. Polymer Nanocomposites with Different Types of Nanofiller. *Nanocomposites - Recent Evol.* (2019). doi:10.5772/intechopen.81329.

[3] Camargo Pedro Henrique Cury, Satyanarayana KesturGundappa&Wypych Fernando. Nanocomposites: Synthesis, Structure, Properties and New Application Opportunities. *Mater. Res.* **12**, 1–39 (2009).doi: 10.1590/S1516-14392009000100002.

[4] Kandulna, R. & Choudhary, R. B. Concentration-dependent behaviors of ZnO-reinforced PVA–ZnO nanocomposites as electron transport materials for OLED application. *Polym. Bull.* **75**, 3089–3107 (2018).doi: 10.1007/s00289-017-2186-9.

[5] Shih, Y. F., Chou, M. Y., Lian, H. Y., Hsu, L. R. & Chen-Wei, S. M. Highly transparent and impact-resistant PMMA nanocomposites reinforced by cellulose nanofibers of pineapple leaves modified

by eco-friendly methods. *Express Polym. Lett.***12**, 844–854 (2018).doi: 10.3144/expresspolymlett.2018.72.
[6] Evchuk, I. Y., Musii, R. I., Makitra, R. G. &Pristanskii, R. E. Solubility of polymethyl methacrylate in organic solvents. *Russ. J. Appl. Chem.***78**, 1576–1580 (2005).doi: 10.1007/s11167-005-0564-9.
[7] Weismann, M. &Panoiu, N. C. Theoretical and computational analysis of second- and third-harmonic generation in periodically patterned graphene and transition-metal dichalcogenide monolayers. Phys. Rev. B 94, (2016).doi: https://doi.org/10.1103/PhysRevB.94.035435.
[8] Ma, H., Shen, Z. & Ben, S. Understanding the exfoliation and dispersion of MoS_2nanosheets in pure water. J. Colloid Interface Sci. 517, 204–212 (2018).doi:10.1016/j.jcis.2017.11.013.
[9] Manzeli, S., Ovchinnikov, D., Pasquier, D., Yazyev, O. V. &Kis, A. 2D transition metal dichalcogenides. Nat. Rev. Mater. 2, (2017).doi:10.1038/natrevmats.2017.33.
[10] Eda, G. et al. Photoluminescence from chemically exfoliated MoS_2. Nano Lett. 11, 5111–5116 (2011).doi:10.1021/nl201874w.
[11] Rezk, A. R. et al. Acoustic-Excitonic Coupling for Dynamic Photoluminescence Manipulation of Quasi-2D MoS_2 Nanoflakes. Adv. Opt. Mater. 3, 888–894 (2015).doi:10.1002/adom.201500034.
[12] Walia, S. et al. Characterization of metal contacts for two-dimensional MoS_2 nanoflakes. Appl. Phys. Lett. 103, 232105 (2013).http://dx.doi.org/10.1063/1.4840317.
[13] Lopez-Sanchez, O., Lembke, D., Kayci, M., Radenovic, A. &Kis, A. Ultrasensitive photo detectors based on monolayer MoS_2. Nat. Nanotechnol. 8, 497–501 (2013).DOI:10.1038/nnano.2013.100.
[14] Yao, Y. et al. Large-scale production of two-dimensional nanosheets. J. Mater. Chem. 22, 13494–13499 (2012).doi:https://doi.org/10.1039/C2JM30587A.
[15] Kaushik, V. et al. Scalable Exfoliation of Bulk MoS_2 to Single- and Few-Layers Using Toroidal Taylor Vortices. Nanomaterials 8, 587 (2018).doi: 10.3390/nano8080587.
[16] Zhang, X. et al. Physical vapor deposition of amorphous MoS_2 nanosheet arrays on carbon cloth for highly reproducible large-area electrocatalysts for the hydrogen evolution reaction. J. Mater. Chem. A 3, 19277–19281 (2015).doi:https://doi.org/10.1039/C5TA05793K.
[17] Feng, X. et al. Novel mixed-solvothermal synthesis of MoS_2nanosheets with controllable morphologies. Cryst. Res. Technol. 48, 363–368 (2013).doi:10.1002/crat.201300003.
[18] Bai, L.-Z. et al. Enhanced Specific Capacitance of Few-Layer MoS_2 Nanosheets via SDBS-Assisted Hydrothermal Method. J. Nanosci. Nanotechnol. 18, 1804–1810 (2018).DOI: 10.1166/jnn.2018.14234.
[19] Deokar, G., Vignaud, D., Arenal, R., Louette, P. &Colomer, J.-F. Synthesis and characterization of MoS_2 nanosheets. Nanotechnology 27, 075604 (2016).doi: 10.1088/0957-4484/27/7/075604.
[20] Ren, W., Zhou, W., Zhang, H. & Cheng, C. ALD TiO_2 -Coated Flower-like MoS_2 Nanosheets on Carbon Cloth as Sodium Ion Battery Anode with Enhanced Cycling Stability and Rate Capability. ACS Appl. Mater. Interfaces 9, 487–495 (2017).doi:10.1021/acsami.6b13179.
[21] Huang, Y., Guo, J., Kang, Y., Ai, Y. & Li, C. M. Two dimensional atomically thin MoS_2 nanosheets and their sensing applications. Nanoscale 7, 19358–19376 (2015).https://doi.org/10.1039/C5NR06144J.
[22] Bo Weng, Xin Zhang, Nan Zhang, Zi-Rong Tang, and Yi-Jun Xu.The Two-Dimensional MoS_2 Nanosheets-Coated Bi2S3 Discoids : Synthesis , Formation Mechanism and Photo catalytic Application. (2015). Langmuir, 2015, 31 (14), pp 4314–4322.doi:10.1021/la504549y.
[23] Perumal Veeramalai, C., Li, F., Xu, H., Kim, T. W. & Guo, T. One pot hydrothermal synthesis of graphene like MoS_2 nanosheets for application in high performance lithium ion batteries. RSC Adv. 5, 57666–57670 (2015).https://doi.org/10.1039/C5RA07478A.
[24] Yang, X. et al. Antibacterial activity of two-dimensional MoS_2sheets. Nanoscale 6, 10126–10133 (2014).https://doi.org/10.1039/C4NR01965B.

[25] Rajesh kumarAnbazhagan, Hsing-Ju Wang, Hsieh-Chih Tsai, Ru-Jong Jeng.Highly concentrated MoS_2 nanosheets in water achieved by thioglycolic acid as stabilizer and used as biomarkers.RSC Adv., 2014,4, 42936-42941.doi:10.1039/C4RA07512A.

[26] Wang, X., Xing, W., Feng, X., Song, L. & Hu, Y. MoS_2/Polymer Nanocomposites: Preparation, Properties, and Applications. Polym. Rev.57, 440–466 (2017).doi:10.1080/15583724.2017.1309662.

[27] Liang, G. *et al.* Optical limiting properties of a few-layer MoS_2 /PMMA composite under excitation of ultrafast laser pulses. *J. Mater. Chem. C***7**, 495–502.(2019).doi:10.1039/C8TC04200D.

[28] Shinde, S. M., Kalita, G. &Tanemura, M. Fabrication of poly(methyl methacrylate)-MoS_2/graphene heterostructure for memory device application. *J. Appl. Phys.***116**, (2014).doi: 10.1063/1.4903552.

[29] Burman, D., Sharma, A. & Guha, P. K. Flexible Large MoS_2 Film Based Ammonia Sensor *.IEEE Sensors Lett.***2**, 1–4 (2018).doi:10.1109/LSENS.2018.2817651.

[30] Ma,D.*et al.* A universal etching-free transfer of MoS_2 films for applications in photodetectors. *Nano Res.***8**, 3662–3672 (2015).doi:10.1007/s12274-015-0866-z.

[31] Kim, J. H. *et al.* Thickness-dependent electron mobility of single and few-layer MoS_2 thin-film transistors. *AIP Adv.***6,** 2–7 (2016).Doi: https://doi.org/10.1063/1.4953809.

[32] Gao, D. et al. Ferromagnetism in freestanding MoS_2nanosheets. *Nanoscale Res. Lett.* **8**, 1–8 (2013).doi:https://doi.org/10.1186/1556-276X-8-129.

[33] Liu, H. & Chi, D. Dispersive growth and laser-induced rippling of large-Area single layer MoS_2 nanosheets by CVD on c-plane sapphire substrate. *Sci. Rep.***5,** (2015).https://doi.org/10.1038/srep11756.

[34] Fan, X. *et al.* MoS_2actuators: reversible mechanical responses of MoS_2-polymer nanocomposites to photons. *Nanotechnology***26,** (2015). doi: 10.1088/0957-84/26/26/261001.

[35] Visic, B. *et al.* Optical properties of exfoliated MoS_2coaxial nanotubes - analogues of graphene. *Nanoscale Res. Lett.***6,** 1–6 (2011).doi:https://doi.org/10.1186/1556-276X-6-593.

[36] Muscuso, L., Cravanzola, S., Cesano, F., Scarano, D. &Zecchina, A. Optical, vibrational, and structural properties of MoS_2nanoparticles obtained by exfoliation and fragmentation via ultrasound cavitation in isopropyl alcohol. *J. Phys. Chem. C***119**, 3791–3801 (2015).

[37] Zhou, B. et al. Evolution of Electronic Structure in Atomically Thin Sheets of WS_2 and WSe_2.ACS Nano7, 1, 791–797 (2013).doi: 10.1021/nn305275h.

[38] B. Bijoy, Synthesis and characterization of nanocrystalline ZnS thin films and quantum dots by chemical process, PhD thesis Guwahati University, pp. 210–215 (2015). *http://shodhganga.inflibnet.ac.in:8080/jspui/bitstream/10603/50845/11/11_chapter%205.pdf.*

[39] Aziz, S. B., Gh, O., Brza, M. A., Azawy, A. K. & Tahir, D. A. Results in Physics Effect of carbon nano-dots (CNDs) on structural and optical properties of PMMA polymer composite. *Results Phys.***15**, 102776 (2019).doi: 10.1016/j.rinp.2019.102776.

[40] Harvey, A. *et al.* Luminescent transition metal dichalcogenide nanosheets through one-step liquid phase exfoliation. *2D Mater.***3,** 035014 (2016).doi:10.1088/2053-1583/3/3/035014.

[41] Mak, K. F., Lee, C., Hone, J., Shan, J. & Heinz, T. F. *Atomically thin MoS_2: A new direct-gap semiconductor.* doi:10.1109/TIM.2014.2298672.

[42] Splendiani, A. *et al.* Emerging photoluminescence in monolayer MoS_2. *Nano Lett.***10,** 1271–1275 (2010).https://doi.org/10.1021/nl903868w.

[43] Oliver, W. C. & Pharr, G. M. *An improved technique for determining hardness and elastic modulus (Young's modulus)* (2014).doi:10.1557/JMR.1992.1564.

[44] Gonalves, G. *et al.* Graphene oxide modified with PMMA via ATRP as a reinforcement filler. *J. Mater. Chem.***20,** 9927–9934 (2010).doi: 10.1039/c0jm01674h.

Analysis and Simulation of Stability Technique of ALFC in Isolated Area of Power System

Sanitha Michail C

CMR Institute of Technology, Bengaluru

E-mail: sanitha.c@cmrit.ac.in

Abstract. Major function of ALFC loop (automatic load frequency control) in a power system is to regulate the megawatt output and frequency of the generator. Variation in the load demand brings variation in the response of the system frequency. State equations of ALFC loop of an isolated power system, with and without integrator, are examined in the paper. This paper deals with the analysis and simulation of pole placement technique for stability of the system .Pole placement method will help the designers to choose the appropriate design parameters to get the desired response from the system.

Keywords: Automatic load frequency loop, power system area, pole placement, frequency deviation, control parameter

1. Introduction

Frequency and voltage are the two important parameters in power system. It has to be continuously monitored and maintained within the limits. The main aim of an interconnected power system is to deliver the load in both active and reactive power by maintaining the voltage and frequency at desired limit. There are two major control loops in generator to achieve this task. (i) Automatic Voltage Regulator (AVR) (ii) Automatic Load Frequency Control (ALFC). Figure.1 shows the schematic diagram of AVR and ALFC loops. Real power changes mainly depend on changes in rotor angle which in turn affects the frequency. Reactive power changes depend on magnitude of voltage which depends on excitation of voltage. Automatic Load Frequency Control loop controls the megawatt output and frequency of the generator while terminal voltage V is controlled by Automatic Voltage Regulator loop .ALFC loop has a fast primary loop which respond to the frequency signal and a slow secondary loop to maintain the fine adjustment of the frequency and maintain megawatt power interchange with other pool members.

2. ALFC of Isolated Area System

Figure.2 shows the schematic diagram of a speed governing system of a generator with a regulation characteristics R. This system can also be called as the primary control loop. Pressurized steam which controls the turbine is in turn controlled by the position of x_E. Downward or upward movement of point E decreases or increases the flow of steam to the turbine respectively. It can be measured in megawatt change ΔP_V .This flow can be translated into turbine power change ΔP_T. Hydraulic amplifiers are used to control the main valve against the high pressure steam.

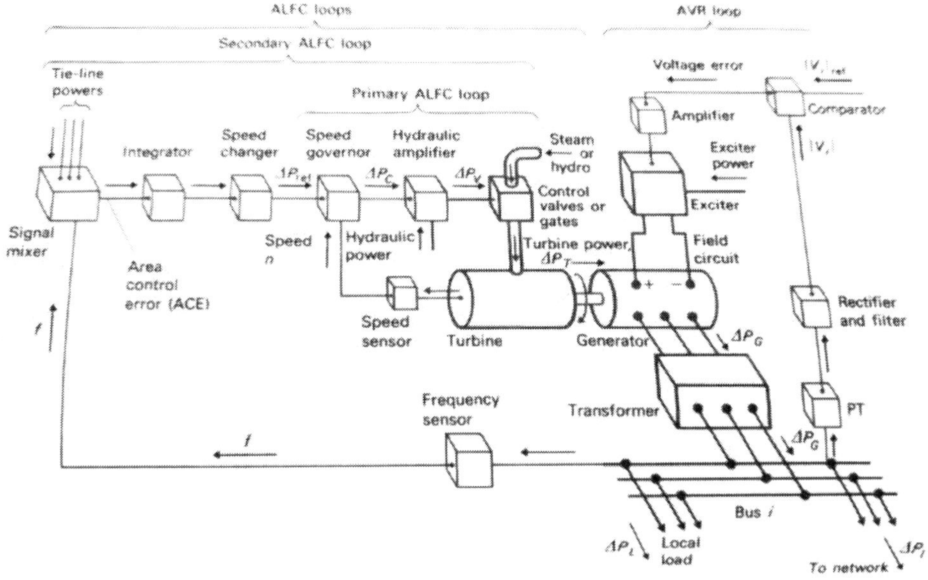

Figure.1 AVR and ALFC loops

Position of x_D of the pilot valve is the input to the hydraulic amplifier and position of x_E of the main piston is the output. Position of pilot valve will be changed (i) if there is a change in the power settings ΔP_{ref} , point A (ii) due to the position changes in main piston, point C (iii) due to the changes of point B due to the variations in speed of the generator.[1]. Figure.3 shows the block diagram of speed governor.

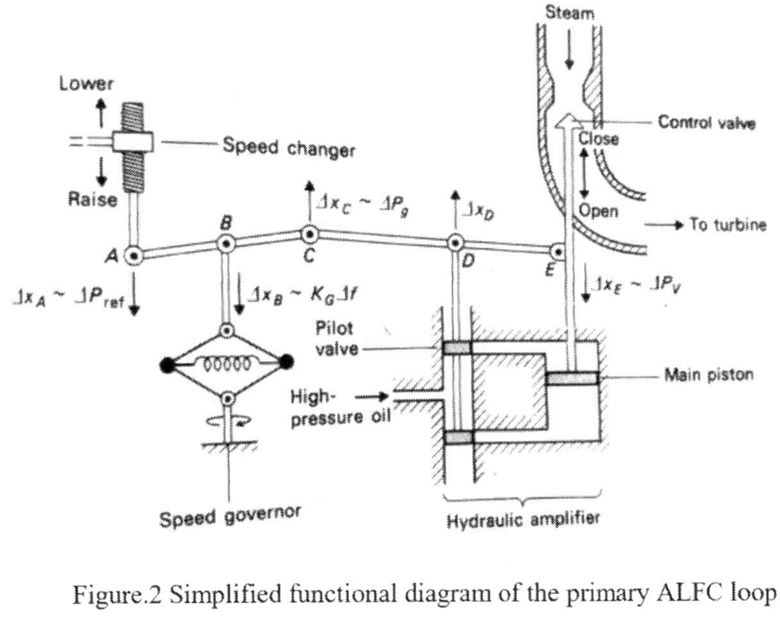

Figure.2 Simplified functional diagram of the primary ALFC loop

$$\Delta P_g(s) = \Delta P_{ref}(s) - (1/R)\Delta f(s)$$

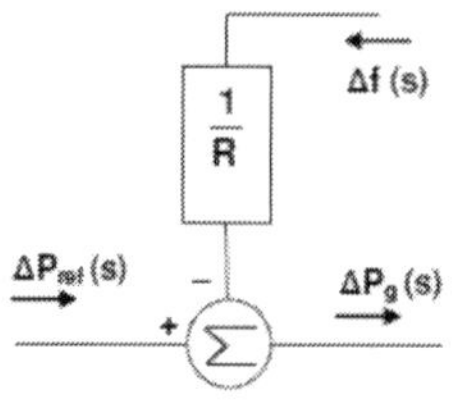

Figure.3 Block diagram of speed governor

Output equation of hydraulic actuator can be written as $\Delta P_v = k_H \int \Delta x_D dt$, k_H depends on cylinder geometrics and fluid pressure. Change in x_D can be written as $\Delta x_D = \Delta P_g - \Delta P_v$

$$\Delta P_v(s) = \left(\frac{k_H}{s}\right)\Delta x_D(s) \qquad (1)$$

$$\Delta x_D(s) = \Delta P_g(s) - \Delta P_v(s) \qquad (2)$$

Solving (1) and (2) equations we get,

$\Delta P_v(s) = \frac{1}{1+sT_H}\Delta P_g(s), T_H = \frac{1}{k_H}$ where T_H is the hydraulic time constant.

Transfer function of the hydraulic valve actuator can be written as $G_H(s) = \frac{\Delta P_v(s)}{\Delta P_g(s)} = \frac{1}{1+sT_H}$

The block diagram of hydraulic actuator is shown in figure.4

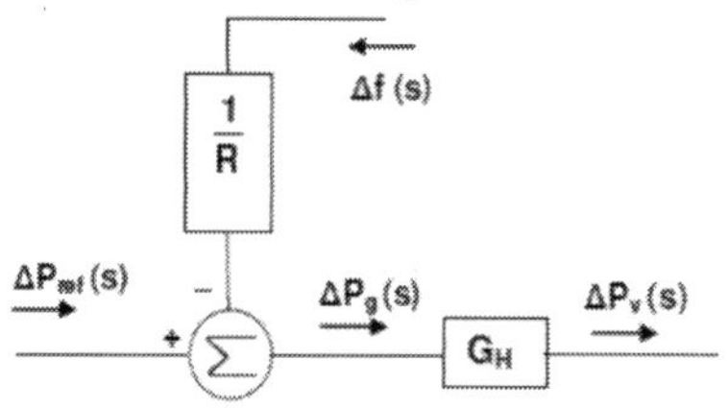

Figure.4 Block diagram of speed governor along with hydraulic actuator

Accelerating power in turbine can be written as $\Delta P_T(s) - \Delta P_G(s)$.

Transfer function of turbine can be written as $\Delta P_T(s) = G_T \Delta P_v(s) = \frac{1}{1+sT_T}\Delta P_v(s)$

Change in generator power [2] depends on the change in load power. $\Delta P_G(s) = \Delta P_D(s)$
Linear model of primary ALFC loop is shown in figure.5

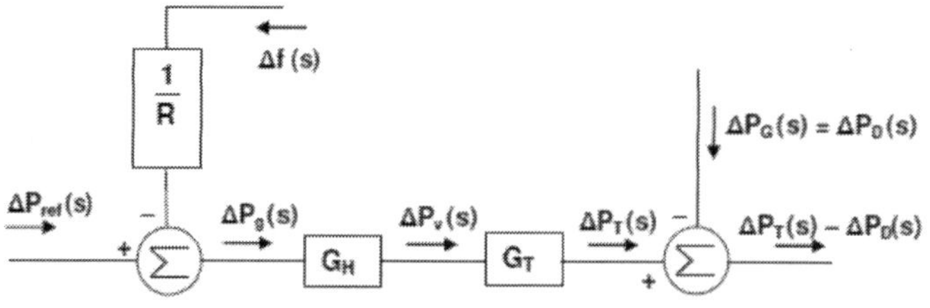

Figure.5 Linear model of primary ALFC loop

Considering the kinetic energy of the rotor proportional to the square of speed and load as a function of voltage and frequency, $W_{kin} = W_{kin}^{o}\left(\frac{f}{f^{o}}\right)^{2} MW \sec$

$$D = \frac{\partial P_D}{\partial f} MW / Hz$$

Thus $\Delta P_T - \Delta P_D = \frac{d}{dt}(W_{kin}) + D\Delta f$ and $f = f^{o} + \Delta f$

$\Delta P_T - \Delta P_D = \frac{2W_{kin}^{o}}{f^{o}}\frac{d}{dt}(\Delta f) + D\Delta f$ MW. Let $H = \frac{W_{kin}^{o}}{P_r}$ MW sec/MW (or sec) then

$$\Delta P_T - \Delta P_D = \frac{2H}{f^{o}}\frac{d}{dt}(\Delta f) + D\Delta f \text{ pu MW}$$

$$\Delta P_T(s) - \Delta P_D(s) = \frac{2H}{f^{o}} s\Delta f(s) + D\Delta f(s)$$

$$\Delta f(s) = G_p(s)[\Delta P_T(s) - \Delta P_D(s)]$$

$G_p(s) = \frac{1}{\frac{2H}{f^{o}}s + D} = \frac{\left(\frac{1}{D}\right)}{1 + s\frac{2H}{f^{o}D}} = \frac{K_p}{1 + sT_p}$ where $K_p = \frac{1}{D}$ and $T_p = \frac{2H}{f^{o}D}$

Also $\frac{K_p}{1 + sT_p}$ can be written as $\frac{1}{2Hs + D}$ .Block diagram of primary ALFC loop is shown in figure.6

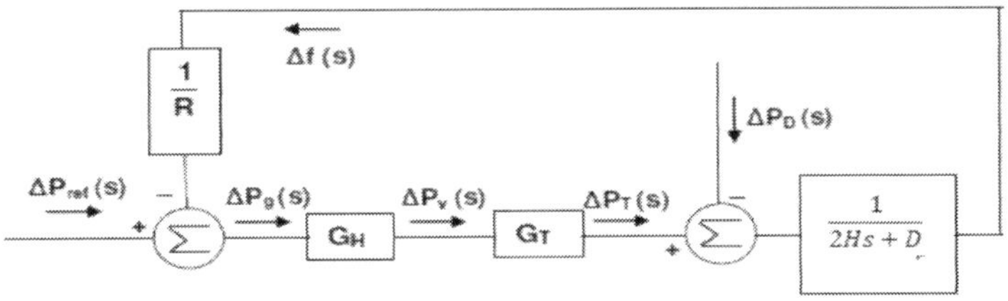

Figure.6 Block diagram of primary ALFC Loop

We can make the speed changer be commanded by a signal obtained by amplifying [4] and integrating the frequency error. $\Delta P_{ref} = -K_I \int \Delta f dt$ where K_I is the integral control.Figure.7 shows the block diagram of complete ALFC loop

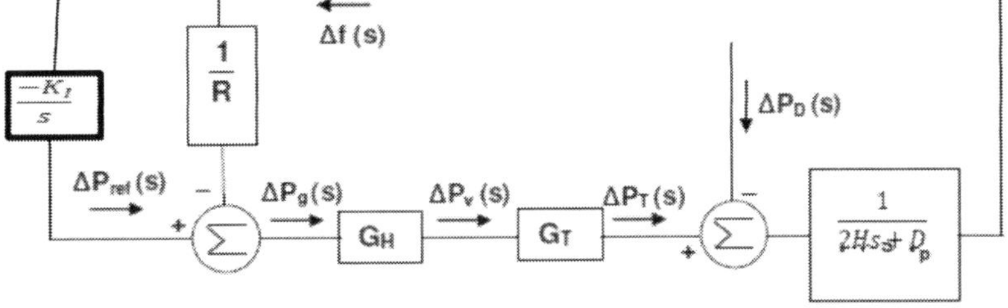

Figure.7 Block diagram of complete ALFC loop

3. State Space model of single area AGC

From the above equations the state equations can be written as

$$\begin{bmatrix} \dot{\Delta f} \\ \dot{\Delta P_T} \\ \dot{\Delta P_V} \\ \dot{\Delta P_{ref}} \end{bmatrix} = \begin{bmatrix} -D/2H & 1/2H & 0 & 0 \\ 0 & -1/T_T & 1/T_T & 0 \\ -1/RT_H & 0 & -1/T_H & 1/T_H \\ -K_I & 0 & 0 & 0 \end{bmatrix} \begin{bmatrix} \Delta f \\ \Delta P_T \\ \Delta P_V \\ \Delta P_{ref} \end{bmatrix} + \begin{bmatrix} -1/2H \\ 0 \\ 0 \\ 0 \end{bmatrix} \Delta P_D$$

Standard form of state space model can be written as

$\dot{X}(t) = AX(t) + Bu(t)$

$y(t) = CX(t) + Du(t)$ where y(t) is the output , u(t) is the input vector and X(t) is the state vector. Here

$X(t) = \begin{bmatrix} \Delta f & \Delta P_T & \Delta P_V & \Delta P_{ref} \end{bmatrix}^T$

$u(t) = \Delta P_D$ and $y(t) = \Delta f$

4. Stability technique-Pole Placement Design

The stability of the power system is determined by the matrix A. If the eigen values of matrix A lie on the LHS of s-plane the system will be stable. For an unstable system it is possible to design a controller to locate the poles of the system on LHS. The control signal used is given by
$u(t) = -KX(t)$ where K is the vector of feedback gain as shown in figure.8

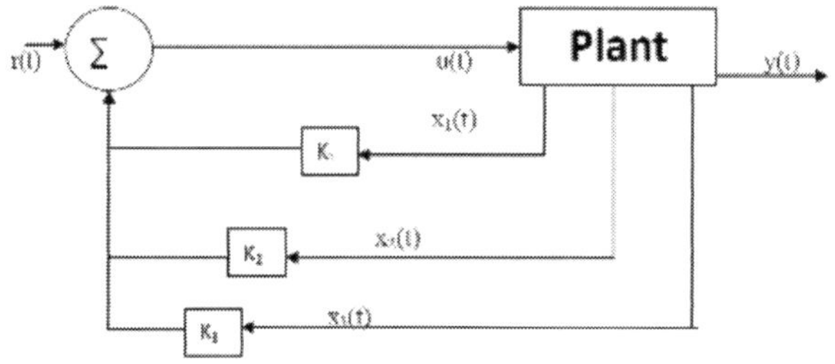

Figure.8 Pole placement design

$\dot{X}(t) = AX(t) + B(-KX(t))$ is the modified state vector.

The characteristic equation for the above model is given by $|sI - A + BK| = |sI - A_p| = 0$

This equation can be expanded into

$$|sI - A + BK| = s^n + (a_{n-1} + k_n)s^{n-1} + \dots\dots..(a_1 + k_2)s + (a_0 + k_1) = 0$$

Consider the desired closed loop pole locations are $-\lambda_1, -\lambda_2, \dots\dots -\lambda_n$, then the desired characteristics equation can be written as

$$d(s) = (s + \lambda_1) + (s + \lambda_1) + \dots..(s + \lambda_n) = s^n + d_{n-1}s^{n-1} + \dots.. + d_1 s + d_0 = 0$$

Equating the two characteristic equations we get

$$a_{n-1} + k_n = d_{n-1} \Rightarrow k_n = d_{n-1} - a_{n-1}$$
$$a_0 + k_1 = d_0 \Rightarrow k_1 = d_0 - a_0$$
$$K_i = d_{i-1} - a_{i-1}$$

5. Results & Discussions

Here we have taken an isolated system without AGC ,T_H=0.3 s,T_T=0.6s,H=5s and D=1.For this[3] system R > 0.0183 for stability.Figure.9 shows the simulink model of a stable primary ALFC loop. Figure.10 shows the frequency deviation of the given system with R=0.05 is stable and figure.11 shows that if R<0.0183 it is unstable.

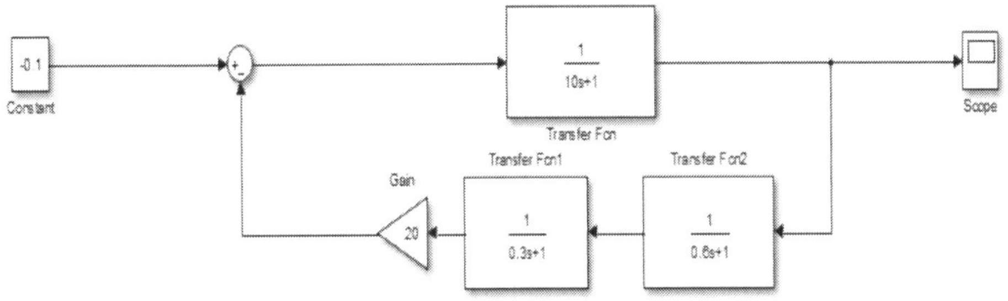

Figure.9 Simulink model of a primary ALFC loop

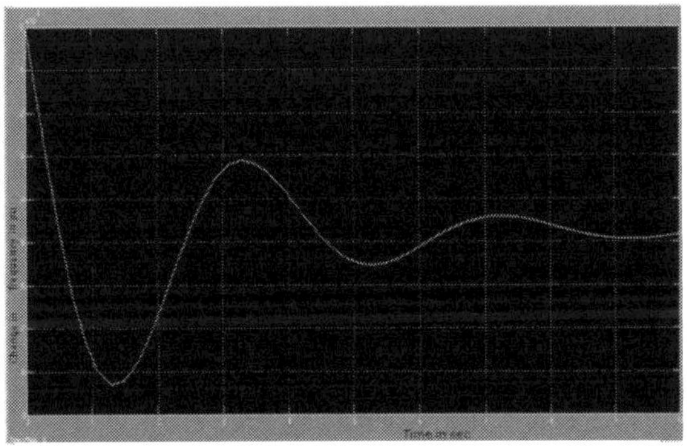

Figure.10 Frequency response of the system for R=0.05

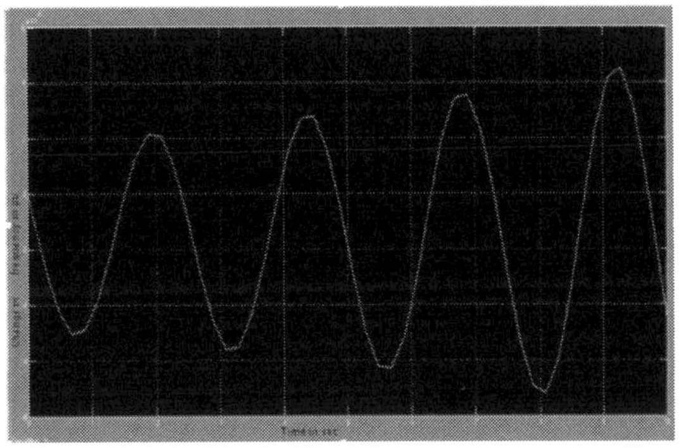

Figure.11 Frequency response of the system for R=0.016

The roots of the characteristics equation of the given system are -4.3216 + 0.0000i, -0.3877 + 1.5968i and -0.3877 - 1.5968i if the gain is chosen as 20. Figure.12 shows the frequency deviation of the complete ALFC system with integral controller K_I =5 is stable . The roots of the characteristics equation of the given system with integral controller are -4.2851 + 0.0000i,-0.2725 + 1.5337i,-0.2725 - 1.5337i and -0.2669 + 0.0000i .

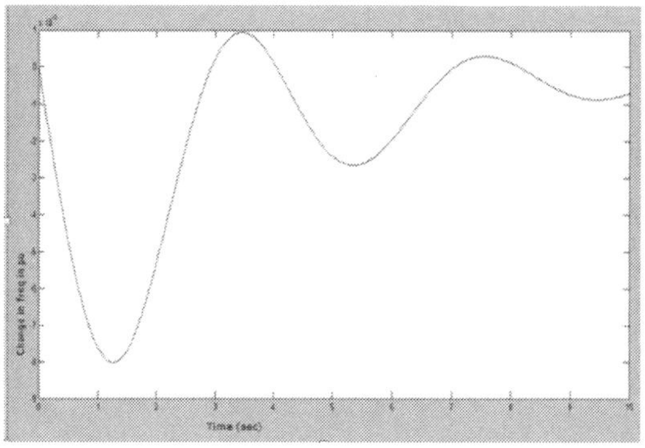

Figure.12 Frequency deviation of complete ALDC loop with integral controller, K_I =5

If we choose K_I =16 system is becoming unstable and roots are -4.1977 + 0.0000i,0.0111 + 1.5153i,0.0111 - 1.5153i and -0.9215 + 0.0000i .Here the pair of complex conjugate roots lying on RHS clearly explains the instability of the system(Figure.13).

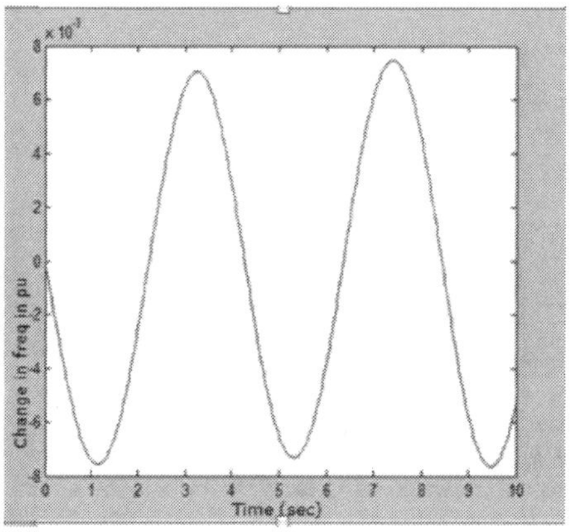

Figure.13 Frequency deviation of complete ALDC loop with integral controller, K_I =16

Pole placement technique will improve the stability of the system. As explained before poles can be placed at the desired location. Consider the conjugate pair of poles to be placed at -4±6i. The roots of the characteristics equation of the complete ALFC system with K_I =16 are -4.0000 + 6.0000i,-4.0000 - 6.0000i,-4.3216 + 0.0000i and -0.9215 + 0.0000i. Figure.14 shows the response of the system after the pole placement where frequency quickly reaches the steady state.

6. Conclusion

Pole placement technique in an isolated area of power system has been explained in the paper to help the designers to choose the design parameters. A detailed discussion of controller poles influence on the closed loop system has been presented in the paper. Once the desired response of the closed loop system is specified, computation of the controller can be calculated automatically. Adequate adaptation of design parameters will help to obtain the desired location of the poles of the system for the realistic problems.

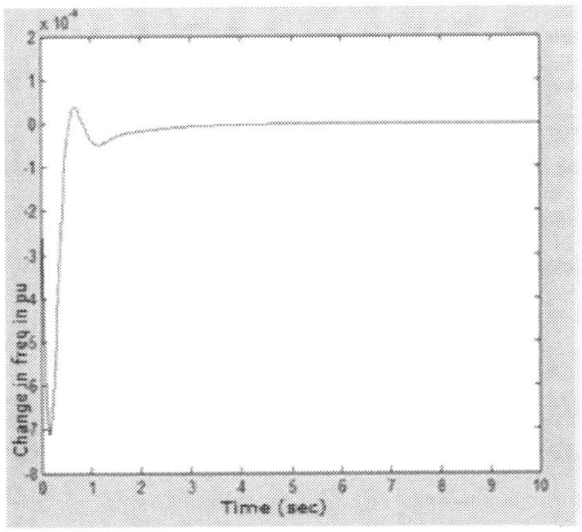

Figure.14 Frequency response after the pole placement

References

[1] Elgerd O L"*Electric Energy Systems Theory An Introduction*"McGraw Hill Education edition,1983

[2] G. Shahgholian, P. Shafaghi and H. Mahdavi-Nasab, "*A comparative analysis and simulation of ALFC in single area power system for different turbines*," 2010 2nd International Conference on Electronic Computer Technology, Kuala Lumpur, Malaysia, 2010, pp. 50-54, doi: 10.1109/ICECTECH.2010.5479992.

[3] Uma Rao.K"*Power System: Operation & Control.*" Wileyindia, www.wileyindia.com/power-system-operation-control.html/.

[4] M S Parvathy Thampi, "*Low Voltage High Efficient Synchronous Buck Converter for Mobile Chargers*,"Journal of Solid State Technology,Volume: 63 Issue: 4 ,Publication Year: 2020

Growth, Nonlinear Optical, Electrical, Mechanical and Dielectric Properties of Zinc Sulphate Doped L-Alanine Single Crystal for Optoelectronic Applications

M R Hareeshkumar[1], M R Jagadeesh[2*], G J Shankaramurthy[1] and B M Prasanna[3]

[1] Department of Physics, UBDT College of Engineering, Davangere, Constituent college of VTU Belagavi, Karnataka, 577004, India
[2] Department of Physics, Jain Institute of Technology, Davangere, Affiliated to VTU Belagavi Karnataka, 577003, India
[3] Department of Chemistry, Jain Institute of Technology, Davangere, Affiliated to VTU Belagavi Karnataka, 577003, India

e-mail: jmrshyagale@gmail.com

Abstract. Zinc sulphate added L-alanine (LAZS) crystals were obtained from solvent evaporation of solution method. The obtained samples were taken for diverse characterizations tool such as, XRD, functional group identification, optical transparency, simultaneous thermal analysis, dielectric, hardness and second harmonic generation studies. Crystal diffraction study proves that LAZS crystal fit into orthorhombic system and having space group $P2_12_12_1$. FTIR confirms presences of various functional groups present in the sample. The thermal study infers that LAZS is steady up to 288 °C. High transmittance and extended transparency of LAZS is noticed from linear optical study. The calculated hardness values were found to be considerable and the dielectric parameters are noticed characteristically followed by NLO study.

Keywords: alanine, zinc sulphate, evaporation, hardness, SHG

1. Introduction

Research on novel hybrid NLO materials is becoming fascinating since their nonlinear optical effects used in optical communication and other significant electro optical applications [1-7]. Organic molecules having conjugated bonds possess high inherent nonlinearity, synthetic flexibility, and noticeable laser damage threshold. On the other side, Inorganic materials are covalent and ionic and the NLO nature is a combined effect. In case of semi organic crystals, it so happens that, when organic materials are mixed with inorganic materials, the resulting hybrid materials share the characteristics of both the above with the improved properties. L-alanine (LA) having the zwitterionic structure (NH_3^+ and COO^-) both in solid state and also in aqueous solution. LAL has orthorhombic nature with cell parameters values of a=6.302Å b=12.343Å c=5.78Å [8-10]. The various combinations of LAL single crystals have been accounted recently in the scientific literature [11-15]. In this context, LAZS crystal was grown by the solvent evaporation method and subjected to different characterizations to explore its potential properties and the results are discussed.

2. Experimental

2.1. *Synthesis and crystal growth*

The crystal under study was harvested using AR grade LA and zinc sulphate in a 1:0.1 molar ratio. The chemicals were methodically liquefied in the considered quantity of water and stirred well. The obtained solution is filtered and kept to crystallize. Because of homogeneous nucleation, seed crystals were created. Further, solvent evaporation led to crystals with the dimensions of 35x15x8 mm^3 were obtained in 38 days and is depicted in Figure 1.

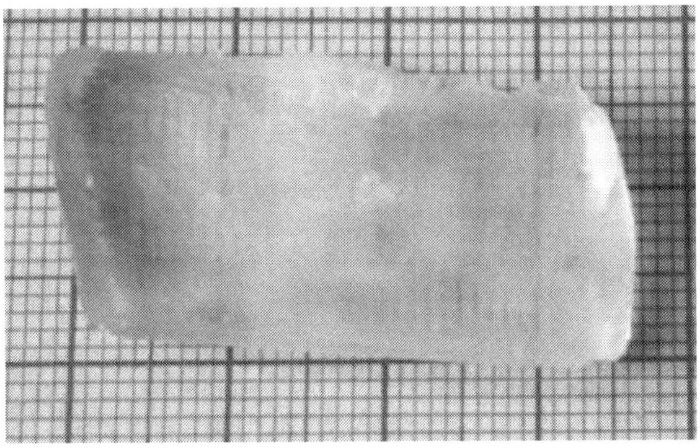

Figure 1. As grown LAZS crystals.

3. Results and discussions

3.1. XRD studies

The diffrcation study of LAZS crystals was carried out with a Bruker Smart Apex Duo diffractometer. The study validate that LAZS fit into orthorhombic crystal family having the space group $P2_12_12_1$ with unit cell parameter a = 5.78 Å, b = 6.09 Å, and c = 12.35 Å, $\alpha = \beta = \gamma = 90°$ with volume V = 430 $Å^3$. These values of LAZS are in concurrent with the reported values of pure LA [9].
Well-grounded sample was examined by powder XRD (Bruker AXS D8). Typical 2θ scan having rate of 2 degree/minute has been made from 10° to 70°. In doped materials, both intensities and 2θ values were shifted due to the active role of dopant in the lattice of the grown crystal but there is no alteration in the crystal system. The obtained pattern is in Figure 2 and it reveals the well crystalline character.

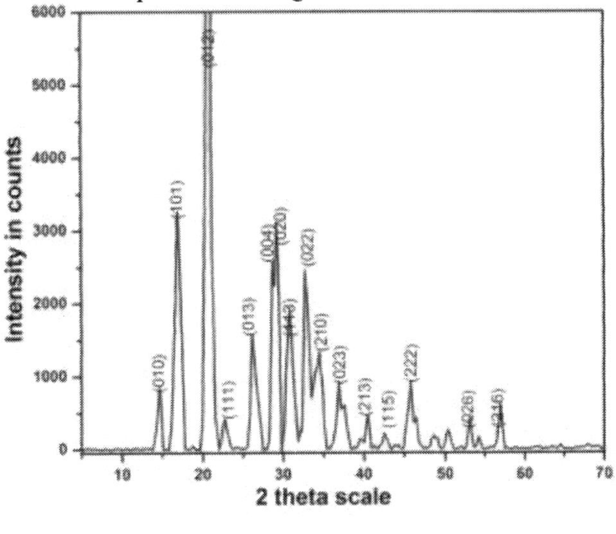

Figure 2. Powder X-ray diffraction patterns of LAZS

3.2. FTIR spectral analysis

The presence of different functional groups in terms of molecular vibration and corresponding chemical bonding were noticed by Avatar 370 FT-IR spectrum spectrometer in the series 4000-400 cm^{-1} and Figure 3. is the resulting spectra.

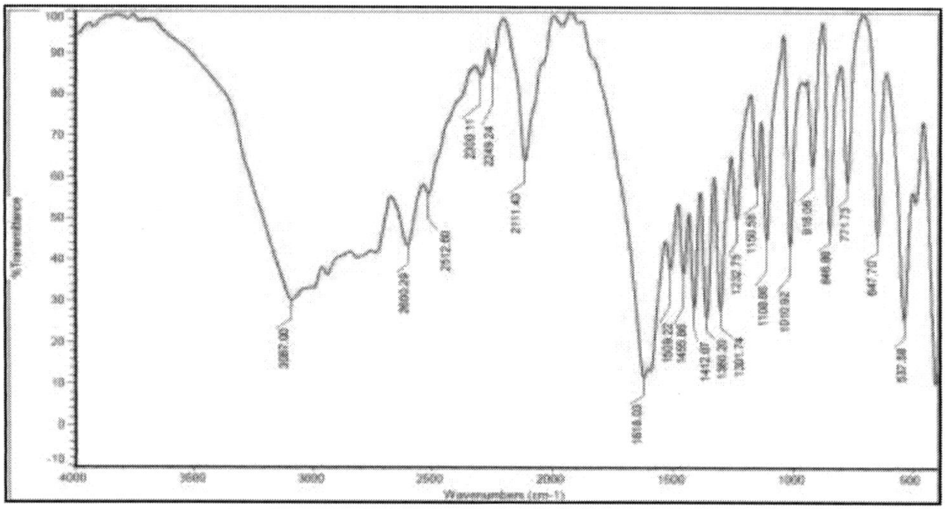

Figure 3. The FTIR spectra of the LAZS crystals.

The wide-ranging peak at 3087 cm^{-1} could be of the NH_3^+ asymmetric vibrations also the amine group of NH_3^+ bending is with the intense peak at 1617.99 cm^{-1}. Asymmetric bending vibrations of NH_3^+ were at 2249 cm^{-1} and 2111 cm^{-1}. The symmetrical stretching of CO_2 is at 1412 cm^{-1}. The peak at 1360 cm^{-1} indicates the C-H deformation in CH_3. NH_3^+ rocking and C-O stretching is attributed to the sharp peak at 1108 cm^{-1}. The C = O bending is at 774 cm^{-1} and the deformation of COO^- is about 648.03 cm^{-1}. FTIR analysis confirms the presence of the predictable groups of LAZS.

3.3. UV-Visible absorption Studies

It gives information about molecular information as it involves the electron movement in σ and π orbital from the lower to elevated states. The crystals with wide and high transmittance value possess improved SHG leads to better performing materials. In the range 200 to 1600 nm, the optical absorbance spectrum was recorded and is depicted in Figure 4. Further, there is no noticeable absorption peak region of study. The cutoff wavelength of the LAZS is found to be 282 nm which is sufficient for many applications. Hence, both fundamental and second harmonic wave transmission could be achieved which suggests for possible potential applications.

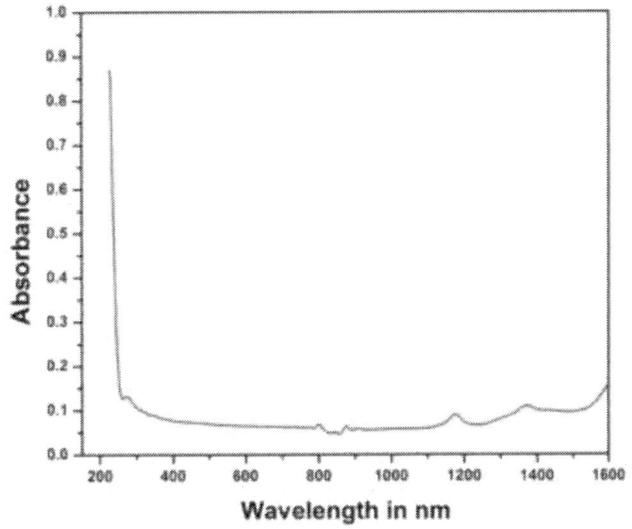

Figure 4. The UV spectra of LAZS crystals.

Generally, optical band gap (Eg) is the threshold for the photons to be absorbed. This calculation helps in understand the possible transitions in the crystal. Here, the absorption coefficient (α) was calculated by

$$\alpha = \frac{1}{d}\, log \frac{1}{T}$$

Here, α was calculated through the equation

$$\alpha hv = Q(hv - Eg)^n$$

Where Q a constant and n is an index, which can have the values with respect to the type of the transitions. Accordingly the equation becomes

$$\alpha = \frac{A(hv - Eg)^{1/2}}{hv}$$

By considering the graph of $(\alpha hv)^2$ versus hv in eV as shown in the Figure 5. and the resulting band gap and is found to be 5.1 eV.

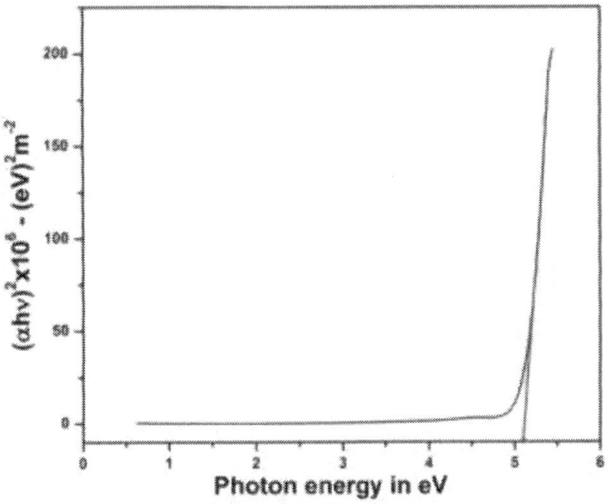

Figure 5. Graph of $(\alpha hv)^2$ versus energy hv.

Since refractive index (n) is a function of wavelength and it has influence on NLO nature. It is a photon energies reliant function showing the crystal internal ability which is required for the solid state devices. The calculated n value of LAZS is 1.98 as obtained from below equation.

$$\frac{n^2 - 1}{n^2 + 2} = 1 - \sqrt{\frac{E_g}{20}}$$

Where, E_g is the band gap

All these UV associated parameters used to compute the electrical susceptibility(χ_c) based on the consideration that, generally the naturally obtainable ones are non-magnetic at optical frequencies and hence, their n value is almost same as relative permittivity of the materials which is given by

$$\chi_c = \varepsilon_r - 1$$

Or

$$\chi_c = n^2 - 1$$

Thus, χ_c value of the crystal is found to be 2.94. If the resulting value of n is greater than unity then, crystal can easily be polarized by exposing to intense light. These optical properties show the internal efficiency of LAZS for the solid state device applications.

3.4 SEM and EDAX studies

The surface character of LAZS was studied by JEOL JSM-6390LV model SEM instrument and the recorded image is depicted in Figure 6. Since organics are generally insulating, to make it conductive and to achieve good emission of secondary electrons the carbon coating was made before exposing the crystal to electron beam. LAZS crystal has almost smooth surface however, few micro cracks.

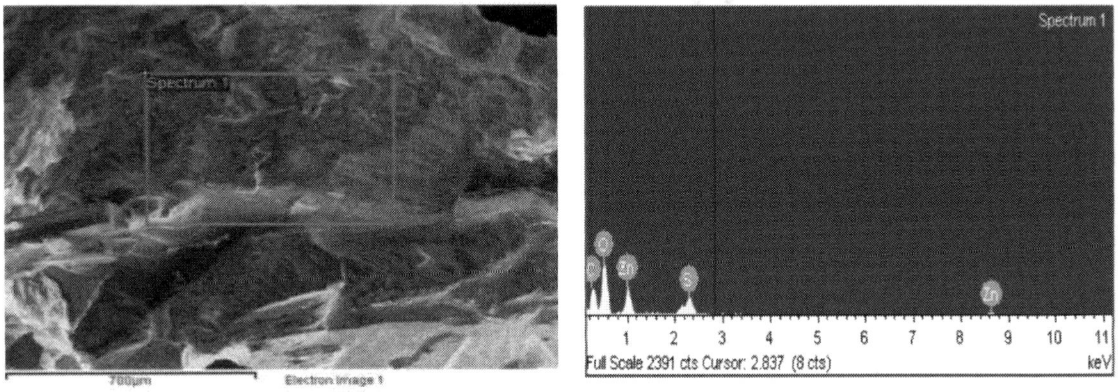

Figure 6. SEM image & EDAX spectrum of LAZS crystals

EDAX technique is essential in identifying the presence of elements in the sample wherein it measures the number of characteristic X-rays produced by the innermost shell of a specific element present in the sample. LAZS EDAX spectrum is depicted in Figure 6. and it confirms the presence of zinc sulphate in the L-alanine.

3.5. Thermal (TG-DTA) studies

These were used to determine the thermal activity of LAZS. The thermal behavior of LAZS was made in presence of nitrogen at the rate of 10°C/min. between ambient temperatures to 800°C. The primary weight of LAZS was 8.4 mg. The TGA plot (Figure 7) signifies the single stage of sample decay.

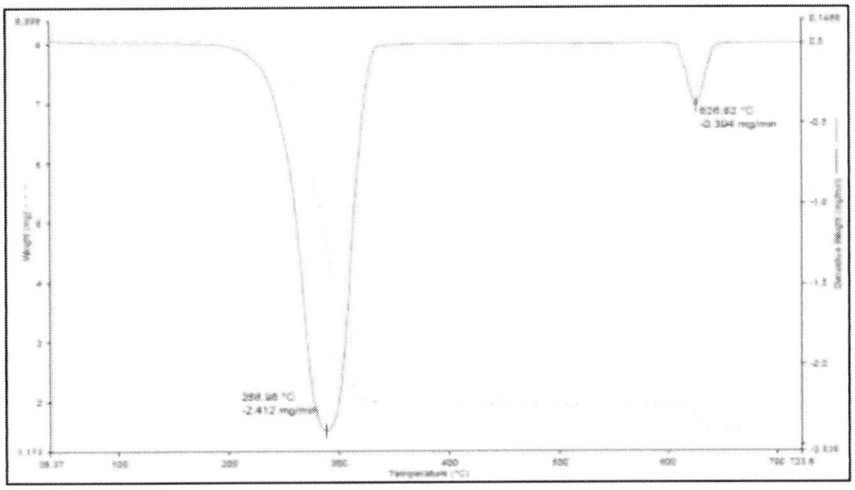

Figure 7. TG-DTA Thermogram of LAZS crystal

In the beginning, there is no weight loss until the temperature exceeds 210°C, after which the sample starts to decay and reaches 289°C. The dissociation of LAZS causes mass loss between 210°C and 289°C. The dissociation of the LAZS structure is seen through strong endothermic peak in the DTA curve. The material's strong crystalline quality and purity are shown by the sharp curve. As a result of the TG-DTA analysis, LAZS is stable and can be used in devices up to 288°C.

3.6. SHG studies

To recognize the NLO property of LAZS, Kurtz powder method was used [16]. Nd:YAG laser (1064 nm) beam having input power of 1.9 mJ/pulse was used for the study. Well-grounded sample of same particle size (125-150 μm) were filled in a capillary and exposed to the laser with KDP and urea as reference. Emitted green light (532 nm) from the LAZS confirmed the SHG from the crystalline sample and it showed the corresponding efficiency of 0.9 and 0.6 times that of pure KDP and urea sample respectively.

3.7. Dielectric studies

The electric field distribution of the sample and its frequency dependence will decide the material applications. Dielectric measurements were carried out with HIOKI-LCR HiTester 3535 from 100Hz to 5MHz. The silver paste is added as parallel plates to the opposite faces of the crystal dielectric parameters and AC conductivity (σ_{ac}) along with frequency dependence were observed at ambient temperature.

Dielectric constant (ε_r) was calculated through the relation ε_r=Ct/ε_oA, and its variation with and applied frequency was noticed along with loss factor tanδ and are shown in Figures 8 and 9 respectively. It is seen that the value of ε_r weakens with frequency increment and finally attains constant at elevated frequencies which is a normal dielectric behavior. At low frequencies all the four polarizations are significant which depend on limpidness and perfection of the crystal. On increasing frequncy, in sequence, the polarization mechanisms will be neutralized since they will become incapable to respond in accordance with change in electric field. At higer frequencies (10^{15}Hz) electronic polarization will have major role. The low ε_r value materials possess the possible potential applications in microelectronics. Further, With increasing frequency, dielectric loss also decreases. The sample's low dielectric loss at high frequency indicates that it has good optical efficiency and minimum defects.

The AC condctivity was measured by using the relation $\sigma_{ac} = \varepsilon_r \varepsilon_o \omega \tan\delta$ where ε_o is the free space permittivity and ω is the applied field's angular frequency and is depicted in the Figure 10. The conductivity increases as the frequency increases, as seen in the graph. This feature is endorsed to the decrease in the electronic polarization at upper frequencies.

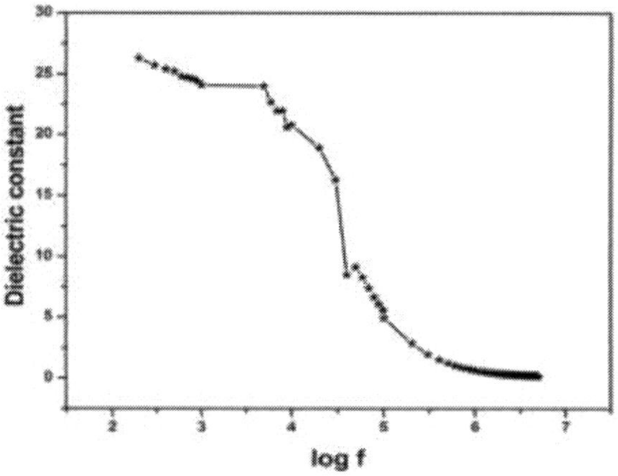

Figure 8. Dependence of dielectric constant on frequency

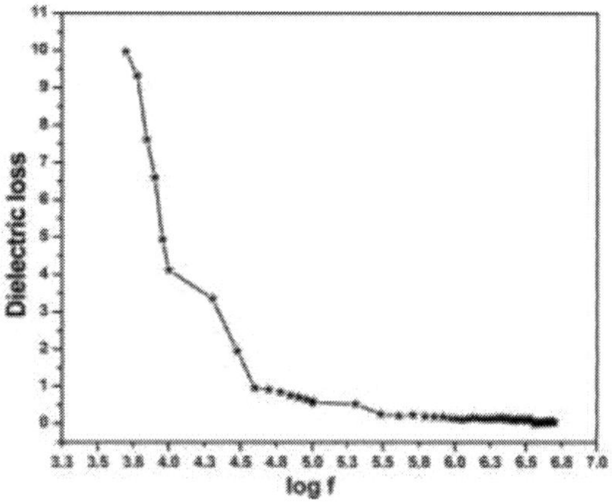

Figure 9. Variation of dielectric loss with frequency

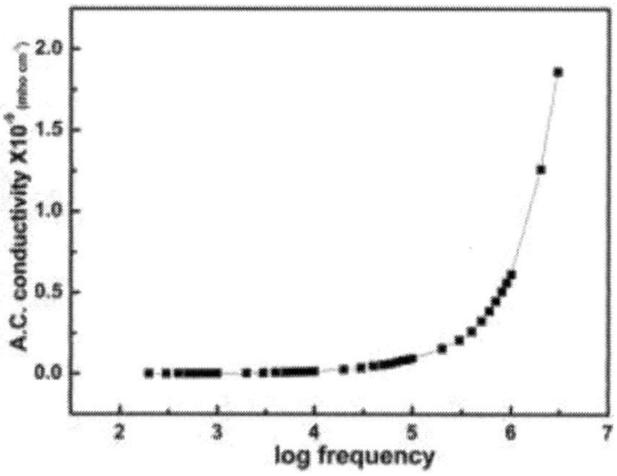

Figure 10. Variation of AC conductivity with frequency

3.8. Microhardness studies

The hardness analysis assists in the comprehension of the crystal's strength and deformation. It's also important to note that the mechanical properties of NLO materials have a strong relationship with their resistance to structural destruction or deformation. Its value is determined by the elastic and plastic properties of the indenter under study, as well as the measurement conditions. The Micro Indentation Hardness Test is ideally suited for crystals, with forces usually below 2N.

The most commonly known pyramid indenter in which opposite sides converge at the apex at an angle (α) of 136°. The ratio of the applied load to the region of the indentation is known as hardness. As a result, the Vicker's hardness number H_V is written as

$$H_V = \frac{Load}{Pyramidal\ area} = \frac{2P\sin(\alpha/2)}{d^2} = \frac{1.8544\,P}{d^2} \quad \text{Kg/mm}^2$$

The applied load P in gm, and the diagonal length d is measured in micrometers. Various weights from 5 to 250 gms may be used for indentation. The size of the indentation impression is determined using a calibrated microscope. The nature of calculated Hv values different loads is shown in the Figure 11.

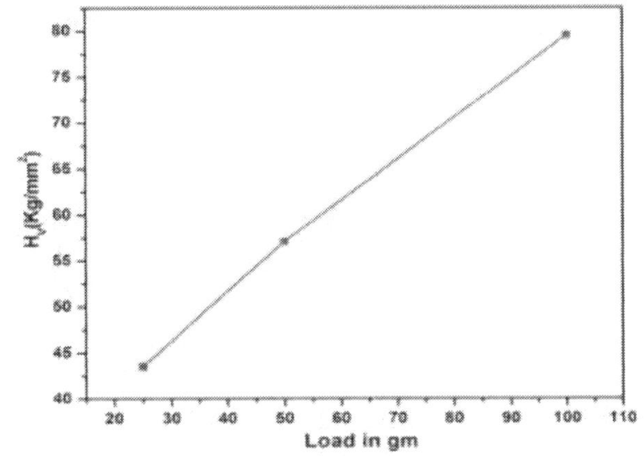

Figure 11. Variation of hardness number with applied load

Here, Hv rises with increasing applied load and cracks were observed for loads greater than 100 gm. According to Mayer's law, the connection between applied load and indentation size is given by

$$P = kd^n \quad \text{or}$$

$$\log P = \log k + n \log d$$

Where, k is a particular material constant. The value of 'n'indicates Mayer's index. Plotting the graphs of log p and log d provides the value of n and which is normally linear in nature, and the resulting slope gives the n value. If n>2, Hv will increase as the load increases, and if n<2, Hv will decrease as the load increases. Moreover, for hard materials, the value 'n' should be between 1 and 1.6, while for soft materials, it should be greater than 1.6 [17]. Based on this, for the grown crystal the resulting graph is shown in Figure 12.

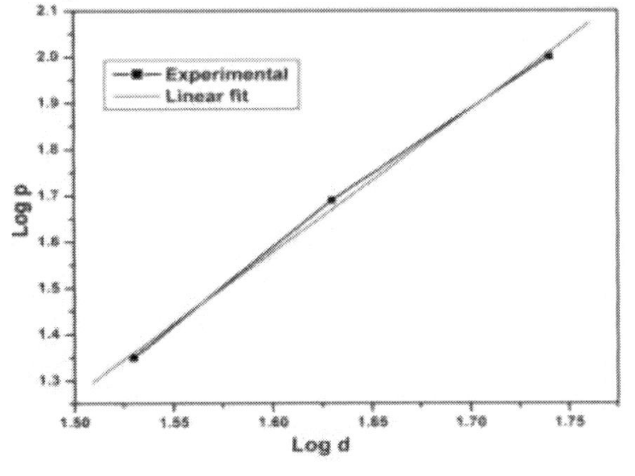

Figure 12. Plot of log p versus log d

The LAZS calculated 'n' value is 2.9 after the linear fitting suggest that it LAZS is soft material. Further, the elastic stiffness constant (C_{11}) gives an indication of the material's peak energy absorption before a fracture occurs. It also expresses the concept of material resistance to deformation when a load is applied to a flat crystal surface. Further, In calculating the C_{11}, the tightness of the bonding and the rate of variance with the position of the atoms are the two most significant factors. The corresponding C11 value for the applied loads on the crystals was obtained from the following equation. According to Wooster's formula given by

$$C_{11} = (H_V)^{\frac{7}{4}}$$

Yield strength (σ_V) is the stress under which a material begins to warp plastically, before deforming elastically.The value of σ_V for the different loads was calculated by using the following equation and are tabulated in the table 2.

$$\sigma_V = \frac{H_V}{2.9}[1-(n-2)]\left(\frac{12.5(n-2)}{1-(n-2)}\right)^{n-2}$$

Table 1. Calculated mechanical parameters.

Load P in gm	Hv (Kg/mm^2)	σ_v (GPa)	C_{11}x10^{14} Pa
25	*43.55*	*12.79*	*7.38*
50	*57.1*	*16.78*	*11.86*
100	*79.5*	*23.36*	*21.16*

4. Conclusions

L-alanine zinc sulphate (LAZS) crystal was grown by slow evaporation method. LAZS fit into orthorhombic system and is stable up to 288 °C as observed from diffraction and thermal studies respectively. UV study illustrates that LAZS has extensive transparency in complete visible and NIR regions with cutoff wavelength of 282 nm. According to dielectric studies LAZS has a normal dielectric nature with minimal defects, The SHG conversion efficiency of LAZS is 0.9 and 0.6 times that of KDP and urea respectively. The hardness of LAZS diminutions with load and also confirms the absence of liquid inclusions. With increased transparency, thermal stability and NLO efficiency of LAZS proposes that, it can be a useful candidate for device applications.

5. References

[1] Pepper D M 1982 'Nonlinear Optical Phase Conjugation' *Optical Engineering* **21**, 156-183
[2] Chemla D S and Zyss J 1987 'Nonlinear optical properties of organic molecules and crystals' Academic Press, *Orlando, New York,* **Vol 1-2**
[3] Dimitriev V G, Gurzadyan G G and Nikogosyan D N 1991 'Handbook of Nonlinear Optical Crystals' *Springer -Verlag, New York*
[4] Zernike F and Midwinter J 1973 '*Applied Non linear optics'* (Wiley, New York)
[5] Robert W and Boyd 1992 *'Nonlinear Optics'* Academic Press, INC. (Harcourt Brace Jovanovich Publishers)
[6] S. Suresh S, Ramanand A, Jayaraman D and Mani P 2012 *Reviews on Advanced Materials Science* **30** 175-183
[7] Yariv A 1975 *Quantum Electronics Second Edition*, (John Wiley and Sons, New York)
[8] Bernal J D 1931 *Zeitschrift für Kristallographie - Crystalline Materials* **78** 363 – 369
[9] Simpson Jr H J and Marsh R E 1966 **20** 550 – 555
[10] Destro R, R.E. Marsh R E and Bianchi R 1988 *J.phys.Chem.* **92** 966-973
[11] Arun K J and Jayalekshmi S 2009 *J. Minerals and Materials Characterization and Engg* **8** 635-646
[12] Jaikumar D, Kalainathan S and Bhagavanarayana G 2009 *J. Cryst.Growth* **312** 120-124
[13] Malinga P and Joseph Arul Pragasam A *Journal of Russian Laser Research* **34** 346-350

[14] Shanthi D, Selvarajan P and Jothi Mani R 2014 *Optik* **125** 2531-2537
[15] Karan C, Prabakaran A R, Prabhaharan M and Vinitha G 2019 *Journal of Electronic Materials* **48** 7915-7922
[16] Kurtz S K and Perry T T 1968 *J. Appl. Phys.* **39**(8) 3798–3813
[17] Onitsch E M 1947 *Mikrosckopie* **2** 131-151

Implementation of BL-PFC Converter Fed BLDC Motor Drive for Application less than 800 W

S.Dhamodharan[1], K.Sebastirani[2], Dhanaseelan.N[3], Guru Prasath T.V[3], Nivedha.S[3], Swathi Harikumar[3]

[1,3]Department of EEE, Sri Shakthi Institute of Engineering and Technology, Coimbatore, Tamil Nadu, INDIA
[2]Department of EEE, Sri Ramakrishna Engineering College, Coimbatore, Tamil Nadu, INDIA

Email: sdhamodharan21@gmail.com

Abstract: This article analyses the performance of a PFC SEPIC-fed BLDC motor drive for low-power applications. The BLDC motor's speed is regulated by changing the DC connection voltage of the VSI that supplies the BLDC motor. The VSI feeding BLDC motor is used to electronically commutate the BLDC motor, which works at a low frequency to minimise switching losses. A bridgeless design of a single-ended primary-inductor converter is proposed that eliminates the need for a diode bridge rectifier. A PFC-based SEPIC is intended to operate in a discontinuous inductor current mode, with a single voltage sensor, to achieve inherent power factor correction at the AC mains. The converter proposed here has a low conduction loss, a low harmonic content, and a power factor that approaches unity. Simulate the circuit by connecting the output voltage to a resistive load. The recommended drive's output is tested experimentally on a built prototype. At the AC mains, improved power quality is achieved under all conditions, and thus the obtained power quality indices are obtained. MATLAB-Simulink is used for all simulation work.

Keywords: BLDC Motor, PFC Converter, Matlab, SEPIC

1. Introduction

The last decade has seen dramatic change in industries, with induction motors being phased out and replaced by brushless DC motors. The rotor of an ac induction motor does not include any magnets; instead, it contains a series of laminations and windings. Although the majority of AC induction motors operate directly from AC power without the use of a controller, if variable speed operation is needed, a VFD must be mounted between the AC power and the motor. This VFD modulates the motor's speed by modulating the frequency with which the motor is powered. Even with a variable frequency drive mounted, some industrial induction motors have a small speed range. Furthermore, they are unsuitable for delivering the rated torque at extremely low speeds. A BLDC motor, on the other hand, replaces the rotor's windings with a set of permanent magnets. These magnets generate an enticing field that interacts with the stator field to generate force. A three-phase voltage source inverter is needed to accomplish electronic commutation of BLDC motors based on the location of the rotor as sensed by Hall-effect position sensors[1].

Converters with enhanced power efficiency are used to correct the power factor of alternating current mains. The literature has identified a variety of non-isolated and isolated PFC converter configurations for improving the power quality of alternating current mains [7]. The sensing specifications associated with the PFC converter's mode of operation are largely responsible for the cost of these PFC converters. There are two modes of operation for a PFC converter: continuous conduction and discontinuous conduction. The continuous conduction mode of operation of a PFC converter reduces the stress on the PFC converter switches while still allowing for supply voltage sensing. A PFC converter operating in DCM, on the other hand, requires only a single voltage sensor to regulate the DC connection voltage, and inherent power factor correction is accomplished at the AC mains, but at the cost of increased stress on the PFC converter switches. As a result, applications requiring low power are limited to this mode of operation. This paper will therefore be focused on the design and implementation of a bridgeless SEPIC for feeding a low-power BLDC motor drive. A PFC-based BLDC drive has been suggested using brushless SEPICs [4]; however, the configuration uses a BL-SEPIC as described in this configuration and work is limited to simulation studies [2]. As a result, this research provides experimental validation for BL-SEPIC, a low-power household application that uses a single low-side driver to power a BLDC motor drive.

2. Literature review

Chen, Y., et al., 2013. A brushless dc (BLDC) motor replaces the mechanical commutator and A dc brush motor's brush gear is connected to an electrical commutating unit. A permanent magnet can be found on the rotor (PM). A motor with a broader speed range and a lower maintenance requirements than a brushless DC motor, is more powerful than an inlet motor. The engine works similarly with a DC brush engine, since the BLDC motor's back electric force may have a trapezoidal wavelength. [1]

V. Bist., et al. The brushless DC engine is called a three-phase synchronous motor with three phase windings on the stator and permanent magnets on the rotor. A three-phase voltage source inverter is required to achieve an electronical switching of the BLDC motor based on the positioning of the rotor sensed by Hall-Effect position sensors. The controller supplies current pulses to the motor windings and thus controls the motor's speed and torque. This control system eliminates the need for the commutator (brushes) used in a large number of traditional electric motors. [3]

Jang, Y., et al., 2011. Brushless DC motors are constructed with wound wire coils and a two-pole electromagnet armature. The commutator, which is a mechanical rotary switch, reverses the current's directionality twice every cycle. This allows current to flow through the armature, causing the electromagnet's poles to wave against the permanent magnets on the motor's exterior. As the poles of the electromagnet in the armature touch the poles of the permanent magnet, the commutator reverses the polarity of the electromagnet in the armature. [5]

Fardoun, A., et al., 2018. Brushless DC motors emerged in the 1960s with the advent of solid state electronics. Brushless PM DC motors did not enter the industry until the 1970s. The delay was caused by the creation of not only PMs with high energy, and electrical controllers that can substitute mechanical switching with electronic switching. [6]

3. Proposed system

The proposed system evaluates the effectiveness of a low-power PFC SEPIC-fed BLDC motor drive. The speed of the BLDC motor is controlled by changing the VSI connection voltage for the BLDC motor. The VSI feeding motor BLDC is used to switch the BLDC motor electronically, which works at low frequencies to reduce switching losses. The bridge-free layout of a single-end primary inductor converter removes the need for a bridge rectifier. A SEPIC based on PFC is expected to act with such an independent voltage sensor in an interruptive inductor current mode to accomplish innate power factor on the AC mains. The converter proposed in Figure.1 has a low driving loss, a good dynamic content, and a nearing unity power factor. Simulate the circuit with resistive load connecting the output voltage. The recommended drive output is tested on a built-in prototype experimentally. At the AC pipeline, increased performance is measured under all conditions, thus acquiring the results of power quality indices.

BLDC motor speed control

The recommended Dc motor drive is a bridgeless SEPI VSI-driven BLDC motor. By adjusting the VSI DC connection voltage, the BLDC motor speed is controlled by a single voltage sensor. This allows the electronic switching of Dc motors during VSI operations at low frequency, which significantly reduces switch losses compared to the PWM. Due to the lack of DBR, bridge-less converters have low diode performance losses leading to maximum efficiency. For PF correction and voltage, a DICM front-end bridge free SEPIC can be used. High-frequency field-effect metal oxide transistors are used on the front of the rectifier for high-frequency operation. The component size is reduced. MOSFETs are used in VSI for low frequency applications. When connected to AC power supplies, the driving keeps superior PF and better efficiency levels, allowing for various variable speeds. [2].

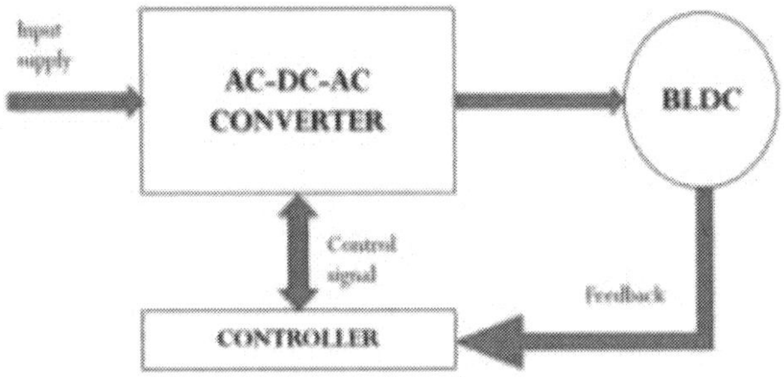

Figure 1: Block diagram of the proposed system

4. Mathematical model

The PFC BL-SEPIC is designed to operate in discontinuous conduction mode, which means that during the switching time, the current flowing through the output inductors (Lo1 and Lo2) is interrupted. A 500W (Pmax) PFC converter will be used with the chosen BLDC motor. For a wide range of speed variations, the DC connection voltage is regulated from a low value of 50 V (Vdc min) to the rated voltage of 200 V (Vdc max), with the supply voltage varying from 170 V (Vs min) to 270 V (Vs max).

For a 3 percent permissible ripple voltage, DC connection condenser value is measured respectively at higher and lower DC connection voltages. As a consequence, the DC connection capacitor is structured to obtain minimal ripple even in the lower DC connection voltage.

$$C_{max} = \frac{I_m}{\omega_L V_m}\tan(\theta) = \frac{(P_{max}\sqrt{2}/V_s)}{\omega_L \sqrt{2}V_s} = \tan(\theta) \qquad (1)$$

Where, the angle of displacement between the fundamental portion of supply voltage and supply current is equal to one. The filter capacitor is chosen to be less than Cmax.

$$L_f = L_f - L_s \qquad (2)$$

$$L_{req} = \frac{1}{4\pi^2 f_s^2 C_f} - 0.035\left(\frac{1}{\omega_L}\right)\left(\frac{V_s^2}{P_{max}}\right) \qquad (3)$$

5. Working

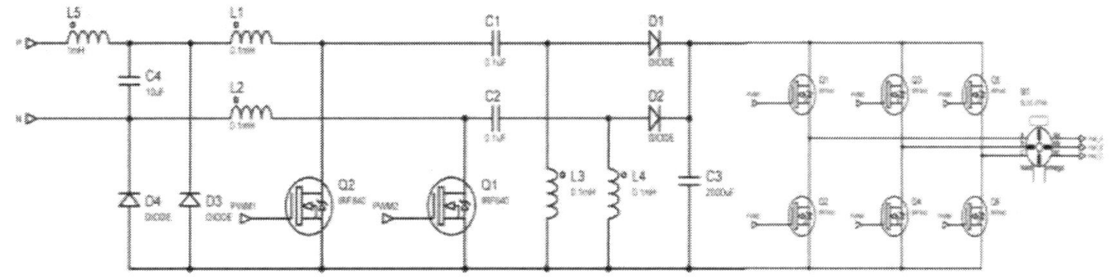

Figure 2: Circuit diagram of the proposed Converter

The proposed method function is divided into two components, each of which corresponds to its application for a full line and a switching cycle. In this project, a bridgeless SEPIC converter fed BLDC motor is proposed. From Figure 2, The bridgeless SEPIC converter is fed an AC voltage as input. The SEPIC transforms the AC voltage from the input into DC voltage. It can be used for both buck and boost operations. If the duty cycle is less than 50%, it operates in buck mode; if the duty cycle is greater than 50%, it operates in boost mode; in the meantime, it is used for PFC. When compared to traditional buck and boost converters, it improves performance. Two inductors, an IGBT, and two capacitors are included in the circuit. The output DC bus voltage is regulated using PWM control. The three phase inverter receives this DC bus voltage. The input DC voltage is converted into AC voltage by this inverter.

The BLDC motor is supplied with this AC voltage. The halleffect sensor is used to provide PWM power to the inverter. The rotor location is determined using the Hall effect. The location of the rotor is detected by the PWM controller, which then energies the next winding in order to rotate the rotor. The speed of a BLDC motor is controlled using PI power. The service cycle of the inverter is adjusted by the PI control based on the set point and feedback.

6. Matlab simulation and result

The output voltage is connected to a resistive load, and the circuit is simulated in Matlab shown in Figure 3. The performance of the suggested drive is supported by real-world results obtained on a prototype. For all conditions, improved power quality is achieved at the AC mains from Figure 4, and thus the obtained power quality indices are achieved. MATLAB-Simulink is used for the entire simulation process. Waveform of the DC link voltage is shown in Figure 5 and the motor speed waveform is shown in Figure 6. The hardware setup image is shown in Figure 7.

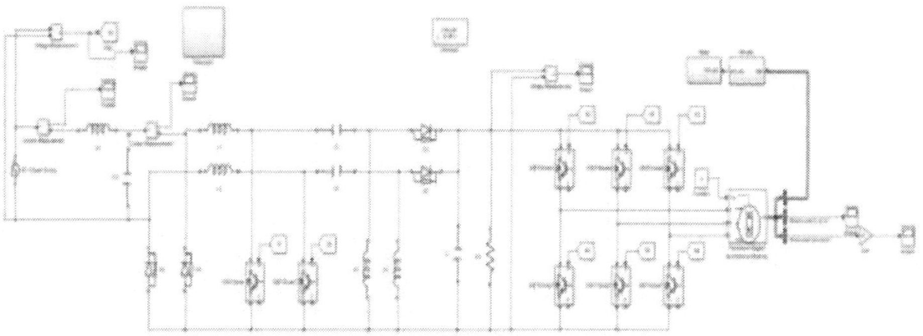

Figure 3. Matlab Simulink model of the proposed system

IOP Conf. Series: Materials Science and Engineering 1166 (2021) 012036 doi:10.1088/1757-899X/1166/1/012036

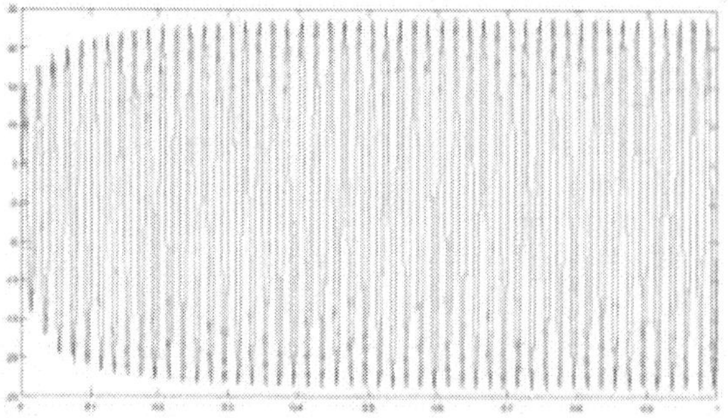

Figure 4: Waveform of Input current of SEPIC PFC Converter

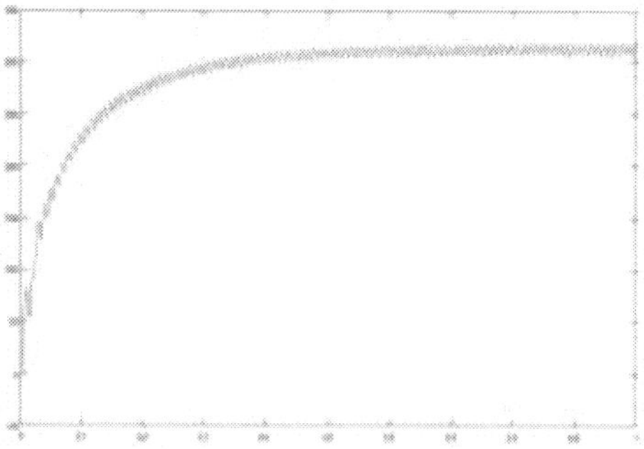

Figure 5. Waveform of DC link voltage output

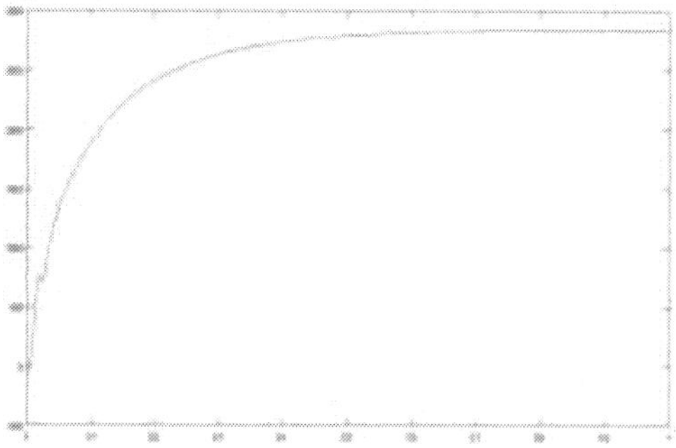

Figure 6. Waveform of Speed of the motor

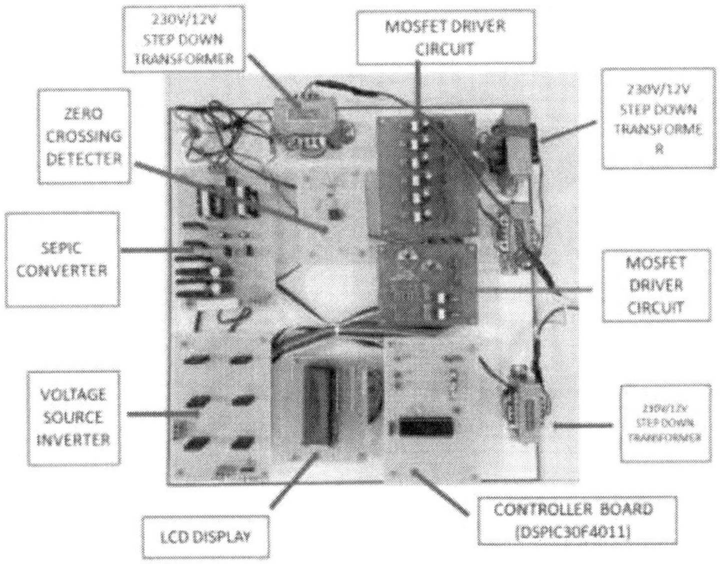

Figure 7: Hardware of the proposed system

7. Conclusion

The BLDC motor's speed has been regulated using the VSI feeding's variable DC connection voltage. This PFC converter has been used to operate a three-phase VSI in lower frequencies switching mode, resulting in lower Losses switching. A DICM front-end was used for DC-link voltage control and power factor correction in the AC network. For speed control and supply voltage fluctuations across a wide range, the output of the proposed drive was found quite satisfactory. A framework of the system drive was built and tested, yielding satisfactory results across the entire speed range and a wide range of supply voltages. The obtained power quality indices were discovered to be within the limitations of the world power quality standard.

References

[1] Y. T. Chen, C. L. Chiu, Y. R. Jhang, Z. H. Tang and R. H. Liang, "A Driver for the Single-Phase Brushless DC Fan Motor With Hybrid Winding Structure," *IEEE Trans. Ind. Elect.,* 2013.

[2] M. R. Sahid, A. H. M. Yatim and T. Taufik, "A new AC-DC converter using bridgeless SEPIC," Proc. of 36th *IEEE Annual Conf. of Ind. Electron. Society (IECON)*, pp.286-290, 7-10 Nov. 2010

[3] V. Bist and B. Singh, "An Adjustable Speed PFC Bridgeless Buck-Boost Converter Fed BLDC Motor Drive", *IEEE Tran. Ind. Electron.*, vol.**61**, no.**6**, pp.2665-2677, June 2014.

[4] B. Singh and V. Bist, "Power Quality Improvement in PFC Bridgeless SEPIC Fed BLDC Motor Drive", *Int. Jr. Emer. Electr. Power Sys. (IJEEPS),* vol.**14**, no. **3**, pp. 285–296, June 2013

[5] Y. Jang and M. M. Jovanovi, "Bridgeless High-Power-Factor Buck Converter," *IEEE Trans. Power Elect.*, vol.**26**, no.**2**, pp.602-611, Feb. 2011.

[6] A. A. Fardoun, E. H. Ismail, A. J. Sabzali, M. A. Al-Saffar, "New Efficient Bridgeless Cuk Rectifiers for PFC Applications," *IEEE Trans. Power Electron.*, July 2012

[7] B. Singh, B. N. Singh, A. Chandra, K. Al-Haddad, A. Pandey and D.p Kothari, "A review of single phase improved power quality AC-DC, converters," *IEEE Trans. Ind. Elect.* Oct. 2003.

Effect of sustainable materials on embodied energy, carbon footprint and cost for a proposed conventional apartment

Abhishek Manjunath[1], Nayana N Patil[1*]

[1] Department of Civil Engineering, Ramaiah University of Applied sciences, Bengaluru, Karnataka, India
[1]abhimaan13@gmail.com
*nayanapatil.ce.et@msruas.ac.in

Abstract. The construction industry accounts for an incredible 36% of worldwide energy usage, and 40% of CO_2 emissions. Therefore, it is required to reduce the impact of construction on the environment. In this study, a few green materials have been selected along with few green techniques and are applied to the apartment and a comparison is provided. An ongoing project consisting of 18 typical floors and basement is selected for the study. Estimation of quantities are done according to the drawings and major materials such as blocks, internal paints, flooring and concrete are replaced with proposed sustainable materials. Embodied energy and carbon footprint analysis is performed for the building components such as blocks, tiles, paints, concrete and plastering. Alternate materials like compressed stabilized earth blocks, clay plaster, wallpaper, terrazzo tiles and blended cement concrete are chosen as replacements for the conventional materials. A comparison is provided with conventional materials with respect to the chosen sustainable materials. The results show 73% reduction in embodied energy and 52% reduction in carbon footprint of the structure. Also, reduction in cost by 30%. Hence, reducing the impact on the environment and making the structure sustainable.

Keywords: Sustainable materials, embodied energy, carbon footprint, compressed stabilized earth blocks, cost analysis

1. Introduction

The construction industry necessitates the extraction of large quantities of materials, which consumes energy resources and releases harmful pollutant emissions into the biosphere. Each material must be extracted, processed, and transported to its final destination. The energy used in these activities is necessary for human growth, but it also jeopardises the biosphere's quality and long-term viability due to unwanted or "second" order impacts. (Hammond, 2008) On a local, regional, or national scale, many of these side effects of energy production and consumption result in resource uncertainty and potential environmental risks. Energy and pollutant emissions like carbon dioxide (CO2) can be considered "embodied" in materials. Thus, embodied energy can be observed as the quantity of energy required to process, and supply to the construction site, the material under consideration. In order to determine the magnitude of this embodied energy, a methodology is required that sums the energy inputs over the major part of the material supply chain or life-cycle. (Imperatives, 1987) In the present context, the emission of energy related pollutants (like CO_2), which is a concern in the context of global warming and climate change, may be viewed over their lifecycle. This gives rise to the notion of 'embodied carbon'. The aim of the present study was to develop an open-access, reliable database of both embodied energy and carbon for (principally) Indian construction materials.

Table 1. Example for difference between green concept and sustainable concept.

Factor	Green concept	Sustainable concept
Clay Plastering	Clay is a naturally abundant Material.	The embodied energy is considerably low.
Wallpapers	Green product, eco-friendly and durable	But not sustainable because it is harvested in an environmental irresponsible way (by depleting the forest) and even expensive

Table 1 shows the difference between green concept and sustainable concept which implies they are similar but not the same.

Materials manufacturing consume maximum energy and large emission of carbon dioxide. This leads to global warming and depletion of non-renewable resources. Therefore, utilization of natural and renewable energy sources in the construction industry to minimize the drastic effect on the environment by the buildings. By comparative study, awareness and motivation about the advantages and benefits of green construction can be achieved.

2. Methodology and Materials

2.1 Methodology.

The main aim of conducting this study is to compare the benefits of alternate sustainable materials in the proposed residential complex, for that the following objectives are considered and followed.

Table 2. Methodology adopted for the study

Objectives	Methodology
Choosing Structure for the analysis	An on-going project consisting of a basement, stilt floor and 18 upper floors
Choosing materials with high impact on environment	Quantifying the materials needed for construction and choosing the major materials such as concrete, blocks, flooring tiles, paints and plaster
Embodied energy, carbon footprint and cost analysis	Embodied energy, cost and carbon footprint of the quantified materials are calculated. This is performed by data collection by referring journals and articles
Comparison between conventional and proposed materials	The results obtained by the analysis is then compared with actual construction parameters

Table 2 shows the methodology applied for the research conducted on conventional and sustainable approach for the selected structure

2.2 Selection of apartment.

The residential apartment chosen for the comparative study is located in Akshayanagar, Bangalore. An 18 storied residential building with a stilt floor and a basement. It has a built - up area of

1000sqm. Each floor has 4 numbers of 3BHK and 2 numbers of 2 BHK apartments. The site has been designed to have 25% garden area included in super built up area.

Table 3. Details of the selected apartment

Salient Features of the Project	
Project Name	Hiranandani Club Meadows
Project type	Commercial Project
Client	House of Hiranandani
Type of Contract	Item rate contract
Project Location	Akshayanagar, Near Hulimavu Lake, Hulimavu, Bannerghatta, Bangalore - 560068
Project Start Date	February 2019
Project Finish Date	July 2021
Project Specifications	The entire project consists of 5 towers and a clubhouse Lake Verandah, Hill Crest, Club Meadows, Queen's Gate and Evita & Torino
Evita & Torino	1 Basement + Ground + 18 Upper Floors

The information is collected from the site office of Hiranandani Club House Meadows. *(Source: House of Hiranandani Clubhouse Meadows office)*

IOP Conf. Series: Materials Science and Engineering 1166 (2021) 012037 doi:10.1088/1757-899X/1166/1/012037

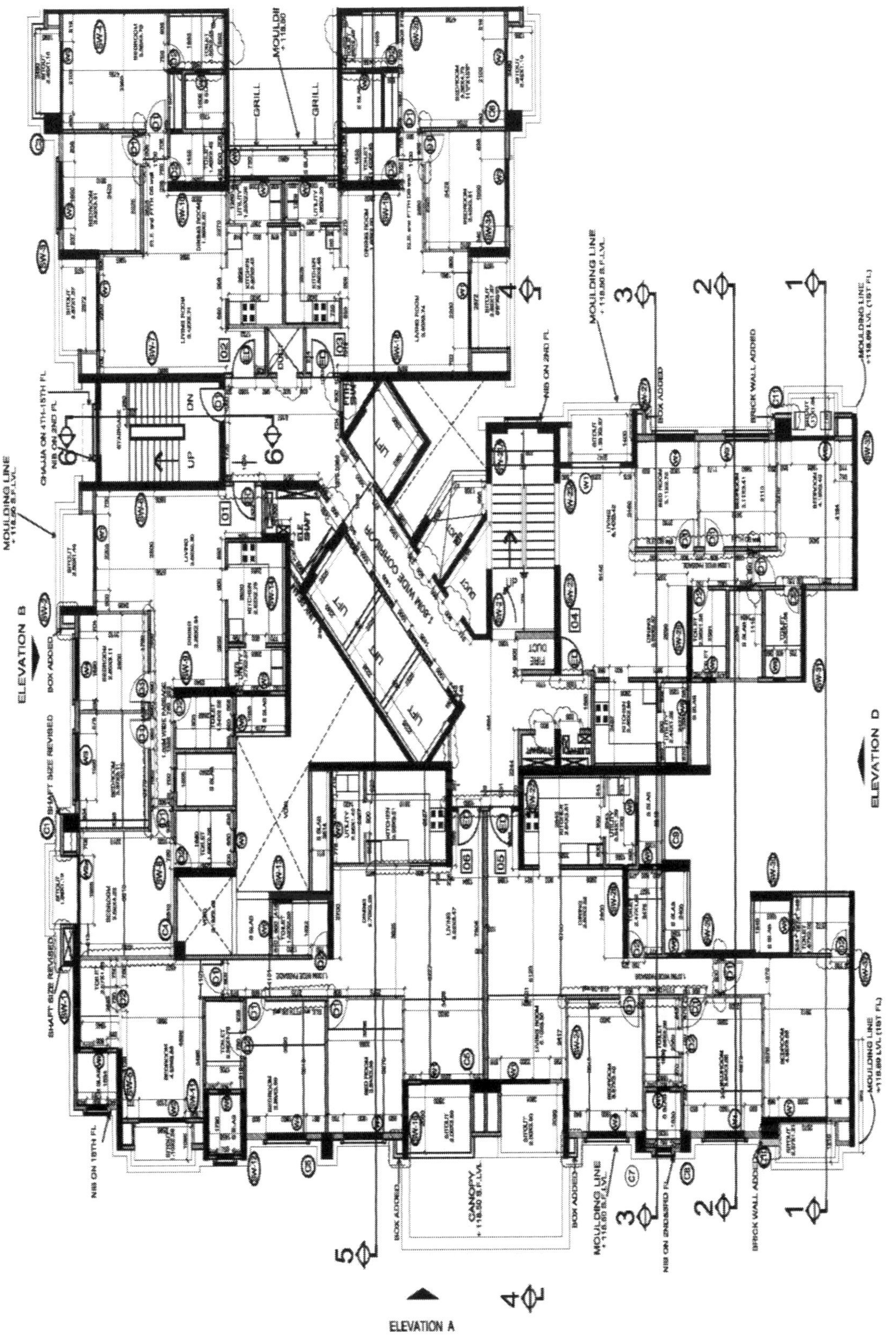

Figure 1. Typical Floor plan
(Source: House of Hiranandani Clubhouse Meadows office)

2.3 Materials used.

Once the values of embodied energy and carbon footing for early versions of the database had been selected, it was possible to apply the data in practical situations. The embodied energy and carbon footing of typical dwellings were analysed by first determining the quantities of material consumed during construction. The selection of material for replacement are based on certain criteria such as embodied energy, local availability, recycle content, functional life period, material cost, maintenance cost, construction waste management, minimum time consumption and toxicity/safety. In this study, materials such as clay, terrazzo tiles, blended cement concrete, wallpapers, compressed stabilized earth blocks (CSEB) and low E glass are considered as the alternatives for the conventional materials. Adopting the selected materials on the chosen structure from the quantity estimation obtained for this building and results are obtained. (Jin lee kim, 2014) (Sustainable construction managementat project level: a modified environmental management system structure, 2008)

2.3.1 Building block. An alternate for concrete blocks are compressed stabilized earth blocks (CSEB) which is low in embodied energy and made of earthly natural materials. This helps in reduction of carbon dioxide and embodied energy. The block is made of mix of fairly dry inorganic subsoil, non-expansive clay and aggregate along with a binder such as cement (Imperatives, 1987).

Table 4. Building blocks comparison between conventional and green building

Item Description	Size inch	Quantity	Unit	Embodied energy per unit	Embodied Energy (MJ/Unit)	Carbon Footprint per Kg	Carbon Footprint (CO2e)Kg	Cost per unit	Cost (Rs.)
Conventional building Block work	4	223109	Per piece	0.81	180718.29	0.06	267730.8	39	8701251
	6	61244		0.96	58794.24	0.07	115751.2	48	2939712
Compressed stabilized earth block	4	223109		0.41	91474.69	0.022	39267.18	26	5800834
	6	61244		0.49	30009.56	0.026	19108.13	33	2021052

(Riza, 2010), Embodied energy and carbon footprint values are taken from the cited paper.

2.3.2 Clay plastering is an old method of finishing surfaces which is proven to be sustainable. The benefits include cooler interior temperature, low embodied energy and naturally occurring material. Clay plaster is a mixture of clay and sand that makes a beautiful, environmentally friendly alternative to conventional plaster and paint. It is natural, non-toxic, durable and beautiful. Unlike most paint, it does not contain VOC's. (Natural Stabilized Earth Panels versus Conventional Façade Systems. Economic and Environmental Impact Assessment, 2018). Table 5 shows the comparison of Plastering material quantity, embodied energy and carbon footprint.

Table 5. Plastering material quantity, embodied energy and carbon footprint comparison

Item Description	Quantity	Unit	Embodied energy per unit	Embodied Energy (MJ/Unit)	Carbon Footprint per KG	Carbon Footprint (CO2e)	Cost per unit	Cost (Rs.)
Cement Plastering	3220	Sqm	94.5	304290	0.18	13910.4	161	518420
Clay Plastering	3220	Sqm	28	90160	0.047	2572.78	96	309120

(Natural Stabilized Earth Panels versus Conventional Façade Systems. Economic and Environmental Impact Assessment, 2018)

2.3.3 Blended cement concrete. The concrete made with 24% fly ash replacement and 36% replacement with GGBS showed highest reduction of 44% in embodied energy and 24% in cost. This can save up costs as well as reduce the impact on environment. comprising OPC that has been partly substituted by supplementary cementitious materials, are used as binders for concrete. Commonly used substitutes include fly ash, a fine waste residue that is collected from the emissions liberated by coal burning power stations, and ground granulated blast furnace slag(GGBS), a waste by-product from steelmaking. According to Flower and Sanjayan use of blended cements results in reduction of CO2 emissions by 13–22%. (An Economic and Embodied Energy Comparison of Geo-polymer, Blended cement and Traditional Concretes, 2014). Table 6 gives the Comparison of concreting materials used in conventional and sustainable construction.

Table 6. Comparison of concreting materials

Item Description	Quantity	Unit	Embodied energy per unit	Embodied Energy (MJ/Unit)	Carbon Footprint per KG	Carbon Footprint (CO2e)	Cost per unit	Cost (Rs.)
Concrete	6603	Cum	3890	25685670	0.25	4333219	5830	38495490
Blended cement concrete 24% fly ash and 36% GGBS	6603	Cum	982	6484146	0.14	2244492	2845	18785535

(An Economic and Embodied Energy Comparison of Geo-polymer, Blended cement and Traditional Concretes, 2014)

2.3.4 Flooring material Terrazzo is a composite material either poured in place or precast or hydraulically pressed as tiles. Terrazzo is used for floor and wall decorative finishes. It consists of marble, quartz, granite, glass or other suitable chips; sprinkled or unsprinkled, and poured with a binder that is cementitious, chemical or a combination of both. Terrazzo is cured, ground and polished to a smooth surface or otherwise finished to produce a uniformly textured surface. When comparing terrazzo to four other flooring types, it was observed, on the one hand, that porcelain stoneware and stoneware presented 65% greater embodied energy values, while in contrast, the values for granite and linoleum were less than 79% and 92%, respectively. Table 7 gives the comparison of flooring materials used in the study.

Table 7. Comparison of flooring material

Item Description	Quantity	Unit	Embodied energy per unit	Embodied Energy (MJ/Unit)	Carbon Footprint per KG	Carbon Footprint (CO2e)	Cost per unit	Cost (Rs.)
Ceramic Tile Flooring	11609	sqm	157	1822613	0.613	227722	611	7093099
Terrazzo Tile Flooring	11609	sqm	74	859066	0.51	82888.26	1076	12491284

(Materials, 2017) (Deshmukh, 2014)

2.3.5 Wallpapers are made of renewable resource such as bamboo and wood. Therefore, having less embodied energy and impact on the environment than paints. Wallpapers are durable, long lasting, and cleanable to meet the needs of different lifestyles and applications, holding up to the wear and tear of children or conditions in high traffic areas. According to a lifecycle analysis, it was established that wallcoverings now can last five times longer than paint under typical usage conditions. Table 8 gives comparison of internal finishing comparison. (Materials, 2017)

Table 8. Comparison of Internal Finishing comparison

Item Description	Quantity	Unit	Embodied energy per unit	Embodied Energy (MJ/Unit)	Carbon Footprint per KG	Carbon Footprint (CO2e)	Cost per unit	Cost (Rs.)
Emulsion Paints (2 coats)	31840	Sqm	39.24	1249401.6	2.54	40437	398	12672320
Wallpapers	15920	Sqm	15	238800	1.87	6847.192	605	9631600

3. Results and Discussion

Selection and adoption of sustainable materials to the selected structure is performed according to standard procedures and rates. It is found that locally available sustainable materials can solve the problem of costs as well as the harmful impact on the environment. The following shows the results of the analysis for the selected structure in aspect of cost, embodied energy and carbon footprint.

3.1 Complete embodied energy and carbon footprint of conventional building

Embodied energy calculation of whole building is calculated by considering energy embodied in MJ of each work at each stage as shown in Table 9. For this, considering embodied energy of each material is necessary. Embodied energy calculation of conventional building at each stage by considering embodied energy of each materials of entire work. Table 10 gives the summary of carbon footprint for the building.

Table 9. Summary of Embodied energy

Summary of embodied energy in MJ		
Description	Conventional Materials	Sustainable Materials
Building Blocks	239512.53	121484.25
Plastering	304290	90160
Concrete	25685670	6484146
Flooring	1822613	859066
Internal Finishing	1249401.6	238800

Table 9 represents the embodied energy in MJ for the conventional materials as well as the selected alternate materials. These values are for the entire structure consisting of basement, stilt floor and 18 typical floors. It can be observed that the embodied energy of sustainable materials are less when compared to conventional materials for the same quantity.

Table 10. Summary of carbon footprint

Summary of carbon footprint in CO_2(eq)		
Description	Conventional Materials	Sustainable Materials
Building Blocks	383481.96	58375.312
Plastering	13910.4	2572.78
Concrete	13910.4	2572.78
Flooring	227722.144	82888.26
Internal Finishing	40436.8	6847.192

Table 10 represents the carbon footprint in CO_2(eq) for the conventional materials as well as the selected alternate materials. These values are for the entire structure. It can be observed that the carbon footprint of sustainable materials are less when compared to conventional materials for the same quantity.

Table 11. Summary of Costs

Summary of costs in INR.		
Description	Conventional Cost	Sustainable Cost
Building Blocks	11640963	7821886
Plastering	518420	309120
Concrete	38495490	18785535
Flooring	7093099	12491284
Internal Finishing	12672320	9631600

Table 11 represents the cost in INR for the conventional materials as well as the selected alternate materials. These values are for the entire structure. It can be observed that the cost of selected sustainable materials are less except for terrazzo tile flooring. But the cost is less when compared entirely with the conventional mode.

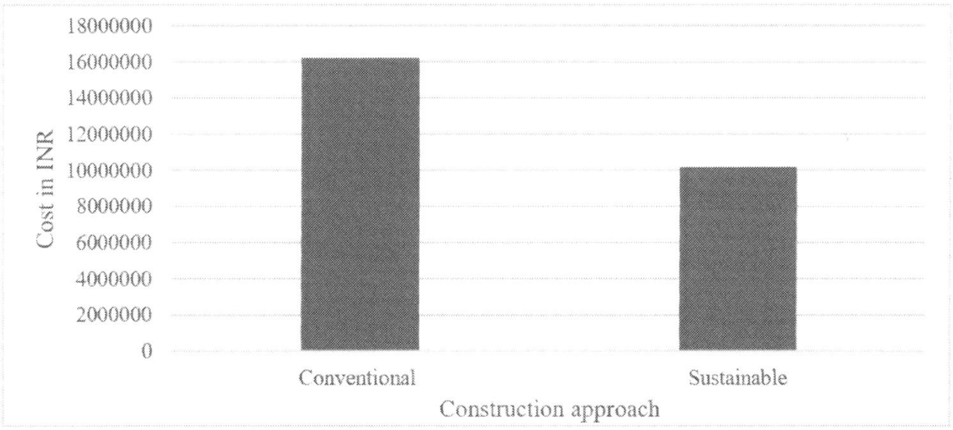

Figure 2. Cost comparison of overall cost for different building approaches

Figure 2 represents the cost comparison between conventional and sustainable concepts showing that sustainable concept is cheaper when compared to conventional.

The embodied energy and carbon footprint comparison for different materials are shown in the Figure 3 and Figure 4.

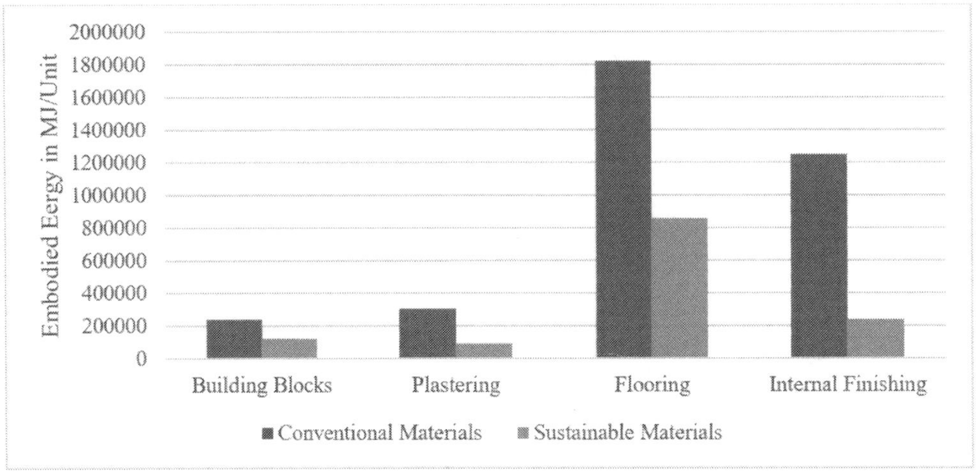

Figure 3. Comparison of embodied energy

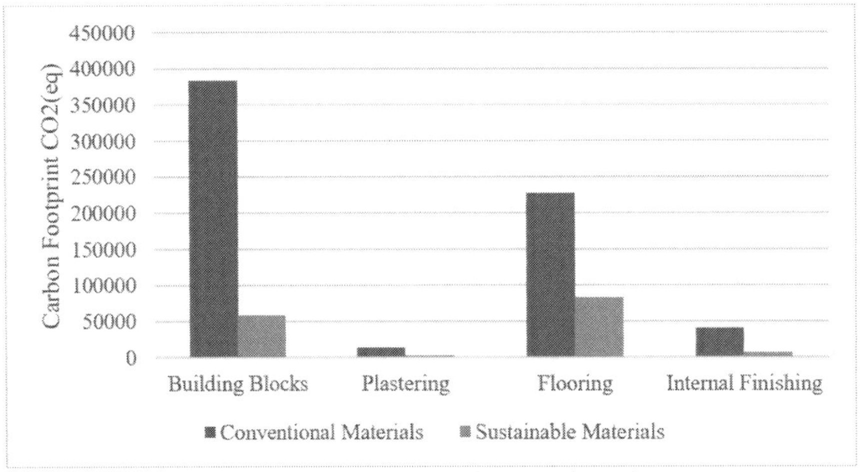

Figure 4. Comparison of carbon footprint

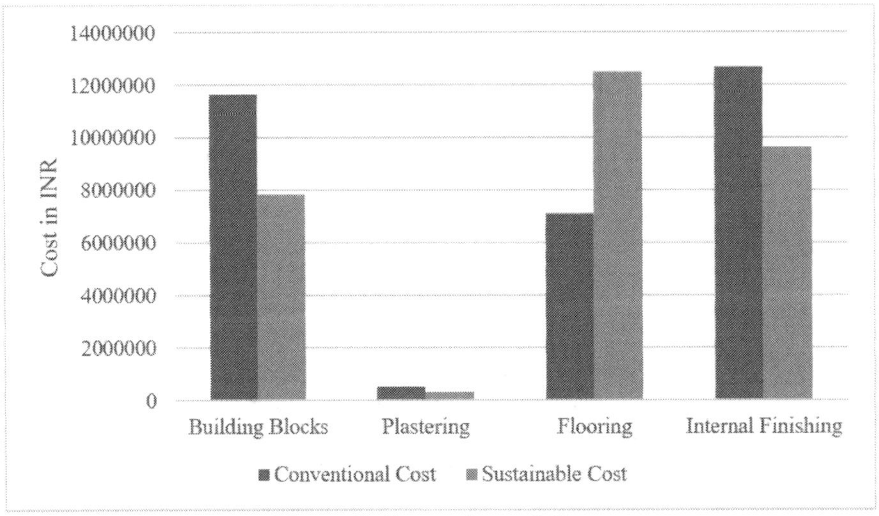

Figure 5. Comparison of Costs

4. Conclusion.

1. The CSEB blocks costs 32% cheaper than concrete blocks. Also the embodied energy and carbon footprint are 49% and 84% reduced when compared to conventional materials.
2. The Blended cement concrete costs 51% less than current market rate of M40 grade concrete. It also exhibits 74% reduction in embodied energy and 48% reduction in carbon emission.
3. Terrazzo tile flooring is 43% costlier than conventional tiles but the embodied energy and carbon footprint of the flooring are drastically less, 52% and 63% respectively.
4. Internal painting is replaced with wallpapers which resulted in cost reduction by 24% and reduction in embodied energy and carbon emission of 80% and 83% respectively.
5. By using clay plaster, the cost is reduced by 40% whereas the embodied energy and carbon emission are reduced by 70% and 81% respectively.
6. The material cost with respect to the selected materials is reduced by 30%.
7. The Embodied energy of the structure is reduced by 73% by adopting the chosen sustainable materials.
8. The Carbon emission of the structure is reduced by 52**%** by adopting the chosen sustainable materials.

5. References

[1] Manjunatha L R, Sandhya anvekar 2014 *An Economic and Embodied Energy Comparison of Geo-polymer, Blended cement and Traditional Concretes*. Researchgate, Journal of Civil Engineering Technology and Research

[2] Rohit Deshmukh, Ashok More. 2014. *Low energy green materials by embodied energy analysis*. International journal of civil and structural engineering research

[3] G. P. Hammond, C. I. Jones. 2008. *Embodied energy and carbon in construction materials*. Institution of civil engineers - energy

[4] *India Construction Materials Database of Embodied Energy and Global Warming Potential METHODOLOGY REPORT 2017*

[5] Jin Lee Kim, Martin Greene, Sunkuk Kim. 2014. *Cost comparative analysis of a new green building code for residential project development*. Journal of construction engineering and management.

[6] Carmen Galán-Marín, Alejandro Martínez-Rocamora, Jaime Solís-Guzmán, Carlos Rivera-Gómez. 2018 *Natural Stabilized Earth Panels versus Conventional Façade Systems. Economic and Environmental Impact Assessment*. Sustainability Journal MDPI.

[7] Venkatarama Reddy. 2009. *Sustainable materials for low carbon buildings*. International journal of low carbon technologies.

[8] Fetra Venny Riza, Ismail Abdul Rahman, Ahmad Mujahid Ahmad Zaidi. 2010. *A brief review of compressed stabilized earth brick*. International conference on science and social research.

[9] Bardhan S. 2011. *Embodied energy analysis of multi storied residential buildings in urban India*. WIT transactions on ecology and the environment

[10] Imperatives. 1987. *Report of the World Commission on Environment and Development: Our common future*. Development and International Cooperation

[11] Urban Persson, Anne Landin, Stefan Olander, 2008. *Sustainable construction management at project level: a modified environmental management system structure*. World sustainable building conference

Implementation of sustainable materials in an existing residential building and comparison

Prajwal S, Akashaya Kumar V H, Nayana N Patil

[1] Department of Civil Engineering, Ramaiah University of Applied sciences, Bengaluru, Karnataka, India

*nayanapatil.ce.et@msruas.ac.in

Abstract. Construction and construction related industries are chiefly responsible for depletion of natural resources and environmental problems, hence it is necessary to solve these issues using sustainable materials and techniques. In the study, a few alternative sustainable materials have been selected and have been implemented in an existing residential building. The effectiveness of sustainable materials in reducing the embodied energy and carbon footprint is enumerated in the selected conventional building constructed in 2014 and re-estimated as per the current SR Book of rates. The materials like building blocks, flooring material, paints, doors and windows and conventional slabs are replaced by their alternatives chosen through a detailed literature survey. To quantify usefulness of green materials in terms of cost and positive environmental impacts like reduced embodied energy and carbon footprint, a similar green residential building is proposed and to conduct comparative study for green versus conventional with all respective aspects. Materials used are brick prepared using mine waste, SHGC glass, filler slab, Bamboo doors and windows and low VOC paint materials. Overall construction cost is also 15.2% lesser than conventional as per the abstract between re-estimated conventional building and the sustainable green building. The overall carbon foot print and the embodied energy of sustainable green building is 39.3% lesser than the conventional one.

Key words: Sustainable, conventional, Green, Residential, Filler slab, Building block.

1. Introduction

The conventional buildings are one of the main factors causing depletion of natural resources, the major cause for environmental issues [1]. The construction industries are responsible for these environmental problems, hence it is necessary to be solved by finding new alternatives for replacing and implementing the green materials to achieve sustainable output [2]. Implementation of sustainable construction will certainly decrease the energy consumption and in turn reduce the carbon emissions. Another way to attain green construction is by utilizing different waste materials for manufacturing building materials [3]. The commonly used waste materials for manufacturing the building materials are fly ash, GGBS, IOT, lime, M-Sand, waste collected from demolished building, mine waste which are locally available [6]. Green and sustainable buildings are similar but in some perception they are little different, the answer is that although the terms can sometimes be used interchangeably, there are some key differences. [4] Green isn't always sustainable but sustainable is always green.

Materials manufacturing consume maximum energy and large emission of carbon dioxide. This is leads to global warming and depletion of non-renewable resources [5]. Therefore utilization of natural and renewable energy sources in the construction industry to minimize the drastic effect on the environment by the buildings. By the comparative study embodied energy and carbon footprint can be minimised along with cost [6]. In the study cost, embodied energy and carbon footprint of the building is reduced.

2. Materials and methodology.

2.1. Methodology. The main aim of conducting the study is to find the usefulness of replaced alternative materials in a existing residential green building. The following objectives are considered and followed.

1. Selection of existing conventional residential building built in 2014 and re-calculation (quantity estimation and rate analysis) of the cost of construction as per the current market rates using SR Book 2018-19 of PWD, Karnataka.
2. All alternative materials and sustainable material are adapted to the proposed green building and again calculation of cost (quantity estimation and rate analysis) has been done as per the current market rates using SR Book 2018-19 of PWD, Karnataka.
3. Comparison of the cost, embodied energy and carbon footprint has done between the selected conventional building versus proposed green building.

2.2. *Selection of residential building*.

The conventional building chosen for the comparative study is located in Whitefield, Bangalore. The building is a 4 storied residential building with a stilt floor. It has a site area of 6000sqft, site area acquired by building is 2400sqft and built up area is about 7208sqft. The selected building details is as shown in below table 1.

Table 1 Selected conventional building data

Project Name	Residential project
Total Area of the Building	2400 sqf , (240sqm)
Built up area	7208sft (720.8sqm)
Levels	Ground floor + 4 upper floors
Typical Floors	1 No of 3BHK in each floors
Total Flats	4 number of flats
Project completed	2013-14
Located	White-field, Bangalore

2.3. *Materials used for replacement tn the praposed structure.*

From the quantity estimation obtained for this building, materials are the important component of the building; hence these materials are replaced with green material. All selected component of green building and with respect to existing residential building are shown in the table 2 below.

Table 2: Materials used in residential building

Building component	Conventional Building material	Alternate Sustainable material
Building block	Normal Clay Bricks	Un-stabilized Bricks of mine waste
Building window envelop	Ordinary Glass	SHGC glass
Building slab	Conventional slab	Filler slab
Flooring material	Marble and Granite	Kota stone (of marble & granite finish)
Paints	VOC paints	Low VOC paints
Wood	Burma/Teak wood	Bamboo

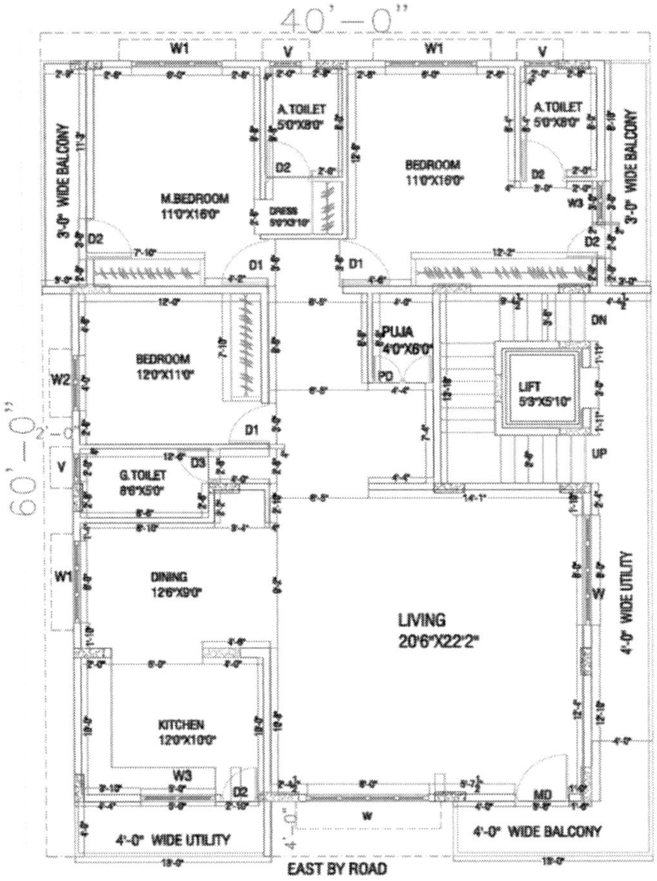

Fig. (1) Typical ground floor, 1st ,2nd and 3rd floor plan of the building

2.3.1. ***Building block***.

The building blocks are prepared using waste materials are a good alternative to normal brick to resolve the problems. This helps in reduction of carbon dioxide in many ways by using mine waste which is locally available in Belgaum and Bellary. [5] The bricks chosen for the study are prepared using red mud, IOT, GGBS and lime waste of proportion 45 to 60%, 15 to 30%, 10%, 3.75% and water 11.25%. And its quantity and cost detail is as shown in table 3 below.

Table 3 Comparison of building blocks between conventional and sustainable building

Building block type	Dimension	Quantity(cum)	Rate per (cum)	Total cost
Conventional building Block work	150mm block work	156.72	5802.7	909399
	100mm block work	26.4	7482.9	197549
Green building Block work	150mm block work	156.72	3570.96	559641
	100mm block work	26.4	4147.9	109505

2.3.2. *Solar heat gain coefficient (SHGC) window:*

SHGC is the fraction of solar radiation admitted through the glass or the window in which it has been used. It is expressed as a number in between 0 and 1. If a window has a SHGC value of 0.5, it transmits 50% of the incident solar radiation through it. When air conditioning is generally not of concern, a higher SHGC in the range of 0.30 to 0.60 can be helpful, as during winter months, the solar heat gained can help warm the house. [6]

2.3.3. *Filler slab:*

It is a technique to replace the concrete in the tension zone. The filler material, thus, is not a structural part of the slab. By reducing the quantity and weight of material, the roof become less expensive, yet retains the strength of the conventional slab. The most popular filler material is the roofing tile. The structure requires less steel and cement and it is also a good heat insulator. It is adopted in the construction of green building in which 50mm of the concrete in the tension zone is replaced by terracotta tiles which again have a U-value half of that of concrete. In table 4 all the description of filler slab and conventional slab materials are mentioned. [4]

Table 4 Filler slab and conventional slab data

Particular	Description
Floor dimension	6.3x6.8m
Service load	2 KN/m^2
Concrete	M-20
Steel	Fe-415
Conventional roof	RCC
Tile	Hollow clay tile
Tile dimension	350x250x65mm

Table 5 Calculation of embodied energy of conventional slab and filler slab

Particulars	Quantities of Conventional Slab (Kg)	Embodied Energy for Materials (MJ/kg)	Embodied Energy for Conventional Slab (MJ)	Quantities of Filler Slab (Kg)	Embodied Energy for Filler Slab (MJ)
Cement	2556.60	5.2	13294.34	2264.45	11775
Fine Aggregate	3854.11	0.10	404.68	1252.74	131.53
Course Aggregate	7708.28	0.08	639.78	2505.51	207.95
Steel	129.88	32	4156.25	117.14	3748.57
Hollow Roof Tiles	N/A	6	N/A	20 Nos	120 MJ
Total embodied energy			**18494.33 MJ**		**15983.07 MJ**

Table 5 shows the results are obtained from the calculation of embodied energy of conventional slab is 18494.33 MJ and filler slab is 15983.07MJ and 13.57% is decreased in filler slab with respect to conventional slab and where it has a difference of 2,511.2646MJ.

Table 6 Calculation of carbon footprint of conventional slab and filler slab

Particulars	Embodied Energy for Conventional Slab (MJ)	Carbon Emission	Carbon Footprint for Conventional Slab (CO_2Eq to Kg)	Embodied Energy for Filler Slab (MJ)	Carbon Footprint for Filler Slab (CO_2Eq to Kg)
Cement	13294.34	0.4048	2018.50	11775	1757.97
Fine Aggregate	404.68	0.0088	3.56	131.53	1.15
Course Aggregate	639.78	0.0088	5.63	207.95	1.83
Steel	4156.25	2.196	9127.13	3748.57	8231.87
Hollow Roof Tiles	N/A	0.0056	N/A	72	0.403
Total carbon footprint			**11,154.84**		**9,993.23**

Table 6 shows the results are obtained from the calculation of carbon footprint of conventional slab is 11,154.84 CO_2/Kg and filler slab is 9,993.23 CO_2/Kg and 10.41% is decreased in filler slab with respect to conventional slab and where it has a difference of 1,161.61 CO_2/Kg.

Table 7 Calculation of cost of conventional slab and filler slab

Particulars	Quantities	Rate per Unit	Cost (Rs)	Quantities	Cost (Rs)
Cement	2556.60 Kg (50 bags)	INR 340/bag	17000	2264.45 Kg (45 bags)	15300
Fine Aggregate	3854.11 Kg (136.10C.ft)	INR 70/ Cu.ft	8166	1252.74 Kg (44.240 Cu.ft)	2654.4
Coarse Aggregate	7708.28 Kg (272.216C.ft)	INR 23/Cu.ft	6261	2505.51 Kg (88.48 Cu.ft)	2035.04
Steel	129.88 Kg	INR 51.8/ Kg	6727.94	117.14 Kg	6068.01
Hollow Roof Tiles	N/A	INR. 20/Tile	N/A	20 Nos	240
Total Cost INR			**38,154.94/-**		**26,297.45/-**

Table 7 shows the results are obtained from the calculation of the total cost of conventional slab is INR 38,154.94/- and filler slab is INR 26,297.45/- and 32% of the total cost is decreased in filler slab with respect to conventional slab.

2.3.4. Flooring material.

Kota stone flooring (with Kota stone of marble finish and granite finish*)* In spite of its advantages over marble and granite, Kota stone is much cheaper than both. This combined with its longevity and other advantages makes Kota stone a truly incredible building material. And it also certified by GRIHA as green certification. Table 8 gives the estimation cost details of flooring materials.

Table 8 Flooring material cost of both the buildings

Building type	Conventional Building	Sustainable building
Total amount of marble and its alternative material (Kota stone MF)	4,13,770.77/-	2,49,734.20/-
Total amount of granite and its alternative material (Kota stone GF)	1,24,683.3/-	72,267/-

Table 8 shows the results are obtained from the calculation of the total cost of conventional flooring is INR 538453.07/- and Kota flooring is INR 3,22,001.2/- and 41% of the total cost is decreased in Kota flooring with respect to conventional flooring.

2.3.5. ***Bamboo doors and window.***

Bamboo replaces timber frames appropriate function, the mat shutters fixed to bamboo frame bamboo boards fixed to the frame which wall can be used as a door[7]. There are some inherent properties of bamboo that makes it a preferred material for longevity and sustainability. Bamboo frames are the sustainable way to reduce the usage of hardwood in interior. Highly compressed bamboo strands give these modern frames the look of a high grade dark wood grain. They are available in different colors and sizes and cost estimation is as shown in Table 9.

Table 9 Cost of green building bamboo doors and windows and conventional building Burma/teak wood doors and windows

Doors and windows	Common Quantity	Price/piece (avg)	Amount INR	Doors and windows	Price/ piece (avg)	Amount INR
Bamboo doors	20	21000	420000	Burma wood doors	35000	700000
Bamboo window	48	15000	720000	Teak wood window	18000	864000
Total amount of doors and windows INR			**11,40,000/-**			**15,64,000/-**

Table 9 shows the results are obtained from the calculation of the total cost of conventional doors and windows is INR 15,64,000/- and green doors and windows is INR 11,40,000/- and 27% of the total cost is decreased in Kota flooring with respect to conventional.

2.3.6. Alternative paints

Low VOC paints are those that contain less VOC (Volatile Organic Compounds) or VOC Solvents than traditional coatings. The VOC solvents act to slow the initial drying by maintaining a "wet-edge" which gives a longer time to work with the product, so to overcome this application methods will also need to be altered to achieve the best finish[6]. High levels of VOC solvents contribute to the formation of pollution and reduce the indoor air quality. In order to minimize the level of VOCs released during construction and to take care of occupant health from regular paints this low VOC paints are used and its cost estimation is as shown in table 10.

Table 10 Cost of internal and external wall painting

Conventional Building	Green Building
Internal painting INR.3,25,913.64/-	Internal painting INR.3,42,683.65/-
External painting INR.123435.08/-	External painting INR.446,14/-
Total cost. INR 4,49,348.7/-	**Total cost. INR 7,88,829.65/-**

Table 10 shows the results are obtained from the calculation of the total cost of conventional paints is INR 4,49,348.7/- and green paints is INR 7,88,829.65/- and 44% of the total cost is increased in low VOC paints with respect to conventional paints.

3. Results and Discussion:

Selection and adoption of sustainable materials to the selected conventional residential building is re-calculated as per the current SR-Book (standard procedures and rates). It is found that locally available sustainable materials can solve the problem of embodied energy, carbon emission and costs where the harmful impact on the environment can be reduced. The following shows the results of the analysis for the conventional with respect to sustainable green building in aspect of cost, embodied energy and carbon footprint.

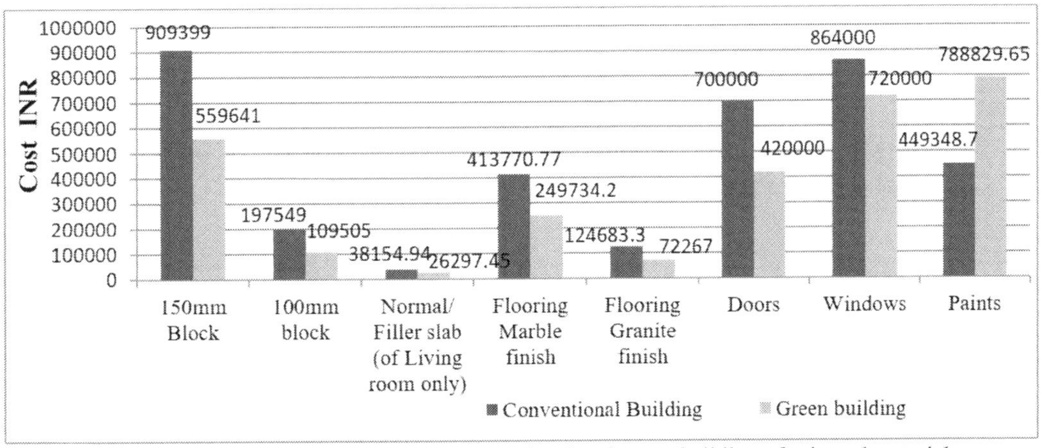

Figure (2) Cost comparisons between conventional and green building of selected materials

The cost between the conventional building and sustainable green building of selected materials is as shown in above Figure 2, where the graphs representations are based on the cost calculations of each selected material either conventional materials or green materials as shown in the tables of cost calculation of materials.

The cost calculation of the building blocks are as mentioned in the Table 3, where the calculation is done using the quantities (units are as per cubic meter) and overall cost of the building blocks of conventional and green buildings, where the Figure 2 shows the green building block work of 150mm block work is 39% lesser in price than conventional, similarly in 100mm block work 45% is lesser.

The cost calculation of the normal slab and filler slab is mentioned in the Table 7, where the calculation is done using the quantities (of cement, fine aggregate, coarse aggregate steel and hollow roof slab) and overall cost of the slab of conventional and green buildings, where the Figure 2 shows that green building filler slab is 32% lesser in price than conventional normal slab.

The cost calculation of the flooring material is as mentioned in the Table 8, where the calculation is done using the quantities (of cement, fine aggregate, coarse aggregate steel and hollow roof slab) and overall cost of the filler slab of conventional and green buildings, where the Figure 2 shows that green building filler slab is 40% lesser in price than conventional flooring.

The cost calculation of the doors and windows materials are as mentioned in the Table 8, where the calculation is done using the quantities (units in numbers) and overall cost of the doors and windows of conventional and green buildings, where the Figure 2 shows that green building bamboo doors and windows is 28% lesser in price than conventional Burma/teak wood.

The cost calculation of the wall paints is as mentioned in the Table 10, where the calculation is done using the quantities and overall cost of the normal paint of conventional and low VOC paints of green buildings, where the Figure 2 shows that green building low VOC paint is 22% greater in price than conventional normal paints.

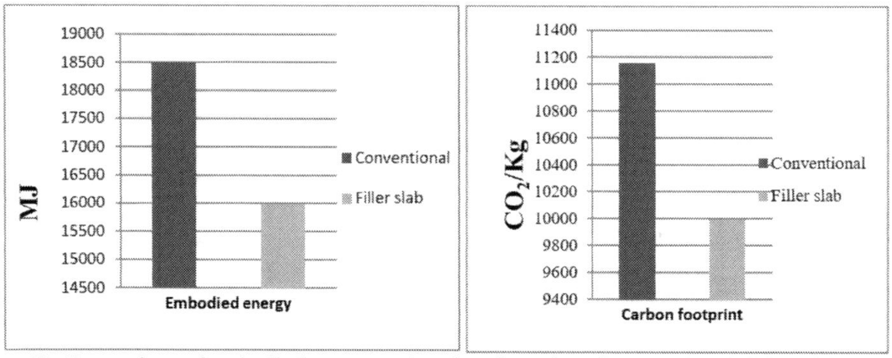

Figure (3) Comparison of embodied energy and carbon footprint of conventional slab and filler slab

The embodied energy and carbon footprint comparison between conventional slab and fillers slab of both the conventional and sustainable green building are as mentioned in the Table 5 and 6, where the Figure 3 is the graph representation of embodied energy and carbon footprint comparison is as shown.

After the selection of necessary sustainable materials, estimating the quantities and its rate analysis are prepared and comparison graphs are represented and an abstract of green building is considered at every stage of work like excavation, PCC, plastering, RCC, flooring, painting and etc. In overall cost estimation comparison between Conventional building and Sustainable green building, and its difference is Rs.19,48,685/- that is 15.2% lesser cost after its complete estimation. In overall cost estimation comparison between Conventional building and Sustainable green building, and its difference is Rs.19,48,685/- that is 15.2% lesser cost after its complete estimation.

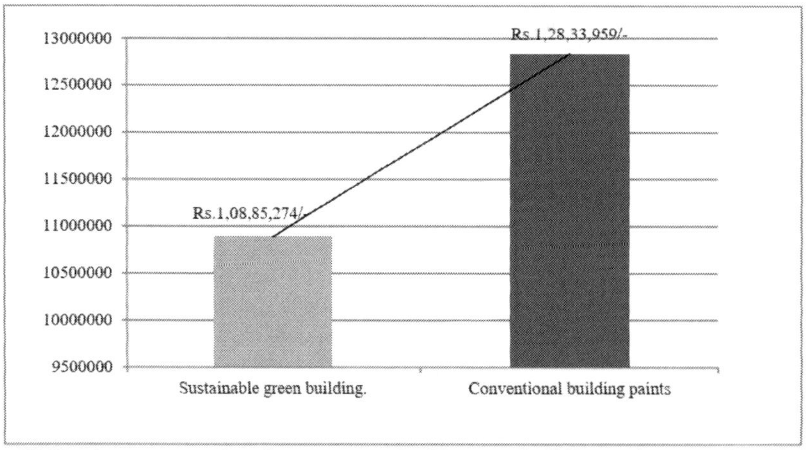

Figure (4) Total cost comparison between sustainable green building and conventional building.

Embodied energy calculation of complete building is calculated by considering the energy embodied in MJ of each work at each stage as shown in Table 11. For this considering embodied energy of each material is necessary. Embodied energy calculation of conventional building at each stage and every by considering embodied energy of each material of entire work.

Embodied energy and carbon foot print comparison between the sustainable green building and the conventional building. And the graph is as per the calculation.

IOP Conf. Series: Materials Science and Engineering 1166 (2021) 012038 doi:10.1088/1757-899X/1166/1/012038

Table 11 Embodied energy calculations

Particulars	Embodied energy of conventional building (in MJ/Unit)	Embodied energy of green building (in MJ/Unit)
PCC	30642.26	29817.76
RCC of slab, column, beam, footing, Cheju, lintels & stairs	34626.98	33856.48
Painting	107712.5	30445
Block work	351605.8	219999.4
Excavation	345375.8	213529.4
Plastering	184600.1	184243.7
Total	**869963.34**	**527648.04**

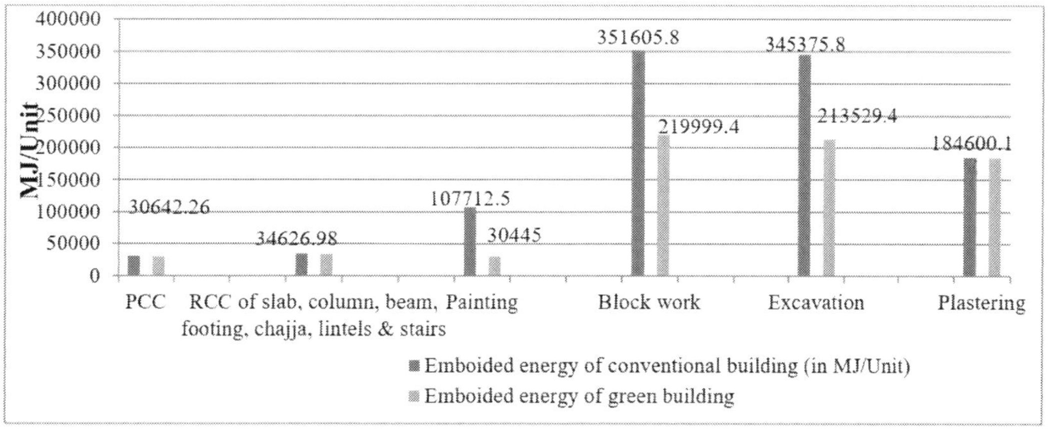

Figure (5) Total embodied comparison between sustainable green building and conventional building.

The calculated embodied energy data of each item in MJ/unit of the conventional building and sustainable green building is shown in figure (9), where total embodied energy of conventional building is 869963.34 MJ/unit and sustainable green building is 527648.04 MJ/units, where the difference is 342315.3 MJ/units, sustainable green building is 40% lesser in consumption of embodied energy.

Table 12 Carbon footprint calculations

Particulars	Carbon footprint of Conventional building (CO_2 /kg)	Carbon footprint of Sustainable building (CO_2 /kg)
Reinforcement Steel	387800.000	387800.000
Concrete	1053346.875	561122.94
Blocks 4inch	267730.800	97032.496
Blocks 6 inch	115751.160	36981.048
Plastering	56006.280	11724.526
Finishing	72771.000	24644.730
Flooring	316798.400	138599.30
External Cement Plastering	455738.400	455738.400
External Painting	14859.000	14859.000
Pavers	282.880	216.320

The calculated carbon footprint data of each item in CO_2/kg of the conventional building and sustainable green building is shown in table 11

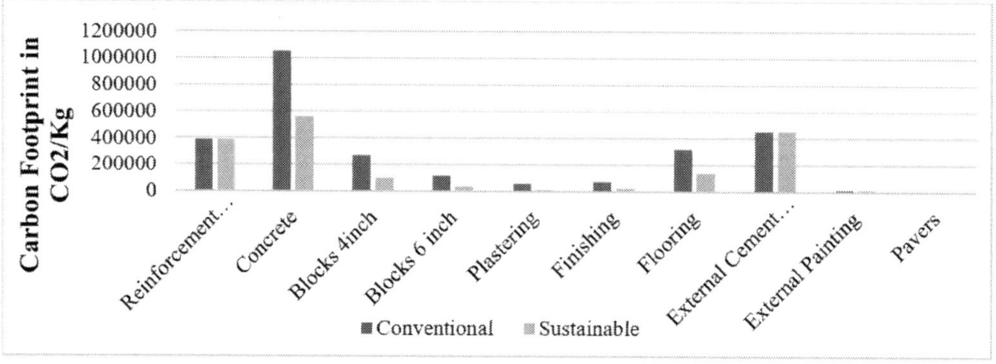

Figure (6) Total carbon footprint comparison between sustainable green building and conventional building

From the results presented in above graphs, it can be observed that the sustainable green building is holds good in every comparison than the conventional building Where here in graph each and every particulars of construction stage has lesser carbon footprint due to the less carbon footprint of each selected alternative building material like kota stone, mine waste bricks, bamboo materials, low VOC paints etc.

4. Conclusion.

1. The building block from the mine waste is 35% to 45% less in price compared to conventional clay brick.

2. The filler slab tiles has reduced the 25% of concrete consumption in slab construction in every floors and also cost is reduced 32% lesser than conventional building slab.

3. In flooring, Kota stone marble and granite finish flooring cost is 40% lesser than conventional flooring and also looks similar.

4. The Bamboo doors and windows cost is 28% lesser than Burma/Teak wood and suits to say sustainably green.

5. Low VOC paint and another partial external elevation and wall painting is replaced by Green wall (vertical gardening) which cost is 22% greater than conventional building wall painting.

6. Overall construction cost is also 15.2% lesser than conventional as per the abstract between re-estimated conventional building and the sustainable green building.

7. The overall building carbon foot print and the embodied energy is 39.3% of sustainable green building is lesser than the conventional one.

References

[1] Anjali Guptha., 2017. Planning, Analysis, Design and Management of Multi-Storeyed Green Building- published thesis of masters in engineering
[2] Gupta, A., 2017. Building a Green Home Using Local Resources and Sustainable Technology in Jammu Region–A Case Study. Energy Procedia, 115, pp.59-69
[3] Robichaud, L.B. and Anantatmula, V.S., 2011. *Greening project management practices for sustainable construction*. Journal of management in engineering, 27(1), pp.48-57
[4] Reffat, R., 2004, February. *Sustainable construction in developing countries*. In Proceedings of First Architectural International Conference, Cairo University, Egypt
[5] Beulah, Clifford, Jashwanth and Prajwal 2018, *Utilization of mine waste for the preparation of mine waste*, thesis in Christ (Deemed to be University) Faculty of Engineering.
[6] Allen (2016) *Twenty-Five Years of Green Chemistry and Green Engineering*: The End of the Beginning, ACS organisation journal.
[7] Tejasheker (2016) *Selection and implementation of Green materials and green techniques for the sustainable development of a residential layout*, thesis in MSRUAS.
[8] Venkatarama Reddy, B.V., 2009. *Sustainable materials for low carbon buildings*. International Journal of Low-Carbon Technologies, 4(3), pp.175-181

Microwave-assisted extraction of chlorophyll and polyphenol with antioxidant activity from *Pandanus amaryllifolius* Roxb. in Vietnam

Nguyen Hong Khoi Nguyen[1]*, Nguyen Thi Diem An[1], Phan Kim Anh[2] and Tran Thanh Truc[3]

[1] Faculty of Environmental and Food Engineering, Nguyen Tat Thanh University, Ho Chi Minh City, Vietnam
[2]Chemical and Environmental Engineering Faculty, Lac Hong University, Vietnam
[3] Food Technology Department College of Agriculture, Can Tho University, Vietnam

*Corresponding author's email: *nhknguyen@ntt.edu.vn*

Abstract. *Pandanus amaryllifolius* Roxb., also called as pandan leaf, is distributed mainly in hot and humid climate areas. Pandan leaf, grown very popular in Vietnam, contains many biological value compounds such as chlorophyll and polyphenol with oxidation prevention activity. Microwave-assisted extraction (MAE) has advantages such as reduced extraction time and high extraction efficiency in comprison to other extraction technologies. The aim of this study was to assess the effect of microwave-assisted extraction conditions on chlorophyll and polyphenol content as well as antioxidant activity of pandan leaf. Research results show that acetone solvents of 90°, pandan powder : acetone ratio of 1:30, microwave capacity of 300W, microwave-assisted time of 2 minutes are parameters for the highest extraction efficiency. With these conditions, chlorophyll a, b and total chlorophyll content are respectively 9.4278 µg/mL, 4.2460 µg/mL, 13.6738 µg/mL, polyphenol content is 2.7577 g/L and DPPH radical collection activity 1409.51 µmol/L.

Keywords: microwave-assisted extraction, *Pandanus amaryllifolius* Roxb., chlorophyll, polyphenol, DPPH, radical-scavenging activity

1. Introduction

Pandanus amaryllifolius Roxb., also known as pandan leaf [6], belongs to the *Pandanaceae* family, *Pandanus* genus. There are about 700 species of which 52 species are found in the Philippines but only *Pandanus amaryllifolius* Roxb. and *Pandanus odoratissimus* Linn. are scented species [7], [8]. *Pandanus* genus is grown primarily in hot and humid climate areas such as Africa and the Pacific. *Pandanus* genus leaves are called with different names such as pandan wangi (Malaysia), Toei hom (Thailand), daun pandan (Indonesia), pandan mabango (Philippines), and sticky leaves (Vietnam).[7]. *Pandanaceae* family trees are tall or brush-like plants [9]. The *pandanus* genera are dust-sized trees that are 1 m to 20 m high, with broad foliage and moderately developed [9]. Pandan leaves are pale green to dark green, with a leaf's length of about 30 cm and a width of about 6 cm [9]. Pandan leaves (*Pandanus amaryllifolius* Roxb.) do not produce fruit because they undergo vegetative propagation [10]. Since their sufficient chlorophyll content, pandan leaves that are also a common green color is used in foods [11].

The important chemical component to create the fragrance of pandan leaves is 2-acetyl-1-pyrroline, the volatile flavoring components include hexanone, 2-hexanone, 3-methyl pyridine, 2-penten-1-ol, nonanone, benzaldehyde and linalool [12]. In some other studies, the chemical composition of pandan leaves also contains maltodextrin [10]. In addition, pandan leaves also contain quercetin [13], carotenoids, tocopherols, tocotrienols, and essential oil content [14]. Besides, pandan leaves comprize phytochemicals such as steroids, carbohydrates, polyphenols, alkaloids, flavonoids, saponins, tannins [15].

Chlorophyll is discriminated by a small reversal of substituents on the B ring from methyl to formyl. Chlorophyll a and chlorophyll b coexist in plants in a 3 : 1 ratio with the dominant methyl (chlorophyll a) variant. Chlorophyll a contains methyl substituents -CH3 which are less stable and have greener properties than chlorophyll b containing formyl substituents -CHO [2]. According to some other studies, plants in shade have a higher proportion of chlorophyll b compared to chlorophyll a. This difference was observed in sunlight of some other types of leaves, as well as when a single species was grown under different light intensities [3].

Polyphenols are a large family of naturally occurring organic compounds that are abundant in plants [4]. Polyphenols are polyhydroxylated phytochemicals and they represent a range of similarly structural compounds. Polyphenols are synthetic by shikimate and acetate/polyketide pathways, and they are plant secondary metabolites. Glucose metabolism produces acetic acid and shikimic acid [5]. The physical and chemical properties of secondary metabolites are distinct so the polyphenol structure is very diverse. Soluble phenolic is present in the vacuoles of plant cells, while insoluble phenolic is found in the cell walls, but the outer layers of plant cells contain higher levels of phenolic than their inner layers [4].

2. Materials and methods

2.1. Material preparation

Pandanus amaryllifolius Roxb. leaves were harvested after 3 months of planting in Long An, Vietnam. Pandan leaves were washed, dried, powdered, sieved and refrigerated at 4°C during the study. Acetone, ethanol, methanol and Na_2CO_3 were originated from China; Folin-Ciocalteu, 2,2-diphenyl-1-picrylhydrazyl (DPPH) and Trolox were bought from Sigma Aldrich, USA; second distilled water was prepared before performing experiments in the laboratory.

2.2. Procedure

To evaluate the effect of the extraction conditions, 2 g of pandan leaf powder (*Pandanus amaryllifolius* Roxb.) was extracted under the investigating conditions - solvents (acetone 90°, methanol 100°, ethanol 96°); microwave powers (100 W, 300 W, 400 W, 600 W, 800 W); pandan powder-to-solvent ratios (1:15, 1:30, 1:45) and microwave-assisted extraction times (30 seconds, 1 minute, 2 minutes, 3 minutes, 4 minutes). The extract was descaled by filtering with Whatman No.1 filter paper and then diluted with suitable solvent used in extraction process to 100 mL. The obtained filtrate is used for further analysis of chlorophyll, polyphenol content and DPPH radical-scavenging activity.

2.3. Analytical method

2.3.1. Determination of the method of chlorophyll. Chlorophyll a and b are absorbed with maximum light at 662 - 665 nm, minimum at 645 - 652 nm, which represents the intensity of pigment or amount of chlorophyll in extract. Based on this property, chlorophyll can be quantified by spectrometric method [1]. For each solvent, the chlorophyll content was calculated by the formula:

Calculation formula (using acetone 90°) [29]

$$\text{Chlorophyll a content } (\mu g/mL) \approx 11.93 \times A_{664} - 1.93 \times A_{647}$$

$$\text{Chlorophyll b content } (\mu g/mL) \approx -5.5 \times A_{664} + 20.36 \times A_{647}$$

$$\text{Total chlorophyll } (\mu g/mL) \approx 6.43 \times A_{664} + 18.43 \times A_{647}$$

Calculation formula (using methanol 100°) [29]

$$\text{Chlorophyll a content } (\mu g/mL) \approx 16.29 \times A_{665} - 8.54 \times A_{652}$$

$$\text{Chlorophyll b content } (\mu g/mL) \approx -13.58 \times A_{665} + 30.66 \times A_{652}$$

$$\text{Total chlorophyll } (\mu g/mL) \approx 2.71 \times A_{665} + 22.12 \times A_{652}$$

Calculation formula (using ethanol 96°) [30]

$$\text{Chlorophyll a content } (\mu g/mL) \approx 11.75 \times A_{662} - 2.53 \times A_{645}$$

$$\text{Chlorophyll b content } (\mu g/mL) \approx -18.61 \times A_{662} - 3.96 \times A_{645}$$

$$\text{Total chlorophyll } (\mu g/mL) \approx 7.79 \times A_{662} + 16.08 \times A_{645}$$

2.3.2. Determination of polyphenol content. The total polyphenol content (TPC) was determined by the Folin-Ciocalteu method, the reaction in blue [28]. The obtained extract was diluted with distilled water, 1 mL of diluted solution was drawn into a test tube, 1 ml of Folin-Ciocalteu reagent was added, waited for 5 minutes, then 1 mL of Na_2CO_3 was added and put in dark place for 30 minutes. Then poured into the cuvette and optically measured at a wavelength of 765 nm.

2.3.3. Determination of the antioxidant activity (DPPH). DPPH (2,2-Diphenyl-picrylhydrazyl) method is used to find antioxidant activity of panda leaf extract.The purple color of DPPH free radicals is changed depending on the antioxidant capacity of the extract, the higher the antioxidant capacity of the extract, the more purple color is lost [19]. The standard curve is contructed by using Trolox. The operation consists of 4 steps as follows. First, extracted solution was diluted with methanol with the appropriate dilution factor. Second, 0.2 mL of diluted solution and 3 mL of DPPH reagent were drawn into a test tube. Third, the reaction was taken place for 30 minutes in the absence of light. Finally, the post-reaction solution was poured into the cuvette for photometric measurement at 515 nm.

2.3.4. Data analysis. All results of the experiment were performed in triplicates. The results are calculated as the average of 3 runs, the values are shown with the standard deviation. IBM SPSS Statistics 20 were used to get data analysis. One-factor variance analysis ANOVA was used to test the mean hypotheses of the sample groups with the probability of error of only 5%..

3. Results and discussion

3.1. Effect of acetone 90°, methanol 100° and ethanol 96° on the extraction process

The solvent kinds have a significant effect on the extracted substance content and antioxidant activity of the extract

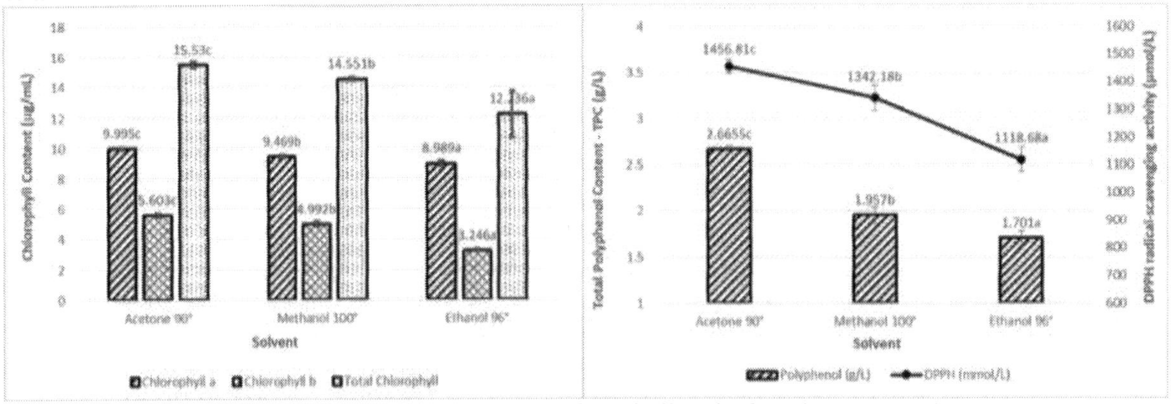

Figure 1. Effect of solvent kinds on chlorophyll, polyphenol content and DPPH radical-scavenging activity.

Figure 1 shows different solvents resulting in different chlorophyll, polyphenol content and DPPH free radical-scavening activity in the extracts. The chlorophyll a (9.995 ± 0.07 µg/mL), chlorophyll b (5.603 ± 0.15 µg/mL); total chlorophyll (15.53 ± 0.21 µg/mL); polyphenol content (2.6655 ± 0.04 g/L) and DPPH radical removal activity (1456.81 ± 26.62 µmol/L) were the highest values when extracted with acetone 90° and the lowest ones when extracted with ethanol 96°. The relative static permittivity or the dielectric constant of a solvent is a relative measure of its chemical polarity. The difference of the relative static permittivity will lead to the variance polarization of the extraction solvent. The strongly polarized solvent has a large dielectric constant; the less polar the solvent, the smaller the dielectric constant. When the solvents are combined, the relative static permittivity and the polarity of the solvent mixture change, the solubility for the solutes is much greater than that of the individual solvents [20]. Therefore, when combining water with different organic solvents, it is possible to increase the efficiency of extraction of solvents on some extracts. This outcome is similar to the regconized result that the acetone-water mixture is the extraction solvent that get the highest yield with a number of polyphenols, as well as flavonoids and anthocyanins [21]. The research of Ritchie, 2008 on extracting chlorophyll from spinach and some algae by acetone, methanol, ethanol with concentrations respectively of 90°, 100° and 100°, acetone 90° also gives the highest extraction efficiency [30]. The study of chlorophyll extraction on the ingredients Lemna minor (duckweed), Potamogeton crispus, and Egeria densa with acetone 80°, acetone 90° and ethanol 95° also shows that acetone 90° for the highest extraction yield [22].

3.2. *Effect of microwave power on the extraction process*

The microwave power produces the necessary heat to be applied to the extracted sample, so the microwave power significantly affects the extraction process.

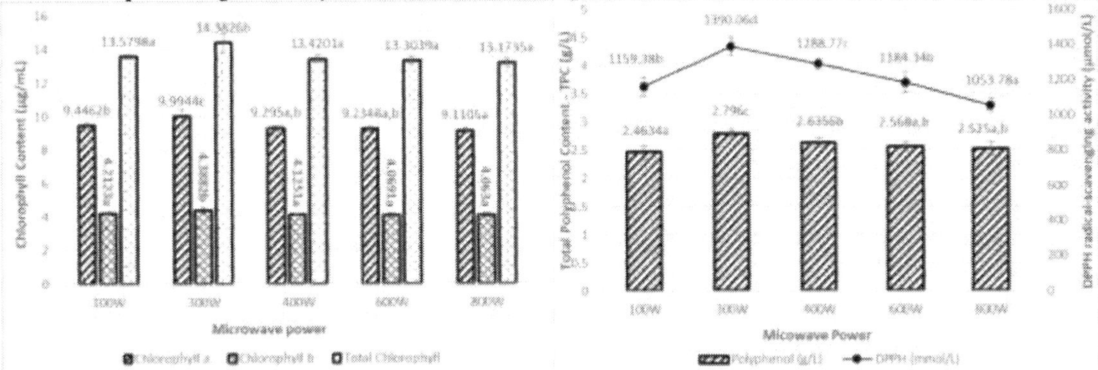

Figure 2. Effect of microwave power on chlorophyll a, b; total chlorophyll; polyphenol content and DPPH radical-scavenging activity.

Figure 2 shows that at 300W the chlorophyll a (9.9944 ± 0.41 µg/mL), chlorophyll b (4.3882 ± 0.162 µg/mL); total chlorophyll (14.3826 ± 0.566 µg/mL); polyphenol content (2.796 ± 0.07 g/L) and DPPH radical removal activity (1390.06 ± 54.19 µmol/L) are the highest values. Each power of microwave oven generates a certain temperature impacting on the sample in the microwave oven. Microwave radiation causes interruption of plant tissue, releasing compounds having in the materials into solvents [23], [24]. Microwave power affects the interaction between particles and solvents, the rate of solvent diffusion into solid samples [18]. Acetone 90° has a boiling point of 57.4°C [25], so at a power of 100 W, the temperature is not sufficient to penetrate all the molecular particles resulting in an extraction efficiency not as high as at 300 W. At the higher microwave power,(400, 600, and 800W) the higher thermal energy causes thermal degradation, reducing chlorophyll, polyphenol content and DPPH radical-scavenging activity. In the study of (Nguyen Thi Hanh, 2016), effect of thermal processing on chlorophyll and vitamin C content in peas (*Pisum sativum*) also shows that when blanching temperature increases to 90°C, chlorophyll content reduces. According to another research, Mustapa N A also uses the microwave power of 300W to extract chlorophyll from *Clinacanthus nutans Lindau* (C. nutans) [27].

3.3. Effect of pandan powder : solvent ratio on the extraction process

The ratio of pandan powder : solvent affects the difference in solute concentration inside the material and outside the environment

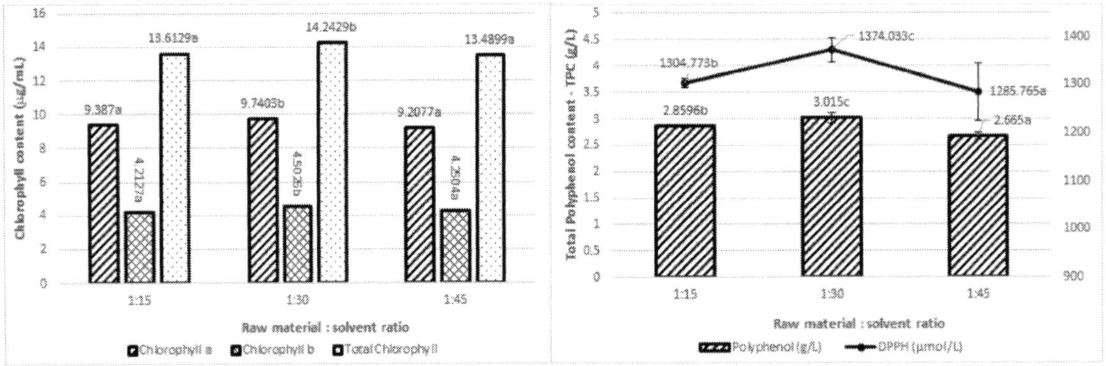

Figure 3. Effect of pandan powder : solvent ratio on chlorophyll a, b; total chlorophyll; polyphenol content and DPPH radical-scavenging activity.

Figure 3 shows that chlorophyll a (9.7403 ± 0.37 µg/mL), chlorophyll b (4.5026 ± 0.19 µg/mL); total chlorophyll (14.2429 ± 0.46 µg/mL); polyphenol content (3.015 ± 0.096 g/L) and DPPH radical removal activity (1374.03 ± 25.26 µmol/L) are the highest values at pandan powder : solvent ratio of 1:30 in the extraction. In microwave-assisted extraction, the material-to-solvent ratio determines extraction effectiveness. If the volume of acetone is less than the number of solid sample, the solvent does not penetrate into the material, microwave radiation cannot affect the molecular particles much, resulting in these particles being limited to rotate, reduces extraction efficiency. If the volume of the solvent is greater than the number of solid samples, the solvent will fully absorb into the particles, microwave radiation will better affect the particles, making the particles move more strongly, the extraction efficiency should be higher. But if the solvent volume is too high, the extracted mixture will be diluted, leading to a longer thermal action time, reducing the chlorophyll, polyphenol content and the DPPH free radical scanning ability [16].

3.4. Effect of microwave-assisted time on the extraction process

Microwave-assisted extraction time generates heat that acts directly on material molecules, so microwave-assisted extraction time is one of the important factors affecting the extraction process.

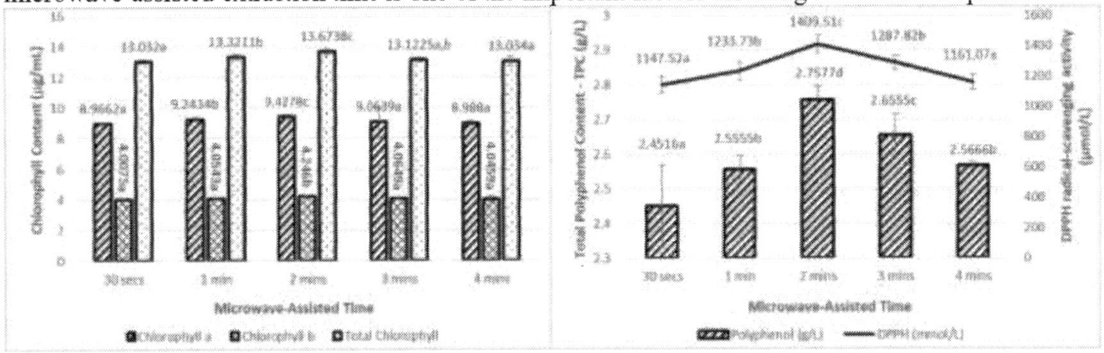

Figure 4. Effect of microwave-assisted time on chlorophyll a, b; total chlorophyll; polyphenol content and DPPH radical-scavenging activity.

Figure 4 shows that chlorophyll a (9.4278 ± 0.13 µg/mL), chlorophyll b (4.246 ± 0.07 µg/mL); total chlorophyll (13.6738 ± 0.15 µg/mL); polyphenol content (2.7577 ± 0.04 g/L) and DPPH radical removal activity (1409.51 ± 55.8 µmol/L) are the highest values obtained in 2 minutes of microwave-assisted extraction time. The power level and extraction time are the main parameters affecting the efficiency of the extraction process [26]. At the time of microwave-assisted extraction of 30 seconds, due to the short time, the microwave energy is low. At microwave-assisted time of 1 minute, radiation is more active than 30 seconds, so the extracted substance content is a bit higher. At the microwave-assisted time of 2 minutes, the extraction time is longer than 30 seconds and 1 minute, so the microwave radiation has a positive impact on the particles, causing friction and breaking down the cell walls, diffusing many soluble substances into solvent. At 3 minutes and 4 minutes, microwave radiation strongly affects the particles, causing more friction, resulting in more impact temperatures, which degrades some extracted compounds. Prolonged microwave support can lead to the decomposition of target compounds along with the overheating of soluble substances [17]. Microwave-assisted extraction time of 2 minutes is also the result of the study of F. Dahmoune when extracting phenolic compounds from pistachio leaves (*Pistacia lentiscus*) [31].

4. Conclusion

Pandan leaf is a very popular plant grown in Vietnam. Factors affecting chlorophyll, and polyphenol extraction, as well as DPPH antioxidant activity of extract were investigated, they are solvent kinds, microwave capacity, pandan powder : acetone ratio, and extraction time used with microwave. The research resulted that the used solvent was acetone 90°; microwave capacity was 300W; pandan powder : acetone ratio was 1:30 and used time with microwave was 2 minutes. These parameters brought highest amounts of chlorophyll a, b; total chlorophyll; polyphenol content and DPPH radical-scavenging activity.

Acknowledgements
This study was done at Nguyen Tat Thanh University Ho Chi Minh City, Viet Nam

References

[1] Palta J P 1990 Leaf chlorophyll content *Remote sensing reviews* **5** 207
[2] Humphrey A M 2004 Chlorophyll as a color and functional ingredient *J. of Food Sci.* **69** C422
[3] Boardman N K 1977 Comparative photosynthesis of sun and shade plants *Annual Review of Plant Physiology* **28** 355
[4] Rajbhar K A R I S H M A, Dawda H I M A N S H U and Mukundan U S H A 2015 Polyphenols: methods of extraction *Sci. Revs. Chem. Commun* **5** 1
[5] Anesini R F C and Ferraro G E 2008 Total polyphenol content and antioxidant capacity of commercially available tea (Camellia sinensis) in Argentina *J. of A. and Food Chem.* **56** 9225
[6] Jimtaisong A and Krisdaphong P 2013 Antioxidant activity of Pandanus amaryllifolius leaf and root extract and its application in topical emulsion *Tropical Journal of Pharmaceutical Research* **12.3** 425
[7] Nonato M G, Takayama H and Garson M J 2008 Pandanus alkaloids: chemistry and biology *The Alkaloids: Chemistry and Biology* **66** 215
[8] Ghasemzadeh A and Jaafar H Z E 2013 Profiling of phenolic compounds and their antioxidant and anticancer activities in pandan (Pandanus amaryllifolius Roxb.) extracts from different locations of Malaysia *BMC complementary and alternative medicine* **13** 1
[9] Routray W and Rayaguru K 2010 Chemical constituents and post-harvest prospects of Pandanus amaryllifolius leaves: a review *Food Reviews International* **26** 230
[10] Jusril M S A N A B, Mamat A S and Aslam M S 2016 Effect of drying methods on the antioxidant properties of Pandanus amaryllifolius *Indian Research Jounal of Pharmacy and Science* **3** 706
[11] Seng K L, Che Man Y B, Tan C P, Osman A, and Hamid N S A 2005 Process optimization of encapsulated pandan (Pandanus amaryllifolius) powder using the spray-drying method *J. Sci. Food Agric.* **85** 1999

[12] Wakte K V, Nadaf A B, Thengane R J and Jawali N 2009 Pandanus amaryllifolius Roxb. cultivated as a spice in coastal regions of India *Genetic resources and crop evolution* **56** 735
[13] Miean K H and Mohamed S 2001 Flavonoid (Myricetin, Quercetin, Kaempferol, Luteolin, and Apigenin) content of edible tropical plants *J. Agric. Food Chem.* **49** 3106
[14] Lee B L, Su J and Ong C N 2004 Monomeric C18 chromatographic method for the liquid chromatographic determination of lipophilic antioxidants in plants *J. Chromatogr. A* **1048** 263
[15] Aini R and Mardiyaningsih A 2016 Pandan leaves extract (Pandanus amaryllifolius Roxb) as a food preservative *Indones. J. Med. Heal.* 7 166
[16] Khaing T A 2011 Evaluation of the antifungal and antioxidant activities of the leaf extract of aloe vera (Aloe barbadensis Miller) *World Academy of Science, Engineering and Technology* **75** 610
[17] Routray W and Orsat V 2012 Microwave-assisted extraction of flavonoids: a review *Food Bioprocess Technol.* **5** 409
[18] Ma W, Lu Y, Dai X, Liu R, Hu R and Pan Y 2009 Determination of anti-tumor constitute mollugin from traditional Chinese medicine Rubia cordifolia: comparative study of classical and microwave extraction techniques *Sep. Sci. Technol.* **44** 995
[19] Bouras O B M, Chadni M, Barbaa F J, Grimia N and Vorobiev E 2015 Optimization of microwave-assisted extraction of polyphenols from Quercus bark *Industrial Crops and Products* **77** 590
[20] Moldoveanu S and David V 2014 *Modern Sample Preparation for Chromatography*, (Amsterdam: Elsevier)
[21] Hismath I, Wan Aida W M and Ho C W 2011 Optimization of extraction conditions for phenolic compounds from neem (Azadirachta indica) leaves *International Food Research Journal* **18** 931
[22] Su S, Zhou Y, Qin J G, Yao W and Ma Z 2010 Optimization of the method for chlorophyll extraction in aquatic plants *Journal of Freshwater Ecology* **25** 531
[23] Michalak K C I, Tuhy Ł and Chojnacka K 2015 Seaweed extract by microwave-assisted extraction as plant growth biostimulant *Open Chem.* **13** 1183
[24] Al-Harahsheh M and Kingman S W 2004 *Microwave-assisted leaching - A review*, Hydrometallurgy **73** 189
[25] Howard K S and McAllister R A 1958 The viscosity of acetone-water solutions up to their normal boiling points *AICHE Journal*l **4** 362
[26] Routray W and Orsat V 2012 Microwave-assisted extraction of flavonoids: a comprehensive overview *Food Bioprocess Technol.* **5** 409
[27] Mustapa A N, Martin A, Mato R B and Cocero M J 2015 Extraction of phytocompounds from the medicinal plant Clinacanthus nutans Lindau by microwave-assisted extraction and supercritical carbon dioxide extraction *Industrial Crops and Products* **74** 83.
[28] Alonso A M, Guillen D A, Barroso C G, Puertas B and Garcia 2002 A determination of antioxidant activity of wine byproducts and its correlation with polyphenolic content *Journal of Agricultural and Food Chemistry* **21** 5832
[29] Ritchie R J 2006 Consistent sets of spectrophotometric chlorophyll equations for acetone, methanol and ethanol solvents *Photosynthesis research* **89** 27
[30] Fabrowska J, Messyasz B, Szyling J, Walkowiak J and Łęska B 2018 Isolation of chlorophylls and carotenoids from freshwater algae using different extraction methods *Phycological Research* **66** 52
[31] Dahmoune F, Spigno G, Moussi K, Remini H, Cherbal A and Madani K 2014 Pistacia lentiscus leaves as a source of phenolic compounds: Microwave-assisted extraction optimized and compared with ultrasound-assisted and conventional solvent extraction Industrial Crops and Products **61** 31

Evaluation of antibacterial properties of some essential oils from Vietnam

D N Do[1*], D P Nguyen[1], T T A Tran[2,3], D Q Nguyen[2,3], N M T Lai[2], T H Huynh[4], V M Do[4], X C Luu[5] and X T Le[2]

[1] Faculty of Chemical Engineering and Food Technology, Nguyen Tat Thanh University, Ho Chi Minh City, Viet Nam.
[2] Department of Chemical Engineering, Ho Chi Minh City University of Technology, Ho Chi Minh City, Vietnam
[3] Biomass Laboratory, Ho Chi Minh City University of Technology (HCMUT), 268 Ly Thuong Kiet Street, District 10, Ho Chi Minh City 700000, Vietnam
[4] Institute of Environmental Technology, Vietnam Academy of Science and Technology, No. 18 Hoang Quoc Viet Road, Cau Giay Dist., Hanoi, Vietnam.
[5] Faculty of Drug, Cosmetic and Food, Dong An Polytechnic, 90, 30-4 Street, Di an City, Binh Duong Province, Vietnam.
*Corresponding author: ddnhat@ntt.edu.vn

Abstract. Natural products have attracted attention from scientists because of their potential to replace antibiotics in the near future. Thanks to Vietnam's richness in biodiversity, this study aimed at evaluating antibacterial activity of eleven essential oils from various plants. These essential oils were collected from various plants across Vietnam such as corn mint, clove, clove basil, eugenol, tea tree, terpinen-4-ol, kaffir lime, Mexican mint, rosemary, *Homalomena occulta,* and betel. By chemical reaction and fractional vacuum distillation method, eugenol and terpinen-4-ol was isolated from clove basil and tea tree oils with the content reaching up to 97.0% and 95.8%, respectively. The antibacterial activity against *Staphylococcus aureus*, *Bacillus subtilis, Escherichia coli* of essential oils was determined by disk diffusion and broth dilution method. As a result, Mexican mint oil and eugenol exhibited high antibacterial properties with low MICs and MBCs (62.5 to 125 μL/L), whereas other essential oils (clove, clove basil, betel, tea tree, terpinen-4-ol) had higher MIC and MBC values ranged from 125 to 500 μL/L.

1. Introduction

From ancient times to the modern era, mankind has been able to adapt and thrive with precious help from plants. Of all six kingdoms on earth, plants, which belong to Plantae, can produce secondary metabolites. These compounds help plants survive in their environment by acting as attractants of pollinators. Chemical defenses against micro-organisms, insects, and predators [1-2] take advantage of this secret weapon from plants; human collects these phytochemicals in the form of essential oils through hydro-distillation. Essential oils possess various biological activities from antimicrobial, antibiotic, insecticidal properties to highly important pharmaceutical activities [3]. They are a major source of antibiotics, perfumes, cosmetics as well as food additives. Fortunately, Vietnam is a wonderful place to cultivate various medicinal plants such as *Ocimum gratissimum* L, *Syzygium aromaticum*, *Rosmarinus officinalis, M. arvensis, Citrus hystrix, Piper betle, Homalomena occulta* (Lour.) Schott, *Melaleuca alternifolia, Coleus amboinicus*. They are well-known for their antibacterial and anti-inflammatory properties and are traditionally used to treat oral diseases [4].

Oral diseases are one of the most harmful diseases globally. According to the Global Burden of Disease (GBD) 2010 study, around 3.9 billion people worldwide live with dental conditions, such as dental caries (tooth decay), periodontal (gum) disease, and oral cancer. Untreated caries in permanent teeth were the most common (35% for all ages combined), whereas severe periodontitis and untreated caries in deciduous teeth were the 6th and 10th most prevalent conditions, 11% and 9%. Severe tooth loss was the 36th most prevalent condition, with a global estimate of 2% [5]. These oral diseases have substantially reduced the life quality of patients due to chronic pain and economic burden. Oral disease affects different life stages of humans from children, adults to older people. Due to a lack of awareness and healthcare service, a large group of people in developing countries have suffered from these dental conditions. Particularly in Vietnam, there is a high number of patients who experience dental caries [6].

Therefore, essential oils extracted from plants are a valuable source to solve this problem. These active ingredients can be added to oral caring products like mouthwashes and toothpaste to treat oral diseases. The purpose of the research is to determine the antibacterial properties of some essential oils in Vietnam and to orient their application to mouthwash products.

2. Materials and Methods

2.1 Source of Essential Oil
Essential oils were collected from various plants like *Ocimum gratissimum L, Syzygium aromaticum, Rosmarinus officinalis, M. arvensis, Citrus hystrix, Piper betle,Homalomena occulta (Lour.) Schott, Melaleuca alternifolia, Coleus amboinicus*. These plants were cultivated in Vietnam..

2.2 Eugenol and terpinen-4-ol isolation
Eugenol isolation procedure: 50 g of clove basil oil was added into the mixture of saturated KOH and alcohol solution (pH = 14). The volume ratio of oil and KOH was 1:1. The reaction was carried out at 45°C and stirred for 30 minutes. After that, the reaction mixture was then inserted separating funnel and allowed to stand for a while to form two layers. The bottom layer was K-eugenolat (aqueous layer) and the top was the other compounds. K-eugenolat was then acidified with HCl (36.5%) until pH = 4. The eugenol was formed at the bottom layer. The oil was then purified by hydro-distillation using the Clevenger-type apparatus. The yield of eugenol was 69.24 %. The percentage of eugenol was analyzed by GC/MS: from 81.1% (from clove basil oil) up to 97.0%.

To isolate terpinen-4-ol from tea tree oil, a vacuum fractional distillation method was applied. This method was suitable for this experiment as it was used for separating elements which had different boiling points. Furthermore, the system was operated at vacuum condition, which allowed the terpinen-4-ol to evaporate at low temperature and prevent it from decomposition. The system was operated at 60 mmHg and 115 °C (top pressure) with 300 mm Hempel column. With 100 mL tea tree oil (39.4% terpinen-4-ol), 31.1 g terpinen-4-ol was obtained. GC-MS analysis revealed the percentage of purified terpinen-4-ol up to 95.8%.

2.3 GC-MS analysis
The analysis of essential oils was performed using a GC-MS G1530A-serial US00002778. An electron ionization system with an ionization energy of 70 eV was used. The column was maintained at 40 °C for 1 min and then programmed to 280 °C at 10 °C/min and held for 3 min. The injector was 250 °C. The carrier gas was helium introduced at a rate of 1.0 ml/min. A volume of 0.1 μl of essential oil was injected manually.

2.4 Tested micro-organism and culture
The essential oils were individually tested against a panel of micro-organisms, including Gram-positive Bacillus subtilis, Staphylococcus aureus, and Gram-negative Escherichia coli. The reference strains of

bacteria were maintained on nutrient slants at 4 °C. All the strains were activated in nutrient agar at 37 °C for 24 h.
The products were weighed in each experiment to calculate the yield and underwent chromatographic analysis to investigate the chemical composition and the 1,8-cineole content. The following formulas calculated the yield of each fraction and the overall 1,8-cineole recovery:

2.5 Disk-diffusion assay
The essential oils were dissolved in DMSO (dimethylsulfoxide) (the ratio of DMSO: essential oil = 4: 1). The antibacterial tests were then carried out by disk diffusion method [7] with bacterial density = 108 CFU/mL. The discs (6 mm in diameter) were impregnated with 10 µL of essential oil placed on the inoculated agar. Negative controls were prepared using DMSO. Gentamicin diluted with DMSO (10 µg per disc) was used as the positive reference standard to determine the sensitivity of bacterial species tested. The inoculated plates were incubated at 37 °C for 16-14 hours. The antibacterial activity was evaluated by measuring the zone of inhibition.

2.6 MIC and MBC determination.
Based on the results of the antimicrobial activity of essential oils in the disk-diffusion assay, different concentrations of essential oils were investigated to determine MIC and MBC values by broth dilution method [8]. Essential oils with high bacterial activity at susceptibility test were collected to evaluate the MIC and MBC values. The bacterial strains B. subtilis, E. coli, S. aureus grew in nutrient broth, and they were incubated at 37°C for 48 hours. The bacterial density was adjusted at 5x105 CFU/mL. Each essential oil was diluted with tween 80 (the ratio of tween 80 and essential oil = 2:1) and water at an initial concentration of 1000 µL/L. Then the solution was diluted with nutrient broth at the concentration of 500, 250, 125, 62.5 (µL/L). Bacterial culture (0.1 mL) was added to each tube containing different concentrations of essential oils. The tube was then incubated at 37 °C for 18 - 24 hours. After that, the turbidity of each tube was observed to recognize the MIC values. The dish spread with 50 µL of bacterial culture in each tube was incubated at 37 °C for 24 hours. After that, the dish without bacterial growth was considered to be MBC value.

3. Results and discussion

3.1. Chemical constituents of nine essential oils
A total of nine essential oils was introduced in this part. They were extracted from various plants like corn mint, kaffir lime, clove, clove basil, Mexican mint, rosemary, tea tree, Homalomena occulta, and betel. The major chemical constituents of these essential oils were analyzed by GC-MS (Table 1).

Table 1. Major chemical constituents of nine essential oils.

Essential oils	**Major components**	**Content (%)**
S. aromaticum	Eugenol	70.7
	Aceteugenol	17.8
	Caryophyllene	10.4
O. gratissimum	Eugenol	81.1
M. alternifolia	Terpinen-4-ol	39.4
	γ–terpinene	25.8
	α-terpinene	15.1
C. amboinicus	Carvacrol	51.6

	3-carene	15.7
	p-methyl cumene	11.4
	Caryophyllene	11.3
C. hystrix	β-pinene	33.5
	Limonene	25.6
	Sabinene	15.5
P. betle	Chavibetol	64.4
	Limonene	11.2
M. arvensis	Menthol	72.6
H. occulta	Linalool	65.9
S. rosmarinus	1R-a-pinene	23.6
	Eucalyptol	14.8
	(-)-Verbenone	13.5

3.2. Disk diffusion assay.

Antibacterial susceptibility of essential oils against B.subtilis, E.coli, and S.aureus was presented in figure 1. Of all eleven essential oils extracted, it was clear that C. amboinicus, eugenol and O. gratissimum oil (inhibition diameters ranged from 15–20.5 mm), inhibited significantly the growth of all the micro-organisms tested when compared with positive control (Gentamicin). M. alternifolia, S. aromaticum and P. betle oil showed moderate antibacterial activity. On the other hand, *S. rosmarinus oil and H. occulta, M. arvensis and C. hystrix* oil were weakly effective against these bacterial strains. Therefore, these 4 essential oils were eliminated in further assays.

Moreover, eugenol and terpinen-4-ol, isolated from O. gratissimum and M. alternifolia oil respectively, exhibited higher zones of inhibition in their purified forms. While O. gratissimum oil (eugenol content: 81.1%) had inhibition zones ranged from 10.5-14.0 mm, eugenol (97%) had slightly higher values from 15.0 to 16.3 mm. Similar to eugenol, zone diameters of terpinen-4-ol (87%) were around 1.2 times as many as M. alternifolia oil (terpinen-4-ol content: 39.4%). These data revealed that antibacterial properties of clove basil and tea tree depended on compounds with significant structures like eugenol and terpinen-4-ol.

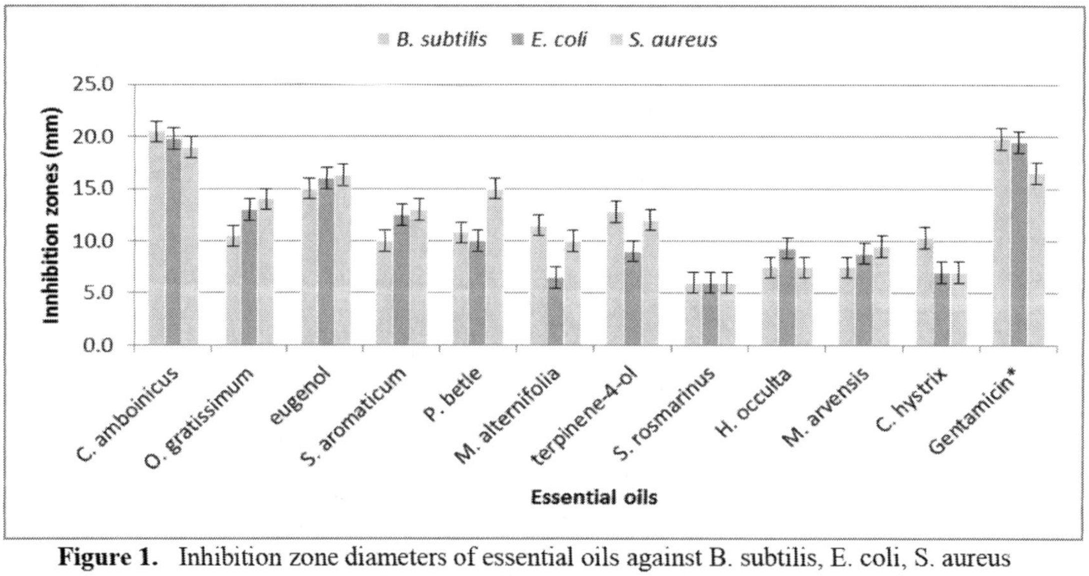

Figure 1. Inhibition zone diameters of essential oils against B. subtilis, E. coli, S. aureus

These plant extracts from the steam distillation process show antibacterial activity at a different level, which results from the chemical structures presented in each essential oil. As defense mechanisms against predators, plants have the ability to synthesize aromatic compounds that fortunately serve as natural antibiotics to humans. These secondary metabolites contain mostly phenols or their oxygen-substituted derivatives. GC-MS analysis revealed the major constituents in C. amboinicus to be dominated by phenolic compounds, specifically carvacrol with 51.58%, which alters the cell wall or disrupt the cell membrane of bacteria [9]. Thus, its structure is probably responsible for the strong growth inhibitory effect observed. P.betle oil also contained 64.4% chavibetol, another phenolic compound. Likewise, eugenol, a major component in O. gratissimum (81.1%) and S. aromaticum (70.67%) has its hydroxyl group on phenol ring, which is thought to get bind to proteins, preventing enzyme action [10]. Beside phenolic compounds, monoterpenoid like terpinen-4-ol found in M. alternifolia (39.4%) contributed to antibacterial activity of this essential oil. However, the selected bacterial strains are almost resistant to *S. rosmarinus, H. occulta, M. arvensis and C. hystrix* oil as these four essential oils have a few to no phenols.

Even though Gram-positive and Gram-negative bacteria are different in their structures, there was no obvious difference between them in the susceptibility test. This result was in agreement with the study by Ouattara [11]. One study found that Gram-positive bacteria to be more sensitive to essential oils than Gram-negative because of the relatively impermeable outer membrane surrounding Gram-negative bacteria [12]. However, another study suggested that there was a time delay in the Gram-negative growth. Therefore, over time, these essential oils would have the same effect on both Gram-positive and Gram-negative bacteria [13].

3.3. MIC and MBC values

After screening for the presence of antibacterial activity, the MICs of selected essential oils were determined by the broth macro-dilution method. Not only presenting qualitative data like the disk diffusion technique, this method also shows quantitative values of antibacterial activity. These MIC values can be serves as a guideline for adjusting an optimal concentration of drugs to treat patients, which reduces the spread of antibiotic resistance in the community.

In order to have accurate results, surfactants were introduced to lower the surface tension between essential oils and broth; thus, allow better solubility of apolar particles in broth (polar solution). Despite the fact that DMSO is more preferable to other solvents in scientific literature, Tween 80 is applied in this experiment due to its low toxicity and safety regarding food and pharmaceutical applications [14]. Furthermore, DMSO at a concentration of 4% and beyond was reported to decrease the growth of bacteria, which unfortunately causes errors in experiments [15].

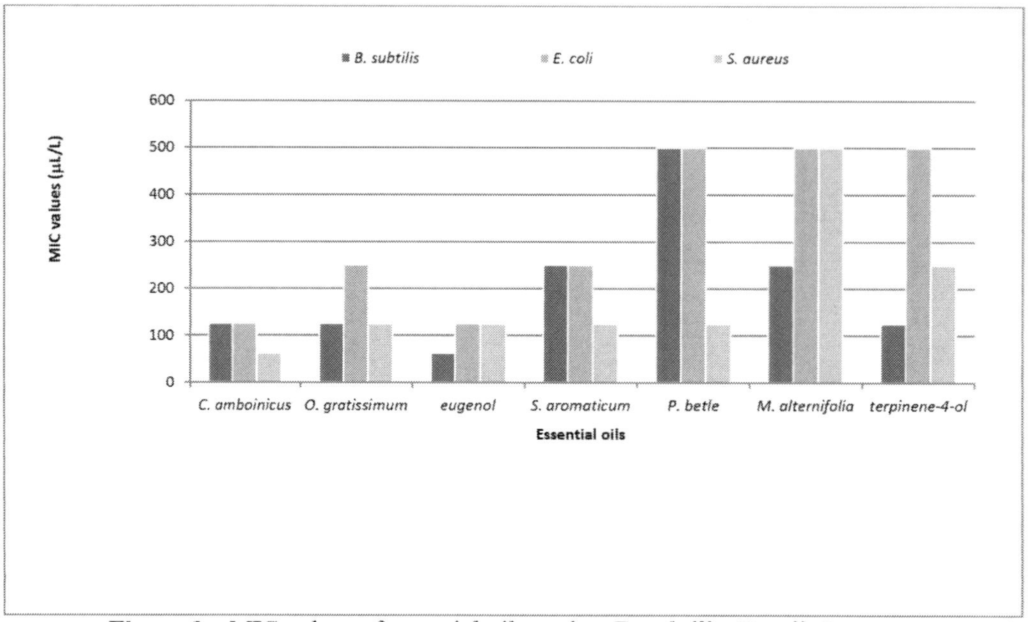

Figure 2. MIC values of essential oils against B. subtilis, E. coli, S. aureus

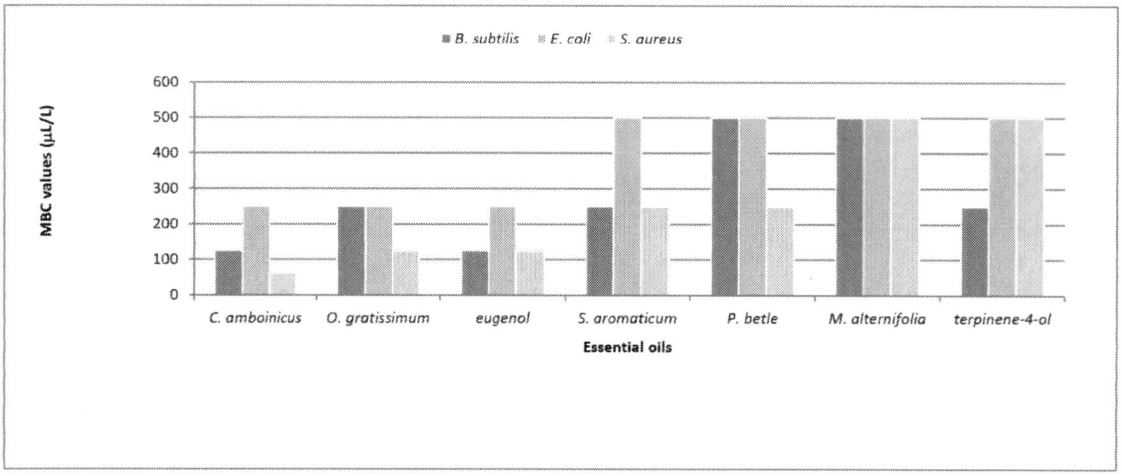

Figure 3. MBC values of essential oils against B. subtilis, E. coli, S. aureus.

As can be seen from figure 2, the MIC range of those essential oils varied from 62.5 µL/L to 500 µL/L. Generally, Mexican mint and eugenol showed low MICs (62.5-125 µL/L). The MIC of Mexican mint (125 µL/L) against *E.coli* was supported by Aguiar and partners as their result shared almost the same value, which was 128 µL/L [9].

Similarly, the MIC values of eugenol against all bacterial strains were in line with Atanasova-Pancevska's research (195µL/L) [16]. Followed by these two strong antibacterial properties, clove, clove basil, and terpinen-4-ol had higher MIC values from 125 to 500 µl/l. Both clove and clove basil had eugenol as a major constituent with 70.7% and 81.1% content, respectively. Lots of studies from Murbach Teles Andrade [17], Fu [18], and Adebolu [19] had MICs higher than these three examined essential oils. This difference can be explained by the compositions of essential oils that depend heavily on the geographical region of plant cultivation, the phenological stage of the plant and its chemotype, and the extraction of the oils.

Based on the disk diffusion method results, it was clear that the bigger the zones of inhibition were, the lower the MICs would be and the higher the antibacterial properties of EOs would exhibit. Besides, MIC values were approximately equal to or as twice as MBCs (figure 3). Therefore, Mexican mint, eugenol, as well as terpinen-4-ol could be used as active ingredients in mouthwash due to their effective antibacterial activities.

4. Conclusions

Of all the essential oils tested, Mexican mint oil, eugenol, and terpinen-4-ol strongly inhibited the growth of B. subtilis, E. coli, and S. aureus. Their application into the mouthwash product will be full of promise as it showed great antibacterial properties with a low concentration of essential oils and good sensory quality.

Acknowledgment
This work was supported by the Vingroup Innovation Foundation (VINIF) under the project code DA120_15062019/year 2019.

References

[1] Chen M S 2008 *Insect Sci.* **15**(2) 101-114
[2] Ha C T, Thai T H, Hien N T, Anh H T, Diep L N, Thuy D T, Nhat D D and Setzer W N 2019 *Nat. Prod. Commun.* **14**(6) 1934578X19860370
[3] Bakkali F, Averbeck S, Averbeck D, Idaomar M 2008 *Food Chem. Toxicol.* **46**(2) 446-475
[4] Palombo E A J E-b c and Medicine A 2011 *Evid.-Based Complementary Altern. Med.* **2011**
[5] Marcenes W, Kassebaum N J, Bernabé E, Flaxman A, Naghavi M, Lopez A and Murray C J 2013 *J. Dent. Res.* **92**(7) 592-597
[6] Khanh L N, Ivey S L, Sokal-Gutierrez K, Barkan H, Ngo K M, Hoang H T, Vuong I and Thai N 2015 *Am. J. Public Health* **105**(12) 2510-2517
[7] Saha S, Dhar T, Sengupta C and Ghosh P 2013 *Czech J. Food Sci.* **31**(2) 195-202
[8] Deshpande S and Kadam D G 2013 *AJPCR* **6**(5) 99-101
[9] Aguiar J J, Sousa C P, Araruna M K, Silva M K, Portelo A C, Lopes J C, Carvalho V R, Figueredo F G, Bitu V C and Coutinho H D 2015 *Eur. J. Integr.* **7**(2) 151-156
[10] Marchese A, Barbieri R, Coppo E, Orhan I E, Daglia M, Nabavi S F, Izadi M, Abdollahi M, Nabavi S M and Ajami M 2017 *Crit. Rev. Microbiol.* **43**(6) 668-689
[11] Ouattara B, Simard R E, Holley R A, Piette G J-P and Bégin A 1997 *Int. J. Food Microbiol.* **37**(2-3) 155-162
[12] Smith-Palmer A, Stewart J and Fyfe L 2001 *Food Microbiol* **18**(4) 463-470
[13] Tassou C C, Nychas G J E and biodegradation 1995 *Int. Biodeterior. Biodeg*radation **36**(3-4) 411-420
[14] Luo J, Li Y and Zhou M 2006 *CJIHAC* **8**(3) 1-4
[15] Wadhwani T, Desai K, Patel D, Lawani D, Bahaley P, Joshi P and Kothari V *Internet J. Microbiol.* 2009 **7**(1)
[16] Atanasova-Pancevska N, Bogdanov J and Kungulovski D 2017 *Open Biological Sciences Journal* **3**(1)
[17] Murbach Teles Andrade B F, Nunes Barbosa L, Bérgamo Alves F C, Albano M, Mores Rall V L, Sforcin J M, Fernandes A A H and Fernandes Júnior A 2016 *J. Essent. Oil Res.* **28**(3) 227-233
[18] Fu Y, Zu Y, Chen L, Shi X, Wang Z, Sun S and Efferth T 2007 *Phytotherapy res*earch **21**(10) 989-994
[19] Adebolu T and Oladimeji S A 2005 *Afr. J. Biotechnol.* **4**(7) 682-684 *genetic diversity* (Australian Centre for International Agricultural Research) p 94 Ayepola O and Adeniyi B 2008 *Journal of Applied Sciences Research* **4**(11) 1410-1413

The green synthesis of iron oxide micro/nano structures for photocatalytic applications

Rajita Ramanarayanan [1, 2], Sindhu Swaminathan[3]

[1] Department of Physics, Government Victoria College, Palakkad, Kerala, India
[2] Department of Physics, S.A.R.B.T.M Government College, Koyilandy, Kerala, India
[3] Department of Nanoscience and Technology, University of Calicut, Kerala, India

rajitaarun2006@gmail.com

Abstract. Iron oxide nanoparticles are continuously drawing researcher's attention because of their unique electronic and physicochemical properties in the field of catalysis, drug delivery and energy harvesting applications. The green synthesis of iron oxide magnetic micro/nano one dimensional structures (Fe_2O_3) using Black tea as a reducing agent was conducted in this report. The cotton was used as a template to obtain needle like structures as characterized using scanning electron microscopy (SEM). Finally the synthesized iron oxide structures showed good photocatalytic activity under solar illumination in the degradation of Methylene Blue dye.

1. INTRODUCTION

Among the metal nanoparticles, the magnetic nanoparticles have assumed great significance in the last decade due to their biomedical applications apart from their catalytic and sensory applications. In magnetic nanoparticles iron is a great interest because it is cheap, easily available and shows more magnetic properties than other metallic elements. [1].The magnetic nanoparticles like iron oxide (Fe_3O_4, Fe_2O_3) structures have been extensively studied due to their exceptional properties in nanoregime. These materials have high specific area and exhibit superparamagnetic properties to be used in targeted drug delivery and chemotherapy applications [2, 3]. Among various synthesis procedures green synthesis of iron oxide nanoparticles is being widely accepted in scientific community since the last decade due to its cost effective and environment friendly approach [4]. The use of black tea extract as reducing agent for the synthesis of iron oxide particles has been previously reported [5]. This work reports the use of black tea extract as reducing agent for the synthesis of iron oxide nano/micro structures with cotton as bio-template. The obtained iron oxide needle like structures was used in the photocatalytic degradation of methylene blue dye to be used for water purification purposes.

2. MATERIALS AND METHODS

2.1 Materials

Ferric nitrate [Fe_2 $(NO_3)_3.9$ H_2O] purchased from Merck Chemicals, black tea and surgical cotton was collected from local market The extract of the tea was prepared (Figure 1) by taking 8.00 g of tea using weight machine in 80 ml of distilled water The solution was heated at 60° C in the magnetic stirrer (IKA Basic) to get the tea extract. After the cooling, the solution was by using whattsman grade one filter paper to remove the tea grains .The filtered black tea extract was used as the reducing agent.

Figure 1. Preparation of extract (a) tea powder (b) Heating (c) Filtering the extract

2.2 Synthesis of iron oxide micro/nano needles

The iron oxide nano/microstructures (Figure 2) were prepared by taking 100 mm of 100 ml Ferric nitrate solution was taken as precursor. The cotton was soaked in black tea extract for 10 minutes so that extract was absorbed in the cotton. The surgical cotton (8cm x 6cm) was added to the precursor solution and stirred well for 10 minutes. The colour of the cotton changes to brown and it was first air dried and then dried at 60 ° C in oven. The dried template was placed inside furnace at 450 ° C for 2 hours to remove the template. On cooling we get a reddish brown powder of iron oxide.

Figure 2. Synthesis (a) precursor solution (b) template in precursor (c) iron oxide template

2.3 Photocatalysis

5 mg of the iron oxide nanopowder was mixed with 50ml of 20 ppm Methylene Blue (MB) dye stirred for two minutes and placed in solar illumination (870-880 Watt/m^2) for 6 hours. After six hours the supernatant solution was taken using a pipette and UV absorbance spectrum was taken and compared with the original dye solution.

2.4 Measurements

The UV-vis absorption measurements were recorded using spectrophotometer (Shimadzu) in the range of 200-900 nm. The solar illumination measured through Solar power meter SP 1300. Fourier transform infrared spectra (FTIR) of the extract were measured using Perkin Elmer spectrometer in the range 4000-450 cm^{-1}. X-ray diffraction data was recorded by using Cu-Kα radiation (1.5406A°) and morphology observed through field emission electron microscope (FE-SEM).

3. RESULTS AND DISCUSSION

3.1 FTIR Measurements

FTIR of the tea extract (Figure 3) corresponds to the functional groups associated with reduction of the metal salt solution. Peak at 3418 corresponds to OH groups (OH stretching) then a small peak at 2926 corresponds to carboxyl group (COO^-). A sharp peak at 1644 corresponds to H-O-H bending. Peaks at 1450, 1390, 1231 corresponds to aromatic C=C and phenolic C-O vibrations. This group corresponds to Gallic acid which is a good reducing agent in agreement with literature reports [6].

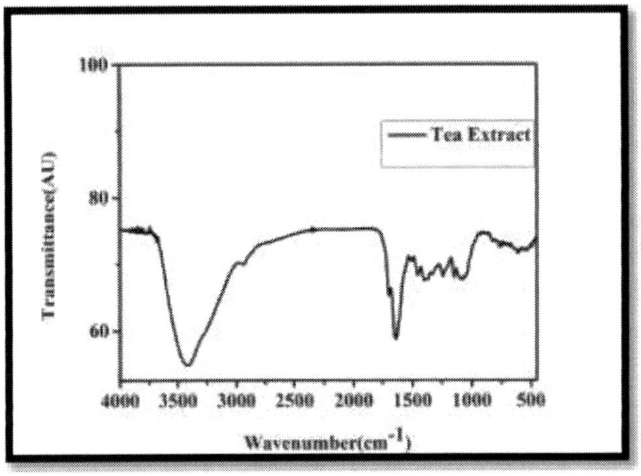

Figure 3 FTIR spectrum of black tea extract

3.2 XRD Measurements

XRD analysis (Figure 4) shows the peaks for iron oxide consistent with that of the JCPDS reference file for Iron Oxide (JCPDS card number 39-1346). The dominant diffraction peaks at 33.33, 35.37, 54, 57, 62.67 theta correspond to (2 2 0), (3 1 1), (4 2 2), (5 1 1) and (4 0 0) crystal planes, respectively of γ-Fe_2O_3.

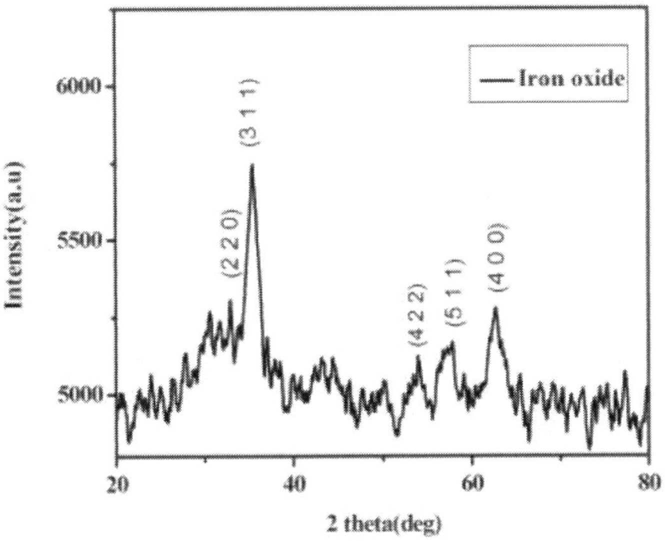

Figure 4 XRD of Iron oxide nano/microstructures

3.3 SEM Analysis

The size and the morphology of the synthesised product by SEM analysis shows needle shaped one dimensional structures as shown in Figure 5. SEM image shows the synthesized Nano needles have breadth in nanometer and length in micrometer range.

Figure 5. SEM analysis

3.4 Photocatalytic study

The dye degradation of MB dye upon addition of the synthesized sample under solar illumination after 6 hours has been plotted in Figure 6. It is seen that the dye degraded to 90.3 % after solar illumination observed from the decrease of the absorbance peak of MB dye at 663.5 nm. The use of very small amount of photocatalyst (5mg) to 50ml of 20ppm MB dye shows good activity of the iron oxide synthesized in this work as a photocatalytic agent coherent with the reported literature of iron oxide nanoparticles [7]. The increased performance could be due to enhanced formation of hydroxyl ion responsible for dye degradation due to increased surface area of needle type one dimensional nano/micro structures.

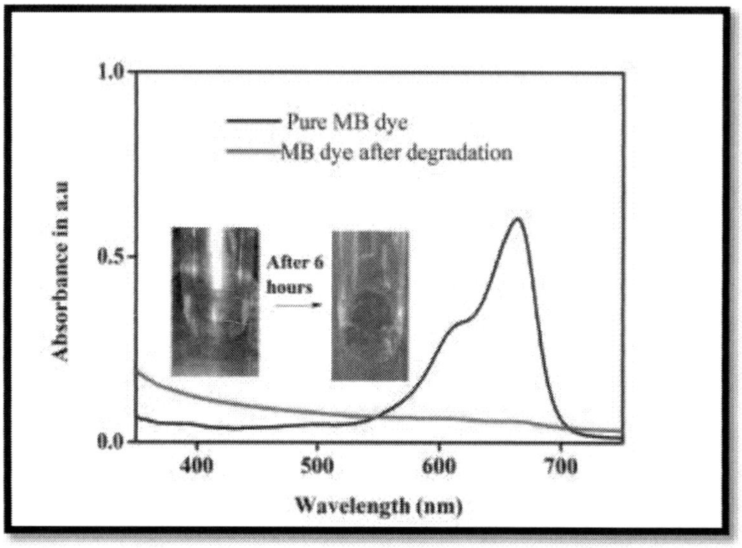

Figure 6. Photo catalytic study with MB dye

4. CONCLUSION

In this study Iron oxide nano/micro one dimensional structures was synthesised by green synthesis by using black tea extract and cotton template. The synthesized sample was characterized using X ray diffraction and scanning electron microscopy (SEM) which shows Fe_2O_3 needle type structures The Black tea extract was studied using Fourier Transform Infrared spectroscopy (FTIR). FTIR of tea extract gives information about functional groups corresponding to Gallic acid in the Black tea extract which is good reducing agent. The obtained iron oxide powder was used for the photocatalytic degradation of methylene blue dye under solar illumination. It was observed that 90.3% degradation was seen after 6 hours of exposure showing good potential for waste water treatment.

ACKNOWLEDGEMENTS

The authors acknowledge Sophisticated Test and Instrumentation Centre (STIC) Cochin University of Science and Technology for SEM analysis. The authors also acknowledge, Department of Physics, University of Calicut for XRD analysis.

References

[1]. Abolfazal A, Mohammed S and Soodabeh D 2012 *Nanoscale Research letters* **7** 144.
[2]. S F Hasany, I Ahamed, Rajan J and A Rehman 2012 *Nanoscience and Nanotechnology* **2(6)** 148
[3]. Alina M P, Simona L I, Carmen S C, Mariana C C, Milhai S and Daniela P 2013 *Journal of Nanomaterials* **2013** 1
[4] Ranjana J, Ranjana D and Nandini S 2016 *International Journal of Innovations in Engineering and Technology* **7** 663
[5] Sevil C and Nursah K 2018 *Materials Focus* **7** 316
[6] Carmen C, Rafael G and M Carmen L 2003 *J. Agric. Food Chem.* **51** 4427
[7] Shahana B, Aarti K and Raja S 2018 *Materials Research Bulletin* **97** 121

Implementation of MIMO OFDM NOMA System using Iterative Algorithm

Anil S, Pappa M and Ramesh C

Dept of ECE, CMR Institute of technology, Bangalore, Karnataka, India

anls18vlsi@cmrit.ac.in, pappa.m@cmrit.ac.in, crameshmail@gmail.com

Abstract: In this modern world, technology has improved a lot and wireless communications are much necessary. MIMO wireless technology is very much essential for providing increased connectivity and spectral efficiency combined with the reliability of advanced connectivity using what were previously considered as interference paths. The multiple access system is considered such that they provide the different methods in communication for multi-users. In this work, the iterative algorithm is used such that it helps in defining random array of signals and assigning the phase to the given measured amplitude. Different types of multiple access systems such as Non-Orthogonal multiple access (NOMA), Orthogonal multiple access (OMA) and Adaptive Orthogonal multiple access (AOMA)which are used to analyze transmission of multiple data stream over power domain in the presence of AWGN and Rayleigh channel. The data transmitted over the channel may vary due to some interference on its path. The signal strength is calculated in the receiver using the fading channel and the other noises that gets added in the channel. MATLAB software tool is used for simulation and the performance analysis in terms of average sum value i.e. the signal strength with respect to SNR for all the above mentioned multiple access techniques is compared. The simulation results clearly exhibits that NOMA outperforms the other two cases and is one of the promising technology for fifth generation wireless communications.

Keywords: MIMO, OMA, NOMA, AOMA, Iterative algorithm

1. Introduction

In the past several years, the way people communicate with each other has been revolutionized by mobile communication that adds mobility to the loop. A lot has been accomplished in a very short time on the wireless journey. Wireless access technology has advanced through leaps and bounds and is close to the fifth generation (5 G). In the past, wireless access technology has pursued a number of progressive routes in an extremely mobile setting aimed at efficiency and performance Today, MIMO technology is a key component of many modern wireless communication technologies including Long Term Evolution (LTE) and WiMAX. MIMO communications systems benefit significantly from spectral performance, power and energy efficiency compared to standard SISO Systems. In wireless propagation, whenever changes in amplitude and phase of radio waves occur, mobile stations transmit these changes through fragments of information. Amplitude or phase can vary conclusively on the receiver side of the mobile applications systems. This leads to deterioration in the system performance and affects the QoS parameters, because the accuracy of the standardized or instantaneous estimated channel has a strong impact on the performance of the receiver. So to improve accuracy of the received signal, channel estimation technique is introduced. Radio stations in mobile communication systems are often the most obsolete or fading channels because of the multi path propagation, due to which inter symbol interference (ISI) mixes with the received signal. To remove the ISI from the signal, many types of acquisition techniques or detection algorithms are used in conjunction with the receiver. These detectors or acquisition techniques should have the prior knowledge on channel impulse response (CIR) which can be provided by separate channel estimator. The channel estimation is based on a known sequence of different bits of specific and repeated transmissions throughout the transmission burst. This enables the channel estimator to estimate CIR (i.e.) channel impulse response from the known transmitted bits for each burst separately and from the corresponding received samples of the signal. The channel estimation category or classification is shown in Fig 1. They are Training based estimations, Blind channel estimations and semi-blind channel estimations.

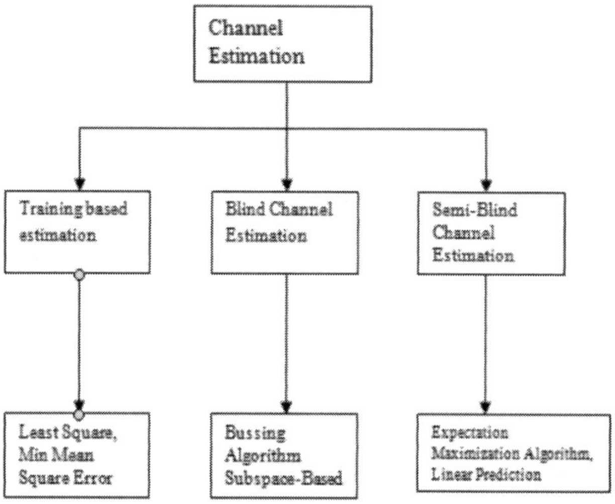

Figure 1. Types of Channel Estimation

Along with the data symbols, the training based estimation is performed through the arrangement of pilots either in block type or the comb type. In Block type arrangement of pilots, one symbol full of pilot subcarriers is transmitted in a synchronous format by aligning in time and transmitting it periodically. This type of estimation is used for slow fading channel. In comb type arrangement of pilot estimation, the pilots are inserted to the OFDM symbols with a periodic frequency bins.

2. Literature Survey

R.S.Ganesh, Dr J.Jayakumari, Akhila I.P[1]has discussed a technique for multiple input multiple output(MIMO) systems, using a set of pilot carriers in the presence of Rayleigh fading .The algorithms that is used for estimating the channel are the conventional Least Square (LS) and Minimum Mean Square (MMSE) estimation .By means of estimated Bit Error Rate (BER) and Mean Square Error(MSE),MIMO-OFDM system performance is analyzed.

Arsla Khan, Soo Young Shin [2], has shown that, performance of typical NOMA systems in the presence of linear precoding techniques provides trade off with complexity. Actually Performance improves at the cost of slight increment in complexity.

Zhiguo Ding, Xianfu Lei,George K. Karagiannidis, Robert Schober, Jinhong Yuan,Vijay K. Bhargav [3], has discussed the extensive research aspects of NOMA, including innovations as well as their emerging applications.

Sergey D. Sosnin, Gang Xiong, and Yongjun Kwak[4] , suggested that when iterative interference cancellation is used along with a reference signal in a data aided manner, improved channel estimation can be achieved. When interference cancellation is done using data aided signals in combination with the reference signal, it improves the overall link performance as well as improved channel estimation in NOMA receiver.

Muyiwa B. Balogun, Fambirai Takawira, Olutayo O. Oyerinde [5], has proposed that with the combination of Orthogonal Frequency Division Multiplexing (OFDM) and Non-Orthogonal Multiple Access (NOMA), spectral efficiency can be enhanced better. An efficient and improved algorithm namely weighted least square based iterative algorithm is proposed and implemented for the channel estimation in NOMA-OFDM systems. By using the improved weighted least square algorithm as the initialization step, an iterative linear minimum mean-square-error (LMMSE) algorithm is developed to estimate the channel. For MIMO OFDM NOMA systems, with the aid of imperfect successive interference cancellation (SIC), this algorithm is implemented for estimating the channel.

3. Multiple Access Systems

In wired or wireless channels, multiple numbers of users are serviced using multiplexing techniques or access methods. On the other hand Communication channels, whether they belong to the category of wireless spectrum segments or cable connections, are tend to be lavish and expensive. So in order to use the available bandwidth efficiently, service providers must allocate users over limited resources to make a profit. Sometimes sharing of these limited channels among many users is also enabled by the access methods to provide the economy of scale necessary for a successful communications business. As shown in figure 2, frequency division multiple access (FDMA), time division multiple access (TDMA), code division multiple access (CDMA), orthogonal frequency division multiple access (OFDMA) and spatial division multiple access (SDMA) are the five different multiple access techniques. This paper concentrates on Orthogonal multiple access (OMA) and Non Orthogonal multiple access (NOMA).These multiple access has different types respectively and OMA is divided in to TDMA, FDMA, CDMA and OFDMA. The NOMA is divided into IDMA LDS, CDMA, Power domain NOMA and MUSA.

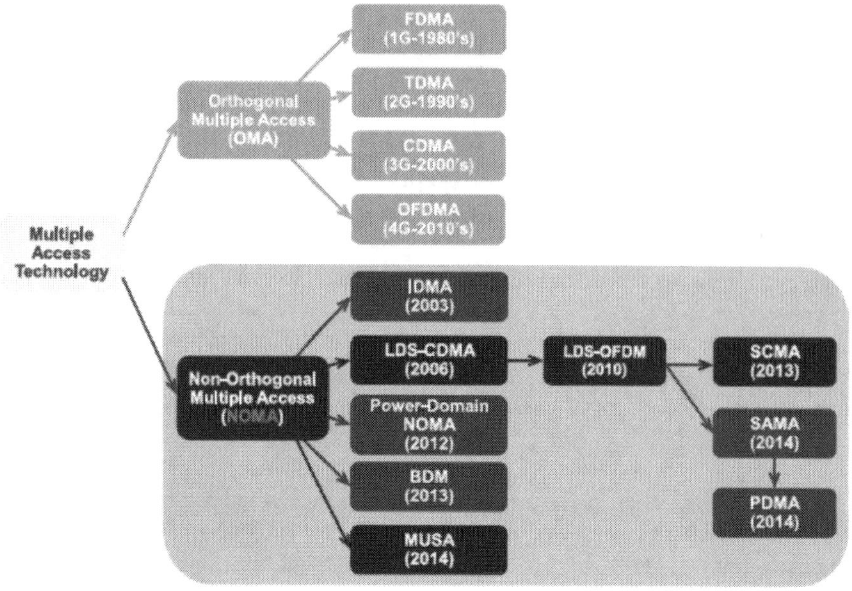

Figure 2. Multiple Access Technologies [6]

Orthogonal Multiple Access: The Orthogonal Multiple Access techniques have several different access technologies that are Time Division Multiplexing Access, Frequency division Multiplexing Access and Code Division Multiplexing Access. None of these can meet the parameters which are better than the NOMA. These schemes are defined in a different ways; in TDMA the signals sent to the user are sent in non overlapping time slots. They need perfect timing synchronization. In FDMA the signals for each users is assigned to a subcarrier. The CDMA uses code to divide the user in the respective channel.

Power Domain NOMA: Power domain NOMA defines multiple user usage in the power domain and examines channel differences between multiplexed users. On the transmitter side the signals from different users are combined or superimposed and the outgoing signals are transmitted at the same frequency to the same users. On the recipient side, to obtain the desired signals, multi-user acquisition algorithms such as SIC(Successive Interference Cancellation) is used as shown in figure 3.

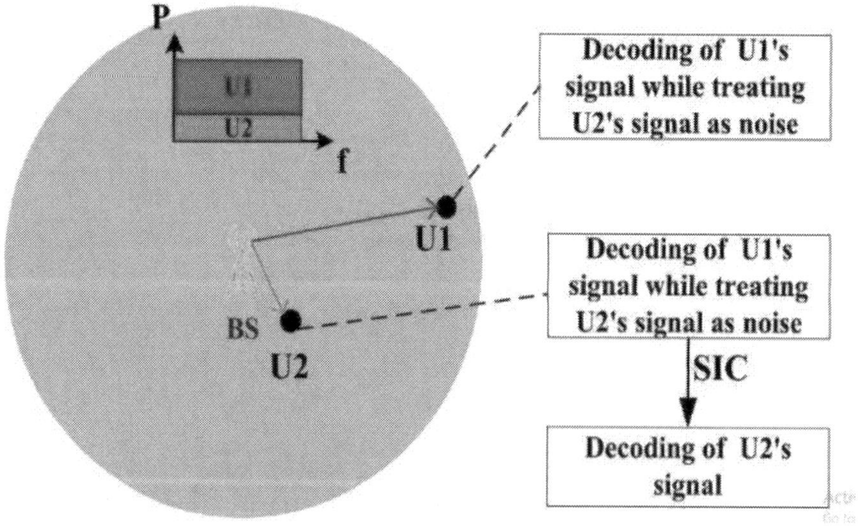

Figure 3. NOMA Technology [6]

4. Iterative Algorithm

This algorithm explains the random array and it assigns phase to the measured amplitudes, thus it has a capability of generating the complex-valued spectrum at zeroth iterate. Let's consider the random array as and hence the complex valued spectrum can be given as $\emptyset^0$ and hence the complex-valued spectrum given as
$S^0 = p^{meas} exp(i\emptyset^0)$.A following sequence can be initiated as shown below: Inverse Fourier Transform is used on the function S^0 to get the zero-th estimate in the group of longitudinal distributionS^0. Impose the time domain constraints by applying information on the distribution. Calculating the FT of domain function, and get a subsequent estimate in the frequency domain S^1 and calculate its phase $\emptyset^1 = arg(S^{r(1)})$. Frequency domain iterative can be updated by imposing the frequency domain constraint $S^1 = p^{meas}exp(i\emptyset^1)$. The feature of iterative method is that, different solutions are obtained from the initial phases and these are not always correct sometimes they are distorted. There is a need in the selection and averaging the solutions given, because at end only one profile is required. There are different types of averaging:
Two step averaging: As shown in figure 4, here the first step would be averaging the Hilbert profile as reference and in second, new profile is being used as reference. The drawback here is we lose a lot of time to attain advantage inX^2.

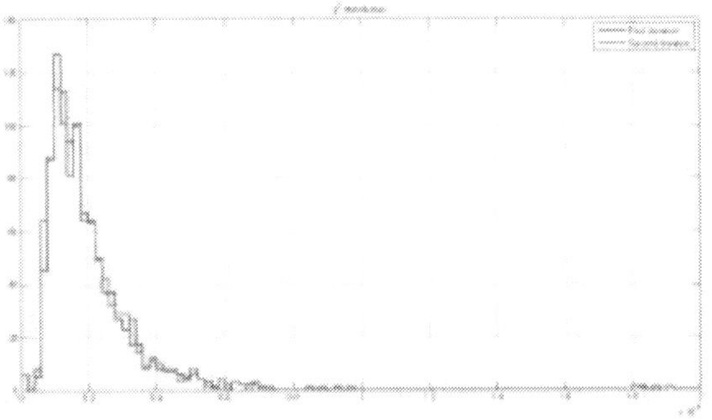

Figure 4. Two step Averaging in Iterative Algorithm

Averaging is a simple technique where weight can be considered as correlation coefficient and uniform averaging also works similar in principle. The linear sampling is used here for reconstruction methods.

5. Methodology

Multiple-input Multiple-output is an antenna technology used in the transmission and reception of more data or signals in multiple paths for wireless radio communication. There can be various MIMO configurations 2*2, 4*4 etc. Pre-coding, spatial multiplexing and diversity coding are the three important categories of MIMO Systems. In MIMO-OFDM systems, channel estimation plays a crucial or vital part in analyzing the performance of the system. To support pre-coding and scheduling, channel state information (CSI) is measured. During its transmission noise gets added as unwanted signal and interfere the desired signal as shown in figure 5. Two channel conditions namely AWGN and Rayleigh are considered here.

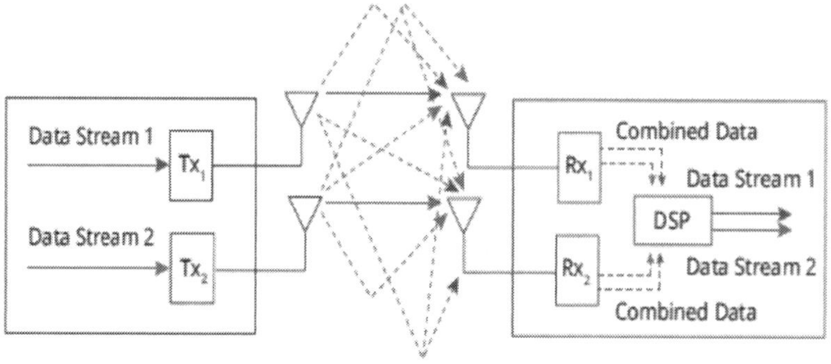

Figure 5: MIMO Principle

In presence of AWGN Channel the output is calculated as

$$y(m) = x(m) + w(m) \quad (1)$$

Where $y(m)$is the output, $x(m)$ is the input, and $w(m)$is the noise.

In presence of Rayleigh fading channel, the output is calculated as

$$y(m) = h(m)x(m) + w(m) \quad (2)$$

Where $h(m)$denotes the fading process.

The multiple access system considered here is NOMA, OMA and AOMA. As shown in the figure 6 the user 1 transmits the signal they note a transmission power when they do so. The AWGN with zero mean and variance is used. With help of the SIC the base station decodes the signal of user 1 and user 2 in two stages. The user 2 signal is treated first by considering the user 1 as the noise. In next stage the receiver subtracts the decoded signal from the received signal and decodes the user 1. As the signal starts propagating the decay of the signal also starts. We know that signal strength becomes week if the distance between the antennas increases. There will also be a phase delay in a signal because it is a wireless channel that is called as echo signal.

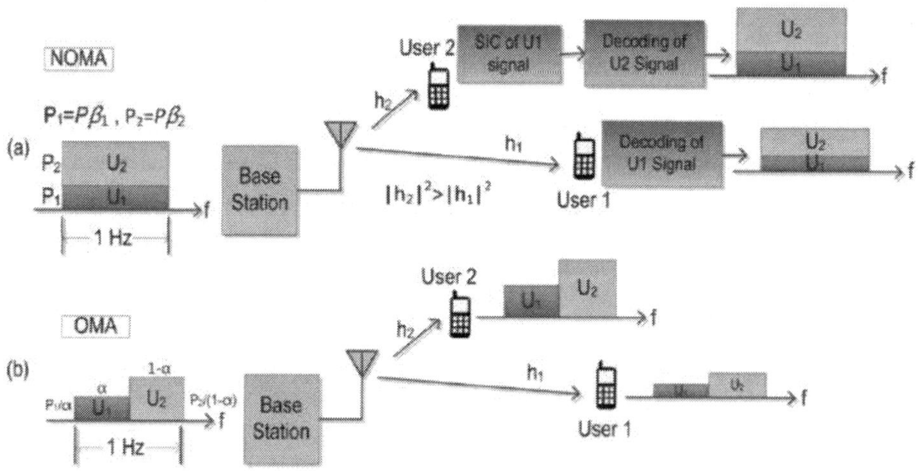

Figure 6. Multiple access scheme for two user scenario (a) NOMA (b) OMA [7]

Here the channel estimation is done with and without NOMA with respect to MIMO.TDMA scheme with fixed and adaptive form for the OMA (without NOMA) is implemented. In NOMA the Power Control is used to reduce the interference of the channel as shown in figure 7. In this case SIC (successive interference cancellation) is used for the interference cancellation.

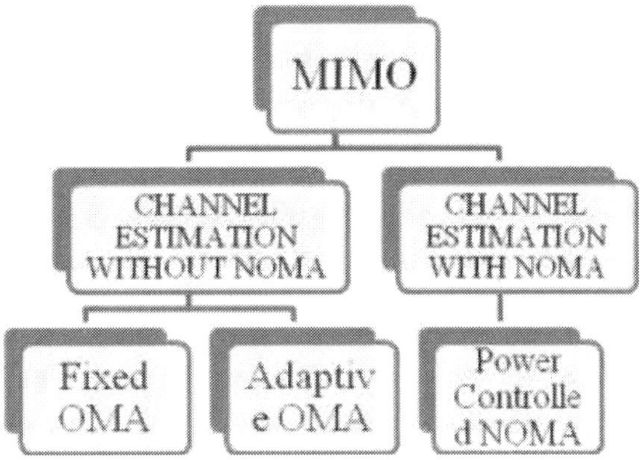

Figure 7. Channel Estimation with and without NOMA

The channel estimation of a MIMO is done through three methods they are NOMA, OMA and AMOA. The parameters such as SNR, and power factors and the time synchronization are initialized. The loss of signals such as fading loss, group loss and Rayleigh gain is calculated.. These all losses are calculated together for the given threshold values. They are calculated with respect to the SNR values and also for different users. The calculated Gaussian g^{th}values are compared with the fading losses; if fading loss is high then the signal strength is calculated and will be given it as the output. If the gth values are high then we calculate the different SNR to different users again as shown in figure 8.

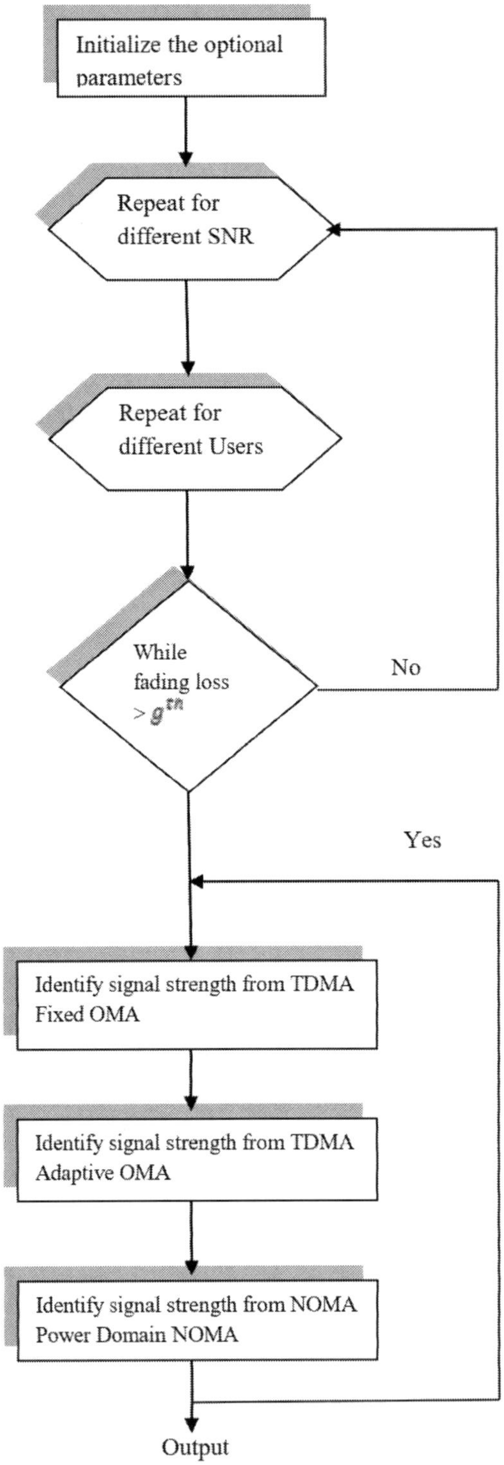

Figure 8: Flow Chart

An OFDM based on NOMA scheme with two users is considered, where the users transmit signals $S_k(m), k \in K = \{1,2\}$on subcarrier $m \in \{1, \dots \dots, M\}$. In order to minimize or remove completely the multiuser interference on each subcarrier, two users are considered here..SNR is calculated as:

$$SNR = {}^{P}/_{N_0} \quad (3)$$

Where P is the transmit power and N_0 is the noise added due to the channel.

In terms of decibels, it is mentioned as,

$$SNR_V = 10^{\frac{SNR_{DB}}{10}} \quad (4)$$

SNR Variation with respect to time in terms of $t_{step} = 0.1\ seconds$ is used in the calculation. It is considered in terms of $t_{step} : t_{step} : 1 - t_{step}$. Here the path loss factor is -3.76 and the Gaussian value is

$$g^{th} = \left(\frac{35}{289}\right)^{PL} \quad (5)$$

The phase is calculated in terms of variations with respect to time and it is denoted as $0{:}\,p_{step}:1$

Uniformly Generated signal is represented by the following equations:

$$a = rand(1,1) \quad (6)$$

$$a = x = \sqrt{a} * \sqrt{{}^{3}/_{2}} \quad (7)$$

Random signal is represented by $b = rand(1,1)$ (8)

Signal transmitted is $signal\ Tx = \left(a * b * \sqrt{\frac{3}{2}}\right) * \text{signal}$

Fading Loss and Group Loss is calculated by,

$$P_{loss} = [norm(a,b)]^{PL\ factor} \quad (9)$$

$$g_{Ls} = P_{loss} + \left(\frac{lognormal}{P_{Lavg*m\ log}}\right) \quad (10)$$

Where $\log_{normal} = 10^{rand*0.8}$

To calculate the overall loss, Rayleigh gain is required and it is calculated using

$$R_{gain} = abs\left(\frac{(rand+li*rand)^2}{2}\right) \quad (11)$$

Combined or Overall loss is calculated using

$$G_{all} = g_{Ls} * R_{gain} \quad (12)$$

5.1 Channel estimation without NOMA

The MIMO-OFDM was introduced for the next generation communication that is the 4G connection using TDMA. The use of TDMA will over-come some challenges which faced by the CDMA, they are limited performance and complexity in detection of multiuser. Here we show different numerical calculations or simulation results for the Adaptive TDMA and Fixed TDMA. The NOMA-OFDM scheme has two user transmitting signals $S1(m), xi \in XI = \{1.2\}$ on subcarrier $m \in \{1, \dots \dots \dots M\}$. These two users are used to reduce completely or minimize the multiuser interference (MUI) on subcarrier used each and also help in reducing the complexity. To separate the superimposed signals of various users, successive interference cancellation technique (SCI) is proposed.

Power 'PL' is assigned to the user 'xi' based on the distance from the base station, hence the signals from the xi users on subcarrier m. They can be represented as $X_{xi}^{n}(m) = \sqrt{PL_{xi}(m)S_{xi}^{n}(m)}$.

The Fixed TDMA is calculated with respect to the strength of signal and the noise in it:

$$OMA = \log_2(1 + G_{all} * P_{sum})/2 \qquad (13)$$

The adaptive TDMA as considered to be having multiple noises which are produced in transmission of a signal given as:g_1, g_2

Phase shift also calculated as $p1_v = P_{sum} * PS_v$

Then the Adaptive OMA in TDMA is given as:

$$rate\ 1 = t_{temp} \qquad (14)$$

$$rate\ 2 = (1 - t_{temp}) * \log_2(\frac{1+p_{2max}}{(1-t\_max)*G(2)}) \qquad (15)$$

$$AOMA(snr_{ix}) = AOMA(snr_{ix}) + sum(rate1 + rate2) \qquad (16)$$

5.2 Channel Estimation with NOMA

The NOMA OFDM is also given by the power tool through the power control technique and SIC. The noise or interference cancellation in the given NOMA is done by strengthening the channel by power control method. Hence when we increase the signal strength as required it can travel to longer distance and also without much disturbance and interference, the signal will reach its destination.

- Step 1:

Identifying Power control parameters, as P_1 from user 1 and P_2 from user 2,

$$P_1 = P_{sum} + P_s \qquad (17)$$

$$P_2 = P_{sum} - p1_v \qquad (18)$$

- Step 2:

Finding maximum relative rate of amplitude:

$$rate11 = \log_2(1 + \frac{(g_1+P_1)}{(g_2*P_2+N_0)}) \qquad (19)$$

Where N_0 is average noise

$$rate12 = \log_2\left(1 + \frac{g_2}{\frac{P_2}{NO}}\right) \tag{20}$$

Multiplying equation (19) and (20) we have:

$$Relative\ Rate\ 1 = rate11 * rate12 \tag{21}$$

Hence for the other user also we calculate the *rate21 and rate22* in same as above by taking:

$$rate21 = \log_2\left(1 + \frac{g_1}{\frac{P_1}{NO}}\right) \tag{22}$$

$$rate22 = \log_2\left(1 + \frac{(g_2+P_2)}{(g_1*P_1+N_0)}\right) \tag{23}$$

Multiplying equation (22) and (23) we have:

$$Relative\ rate\ 2 = rate21 * rate22 \tag{24}$$

- Step 3

NOMA based output signal is calculated based on relative power $R_{1\,max}$ and $R_{2\,max}$.

$If\ R_1max > R_2max$ NOMA signal depends on Relative Rate 1 or else depends on Relative Rate 2.

6. Simulation Results and Discussion

The proposed iterative algorithm is utilized to investigate and analyze the performance of MIMO OFDM systems channel estimation with and without NOMA. SNR ratio variation rate of 5 db is used for the simulation of average sum rate of OMA, AMOA and NOMA.

Table 1. Average sum value (strength of signal) with respect to SNR in db

SNR in terms of dB	Average Sum Rate		
	Without NOMA		With NOMA
	OMA	AOMA	NOMA
0	0.2482	0.2753	0.2864
5	0.4688	0.5213	0.5511
10	0.8387	0.9280	1.0027
15	1.3972	1.5299	1.6921
20	2.1826	2.3535	2.6643

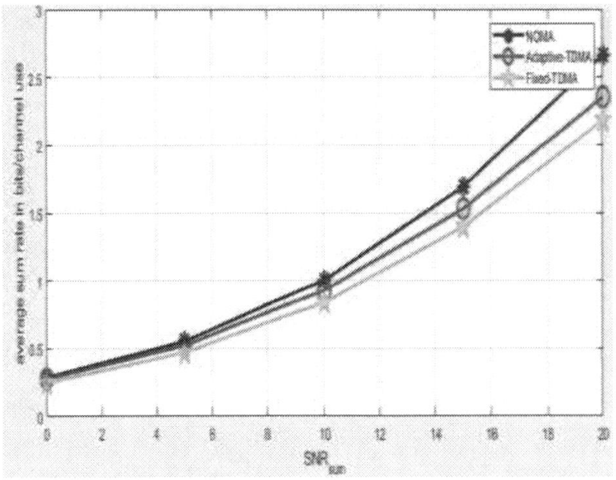

Figure 9: Average sum rate versus SNR

The impact of SNR with average sum-rate using OMA,AOMA and NOMA is shown above in Table 1 and Figure 9. As per the results it is obvious that the MIMO OFDM Systems better in presence of NOMA than without NOMA. Simulations are even plotted in terms of Logarithmic values versus SNR in Table 2 and Figure 10.

Table 2: Logarithmic values of Average sum value (strength of signal) with respect to SNR in db

SNR in terms of dB	Average Sum Rate		
	Without NOMA		With NOMA
	Log OMA	Log AOMA	Log NOMA
0	-9.2616	-9.2160	-9.1985
5	-7.1354	-7.0640	-7.0270
10	-5.1282	-5.0285	-4.9570
15	-3.3171	-3.1956	-3.0730
20	-1.7031	-1.5737	-1.3868

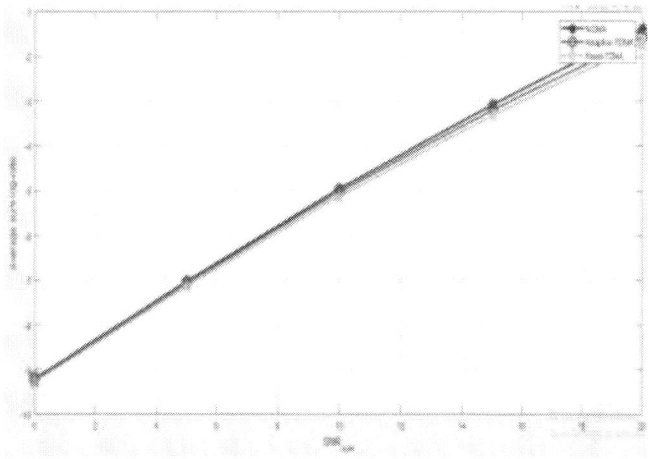

Figure 10: Logarithmic Average sum rate versus SNR

7. Conclusion

In this paper channel estimation problem for OFDM based NOMA system using Iterative algorithm has been presented .Here two users using single antenna are illustrated for transmission of a carrier data. There are various transmissions when we consider MIMO. With estimated channels at the base station, we derived sum rates that are achievable with OMA, AOMA and NOMA. The average sum of a signal is estimated to move in a channel. Simulations results are presented for estimated channel with NOMA and without NOMA and by comparing both, we can conclude that with NOMA higher number of signals can be transmitted at a time which is also described as MIMO-NOMA. The results show that the iterative technique ensures enhanced efficiency and overall system performance. Since NOMA and OMA finds extensive applications in many sectors and also used in the internet generations, this proposed work achieves considerable improvement in signal transmission.

References

[1] Ganesh, R.S., J. Jayakumari, and I.P Akhila. "Channel estimation analysis in MIMO-OFDM wireless systems", 2011 International Conference on Signal Processing Communication Computing and Networking Technologies, 2011.

[2] Arsla Khan & Soo Young Shin, 2017, Linear Pre-coding Techniques for OFDM-Based NOMA over Frequency-Selective Fading Channels

[3] Zhiguo Ding, Xianfu Lei, George K. Karagiannidis, Robert Schober, Jinhong Yuan, Vijay K. Bhargava, "A Survey on Non-Orthogonal Multiple Access for 5G Networks: Research Challenges and Future Trends", IEEE Journal on Selected Areas in Communications, 2017.

[4] Sergey D. Sosnin, Gang Xiong, and Yongjun Kwak, "Iterative Channel Estimation with data-aided Interfernce Cancellation for Non-Orthogonal multiple access, IEEE 88TH Vehicular Technology Conference, 2018

[5] Muyiwa B. Balogun, Fambirai Takawira, Olutayo O. Oyerinde. "Weighted Least Square Based Iterative Channel Estimation for Uplink NOMA-OFDM Systems, IEEE, 2020

[6] A Survey on Non-Orthogonal Multiple Access for 5G Networks: Research Challenges and Future Trends: Zhiguo Ding, Senior Member, IEEE, Xianfu Lei, Senior Member, IEEE, George K. Karagiannidis, Fellow, IEEE, Robert Schober, Fellow, IEEE, Jinhong Yuan, Fellow, IEEE, and Vijay K. Bhargava, Life Fellow, IEEE.

[7] S M Riazul Islam,Nurilla Avazov, Octavia A Dobre, Kyung-Sup Kwak, " Power-Domain Non-Orthogonal Multiple Access (NOMA) in 5G Systems:Potentials and Challenges", IEEE Communications Surveys & Tutorials, 2017

Design, analysis, and optimization of an FSAE upright

Ojas Babannavar[1] **and Aditya M Deshpande**[2]

[1 2]School of Mechanical Engineering, KLE Technological University, Hubli, KA – India

Corresponding author: *aditya@kletech.ac.in*

Abstract. Components of a formula car experience many forces and have to sustain in various conditions. The design of the components must be keeping in mind the safety and efficiency of the component. In Formula racing, the car has to be as light as possible. Cars with powerful engines are faster in the straights, but cars with lighter weight are faster in the corners. Optimizing the components to be light in weight is a critical task and has to be done meticulously. Several conditions and forces have been considered, and the simulation has to be as close to the real-life conditions as possible for accurate results. This paper records the detailed study, design, and analysis done for the weight reduction of the component. Data from previous years' performance of the car was also considered during the analysis. Based on the data, the material selection and designing of the component was made in Solidworks. The Structural and Fatigue analysis were done in ANSYS Workbench; based on these results, the component's optimization was done.

1. Introduction

Formula Student is an engineering design competition which is held all over the world in different countries. It is the right platform for engineering students around the world to learn and showcase their engineering skills. Each year Formula Student teams design and build open-wheel, open-cockpit race cars and compete in these events worldwide.

The upright or knuckle is an essential part of the vehicle as it holds the wheel assembly together. The design and manufacturing of this part have to be up to the mark. The component faces many forces and loads. Some of the loads and forces cannot be predicted or calculated, like the impact of an unexpected bump. Being a Formula car, this runs on race tracks with very few undulations and bumps. Nevertheless, it has to be kept in mind that the car travel speed is very high and is subject to high g-forces.

IOP Conf. Series: Materials Science and Engineering 1166 (2021) 012043 doi:10.1088/1757-899X/1166/1/012043

2. Methodology

2.1. Iterative design process:

An iterative design process is followed for the optimization of the upright design. As this paper's focus is away from generative design, the iterative design process is a good alternative to optimize the design of any mechanical component. The overview of the process is explained in the flowchart, as shown in figure 1.

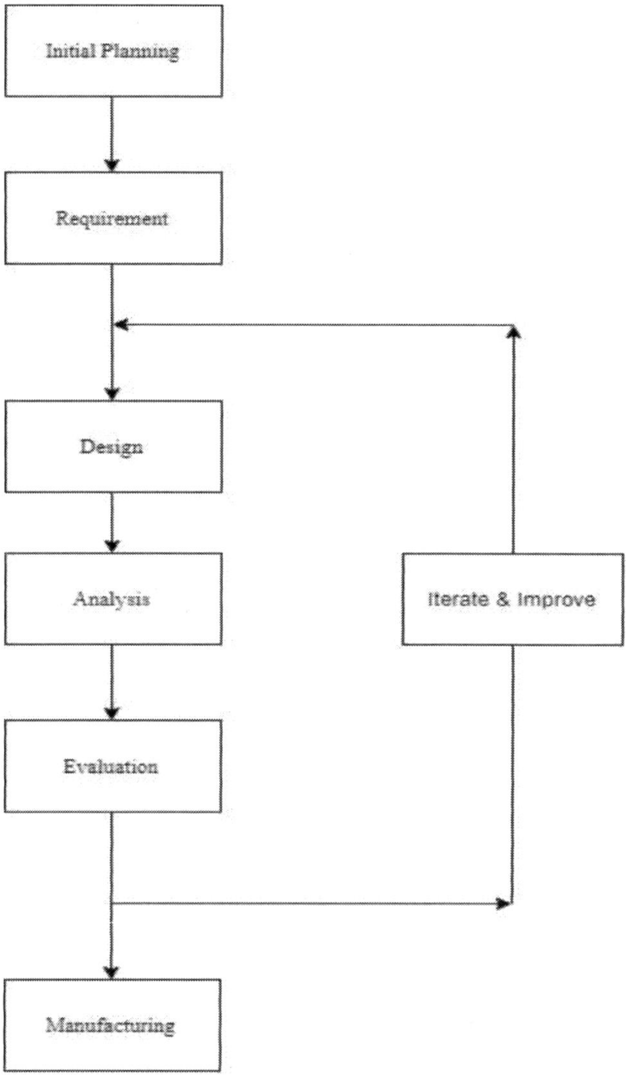

Figure 1. Iterative design process

The above procedure has yielded promising results, and when executed by using basic principles and the proper engineering practices, it will lead to an optimized and improved design. Whenever there is a lack of access to generative design and subsequent 3D printing, metal 3D printing, this method can very well be used to optimize the design.

2.2. Data acquisition

A significant part of the design process is based on the data collected during the competition the previous year. Data was collected in dynamic events like Autocross and Endurance Race. This data was processed and studied, and the inferences from this were taken as input parameters for the design. Vehicle speed, g-forces experienced, and the path on the track was some parameters that were derived from the acquired data as shown in table 1. Load calculations were done considering the car's g-forces, which we derived from the data acquired.

Table 1. Data acquisition of the car's performance on the track.

	Autocross	**Endurance**
Max. Speed Achieved (Km/Hr)	67	129
Average G Force Experienced (G)	2.3	2.7

2.3. Analytical calculations

The following parameters offer significant loads and conditions on an upright. Upright affects the vehicle dynamics greatly, and all these parameters are calculated based on vehicle dynamics of the car and upright under study [1][2].

2.3.1 *Braking force:* Force on front callipers = Braking torque / Radius of rotor

$$= T_{br} / R$$
$$= (383.73 \times 10^3) / 62.5$$
$$= \mathbf{6124.98\ N}$$

2.3.2 *Steering force:* Reaction at each tire $= (R_A/ 2) + (6 \times 9.81)$

$$= (1617/2) + 58.86$$
$$= 867.36\ N$$

Therefore, force applied by tie-rod $= 0.7 \times 867.36$

$$= \mathbf{607.15\ N}$$

2.3.3 *Cornering force:*% of load transfer, LLT = (g × height of CG) / Track Width

$$= (2.7 \times 0.254) / 1.219$$
$$= 0.31\%$$

% load on inner wheels = 0.5 – 0.31 = 0.19 = 19%

% load on outer wheels = 0.5 + 0.31 = 0.81 = 81%

Load transferred on outer wheels = (81/100) × 270 = 219.38 kg

Load transferred on front outer wheel = (219.38 / 2) = 109.6 kg

$$= \mathbf{1096\ N}$$

2.3.4 *Lateral Force:* $F_L = g \times 9.81 \times$ mass on single wheel

$$= 2.7 \times 9.81 \times 75$$
$$= \mathbf{1986.525\ N}$$

2.3.5 *Vertical Load:* F_V = Mass on wheel × 9.81

$$= 75 \times 9.81$$
$$= \mathbf{735.75\ N}$$

3. 3D Model

3.1. Geometry

The 3D Modelling was done in Solidworks 2020. As the design philosophy was to mainly optimize the upright from the 2019 vehicle, the design was started with the 2019 model as the base geometry. A fundamental structure was designed, and then based on the boundary conditions and mounting points on the upright, optimization was carried out. The 2019 upright was for a 13-inch wheel. The new 2020 design is for the 10-inch wheel.

48.61% reduction in weight was achieved in the new design compared to the old design of the upright. FEM analysis of the upright was done to justify the safety of weight reduction. The old and new designs of the upright and its mass properties are shown in figure 2 and figure 3.

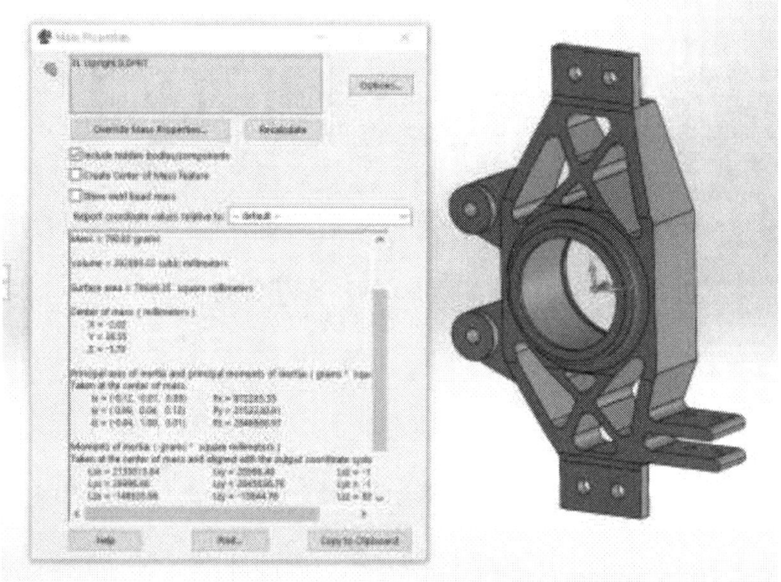

Figure 2. 2019 Upright design and it's mass Properties

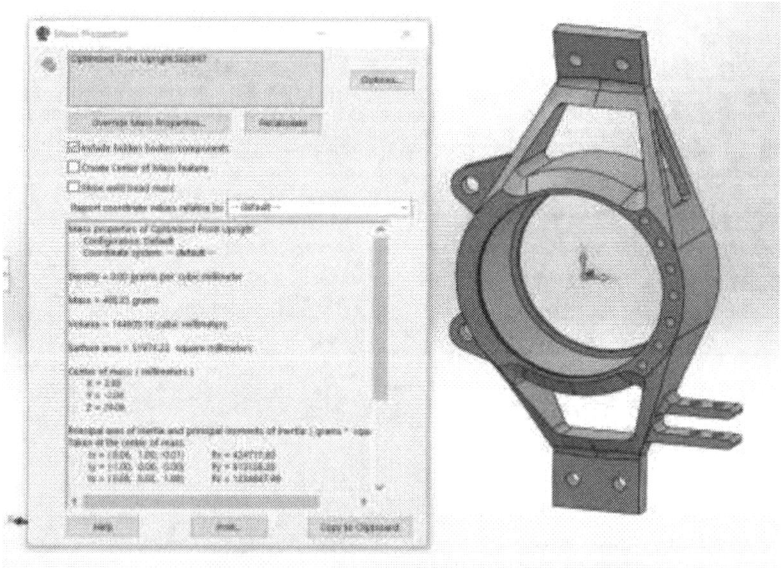

Figure 3. 2020 Upright design and it's mass properties

3.2. Design process and manufacturability

Numerous iterations were made to obtain the design shown below based on the results obtained on static analysis for Von Mises stress, stress flow, and maximum deformation. The uprights designed for race cars are required to have a high stiffness to resist the camber changes in the wheel assembly.

Manufacturability is one of the main factors considered while designing the upright. Generative design and additive manufacturing are the most preferred and most efficient ways to reduce the weight of any mechanical component as it gives much flexibility in designing and manufacturing. However, it has some drawbacks, like economic feasibility and manufacturing accessibility. These points play a vital role in a formula student team while designing and manufacturing a component. As this technology was not feasible and not accessible easily, the authors concentrated on a design that can be manufactured using traditional techniques.

4. Analysis

The analysis was done in Ansys Workbench 18.2. The details are discussed in the following sections.

4.1. Pre-processing:

4.1.1. Meshing: A tetragonal mesh was generated on the model, as shown in figure 4 below. The element size was kept as low as possible considering the computational power available for the analysis. The lower the element size higher the accuracy of the result [3]. The mesh parameters are as follows;

- Size Function = Proximity and Curvature
- Element Size = 12.21 mm
- Nodes = 706656
- Elements = 464159

More significant the number of nodes, the better the accuracy of the result. Keeping this in mind and considering the available computational power, the fine mesh was generated.

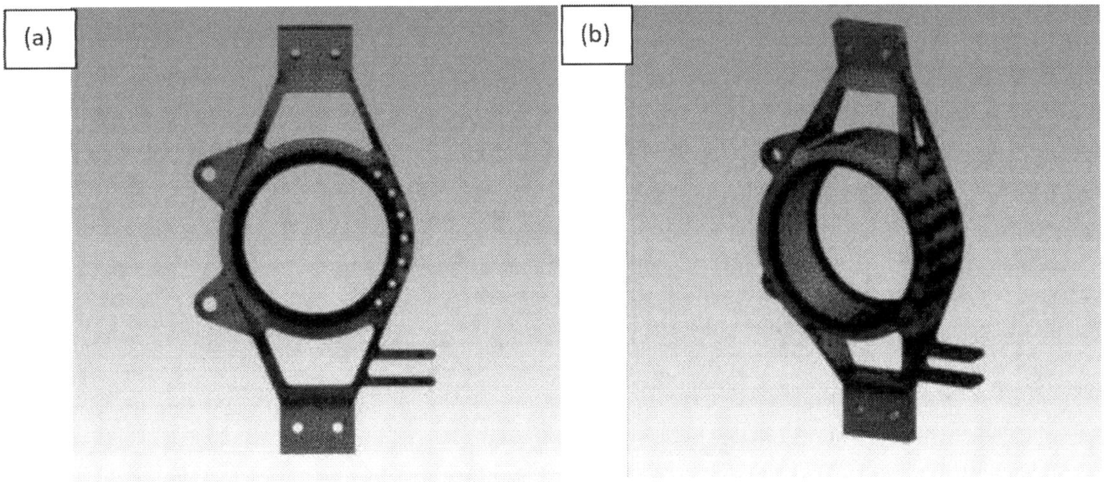

Figure 4. Generated mesh (a) Front View; (b) Isometric View

4.1.2. Boundary conditions: The boundary conditions and loading parameters are as calculated in the above sections of the paper. For the analysis of this design, the maximum loading condition in one of the dynamic events was considered. Autocross consists of different maneuvers like chicanes, slaloms, and hairpin bends. For analysis, the slalom part of the autocross is considered as the magnitude of the force is applied cyclically. Fatigue analysis is done concerning those parameters. The loading conditions and the forces applied are as shown in below figure 5.

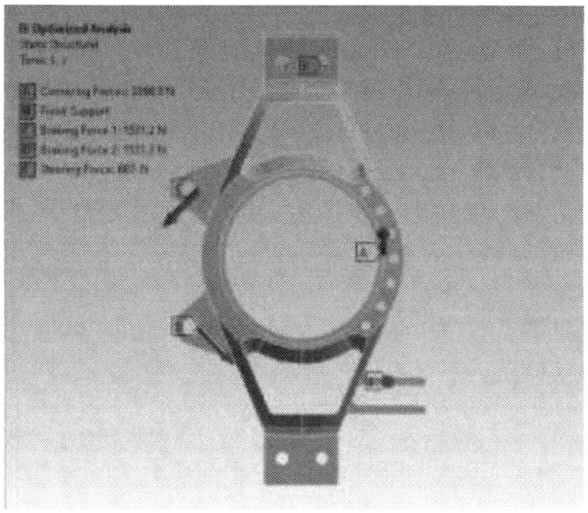

Figure 5. Boundary conditions applied to the upright are as shown in the analytical calculations part.

4.1.3. Material selection and its properties: The material used for this design was Al 7075 T6. The material was selected for its low weight and high strength properties [4]. This material is aerospace-grade aluminium and is used to manufacture aerospace parts. The specifications and properties of the material are shown in figures 6(a) & 6(b).

Properties of Outline Row 3: Aluminum Alloy

	A	B	C	D	E
1	Property	Value	Unit		
2	Material Field Variables	Table			
3	Density	2810	kg m^-3		
4	Isotropic Secant Coefficient of Thermal Expansion				
5	Coefficient of Thermal Expansion	2.3E-05	C^-1		
6	Isotropic Elasticity				
7	Derive from	Young's Modulu...			
8	Young's Modulus	71700	MPa		
9	Poisson's Ratio	0.33			
10	Bulk Modulus	7.0294E+10	Pa		
11	Shear Modulus	2.6955E+10	Pa		
12	Alternating Stress R-Ratio	Tabular			
16	Tensile Yield Strength	503	MPa		
17	Compressive Yield Strength	2.8E+08	Pa		
18	Tensile Ultimate Strength	572	MPa		
19	Compressive Ultimate Strength	0	Pa		

Figure 6(a) Aluminum 7075 T6 material properties

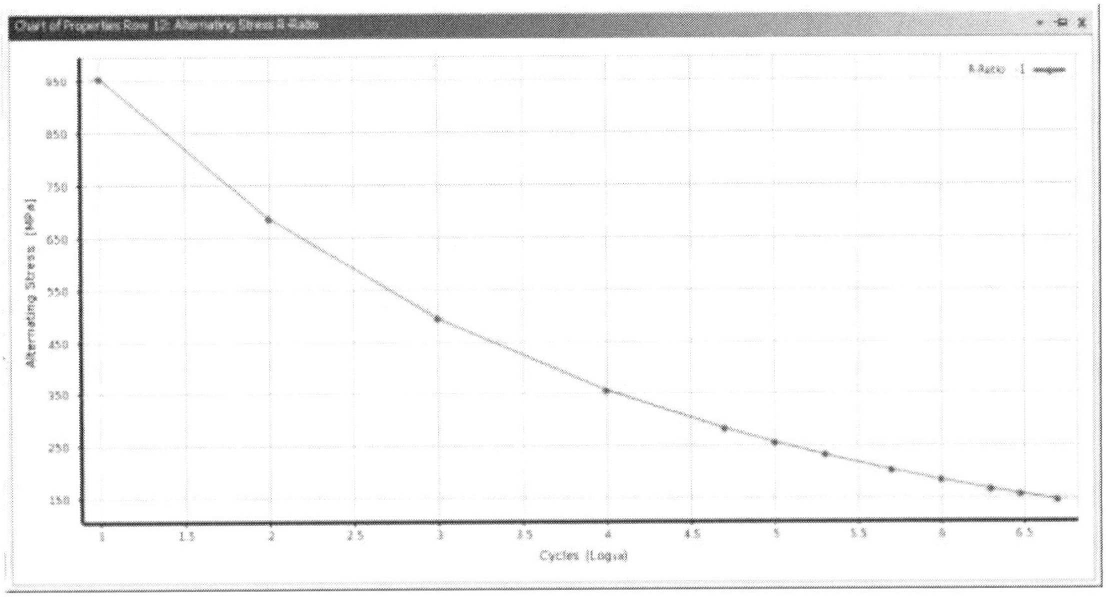

Figure 6(b) SN Curve of Aluminum 7075 T6

4.2. Post-processing:

Static Structural analysis and fatigue analysis was carried out on the upright. The loading conditions for the analysis are as calculated and discussed above. Below are the analysis parameters and the performance results of the component.

The Equivalent or Von Mises Stress and the Total Deformation of the upright are as shown in figures 7(a) and 7(b).

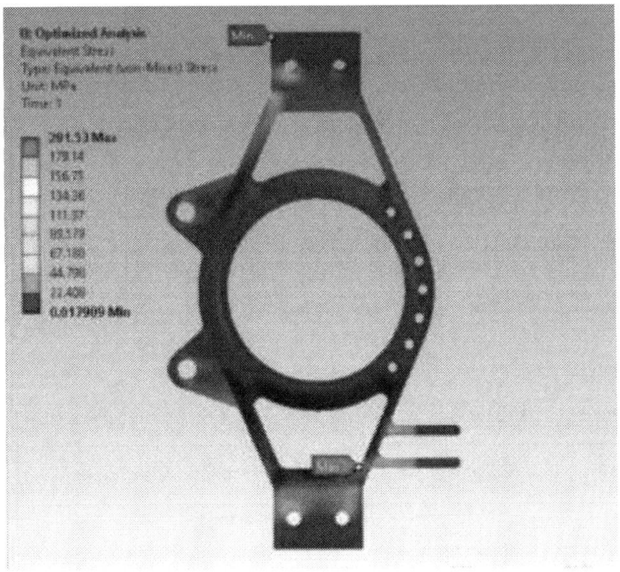

Figure 7(a). Equivalent stress or Von Mises stress

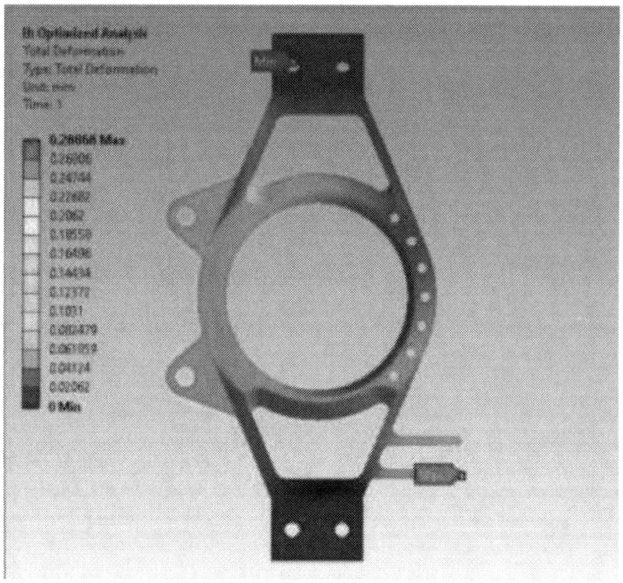

Figure 7(b). Total deformation

As discussed earlier, the analysis is being done to keep in mind the maximum cyclic loading the upright will experience in the dynamic events in autocross. Autocross is a form of competition in which cars are driven around an obstacle course, typically marked out by cones. Slalom is a part of this autocross track which is to move or race in a winding path, avoiding obstacles [5].

Considering the cyclic loading in the above situation, fatigue analysis was done, and the results are as shown in figures 8(a) and 8(b). The fatigue analysis parameters are as follows [6][7];

- Loading Type = Fully Reversed ; Analysis Type = Stress Life
- Mean Stress Theory = Goodman ; Stress Component = Equivalent (Von Mises)

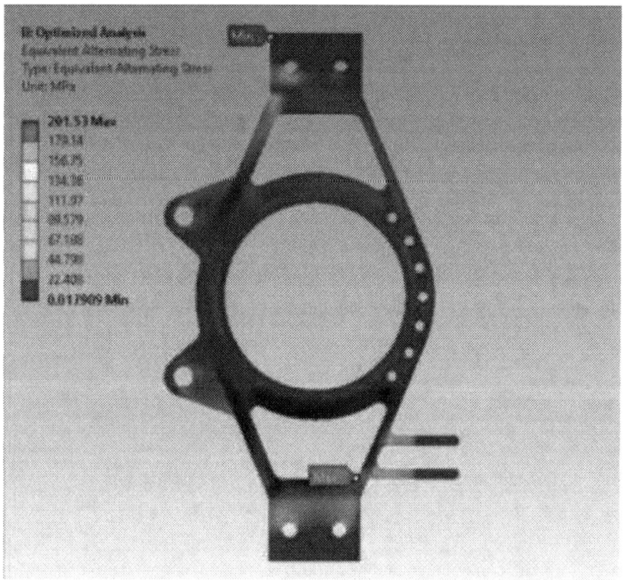

Figure 8(a): Equivalent alternating stress

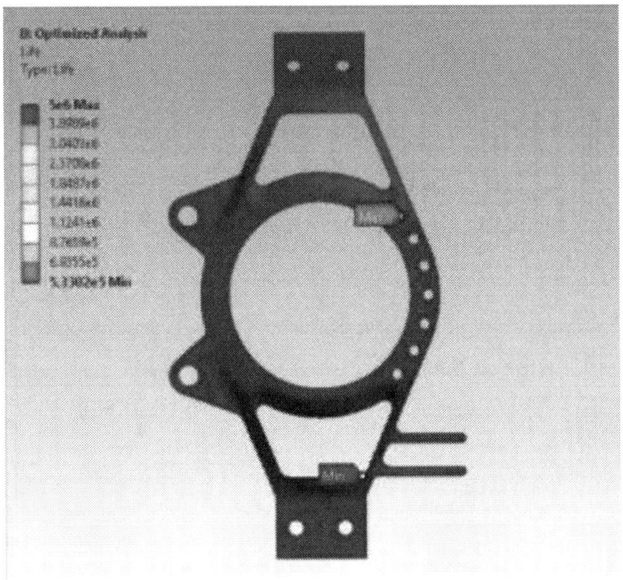

Figure 8(b): Life or Number of cycles

5. Results

The results from the above analysis are as shown in below table 2. It is essential to keep in mind the material properties, especially the maximum yield strength of Aluminum 7075 T6, to interpret the results better.

Table 2. Results from the above analysis.

Parameter	Value
Max. Equivalent Stress (Von Mises)	201.53 MPa
Max. Equivalent Alternating Stress	201.53 MPa
Min. Life or Number of Cycles	5.3302×10^5

6. Conclusion

The iterative design process was followed while designing this upright. The previous design of the upright was used as a benchmark to improve the new design of the upright. Drastic improvements were made, and optimization was done. After a couple of iterations of design and analysis, an optimized design was achieved. The new design was 48.61% lighter than the previous design. This component is part of wheel assembly on all four wheels. Hence, this sums up to a total weight reduction of 48.61% in the wheel assembly sub-system of the car.

The structural and fatigue analysis of the upright was conducted, and it was observed that the component sustained the static and cyclic loading without any premature failure. Based on the above results, it can be concluded that the component is safe to be used in its application. Design optimizations made to the design based on the analysis results of the older designs have contributed to the weight reduction of the component without compromising the performance and safety of the upright. It was also observed that the stress life of the upright could be greatly enhanced by improving the geometry of the upright based on the stress-induced and stress flow. Considering the result of the study conducted, it can be concluded that major improvements in terms of fatigue life enhancement, performance prediction, vehicle reliability, weight reduction, and other design parameters can be achieved by carrying out the static and fatigue analysis of the components.

7. References

[1] Milliken WF, Milliken DL. 1995 *Race car vehicle dynamics* (Warrendale, PA: Society of Automotive Engineers)
[2] Bhavikatti S S, Rajashekarappa K G 1994 *Engineering Mechanics* (New Age International)
[3] Gokhale N S 2008 *Practical finite element analysis* (Finite to infinite)
[4] Bauccio M 1994 *ASM Engineered Materials Reference Book*, 2nd Edition, (ASM International Materials Park OH)
[5] Balkwill J 2017 *Performance vehicle dynamics: engineering and applications* (Butterworth Heinemann)
[6] Khurmi RS, Gupta J K 2005 *A textbook of machine design* (Eurasia)
[7] Bhandari V B 2010 *Design of machine elements* (Tata McGraw-Hill Education)

IOP Conf. Series: Materials Science and Engineering 1166 (2021) 012044 doi:10.1088/1757-899X/1166/1/012044

On-chip Analysis of Etched Drain based Cyl. GAA TFET with Elevated Density Strip

Arya Dutt[1], Sanjana Tiwari[1], Mayuresh Joshi[1], Prakhar Nigam[1], Ribu Mathew[1], Ankur Beohar[1*]

[1]School of Electrical & Electronics Engineering (SEEE) VIT Bhopal University Bhopal, India

ankurbeohar16@gmail.com*

Abstract. This paper comprises of, designing and analysis of novel 3D Gate All Around Cylindrical tunnel field effect transistor (TFET). The device designing incorporates hetero-substrate (HeS) material with inclusion of etched drain (ED) and elevated density strip (EDS) at source-to-channel junction for reduction in tunneling barrier width resulting in better ON-current (I_{ON}). For analysis purpose of device, investigation is carried out in terms of drain current profile, subthreshold swing (SS) and parasitic capacitances. The device has been recorded with steepest SS=35mV/Dec, robust OFF-current (I_{OFF})=1.51X10^{-19} A/μm, high I_{ON}=1.52X10^{-5} A/μm. Designing and analysis of the proffered structure has been executed using Technology Computer-Aided Design (TCAD) 3D device computation software.

Keywords. Gate all around (GAA), Etched Drain (ED), Elevated density Strip (EDS), Hetero-Substrate (HeS) Material, Tunnel Field Effect Transistor (TFET)

1. INTRODUCTION

For electronics market, MOSFETs has been the most prior choice in terms of manufacturing due to its unbeatable performance. However, during the excessive downscaling in nanoscale technology, short channel effects (SCEs) have been the only cause of degradation in performance, most prominently in terms of current profiles [1-4]. Compared to MOSFETs, TFET has an upper with regard to various parameters such as low I_{OFF}, SS <60mV/dec as well as higher switching speed. For better performance and overcoming SCEs, TFETs belonging to Band-to-Band tunneling (BTBT) phenomenon can be most emerging alternative. Instead of various advantages, conventional TFET faces major degrading issues such as low I_{ON} and ambipolar behavior [5-12]. Tunneling capability and ON-state current (I_{ON}) for proffered structure can be anticipate by utilizing Wentzel–Kramers–Brillouin (WKB) approximation [5].

$$I_{ON}\,\alpha\,T_{WKB} \approx \exp\left(\frac{-4\lambda\sqrt{2m*}\sqrt[3]{Eg}}{3qh(Eg+\Delta\emptyset)}\right) \qquad (1)$$

IOP Conf. Series: Materials Science and Engineering 1166 (2021) 012044 doi:10.1088/1757-899X/1166/1/012044

Here, m* represents mass of carrier, Eg is semiconductor material energy band gap, λ is geometrical-dependent tunneling length, q depicts the electron majority charges and $\Delta\emptyset$ is difference in energy for source region's valence band and channel's conduction band. WKB approximation recommends I_{ON} can be ameliorated by minimizing λ, m* and E_g. Few pioneering examples for addressing performance issues of TFET are Dual-material Gate TFET [13], Drain Underlap GAA TFET[14],Double-gate TFET [15-17], hetero-gate TFET[18], Dual-material Tri Gate TFET[19], SOI based TFET[20], Fin-TFET [21]. Although, the mentioned devices depicted elevated performance, still some crucial parameters such as low ambipolar behavior resulting in high leakage current, thermal issues and low SS overall affecting the device performance. In this work, hetero-substrate (HeS) material cylindrical gate all around (GAA) etched drain (ED) with elevated density strip (EDS) TFET is proposed. The proffered structure consists of twain substrate materials with distinct band-gaps and a etched drain at oxide level situated at drain side. Further, the proposed device is integrated with elevated density strip (EDS) layer inserted across the source-channel junction. Analyzed simulated characteristics depicts, in comparison to a regular TFET device with different referred devices, the proposed device with hetero substrate depicts better performance in terms of immunity towards SCE's especially I_{OFF}, high speed and better electro-static gate controllability. The proffered device will be designed and analyzed with the supporting tool tcad synopsys.

2. Device Modeling And Simulation Environment

The proffered idea in presented work is a hetero substrate material cylindrical GAA TFET with an EDS inserted over the junction formed across source and channel region of device. Fig 1(a) depicts the 3D structure of proffered structure whereas Fig. 1(b) displays 2D cutline illustration for the proffered structure. Furthermore, the proffered device has been calibrated with [14]. The hetero substrate encompasses of two material regions with different band-gap: (i) region-I low band-gap material (Si-Ge) for source and channel region, and (ii) region-II high band-gap material (GaAs) for drain region [22]. The region of hetero substrate with SiGe material is placed for source and channel for reducing the fringing field from source to gate. Premise is critical as a reduction in fringing field increases the tunneling across the source to gate, thereby higher I_{ON}. Whereas, the placement of high band-gap material for the drain side results in high tunneling barrier width effect and thereby low I_{OFF} or leakage current [14]. Furthermore, use of SiO_2 as oxide material under gate will reduce the fringing field effect at source-to-channel region resulting in high I_{ON} and better SS [7]. The device modeling parameters used in TCAD Synopsys are: Etched Drain Thickness (t_{ED})=5nm, Oxide thickness (t_{oxi}) =2nm, channel length (L_{chnl}) = 30nm, radii(R) = 7nm, gate work function =4.8 eV(Tungsten diSilicide), P^{++} doped (elevated doped) ESD for conc.=$10^{20}/cm^3$, P^+ channel doped for conc = $10^{15}/cm^3$, P^+ source doped for conc.= $10^{18}/cm^3$, N^+ drain doped for conc.=$5x10^{17}/cm^3$. Extended length for source (L_{sxted}) and drain (L_{dxted}) = 40nm. Non-local BTBT model, trap assisted tunneling (TAT) model to measure effect of I_{OFF}, Shockley Read Hall (SRH) model for fixed minority carriers lifetime [23].

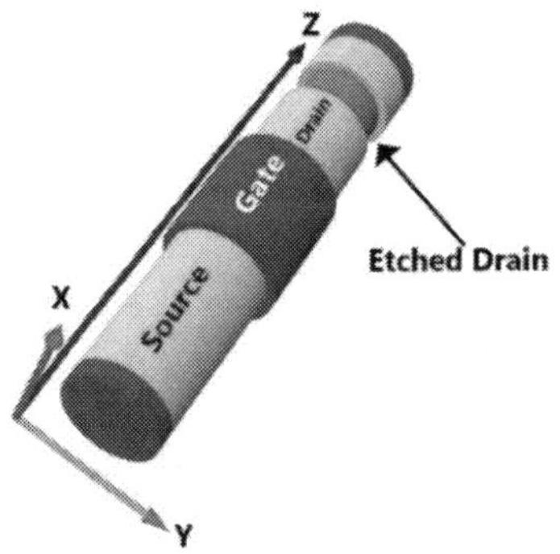

(a)

IOP Conf. Series: Materials Science and Engineering 1166 (2021) 012044 doi:10.1088/1757-899X/1166/1/012044

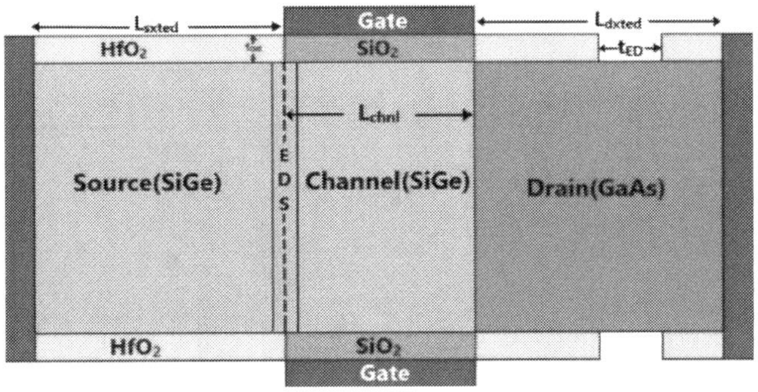

(b)

Figure 1(a) 3D illustration, and (b) 2D Cutline illustration of Cyl. GAA hetero-substrate TFET

3. Results And Discussion

The simulated DC plots for HeS-EDS and other referred devices are presented in Fig. 2. Presented characteristics demonstrates, it is observed that the robust I_{OFF} is $1.51X10^{-19}$ A/μm at V_{gs} = 1.5V for hetero substrate-EDS TFET as high band-gap substrate material is present at drain region it produces high fringing field effects within drain channel junction and causes increase in tunnel barrier width, which weakened electric field at drain side. Whereas the incorporation of low band-gap material at source with EDS and channel deteriorate the fringing field effects and elevate the probability of tunneling charges carriers by reducing barrier width at source-to-channel junction and stimulate high electric field to conduct high I_{ON} as compare to other [7],[14] structure.

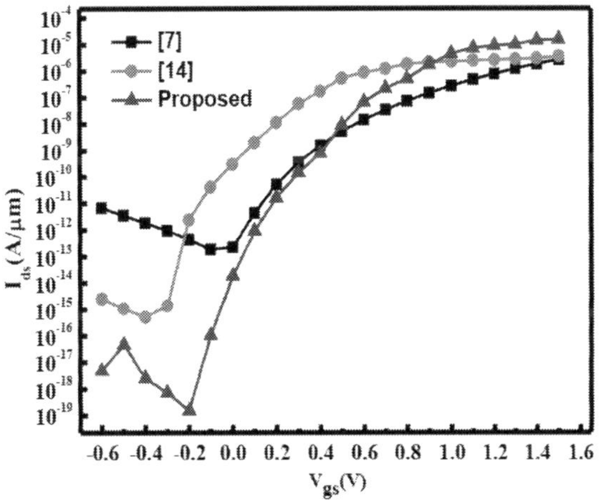

Figure 2: represents I_{ds} vs V_{gs} simulated transfer plots at V_{ds}= 1 V for Cyl. GAA TFET at hetero-substrate with EDS with [7] [14]

Thus, the performance attributes for the proffered structure are robust and optimized for the improved reliability as compared to other reported work [7][14]. Here ON current has been computed using WKB approximation as referred in eq (1). Moreover, the obtained result is very prominent for the scaled L_{chnl} ~ 30 nm and to obtain the steepest subthreshold swing as 35 mV/decade.

Fig. 3 portrays, analyzed electric field characteristics for all the considered devices. Elevated electric field for proffered device due to presence of EDS at source-to-channel region enhances the density of charges for the particular region which signifies it's current driving capability as compared to [7][14] structures [19]. Current driving capability can be resulted as high tunneling probability at source-to-

IOP Conf. Series: Materials Science and Engineering 1166 (2021) 012044 doi:10.1088/1757-899X/1166/1/012044

channel edge. Furthermore, at drain region, due to incorporation of GaAs (high band-gap material) based substrate junction tunneling width will be broaden resulting in steepest electric field for the drain area.

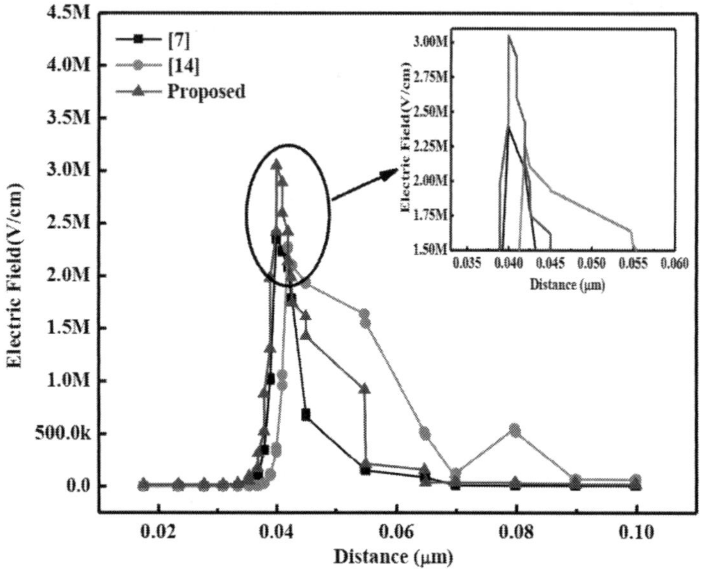

Figure 3: Simulated electric field characteristics for proffered structure with referred work

Fig 4. depicts transconductance (g_m) for all the structures, g_m plays a vital role it terms of driving capability of the current. Furthermore, due to presence of EDS and SiGe (Low band-gap material) as substrate for source and channel areas the charges collected at the source-to-channel edge aids for enhancement of optimum g_m as compared to [7] and [14].

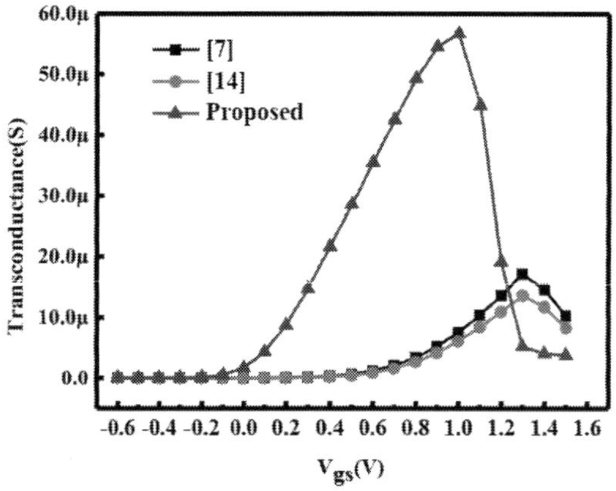

Figure 4: Computed transconductance characteristics for proffered structure with referred work

Fig. 5 and 6 shows Cgd and Cgs at drain voltage (V_{ds}) = 1V and channel length (L_{chnl}) = 30nm as an operation of V_{GT} for all considered structures.

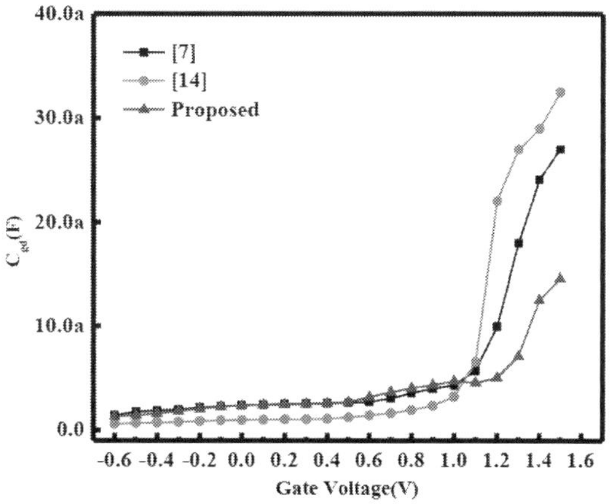

Figure 5: Simulated Characteristics for C_{gd} vs V_{gs} for proposed structure with referred structures

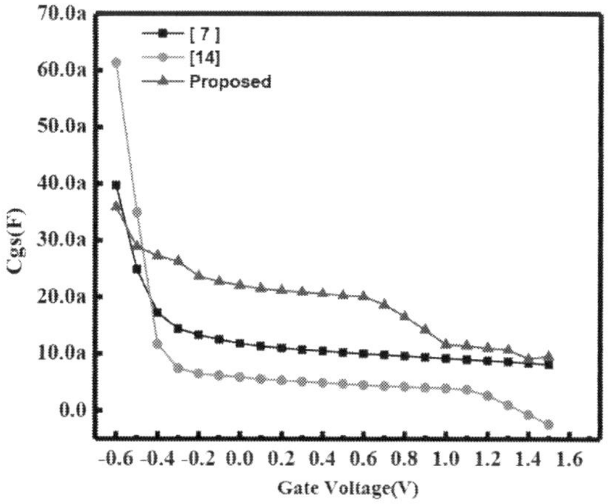

Figure 6: Simulated Characteristics for C_{gs} vs V_{gs} for proposed structure with referred structures

A low C_{gd} and a high C_{gs} are examined in the results for the hetero substrate-EDS in contrast to [7][14], where a crucial aspect plays into significant the switching speed of the proffered TFET structure using the computation delay = $C_{gg}V_{DD}/I_{ON}$, Where $C_{gg} = C_{gd}+ C_{gs}$, In which $C_{gd} = C_{dif} + C_{dov} + C_{of} + C_{gdinv}$ and $C_{gs} = C_{of} + C_{sif}$. Here C_{dov} is overlap capacitance for drain side, C_{sif} and C_{dif} are the capacitance for interior fringing fields and C_{of} is the capacitance for exterior fringing fields [7]. It has been examined that C_{gd} increases corresponding to rise in gate-to-source voltage (V_{gs}) because of the accumulation of a charge depletion layer from drain-to-source region and reduces drain-to-channel barrier width. Moreover, due to broadening of the depletion width the linking within the source and channel region weakens that results in drop in C_{gs} with respect to V_{gs}.. Table 1 represents the summarized relative inspection of all the work compared with the proposed work as well as some experimental work on the basis of device parameters.

Table 1. Relative inspection of proposed device with other referred work and experimental work

Parameters	[7]	[14]	[8]	[21]	Proposed
I_{ON}	2.62μ	3.52μ	12.1μ	0.139μ	15.2μ
I_{OFF}	186f	0.514f	5.4n	4.96p	0.151a
I_{ON}/I_{OFF}	14M	14.5G	2.24k	28k	100T
C_{gs}(aF)	8.13	2.35	-	-	9.6
C_{gd}(aF)	27.6	35.2	-	-	14.6
SS(mV/Dec)	54.4	52.6	52.8	81	35

4.CONCLUSION

The paper presents, a novel idea of Cyl. GAA Hetero-substrate TFET having an EDS at source-to-channel end and ED at drain side. The simulated investigation of the impact of spacer technology on the proffered device is made. The device has been computed and tested regarding diverse material differences and detected robust I_{OFF} of $1.51X10^{-19}$ A/μm at $V_{GS} = 1.5$ V and SS of 35 mV/decade and I_{ON}/I_{OFF} is more than 10^{14} for threshold voltage of 0.5 V. Furthermore, a low C_{gd} of 14.6 aF and high C_{gs} of 9.6 aF at $V_{ds} = 1.0$ V is observed. Therefore, the proffered structure will be valuable for nanoscale and for system-on-chip (SoC) applications. For future work, the proffered structure can be implemented at circuit level and investigated.

ACKNOWLEDGMENT

Authors appreciate the technical facilities furnished by VLSI Circuit & System Design Research Group, IIT Indore, Indore.

REFERENCES

[1]. Mertens H., Ritzenthaler R, HikavyyA., Kim M.S., Tao Z., Wostyn K. 2016 Gate-All-Around MOSFETs based on Vertically Stacked Horizontal Si Nanowires in a Replacement Metal Gate Process on Bulk Si Substrates *Symposium on VLSI Technology Digest* 1-2

[2]. Kim S., Kim M., Ryu D., *et.al.* 2020 Investigation of Electrical Characteristic Behavior Induced by Channel-Release Process in Stacked Nanosheet Gate-All-Around MOSFETs *IEEE Trans. Electron Devices* **67** 6 2648-2652

[3]. Saxena R.S., Kumar M.J. 2009 Stepped Oxide Hetero-Material Gate Trench Power MOSFET for Improved Performance *IEEE Trans. Electron Devices* **56** 6 1355-1359

[4]. Jaafar H., Aouaj A., Bouziane A., Iñiguez B. 2018 An analytical drain current model for cylindrical gate DMG-GC-DOT MOSFET *International Journal of Electronics Letters* **7** 4 1-15

[5]. Beohar A., Vishvakarma S.K. 2016 Performance enhancement of asymmetrical underlap 3D-cylindrical GAA-TFET with low spacer width *IET Micro & Nano Letters* **11** 8 443–445

[6]. Jhan Y.R., Wu Y.C., Hung M.F. 2013 Performance enhancement of nanowire tunnel field-effect transistor with asymmetry-gate based on different screening length *IEEE Electron Device Lett.* **34** 12 1482–1484

[7]. Vishvakarma S.K., Beohar A., Vijayvargiya V., Trivedi P. 2017 Analysis of DC and analog/RF performance on Cyl-GAA-TFET using distinct device geometry *Journal of Semiconductor* **38** 7

[8]. Choi W. Y., Park B-G., Lee J. D., Liu T-J. K. 2007 Tunneling Field-Effect Transistors (TFETs) With Subthreshold Swing (SS) Less Than 60 mV/dec *IEEE Electron Device Lett.* **28** 8 743-745

[9]. Chattopadhyay A., Malik 2011 A. Impact of a spacer dielectric and a gate overlap/underlap on the device performance of a tunnel field-effect transistor *IEEE Trans. Electron Devices* **58** 3 677–683

[10]. Wu J., Min J., Taur Y. 2015 Short-channel effects in tunnel FETs *IEEE Trans. Electron Devices* **62** 9 3019–3024
[11]. Kim S.H., Aggarwal S., Jacobson Z.A., Matheu P., Hu C., Liu T-J. K. 2010 Tunnel field effect transistor with raised germanium source *IEEE Electron Device Lett.* **31** 10 1107–1109
[12]. Pandey C.K., Dash D., Chaudhury S. 2020 Improvement in Analog/RF Performances of SOI TFET Using Dielectric Pocket *International Journal of Electronics* **107** 11 1-17
[13]. Saurabh S., Kumar M.J. 2011 Novel Attributes of a Dual Material Gate Nanoscale Tunnel Field-Effect Transistor *IEEE Trans. Electron Devices* **58** 2 404-410
[14]. Beohar A., Yadav N., Vishvakarma S.K. 2017 Analysis of Trap Assisted Tunneling in Asymmetrical Underlap 3D-Cylindrical GAA-TFET based on Hetero-Spacer Engineering for Improved Device Reliability *IET Micro & Nano Letters* **12** 12 982-986
[15]. Saurabh S., Kumar M.J. 2009 Impact of Strain on Drain Current and Threshold Voltage of Nanoscale Double Gate Tunnel Field Effect Transistor: Theoretical Investigation and Analysis *Japanese Journal of Applied Physics* **48** 064503
[16]. Narendiran A., Akhila K., Bindu B. 2015 A Physics-Based Model of Double-Gate Tunnel FET for Circuit Simulation *IETE Journal of Research* **62** 3 1-7
[17]. Vijayvargiya V., Vishvakarma S.K. 2014 Effect of drain doping profile on double gate tunnel field-effect transistor and its influence on device RF performance *IEEE Trans. Nanotechnology* **13** 5 974–981
[18]. Choi W.Y., Lee W. 2010 Hetero-gate-dielectric tunneling field-effect transistors *IEEE Trans. Electron Devices* **57** 9 2317–2319
[19]. Saha P., Sarkhel S., Sarkar S.K. 2018 3D Modelling and Performance Analysis of Dual Material Tri-Gate Tunnel Field Effect Transistor *IETE Technical Review* **36** 2 1-13
[20]. Goswami P.P., Bhowmick B. 2019 Optimization of electrical parameters of pocket doped SOI TFET with L shaped gate *Springer Journal of Silicon*, **12** 693–700
[21]. Kim J. H., Kim H. W., Kim G., Kim S., Park B-G. 2019 Demonstration of Fin-Tunnel Field-Effect Transistor with Elevated Drain *Micromachines* **30** 1-10
[22]. Beohar A., Yadav N., Shah A.P., Vishvakarma S.K. 2018 Analog/RF characteristics of a 3D Cyl underlap GAA-TFET based on a Ge-Source using fringing field engineering for low power applications *Springer Journal of Computational Electronics* **17** 4 1650-1657
[23]. Sentaurus device manual 2017.1, https://www.synopsys.com/silicon/tcad/sentausdevice.html.

IOP Conf. Series: Materials Science and Engineering 1166 (2021) 012045 doi:10.1088/1757-899X/1166/1/012045

In-silico Investigation of Cyl. Gate all Around (GAA) Tunnel Field Effect Transistor (TFET) Biosensor

Sanjana Tiwari[1], Arya Dutt[1], Mayuresh Joshi[1], Prakhar Nigam[1], Ankur Beohar[1*], Ribu Mathew[1]

[1]School of Electrical & Electronics Engineering (SEEE) VIT Bhopal University Bhopal, India

[*]Corresponding author: *ankurbeohar16@gmail.com*

Abstract. For some time now, the advancement of low power and high sensitivity biosensors has been the center of attention for *in-situ* detection and monitoring which form an integral part of portable health monitoring systems. This paper elucidates the design-optimization of a cylindrical (cyl.) gate all around (GAA) tunnel field effect transistor (TFET) biosensor with retrograde doping using numerical modeling. The device consists of n^+ heavily doped SiGe substrate and two insulated gates i.e. primary gate (PrG) and biasing gate (BG) with suitable work functions. Sensitivity of the biosensor is investigated by varying dielectric constant (k) and charge density (ρ) in the active region of device. TFET biosensor design and simulation is performed using TCAD Synopsys software. Computations are carried out for various conditions of dielectric constant (k) and charge density (ρ) for analyzing the sensor sensitivity. Simulation results show that for k=10 and $\rho = 3.0 \times 10^{12} cm^{-2}$, there is low leakage current (I_{OFF}) = 1.0×10^{-16} A/μm, and high ON current (I_{ON}) = 1.0×10^{-6} A/μm. Results obtained in this work as useful as it will act as a design guideline for developing TFET biosensors for various biological applications.

Keywords. Biosensor, TFET, Gate all around (GAA), short channel effect, leakage current

1. Introduction

Presently, the center of attention have been the enhancement of biosensors due to a surge in new strains of bacteria and virus induced diseases like COVID 19. For *in-situ* detection of bio-molecules, the desired characteristics of a biosensor includes increase in susceptibility, rapid latency, immense span of measurement, real time detection, portability, low power consumption, to cite a few. Literature encompasses examples of devices for detecting bio-molecules with various technologies [1-9].

One of the solutions that has been highly explored in recent times is field effect transistor (FET) based biosensor. Even though, FET biosensors depict desired characteristics, performance of such sensors is limited due to the short channel effects(SCEs). In contrast to a conventional FET device, a Tunnel Field Effect Transistor (TFET) with gate all around (GAA) structure depicts lower subthreshold swing (SS) with reduced SCEs. Some innovational solutions for the performance improvement of GAA TFET include source pocket Silicon (Si) TFET [10], 2-D tunneling FET with localized charges [11], double gate TFET with drain underlap [12-14], cylindrical TFET with reliability issues [15], hetero gate dielectric TFET [16], asymmetrical underlap 3D-cylindrical GAA-TFET with low spacer width [17]. Although Si based GAA TFET depicts high performances, it has limitations such as low thermal stability, high OFF current (I_{OFF}) and high SS due to ambipolar behaviour.

In this paper, we have performed the design-optimization of a cylindrical (cyl.) gate all around (GAA) tunnel field effect transistor (TFET) biosensor. Furthermore, detailed investigation is performed to comprehend the impact of multiple specification on sensor performance i.e. sensitivity. The paper has the following sections: the device details and modelling-simulation environment are explained in section 2. Subsequently, results obtained through numerical simulation are summarized in section 3, and the conclusion of the computations is detailed in segment 4.

2. Device Details and Modelling Parameters

The device under study is a cyl. GAA TFET with retrograde doping. 3D structure and the transversal view of the proffered structure are shown in Figure 1(a) and (b) respectively. Retrograde doping is incorporated in subsidiary band gap substrate (SiGe) at the source to channel edge. Parameters of device and simulations environment are as follows: p-channel doping concentration in the core region=$10^{15}cm^{-3}$ and outer region=$10^{14}cm^{-3}$, source-region doping concentration (p-type)=$1x10^{18}cm^{-3}$, drain-region doping concentration (n-type) = $5x10^{20}cm^{-3}$, oxide thickness (t_{oxide}) = 5nm , channel length ($L_{channel}$) = 20nm , gate work function of control gate (CG) = 4.1eV , gate work function of polarity gate (PG) =4.53eV, extended source length (L_{extss}) and drain length (L_{extdd}) = 40nm. The device modelling and simulation has been performed using TCAD Synopsys software.

The device substrate material is Silicon-Germanium (Si-Ge) which has better SS due to narrow bandgap. The RD doping used exemplify a high doping region (inner-region) that incorporates 25% of the channel region with remaining portion relatively less than the inner region [18]. Premise results in reduced SCEs, and improves I_{OFF} along with SS. The primary gate (PrG) and biasing gate (BG) have been modeled with appropriate work function as well as polarity controlled voltage. By etching an adequate segment of PrG dielectric section Nano gap cavity is incorporated inside the CG dielectric about the BG edge. This region acts as the active site for biomolecule sensing. As a effect of low-k dielectric (SiO_2) a reduction in the influence of fringing field is observed. Simulation models like band-to-band tunneling (BTBT) model, trap assisted tunnelling (TAT) which represents the supremacy of the I_{OFF}, Shockley Read Hall (SRH) for creation and rebinding of the hole-electron pairs in the semiconductors, Wentzel Kramer's Brillouin (WKB) approximation and Kane's models have been incorporated for device modeling and simulation [19].

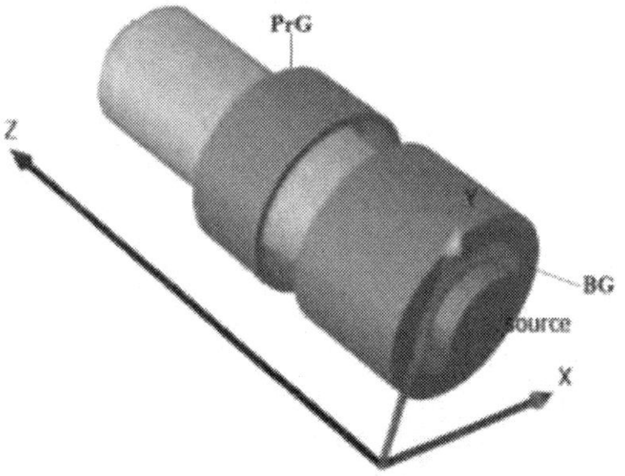

(a)

IOP Conf. Series: Materials Science and Engineering 1166 (2021) 012045 doi:10.1088/1757-899X/1166/1/012045

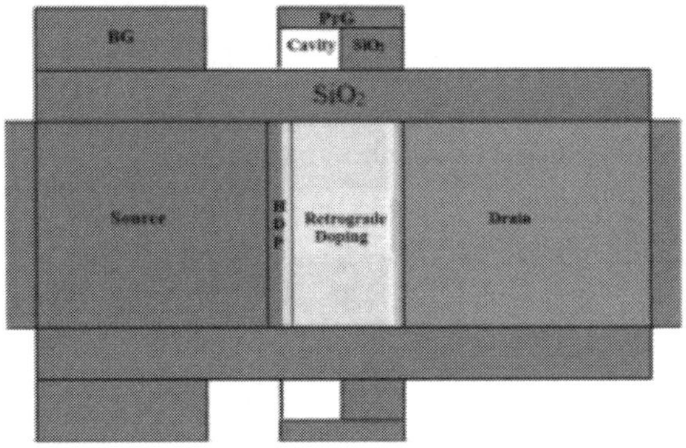

(b)

Figure 1: (a) 3-D view of Cylindrical (Cyl) Gate all around (GAA) highly doped cavity retrograde doped TFET, (b) transversal view of the device

3. Results and Discussion

This segment, analyzes the influence of various parameters on the receptiveness of the retrograde doping based TFET biosensor. The target biomolecule concentration variation is mimicked as change in dielectric constant (k). The condition k=1 depicts inadequacy of biomolecules, whereas k>1 represents the occurrence of biomolecules in the active region of the biosensor. The tunnel barrier reduces as a result of increasing k which intensifies band bending. Rising barrier width results in elevated negative charge concentration. A lower measure of surface potential is observed beneath the cavity sector when biomolecules are inadequate (k=1). On the other hand, for the availability of biomolecules (k>1), the surface potential attains a greater value beneath the cavity sector. In this work, dielectric constants are varied as k=1, 5 and 10, and charge densities are imparted as $\rho = -5x10^{-11}$, $-1x10^{-12}$ and $-3x10^{-12}$ cm^{-2}) depicting the variation in the biomolecules [20]. The highly doped pocket placed at the source edge results in a tunnel gap which induces a restraint between conduction and valence bandwidth resulting in a tunneling event at source-channel interface as a result of low band gap substrate. Furthermore, tunnel barrier width at the drain-channel interface increases as a result of high band gap substrate at the drain side.

The variation in drain current relies on the parameters k and ρ for different gate source voltage (V_{GS}) is shown in Fig.2. With the rise in accumulation of biomolecules the switching current (i.e I_{ON}) of the device increases. Furthermore the inclusion of retrograde doping in the device results in lower leakage current (I_{OFF}). It also resolves the concern of a higher SS as in conventional MOSFET's.

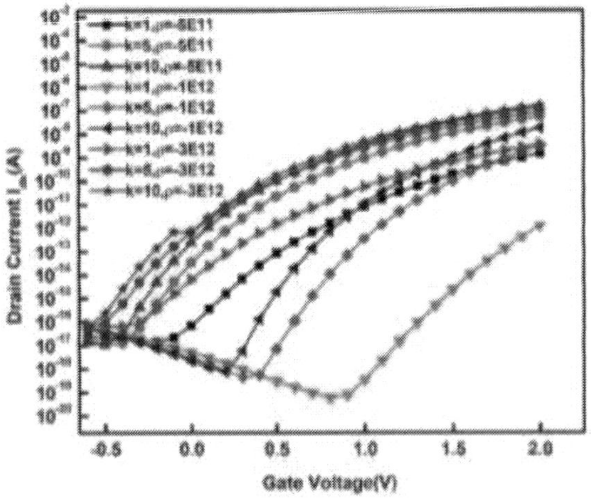

Figure 2: Drain current as a function of parameters k and ρ for different V_{GS}

The variation in electric field (EF) relies on the parameters k and ρ along the longitudinal direction (length of the device) is shown in Fig.3. With the rise in k the gate-channel coupling elevates the surface potential. On account of accumulation of the biomolecules which promotes to an improved flow of charges results in higher EF, and thereby higher tunneling across the valence and conduction bands. With increase in the parameter k values, at the source-channel intersection, the EF improves effectively, also EF reduces as the tunneling barrier increases.

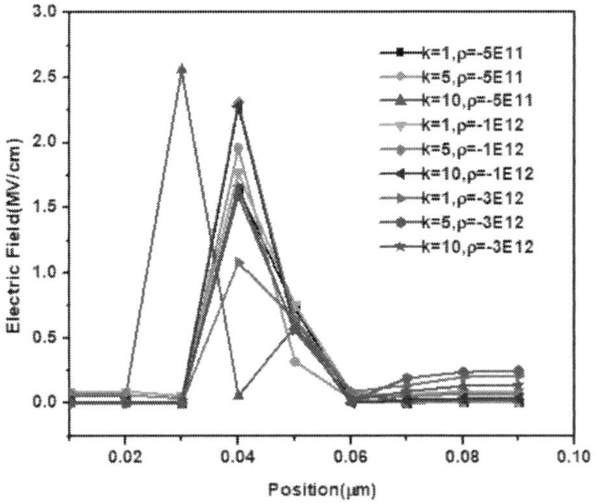

Figure 3: Electric field as a function of parameters k and ρ along sensor length

The variation in current density relies on the parameters k and ρ alongside the longitudinal direction (length of the device) is shown in Fig.4. The current density depends on various parameters such as bandgap of the material, also the inclusion of the WKB approximation supports for higher current density. Furthermore it is witnessed that as the concentration of biomolecules rises, there is an enhancement in the density of charges resulting in a higher current density.

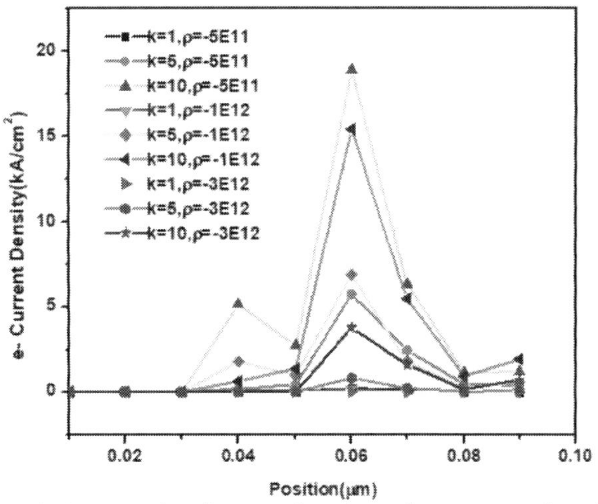

Figure 4: Current density as a function of parameters k and ρ along sensor length

Under the cavity region at source-channel intersection as the electric field reduces, an increase in charge concentration is observed leading to reduced I_{ON}. As dielectric constant increases the I_{ON}/I_{OFF}

IOP Conf. Series: Materials Science and Engineering 1166 (2021) 012045 doi:10.1088/1757-899X/1166/1/012045

ratio tends to increase. Moreover it reduces as the negative charge density increases. Sensitivity is defined by eq.(1)

$$S_{drain}=\left(\frac{\left(I_{drain}^{biomolecule}-I_{drain}^{air}\right)}{I_{drain}^{air}}\right) \quad (1)$$

I_{drain}^{air} is the drain current with k=1 representing biomolecules are absent and $I_{drain}^{biomolecule}$ is the drain current with k > 1 representing the occurrence of biomolecules in the cavity section and as the dielectric constant rises the sensitivity of the drain current also rises. Furthermore, the ability of sensing the biomolecular element in the cavity is represented by a significant repositioning in the drain current sensitivity across gate-to-source voltage (V_{gs}). With variations in the dielectric constant and charge density at low V_{ds} an enhancement of drain current sensitivity is observed, for a biosensor based application [21-24].

The variation in device performance relies on the parameters k and ρ for different V_{GS} values are summarized in Table 1. Device capacitances (C_{gs}) and (C_{gd}) as a function of parameters k and ρ for various V_{GS} is shown in Figure 5 and Figure 6 respectively. The parasitic capacitances play a crucial role in sensor's switching capability as well as leakage current. The parameter C_{gd} is calculated by eq.(2)

$$C_{gd} = C_{of} + C_{dif} + C_{gdinv} \quad (2)$$

where the symbol C_{of} represents outer fringing capacitance and C_{dif} represents drain region inner fringing field capacitance. The symbol C_{gdinv} represents inverted gate-drain capacitance [17].

In TFET, the parameter C_{gd} should be less for better performance. As depicted in Figure 5 the steepest C_{gd} is observed for k=1 and ρ = -3.0x10^{-12} cm^{-2}. As mentioned earlier k=1 signifies absence of biomolecules in the cavity section under gate due to which the barrier width increases hindering the tunneling channel-to-drain section.

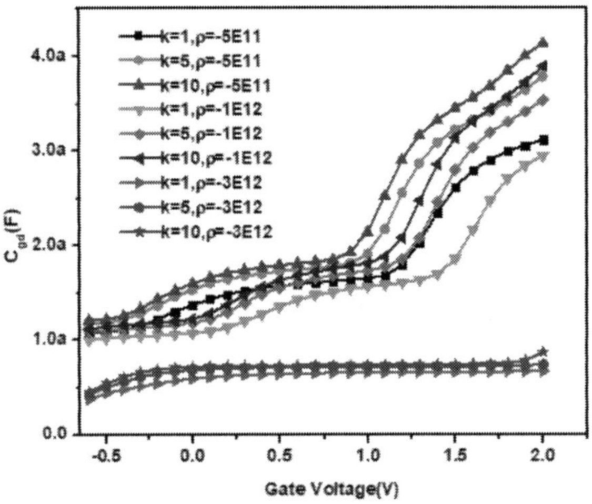

Figure 5: Capacitance (C_{gd}) as a function of parameters k and ρ for various V_{GS} values

The parameter C_{gs} can be obtained by eq.(3)

$$C_{gs} = C_{of} + C_{sif} \quad (3)$$

where, C_{of} represents outer fringing capacitance and C_{sif} depicts source region inner fringing field capacitance [17]. Due to the presence of SiGe HDF layer it results in a higher C_{gs} value.

Table 1. Change in device performance as a function of parameters k and ρ

Parameters	ρ = -5.0x10^{-11} cm^{-2}			ρ = -1.0x10^{-12} cm^{-2}			ρ = -3.0x10^{-12} cm^{-2}		
	K=1	K=5	K = 10	K=1	K=5	K = 10	K=1	K=5	K = 10
I_{ON} (A)	3.68n	90n	216n	1.5n	57n	150n	1200n	3.21n	18.3n
I_{OFF} (A)	81.9a	93.5a	63.8a	71.7a	58.4a	33.7a	.093a	.858a	.786a
I_{ON}/ I_{OFF}	.0449G	.963G	3.38G	.021G	.976G	.445G	.0128G	3.74G	23.3
EF(V/µm)	1.65M	1.96M	.586M	1.08M	1.59M	1.66M	1.77M	2.31M	2.28M
ECD (A/cm^2)	.216k	5730k	18.9k	.184k	6.88k	15.4k	.001k	.824k	3.77k
C_{gs} (F)	2.57a	3.79a	4.31a	2.66a	3.89a	4.41a	4.59a	6.41a	7.21a
C_{gd} (F)	1.08a	1.17a	1.21a	1.0a	1.09a	1.12a	.379a	.428a	.461a

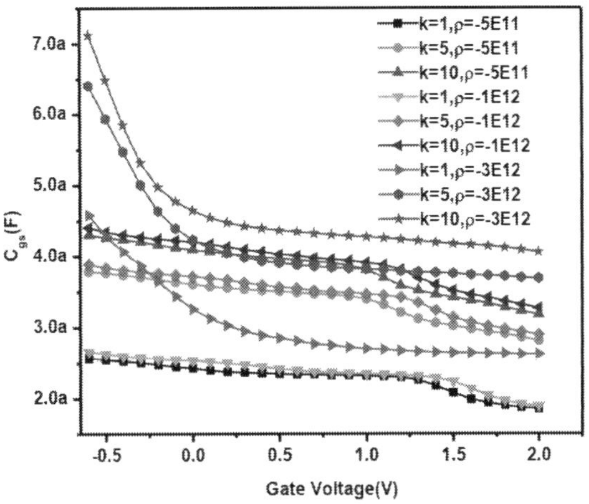

Figure 6: Capacitance (C_{gs}) as a function of the parameter k and ρ for different V_{GS} values

4. Conclusion

This paper elucidates the numerical investigation of a cyl. gate all around (GAA) TFET biosensor. Sensitivity of the biosensor is analysed using device simulation software. The variation in the concentration of biomolecules in the active region of the biosensor is mimicked by varying the dielectric constant (k) and charge density (ρ). The sensor performance has been investigated through the DC and transient analyses. Parameters which have been observed in this study include I_{ON}, I_{OFF}, C_{gs}, C_{gd}, EF Intensity and electron current density for variation in dielectric constant and charge density depicting biomolecule concentration change. Simulation results show that in the cyl. GAA retrograde doped highly doped pocket TFET biosensor, for k=10 and $\rho = 3.0\text{x}10^{12}\text{cm}^{-2}$ the truncated OFF current (leakage current) is ($I_{OFF} = 1.0\text{x}10^{-16}$A/µm), and high ON current is ($I_{ON} = 1.0\text{x}10^{-6}$ A/µm). In addition, it is observed that the presence of highly doped pocket between the source and source ended channel enhances the device efficacy due to reduced influence of fringing electric fields of gate towards the source. Premise is primarily due to tunnelling of barrier width. $C_{gd} = 5.0\text{x}10^{-19}$F and $C_{gs} = 7.0\text{x}10^{-18}$F at drain voltage of $V_{ds} = 1$V have been observed in numerical simulation.

Acknowledgment

Authors would like to give thanks to Nanoscale Devices, VLSI Circuit and System Design Research Group, IIT Indore, Indore for technical support.

References

[1] Mathew R and Sankar A.R 2018 A review on surface stress based miniaturized piezoresistive SU-8 polymeric cantilever platform sensors *Nano-Micro Letters Springer publishers* **10** 35

[2] Mathew R and Sankar A.R 2015 Design of a triangular platform piezoresistive affinity microcantilever sensor for biochemical sensing applications *J. Physics D: Applied physics IOP publishers* **48** 205402

[3] Mathew R and Sankar A.R 2018 Impact of isolation and immobilization layers on the electro-mechanical response of piezoresistive nano cantilever sensors *Journal of Nano science and Nano technology ASP* **18** 1636-1647

[4] Mathew R and Sankar A.R 2019 Influence of surface layer properties on the thermo-electro-mechanical characteristics of a MEMS/NEMS piezoresistive cantilever surface stress sensor *Materials Research Express, IOP publishers* **6** 086304

[5] Mathew R and Sankar A.R 2020 Temperature drift-aware material selection of composite piezoresistive micro-cantilevers using Ashby's methodology *Microsystem Technologies, Springer publications* 1-14

[6] Mathew R and Sankar A.R 2018 Optimization of a Nano-Cantilever Biosensor for Reduced Self-heating Effects and Improved Performance Metrics *Journal of Micromechanics and Microengineering, IOP publishers* **28** 085012

[7] Vijh, M., Gupta, R. S., and Pandey, S. 2019 Investigation of Tunnel Field Effect Transistor for Biosensing Applications *PhotonIcs & Electromagnetics Research Symposium - Spring (PIERS-Spring)* 229-233

[8] Kanungo S., Chattopadhyay S, *et al.* 2015 Comparative Performance Analysis of the Dielectrically Modulated Full- Gate and Short-Gate Tunnel FET based biosensors *IEEE, Transactions on Electron Devices* **62** 3

[9] Kumar, S., Singh, Y., & Singh, B. 2020 Extended Source Double-Gate Tunnel FET Based Biosensor with Dual Sensing Capabilities *Silicon*

[10] Chang H, Adams B, Chien P *et al.* 2013 Improved Subthreshold and Output Characteristics of Source- Pocket Si Tunnel FET by the Application of Laser Annealing *IEEE Transactions on Electron Devices* **60** 1 92-96

[11] Vishnoi R and Jagadesh M K 2014 Two Dimensional Analytical Model for the Threshold Voltage of a Tunneling FET with Localized Charges *IEEE Trans. on Electron Devices* **61** 3054-3059

[12] Xu P, Lou H, Zhang L *et al.* 2017 Compact Model for Double-Gate Tunnel FETs With Gate–Drain Underlap *IEEE Transactions on Electron Devices* **64** 12 5242-5248

[13] Pon A. Tulasi K.S.V.P., Ramesh R. 2019 Effect of interface trap charges on the performance of asymmetric dielectric modulated dual short gate tunnel FET *AEÜ International Journal of Electronics and Communication* **102** 1-8

[14] Priya G L, Balamurugan N.B 2019 New dual material double gate junctionless tunnel FET: Subthreshold modeling and simulation *AEÜ International Journal of Electronics and Communication* **99** 130-138

[15] Beohar A, Yadav N and Vishvakarma S.K 2017 Analysis of Trap Assisted Tunneling in Asymmetrical Underlap 3D-Cylindrical GAA-TFET based on Hetero-Spacer Engineering for Improved Device Reliability *IET Micro & Nano Letters* **12** 12 982-986

[16] Rahimian M, Fathipour M 2017 Improvement of electrical performance in junctionless nanowire TFET using hetero-gate-dielectric *Materials Science in Semiconductor Processing* **63**

[17] Beohar A., Vishvakarma S.K. 2016 Performance enhancement of asymmetrical underlap 3D-cylindrical GAA-TFET with low spacer width *IET Micro & Nano Letters* **11** 8 443–445

[18] Gupta D and Vishvakarma S K 2016 Improvement of Short Channel Performance of Junction-Free Charge Trapping 3-D NAND Flash Memory *Micro & Nano Letters* **12** 1

[19] Sentaurusdevicemanual2017.1,"https://www.synopsys.com/silicon/tcad/sentausdevice.html.

[20] Venkateshm P., Nigam K., Sharma D. *et al.* 2017 A dielectrically modulated electrically doped tunnel FET for application of label free biosensor *Superlattices Microstructures* **109** 470–479

[21] Choi J.M., Han J.W., Choi S.J., *et al.* 2010 Analytical modeling of a nanogap embedded FET for application as a biosensor *IEEE Trans. Electron Devices* **57** 12 3477–3484

[22] Narang R., Saxena M., Gupta R.S., *et al.* 2012 Dielectric modulated tunnel field-effect transistor-A biomolecule sensor *IEEE Electron Device Letter* **33** 2 266–268

[23] Sarkar D., Banerjee K. 2012 Proposal for tunnel field-effect transistor as ultra-sensitive and label free biosensors *Applied Physics Letters* **100** 143108-1–143108-4

[24] Narang R., Reddy K.V.S., Saxena M., *et al. 2012* A dielectric modulated tunnel FET based biosensor for label free detection: analytical modelling study and sensitivity analysis *IEEE Trans. Electron Devices* **59** 10 2809–2817

Vehicle Recognition using extensions of Pattern Descriptors

V Keerthi Kiran[1,3*], Sonali Dash[1,4], Priyadarsan Parida[2,5]

[1]Department of Electronics and Communication Engineering, Raghu Institution of Technology, Visakhapatnam, India-531162
[2]Department of Electronics and Communication Engineering, GIET University, Gunupur, India -765022
[3]*v.keerthikiran@giet.edu*
[4]*sonali.isan@gmail.com*
[5]*priyadarsanparida@giet.edu*

Abstract. Vehicle identification and classification for still images are incredibly useful and can be extended to a range of traffic surveillance operations. Reliable and accurate recognition of vehicles is however a challenging issue due to changes in vehicle appearance and illumination difference in real time scene. In this paper, we present a simple and effective way of vehicle recognition technique based on vehicle's local texture features extraction and classification. The local features are extracted individually using the Local Binary Pattern (LBP), Median Binary Pattern (MBP), Gradient directional pattern (GDP), and Local Arc Pattern (LAP) descriptors and feed into Support Vector Machine (SVM) for classification. We also focus on vehicle classification using various color spaces like RGB, HSV, YCbCr for the texture descriptors extraction. The primary focus is to observe the effect of colour information on vehicle classification efficiency across different colour spaces. Initially, experiments are conducted for the classification of gray-level vehicle images of five different classes from the CompCars dataset. Then experiments are extended to different color spaces for the same dataset for color texture classification. The integration of different colour details increases the efficiency of vehicle classification, according to the experimental results.
Keywords: Support Vector Machine (SVM), Local Binary Pattern (LBP), Median Binary Pattern (MBP), Gradient directional pattern (GDP), and Local Arc Pattern (LAP)

1. Introduction

In a wide variety of computer vision applications, such as real-time traffic monitoring, transportation, traffic management, and the ability to distinguish vehicles in surveillance images is essential [1]. For example, video surveillance systems provide fast and accurate data for improved protection and traffic flow such as lane crossings, parked vehicles, traffic congestion, vehicle count as well as identification of the number plate, kind of vehicle, speed, and direction of the vehicles. The main objective of vehicle detection is usually to recognize potential vehicle positions in a given image for further processing tasks, marking them as a region of interest (ROI). In contrast, accurate detection of automobiles is a difficult and complicated task [2]. The challenge stems from the wide range of vehicle appearances (e.g., size, shape, and orientation) as well as reduced visibility due to camera noise or adverse weather conditions (e.g., rain, fog). Besides, the disparities in lighting, occlusion, and an unconstrained background intensify the complexity in comparison with general object detection.

Despite of these challenges, the research community and the transport sector have led to the advancement of various kinds of vehicle monitoring and detection approaches to increase the reliability of the traffic control systems and applications mentioned above. As a result, many new approaches have been suggested for the identification of vehicles in the literature that mostly focused on handcrafted or learning features [3]. Su et al. [4] suggested a method that uses rotation-invariant HOG-like features, which is also one of these methods. In [5] also most popular handcrafted features like Gabor filters, SIFT, and direction gradient have been used. Although these methods are resistant to intra-class dissimilarities, they appear to be susceptible to inter-class distinctions.

Some researchers are exploring learning feature-based approaches in vehicle detection tasks due to their excellent learning capacity. The majority of learning features based methods transform vehicle recognition into a classification problem by constructing vehicle proposals and then categorizing each one separately using classification algorithms such as support vector machines (SVM) or convolutional neural networks (CNN). The majority of current research focuses on categorizing vehicles into major groups, such as motorcycles, cars, buses, and trucks, however, but this lacks enough capabilities to meet user's needs. Many researchers have used frontal images to investigate the identification and recognition of vehicle logos to provide access to details concerning the vehicle manufacturer[6].

Yong Tang et al., [7] used Haar-like features and AdaBoost algorithms to extract features and create classifiers for vehicle recognition for on-road images. They have used histogram intersections to calculate similarities between various Local Gabor Binary Pattern (LGBP) and were used to determine the Euclidean distance in the closest neighborhood. When their model is validated with seven classes and 227 samples, a recognition rate of 92% is published. M. Hassaballah et al.,[8] have suggested an efficient vehicle identification system using clustering forests. Whereas visually identifiable codebooks are constructed using clustering forests based on the Chi-square for the features generated by the Local Binary Pattern (LBP). On the UIUC car dataset, they achieved detection rates of 99.6% for single-scale parts and 98.9% for multiscale parts, respectively. Wenjin Chu et al., [9] have developed a facial expression recognition algorithm that uses Gradient Direction Pattern (GDP), LBP, and Sparse Representation Classification (SRC). They have achieved recognition rates of 50% and 67.14% for GDP + SRC and LBP + SRC respectively and suggested that combining these pattern descriptors improves the recognition rate. To improve the performance of the face recognition system, Niaraki et al., [10] suggested a feature extraction algorithm built on the Co-occurrence Matrix of Local Median Binary Pattern (CMLMBP). The co-occurrence matrix of the processed image is determined after the Local Median Binary Pattern of the input image has been calculated. They have reported a recognition rate of 96.25% and 100%, with the ORL and Yale databases respectively. The continuation of the article is set out as follows: Literature review is discussed in section 2. Section 3 describes the proposed methodology. Experimental results are discussed in section 4. Finally, in Section 5, the conclusions are addressed.

2. Literature review

Two competing goals are expected to achieve a good texture feature, a key component of the texture classification. Local Binary Pattern (LBP) methods have been identified as one of the most prominent and commonly studied texture patterns, with a wide range of LBP variants being suggested to a variety of different applications.

2.1 Local Binary Patterns (LBP)

LBP is a type of texture descriptor that is both simple and powerful [11]. The 3 x 3 window covers each pixel of a grayscale image. The neighboring pixels are modified according to the center pixel. Where the value of neighboring pixels greater than the center pixel is given a value of 1, while the values of pixels with the value below the center pixel are 0. The thresholded neighboring pixels are then combined to form a binary code that defines the texture of the concerned pixel. As the image is divided into a grid, a histogram of binary patterns for each pixel inside a patch is determined. As a result, a histogram is generated for each grid cell, which is then used to allocate a terrain class to the cell. As an 8-bit binary sequence can have 256 values, we can classify it using a histogram of 256 dimensions. Figure 1. shows a 3x3 pixel pattern of an image.

45	26	85
33	64	87
89	75	22

0	0	1
0		1
1	1	0

Figure 1. LBP code generation for a 3 × 3 pixel neighborhood

Binary Pattern = 00110110

2.2. Median Binary Patterns (MBP)pp0

Instead of using the central pixel always, the median binary pattern (MBP) operator maps a localized binary pattern to a threshold of pixels against the median value within the neighborhood to increase the sensitivity to microstructure and noise [12]. The median value of the patch is 120 for the example shown in Figure 2. The following equation is used to measure the MBP at pixel (i, j):

$$\mathrm{MBP}(i,j) = \sum_{k=0}^{L-1} 2^k H(b_k - \tau) \quad (1)$$

where L is the patch size (i.e., L = 9 for a patch of 3×3), and τ is the median of the local patch. When the threshold is set to the patch's median, the resulting binary pattern must contain at least five one bits, and only 256 binary patterns are possible. In this case, the MBP only produces a 256 binary pattern subset rather than the entire 9-bit spectrum of binary patterns (i.e. [0, 511]). By measuring the distribution of these patterns over the image using the MBP histogram, the texture descriptor is generated. The complete image is then modified to a 256 x 1 vector, which represents the MBP histogram that uses the median inside the local image patch.

127	130	135
120	127	88
90	30	43

1	1	1
1	1	0
0	0	0

Figure 2. MBP code generation for a 3 × 3 patch

Here L=9; $\tau = 120$

Output binary pattern MBP = $100011110_2 \equiv 286$.

2.3. Gradient directional pattern (GDP)

The LBP operator determines the threshold based on the center pixel's value. As a result, the LBP codes are vulnerable to noise and lighting fluctuations, as even small adjustments will cause their value to move concerning the center pixel. The direction of each pixel's gradient vector in an image is calculated first with the following equation:

$$\alpha(x, y) = tan^{-1}\left(\frac{G_x}{G_y}\right) \quad (2)$$

The gradient path angle (x, y) for pixels is represented here by $\alpha(x, y)$, and G_x and $G_y$are the two gradient vector elements that can be achieved with the implementation of the Sobel operator on the source image[13]. To obtain the values of G_x and G_y, the Sobel operator convolves the image with a horizontal and vertical kernel, as seen in Figure 3.

-1	-2	-1
0	0	0
1	2	1

a

-1	0	1
-2	0	2
-1	0	1

b

Figure.3. Sobel kernels (a) Horizontal kernel (b) Vertical kernel

The GDP operator selects a 3x3 neighborhood around each pixel of the image after measurement of gradient directions and quantizes the adjacent gradient directional angles to the central pixel direction angle through a threshold τ. In this case, Neighboring pixels with a gradient angle around the center angle are quantized to 1, while the rest are quantized to 0. The center pixel is then given the binary pattern value that has occurred.

2.4. *Local Arc Pattern (LAP)*

Local Arc Pattern (LAP) is a pattern that addresses nearly all of the cost concerns and shortcomings associated with LBP[14]. It computes two different patterns from the local 5x5 pixel region, that represent the local pattern at the center pixel. The LAP technique is a variant of the GDP. In all possible directions, the local pattern at a pixel describes variations in the grey color intensities of its neighbors.

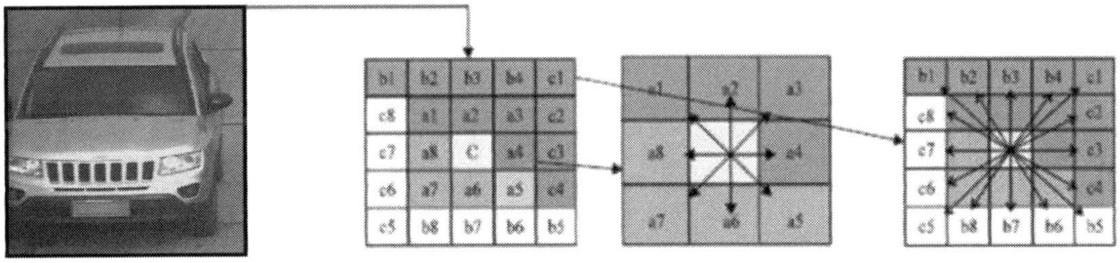

(a) (b) (c) (d)

Figure 4. Local pixels notation (a) Vehicle image, (b) Local 5x5 pixels region, (c) Pixels used for pattern-1, and (d) Pixels used for pattern-2

The LAP pattern of the center pixel of the region, "C" is shown in Figure. 4, can be calculated with a 5 x 5 pixel local region. The LAP Binary patterns are based on grey color-intensity values of a(1-4), b(1-4), and c(1-4). One 4-bit binary pattern (i.e., Pattern-1 (P1)) and one 8-bit binary pattern (i.e., Pattern-2 (P2)) and make up a LAP pattern. P1 is determined from the grey color intensity values of a(1-8), while P2 is calculated from the grey color intensity values of b(1-8), and c(1-8).

P1 is limited to 24 = 16-bit combinations, while P2 is limited to 28 = 256-bit combinations. A bin is generated for each combination to count the number of times the combination occurs inside a given block. The LAP histogram for a block is combined into 16 bins for P1 and 256 bins for P2, respectively. As a result, each block's feature vector length is 16+256 (272) for the proposed method. Figure 4(c) illustrates how a(1-8) in a 4-bit binary pattern reflects the grey color attribute of the pixels. As shown in figure 4(d), the 8-bit binary pattern, b(1-8)and c(1-8) represent the corresponding grey color value of the pixels. Figure 5. illustrates how to acquire LAP patterns from a 5x5 pixel region. The final feature vector of an image is formed by concatenating the histograms of all blocks in the image.

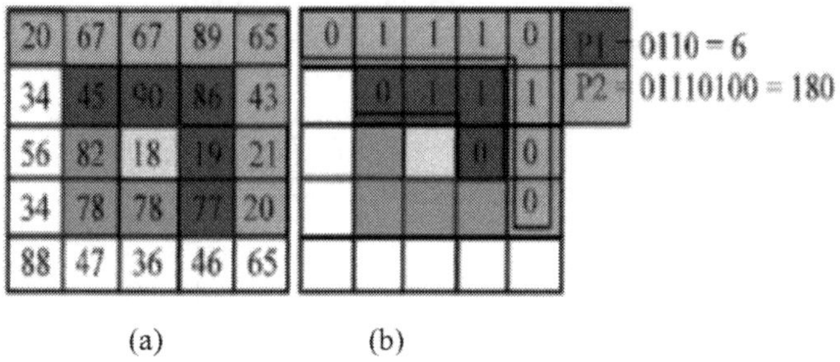

Figure 5. Illustration of LAP code generation (a) 5x5 pixels local region, (b) LAP representation of pixel 18

3. Proposed model

The aim of this work is to compare various texture descriptors in extracting the features of vehicles which in turn makes an effect on the recognition rate. Investigations are also carried how the various color spaces make out an impact on the texture descriptors for improving the classification accuracy. The proposed model is shown in Figure 6.

- Initially, in the pre-processing stage, two data sets are created out of which one dataset contains 70% training images and 30% of images are used for validation and another dataset consists of 50% training images, and 50% of images are used for validation. Then the images are resized to an RGB scale of 256x256x3 before they are applied to various descriptors.
- In the next step, for the study of various color spaces, the dataset is converted into the gray level, RGB, YCbCr, and HSV color spaces.
- The LBP, MBP, GDP, and LAP descriptors process the different color space images and extract the texture features.
- The features considered here are absolute mean, mean square error, standard deviation, and entropy.
- The extracted features are feed to SVM with different kernel functions like Radial Basis Function (RBF) and polynomial function for classification.

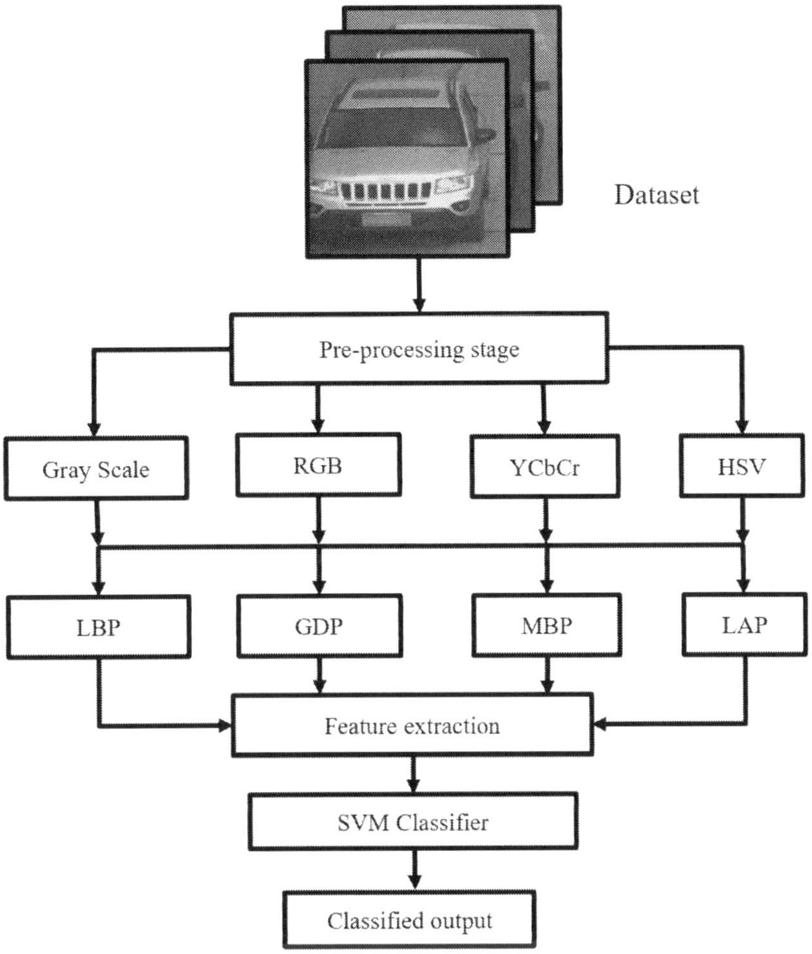

Figure 6. The block diagram of the proposed model

Figures 7, 8, 9, and 10 shows the generated images for a class of CompCars dataset [15] by the four pattern descriptors (i.e., LBP, MBP, GDP, and LAP) for various color spaces.

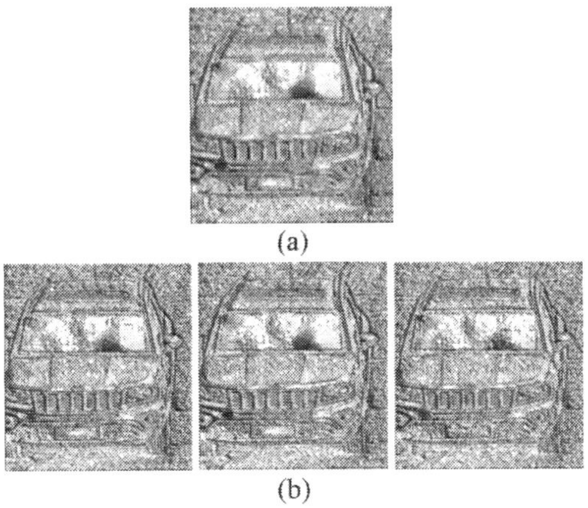

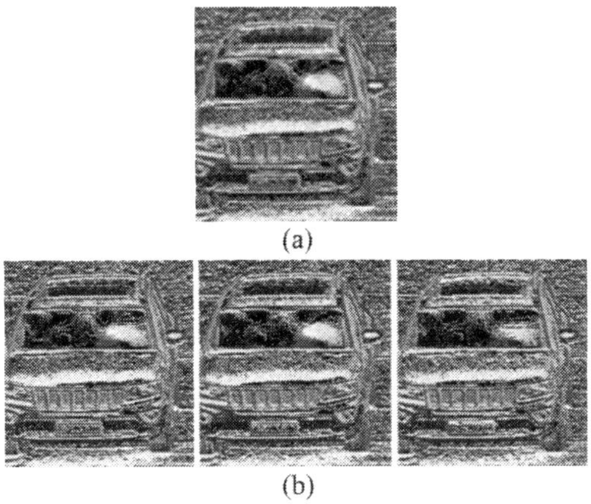

(c)

(d)

Figure 7. LBP descriptor generated images of class 'Jeep' (a) Gray level image (b) RGB (c) YCbCr (d) HSV

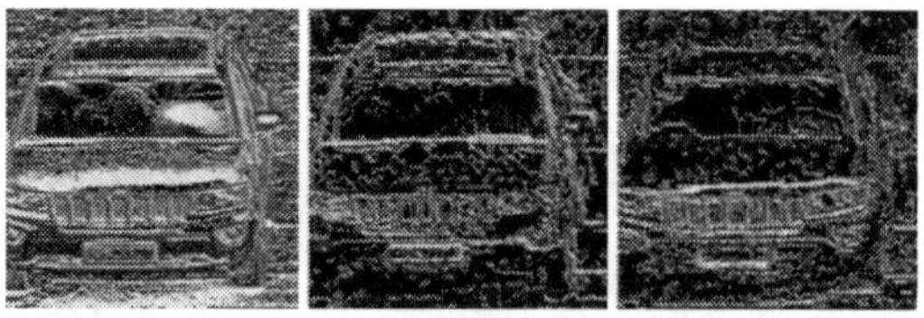

(c)

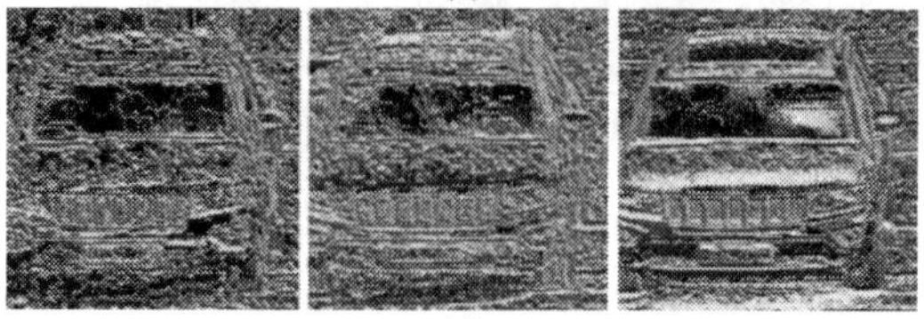

(d)

Figure 8. MBP descriptor generated images of class 'Jeep' (a) Gray level image (b) RGB (c) YCbCr (d) HSV

(a)

(b)

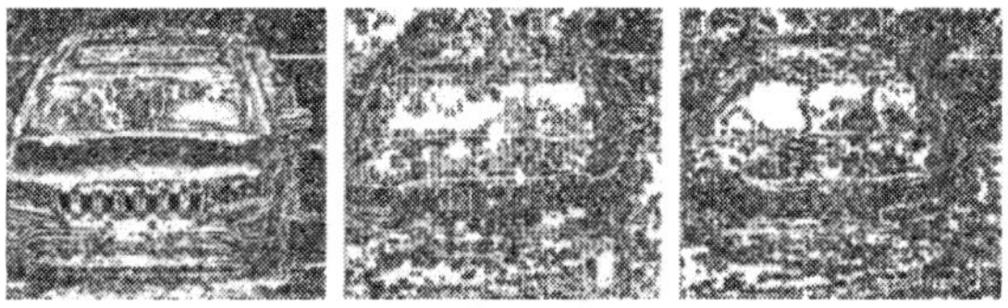

(c)

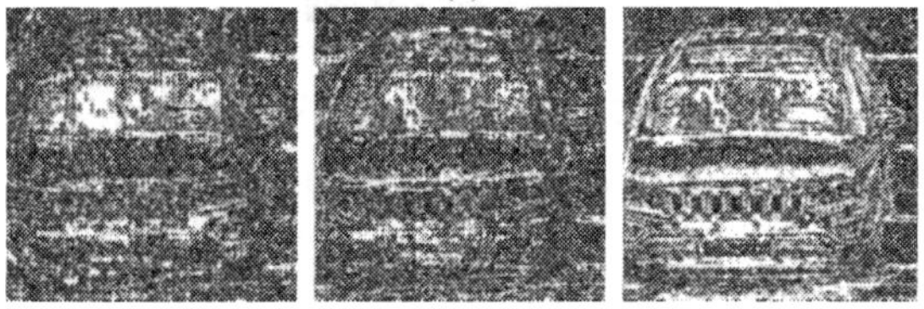

(d)

Figure 9. GDP descriptor generated images of class 'Jeep' (a) Gray level image (b) RGB (c) YCbCr (d) HSV

(a)

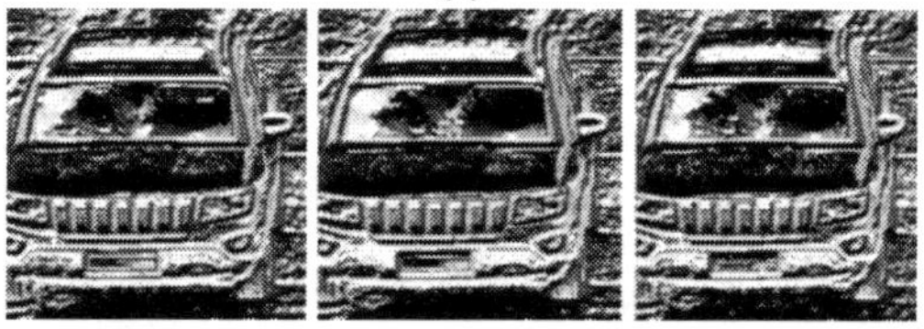

(b)

(c)

(d)

Figure 10. LAP descriptor generated images of class 'Jeep' (a) Gray level image (b) RGB (c) YCbCr (d) HSV

4. Experimental results

The proposed model is tested on the CompCars dataset which is categorized into Group-I and Group - II datasets. The Group-I dataset consists of 70% images for training and 30% for validation. Whereas

Group -II dataset contains 50% images for training and 50% for validation. In each dataset, 50 images of five different classes are considered. The experiments are conducted on MATLAB software.

Table 1 shows the classification results for various descriptors of Gray level color space images. For the Group-I dataset, the LBP and LAP descriptors achieved the highest accuracy of 37.60% with SVM using RBF kernel function with a box constraint of 1 and polynomial function of order 3 respectively. The LAP descriptors achieved the highest accuracy of 44.0% with SVM using the polynomial function of order 3 in the Group -II dataset.

Table 1. Classification accuracy of LBP, MBP, GDP, and LAP for Gray level color space images

Kernel	Group -I dataset				Group -II dataset			
	LBP	MBP	GDP	LAP	LBP	MBP	GDP	LAP
	Classification Accuracy	Classification Accuracy	Classification Accuracy	Classification Accuracy	Classification Accuracy	Classification Accuracy	Classification Accuracy	Classification Accuracy
POLY 2	33.60%	34.40%	36.80%	32.00%	36.00%	28.00%	36.00%	38.67%
POLY 3	35.20%	32.00%	33.60%	37.60%	30.67%	34.67%	30.67%	44.00%
RBF 1	37.60%	20.00%	28.00%	33.60%	34.67%	17.33%	36.00%	36.00%
RBF 2	36.80%	20.00%	28.00%	34.40%	30.67%	17.33%	36.00%	32.00%

Table 2 shows the classification results for various descriptors of Gray level color space images. For the Group-I dataset, the LAP descriptor achieved the highest accuracy of 43.2% with SVM using the polynomial function of order 3. The LAP descriptors achieved the highest accuracy of 48.0% with SVM using the polynomial function of order 2.

Table 2. Classification accuracy of LBP, MBP, GDP, and LAP for R G B color space images

Kernel	Group -I dataset				Group -II dataset			
	LBP	MBP	GDP	LAP	LBP	MBP	GDP	LAP
	Classification Accuracy	Classification Accuracy	Classification Accuracy	Classification Accuracy	Classification Accuracy	Classification Accuracy	Classification Accuracy	Classification Accuracy
POLY 2	40.00%	35.20%	40.00%	38.40%	38.67%	37.33%	36.00%	48.00%
POLY 3	39.20%	40.80%	28.00%	43.20%	34.67%	36.00%	32.00%	46.67%
RBF 1	30.40%	29.60%	28.00%	35.20%	30.67%	36.00%	30.67%	33.33%
RBF 2	28.80%	27.20%	30.40%	35.20%	32.00%	32.00%	30.67%	32.00%

Table 3 shows the classification results for various descriptors of Gray level color space images. For the Group-I dataset, the LAP descriptor achieved the highest accuracy of 52.0% with SVM RBF kernel function with a box constraint of 1 and 2. The LAP descriptors achieved the highest accuracy of 48.0% with SVM using a box constraint of 2.

Table 3. Classification accuracy of LBP, MBP, GDP, and LAP for HSV color space images

Kernel	Group -I dataset				Group -II dataset			
	LBP	MBP	GDP	LAP	LBP	MBP	GDP	LAP
	Classification Accuracy	Classification Accuracy	Classification Accuracy	Classification Accuracy	Classification Accuracy	Classification Accuracy	Classification Accuracy	Classification Accuracy
POLY 2	45.00%	42.40%	44.80%	48.80%	38.67%	37.33%	34.67%	42.67%
POLY3	31.00%	42.40%	39.20%	42.40%	38.67%	40.00%	38.67%	42.67%
RBF 1	38.40%	47.20%	36.00%	52.00%	42.67%	44.00%	36.00%	45.33%
RBF 2	37.60%	46.40%	35.20%	52.00%	44.00%	44.00%	36.00%	48.00%

Table 4 shows the classification results for various descriptors of Gray level color space images. For the Group-I dataset, the LAP descriptor achieved the highest accuracy of 46.4% with SVM using the polynomial function of order 2. The MBP and LAP descriptors achieved the highest accuracy of 40.0% with SVM using the polynomial function of order 2.

Table 4. Classification accuracy of LBP, MBP, GDP, and LAP for YCbCr color space images

Kernel	Group -I dataset				Group -II dataset			
	LBP	MBP	GDP	LAP	LBP	MBP	GDP	LAP
	Classification Accuracy	Classification Accuracy	Classification Accuracy	Classification Accuracy	Classification Accuracy	Classification Accuracy	Classification Accuracy	Classification Accuracy
POLY 2	42.40%	39.20%	37.60%	46.40%	38.67%	40.00%	33.33%	40.00%
POLY 3	39.20%	35.20%	36.00%	36.00%	36.00%	37.33%	36.00%	33.33%
RBF 1	38.40%	39.20%	37.60%	36.80%	38.67%	38.67%	32.00%	30.67%
RBF 2	36.80%	39.20%	36.80%	36.00%	33.33%	37.33%	33.33%	32.00%

5. Conclusion

In the field of intelligent transportation systems, accurate and robust vehicle identification and recognition remain a difficult challenge. In this paper, vehicle recognition is performed using the features generated by the pattern descriptors. The experimental results show that the LAP descriptor outperforms the other descriptors and the classification accuracy is improved by considering RGB, HSV, and YCbCr color spaces. The improvement of the classification results using deep learning is the future work to be considered.

References:

[1] Keerthi Kiran V, Parida P, and Dash S 2021 *Adv. in Intell. Sys. and Comput., Springer, Cham* **1180**
[2] Liu W, Liao S, and Hu W 2019 *Neuro. Comput.* **347**, pp 24-33
[3] Wang X, 2013 *Pattern Recog Letters* **34(1)**, pp.3-19
[4] Su A, Sun X, Zhang Y, and Yu Q 2016 *IET Comput Vision*, **10(7)**, pp.634-40
[5] Awad A I and Hassaballah M 2016 *Stud. in Comput. Intell. Springer* Int. P., Cham
[6] Yu Y, Li H, Wang J, Min H, Jia W, Yu J, and Chen C 2020 *IEEE Trans on Intell Trans Sys*
[7] Tang Y, Zhang C, Gu R, Li P, and Yang B 2017 *Multimedia Tools and Applic* **76,** 5817–32
[8] Hassaballah M, Kenk M A, and El-Henawy I M 2020 *Pattern Anal and Applic,* **23,** 1505–21
[9] Chu W, Ying Z, and Xia X 2013 *IEEE Int. Conf. on Green Comput. and Commun. and IEEE Internet of Things and IEEE Cy., Phys. and Soc. Comput.* pp. 1458-62
[10] Niaraki R J and Shahbahrami A 2019 *4th Int. Conf. on Pattern Recognition and Image Analysis (IPRIA)*, pp. 141-4
[11] Ojala T, Pietikäinen M, and Mäenpää T 2002 *IEEE Trans on Pattern Analysis and Machine Intelligence*, **24**, pp. 971-87
[12] Hafiane A, Seetharaman G, and Zavidovique B 2007 *Kamel M.,Campilho A. (eds) Image Analysis and Recognition. Lecture Notes in Comp. Science,* Springer **4633**
[13] Ahmed F, 2012 *Elect. Letters* P. 1203 – 4
[14] Islam M S and Auwatanamongkol S 2013 *Asian J. of Info. Tech., Trends Appl. Sci. Res*, **9**, pp.113-20
[15] Yang L, Luo P, Loy C C, and Tang X 2015 *Proc. of the IEEE conf. on Comp. Vision and Pattern Recognition* pp. 3973-81

IOP Conf. Series: Materials Science and Engineering 1166 (2021) 012047 doi:10.1088/1757-899X/1166/1/012047

Influence of Zinc Doping on the Structural, Morphological, Optical and Electrical Properties of Copper Oxide Thin Films Prepared by Jet-Nebulizer Spray Pyrolysis Technique

V Jagadeesan[1*], A Rani Sangeetha[2], P Anitha[3]

[1 2 3]Department of Electronics, PSG College of Arts & Science, Coimbatore-14, Tamil Nadu, India

*E-mail: vjpsgcas1@gmail.com

Abstract. This research details the deposition and characterization of adding zinc (Zn) impurities with different doping concentrations (1%, 3%, and 5%) into copper oxide (CuO:Zn) by using a jet nebulizer spray pyrolysis system. The impact of zinc doping on the different copper oxide properties was investigated. The structural studies confirm that pure copper oxide and zinc doped copper oxide thin films are matched with monoclinic structure. High resolution schottky field emission scanning electron microscopy and Energy dispersive spectrum analysis identifies the surface structure and chemical composition of pure CuO and CuO: Zn thin films. EDS spectra reveal the occupancy of Cu, O, and Zn elements in the pure CuO and CuO:Zn thin films. The highest transmittance was observed as 66.91% for 5% Zn concentration. The absorption studies clearly indicated that the absorbance of the films was decreased when Zn doping concentration increase. The energy band gap value (Eg) is calculated as 2.01eV for pure CuO and the value of Eg for CuO:Zn thin films are varied from 2.07 to 2.26 eV. Keithley high resistance and low current electrometer was used to measure direct current conductivity of both pure CuO and CuO: Zn thin films and maximum conductivity value was observed for CuO: Zn (5%) thin film.

1. Introduction

Copper oxide consider as an important p-type semiconductor with narrow bandgap, categorized into nanostructures of metal oxide group. Monoclinic structured based CuO contains various characteristics such as chemical, physical and thermal properties [1]. It can be act as solar absorber due to its high solar absorption capability [2]. Furthermore, CuO involved in the fabrication of solar photovoltaic cells since it has suitable optical properties [3]. CuO provides favourable applications in the field of sensors, detectors, capacitors and energy based materials [3-8]. In recent years, influence of various metal oxides such as cadmium (Cd), nickel (Ni), manganese (Mn), zinc (Zn), cobalt (Co) etc., on the characteristics of CuO have been investigated by different researchers. Batuary et al. investigated the impact of Ni doping with CuO and reported on the significant modification of optical and electrical properties of the CuO films [9]. Bayansal et al. examine that the CuO nanostrucres grain size was decreased with increasing Co doping concentration [10]. Rejith et at. reported that the energy band gap is decreased while doping with Cd in nanostructured CuO thin films [11]. Among different transition metals, CuO: Zn thin films have been reported very few. It can be a useful material for optoelectronic application.

Both physical and chemical deposition techniques were involved in the formation of CuO thin films such as pyrolysis deposition, oxidation process, sol-gel, jet nebulizer spray pyrolysis [12-19]. In present work pure CuO and CuO:Zn thin films were deposited via low cost jet-nebulizer spray pyrolysis system (JNSP). It is examined that pure CuO thin film deposited at the fixed temperature 450°C with 0.2M molar concentration have preferred properties. Similarly, CuO:Zn thin films are deposited with Zn concentration 1%, 3% and 5% at the same temperature. In this work the influence of adding Zn impurities with CuO has been studied using properties analysis. To the best of our knowledge CuO: Zn thin films had not yet been reported deposited using JNSP system.

2. Experimental procedures

2.1. Deposition of pure CuO and CuO:Zn

Jet- nebulizer spray pyrolysis system is very simple and cost effective. This system does not require sophisticated apparatus. It consists of only nebulizer with "L" shaped glass tube, hot furnace with temperature controller and compressor. Ultrasonic cleaned glasses were utilized as a base substrate to deposit pure CuO and CuO:Zn. Substrate cleaning is an important method in all kind of thin and thick film deposition. Initially, base glass substrates were soaked in the well blended double distilled water and nitric acid solution for 20 min. After 20 min, these substrates were taken from the solution and clean up with double distilled water. After primary cleaning method, ultrasonic bath cleaner in 30 min timer operation used to clean substrates with acetone solution. This ultrasonic bath cleaning method protects the glass surface from the damage caused by strong alkali and acid. Consequently, dust and alkali-free substrates were produced and taken from the ultrasonic bath. Finally, these substrates are preserved at room temperature

Cupric nitrate trihydrate [$Cu(NO_3)_2 . 3H_2O$] taken as a starting material to produce pure CuO thin film. Zinc acetate dihydrate [$(CH_3COO)_2 Zn. 2H_2O$] taken as a dopant to provide CuO:Zn thin films. To prepare pure CuO, cupric nitrate was blended with double distilled water and magnetically stirred for 60min. similarly, cupric nitrate and zinc acetate were mixed well with double distilled water to prepare CuO:Zn. The quantity of cupric nitrate and zinc acetate were regulated in atomic weight percentage and 1%, 3%, and 5% of Zn concentration taken to deposit CuO:Zn thin films. Pure CuO and CuO:Zn precursor solution was sprayed onto ultrasonically cleaned glass substrate at a fixed temperature of 450°C with 0.2M mole concentration and 4ml volume of solution which was kept on a hot furnace. The hot furnace substrate temperature regulated and fixed at 450°C using a thermocouple. The nozzle of the glass tube attached with the jet nebulizer was fixed at a 5 cm distance from the glass substrate. The air compressor's flow rate has been fixed at 2 bar. After sprayed, all deposited films were kept on the hot furnace plate for 5 min and then cooled naturally at ambient temperature.

2.2. Characterization approach

The crystal structural studies were examining using Shimadzu X-ray diffractometer (XRD) in the 2-theta range from 20° to 60°. XRD has been used to determine the crystalline phase and to study the structural properties. The surface structure and chemical composition were carried out by High-resolution Schottky Field Emission Scanning Electron Microscopy and Energy Dispersive Spectrum. The optical studies of the films were examined using an Ultraviolet (UV) -visible spectrophotometer (JASCO-570). Keithley high resistance and low current electrometer (6517-B) was used to measure direct current conductivity of both pure CuO and CuO: Zn thin films.

3. Result and analysis

3.1. Structural analysis

Figure 1 shows typical structural analysis of XRD patterns for pure CuO and CuO:Zn thin films with different Zn concentrations. The XRD profiles of all the films observed at 2-theta between 20 ° and 60°. The obtained XRD data of pure CuO thin film reveals that the film was a crystal in a monoclinic system (JCPDS No. 89-5899) [20]. The strongest peaks were observed with diffraction angle 2-theta and favorable orientation planes of 35.5° (-111), 38.7° (111), and 53.5° (020). Similar to pure CuO, XRD data for CuO: Zn thin films were also monoclinic crystal structures. Despite this, no other peaks compatible with Zn elements is recognized by XRD data. The reason is XRD system might be challenging to find a small amount of Zn impurities that may present in the cluster form in CuO:Zn. Meantime, increasing doping concentration causes reduced peak intensity in diffraction angle of 53.5° and disappeared completely for the 5% CuO: Zn thin film. The structural analysis revealed that the peak intensity of all the films was decreased with increasing Zn doping concentration.

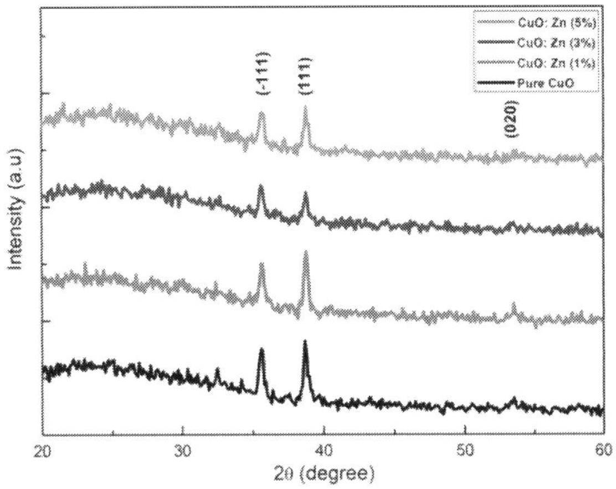

Figure 1. X-ray diffraction peaks of pure CuO and CuO:Zn thin films

The average crystallite size estimated from the following Debye-Scherrer's equation with the help of full width half maximum (FWHM) value which strengthens the structural analysis [21]

$$D=\frac{0.9\lambda}{\beta\cos(\theta)} \tag{1}$$

Where β FWHM of the diffraction peak at diffraction angle θ, λ is the X-rays radiation wavelength, and D is the estimated average crystallite size. The following equations are used to calculate micro-strain (ε) and dislocation density (δ) values [22] which is listed in the table 1.

$$\varepsilon=\frac{\beta\cos\theta}{4} \tag{2}$$

$$\alpha=\frac{1}{D^2} \tag{3}$$

The comparison of average crystallite size and FWHM are also illustrated in the figure 2. The result shows that the average crystallite size is increased with increase of Zn impurities. Accordingly, adding Zn impurities into CuO shows an important role in the crystal size regulating process. The comparison of micro-strain variations and average crystallite size for pure CuO and CuO:Zn is illustrated in the figure 3. The average crystallite size is varied from 17 to 28 nm for pure CuO and CuO:Zn thin film which is listed in the table 1. The comparison plot clearly indicate that the strain values are degreased while increase in the average crystallite size. Similarly, dislocation density values are degreases with increase of Zn impurities.

Table 1. Variation of structural parameters of pure CuO and CuO:Zn thin films

Thin films	2θ (degree)	FWHM (Degree)	Crystallite size (D) (nm)	Micro strain (ε)	Dislocation density (δ) ($\times 10^{15}$ lines m^{-2})
Pure CuO	38.7131	0.4809	17	0.0019	3.25719
CuO: Zn (1%)	38.7516	0.3942	21	0.0016	2.18919
CuO: Zn (3%)	38.7583	0.3533	23	0.0012	1.64718
CuO: Zn (5%)	38.8094	0.3188	28	0.0010	1.38985

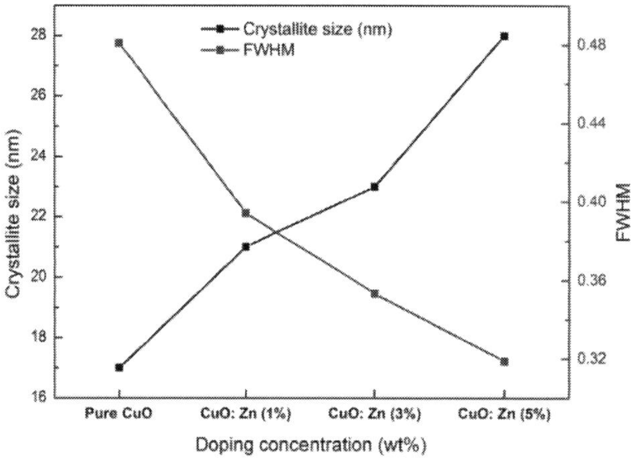

Figure 2. Comparison of crystallite size and FWHM for Pure CuO and Zn doped thin films

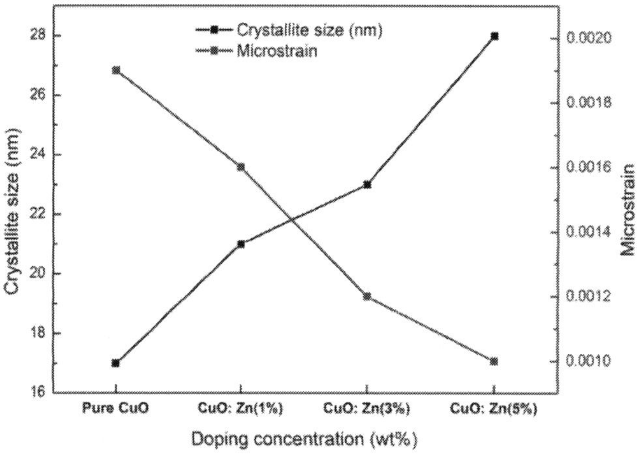

Figure 3. Comparison of crystallite size and micro strain for Pure CuO and Zn doped thin films

3.2 Surface morphological analysis

FESEM is a useful technique to determine the surface structure of thin films. The surface structure images of pure CuO and CuO:Zn thin films with various concentrations are shown in figure 4. Figure 4 (a) shows spherical shape morphology of pure CuO thin and Figure (b)-(d) exhibits the flower-like structure which confirm the presence of Zn impurities in the CuO:Zn [23]. The morphology of the figure (b)-(d) was observed as flower kind of structure. However, the density of the flowers is seen to change, significantly and the aggregation is spotted only for CuO: Zn (1%). Further, elementary composition analysis is carried out to examine the 'Cu', 'O' and 'Zn' in the synthesized materials using Energy Dispersive Spectra (EDS). CuO: Zn (5%) sample was examined to identify the presence of chemical composition in the synthesized material. Figure 5 shows EDS spectra which reveal the occupancy of Cu, O, and Zn elements in the pure CuO and CuO:Zn thin films.

Figure 4. Surface structure images of **(a)** Pure **(b)** 1% **(c)** 3% and **(d)** 5% CuO: Zn

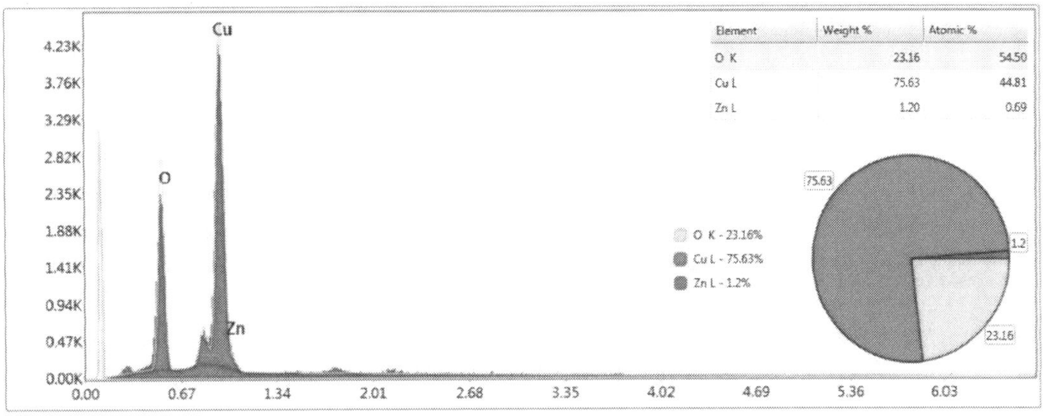

Figure 5. EDS spectra and data of elemental analysis for Zn doped CuO thin film

3.3. Optical analysis

The optical analysis of pure CuO and CuO:Zn thin films deposited by JN spray pyrolysis system were studied. The influence of adding Zn impurities on CuO were investigated. Figure 6 shows the optical transmittance studies exhibited the amount of energy transmitted by the material sprayed on the substrate. The pure CuO thin film exhibited an optical transparency of 49.83%. When increasing Zn doping concentration, the optical transmittance increases from 53.19% to 66.91. CuO: Zn thin films exhibit high transparency in the near-infrared (NIR) region. The maximum transparency is 66.91% for CuO:Zn 5%. As a result it is well know that adding Zn impurities were strongly influences on the optical properties of CuO thin films. The absorption spectra of pure CuO and CuO: Zn thin films deposited by JN spray pyrolysis as shown in figure 7. The spectra reveal that absorption of the entire films exhibit high ultra violet field. From the absorption spectral analysis, absorption capability of the films was reduced when increasing the percentage of Zn impurities. The energy bandgap of the pure CuO and CuO:Zn thin films were evaluated from the absorption values. As CuO is a direct band gap material, the absorption co efficient can be relevant to the energy bandgap (E_g) through Jan tauc's relationship by using following formula [24]

$$\alpha h\nu=\beta(h\nu-E_g)^n \quad (4)$$

β referred as energy-independent constant, $h\nu$ referred as incident photon energy and n referred as constant that determines the type of optical transition. α is the absorption coefficient which can be calculated using below formula [24]

$$\alpha=\frac{2.303A}{t} \quad (5)$$

Where A and t referred as absorbance value and thickness of the CuO films. The relationship between $(\alpha h\nu)^2$ and $h\nu$ showed in the fig. The energy bandgap (E_g) value is estimated from the value of intercept of the straight line at α=0 by plotting $(\alpha h\nu)^2$ against $h\nu$ [25]. According to Jan tauc plot energy band gap value is calculated 2.01eV for pure CuO and value of E_g for Zn: CuO thin films are varied from 2.07 to 2.26 eV. From the figure 8, it is seen that the value of E_g raising with the increase of Zn impurities.

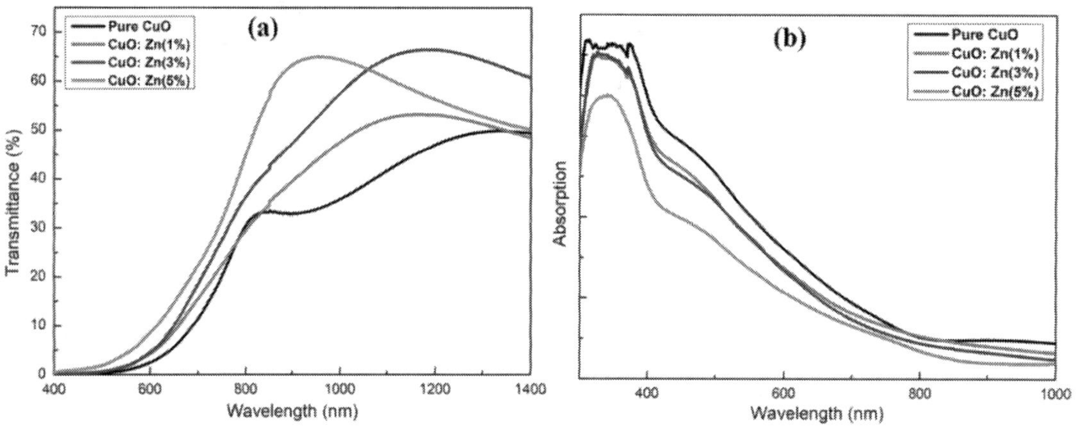

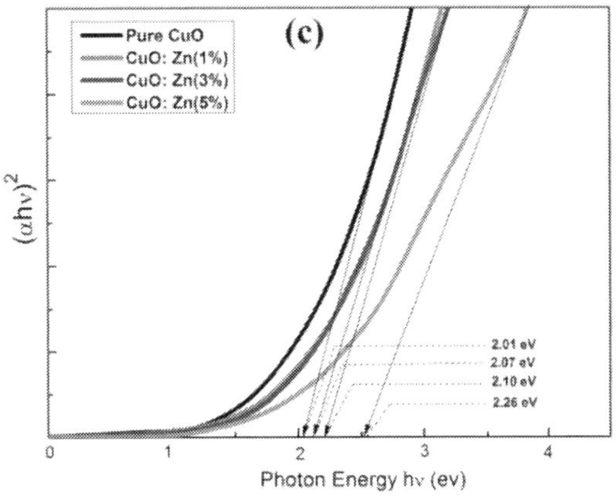

Figure 6. **(a)** Optical transmittance spectra **(b)** Absorption spectra **(c)** Energy band gap of pure CuO and CuO:Zn thin films

3.4 Direct Current (DC) Electrical analysis

The direct current electrical conductivity for pure CuO and CuO:Zn thin films were evaluated using Keithley high resistance and low current electrometer under ambient temperature. Current Vs voltage (I-V) characteristics of pure CuO and CuO:Zn thin films exhibited in figure 7 (a). The current values were measured by varying applied DC voltages. From figure 7 (a), it is noted that an increase in applied voltage leads to a gradual increase in the current value. Similarly, an increase in Zn impurity percentage leads to an increase in the current value. The results show that CuO: Zn (5%) thin films have the highest current value. The DC conductivity for pure CuO and CuO:Zn thin films were determined using the below formula [26]

$$\sigma = \left(\frac{V}{I}\right) X \left(\frac{d}{A}\right) \tag{6}$$

Where 'V' referred as applied voltage, 'I' referred as current, 'd' referred as distance and 'A' referred as cross sectional area of the film. Average conductivity of pure CuO and CuO: Zn thin films depicted in the figure 7 (b). The studies showed that the conductivity value is increased when increasing Zn doping concentration. CuO:Zn (5%) thin film shows the highest conductivity of 7.87 x 10^{-10} Siemens per centimeter (S/cm). The results reveal that adding Zn impurities with CuO tends to increase the conductivity value.

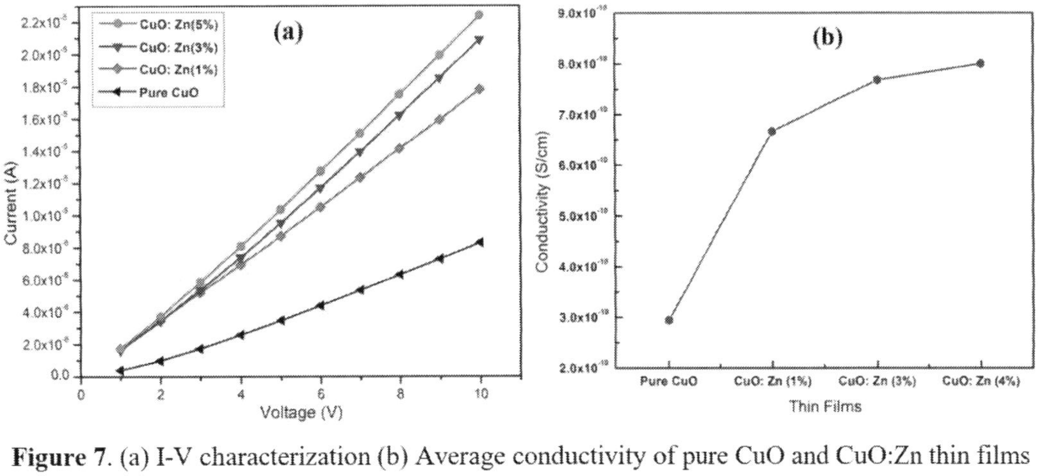

Figure 7. (a) I-V characterization (b) Average conductivity of pure CuO and CuO:Zn thin films

4. Conclusion

In this paper, we reported on the deposition of pure CuO and CuO: Zn thin films with varied Zn impurities (1%, 3%, and 5%) via the JNSP system. The characteristic analysis of pure CuO and CuO: Zn thin films evaluated to study the impact of Zn incorporation with CuO. XRD results reveal that all the films are crystal in a monoclinic structure and agree with JCPDS No. 89-5899. From XRD results, no corresponding peak is observed related to Zn impurity and Zn-based oxide compounds at even CuO:Zn (5%). The intensity of the peaks gets reduced and crystallite size was increased when increasing the Zn doping concentration. FESEM results show spherical-shaped structure for pure CuO and flower-like structure for CuO: Zn thin films. It establishes the presence of Zn impurities in CuO: Zn thin films. EDS analysis certifies the presence of 'Cu', 'O', and 'Zn' elements in the CuO: Zn thin film. The pure CuO thin film exhibited optical transparency of 49.83%. When increasing Zn doping concentration, the optical transmittance increases from 53.19% to 66.91%. The highest transparency is 66.91% for 5% CuO:Zn. The absorption spectra indicate that the absorbance of the films was decreased when Zn doping concentration increase. In agreement with Jan tauc plot, the value of energy bandgap value is calculated at 2.01eV for pure CuO, and the value of Eg for Zn: CuO thin films are varied from 2.07 to 2.26 eV. The DC electrical conductivity of pure CuO and CuO: Zn thin films were measured. It is observed that the current value is increased gradually with increasing applied voltage and the current value is increased when increasing the Zn doping concentration. CuO:Zn (5%) thin film shows the highest conductivity of 7.87×10^{-10} Siemens per centimeter (S/cm). The results reveal that adding Zn impurities with CuO tends to increase the conductivity value. At last, we presume that adding Zn impurities assumed an imperative part in the modification of structural, optical, morphological, and electrical properties of CuO thin films.

References

[1] Tran T H and Nguyen V T 2014 *Int. Sch. Research Not.* 856592

[2] Yoon K H, Choi W J and Kang D H 2000 *Thin Solid Films* **372**, 250

[3] Oral A Y, Mensur E, Aslan M H and Basaran E, 2004 *Mater .Chem. Phys.* **83**, 140

[4] Chandrasekaran S 2013 *Sol. Energ. Mat. Sol. C*, **109**, 220–226

[5] Wang S B B, Hsiao C H H, Chang S J J, Lam K T T 2011 *Sens, Actuators A:Phys.* **171** 207-211

[6] Choi K J, Jang H W 2010 *Sensors* **10** 4083–4099

[7] Zhang X, Shi W, Zhu J, Kharistal D, Zhao W, Lalia B 2011 *ACS Nano* **5** 2013–2019

[8] Rossi C, Zhang K, Esteve D, Alphonse P, Tailhades P, Vahlas C 2007 *J.Microelectromech. Syst.* **16** 919–931.

[9] Baturay S, Tombak A, Kaya D, Ocak Y S, Tokus M, Aydemir M, Kilicoglu T 2016 *J. Sol Gel Sci. Technol.* **78** 422–429

[10] Bayansal F, Taskopru T, Sahin B 2014 *Metallurg. Mater. Trans. A* **45(A)** 3670–3674

[11] Rejith S G 2013 *Adv. Appl. Sci. Res.*, **4(2)**:103-10

[12] Markworth P, Liu X, Dai J, Fan W, Marks T J, Chang R P, 2001 , *J. Mater. Res.* **16** 8 2408–2414

[13] Mahalingam T, Chitra J, Chu J, Velumani S, Sebastian P 2005 *Sol. Energy Mater. Sol. Cells* **88** 209–216

[14] Mageshwari K, Sathyamoorthy R 2013 *Mat. Sci. Semicond. Proc.* **16 (2)** 337-343.

[15] Oral A, Menşur E, Aslan M, Başaran E 2004 *Mater. Chem. Phys.* **83** 140–144.

[16] Ogwu A, Darma T, Bouquerel E, 2007 *J. Achiev. Mater. Manuf. Eng.* **24** 172–177.

[17] Saravanan V, Shankar P, Mani G K, Rayappan J B B 2015 *J. Anal. Appl. Pyrolysis* **111** 272–277.

[18] Khojier K, Savaloni H, Sadeghi Z 2014 *J. Theor. Appl. Phys.* **8** 1–8

[19] Mariappan R, Ponnuswamy V, Suresh R, Suresh P, Chandra Bose A, Ragavendar M, 2014 *J. Alloys Compd.* **582** 387–391

[20] Jagadeesan V, Subramaniam V, 2019 *J Mater Sci: Mater Electron.* **30**, 1571–1578

[21] Patterson A L 1939 *Phys. Rev.* **56** 978–982

[22] Shkir M, AlFaify S 2017 *Sci. Reports*, **7** 16091

[23] Singh, Subhash. 2013 *Journal of Nanoengineering and Nanomanufacturing*. **3 (4)**.

[24] Rafea M A, Roushdy N 2009 *J. Phy. D-Appl. Physics* **42 (1)** 15413

[25] Prakash chand, Manisha, Praveen kumar 2018 *optic* **156** 743-753

[26] Balaji M, Chandrasekaran J, Raja M 2016 *Mater. Sci. Semicond. Process.* **43** 104

Lossless Compression of Satellite Images using a Versatile Hybrid Algorithm

V Swetha[1], G Premjyoti Patil[2], B Shantakumar Patil[3]

[1]Department of Electronics and Communication Engineering, School of Engineering and Technology, CMR University, Off Bagalur Main Rd, near Kempegowda International Airport, Mitganahalli, Hennur Gardens, Chagalahatti, Karnataka, 562149, India
[2]Research Scholar, Visvesvaraya Technological University, Belagavi, Karnataka, 590018, India
[3]Department of Computer Science and Engineering, Sai Vidya Institute of Technology, Rajanukunte, via Yelahanka, Bengaluru, Karnataka, 560064, India

[1]Corresponding author: vuraswetha@gmail.com

Abstract. The remote sensing images taken from various satellites have high resolution and occupy a lot of storage space. It, therefore, becomes difficult to send a greater number of images to the base station. Image compression becomes inevitable in such cases. Satellite image compression decreases repetition in representation of data so that the storage cost and transmission reduces and compensates for the limited on-board resources, in terms of large-scale memory and downlink bandwidth. The aim of this paper is to perform lossless image compression on satellite imagery using a hybrid (DWT-RLE) algorithm to obtain higher compression ratios. This compression is implemented using the software tool MATLAB.

1. Introduction

Remote sensing is the art of acquiring characteristics of an object or an area without being in physical contact with it. Satellites incorporate remote sensing using radio waves for transmission and reception of data. They have a plethora of applications, including imaging, communication, weather forecast, space exploration, military, etc. Some of the imaging satellites are SPOT, IKONOS, Cartosat, Worldview, and QuickBird. Transmission of large panchromatic, multispectral, and hyperspectral images increases the data volume making image compression inevitable. Image compression is broadly classified as lossy and lossless. In lossy image compression, we lose some information, but in lossless image compression, compression can be obtained without any loss of data. Image compression thus becomes an important step in any image processing application.

Wavelet transform has emerged as an essential tool of choice for image compression. The real-world data represents slowly changing trends or oscillations with abrupt transitions. Fourier transform represents data as a sum of sine waves that are continuous and not localized in time or space. It thus cannot represent abrupt changes giving rise to wavelets that accurately can measure abrupt changes. A wavelet is a localized wave with zero mean. The two main components are shifting and scaling. Depending on these two parameters, the wavelet can shrink or expand and shift over time. The compressed wavelet can capture abrupt changes, whereas a stretched wavelet captures slow changes in a signal. The oscillations in a wavelet depend on its vanishing moments.

FIR wavelet filter's length depends on the support size. Different types of wavelets like Morlet, Haar, Mexican Hat, Daubechies, Gaussian, Symlets, Coiflets, Biorthogonal, etc., have different properties. Wavelets with fewer vanishing moments and have less coefficients can be used for compression where smooth reconstruction of an image can be obtained.

Run-Length Encoding (RLE) is a lossless compression methodology which is a relatively uncomplicated form of data compression. The data here is repeated in succession. Instead of representing all the elements, it is stored as a single data value and count. Let us consider a black and white image which has a greater number of white pixels than black. There will be many long runs of white pixels in the blank space and short runs of black pixels within the text.

WWWWWWWWWBWWWWWWWWWWWBBBWWWWWWWWWWWWWWWWWWWW WBB WWWW WWWWWWWWW

After employing RLE to the above line, we get the below code: 9W B 11W 3B 20W 2B 13W.

The paper presents the survey on the various compression methods which provides good scalability and efficiency. [1-20]

The rest of the paper is categorized as follows. Literature Survey done is shown in Section 2. Section 3 explains the proposed Lossless Image Compression Algorithm. Experiment results are presented in Section 4. Concluding remarks are specified in Section 5.

2. Literature Survey

D. Gowri Sankar Reddy [21] developed a wavelet thresholding method to compress LANDSAT images. The image is first encoded with the Discrete wavelet transform using db4 wavelet. After decomposition into frequency bands, the LL coefficients are the approximate coefficients where most of the energy is concentrated, and much of the information is available. The other LH, HL, HH, which are the detail coefficients, have the less significant information. Image compression is achieved by truncating the detail coefficients. By varying threshold, the bit rate can be obtained. Using the proposed method, better PSNR (Peak Signal-to-noise ratio) was obtained for low bit-rate applications. This paper uses only one wavelet, and the thresholding holds good only for LANDSAT images. The proposed method cannot provide better PSNR for high bit-rate applications.

Multispectral images acquire data within the definite wavelength ranges across the band of the electromagnetic spectrum. It requires a lot of memory for storage. Dr. S. Prashantha [22] proposed a multispectral image compression algorithm using Daubechies wavelets, Symlet wavelets, Coiflet wavelet, and Bi-orthogonal wavelet. DWT is applied to the image taking a single band at a time. Depending on the application, decomposition level is selected. After applying the wavelet filters, the image undergoes thresholding where the compression takes place. Db1 and Biorthogonal wavelets obtain higher compression ratios compared to other wavelets. The algorithm proposed uses only multispectral images and neither panchromatic nor hyperspectral images. Though many wavelets are compared using DWT, it does not include any encoding technique to increase the compression ratio.

Sulaiman Khan [23] presented an approach of Haar wavelet transform, Discrete Cosine Transform, and Run-Length Encoding techniques for advanced manufacturing processes with higher image compression ratios. The images are segmented into 8X8 blocks and then decoded to construct the original image. An input image is given to the compression system, which converts the image to a bitstream. Later, a compression technique is applied. To achieve the decompressed resultant image, reconstruction of the original image is performed. The proposed algorithm works with a compression rate of 97.7%. The results showed that the DHT algorithm has a higher compression ratio and peak signal to noise ratio than other algorithms. It is found to be more efficient, less complex, cost-effective, and easy to implement. The main disadvantage of this paper is that it uses only Haar wavelet, which gives less compression ratio. Using DCT, the image must undergo segmentation into 8X8 blocks before it undergoes compression making the process more complex.

3. Proposed Algorithm

The proposed hybrid algorithm, which combines Discrete Wavelet Transform (DWT) and Run-Length Encoding (RLE), overcomes the above disadvantages. Using DWT, the image can be used as a whole and need not be segmented as in Discrete Cosine Transform (DCT). This algorithm can compress both panchromatic and multispectral images of different resolutions, image formats, intensities, and wavelengths. The algorithm has been designed to use and compare five types of wavelets: Haar, Daubechies, Symlets, Coiflets, and Biorthogonal. It also can analyze various levels of decomposition. Apart from DWT, it uses a simple encoding technique, RLE, to enhance the compression ratio. When the wavelet coefficients are segregated using a given threshold value, the coefficients below the value are truncated to zero. Run-Length Encoding becomes an efficient coding technique when there is a lot of repetitive data.

The data set in this proposed method are remote sensing images taken from the satellites CARTOSAT2, CARTOSAT3, INSAT3DR, CHANDRAYAAN2, and MARSORBITER. This consists of ten images, out of which five are panchromatic (Figure 1), and five are multispectral (Figure 2). The images are shown below.

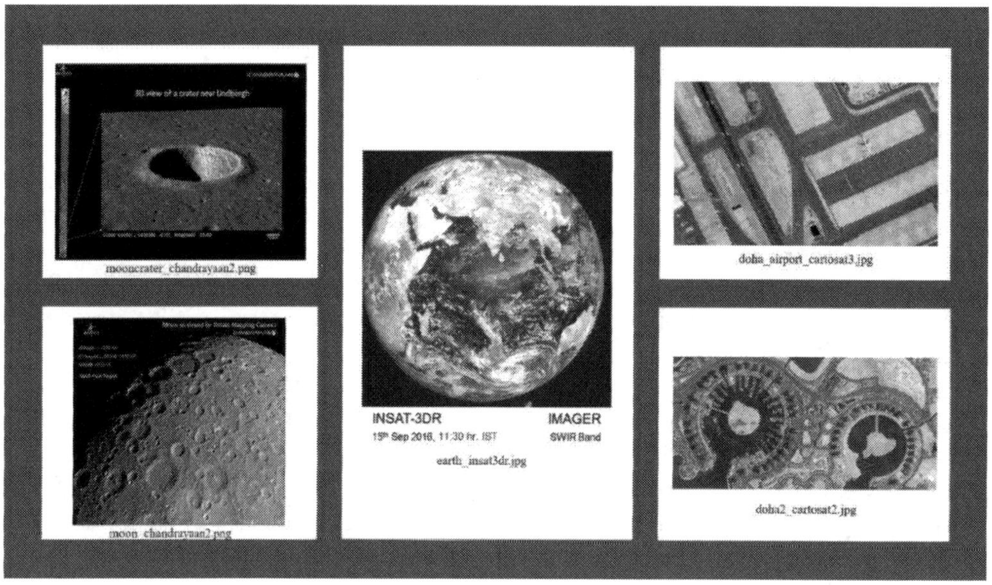

Figure 1. Panchromatic images

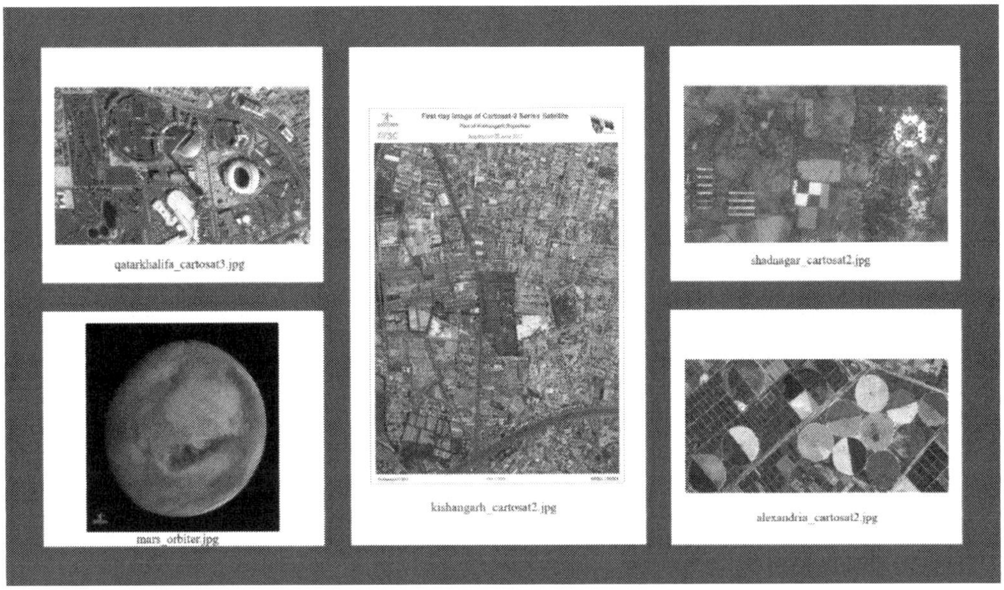

Figure 2. Multispectral images

The proposed algorithm for lossless image compression (Figure 3) consists of the following steps.

1. The satellite image is taken as the input.
2. Discrete Wavelet Transform (DWT) is applied to the image.
3. The resultant coefficients are segregated depending on the threshold value.
4. The coefficients are then quantified.
5. Run-Length Encoding is applied to the quantified coefficients to obtain the compressed output bitstream.

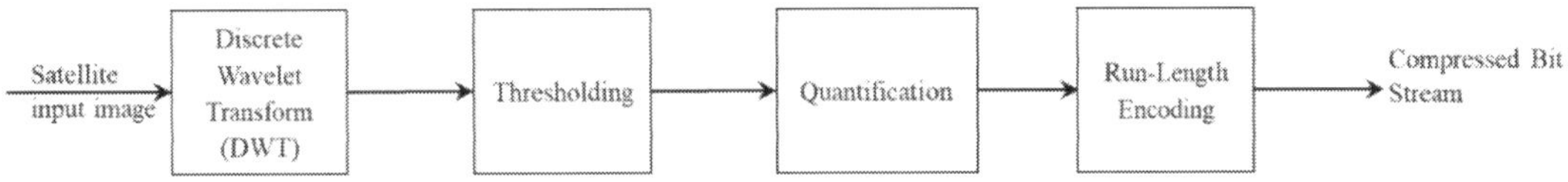

Figure 3. Block diagram of DWT-RLE

The satellite image is taken as the input. If it is a panchromatic image, it is directly wavelet decomposed. But, if the image is multispectral, it is converted to grayscale, and then wavelet decomposed. In the wavelet transform, a choice of five wavelets (Haar, Daubechies, Symlets, Coiflets, and Biorthogonal) are given. After selecting the type of wavelet, an option of decomposition level is shown. If 1 is given, Level 1 decomposition is performed on that image (Figure 4). We get four sub-bands where LL has the approximate coefficients and HL, LH and HH have the detail coefficients. Most of the information is stored in the approximate coefficients, and detail coefficients have less information. The LL band can be further divided depending on the decomposition level.

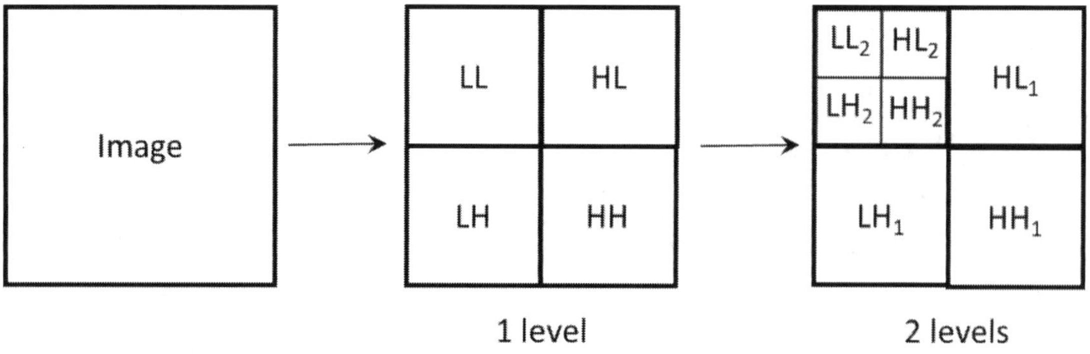

Figure 4. Wavelet Decomposition levels

The flow of the proposed hybrid algorithm is implemented as follows. The panchromatic image of Doha taken by the satellite CARTOSAT2 is taken as the input. (Figure 5).

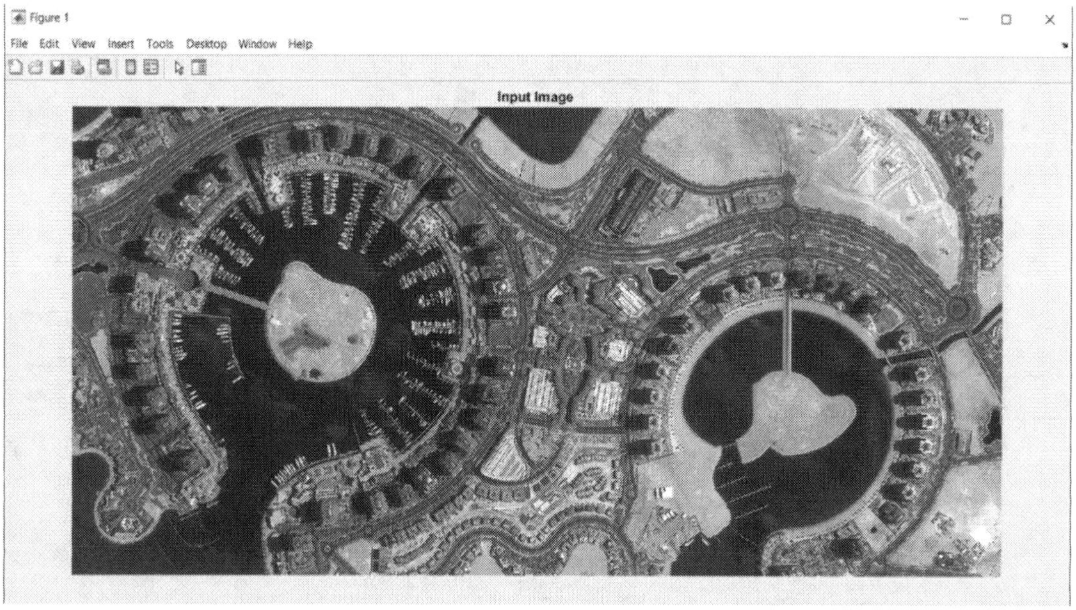

Figure 5. Input image

IOP Conf. Series: Materials Science and Engineering 1166 (2021) 012048 doi:10.1088/1757-899X/1166/1/012048

To implement DWT, an option is provided to choose a wavelet from the five wavelets: Haar, Daubechies, Symlets, Coiflets, and Biorthogonal in the Command Window of MATLAB. Here we have given option 2 and chosen Daubechies wavelet. Next, the decomposition level must be entered. It is entered as Level 1. The flow of inputs can be viewed in the following Figure 6.

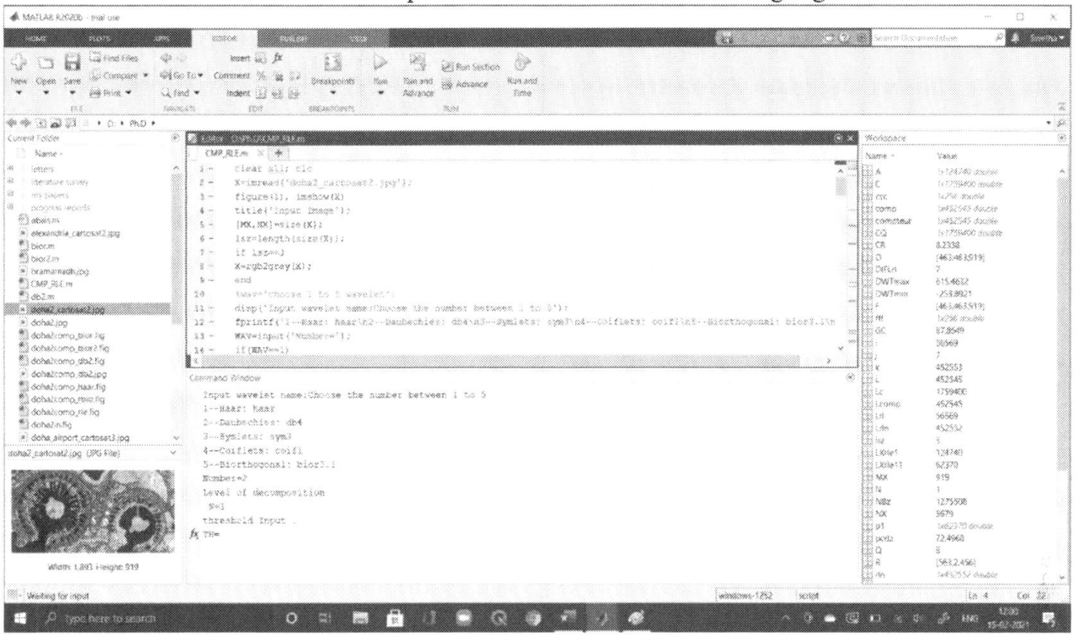

Figure 6. Inputs for wavelet, decomposition level, and threshold in MATLAB tool

After performing Level 1 decomposition, we get the following image (Figure 7).

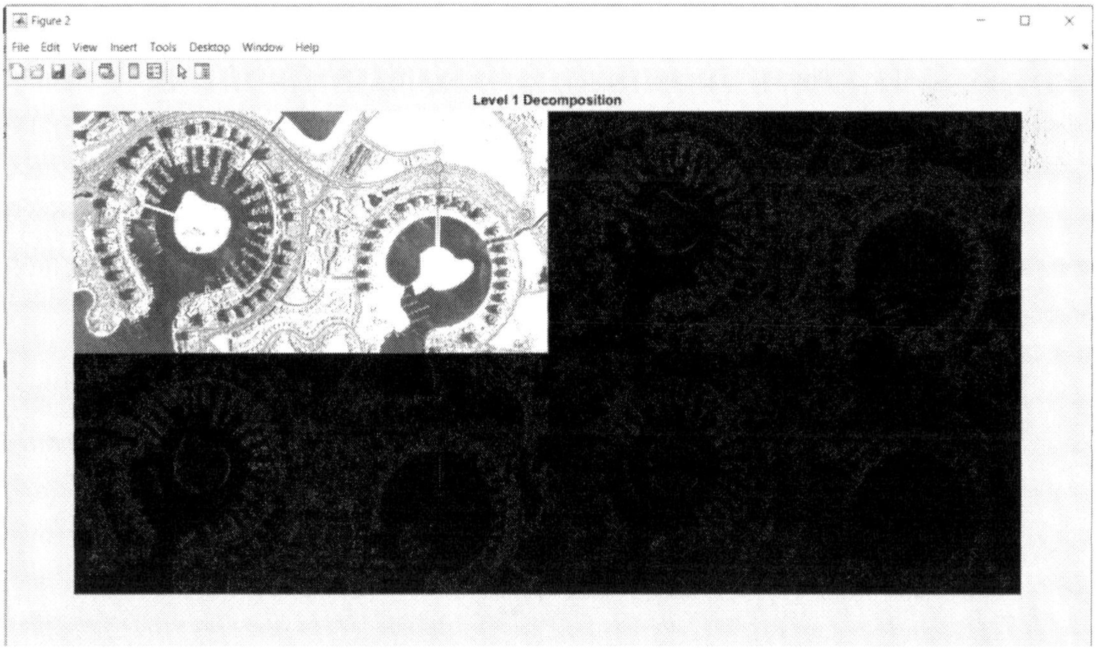

Figure 7. Level 1 decomposition image

The coefficients are then subjected to an input threshold value (given as 50), where the coefficients below the threshold value are truncated. The following equation gives the equation for thresholding.

if abs(C[i])<TH, then C[i]=0 (1)

where C[i]= wavelet decomposed vector

TH=Threshold value

The resultant coefficients are Run-Length Encoded, where the data repeated in succession is stored as a single data value with its count. As there is no loss of information, the output bitstream is thus compressed in a lossless way. With the help of the compressed output bitstream, the Compression ratio is calculated by the given formula.

Compression ratio=Original Image Size/Compressed Image Size (2)

A compression ratio of 8.2338 is obtained for Level 1 Decomposition of Daubechies Wavelet. The compression ratio increases when the level of decomposition increases.

4. Results and Discussion

This paper aims to achieve and analyze high compression ratios using different wavelets and Encoding using the Run-Length Encoding technique. The ten satellite images, both panchromatic and multispectral, are wavelet decomposed using the five wavelets: Haar, Daubechies, Symlets, Coiflets, and Biorthogonal. The compression ratios are analyzed with the three decomposition levels: Level 1, Level 2, and Level 3. The threshold value is taken constant throughout all the comparisons. For example, we take a sample image of a crater on the moon taken by Chandrayaan2. As it is panchromatic, it is directly subjected to Wavelet Decomposition. Firstly, the wavelet is chosen from the given five wavelets and then input the decomposition level. The coefficients are Run-Length Encoded to obtain the compressed output bit stream depending on the given threshold value. The compression ratio is lastly calculated.

If we choose the Daubechies wavelet, Level 1 decomposition, threshold value as 50, the compression ratio is calculated as 18.97. If the Level 2 decomposition is selected with the same wavelet and threshold value, the compression ratio obtained is 41.19. If Level 3 is chosen, the compression ratio would be 48.87 (Table 1). The following table constitutes the comparison of compression ratios of the ten satellite images using five wavelets and three decomposition levels.

Table 1. Comparison of Compression ratios of different images with various wavelets

Images	Wavelets														
	Haar			Daubechies			Symlets			Coiflets			Biorthogonal		
	Decomposition levels														
	1	2	3	1	2	3	1	2	3	1	2	3	1	2	3
doha2_cartosat2.jpg	7.47	10.58	11.20	8.23	12.37	13.20	8.15	12.22	13.01	7.96	11.76	12.52	8.60	11.23	11.36
earth_insat3dr.jpg	9.39	12.46	13.31	10.32	14.56	15.51	10.37	14.48	15.50	10.15	14.08	15.15	10.13	12.17	12.37
moon_chandrayaan2.png	11.09	15.44	15.90	12.09	17.59	17.83	11.98	17.49	17.76	11.68	16.85	17.21	12.84	14.32	13.89
mooncrater_chandrayaan2.png	18.61	38.46	46.24	18.97	41.19	48.87	18.94	40.61	48.68	18.93	40.41	48.86	19.63	33.41	32.48
doha_airport_cartosat3.jpg	10.07	31.53	55.34	10.18	31.53	67.67	10.17	33.53	66.79	10.14	33.20	64.37	10.32	20.93	21.53
alexandria_cartosat2.jpg	7.72	12.11	13.33	8.85	15.26	17.07	8.69	14.79	16.56	8.43	14.07	15.68	9.20	13.02	13.18
kishangarh_cartosat2.jpg	9.86	13.98	14.34	11.02	16.85	17.41	10.89	16.48	17.04	10.74	16.03	16.54	11.79	14.83	14.72
mars_orbiter,jpg	18.17	67.32	151.2	18.24	68.35	158.2	18.22	68.26	160.4	18.21	68.05	158.6	18.22	58.59	90.41
qatarkhalifa_cartosat3.jpg	10.12	23.56	28.50	10.42	29.72	38.58	10.42	29.29	37.89	10.39	28.04	35.87	10.42	29.96	37.89
shadnagar_cartosat2.jpg	10.14	39.45	99.38	10.08	39.69	131.0	10.11	40.06	128.3	10.11	40.04	121.9	10.12	40.52	112.9

The above table is plotted into graphs for all the ten images, panchromatic and multispectral, to analyze and compare the various compression ratios obtained for different wavelets and decomposition levels. The five wavelets and the decomposition levels are shown on the X-axis and the Compression Ratio on the Y- axis in Figure 8.

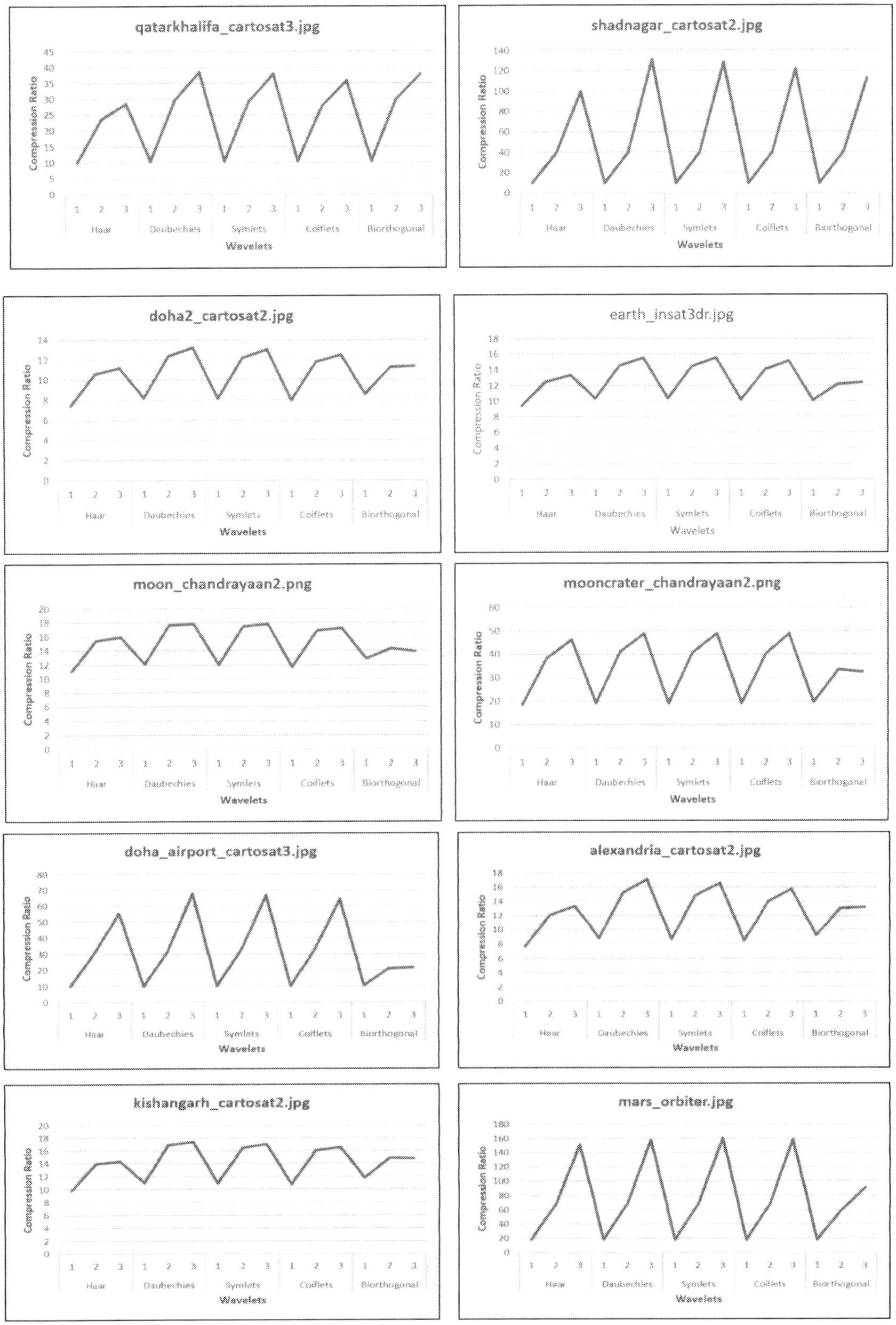

Figure 8. A plot of comparison of compression ratios

5. Conclusion

The ten satellite panchromatic and multispectral images are subject to compression using the hybrid Discrete Wavelet Transform (DWT) and Run-Length Encoding (RLE) algorithm. The images are analyzed using the five wavelets: Haar, Daubechies, Symlets, Coiflets, and Biorthogonal. Levels 1, 2, and 3 decompositions are performed on the images while keeping the threshold value constant. It can be observed that higher compression ratios are achieved using Daubechies and Biorthogonal Wavelets than other wavelets. Run-Length Encoding enhances the compression by replacing the repetitive long runs of truncated data with only a single value and a count.

REFERENCES

[1] Ajay Sudhir Bale, Suhaas V. Reddy, Shivashankar A. Huddar, Electromechanical characterization of Nitinol based RF MEMS switch, *Materials Today: Proceedings*, Volume **27**, Part 1, 2020, Pages 443-445, ISSN 2214-7853, https://doi.org/10.1016/j.matpr.2019.11.263

[2] Ajay Sudhir Bale, J. Aditya Khatokar, Shantanu Singh, G. Bharath, M.S. Kiran Mohan, Suhaas V. Reddy, T.Y. Satheesha, Shivashankar A. Huddar, Nanosciences fostering cross domain engineering applications, *Materials Today: Proceedings*, 2020, ISSN 2214-7853, https://doi.org/10.1016/j.matpr.2020.09.076.

[3] J. Aditya Khatokar, N. Vinay, Ajay Sudhir Bale, M.A. Nayana, R. Harini, V. Suhaas Reddy, N. Soundarya, T.Y. Satheesha, A. Shivashankar Huddar, A study on improved methods in Micro- electromechanical systems technology, *Materials Today: Proceedings*, 2020, ISSN 2214-7853, https://doi.org/10.1016/j.matpr.2020.10.993.

[4] S. A. Huddar, B. G. Sheeparamatti and A. S. Bale, Study of pull-in voltage of a perforated SMA based MEMS Switch, *2017 International conference on Microelectronic Devices, Circuits and Systems (ICMDCS)*, Vellore, India, 2017, pp. 1-4, doi: 10.1109/ICMDCS.2017.8211584.

[5] Ajay Sudhir Bale et al *2020 IOP Conf. Ser.: Mater. Sci. Eng.* 872 012008.

[6] Venkatesh M S, Manoj Patil, Ajay Sudhir Bale, Srujan Ingalgeri. Design of Remotely Monitorable Low Power Phototherapy Unit for Treatment of Neonatal Hyperbilirubinemia, *National Conference at Bapuji Engineering College*, Davangere, India.

[7] Aditya Khatokar J., Mounisha M., Nayana M.A., Ajay Sudhir Bale, Bhavana S. Battery Management System: A Survey. *Journal of Industrial Safety Engineering*. 2020; **7**(1): 29–35p.

[8] Kishan Das Menon H, Aditya Khatokar J, Ajay Sudhir Bale. Enhanced Railway Operations Using Automated Locomotive Simulator. *Trends in Transport Engineering and Applications*.2020; **7**(1): 17–23p.

[9] Ajay Sudhir Bale, Hosamani Ummar Farooq N, Shivashankar Huddar, Automated Diesel transfer system using PLC. *Journal of Industrial Safety Engineering*. 2019; **6**(1): 8–14p.

[10] Aditya Khatokar J, Nayana M A, Soundarya N, Meghana N, Bhavana S, Sunkireddy Umarani, Ajay Sudhir Bale. Electric Vehicles: Transition to Green Zone. *Trends in Transport Engineering and Applications*. 2020; **7**(2): 12–17p.

[11] Raksha K.P., Rajani Alagawadi, Nisha N., Deeksha R., Ajay Sudhir Bale. Advancement of Nanotechnology in Batteries. *International Journal of Energetic Materials*. 2020; **6**(2): 18–24p.

[12] Vinay N., Aditya Khatokar J., Ajay Sudhir Bale. Analysis on Synthesis of Quantum Dots with Their Applications on Photochemistry. *International Journal of Photochemistry*. 2020; **6**(1): 1–11p.

[13] Ajay Sudhir Bale, Bharath G, Kiran Mohan M S, Shantanu Singh, Aditya Khatokar J. Thin-Films: Study of Medical, Display and Environmental Applications. *International Journal of*

Energetic Materials. 2020; **6**(1): 1–6p.
[14] Aditya Khatokar J., Nayana M.A., Ajay Sudhir Bale, Meghana N., Sunkireddy Umarani. A Survey on High Frequency Radios and their Applications. *Journal of Industrial Safety Engineering*. 2020; **7**(1):7–12p.
[15] Harish Koujalgi, Ajay Sudhir Bale. Biometric Based Automatic Ticket Vending Machine for Indian Railways. *International Research Journal of Engineering and Technology (IRJET)*. Volume: **04** Issue: 07 July -2017. e-ISSN: 2395-0056, p-ISSN: 2395-0072.
[16] Ajay Sudhir Bale, Harish Koujalgi. Quality Factor analysis for Nitinol based RF MEMS Resonator. *International Research Journal of Engineering and Technology (IRJET)*. Volume: **04** Issue: 07 July-2017. e-ISSN: 2395-0056, p-ISSN: 2395-0072.
[17] A Recent trend in Microgrid, *First international Conference (IEEE Technically sponsored) Advance Computing and Innovative Technologies in Engineering (ICACITE 2021)*, on 07th March 2021 at Galgotias University, Greater Noida, Uttar Pradesh.
[18] Aditya Khatokar J, Nayana M A, Kishan Das Menon H, Janardhan V, Ajay Sudhir Bale. A Study on Various Approaches in Remote Sensing. *Journal of Telecommunication, Switching Systems and Networks*. 2020; **7**(2): 32–37p.
[19] Ajay Sudhir Bale, J. Aditya Khatokar, M.S. Kiran Mohan, G. Bharath, Shantanu Singh, J. Roshini, Suhaas V. Reddy, Shivashankar A. Huddar, N. Vinay, Nanotechnology as a tool for treating cancerous tumors, *Materials Today: Proceedings*, 2021, ISSN 2214-7853, https://doi.org/10.1016/j.matpr.2020.12.1175.
[20] J. Aditya Khatokar, N. Vinay, B. Sanjay et al., Carbon Nanodots: Chemiluminescence, fluorescence and photoluminescence properties, *Materials Today: Proceedings*, Volume **43**, Part 6, 2021, Pages 3928-3932, ISSN 2214-7853 https://doi.org/10.1016/j.matpr.2021.02.582.
[21] D. Gowri Sankar Reddy, "Satellite Image Compression using Discrete wavelet Transform", *IOSR Journal of Engineering (IOSRJEN)*. ISSN (e): 2250-3021, ISSN (p): 2278-8719 Vol. **08**, Issue 01, V2, PP 53-59, 2018.
[22] Dr. H S Prashantha, Nagesh K N, Karthik H, Ankith Shetty R, Likitha C, "Multispectral Image Compression Using Wavelet Transforms", *International Journal of Engineering Applied Sciences and Technology*, Vol. **5**, Issue 3, ISSN No. 2455-2143, Pages 327-330, 2020.
[23] Sulaiman Khan, Shah Nazir, Anwar Hussain, Amjad Ali and Ayaz Ullah, "An efficient JPEG image compression based on Haar wavelet transform, discrete cosine transform, and run length encoding techniques for advanced manufacturing processes", *Measurement and Control*, Vol. **52**(9-10) 1532– 1544, 2019.
[24] Faisel Ghazi Mohammed1 and Hind Moutaz Al-Dabbas, "The Effect of Wavelet Coefficient Reduction on Image Compression Using DWT and Daubechies Wavelet Transform", *Science International*, 2018.
[25] Mamun, M. A., Hossain, M. A., Mondal, M. N. I., & Aktar, M, "Satellite image compression using integer wavelet regression", *2017 International Conference on Electrical, Computer and Communication Engineering (ECCE)*, doi:10.1109/ecace.2017.7912944, 2017.
[26] Ravi, R. V., & Subramaniam, K., "Image compression using optimized wavelet filter derived from grey wolf algorithm", *Journal of Ambient Intelligence and Humanized Computing*. DOI:10.1007/s12652-020-02290-7, 2020.
[27] K. Pranitha, Dr. G. Kavya, "Data Compression with High Peak Signal to Noise Ratio Using Bisectional Cylindrical Wavelet Transform for a Satellite Image", *International Journal of Engineering & Technology*, **7** (4.6) 469-475, 2018.
[28] Shihab, H. S., Shafie, S., Ramli, A. R., & Ahmad, F., "Enhancement of Satellite Image Compression Using a Hybrid (DWT–DCT) Algorithm", *Sensing and Imaging*, **18**(1), doi:10.1007/s11220-017- 0183-6, 2017.

[29] Geetha, V., Anbumani, V., Murugesan, G., & Gomathi, S., "Hybrid optimal algorithm-based 2D discrete wavelet transform for image compression using fractional KCA", *Multimedia Systems*. DOI:10.1007/s00530-020-00681-6, 2020.

[30] B. Dheepa, R. Nithya, N. Nishavithri, K. Vinoth, K. Balaji, "Directional lifting wavelet transform based SAR Image Compression", *International Journal of Advanced Trends in Computer Science and Engineering*, 2020.

[31] Bo Li, Rui Yang, and Hongxu Jiang, "Remote-Sensing Image Compression Using Two-Dimensional Oriented Wavelet Transform", *IEEE Transactions on Geoscience and Remote Sensing*, VOL. **49**, NO. 1, 2011.

[32] Memane, T., & Ruikar, S. D., "Selection of wavelet for satellite image compression using picture quality measures", *International Conference on Communication and Signal Processing* DOI:10.1109/iccsp.2014.6949997, 2014.

[33] V. Sunil Kumar, M. Indra Sena Reddy, "Image Compression Techniques by using Wavelet Transform", *Journal of Information Engineering and Applications*, ISSN 2224-5782 (print) ISSN 2225-0506 (online), Vol **2**, No.5, 2012.

[34] A Indradjad, A S Nasution, H Gunawan and A Widipaminto., "A comparison of Satellite Image Compression methods in the Wavelet Domain", *The 4th International Conference of Indonesian Society for Remote Sensing, IOP Conf. Series: Earth and Environmental Science* 280, 012031. DOI:10.1088/1755-1315/280/1/012031, 2019.

[35] Delaunay, X., Chabert, M., Charvillat, V., & Morin, G., "Satellite image compression by post- transforms in the wavelet domain", *Signal Processing*, **90**(2), 599-610. DOI: 10.1016/j.sigpro.2009.07.024, 2010.

Development of an integrated system for force measurement on end effectors of serial manipulators (IRB1410) using LabVIEW and data dashboard.

Vasanth kumar.CH[1], Raghavendra Rao.V[2] and Ann C V Medona[3]

[1]Department of AI and DS, EIE, Saveetha Engineering College, Chennai, Tamil Nadu, India
[2]Department of Mechanical Engineering, SRM institute of Science and Technology, Chennai, Tamil Nadu, India
[3]Department of AI and DS, Saveetha Engineering College, Chennai, Tamil Nadu, India

E-mail: *sreevasanthkumarchoda@gmail.com, Raghavev@srmist.edu.in.*

Abstract. A data acquisition method to obtain force measurement on an end effector of a serial manipulator is carried out using LabVIEW Data Dashboard to obtain results wirelessly developing an integrated, system between the serial manipulator, a computer and a remote device. A strut array was manufactured with a 4x4 grid and 8 strain gauges were attached on the parallel sides of the 4 rectangular struts based on strain analysis run theoretically using ANSYS to locate position of maximum strain. Two full bridge circuits from the array were then connected to a NI 9219 analog I/O module set up on a NI cDAQ 9174 chassis to measure the force on the struts whenever a load is applied. Tabular results of strain were obtained which varied linearly with uniformly appreciating loads, graphically displayed using LabVIEW GUI. LabVIEW is further connected wirelessly to the NI Data Dashboard app on a mobile via a shared variable engine which generates the results as a shared variable which can then be displayed on other remote devices. Therefore we can say that an integrated wireless system to remote monitor force applied on an end effector of the serial manipulator IRB1410 has been developed successfully.

Keywords- LabVIEW Data Dashboard, Shared Variable, End Effector, Strut array.

1. Introduction.

With further progress in robotics research, scientists are concerned more and more with the field of compliance robotics and wireless and seamless connectivity to other machines and Central Processing Unit (CPU) mainframes to better control robots/serial manipulator and end effectors remotely and accurately to develop systems that are more flexible and less error prone by eliminating redundancies and to improve worker safety. The best way to improve on this is to find a method to wirelessly provide and receive sensor-based intelligence from the robot in real time and to connect it in a closed feedback loop to continuously update and eliminate errors. Remote access of equipment has always been a priority, an example of which can be seen in the paper by Jagadeesh et al. [1] and which is why wireless sensor networks can play a big role in this field. Robot arm systems have always been complex as evidenced in the paper by Chandra Mouli.et.al. [2] Which describes the construction and analysis of a robotic arm. Remote monitoring and control of such a system has a lot of difficulties. In this case setting up sensor arrays such as Santoso et al. [3] has done in a robotic arm can greatly improve efficiency. The setting up of DAQ and a wireless sensor network for which several previous studies have been carried out [4-11] is an important part of realizing this idea for compliance. Even microcontrollers such as Arduino and Xbee radio modules can be utilized to facilitate wireless sensor networks for remote viewing [5-6], [9-11] which is more cost effective but more error prone as compared to DAQ systems on which several studies have been undertaken [3, 7-8]. The main objective of the project is to run an analysis and receive values wirelessly from an array of sensors present on a robotic gripper with the help of an example metallic strut array to increase compliance methods and to present these results directly on the cellular device of the user instantaneously using web publishing tools or data dashboard and by presenting these results graphically using the GUI interface LabVIEW.

2. Materials and Methods

2.1 Materials and construction

The strut array was constructed using the Aluminium alloy Al H11 which has specifications as shown in table 1.

Table 1. Al H11 Specifications

Young's Modulus MP a	Poisson's Ratio	Bulk Modulus M P a	Shear Modulus M P a
69000	0.33	67647	25940

Strain gauges were affixed on the struts with the following specifications:
1. Gauge Factor: 2:1 ±1%
2. Gauge Resistance: 350 ± 0:3
3. Gauge Length: 3mm

The basic design of the strut array system is made using SolidWorks. It is a 4 x 4 array with each strut having equivalent dimensions of 7 x 5 x 35mm. Each strut is spaced 20mm from each other to form a perfect symmetrical rectangular array. Each strut has two strain gauges attached on parallel sides. The design of a single strut is as shown in Figure 1 (ANSYS). The actual strut system is shown in figure 2. Jumper wires were used for the connection wiring.

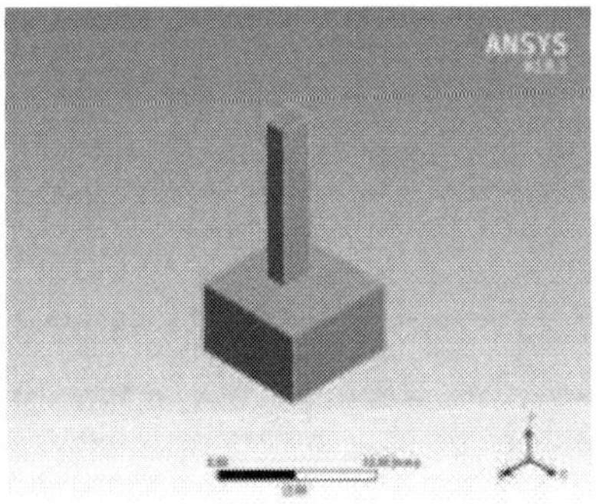

Figure 1. Design of the strut in ANSYS

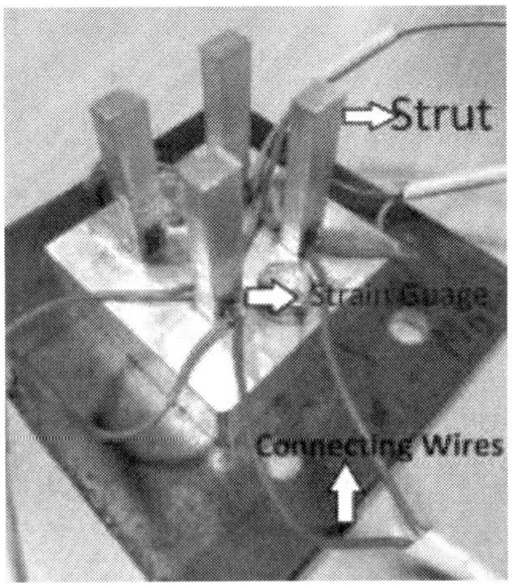

Figure 2. Physical model of strut with mounted sensor array.

2.2 Connecting to DAQ and electronic circuitry:

The sensor array was connected to the analog I/O module NI 9219 which was fixed on the chassis cDAQ 9174 using two full bridge circuits which were connected to two channels on the I/O module. There are 4 channels on the NI 9219 as shown in figure 3. The full bridge circuit connections are as shown in figure 4 with each terminal in a channel connected to a respective connection from the full bridge circuit which comprises 4 strain gauges.

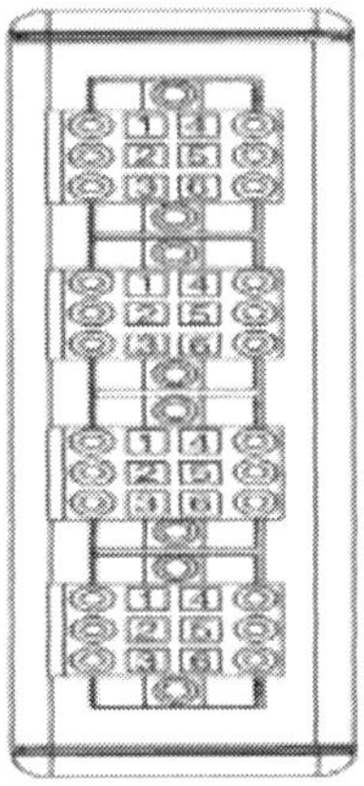

Figure 3. NI 9219 Channels for interfacing analog inputs .

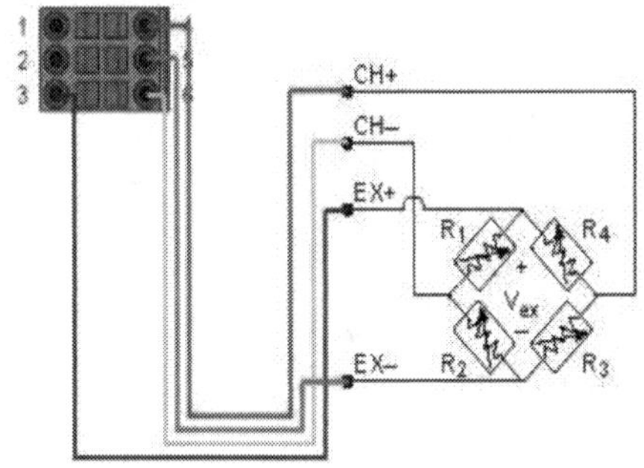

Figure 4. Full bridge circuit connection from sensor array to NI 9219.

The next step is to set up the automatic VI block diagram on LabVIEW to read the values that are inputted from the DAQ module to LabVIEW GUI as shown in figure 5, which is then stored in an excel document and the values can be shown in strain .i.e, mm/mm format by choosing the strain module from the DAQ assistant. This block diagram that runs to read the values dumped by the DAQ module should be stored as a read only shared variable in the Shared Variable Engine (SVE) to be deployed in real time to be used in wireless transfer.

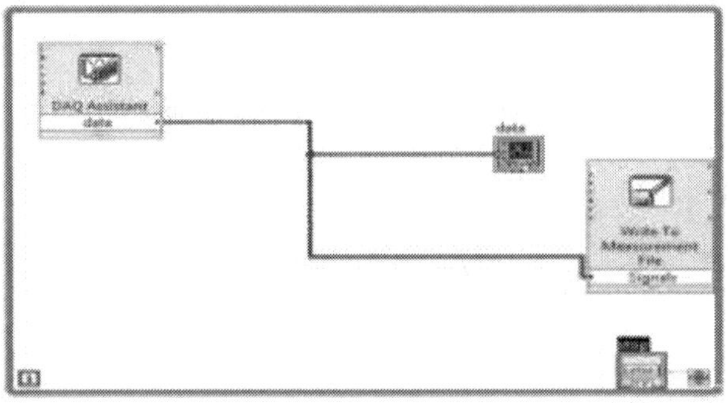

Figure 5. Block diagram to read DAQ analog readings

2.3 Wireless transfer method using data dashboard:

1. The first step is to open the app NI Data dashboard on the cellular device which has an opening panel as shown in figure 6.
2. Click on the '+' sign in the upper corner to open a new dashboard.
3. Next place an indicator or control such as a waveform chart and numeric indicator in the dashboard by selecting from the indicators and controls palette as shown in figure 7.
4. The next step is to deploy the shared variable in LabVIEW in real time. A trial shared variable program which shows a graph of random numbers is shown in figure 8. The difference between this trial program and the actual deployment is that in the trial program the shared variable is in write mode which means the random number function can change the values stored in the SVE whereas in the original program, the shared variable is in read only mode which only reads the values coming from the DAQ and stores it in SVE and continuously updates it.
5. Now the next step that was taken was to set up an internet hotspot connecting the cellular/remote device and the computer.
6. Next select the data link icon present under the placed indicator in data dashboard which opens the connect dialog box.

7. In the connect dialog box select the shared variables option and enter the computer's IP address which establishes the connection between LabVIEW and dashboard and connects to the project library.
8. Open the library which has the required shared variable and select it.
9. Finally press the play button on the upper right-hand corner of the panel on the dashboard to view the results remotely on the mobile as shown in figure 9.

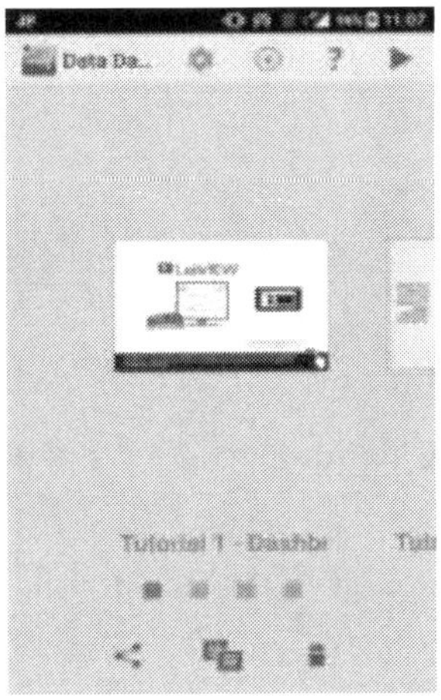

Figure 6. Data Dashboard panel on an android platform

3. Results and Discussions

3.1 Standardized testing with cantilever beam

Before we analyse the strut array we have to first determine if there are any errors in the sensors (strain gauges) we use using a standardized testing base comparing theoretical and experimental data and calculating error percentage. If the error percentage is within 10%, then the strain gauges are good to use and can be used for further testing. Values of the strain value are obtained by a quarter bridge circuit where the strain gauge is placed at the centres of the cantilever which is 181mm from the support. The weight put at the end of the cantilever is 750grams or 7.5N downwards for which the strain values are as shown in table 2.

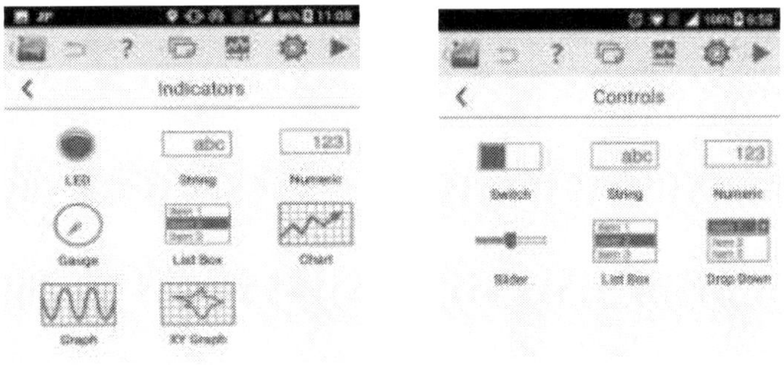

(a) (b)

Figure 7. (a) Indicators and (b) Controls palette In the android platform

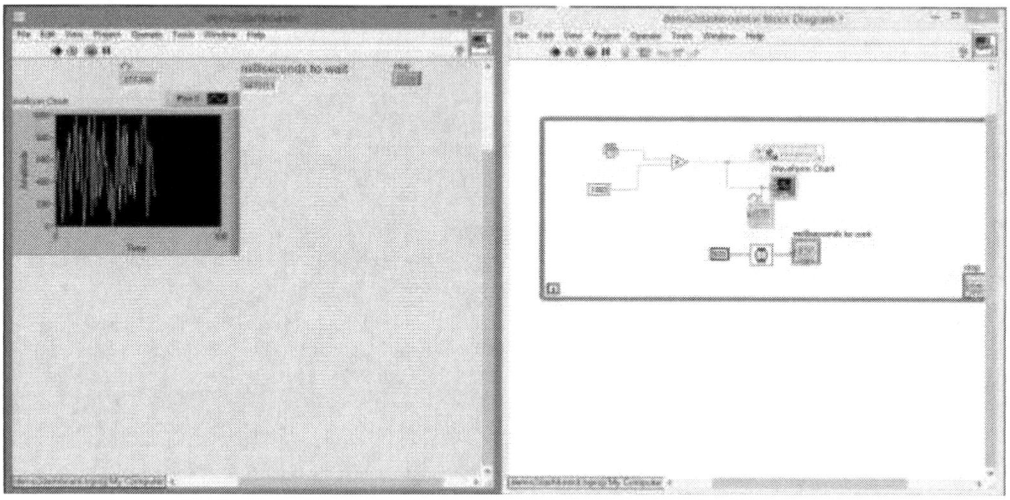

Figure 8. Shared Variable with the data dash board using LabVIEW.

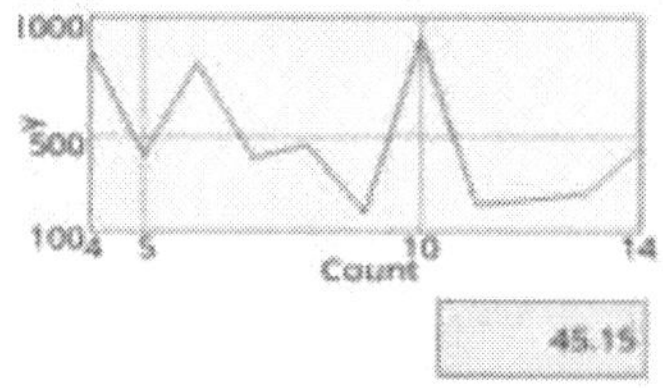

Figure 9. Real time results on dashboard after establishing connection

Table 2. Experimental strain values using cantilever beam as base

S No.	Strain Values (mm/mm)
1	0.12361 $x10^{-3}$
2	0.12361 $x10^{-3}$
3	0.12361 $x10^{-3}$
4	0.12361 $x10^{-3}$
5	0.12361 $x10^{-3}$

Average = 0.12361 $x10^{-3}$ (1)

The theoretical strain at the same point for cantilever beam for the same load of 7.5N comes at $0.134x10^{-3}$ mm/mm.

To calculate the error % see equation 1.

$$[(0.000134 - 0.000123) \div 0.000134]x\,100 = 8.2\% \quad (2)$$

Since the error percent is less than 10% it is safe for us to proceed with this model to calculate the strain on the given strut.

3.2 Theoretical strain analysis on system of struts using ANSYS

A stress and strain analysis is run on the struts to obtain the optimum location to position the strain gauges on the respective struts. A theoretical strain table with respect to length of strut (table 3) and graph (figure 10) are obtained. It is taken for a load of 10N on each strut on ANSYS. The results are obtained as shown in figure 11 (ANSYS) after the boundary conditions are set and figure is appropriately meshed in ANSYS.

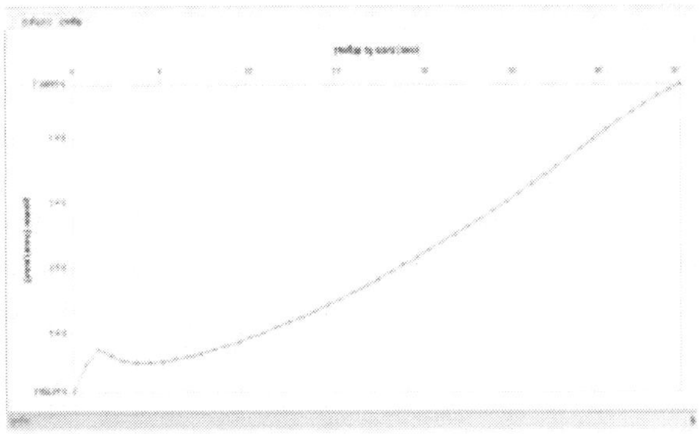

Figure 10. Length of strut versus strain

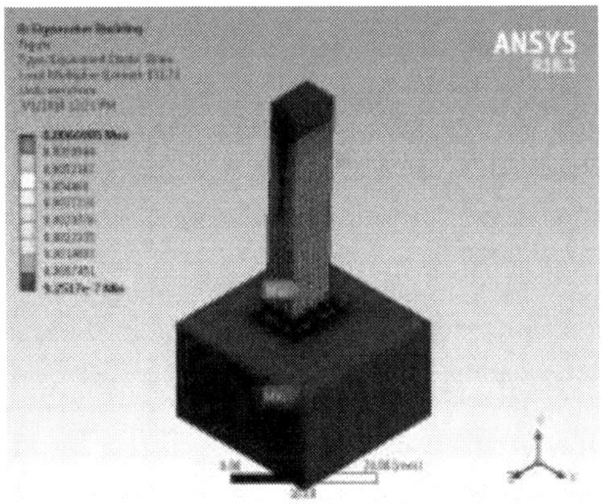

Figure 11. Strain distribution on the strut after analysis

Table 3. Theoretical strain values

Sl no	**Length**[mm]	**Value**[mm/mm]
1	0	4.8875×10^{-3}
2	0.72917	4.4589×10^{-3}
3	1.4583	4.2378×10^{-3}
4	2.1875	4.3403×10^{-3}
5	2.9167	4.4221×10^{-3}
6	3.6458	4.4357×10^{-3}
7	4.375	4.4428×10^{-3}
8	5.1042	4.4188×10^{-3}
9	5.8333	4.3865×10^{-3}
10	6.5625	4.3444×10^{-3}

The max strain is experienced at 4.375mm of the strut from the base, hence the strain gauge is placed at this distance from the base of strut.

3.3 Experimental analysis using NI-cDAQ 9174 and NI9219

Using the NI 9219, strain values were recorded according to the procedure in experimental setup using a strain module on the DAQ assistant (figure 5) for 5 different weights ranging from 1kg to 5kg thereby getting a range of strain values for each load without too much variation from average value as shown in table 4. These results were then displayed in graphical format on LabVIEW GUI which was then displayed in real time on Data Dashboard as shown in figure 9.

Table 4. Experimental strain values from DAQ

LOAD(Kg)	**RANGE OF STRAIN VALUES FOR THE GIVEN LOAD AS PER NI-cDAQ9174**				
1.00	0.000336	0.000338	0.000342	0.000338	0.000338
2.00	0.000357	0.000357	0.000357	0.000357	0.000357
3.00	0.000450	0.000452	0.000449	0.000451	0.000451
4.00	0.000542	0.000543	0.000544	0.000544	0.000545
5.00	0.000546	0.000545	0.000545	0.000543	0.000547

The results that have been measured provide a correct estimate on the forces applied on the serial manipulator IRB 1410 by using strain gauge and a DAQ and remotely view and monitor it. This study further enhances our understanding of remote monitoring of robots used in industries. i.e., compliance robotics. A number of sensor arrays in a variety of configurations and types can be used in this setup to completely monitor a working robotic arm/serial manipulator from afar and many further enhancements to this idea can also be made. A closed feedback system can be put in place so that the gripper/robot itself can be controlled based on the results obtained from the various sensor arrays. A method to implement pseudo force to test out the robot must be forthcoming in order to safely manipulate and run tests on it without actually loading the robot. A web publishing method can be established in order to view and control all data displayed on the GUI in any network by creating a Hyper Text Transfer protocol (HTTP) which should let the user view and control the data with the help of certain plug-ins on any PC or device which has an internet connection. Different materials can be used for testing such as Teflon to replace the Aluminium H11 alloy to test further effectiveness. Strain or other sensor tests for bending and vibration impacts can also be carried out apart from buckling alone.

4. Conclusions

Theoretical strain analysis was carried out on the strut and strain gauges were also tested using standardised cantilever beam with an error percentage less than 10% thus the strut system and the strain gauges are all within acceptable limits and will show the desired results. The strain values were obtained from DAQ. Accurate strain results were obtained for the sensor array via NI 9219 Analog module and were displayed in the GUI, i.e. LabVIEW. Moreover, the results were displayed in real time in a remote device such as a mobile phone using the NI Data Dashboard app thus facilitating remote control. Thus, all the requirements to prove this concept of wireless data transfer and control on a remote device from a sensor array affixed on a robotic gripper has been affirmed.

References

[1] Jagadeesh, R., Ranganatha, R., and Siddaraju, C. (2018). "*Design and implementation of an architecture for a remotely operated plc laboratory using labview.*" **8**, 419–426.

[2] Chandra Mouli, C. (2013). "*Design and implementation of robot arm control using labview and arm controller.*" **6**, 80–84.

[3] Santoso, D., Maryanto, S., Nadhir, A., and Sugiharto, T. (2017). "*A simple and low cost data acquisition system with multi-nodes facility for geophone array sensors.*" **12**, 2109–2114.

[4] Engelbrecht, N. and Penzhorn, W. T. (2005). "Secure authentication protocols used for low power wireless sensor networks." *Proceedings of the IEEE International Symposium on Industrial Electronics*, 2005. ISIE 2005, Vol. 4. 1777–1782.

[5] Harikrishnan, R. (2015). "An integrated xbee arduino and differential evolution approach for localization in wireless sensor networks." *Procedia Computer Science*, **48**, 447 – 453. *International Conference on Computer, Communication and Convergence (ICCC 2015).*

[6] Kabir, A. F. M. S., Shorif,M. A., Li, H., and Yu, Q. (2014). "A study of secured wireless sensor networks with xbee and arduino." The 2014 2nd *International Conference on Systems and Informatics (ICSAI 2014)*. 492–496.
[7] Liu, Q., Xu, Z., and Li, Z. (2017). "*Implementation of hardware tcp/ip stack for daq systems with flexible data channel.*" 53.
[8] Mukaro, R. and Francis Carelse, X. (2000). "*A low-cost microcontroller-based data acquisition system for solar radiation monitoring.*" **48**, 1232 – 1238.
[9] Sung, W. T., Chen, J. H., Hsiao, C. L., and Lin, J. S. (2014). "Multi-sensors data fusion based on arduino board and xbee module technology." 2014 *International Symposium on Computer, Consumer and Control*, 422–425.
[10] Yussoff, Y., Abidin, H. Z., Rahman, R. A., and Yahaya, F. H. (2010). "Development of a pic-based wireless sensor node utilizing xbee technology." 2010 2nd *IEEE International Conference on Information Management and Engineering*. 116–120.
[11] Nikhade, S. and Agashe, A. (2015). "*Wireless sensor network communication terminal based on embedded linux and xbee.*" 1468–1473.
[12] Segura, F.; Bartolucci, V.; Andújar, J.M. Hardware/Software Data Acquisition System for Real Time Cell Temperature Monitoring in Air-Cooled Polymer Electrolyte Fuel Cells. *Sensors* 2017, **17**, 1600.
[13] Kim, H.S., Han, J.S and Lee.Y.H. (2012)."Scalable network joining mechanism in wireless sensor networks. "2012 IEEET optical Conference on Wireless Sensors and Sensor Networks. 45–48.
[14] Kulakov,A., Davcev.D and Trajkovski.G. (2005)."*Application of wavelet neural networks in wireless sensor networks.*"
[15] Escobar, R.F.; Adam-Medina, M.; García-Beltrán, C.D.; Olivares-Peregrino, V.H.; Juárez-Romero, D.; Guerrero-Ramírez, G.V. Monitoring and Control Interface Based on Virtual Sensors. *Sensors* 2014, **14**, 20645-20666.
[16] Hercog, D.; Gergič, B. A Flexible Microcontroller-Based Data Acquisition Device. *Sensors* 2014, **14,** 9755-9775.
[17] Chen, C.-Y.; Liu, C.-Y.; Kuo, C.-C.; Yang, C.-F. Web-Based Remote Control of a Building's Electrical Power, Green Power Generation and Environmental System Using a Distributive Microcontroller. *Micromachines* 2017, **8**, 241.
[18] Shahzad, A.; Landry, R.; Lee, M.; Xiong, N.; Lee, J.; Lee, C. A New Cellular Architecture for Information Retrieval from Sensor Networks through Embedded Service and Security Protocols. *Sensors* 2016, **16**, 821.
[19] Shahzad, A.; Lee, M.; Xiong, N.N.; Jeong, G.; Lee, Y.-K.; Choi, J.-Y.; Mahesar, A.W.; Ahmad, I. A Secure, Intelligent, and Smart-Sensing Approach for Industrial System Automation and Transmission over Unsecured Wireless Networks. *Sensors* 2016, **16**, 322.
[20] Chaos, D.; Chacón, J.; Lopez-Orozco, J.A.; Dormido, S. Virtual and Remote Robotic Laboratory Using EJS, MATLAB and LabVIEW. *Sensors* 2013, **13**, 2595-2612.

Influence of number of studs on compliance of a flight intersection joint

M Umakanth[1, 2], P Uday Kumar[1], V. Narayanamurthy[1, 3] and S Korla[2]

[1] Research Centre Imarat, Hyderabad, Telangana - 500069.
[2] Department of Mechanical Engineering, NIT-Warangal, Telangana - 506004.
[3] Corresponding Authors' Email: v.narayanamurthy@rcilab.in

Abstract: Flight intersection joints are temporary joints used to integrate one airframe section with another. They are characterized by their capacity to resist the rotational flexibility when subjected to an external bending moment, commonly termed as the joint rotational compliance (JRC). Currently, JRC is quantified mostly by extensive experiments on the flight intersection joints. The influence of number of studs in a stud-pocket type intersection joint on JRC is not yet well understood in literature which is addressed in this paper through numerical simulations. The variation in joint compliance and its dependence on number of studs are brought out through the analysis of circumferential distribution of joint displacement, JRC and joint rotational stiffness.

1. Introduction

Slender flights vehicles are made up of separate airframe sections as shown in Figure 1, upon assembly of airborne subsystems within the individual sections, they are joined together using a class of intersection joint. There are different types of flight intersection joints [1-9]. The selection of a type of joint depends on the joint compliance to be achieved, ease of integration, and requirement of a minimal occupied volume. A flight intersection joint (FIJ) is characterized by its resistance to the rotational flexibility when subjected to an external bending moment. This is called as the joint rotational compliance (JRC) and expressed in rad/Nm. This JRC helps in modelling the structure of the flight vehicle as a flexural beam with rotational springs at joints, to predict the natural structural mode shape and modal frequencies.

The JRC is often quantified through extensive experiments and sometimes through empirical relations whose accuracies are always questionable. A stud-nut-slot type of intersection joint which is conformal with cylindrical airframe geometry as shown in Figure-1 is quite popular. Several researchers have investigated the performance of flight intersection joints through numerical and experimental methods [1-13]. Kaplan [10] has determined the flexibility coefficients i.e. a measure of joint rotational stiffness (JRS, an inverse of JRC is another form of representation) for flight structural joint assemblies and reported the challenges in experimental and theoretical determination of the flexibility coefficients and the importance of these values in flight dynamics. Kumar *et al.* [11] and Gunda and Krishna [12] conducted ground resonance test and predicted the free vibration response of the launch vehicles through finite element method (FEA) using beam elements representing the airframes and rotational spring elements representing the FIJs. They observed significant discrepancies in the predictions as compared to the experimental data caused by the assumption of JRC values adopted in FEA and stressed upon the need for an accurate quantification of joint flexibility or the stiffness of the FIJs.

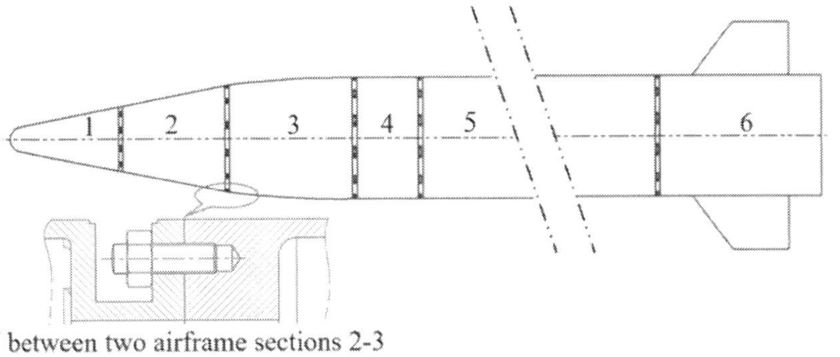

Figure 1. Typical flight vehicle with five intersection joints connecting six airframe sections

Alley and Gerringer [4] and Leadbetter et al. [5] investigated the prediction and measurement of natural vibrations on multi stage launch vehicles where they conducted experimental measurements of free vibration response of launch vehicles and observed considerable difference with the predicted vibration responses. They reported that invariably, computed mode data on launch vehicles which disregard local influences at FIJs will result in frequencies higher than actual; therefore, any reasonable approximation to stiffness of FIJ will move the computed results close to reality. Therefore, Alley and Leadbetter [3] provided a guide to joint evaluation which should provide a simple means for approximating the JRCs in a typical launch vehicle through an empirical relation. But this is based on 12 experiments and its accuracy is uncertain for different diameters and type of FIJs.

The behaviour of FIJs are also modelled using analytical approaches similar to that presented by Wojnar et al [13] for the tubular flange joints. Further the researchers have studied the joint compliances experimentally and stressed the importance of JRC in predicting the dynamic response of a flight vehicle. Although the importance of JRC in a FIJ and its experimental determination are well elaborated, prediction of JRC through numerical methods is not attempted. This paper presents the influence of number of studs on compliance using finite element analysis code ANSYS Workbench [14]. The details of numerical modelling, evaluation of joint opening and computation of JRC are discussed in the paper.

2. Numerical modelling and analysis

The geometry and dimensions of a typical stud-nut type of a flight intersection joint considered in this study is shown in Figure 2 with 8 numbers of M10 studs. The section airframes are made of aluminium alloy with elastic modulus E =70 GPa and Poisson's ratio ν = 0.32. The stud and nut are made of high strength alloy steel with E = 200 GPa and ν = 0.3. The entire geometry is discretised with 20 node hexahedral element with an element size of minimum 1 mm as shown in Figure 3(a). The interface between the external cylindrical surface of stud with 1) receiving internal threaded hole surface in front bulkhead of Section-2 airframe, and 2) inner threaded surface of nut are established with bonded contact to simulate the screwed joint as shown in Figure 3(b).

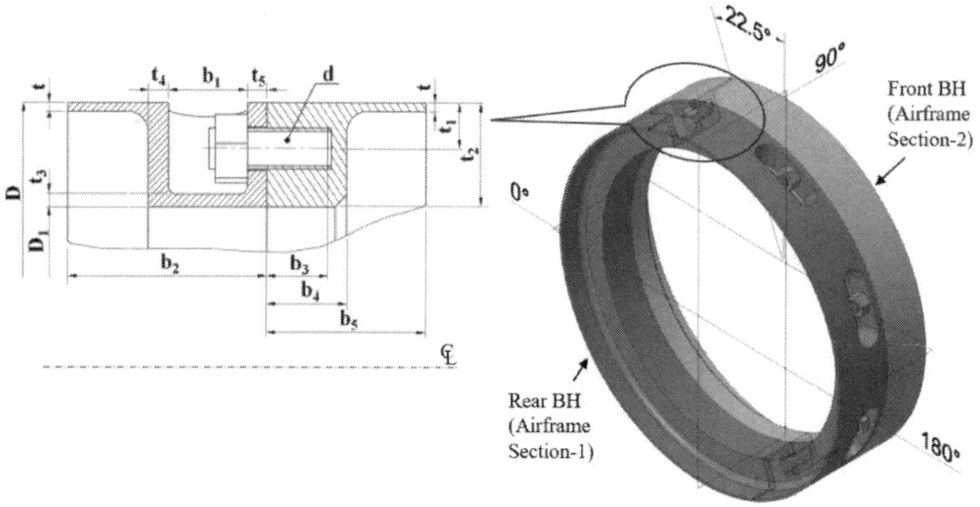

Figure 2: Details of a stud-pocket type flight intersection joint connecting airframe Sections-1 and 2

Frictional contacts with a friction coefficient of 0.5 is established between 1) the butting surface of nut against the flat vertical face in the slot of rear bulkhead in Section-1 airframe; and 2) the two contacting flat end surfaces of both the bulkheads.

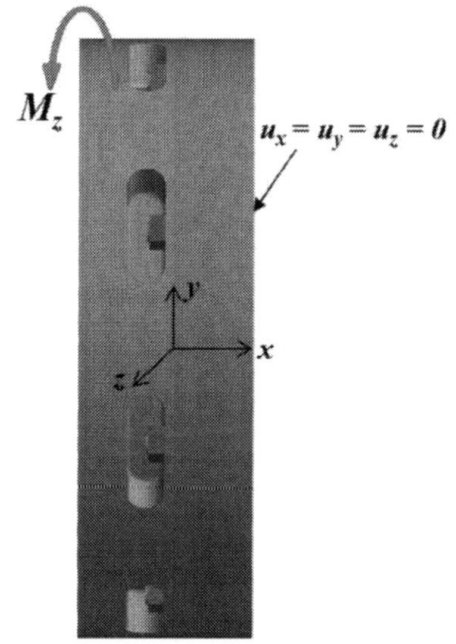

(a) FEA model

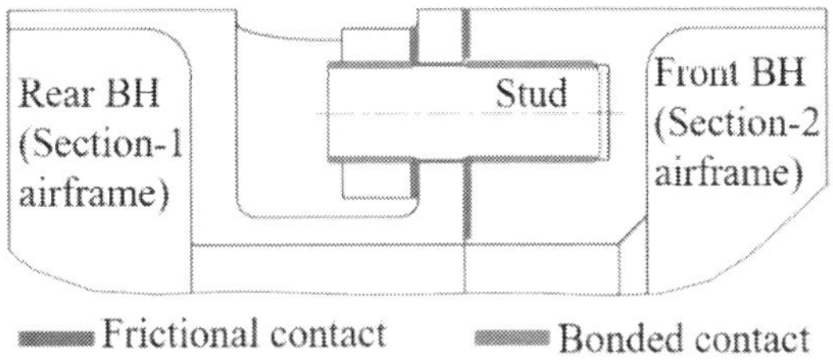

(b) Contacts in FEA model

Figure 3: Contacts, model for FEA with loads and boundary conditions

A pretension of 20000N [15] is applied on the threaded surface of stud and an equivalent compressive force is applied on the flat butting surface of nut on the stud-pocket to simulate the effect of pre-tightening torque of 40 Nm. First yield bending moment M_y is calculated for the geometry of airframe and the external bending moment M is proportionally applied in steps until M_y capacity is reached. The FE model with boundary conditions is shown in Figure 3(a). The axial displacements are measured from the analysis results. The joint compliance and the opening is evaluated to find the variation of JRC and its dependence on the number of studs.

3. Results and discussions

The distribution of axial displacement u_x around the circumference of the joint in degrees with 0° starting from the left reference orientation of the airframe through the top reference (90^0), the right reference (180^0) and the bottom reference (270^0) to the starting reference is shown in Figure 4. From Figure 4 and 5, it can be inferred that the maximum axial opening of the joint is 9.5 mm at 90° for 4 number of M10 studs. The opening reduces to 4.5 mm by increasing the number of studs from 4 to 6 numbers. The opening remains almost same for the joint with 8 and 10 number of studs. Enhancing the number of studs to 12 numbers reduced the opening by 2.5 mm. Further, it can be found that the joint openings are almost zero exactly at all stud locations. The bottom segment of the joint from 180^0 to 360^0 experiences a compressive action due to the hogging moment applied at the joint, tends to close the joint further and experiences almost zero opening. The joint's axial openings at RT are increasing linearly with increase in M/M_y and reaches maximum at $M/M_y = 1$ as seen in Figure 5.

The opening slope θ is calculated based on axial displacement with respect to neutral axis using airframe geometry. The JRC is calculated as ratio of θ / M in rad/Nm. The inverse of JRC is the joint rotational stiffness called as JRS. A combined plot showing JRC, JRS and maximum axial opening as a function of number of studs is shown in Figure 6. The JRC reduces from 1.3 to 0.35 rad/MNm, JRS increases from 0.7 to 2.85 MN/rad and the axial opening reduces from 9.5 mm to 2.4 mm as the number of studs are increased from 4 to 12. Although there is no appreciable difference noticed between 8 and 10 numbers of M10 studs on all three parameters studied here, the 12 number of studs provides least joint compliance with highest rotational stiffness in this stud-nut type of flight intersection joint.

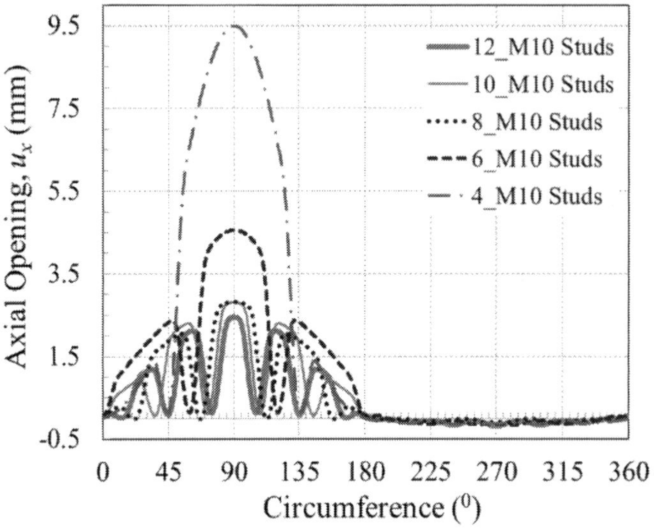

Figure 4: Axial opening along circumference starting from RL at M/M_y =1 for varying number of M10 studs

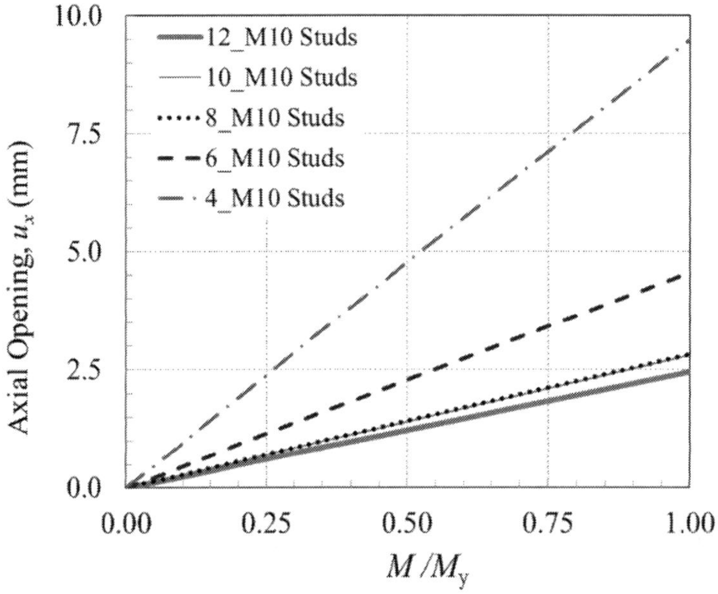

Figure 5: Variation of axial deformation at RT with applied moment for varying number of studs

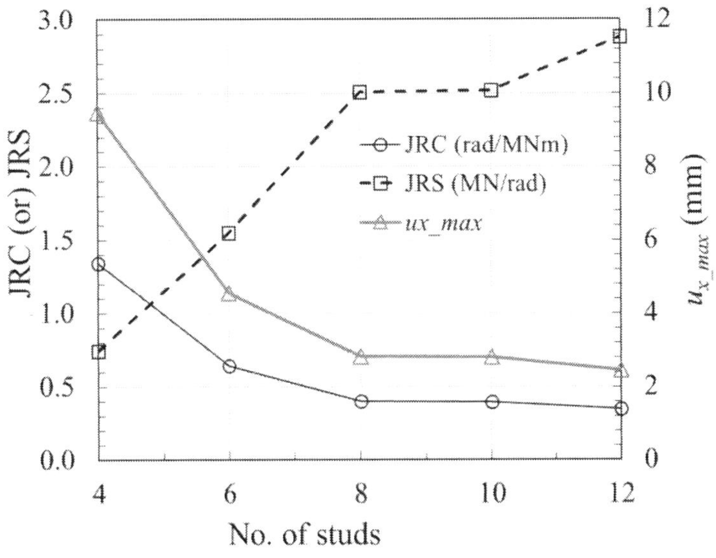

Figure 6: Effect of number of M10 studs on JRS, JRC and maximum axial opening

4. Conclusion

The flight airframe sections are joined to one another through a typical stud-nut type of intersection joint. This joint is characterized by the value of its joint rotational compliance, JRC. A joint need to exhibit a least JRC to simulate a flight structure with minimal flexibility at temporary joints and to have a continuity in natural mode shapes without an appreciable slope change. This paper numerically investigated the influence of number of studs on joint compliance and shown that a stud-nut type of joint with 8 to 12 numbers of M10 studs exhibited good performance in terms of JRC and JRS. The decrease in JRC is very significant from 4 to 8 number of studs. It is recommended to have 12 number of M10 studs to achieve a least joint rotational compliance.

5. References

[1] Masuram U, Narayanamurthy V and Korla S 2021 A review of flight intersection joints. *Journal of Int., Reviews in Aerospace Engineering*, In Press.

[2] Umakanth M, Punna U, Narayanamurthy V and Korla S 2021 Predicting the Rotational Compliance of a Flight Inter-section Joint. *NMIMS Journal of Engineering and Technology Review*, 3(1), 62-73.

[3] Alley V L and Leadbetter S A 1963 The prediction and measurement of natural vibrations on multi stage launch vehicles, *AIAA Journal*, 1(2), 374-79. https://doi.org/10.2514/3.1540.

[4] Alley V L and Gerringer A H 1962 A matrix method for the determination of the natural vibrations of free-free unsymmetrical beams with applications to launch vehicles, NASA TN D-1247.

[5] Leadbetter S A, Alley V L, Robert W H and Gerringer A H 1962 An experimental and analytical investigation of the natural frequencies and mode shapes of a four stage solid propellant rocket vehicle, NASA Technical Note D – 1354.

[6] Maloney J G and Shelton M T Underhill, D.A. Structural dynamic properties of tactical missile joints -Phase 1, General Dynamics, Pomona Division Report No.CR-6-348-945-001, (June 1970).

[7] Maloney J G and Shelton M T Structural dynamic properties of tactical missile joints- Phase 2, General Dynamics, Pomona Division Report No. CR-6-348-945-002, (Sept 1971).

[8] Henson G M and Hornish B A 2010 An evaluation of common analysis methods for bolted joints in launch vehicles, *AIAA Journal*, 3022, 1-27.

[9] Gharouni S M, Panahiha H M and Jam J E 2014 Space shuttle solid rocket motor (SRM) field joint: review paper, Met. Mater. Eng., 20 (3), 155-63. https://doi.10.5937/metmateng1403155G.

[10] Kaplan S M 1971 Flexibility coefficients for structural joint assemblies, *J. of Spacecrafts and Rockets*, 8(1), pp 76-77. https://doi.org/10.2514/3.30223.

[11] Kumar B, Rao NVM, Gupta RK and Mohan JR 2013 Prediction and measurement of free vibration response of multistage launch vehicles, *Proc. of 1st Indian Conference on Applied Mechanics (INCAM-2013),* IIT Madras, pp 1-6.

[12] Gunda J B, and Krishna Y 2014 Influence of joint flexibility on vibration analysis of free-free beams. Nonlinear Engineering, 3(4), pp 237–46. https://doi.org/10.1515/nleng-2014-0012.

[13] Wojnar A and Kozłowski A 2006 Mechanical model for assessment of the stiffness of bolted flanged joint, *Proc. of the 11th Int. Conf. on Metal Structures, Taylor & Francis,* Rzeszów, Poland, pp 188-89.

[14] ANSYS Workbench User's Manual. Version 18.1. ANSYS Corporation Inc., USA. 2019.

[15] Bickford J H 2008 *Introduction to the design and behaviour of bolted joint*, CRC Press, USA.

Chemical composition, antibacterial and antioxidant activities of lemongrass (*Cymbopogon citratus*) essential oil and its fractions obtained by vacuum distillation

D N Do[1*], H T T Nguyen[1], T H Huynh[2], N P Nguyen[2] and X C Luu[3]

[1] Faculty of Environmental and Food Engineering, Nguyen Tat Thanh University, Ho Chi Minh City, Viet Nam.

[2] Institute of Environmental Technology, Vietnam Academy of Science and Technology, Cau Giay Dist., Hanoi, Vietnam.

[3] Faculty of Drug, Cosmetic and Food, Dong An Polytechnic, Di an City, Binh Duong Province, Vietnam.

*Corresponding author: ddnhat@ntt.edu.vn

Abstract. Lemongrass (*Cymbopogon citratus*) essential oil (raw oil) is a natural product with many biological activities and is commonly used in the food and cosmetic industries. In this study, fractions of lemongrass essential oil were segregated from raw oil by vacuum fractionation. These fractions were then examined for their constitution, antioxidant and antimicrobial activities. Gas chromatography-mass spectrometry results indicated a difference in constitution and content of compounds among the fractions and raw oil. β -myrcene was the main ingredient in two fractions, F1and F2, with content 2-4 times higher than the original content in raw oil. At the same time, F4 was the primary fraction for citral recovery, with the highest citral content of 83.53%. Fractions F3, F4, and undistilled oil (Unoil) showed better antioxidation than raw essential oil. On the other hand, the antimicrobial experiments indicated that F1, F2, and F4 had the highest activity in both cases against Staphylococcus aureus and Escherichia coli bacteria. These outcomes could increase the applicability of lemongrass essential oil in many fields, especially in medical and food applications.

1. Introduction

In recent years, consumers have become increasingly interested in natural and eco-friendly products, and they are increasingly looking for healthier and safer products [1]. Pharmaceutical companies are increasingly interested in natural sources of materials to replace synthetic compounds gradually [2]. In the food industry, synthetic additives and preservatives are increasingly limited, so there is a tendency to use additives and antibacterial substances derived from nature that is safe for health growing healthy [3]. In natural alternatives, essential oil is a potential alternative source.

Lemongrass (*Cymbopogon citratus*) is a herb mainly planted in subtropical and tropical areas. The lemongrass essential oil is commonly applied to produce flavors, detergents, cosmetics, and pharmaceuticals. The investigation has indicated that diverse chemical composites of lemongrass oil present antioxidant, antimicrobial, antifungal, anti-inflammatory, antiparasitic, and insect-repellent activities. In the food sector, lemongrass oil is normally acknowledged as safe and can be applied as a replacement food additive to synthetic types [4]. Its significant antibacterial and antioxidant features can be used in food preservation as a substitute for synthetic components, meeting today's consumer needs.

The efficiency of the biological ability of essential oils is based mainly on the chemical constitution, which depends on many different factors such as the growing conditions, harvest time, and extraction method [5-6]. In addition to the attendance of components of essential oils, the content of each ingredient also plays a critical role in their biological activity. Investigates have shown that even minor components participate in the biological function of essential oil owing to the synergistic activity. Studying the bioactive ability of different essential oil fractions can create new applications and enhance the value of essential oils.

There was much attention to the bioactive ability of lemongrass essential oil for commercial applications in various fields. The purpose of the study was to fractionate lemongrass essential oil by vacuum fractionation. Essential oil fractions, which are obtained after distillation, were determined by their physical and chemical properties. The effects of the different chemical compounds in each fraction on their antimicrobial and antioxidant activity were determined.

2. Materials and Methods

2.1 Source of lemongrass essential oil
The crude lemongrass oil was produced by steam distillation in Phong Dien district, Thua Thien Hue province, Vietnam, in October 2020. It was supplied by Suwina Company, Vietnam. Chemicals used in the study were ethanol (Merck), sodium hydroxide, 2,2-Diphenyl-1-picrylhydrazyl (DPPH) (Sigma-Aldrich) analytical grade.

2.2 Essential oil fractionation
The feed section was a 250-mL flask containing 100mL of raw essential oil heated by an electrical heater and fractionated on a Vigreux column with heights of 300 mm. The vapor left the distillation column was then condensed by a Liebig condenser using the 5 °C water as a cooling stream through a chiller. Finally, the condensed liquid was collected in a multi-limb vacuum receiver and remaining in the flasks until the process finished. Fractions F1-F4 were collected, respectively, based on the difference in vapor temperature at the top of the column. The internal vacuum pressure of the whole system could be reduced up to 15 mmHg and was maintained owing to a vacuum pump connecting to the receiver. The fractionation was terminated when no vapor left the distillation column.

2.3 Essential oil component analysis
The chromatography was performed on an Agilent G1530A Gas Chromatography system. The system coupled with an HP-5MS column used He as a carrier gas with a 1 mL/min flow rate. An Agilent Mass Spectrum, Model 5973N MSD, was used as the system detector. The ionization energy was 70 eV, while the ionization source temperature of 220 °C and the mass scan ranged between 40 and 400 amu. The column temperature was initially set at 60 °C and increased at a rate of 10 °C/min up to 280 °C. The injection volume was 1,0 µL with a split ratio of 1:50. The scan regime in the GC/MS was 1 second per turn.

2.4 Antioxidant activity testing
The lemongrass essential oil's antioxidant capacity was estimated as described by Brand-Williams et al. [7] with some modifications. The 1,1-diphenyl-2-picrylhydrazyl (DPPH) radical solution was performed with 3.9 mg of radical and 100 mL of ethanol, then stored at 4 °C for 24 hours. This solution was diluted with ethanol to get an absorbance of 1.000 ± 0.02 at 517 nm in a Shimadzu UV-1800 Spectrophotometer (Shimadzu Corp., Japan). Aliquots of 600 µL of each oil solution with ethanol were put into a cuvette, and 2.4 mL DPPH solution was added. The mix was shaken and left in the dark for 30 min at room temperature. After that, the absorbance was determined at 517 nm in the Shimadzu UV-1800.

$$\% \text{ Inhibition} = [(Ab-As)/Ab] \times 100\%$$

Ab refers to the absorbance of the control, and As refers to the absorbance of the test oil.

2.5 Antimicrobial activity testing

Escherichia coli (ATCC 25922) and Staphylococcus aureus (ATCC 25923 TM) were the bacterial strains used in this investigation. Fresh cultures of microorganisms were cultivated on MHB and were incubated at 37 °C for 24 h. Lemongrass essential oil and its fractions were tested for antibacterial activity by the disk diffusion technique [8]. A 100 µL bacterial culture was spread homogeneously onto the surface of the MHA plates. Filter paper disks with a diameter of 6 mm were loaded with 10 µL of each lemongrass essential oil fraction. After 18-24 h of incubation at 37 °C, the antibacterial activity was recognized by measuring the diameter (mm) of the inhibition zones.

3. Results and discussion

3.1 Composition of the raw lemongrass essential oil and its fractions

Fractions F1, F2, F3, F4, and Unoil were recovered during fractional vacuum distillation. Table 1 showed the composition of compounds presented in the crude lemongrass essential oil and the fractions produced after the separation process. The earliest components recovered at the top of the column were 5-hepten-2-one and β-myrcene. Most of the β-myrcene was recovered in the F1 and F2 fractions during distillation. The content of β-myrcene presented in these fractions is much higher than in the raw oil (from 12.15% in crude oil to 45.21, 26.53% at the F1, F2 fractions, respectively). With this high β-myrcene content, the F1, F2 fractions have potential applications in cosmetics and food. The most abundant compound in raw lemongrass oil was citral. After the separation process, citral concentrated mainly on the F3 and F4 fractions. In which F4 was the fraction with the highest citral content (83.53%). The rest of the essential oil in the flask after the fractional distillation (un-distilled essential oil) contained less volatile constituents such as geranyl acetate (5.61%), trans-α-bergamotene (95.46%), m-camphorene (9.69%), p-camphorene (4.70%). In general, the differences in the composition of fractions (F1, F2, F3, F4, Unoil) compared with raw lemongrass essential oil could lead to differences in their bioavailability.

Table 1. The chemical composition of the raw lemongrass essential oil and its fractions.

Component	Raw oil (wt%)	F1 (wt%) 50-57°C	F2 (wt%) 58-60°C	F3 (wt%) 61-98°C	F4 (wt%) 100-110°C	Unoil (wt%)
2-Pentanone	0.30	0.15	-	-	-	-
5-Hepten-2-one	3.73	14.55	11.13	4.04	0.29	-
β-Myrcene	12.15	45.21	26.53	8.11	0.60	-
2,3-Dehydro-1,8-cineole	0.26	0.16	-	-	-	-
trans-β-Ocimene	1.23	5.83	3.58	1.01	-	-
β-cis-Ocimene	0.78	3.78	2.54	-	-	-
Linalool	1.36	2.32	3.84	3.48	0.93	-
Perillen	0.54	1.85	1.89	0.65	-	-
Citronellal	0.69	1.40	2.92	2.36	-	-
Isoneral	1.75	1.16	2.72	3.18	2.22	-
3,7-Dimethyl-3,6-octadienal	2.92	1.69	4.39	5.78	4.25	2.99
2-Cyclohexen-1-ol	0.28	0.13	0.17	0.43	0.31	-
β-Citral	29.4	7.43	19.55	34.77	37.85	11.08

Geraniol	3.49	0.52	0.95	2.61	4.17	3.56
α-Citral	33.85	6.36	15.41	29.91	45.68	33.79
Geranyl acetate	1.37	0.18	0.21	0.88	0.16	5.61
Caryophyllene	0.47	0.12	0.06	0.46	0.56	1.13
trans-α-Bergamotene	0.38	-	-	-	-	5.46
Octadecane	0.35	-	-	-	2.00	1.05
Naphthalene	0.149	-	-	-	-	1.29
Columbin	0.22	-	-	-	-	0.76
Trichloroacetic acid	0.22	-	-	-	-	0.62
m-Camphorene	0.35	-	-	-	-	9.69
p-Camphorene	0.18	-	-	-	-	4.70

-: Not detected.

3.2 Antioxidant activity

The results of antioxidant ability of raw lemongrass oils and their fractions with DPPH free radical scavenging experiment were indicated in Figure 1. There was a difference in percentage inhibition between raw lemongrass and fractions (F1, F2, F3, F4, Unoil) achieved after distillation. The results displayed that F3, F4, and Unoil had better antioxidant ability than the raw lemongrass essential oil. Meanwhile, F1, F2 showed much lower antioxidant activity than raw essential oil. Previous investigations have demonstrated that citral, the prime terpenoid component of lemongrass essential oil, has excellent antioxidant activity [12]. According to the chemical profile in Table 1, citral was present in a large percentage of the composition of fractions F3 (64.68%), F4 (83.53%), and Unoil (44.87%) fractions. In comparison, citral was present in small proportions in the F1 (13.79%) and F2 (34.96%) fractions. The F3 fraction showed the highest antioxidant activity (57.6%) even though the citral content was lower than the F4 fraction, which may be owing to the synergistic effect of all its constituents, according to some previous studies [13]. In the case of Unoil, although the citral content was not high (44.87%), its antioxidant activity was higher than that of raw essential oil; this may be due to the presence of phenolics compounds with high antioxidant activity such as m-camphorene, p-camphorene, geranyl acetate. In some other studies, this trend was also observed that the residues with higher phenolic content indicated higher antioxidant ability [14].

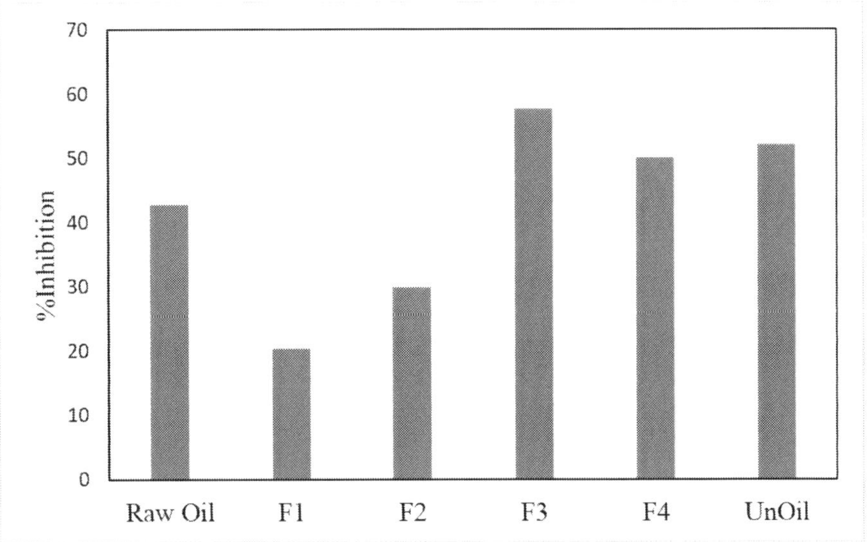

Figure 1. Antioxidant activity (% inhibition of lemongrass essential oil and its fractions)

3.3 Antibacterial activity

The antibacterial ability of the lemongrass essential oil and fractions (F1, F2, F3, F4, Unoil) was investigated through preliminary assays of the disk diffusion experiment. The antimicrobial ability was examined against gram-positive bacteria (Staphylococcus aureus) and gram-negative bacteria (Escherichia coli). The results showed that raw lemongrass essential oil and F1, F2, F3, F4, and Unoil fractions all have antibacterial effects against both Staphylococcus aureus and Escherichia coli. Gram-positive organism (Staphylococcus aureus) with inhibition zones ranging from 10–38 mm was found more sensitive to lemongrass essential oil and its fractions as compared to the gram-negative organism (Escherichia coli) with inhibition zones ranging from 2–26 mm. The diameter of the inhibitory zone (Table 2) for Staphylococcus aureus is shown in the order F2> F1> F4> Raw oil> F3> Unoil. In case of Escherichia coli, diameter of inhibitory zone (Table 2) in order F2> F4> Raw oil F1> F3> Unoil. In general, F1 and F2 fractions had good antibacterial activity; this may be due to the presence of β-myrcene in high content in these fractions; some studies have recognized that β-Myrcene enhances the antibacterial effect when combined with citral [15]. The F4 fraction with the highest citral content showed better antibacterial activity when compared to the raw lemongrass essential oil, which was consistent with the good antimicrobial role of citral in several previous studies [16]. For the residue after distillation, the least volatile fraction showed a lower antibacterial effect than the raw essential oil with both Staphylococcus aureus and Escherichia coli organisms. In summary, lemongrass essential oil and its fractions after the distillation had shown different antibacterial against both gram-negative and positive-gram bacteria. The positive-gram effect is better than compared to gram-negative organisms. The antibacterial ability of lemongrass essential oil and its fractions depended on three prime compounds: α-citral, β-citral, and β-myrcene. This was consistent with some previous reports [17].

Table 2. The diameter of inhibition zone of lemongrass essential oil and its fractions

Oil Fraction	**Diameter of Inhibition Zone (mm)**	
	Staphylococcus aureus	***Escherichia coli***
Raw oil	24.33 ± 0.36	12.05 ± 0.41

F1	34.26 ± 0.66	10.37 ± 0.48
F2	36.18 ± 0.52	26.24 ± 0.16
F3	18.76 ± 0.21	6.51 ± 0.39
F4	32.05 ± 0.44	22.45 ± 0.46
Unoil	10.91 ± 0.29	2.08 ± 0.57

* Data presented as mean of triplicate tests ± SD

4. Conclusions

Lemongrass essential oil (raw oil), its fractions (F1, F2, F3, F4), and the undistilled oil (Unoil) obtained by the fractional distillation showed differences in composition content as well as in their biological activities. This result showed that the vacuum distillation method was appropriate for producing different essential oil fractions, having better biological activities than the raw essential oil. Fractions F3, F4, and Unoil indicated better antioxidation than raw essential oil. On the other hand, fractions F1, F2, and F4 indicated better antibacterial ability than the original raw essential oil. These properties were related to the presence of various compounds in the essential oil composition. The presence of three main components α-citral, β-citral, and β-myrcene played an important role. These results contributed to the increased applicability of lemongrass essential oil. Fractions with high antimicrobial ability have the potential for medical applications as natural antimicrobial agents. The fractions can be used for their antioxidant ability in food applications, such as natural preservatives.

Acknowledgment

This work was supported by the Vingroup Innovation Foundation (VINIF) under the project code DA120_15062019/year 2019.

References

[1] Barbarossa C and De Pelsmacker P 2016 *J. Bus. Ethics* **134**(2) 229-247
[2] Atanasov A G, Waltenberger B, Pferschy-Wenzig E-M, Linder T, Wawrosch C, Uhrin P, Temml V, Wang L, Schwaiger S and Heiss E H 2015 *Biotechnol. Adv.* **33**(8) 1582-1614
[3] Bondi M, Lauková A, de Niederhausern S, Messi P and Papadopoulou C 2017 *Natural preservatives to improve food quality and safety* Hindawi
[4] Hyldgaard M, Mygind T and Meyer R L 2012 *Front. microbiol.* **3** 12
[5] Turek C, Stintzing F C 2013 *Compr. Rev. Food Sci. Food Saf.* **12**(1) 40-53
[6] Bakkali F, Averbeck S, Averbeck D, Idaomar M 2008 *Food Chem. Toxicol.* **46**(2) 446-475
[7] Brand-Williams W, Cuvelier M-E, Berset C 1995 *LWT - Food Sci Technol* **28**(1) 25-30
[8] Ha C T, Thai T H, Hien N T, Anh H T, Diep L N, Thuy D T, Nhat D D and Setzer W N 2019 *Nat. Prod. Commun.* **14**(6) 1934578X19860370
[9] Falcao M A, Fianco A L, Lucas A M, Pereira M A, Torres F C, Vargas R M and Cassel E 2012 *Phytochem Rev* **11**(4) 405-412
[10] Kpoviessi S, Bero J, Agbani P, Gbaguidi F, Kpadonou-Kpoviessi B, Sinsin B, Accrombessi G, Frédérich M, Moudachirou M and Quetin-Leclercq J 2014 *J. Ethnopharmacol.* **151**(1) 652-659
[11] Verma R K, Verma R S, Chauhan A and Bisht A 2015 *J. Essent. Oil Res.* **27**(3) 197-203
[12] Baschieri A, Ajvazi M D, Tonfack J L F, Valgimigli L and Amorati R 2017 *Food Chem.* **232** 656-663
[13] Naik M I, Fomda B A, Jaykumar E and Bhat J A 2010 *Asian Pac J Trop Dis* **3**(7) 535-538
[14] Olmedo R, Nepote V and Grosso N R 2014 *Food Chem* **156** 212-219

[15] Premathilake U, Wathugala D and Dharmadasa R 2018 *Int. J. Minor Fruits, Med. Aromatic Plants* **4** 13-19
[16] Tyagi A K, Malik A 2012 *Innov Food Sci Emerg Technol* **13** 169-177
[17] Onawunmi G O, Yisak W-A and Ogunlana E 1984 *J. Ethnopharmacol.* **12**(3) 279-286

Ab-initio Modeling of Functionalized 2D-Stanene nanostructure in context of FET based Toxic Gas Sensor

Swati Verma[1], Arun Kumar[2], Hemant Kumar[3], Rahul Baghel[4], Latika Pinjarkar[5], Mohan L Verma[6]

[1]Department of Electronics & Tele-communication Engineering, FET-SSGI, Shri Shankaracharya Technical Campus, Chhattisgarh, India
[2]Department of Electronics & Tele-communication Engineering, Bhilai Institute of Technology, Durg, Chhattisgarh, India
[3]Center for Basic Sciences, Pt. Ravishankar Shukla University Raipur, Chattisgarh, India
[4]Department of Electrical & Electronics Engineering .FET-SSGI, Shri Shankaracharya Technical Campus Junwani, Chhattisgarh, India
[5]Department of Information & Technology Engineering FET-SSGI, Shri Shankaracharya Technical Campus Junwani, Chhattisgarh, India
[6] Department of Applied Physics FET-SSGI, Shri Shankaracharya Technical Campus Junwani, Chhattisgarh, India

Abstract. Detecting toxic gases in the environment cannot be minimized in today's highly polluted world and the reality of global warming. Carbon monoxide is a highly toxic air pollutant and can cause serious health problems. Sensing materials plays a critical role in the performance of gas sensors. Nanostructure materials exhibit better performance as sensing channels in application for developing novel FET-based sensing devices. In the proposed work, the ab-initio study is performed to explore the performance of 2D-Stanene nano sheet as a sensing substrate for application of field-effect transistors as gas sensor. We explored CO gas-adsorption behaviour on pristine, hole-defective, and Pt-doped Stanene sheet and analyzed structural, charge-density, electronic and transport properties by first principles calculations and exploit its potential for high-performance gas sensing. Here in, to get the measure of sensing properties of Stanene nano sheet after the adsorption of gas molecule, first the molecular model of all the adsorption cases are modelled by considering all possible anchoring sites and optimized, and then calculations were performed based on density functional theory (DFT).

1. Introduction

From the beginning to the end of the day, we continuously sense something and our mind responds accordingly. We can see, the life of almost every individual across the world depend upon some automatic gadgets, in which sensor plays an important role[1].

It is interesting to see that sensor is a tool that identifies the existence and behavior of the physical environment and provides useful information in the form of signals. Due to this property of the sensor, we can study it at the nanoscale level[2]. Also, various technologies have been developed such as the nano-electromechanical system, nano-opto-electromechanical system, nano-photonics and the combination of nanotechnology with microtechnology[3].

Nano-sensors have a wide variety of applications in the field of healthcare, defense, homeland security, environmental monitoring and light-sensing electronics[4]. We can create more efficient and effective sensor networks through Tin(Sn) material because of its two-dimensional nature and excellent electronic and mechanical properties same as Graphene[5]. Sensing materials play a critical role in the performance of gas sensors.2D nano-materials show much better performance as sensing channels in application for, sensing channels in application for developing novel FET-based sensing devices [6-8].

We first modeled a stanene sheet to be used as sensing substrate, to be used as channel of Field-effect transistors (FETs).Field-effect transistors (FETs) are the kind of transistors that control the flow of current through the electric field. Field effect transistors (FETs) work on a principle that uses the field produced by the gate in a way that makes the channel conduct more or less. Here it consists of terminal, drain and gate where gate manipulates the channel conductivity [9]. The source is a terminal through which carrier (holes or electrons) enter the channel and drain allow the carriers to leave the channel. In bipolar transistors, the current that can be carried between the collector and emitter is modulated by the base current [10].

The working principle of FET sensor is to monitor the conductance change in the gate channel, before and after the adsorption of target molecules .Motivated by the applicability of 2D Graphene for gas sensing applications [11-13], we investigated the gas sensing properties of a monolayer of tin as stanene, in analogy with graphene.

In this work, first-principles methods based on DFT are employed to study the electronic and transport properties of the pristine, defected and Pt-doped Stanene to exploit its performance as a toxic gas sensor. It has been reported that functionalization, introducing dopants, and defects can tune the electronic and magnetic properties of the various nanomaterials[14-16]. Toxic gases are harmful gases and their reckless intake can cause harm to the respiratory system of the body. Carbon monoxide is an odorless toxic gas that is formed due to incomplete combustion of carbon and oxygen. Carbon monoxide is made up of one carbon atom and one oxygen atom with a bond order of 3, where it has one sigma bond and two pi bonds. It is also known as the invisible killer due to its poisonous nature and colorless character [17]

2. Computational Details

We exhibit the whole study of based on Density Functional Theory (DFT) which normally using electron density as a function of space and time. Here, We optimized and concluded all the results using Kohn-Sham coherent density functional method which further implements on computational package using open source software Spanish Initiative for Electronic Simulations with Thousands of Atoms (SIESTA). The SIESTA uses a basis set of strictly-localized atomic orbitals and employs norm-conserving pseudo potentials .All geometry optimizations and electronic structural calculations have been performed by using the SIESTA (Spanish Initiative for Electronic Simulations with Thousands of Atoms) code [18].

Quantum transport theory as implemented in GOLLUM code these codes are based on the density functional theory (DFT)[19].The exchange and correlation energies are treated within the generalized gradient approximation (GGA) [20] according to the Perdew-Burke-Ernzerhof (PBE) parameterization.

Nowadays, Researcher follows the GGA approximation because there are so many advantages over LDA. Local Density Approximation (LDA) have such limitations it does not account for van der Waals bonding, not suitable for strongly correlated system and gives a poor description of hydrogen bonding and over binding error for solids and molecules significantly reduces[21].

IOP Conf. Series: Materials Science and Engineering 1166 (2021) 012052 doi:10.1088/1757-899X/1166/1/012052

Numerical atomic orbital's (NAOs) with double zeta polarization (DZP) basis set have been used for geometry optimization. Minimization of energy has been carried out using standard conjugate-gradient (CG) technique until the forces on each atom are less than 0.005 eV/Å, the mesh cut-off energy is taken equal to 250 Ry.

3. Results and discussion

3.1 Structural Analysis

To explore the interaction of CO gas molecule with defect fuctionalized stanene systems, we firstly modeled and optimized the atomic structures of gas molecules and all Stanene systems .We have considered three sensing substrate system

1. Sn - the pristine Stanene nanostructure, consisting of 40 Tin (Sn) atoms.
2. Sn_V- defective Sn system,by creating a vacancy at center of Stanene nanosheet.
3. Sn_Pt- doped Sn system,by replacing one Sn atom with dopant Pt atom at center of Stanene nanosheet.

We have designed pristine 2D Stanene nano-structure sheet by repeating the unit cell 5 times along Z direction making it a nanosheet of 40 Tin atoms. After initial relaxation of the pristine sheet the relaxed lattice constant of a unit cell is 4.647 Å, and the length of Sn-Sn bond is 0.279 nm .These computed parameters are in good agreement with the previous reported results. After doping of Pt, we have analyzed the change in atomic structure, bond length between Sn-Sn increases to 0.282 nm and the bond between Sn-Pt decreases to 0.265 nm, which supports the strong interaction between the dopant atom and the sheet. In defective Sn system, we have created a vacant space due to creation of vacancy or hole at the center, the bond length between Sn-Sn decreases to a value of 0.275nm, as compared to pure stanene sheet.

In figure 1(a) represents the unit cell of Stanene. In figure 1 (b), a pure stanene sheet is formed due to the repetition of the unit cell of stanene, appearing as a honeycomb-like structure in the 2D plane. In figure 1(c) one atom of Platinum is doped in the center of the pure Stanene sheet because it provides a and figure 1 (d) illustrates the two-dimension Sn sheet with a vacant space formed because of the removal of Sn atom which was at the same place where Pt was doped.

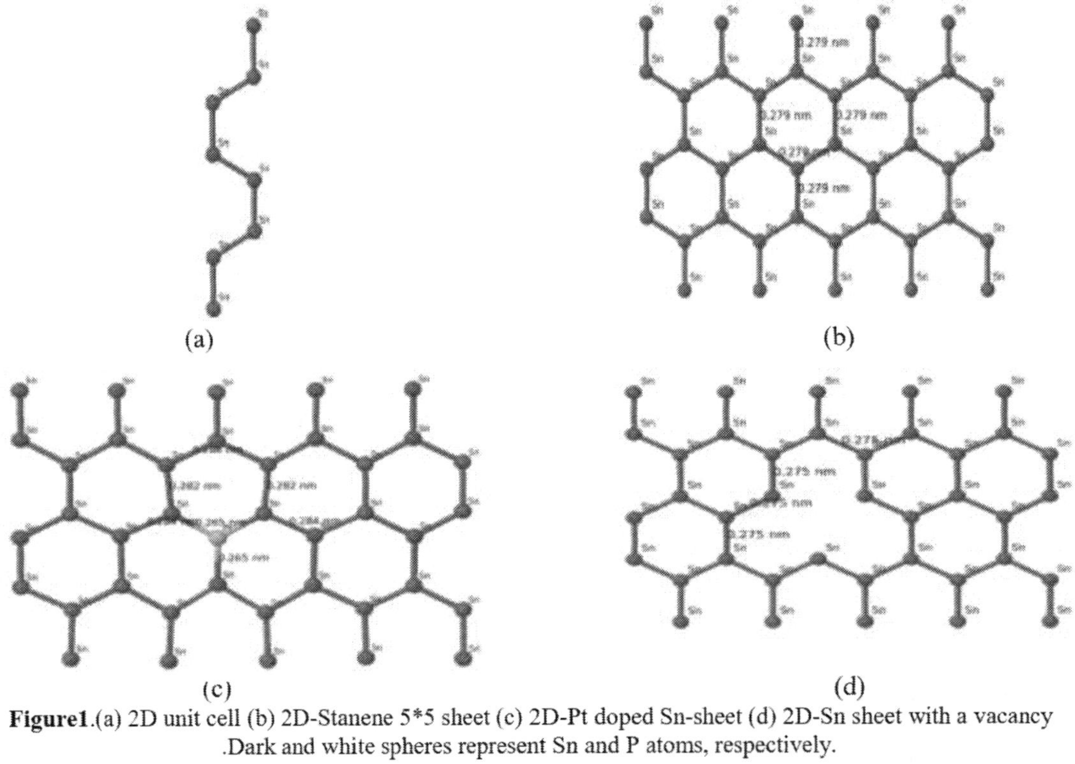

Figure1.(a) 2D unit cell (b) 2D-Stanene 5*5 sheet (c) 2D-Pt doped Sn-sheet (d) 2D-Sn sheet with a vacancy .Dark and white spheres represent Sn and P atoms, respectively.

After the optimization of CO molecule, the bond length of CO is 0.116 nm, which is in accordance with the published results.In the case of CO adsorption, there are six possible coordination configurations depending on the different atoms of CO molecule above three adsorption sites.

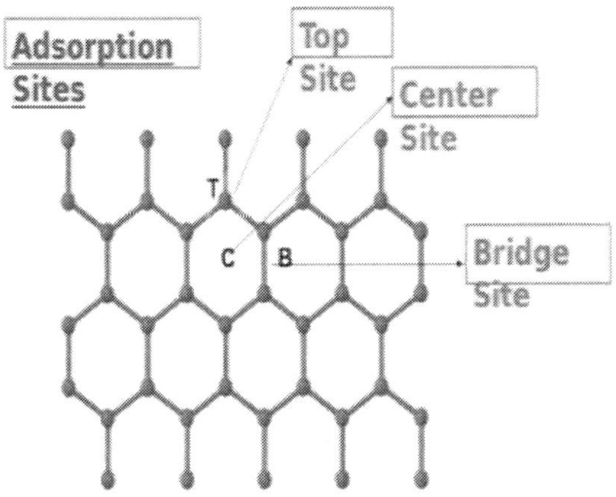

(a)

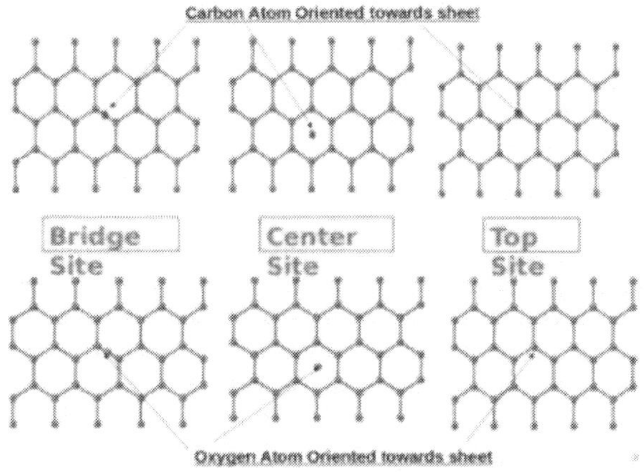

(b)

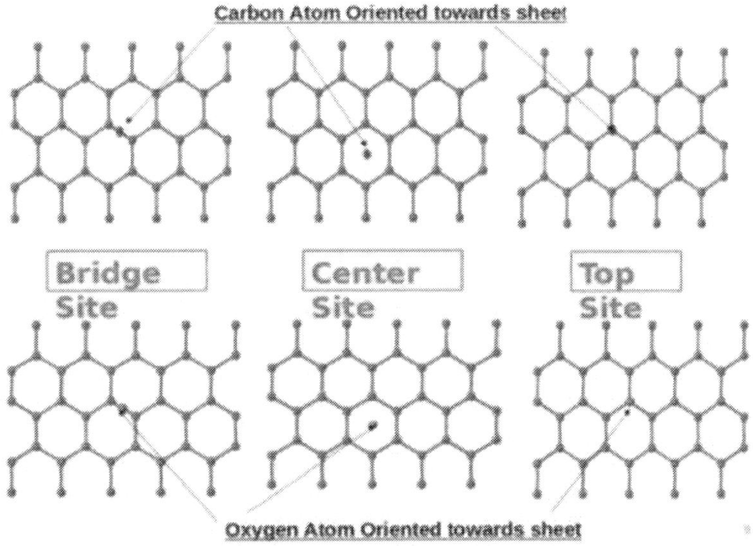

(c)

Figure 2.(a) Different absorption sites of Stanene sheet (b,c) CO molecules interact with Sn sheet and Sn_Pt sheet with two different orientation at 3 location, respectively

We allow CO molecules to interact with the Stanene sheet in all possible ways and found that there are three different absorption sites of interaction, which are, i)T-site, on the top of Sn atom (T), ii)C-site, center of the Sn hexagon (C) in which the molecule is placed in the hollow space region of a honeycomb-like structure of the sheet., and iii)B-site, the bridge point of a Sn-Sn bond (B),where the molecule is placed above the bond of the two atoms of Sn-Sn molecules of the sheet. Figure2 (a) represents the three different adsorption sites. In case of Sn_V sheet, we have placed the molecule on the vacant site.

The CO molecule has a diatomic structure. Due to the presence of three absorption sites, we can get six different coordination configurations depending upon the orientation of carbon and oxygen. In figure 3(b) the CO molecules interact with three different absorption sites, the top row figure shows the carbon atom orientation, forming a bond with a pure Sn sheet. The down row shows the oxygen atom orientation, forming a bond with a pure Sn sheet. In figure 3(c) the CO molecules interact with three different absorption sites, the top row figure shows the carbon atom orientation, making a bond with pure Platinum, doped in Sn sheet. The down row shows the oxygen atom orientation, forming a bond with the Platinum, doped in the Sn sheet.

The side view of CO molecules that interacts with three different sheets with pure Sn sheet, Sn sheet doped with Pt and Sn sheet with vacant space is represent in below figure. When the CO molecules interact with the pure Sn sheet, it makes two bonds between Sn-O with a bond length of 0.216 nm and the length of bond between C-O is 0.141 nm. In figure 3(b) oxygen atom of the CO molecule is bonded with Platinum which is doped in Sn sheet, forming the bond length of 0.248 nm between Pt-O and the bond length between C-O is 0.12nm. Figure 3(c) represents the vacant Sn sheet where the CO molecule formed a bond with Sn, having a bond length of 0.258 nm and C-O molecules having a bond length of 0.127nm.

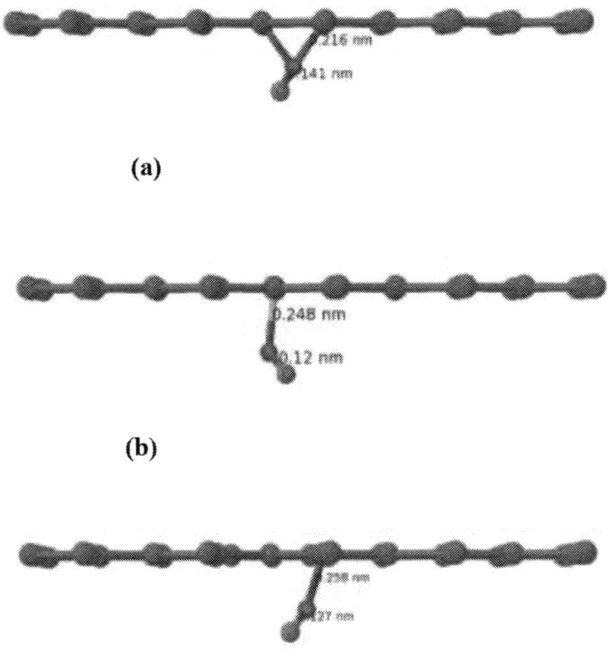

Figure 3. Side view of interaction of CO molecules with (a) Pure Sn sheet (b) Sn_Pt sheet (c) Sn_V sheet

3.2 Charge Density Analysis

Charge Density Analysis plays an important role in determining electronic properties. It gives a clear description of electron density by explaining the interaction between the atoms of the molecules and the process of bond formation which helps in predicting the most reactive position for any electron charge distribution of a molecule. It also provides quantitative information about the different regions of charge distribution for a molecule. In the charge scale, we can observe a higher electron density at the inner-most region (magenta) with a value of +1.0239 and the lowest electron density at the outermost region (red).

The values of the charge density of colors are given in the center of figure 4. In figure 4(a), the pure Sn sheet forms a concentric circle in the center of the hexagonal structure of the sheet and the presence of magenta color, that is visible in the Sn-Sn bond formation region, shows high electron density and indicates the formation of a strong covalent bond.

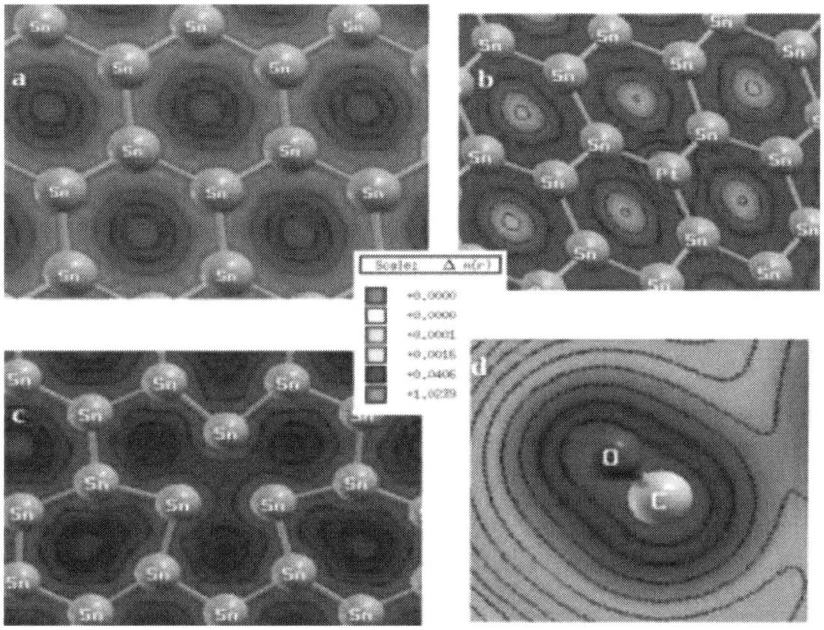

Figure 4. The Charge Density Plot of (a) Pure Sn sheet (b) Platinum doped Sn sheet (c) Sn sheet with one atom vacant (d) CO molecules

Figure 4(b), a higher charge density is observed around Platinum atoms of the platinum doped Sn sheets that means the ionization energy of Platinum (Pt) is higher than Tin(Sn) atom. The formation of concentric circles around the platinum atoms represents the ionic nature and indicates maximum charge concentration, which forms magenta circular structures.

In between the Sn-Sn vacant sheet, in figure 4(c), there are concentric circles around the central region and deformed dumb-bell shapes of the triangular shape formed around the atoms. Hence there is a partially covalent and ionic bond. From figure 4 (d), it is visible that the CO molecules bond forms the dumb-bell shape which indicates the sharing of equal charge between atoms and the charge density in most of the parts are of blue color that shows the presence of the covalent bond.

3.3 Electron Properties Analysis

The forbidden energy gap is the energy difference between the HOMO of the valance band and LUMO of the conduction band in the density of states (DOS) plotting. In any system, the conduction band provides enough energy to the electrons which allows them to move freely and this movement of electrons creates an electric current. The transition of an electron from HOMO of the valance band to the LUMO of the conduction band is very important to understand the Density of States (DOS) plot and Projected Density of states (PDOS) plot.DOS plots show the concentration of states for a particular energy level that an electron can occupy.

In Figure 5(a), the DOS plot shows the variation for all the different types of sheets and their concentration of states is demonstrated, where the pure Sn sheet has lower peaks in the forbidden gap region as compared to the vacant Sn (Sn_V) sheet and doped Platinum Sn (Sn_Pt) sheet. On the other hand, the vacant Sn sheet has higher peaks in the forbidden gap region. On the same figure, continuous waves are visible, which means that there is no energy gap in all three sheets. At the forbidden energy region, the sequence for the formation of lower peaks is Sn<Sn_Pt<Sn_V.

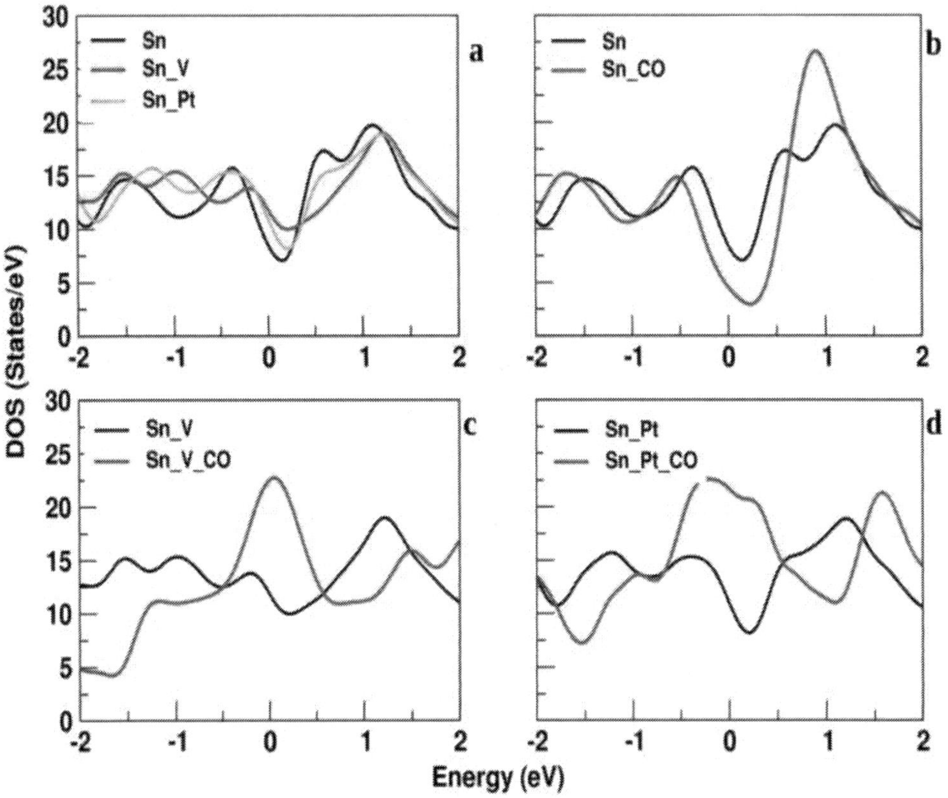

Figure 5. Density of States plotting (a) with all different three Sn, Sn_Pt and Sn_V sheet, (b,c,d) with or without interaction of CO molecules with Sn sheet, Sn_V sheet and Sn_Pt sheet, respectively

In the LUMO or conduction region, higher peaks are formed with the sequence of Sn>Sn_Pt>Sn_V. In Figure 5(b) graph represents the DOS plot of pure Sn sheet and pure Sn sheet interacting with CO (Sn_CO) molecules. In the LUMO or conduction region, the Sn_CO sheet shows high absorbance as compared to the pure SN sheet. In the forbidden energy gap region, the Sn_CO has a lower peak as compared to the pure Sn sheet, without interacting with CO molecules which means that the concentration of electron gets decreases in the region where we interact the CO molecules with the pure Sn sheet.

In Figure 5(c), the graph represents the DOS plot of the vacant Sn sheet which does not interact with CO molecules (Sn_V) versus the vacant Sn sheet which interacts with CO molecules (Sn_V_CO). The forbidden energy gap region of the Sn_V_CO sheet has a higher peak as compared to the Sn_V sheet, which means that the concentration of electron gets increase in the region when we interact with the CO molecules with pure SN sheet. In the LUMO or conduction region, Sn_CO interaction with the sheet shows high absorbance as compared to the pure SN sheet.

In Figure 5, the lower right graph represents the DOS plot of the Platinum doped Sn (Sn_Pt) sheet and platinum doped Sn sheet interaction with CO (Sn_Pt_CO) molecules. While in the forbidden energy gap region, the Sn_Pt show a lower concentration of electrons and in the case of Sn_Pt_CO, it shows a higher concentration of electrons and their antagonistic behavior is visible. The absorbance of CO molecules through the LUMO band of the Sn_Pt_CO sheet shows high energy after 1.5 eV but in the case of Sn_Pt sheet, the LUMO band shows the high peak at 1 eV that denotes the acceptance of electrons in orbital shifting, which means that high energy is required for the bond formation of Planium and CO molecules. PDOS plots depict the energy of any particular orbitals and the concentration of states that an electron can occupy.

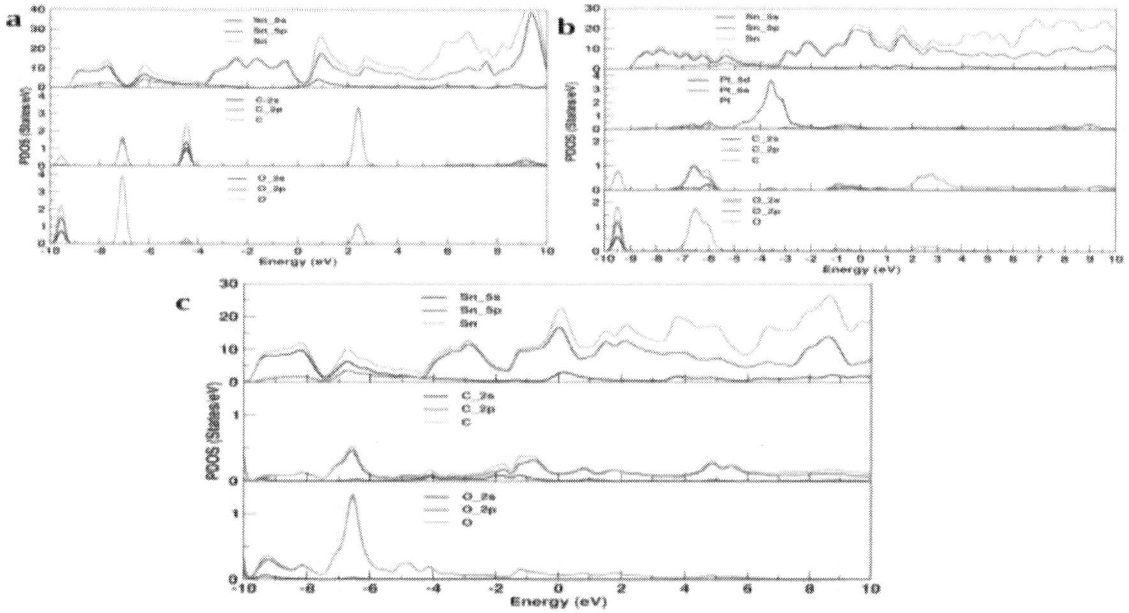

Figure 6. Projected Density of States with interaction of CO molecule with Sn, Sn_Pt and Sn_V sheet.

In all the three possibilities with the CO molecules (in Figure 6) interaction with sheets of pure Sn sheet, platinum doped Sn (Sn_Pt) sheet and vacancy Sn sheet there PDOS are shown in figure 6(a,b,c). In all the cases, the Carbon atom of the CO molecule has shown high absorption in the conduction band as compared to the oxygen atom when they interact with the sheets.

3.4 Electron Transport Analysis

To study the transport properties of our system, we used the GOLLUM code which is based on the equilibrium transport theory. GOLLUM code gives the electronic and thermal transport properties of nanostructured materials. GOLLUM code generates Hamiltonian through Density Functional Theory(DFT) which corresponds further to generate the S-matrix which gives the physical quantities such as current by using additional theory in DFT that is equilibrium Green Function (EGF).

The GOLLUM code helps to analyzed and study the first principle behavior of any nanostructure material through their I-V characteristics by using the DFT-EGF approach.
For transport calculation, the system is divided into three parts that include the left electrode, nanostructured material (scattering region) and right electrode, which are attached in a consecutive matter in series to fulfill the valency of each atom. The electrodes which are used for transport analysis are periodic in structure while the contact nanostructured material is non-periodic in structure.

Figure 7(a), shows the comparative plotting of three different sheets that include pure Sn sheet, platinum doped Sn (Sn_Pt) sheet and vacant Sn(Sn_V) sheet and these sheets are represented in terms of their I-V characteristics. From the graph, it is clear that before the supply of 1V, the output current in the vacant Sn(Sn_V) sheet shows high peaks as compared to the other two different sheets. After the supply of 1V, the Platinum doped Sn (Sn_Pt) sheet shows an efficient current with a magnitude scale of 10-5 Ampere. It can be easily concluded that by adding the functionality to the Sn sheet, in terms of adding vacancy and dopant Pt atom, we can modulate the conductivity of the sheet. Here, considerable rise in the current as shown in figure 7(a) indicates the increase in conductivity of our considered functionalized sheets.

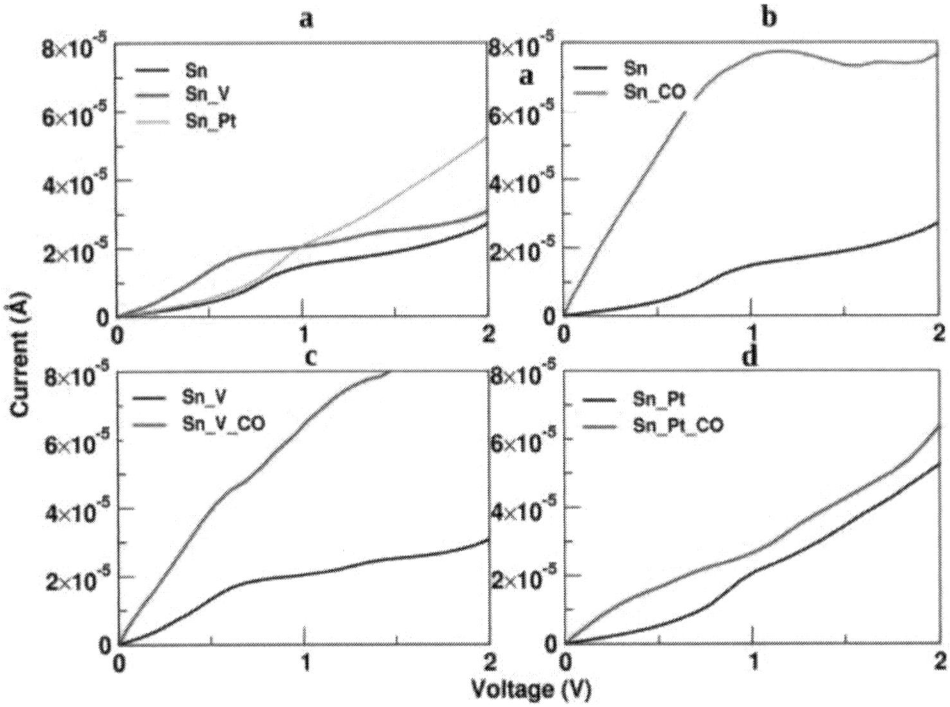

Figure 7. I-V Characteristics with/ without interaction of CO molecules with Sn, Sn_Pt and Sn_V sheet.

For exploring the adsorption of CO molecule with pure Sn sheet, platinum doped Sn (Sn_Pt) sheet and vacant Sn(Sn_V) sheet, we have obtained I-V curves which are plotted in figure 7(b),(c),and (d) respectively. For all the three considered systems, we have compared the I-V curves both in absence and presence of the adsorbed CO gas molecule. The sharp rise in value of current clearly indicates the strong interaction between the adsorbed CO gas with the sensing sheets

4. Conclusion

Structural analysis exhibits the variation in adsorption distance relative to system. High peaks in DOS represents to charge donating nature of adsorbed molecule. Adsorption bond formation nature is found highly with C atom of CO molecule.

Covalent bonds are formed in Sn hexagonal rings/between C & Sn and partially ionic and covalent between Pt & Sn.I-V curves exhibit metallic nature with high rise in current with the adsorption of CO with all considered Sn sheets.

Acknowledgment

Authors are grateful to Chhattisgarh Swami Vivekanand Technical University, Bhilai to grant financial assistance through TEQIP Project CSVTU/CRP/TEQIP-III/86, dated 05/09/2019, to the management of Shri Shankaracharya Technical Campus-SSGI to avail research facilities.

References

[1] Islam, T., Mukhopadhyay, S. C., & Suryadevara, N. K. (2017) "Smart Sensors and Internet of Things": A Postgraduate Paper. *IEEE Sensors Journal, **17(3)**, 577–584.*

[2] Kanoun, O., & Trankler, H.-R. (2004). "Sensor Technology Advances and Future Trends. *IEEE Transactions on Instrumentation and Measurement", **53(6)**, 1497–1501.*

[3] Abdel-Karim, R., Reda, Y., & Abdel-Fattah, A. (2020), "Review—Nanostructured Materials-Based Nanosensors. *Journal of The Electrochemical Society", 167(3)*

[4] Patel, S., Park, H., Bonato, P., Chan, L., & Rodgers, M. (2012) "A review of wearable sensors and systems with application in rehabilitation." *Journal of NeuroEngineering and Rehabilitation 9(1), 21.*

[5] Chen , Z.-W., Shek, C.-H., Wu, C. M. L., & Lai, J. K. L. *(2013),* " Recent research situation in tin dioxide nanomaterials: synthesis, microstructures, and properties. *Frontiers of Materials Science, 7(3), 203–226.*

[6] Choudhuri, I., Kumar, S., Mahata, A., Rawat, K. S., & Pathak, B. (2016). "Transition-metal embedded carbon nitride monolayers: high-temperature ferromagnetism and half-metallicity". *Nanoscale, (2016), **8(29)**, 14117–14126.*

[7] Şahin, H., Cahangirov, S., Topsakal, M., Bekaroglu, E., Akturk, E., Senger, R. T., & Ciraci, S. *(2009)* "Monolayer honeycomb structures of group-IV elements and III-V binary compounds: First-principles calculations". *Physical Review B80(15).*

[8] Miró, P., Audiffred, M., & Heine, T. *(2014), "An atlas of two-dimensional materials. Chem. Soc. Rev.", 43(18), 6537–6554.*

[9] Fortunato, E., Correia, N., Barquinha, P., Costa, C., Pereira, L., Gonçalves, G., & Martins, R. *(2009 "Paper field effect transistor. Zinc Oxide Materials and Devices IV").*

[10] Wise, C. T. A., Suresh, G. R., Palanivelen, M., & Saraswathi, S. "Design of Pentacene based Organic Field Effect Transistor for low-frequency Operational Transconductance Amplifier". *Journal of Circuits, Systems and Computers.*

[11] Chen, G., Paronyan, T. M., & Harutyunyan, A. R. *(2012), "Sub-ppt gas detection with pristine graphene". Applied Physics Letters, **101(5)**, 053119.*

[12] Zhu, F., Chen, W., Xu, Y., Gao, C., Guan, D., Liu, C., Jia, J. *(2015),* 'Epitaxial growth of two-dimensional stanene". *Nature Materials, **14(10)**, 1020–1025.*

[13] Xu, Y., Gan, Z., & Zhang, S.-C. *(2014),* "Enhanced Thermoelectric Performance and Anomalous Seebeck Effects in Topological Insulators*". Physical Review Letters, 112(22).*

[14] Li, N., Wu, Y.-J., & Liu, Z.-W. *(2018),* "Determination of baryon-baryon elastic scattering phase shift from finite volume spectra in elongated boxes". *Physical Review D, 97(1).*

[15] Wu, S.-C., Shan, G., & Yan, B. *,(2014),* "Prediction of Near-Room-Temperature Quantum Anomalous Hall Effect on Honeycomb Materials". *Physical Review Letters, 113(25).*

[16] Lu, G., Ocola, L. E., & Chen, J. *(2009),* "Reduced graphene oxide for room-temperature gas sensors". *Nanotechnology, **20(44)**, 445502.*

[17] Rose , J. J., Wang, L., Xu, Q., McTiernan, C. F., Shiva, S., Tejero, J., & Gladwin, M. T.. *(2017),* "Carbon Monoxide Poisoning: Pathogenesis, Management, and Future Directions of Therapy. *American Journal of Respiratory and Critical Care Medicine", **195(5)**, 596–606.*

[18] http://www.uam.es/siesta

[19] Troullier, N., & Martins, J. L. *(1991),* "Efficient pseudopotentials for plane-wave calculations". *Physical Review B, **43(3)**, 1993–2006.*

[20] Perdew, J. P., Burke, K., & Ernzerhof, M. *(1996),* "Generalized Gradient Approximation Made Simple". *Physical Review Letters,* **77(18**), *3865–3868.*

[21] Patton, D. C., Pederson, M. R., & Porezag, D. V. *(1998),* "The Generalized-Gradient Approximation to Density Functional Theory and Bonding. *Frontiers in Materials Modelling and Design", 37–50.*

Comparative Analysis of Solution Methods for Swing Equation in Power System for Transient Stability Studies

Sanitha Michail C

CMR Institute of Technology, Bengaluru

E-mail:sanitha.c@cmrit.ac.in

Abstract. In power system swing equation plays a remarkable role in investigating the transient stability. Disturbances in system creates imbalance and it may lead to lose the synchronism. Solution of swing equation will help the operators to analyze the stability of the system during the disturbances and initiate the protective mechanism to save the system. The non linear differential swing equation is complex in nature. Commonly used numerical methods to solve swing equations like Point by Point, Modified Euler's and Runge Kutta have explained in the paper. Along with numerical methods, two other methods, evolutionary algorithm based Particle Swarm Optimization method and analytical based Maximum Lyapunov Exponent method has also been investigated in this paper.

Keywords: *Transient stability; numerical methods; analytical method; swing equation;, stability of system*

1. Introduction

Power system stability studies mainly focuses on the dynamics of power system during disturbances. If the system is able to come back to normal operation (without loosing synchronism) when subjected to a disturbance then it is a stable system. In an interconnected power system, synchronous generators are working in synchronism with the rest of the system to generate power [1]. Power system stability can be divided into transient stability, dynamic stability and steady state stability. Dynamic stability investigates whether system can retain stability under small continuous disturbances. Transient stability focuses on stability of the system when it is subjected to major disturbances. Due to this large disturbance, power angle of the synchronous machine changes because of the deceleration/acceleration of the rotor shaft. Steady state stability deals with the response of the system for small and gradual changes in the power system operating conditions.

Due to the increasing demand of power and sustainable energy to load centers, the importance to the stability of power system is increasing. Transmission line outages, different types of faults, sudden loss of generators and other disturbances are creating instability in the system. The transient period of the system is expressed by swing equation and it is a differential equation with non linear nature. Swing curve is the graphical representation of this equation and stability of system can be determined. This will help to analyze machine is in equilibrium or not. Commonly used numerical methods to solve this equation are Point by Point, Modified Euler's and Runge Kutta. These methods too have their own restrictions. Evolutionary algorithms can be applied to solve this equation. One such algorithm, Particle swarm Optimization, is discussed in the paper. Also Analytical solution methods using Lyapunov exponent has investigated in the paper.

2. Swing Equation

The relative motion of the rotor of synchronous machine and the magnetic field of stator as a function of time is expressed by swing equation. $T_a = T_s - T_e$ where T_a is the accelerating torque, T_s is the shaft torque and T_e is the electromagnetic torque.
Accelerating power can be written as

$$P_a = T_a\omega = I\alpha\omega = M\alpha \tag{1}$$

ω is the angular velocity , α is the angular acceleration and I is the moment of inertia in kg-m^2.

$$M = I\omega \tag{2}$$

M is the angular momentum.
Angular acceleration α can be written as

$$\alpha = \frac{d^2\theta}{dt^2} \tag{3}$$

Rotor angle (θ) changes continuously with respect to time during disturbances in the system as follows

$$\theta = \omega_s t + \delta \ ; \tag{4}$$

δ is the angular displacement .The relation between θ and δ is shown in fig.1.

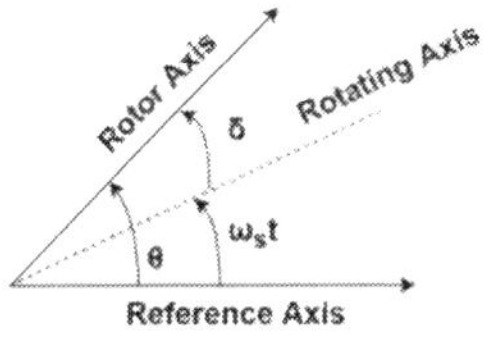

Fig. 1.Rotor position with respect to reference axis

$$\frac{d\theta}{dt} = \omega_s + \frac{d\delta}{dt} \ ; \ \frac{d^2\theta}{dt^2} = \frac{d^2\delta}{dt^2} \tag{5}$$

$$M\frac{d^2\delta}{dt^2} = P_a = P_s - P_e \tag{6}$$

is the swing equation.

3. Numerical methods

Commonly used numerical methods to solve the swing equation are (i)Point by point (ii)Modified Euler's (iii) Runge Kutta

3.1. Point by Point method

Point by point method is used to calculate the rotor position during the disturbances for a small interval of time. It is based on certain assumptions (i) angular velocity calculated at the middle of the time period is constant throughout the period (ii) acceleration power calculated at the beginning of time period remains constant from the centre of previous time period to the centre of time period under consideration.

$$P_{a(0+)} = P_s - P_{e(0+)} \tag{7}$$

$$\frac{d^2\delta}{dt^2} = \alpha_{(0+)} = \frac{P_{a(0+)}}{M} \tag{8}$$

For the 1st time period, change in angular velocity

$$\Delta\omega_1 = \alpha_{(0+)}\Delta t \tag{9}$$

$$\omega_1 = \omega_0 + \Delta\omega_1 = \omega_0 + \alpha_{(0+)}\Delta t \tag{10}$$

here ω_0 is the relative angular velocity at t=0.
Change in rotor angle

$$\Delta\delta_1 = \Delta\omega_1\Delta t \tag{11}$$

$$\delta_1 = \delta_0 + \Delta\delta_1 = \delta_0 + \Delta\omega_1\Delta t = \delta_0 + \alpha_{(0+)}(\Delta t)^2 \tag{12}$$

There are three possibilities for the discontinuity that occur due to switching operation or removal of fault(i) discontinuity at the beginning of the time period (ii) discontinuity at the centre of the time period (iii) discontinuity at some time other than (i) and (ii)

For (i) use the average value of accelerating power before and after the discontinuity (ii) incremental value is taken same as at the beginning of the time period (iii) weighted average of the accelerating power before and after the discontinuity is considered.

3.2. Modified Euler's method

Dependent variables are predicted by Modified Euler's method at the end of the time period by using the derivatives at the beginning of the time period. Derivatives at the end of the time period are computed using these predicted values. Average of these derivatives is used to update the value of the variable. Numerical errors are expected as terms with higher orders of Taylor's series are omitted.
Considering the initial values as δ_0, ω_0 and step size as Δt

$$\frac{d\delta}{dt} = \omega \tag{13}$$

$$\frac{d\omega}{dt} = \frac{P_a}{M} = \frac{P_m - P_{max}\sin\delta}{M} \tag{14}$$

$$\left.\frac{d\delta}{dt}\right|_0 = D_1 = \omega_0 \tag{15}$$

$$\left.\frac{d\omega}{dt}\right|_0 = D_2 = \frac{P_m - P_{max}\sin\delta}{M} \tag{16}$$

$$\delta^p = \delta_o + D_1\Delta t \tag{17}$$

$$\omega^p = \omega_o + D_2\Delta t \tag{18}$$

$$\left.\frac{d\delta}{dt}\right|_p = D_{1p} = \omega^p \tag{19}$$

$$\left.\frac{d\omega}{dt}\right|_p = D_{2p} = \frac{P_m - P_{\max}\sin\delta^p}{M} \tag{20}$$

$$\delta_1 = \delta_0 + \left(\frac{D_1 + D_{1p}}{2}\right)\Delta t \tag{21}$$

$$\omega_1 = \omega_0 + \left(\frac{D_2 + D_{2p}}{2}\right)\Delta t \tag{22}$$

3.3. Runge Kutta method

Runge Kutta method uses a set of formulae and approximations to truncate the Taylor series expansion

$$k_1 = \omega_o\Delta t \tag{23}$$

$$k_2 = \left(\omega_o + \frac{l_1}{2}\right)\Delta t \tag{24}$$

$$k_3 = \left(\omega_o + \frac{l_2}{2}\right)\Delta t \tag{25}$$

$$k_4 = (\omega_o + l_3)\Delta t \tag{26}$$

$$l_1 = \left[\frac{P_m - P_{\max}\sin\delta_0}{M}\right]\Delta t \tag{27}$$

$$l_2 = \left[\frac{P_m - P_{\max}\sin\left(\delta_0 + \frac{k_1}{2}\right)}{M}\right]\Delta t \tag{28}$$

$$l_3 = \left[\frac{P_m - P_{\max}\sin\left(\delta_0 + \frac{k_2}{2}\right)}{M}\right]\Delta t \tag{29}$$

$$l_4 = \left[\frac{P_m - P_{\max}\sin(\delta_0 + k_3)}{M}\right]\Delta t \tag{30}$$

$$\delta_1 = \delta_0 + \frac{1}{6}[k_1 + 2k_2 + 2k_3 + k_4] \tag{31}$$

$$\omega_1 = \omega_0 + \frac{1}{6}[l_1 + 2l_2 + 2l_3 + l_4] \tag{32}$$

As per [2] for a 50Hz, synchronous generator with $M=5.77*10^{-4}$ pu, $|E_g|=1.2$ pu, $x'_d=0.3$ pu, $x_{T1}=0.2$ pu, $x_{L1}= x_{L2}=0.4$ pu is connected to infinite bus bar as given in fig.2. At the center of one of the transmission line a 3 phase fault has occurred and two situations are considered (i) fault is cleared in 2.5 cycles (ii) fault is cleared in 6.25 cycles. Fig 3 shows the plot of swing curve by point by point method for (i) sustained fault (ii) fault cleared in 0.05sec (iii) fault cleared in 0.125 sec using matlab. Fig 4 shows the plot of swing curve by all the three numerical methods for the fault cleared by 6.25 cycles

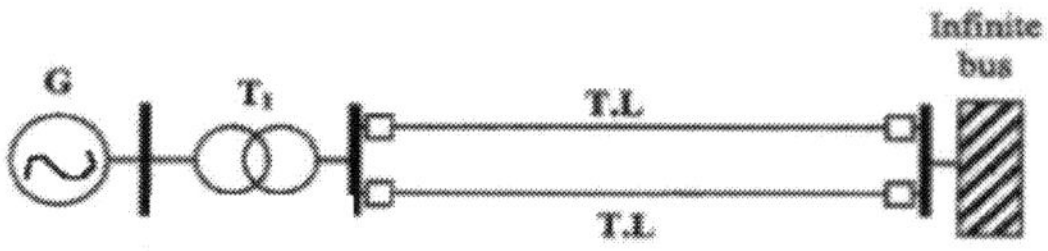

Fig. 2. Synchronous machine connected to infinite bus bars

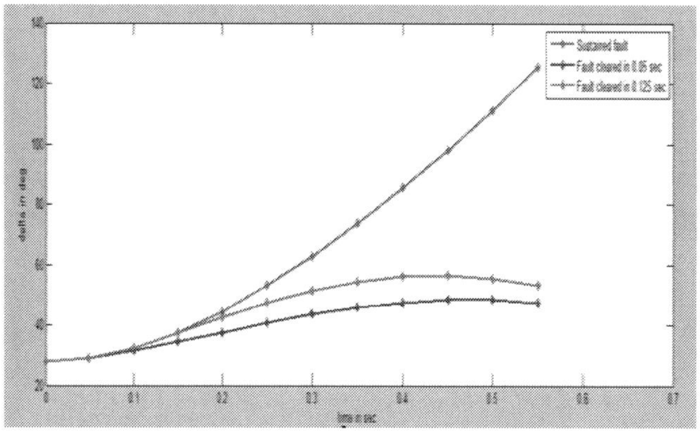

Fig. 3. Swing curve by Point by Point method

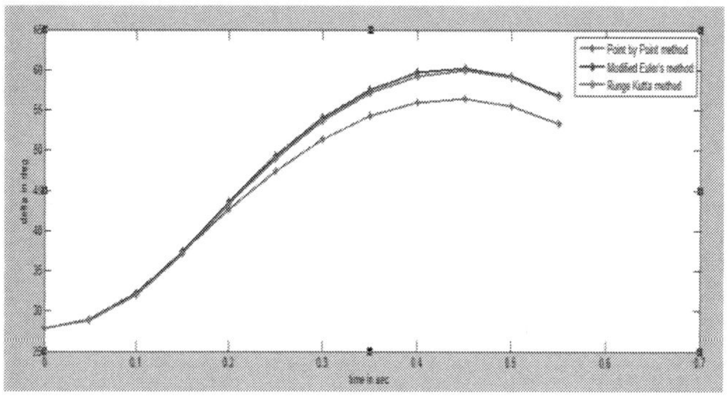

Fig. 4. Swing curve by Point by Point, Modified Euler's & Runge Kutta

4. Evolutionary Algorithm: Particle Swarm Optimization

An evolutionary algorithm, Particle Swarm Optimization [3] can be used in approximation function to find best coefficients. Based on Fourier series swing equation can be approximated [3] as

$$\delta(t) \approx \Delta_{appx}(t) =$$

$$a_o + \sum_{m=1}^{n}\left[a_m \cos\left(\frac{m\pi(t-t_o)}{L}\right) + b_m \sin\left(\frac{m\pi(t-t_o)}{L}\right)\right] \quad (33)$$

$$\delta''(t) \approx \Delta''_{appx}(t) =$$

$$\sum_{m=1}^{n}\left[-\left(\frac{m\pi}{L}\right)^2 a_m \cos\left(\frac{m\pi(t-t_o)}{L}\right)-\left(\frac{m\pi}{L}\right)^2 b_m \sin\left(\frac{m\pi(t-t_o)}{L}\right)\right] \quad (34)$$

$$L = t_n - t_o \quad (35)$$

Objective function is the weighted-residual functional (WRF) in the optimization model. It is minimized to

$$Minimize, WRF = \int |W(t)| \times |R(t)| dt \quad , \quad (36)$$

$R(t)$ is the residual function and $W(t)$ is the weight function.

$$R(t) = f(t, \Delta_{appx}(t), \Delta''_{appx}(t)) \quad (37)$$

Best coefficients can be computed by [3]

$$V_i(t+1) = V_i(t) + c_1(P_{best,i} - X_i(t)) + c_2(G_{best} - X_i(t)) \quad (38)$$

$$X_i(t+1) = X_i(t) + V_i(t+1) \quad (39)$$

c_1 & c_2 : Acceleration constants –positive

G_{best} : Global best position

P_{best} : Personal best position

V_i : Vector for velocity

X_i : Particle position

As per the PSO algorithm given below [3], swing equation has been solved and the curve is plotted when there is a fault at the centre of one of the transmission line and fault cleared in 2.5 cycles and 6.5 cycles in fig.5

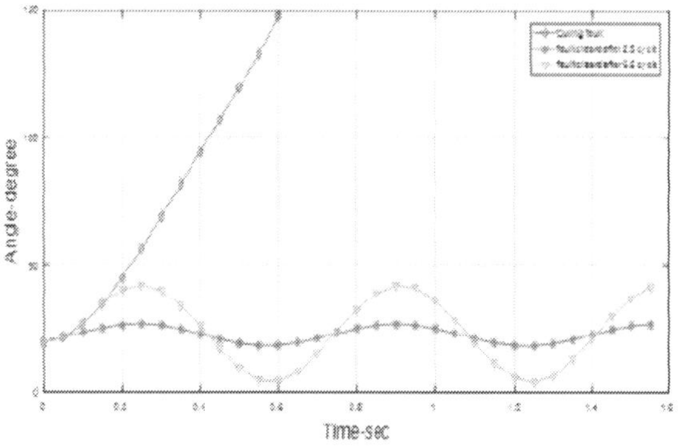

Fig. 5. Plot of swing curve by PSO [3]

Algorithm 1: PSO algorithm

Input: swarm size, initial particle positions, initial velocities
Output: optimal a_0, a_m, b_m
while $t < max\_iterations$ **do**
update particle positions:
$X_i(t) = X_i(t-1) + V_i(t-1)$;
evaluate objective function "weighted-residual functional" W RF for each particle;
find the global best particle G_{best} with minimum objective function value $J(G_{best})$;
if $J(G_{best})$ < tolerance **then**
STOP;
else
find the best position $P_{best,i}$ with minimum objective value for each particle X_i;
if $J(X_i) < J(P_{best,i})$ **then**
- $P_{best,i} = X_i$;
update the velocity V_i for each particle:
— $V_i(t) = V_i(t-1) + c_1(P_{best,i}(t) - X_i(t) + c_2(G_{best} - X_i(t))$;

5. Maximum Lyapunov Exponent

For a system, the Maximum Lyapunov Exponent (MLE)[4] is estimated using phase angles at the generator buses by Phase measurement units (PMU).Stability of the system is indicated by the magnitude and polarity of MLE. It is independent of system model.

$$\dot{\delta}_i^k = \omega_o \Delta\omega_i^k \tag{40}$$

$$\Delta\dot{\omega}_i^k = h_i(\delta^k, \Delta\omega_i^k) \tag{41}$$

This is state space representation of ith synchronous machine of a power system with m number of generators at k^{th} instant of time and ω_o is the synchronous speed in rad/sec, δ_i^k is the internal voltage of ith generator in rad, $\Delta\omega_i^k$ is the per unit speed deviation from synchronous speed of the ith generator, h_i is the dynamics function of the ith generator and $\delta^k = (\delta_1^k.....\delta_m^k)^T$. $X^k = (\delta_1^k....\delta_m^k, \Delta\omega_1^k....\Delta\omega_m^k)$ is the state vector of the power system at k^{th} instant of time. The matrix $Y = (X^1....X^N)^T$ represents time series data of state variables by phase measurements unite. Lyapunov exponents (LE) are the average exponential rate of divergence of nearby trajectories initially separated by small distance in phase plane. The $X^{k'}$ is the nearest neighbour at k^{th} point, $d^k(0) = \min_{k=1:N, |k-k'|>10} \|X^k - X^{k'}\|$, $d^k(0)$ is the initial distance between k^{th} point and nearest neighbour and $d^k(j) = \|X^{k+j} - X^{k'+j}\|$ is the distance after j^{th} time step.

In MLE, $\langle In\ d^k(j)\rangle \approx In\ d^k(0)+\lambda_1(j\Delta t)$, (42)

Δt is the sampling time and λ_1 is the maximum lyapunov exponent.

Fig.6 shows the one line diagram[4] of 9 bus 3 machine system and 3 phase fault has occurred in the transmission line connected between bus 6 and 7 .The fault has occurred at 1 sec and compared the results of the system if the fault is cleared at 1.29 sec and 1.3 sec[3] as shown in fig.7 and fig.8.

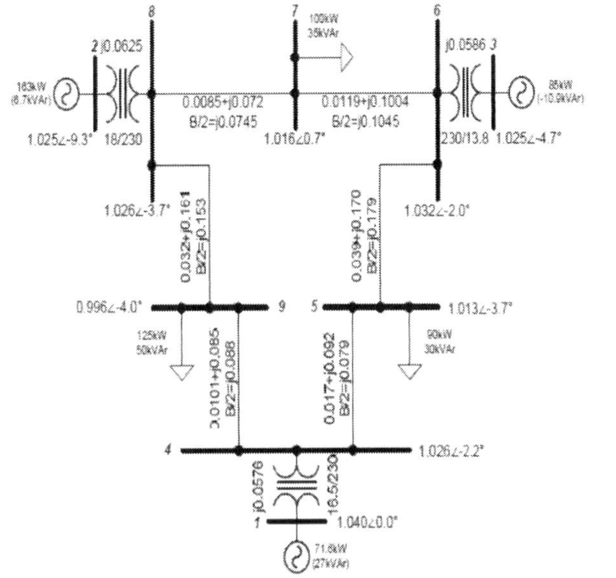

Fig. 6.Nine bus three machine system

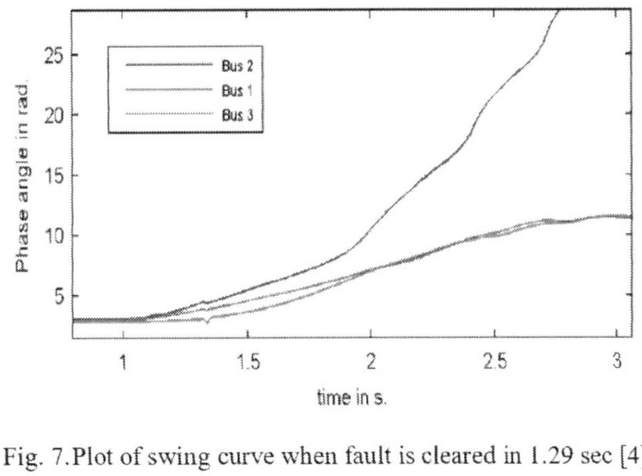

Fig. 7.Plot of swing curve when fault is cleared in 1.29 sec [4]

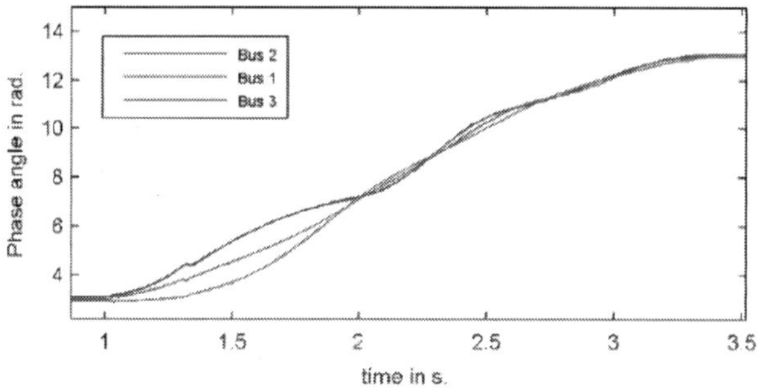

Fig. 8.Plot of swing curve when fault is cleared in 1.3 sec [4]

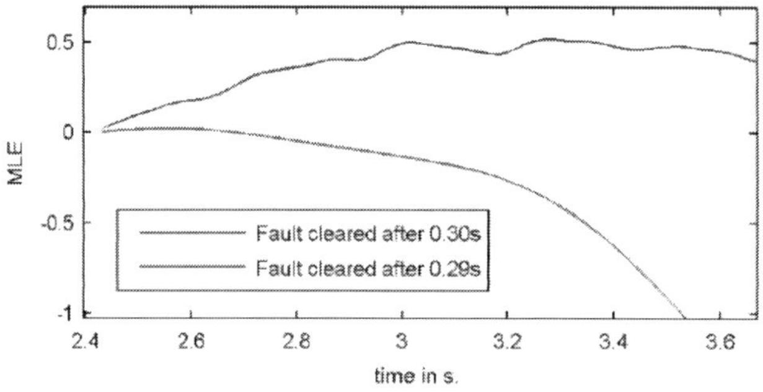

Fig. 9.MLE for the fault cleared in 1.29 sec and 1.3 sec[4]

MLE is negative when fault is cleared at 1.29 sec showing system regaining the stability and it is positive when fault is cleared at 1.3 sec making the system unstable as in fig.9.

6. Conclusions

Location and type of fault will greatly influence the transient stability of power system. Swing equation, a non linear differential equation plays a significant role in investigating the transient stability of power system. Solution of swing equation will help the power system operators to initiate the required emergency operations to save the system. Different methods to solve the swing equation which includes numerical methods like Point by Point, Modified Euler's, Runge Kutta and evolutionary algorithm – Particle swarm optimization method and Lyapunov exponent based MLE method are explained in the paper. These methods will help the power system engineers to study and analyze the system under transient condition.

References

[1] C Sanitha Michail "A Review On Wind Forecasting Methods " *Solid State Technology*",Volume: 63 Issue: 4,Publication Year: 2020

[2] K.Uma Rao, "*Computer Techniques and Models in Power Systems* ",I.K International Publishing House Private Limited,2007

[3] A. Zaidi and Q. Cheng, "*An Approximation Solution of the Swing Equation Using Particle Swarm Optimization*," 2018 IEEE Conference on Technologies for Sustainability (SusTech), Long Beach, CA, USA, 2018, pp. 1-5, doi: 10.1109/SusTech.2018.8671355

[4] P. Banerjee, S. C. Srivastava and K. N. Srivastava, "*A Lyapunov exponent based method for online transient stability assessment*," 2014 Eighteenth National Power Systems Conference (NPSC), Guwahati, India, 2014, pp. 1-6, doi: 10.1109/NPSC.2014.7103871.

IOP Conf. Series: Materials Science and Engineering 1166 (2021) 012054 doi:10.1088/1757-899X/1166/1/012054

Fabrication of titanium dioxide nanotubes with reduced bandgaps by varying electrolytic concentrations

C U Bhadra[1*], Dr D Henry Raja[1] and Dr D Jonas Davidson[2]

[1]Department of Physics, Scott Christian College, Nagercoil 629003, Tamil Nadu, India (affiliated Manonmaniam Sundharanar University Tirunelveli)
[2]Scientist, CSIR-CECRI, Karaikudi, 630003, Tamil Nadu, India
bhadrasenan@gmail.com

Abstract. The main objective of this work is to fabricate titanium dioxide nanotubes of reduced bandgaps for hydrogen generation. Nano scaled material has many peculiar properties than the bulk ones and has led to great break in various fields of science and technology. Titanium dioxide nanotubes have great potential because of its functional properties such as large surface to volume ratio. The TiO_2 nanotubes are grown by electrochemical anodization method at constant DC voltage of 12V for 1 hour. For sample A 0.25 wt% of potassium fluoride dissolved in 99% of ethylene glycol with molarity 0.004M is used as electrolyte. For samples B,C and D the electrolytic solution was prepared by the dissolution of 0.004M of ethylene glycol with copper nitrate($Cu(NO_3)_2 3H_2O$) solution of different concentration (0.002mM,0.003M,0.004M).Titanium dioxide nanotubes formed were studied by Field Emission Scanning Microscope and Photoluminescence Spectroscopy and the results were discussed.

1. Introduction

Because materials exhibit unexpected features when constructed at the nanoscale scale, nanoscience and nanotechnology have sparked a lot of attention in recent years. Nanotubes have enormous technological and biological applications, because of its high surface-to-volume ratio and size dependent properties [1–5].TiO_2 is the most investigated single-crystalline system in the surface science of metal oxides. TiO_2 is rich in properties, is used for Hydrogen Generation in Photo Electrochemical Cell, as a photo catalyst, in solar cell for electric energy production, as a gas sensors, in paint and cosmetic products (as a white pigment), for corrosion protective coating, as an optical coating, in ceramics and in electrochromic devices [6, 7].Titanium dioxide has three main crystalline forms: brookite, anatase and rutile. The rutile phase of titanium dioxide is thought to be the most thermodynamically stable, although anatase has better photoactivity and electrochromic characteristics. Because of its chemical stability, photoactivity, and advantageous relative bandgap location, titanium dioxide is a popular photo catalyst and solar cell material. TiO_2 is a wide-bandgap semiconductor (rutile 3.0 eV and anatase 3.2 eV) so it has to be doped to be photoactive in the visible range [8-10].

IOP Conf. Series: Materials Science and Engineering 1166 (2021) 012054 doi:10.1088/1757-899X/1166/1/012054

2. Experimental Techniques

The number of researchers working on the synthesis and use of diverse nanostructured TiO_2 materials has expanded dramatically in recent years. TiO_2 nanomaterials, such as nanoparticles, Nano rods, Nanowires, Nano sheets, Nano fibers, and Nano tubes, are synthesized using a variety of methods, including the sol-gel method, hydrothermal procedures, and electrochemical anodization. [11-18].

2.1. Electrochemical Anodization

In Electrochemical Anodization Titanium foil (from Sigma-Aldrich) of 0.25mm thickness and 99.70% purity is used as a working electrode i.e., anode and Platinum Pt as co-electrode (cathode) with suitable electrolyte taken in a pure Teflon class beaker. The anodization voltage is applied using a DC eliminator at a particular anodization time. Here four different samples were prepared, for sample A 0.25 wt% of potassium fluoride dissolved in 99% of ethylene glycol with molarity 0.004M is used as electrolyte. For samples B, C and D the electrolyte was prepared by the dissolution of 0.004M of ethylene glycol with copper(II) nitrate solution ($Cu(NO_3)_2$ $3H_2O$) of different concentrations (0.002M,0.003M,0.004M)[19-24].The anodization voltage of 12 V is applied by the DC eliminator connected between the electrodes for 1 hour and then annealed at 480^0C for one hour. The doped metal ions enhance the photo-catalytic activities by reducing electron–hole pair recombinations and/or reducing the bandgap. In some reports researchers have successfully improved the activity of TiO_2 by doping metal ions like Au and Pt but they are very expensive and rare elements.

Table 1: Experimental details carried out

Samples	Anodization Solutions	Conditions
A	0.004M of ethylene glycol and KF	12V 1hour
B	0.004M of ethylene glycol and KF+ 0.002M of copper nitrate solution	12V 1hour
C	0.004M of ethylene glycol and KF+ 0.003M of copper nitrate solution	12V 1hour
D	0.004M of ethylene glycol and KF+ 0.004M of copper nitrate solution	12V 1hour

3. Results and Discussion

Photoluminescence (PL) spectrum is recorded at excitation wavelength 320nm by using Photoluminescence Spectrometer. The surface morphology of titanium foil is measured using Field emission Scanning Electron Microscope.

By using potentiostatic anodization of two electrode configurations, self-aligned TiO_2 nanotube arrays are prepared and characterized. The first graph in the figure shows room-temperature photoluminescence (PL) emission spectra of undoped (sample A) TiO_2 nanotubes array on titanium foil recorded in the range of 300-600nm. A sharp PL emission peak can be clearly seen at 372 nm corresponding to the energy band gap of (3.33 eV).The band gap energy of bulk TiO_2 anatase phase is. 3.2 eV. Two humps were also observed at wavelength 496 nm and 521 nm which may be associated to the free excitations however, a small hump at 377nm and 435nm may be due to the formation of oxygen vacancies at the surface of titania nanotube arrays.

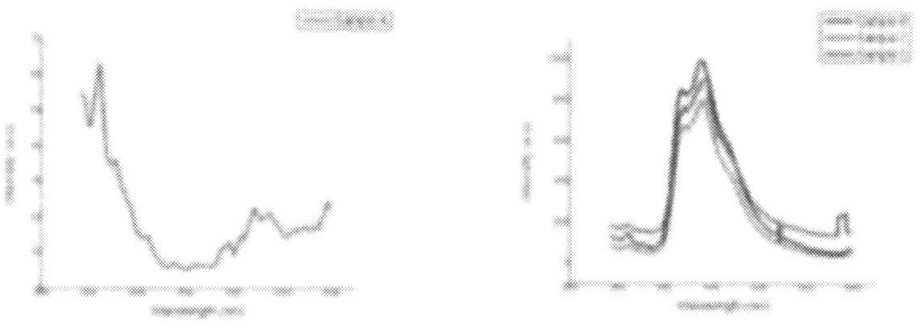

Figure 1. PL spectra of pure and doped TiO_2 nanotube arrays

The second graph in the figure shows room-temperature photoluminescence (PL) emission spectrum of copper nitrate solution dissoluted TiO_2 nanotubes array on titanium foil recorded in the range of 300-600nm of three samples B,C and D. For sample B sharp PL emission peak can be clearly seen at 438 nm corresponding to the energy bandgap of (2.83eV). A hump was also observed at wavelength 358 and 421 nm which may be associated to the free excitations however; a small hump at 528nm may be due to the formation of oxygen vacancies at the surface of titania nanotube arrays [19, 20]. Similarly for sample C and D sharp PL emission peak can be clearly seen at 442 nm and 446nm respectively which corresponds to the band gap energy of (2.81 and 2.78eV). Two humps were observed at wavelength 358 and 421 nm which may be associated to the free excitations however; a small hump at 528nm may be due to the formation of oxygen vacancies at the surface of titania nanotube arrays. It has been reported that TiO_2 nanotubes recorded a maximum peak at 383nm and also TiO_2 doped with copper reduces the bandgap and increases the photocatalytic activities [25, 26].

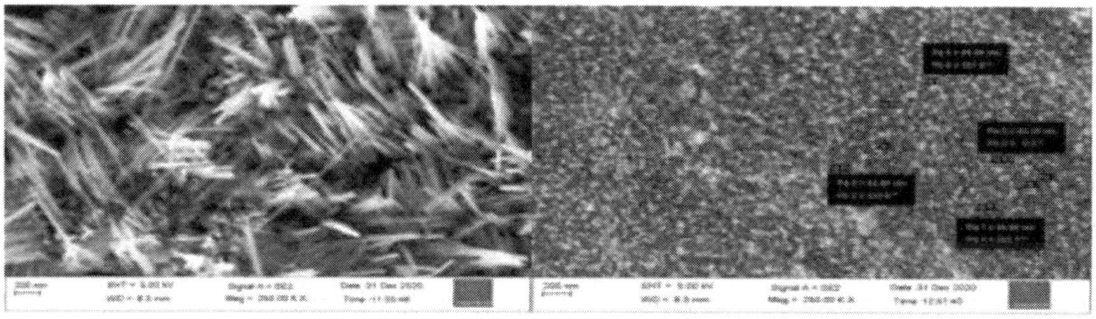

Figure 2. Morphology and top view of TiO_2 nano tube arrays

The extra high tension of 5.00kV is supplied to the sample and magnified at 175.00kX.The left side image shows the titanium dioxide nanotubes of tube length 200nm and the right side image shows the top view of TiO_2 nanotubes with outer radius 43.6nm fabricated by using0.25 wt% of potassium fluoride dissolved in 99% of ethylene glycol with molarity 0.004M as electrolyte.

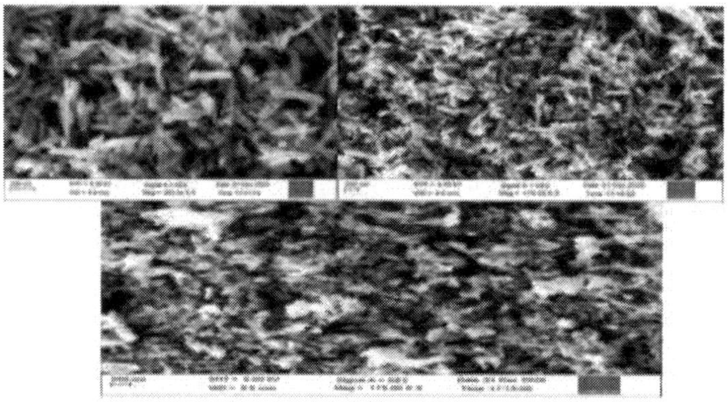

Figure 3. Typical surface morphology copper dissoluted TiO_2 nanotube arrays

The titanium dioxide nanotubes of tube length 200nm are fabricated using 0.25wt% of potassium fluoride dissolved in 99% of ethylene glycol with molarity 0.004M dissoluted with copper nitrate solution of (0.002M, 0.003M,0.004M) as electrolyte. Here the density of tube alignment decreases because of the change in morphology of the tubes. The images were magnified at 362kV has lesser density than images magnified at 175kV each with tubelength 200nm.

4. Conclusion

By anodic oxidation of TiO_2 foil, nanotubes have been fabricated and characterized. For sample A sharp PL emission peak observed at 372 nm corresponding to the energy bandgap of 3.3eV TiO_2 anatase phase. For samples B, C and D the band gap energy decreases as 2.83, 2.81, 2.78eV respectively. In PL spectra, small humps were also observed may be associated to the free excitations and due to the formation of oxygen vacancy at the surface of titania nanotube arrays. By FESEM measurement, change in the morphology of the tubes and tube length has been observed. Finally, the prepared samples B, C and D TiO_2 nanotubes with reduced bandgaps are suitable Hydrogen Generation.

Acknowledgement
The above work was originally carried out by me in Research Department of Physics, Scott Christian College, Nagercoil 3, Tamil Nadu, India without an funding source.

Conflicts of Interest
The authors hereby declare that we have NO financial support from any organization or entity to carry out the work in this manuscript.

References

[1] O Varghese, M Paulose, T LaTempa and C Grimes 2009 *Nano Lett.* **9** 731.
[2] S Roy, O Varghese, M Paulose and C. Grimes 2010 *ACS Nano* *4* 1259.
[3] T Kang, A Smith, B Taylor and M. Durstock 2009 *Nano Lett.* *9 601.*
[4] O Varghese, M Paulose and C Grimes 2009 *Nat. Nanotechnol.4 592.*
[5] K Shankar, J Bandara, M Paulose, H Wietasch, O Varghese, G Mor, T LaTempa, M.Thelakkat and C. Grimes 2008 *Nano Lett* *8* 1654.
[6] U Diebold 2003 *Surf. Sci. Rep.**48*** 53.

[7] A Mills, R H Davies and D Worsley 1993 *Chem. Soc. Rev.* ***22*** 417.
[8] O Legrini, E Oliveros and A. M. Braun 1993 *Chem. Rev.*, ***93***, 671.
[9] A Heller 1995 *Acc. Chem. Res.**28*** 503.
[10] A L Linsebigler, G Lu, J T Yates and Jr *Chem. Rev.* 1995 ***95*** 735.
[11] M. Anpo, *Catal. Surv. Jpn.* 1997, *1*, 169.
[12] P C Maness, S Smolinski, D M Blake, Z Huang, E J Wolfrum and W. A. Jacoby 1999 *Appl. Environ.Microbial* ***65*** 4094.
[13] A Hagen, A Barkschat, J K Dohrmann and H Tributsch 2003 *Sol. Energy Mater. Sol. Cells 77* 1.
[14] M Anpo, S Dohshi, M Kitano, Y Hu, M Takeuchi and M Matsuoka 2005 *Annu. Rev. Mater. Res.**35*** 1.
[15] A Fujishima and K Honda 1972 *Nature* ***238*** 37.
[16] A Fujishima, K Kohayakawa and K Honda *Bull. Chem. Soc. Jpn.* 1975 *48* 1041.
[17] B O'Regan, M Graetzel 1991 *Nature* 353 737.
[18] M Misra, R N Pandey and O N Srivastava 1997 *Int. J. Hydrogen Energy* ***22*** 501
[19] S Sakthivel, M V Shankar, M Palanichamy, B Arabindoo, D W Bahnemann and V.Murugesan, 2004 Enhancement of Photocatalytic activity by metal deposition: characterization and photonic efficiency of Pt, Au and Pd depositedonTiO_2 catalyst *WaterRes.***38** 3001–08
[20] B Xin, L Jing, Z Ren, B Wang and H Fu 2005 Effects of simultaneously doped and deposited Ag on the Photocatalytic activity and surface states of TiO_2 *J.Phys.Chem.B* **109** 2805–09.
[21] L Zang, C Lange, I Abraham, S Storck, W F Maier and H Kisch 1998 Amorphous micro porous titania modified with Platinum (IV)chloride–a new type of hybrid photo catalyst for visible light detoxification, *J.Phys. Chem.* **B102** 10765–71.
[22] V Subramanian, E Wolf and P V Kamat 2001 Semiconductor–metal composite nanostructures. To what extent metal nano particles (Au, Pt, Ir)improve the photocatalytic activity of TiO_2 films, *J.Phys.Chem.***B105** 11439–46.
[23] R Chand, E Obuchi, K Katoh, H N Luitel and K.Nakano 2011 Enhanced photocatalytic activity of TiO_2/SiO_2 by the Influence of Cu-doping under reducing calcination atmosphere *Catal. Commun.* **13** 49–53.
[24] H Li, J Wang, M Liu, H Wang, P Su, J Wu and J Li, 2014 A Nano porous oxide inter layer makes a better Pt catalyst on a metallic substrate: nano flowers on a nano tube bed *NanoRes.*71007-17.
[25] Padmakar G Chavan, Sugat V Shende and Dilip S Joag, Mahendra 2011 A More, *Ultramicroscopy* **111** 415–420.
[26] Jung-Chul Lee Kyung-Soo Park Tae-GeunKim Heon-Jin Choi and Yun-Mo Sung 2006 *Nanotechnology* **17** 4317–21.

Dielectric and Magnetic Studies La^{+3} doped Cobalt Nano Ferrite

N. Ragini[1], Syeda Ayesha[2], K. P Komal[3,5], Tanveer Fatima[4], Harsha G. A[5], Shilpa S[6] and S. Abdul Khader[1,7*]

[1]Department of Physics, G.F.G College, Kuvempu Nagara, Mysore-570023. India.
[2]Department of Chemistry, G.F.G College, Kuvempu Nagara, Mysore-570023, India
[3]Department of Bio-Chemistry, Govt.Science College, Chitradurga-577501. Karnataka.
[4]Department of Physics, Gulbarga University, Kalaburagi-585106, Karnataka, India.
[5]Department of Chemistry, Govt.Science College, Chitradurga-577501, Karnataka.
[6]Department of Physics, G.F.G College, Kadugugi, Bengaluru-560067. Karnataka.
[7]Department of Physics, Govt.Science College, Chitradurga-577501, Karnataka. India.

***Author for Correspondence: khadersku@gmail.com.**

Abstract. La^{3+} substituted cobalt ferrite nano-particles with the chemical composition $La_xCoFe_{2-x}O_4$ (here,x=0, 0.05, 0.1, 0.15 and 0.2) were synthesized using auto-combustion method. Structural and morphological studies were probed using X-ray diffraction (XRD) and Field Emission Scanning Electron Microscopy (FESEM). Formation of rare-earth doped cobalt ferrite was confirmed using the X-ray diffraction studies. Average crystallite size (D) was found to be varying from 43.71nm to 55.37nm with the increase in La^{3+} concentration. Frequency dependent dielectric properties of synthesized samples were measured from 100 Hz to 1 MHz at room temperature using a precision HIOKI make LCR HI-TESTER. Dielectric dispersion was observed at lower frequencies for the synthesized samples. Magnetic properties were studies using Vibrating Sample Magnetometer (VSM), all the synthesized samples exhibited well defined magnetic hysteresis curves, there by confirming the magnetic behaviour of the sintered samples.

1. Introduction

Materials in the nano-scale regime exhibit remarkably new properties compared to materials at their bulk counterpart. Nanomaterials appear in every industry from agriculture to service sector. Numerous applications of nano-technology find in wide range of industries enlarges day to day and a few of applications are transformer inductors, humidity sensors, drug delivery systems, electronic components, multilayer chip inductors, in enhanced photo catalytic performance, magnetic resonance imaging etc. [1-5]. Among the ferrites, spinel ferrites with nano-structure are most important. Ferrites comes under magnetic materials family with the general spinel structure bearing the general chemical formula AB_2O_4 (A^{2+} B_2^{3+} O_4^{2-}) where A is divalent metal ion e.g. Fe, Mg, Zn, Ni, Mn, Cu, Co, etc., and B is trivalent cation e.g. Fe^{3+}, Al^{3+} etc., and O=oxygen anion e.g. O^{2-} Metal cations, A and B are fitted at A-site (tetrahedral site) and B-site (octahedral site) and oxygen has face centred cubic(FCC) close packing structure [6]. Spinel ferrites are familiar dielectric materials with high resistivity and low hysteresis loss. An account of their peculiar magnetic, dielectric and electrical properties of spinel ferrites, they are used in high density information storage system, medical diagnosis, transformer cores, inductor coils, microwave devices, telecommunication, microwave absorbers, magnetic recording, biosensors, power transformers, radio electronics etc. [5,7-10].

Cobalt ferrites are a well-known fact that they are hard magnetic materials, which is studied in detail due to its high coercivity, high chemical stability, high saturation magnetization, mechanical hardness, high resistivity and good electrical insulator [11]. Various research studies explain the cobalt ferrites doped with divalent and trivalent metal ions such as Mn^{+2}, Mg^{+2}, Zn^{+2}, Ni^{+2}, La^{+3}, Ga^{+3} etc. are received increasingly attentiveness in many applications, especially for sensors and high frequency applications. Properties of cobalt ferrites can be altered such as by incorporating a dopant into its structure, sintering temperature and preparation methods. Consequently, based on their structural, optical, magnetic and electrical properties they can be used in many applications [12-14].

In this work, an in-expensive auto-combustion method is chosen as it provides good crystallinity, less power consumption, high purity, ease of sample preparation, stability etc. compared to other techniques like co-precipitation method, sol-gel, micro emulsion route, hydrothermal method [16-20]. Herein, La^{+3} is elected as dopant and substituted it into cobalt ferrite system. Effect of lanthanum (La^{+3}) substitution in $CoFe_2O_4$ can be studied on its structural, morphological, dielectric, electrical and magnetic properties.

2. Experimental

Lanthanum (La^{3+}) doped nano cobalt ferrite powders were synthesized using auto-combustion method. This method is the simplest, reliable method for obtaining high purity nano-materials. To prepare $La_xCoFe_{2-x}O_4$ nano-powders, the precursors selected are, $La(No_3)_3.6H_2O$ (Lanthanum nitrate), $Co(No_3)_2.6H_2O$ (Cobalt nitrate), $Fe(No_3)_3.9H_2O$ (Ferric nitrate) , $C_6H_8O_7$ (Citric Acid). All the chemicals were procured from Sigma Aldrich; all are of 99.8% purity. Calculated stoichiometry molar amounts of lanthanum, cobalt, and ferric nitrates are added with distilled water and mixed thoroughly using a magnetic stirrer until metallic nitrates are dissolved completely at RT. While aqueous solution of citric acid was prepared and added slowly to the aqueous solution of corresponding metallic nitartes. Redox mixture was taken in a silica crucible and allowed in to a muffle furnace, which was already pre-heated to a temperature of 500^0C. Because of the auto-combustion reaction, aqueous mixture of metallic nitrates and organic fuel finally yields porous and fluffy voluminous powder. Obtained voluminous product was well grinded to get magnetic oxide nano powder. As obtained ash was sintered at 900^0C for minimum 4 hours to achieve higher rate of crystallization and finally grinded so that La^{+3} doped $CoFe_2O_4$ ferrite nano powders can be obtained.

3. Results and Discussion

3.1 Phase confirmation studies

Phase confirmation of the prepared samples is done using, Bruker AXE D8 Advanced diffractometer. Diffractogram of $La_xCoFe_{2-x}O_4$ (x=0, 0.05, 0.1, 0.15 and 0.20) powder nano-particles are shown in Fig 1. Obtained XRD data of La^{+3} substituted $CoFe_2O_4$ are compared with standard data (JCPDS PDF card No.22-1086).

Average crystallite size(D) was calculated using the Scherer – Debye Eq. (1) [20].

$$D = 0.9\,\lambda/\beta\cos\theta \ldots\ldots\ldots\ldots\ldots\ldots \quad (1)$$

Lattice parameter or lattice constant(a=b=c) is measured using Bragg's equation for all LCF samples by the maximum intense peak (311) using Eq. (2) [20].

$$a = d_{311}\sqrt{h^2+k^2+l^2} \ldots\ldots\ldots\ldots\ldots\ldots \quad (2)$$

Where d_{311} is the inter-planar spacing for the (311) reflection plane and (h,k,l) are the corresponding miller indices.

Average crystallite size was found to be varying from 43.71nm to 55.37nm along with the increase in the La^{3+}concentration from 0.05 to 0.2. This might be because of the distortion caused in the structure with the substitution of rare earth La^{+3} ions.
Lattice constant increased from 8.37Å to 8.48Å with increase of the dopant La^{+3} concentration, this is due to that La^{3+} has larger ionic radii (1.031Å) replaces the Fe^{3+}(0.64Å) ions having lower ionic radii in La-doped $CoFe_2O_4$ system. This kind of behaviour being caused by cation distribution between the two sites, which requires more energy so that La^{3+} ions to enter the spinel lattice to complete crystallization.

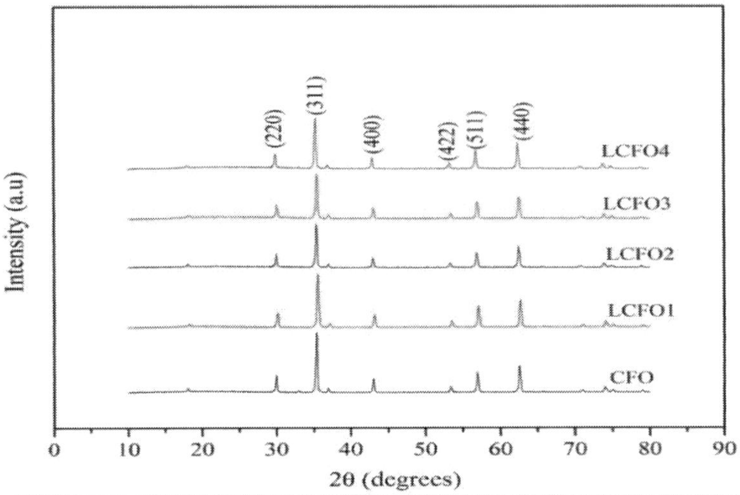

Figure 1. XRD Diffractogram for the Pure and La^{+3} doped $CoFe_2O_4$

3.2 Morphological Studies

Surface morphology of the synthesized La+3 doped Cobalt ferrite (LCF) nano-particles were analysed using FESEM (FEG-Quanta-250), indicates that homogeneous nature for the obtained and samples of nano-powders are shown in Fig 2. Obtained micrographs shows that the morphology of the particles regular distribution of nano-particles. FESEM micrographs further suggests that the samples consist of cubical shaped nano-particles with agglomeration. As the doping concentration of La^{+3} increases, the grains become more and more dense.

3.3 *Dielectric Studies*

Variation of dielectric constant (ε'), and its imaginary counter-part (ε") and ac conductivity (σ_{ac}) with frequency at room temperature for the pure cobalt ferrite and La^{+3} substituted cobalt ferrite is shown in Fig.3 (a) to Fig.3 (d), it is clear that ε' decreases steeply at lower frequencies and remains constant at higher frequencies. Variation of dielectric constant with applied frequency is due to charge transport relaxation. This dielectric dispersion is attributed to Maxwell and Wagner type of interfacial polarization which is in agreement with Koop's theory, as the dielectric constant is a combined effect of dipolar, electronic, ionic and interfacial polarizations [20-24].
In order to understand the conduction mechanism and type of polarons responsible for conduction, ac conductivity, σ_{ac} were estimated as per $\sigma_{ac} = \omega\, \varepsilon_0\, \varepsilon''$, with ε_0 is the permittivity of free space and $\omega = 2\Pi f$. Substitution of La^{+3} ions in place of Fe^{+3} ions, limits the degree of conduction by blocking as per Verwey's hopping mechanism [24].

Variation of σ_{ac} with frequency, ω is shown in Fig.3(c) for the proposed samples. Obtained curves are linear except at very low frequencies, this indicates that the conduction occurs by the hopping of charge carriers between the localized states which confirms the small-polaron type of conduction [25-26].

Figure. 2. Micrographs of the LCF samples with various doping concentrations.

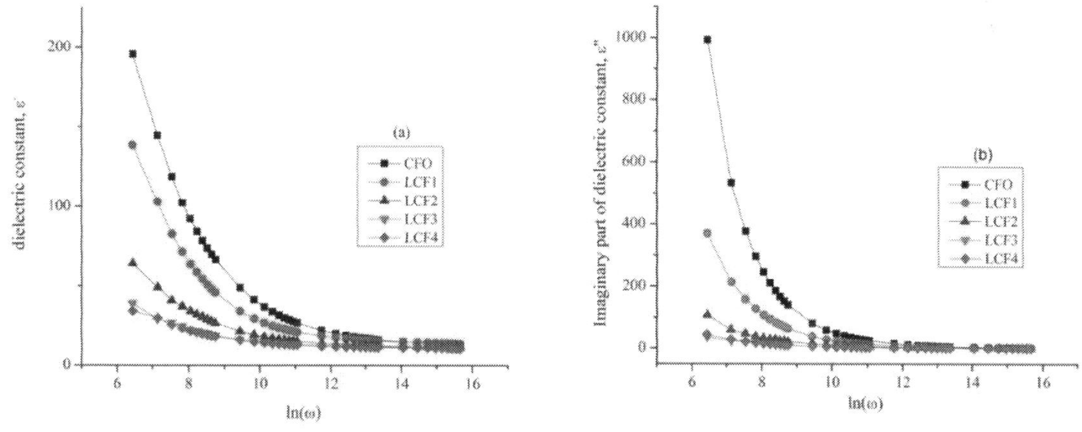

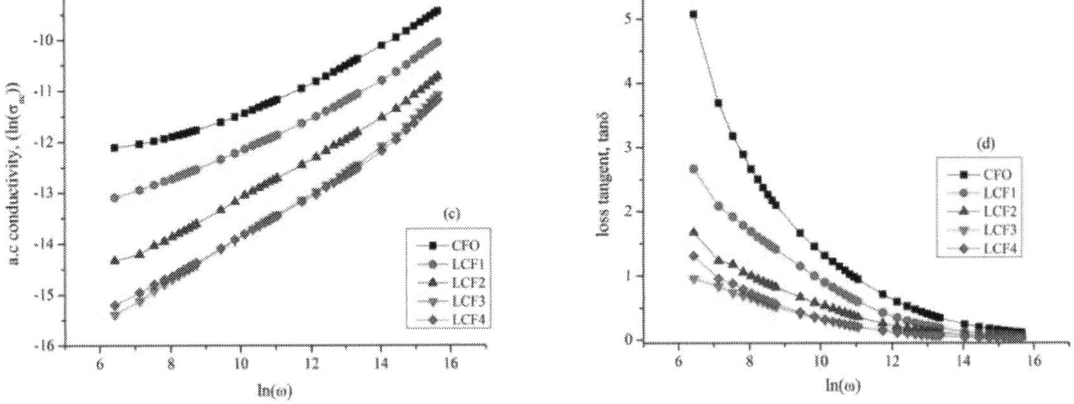

Figure3. Variation of ε^{I}, ε^{II} , $\ln(\sigma_{ac})$ with frequency $\ln(\omega)$ for the proposed samples

4. Magnetic Studies

Magnetic hysteresis loops, at RT were recorded for the proposed samples. M-H loops for synthesized samples sintered at 900^0C are shown in Fig.4. Synthesized and sintered samples exhibit magnetic hysteresis, which confirms the magnetic nature of the samples. Magnetic parameters extracted from M-H loops for pure as well as doped samples were mentioned in the Table.1. Compared to pure $CoFe_2O_4$ sample, remanent magnetization, M_r, decreases with the increase of La^{+3} concentration, it is because of the decrease in Fe^{+3}-Fe^{+3} (B-B) interactions resulting from doping with La^{+3} [29-30]. It is observed that for all sintered samples, coercive field, H_c remains almost same, except for $La_{0.15}CoFe_{1.85}O_4$. It is because of the known fact that, coercivity depends on factors such as magneto-crystalline anisotropy, micro strain, size distribution and magnetic domain size [22-28].

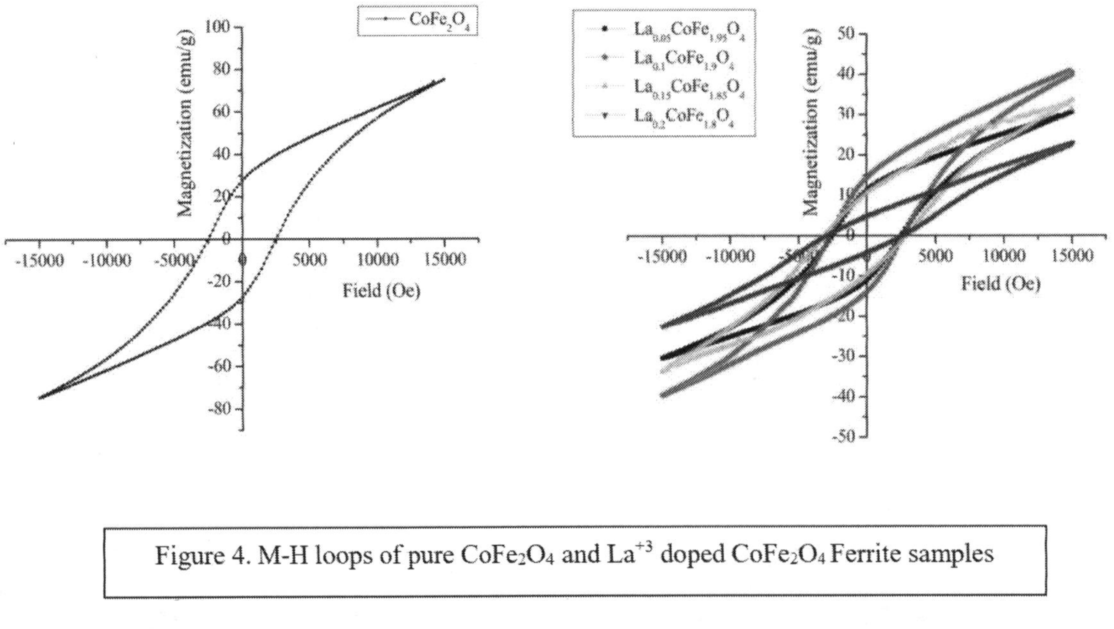

Figure 4. M-H loops of pure $CoFe_2O_4$ and La^{+3} doped $CoFe_2O_4$ Ferrite samples

Table 1. Magnetic Parameters extracted from M-H loops at Room temperature.

Sample	Ms (emu/g)	Mr (emu/g)	Hc (Oe)
$CoFe_2O_4$	75.04	27.87	2548
$La_{0.05}CoFe_{1.95}O_4$	30.64	11.29	2510
$La_{0.1}CoFe_{1.9}O_4$	42.37	14.35	2580
$La_{0.15}CoFe_{1.85}O_4$	33.58	10.53	2900
$La_{0.2}CoFe_{1.8}O_4$	22.97	4.74	2600

5. Conclusions

All the synthesized samples have single phase spinel cubic structure, without any secondary phases indicating the absence of any impurities. From the dielectric measurements, it is observed that the synthesized samples exhibited dielectric dispersion. Linear nature of AC conductivity measurements, confirms the conduction mechanism in the proposed samples is because of small-polarons. Pure and rare-earth La^{+3} doped samples show appreciable magnetic hysteresis with good value of saturation magnetization, remanence at room temperature.

References

[1] P.Kumar, S.K.Sharma,M.Knobel,M.Singh, Jagdish Chand, M. Singh, *J. Electro-ceramics*, **27**(2), (2011) pp.51-55.

[2] T. Ramaprasad, R.J. Kumar, U. Naresh, M.Prakash,D. Kothandan and K.Chandrababu Naidu, *Mater.Res.Express* **5** (2018),095025.

[3] P.Kumar, S.K.Sharma,M.Knobel,M.Singh, Jagdish Chand, M. Singh, *Journal of Alloys and Compounds* **508** (2010)115-118.

[4] Simona Burianova, Jana Poltievora Vejpravova, Peter Holec, Jiri Plocek and Daniel Niznansky, *Journal of Applied Physics,* Vol.**110** (2011),073902.

[5] K.C. B Naidu and W.Madhuri, *Jour. Materials Chemistry and Physics* **187** (2017) 164-176.

[6] S. Sun, H.Zeg,D.Robinson,S.Raoux, P.Rice, S.wang. *J.Appl.Chem.Soc,* **126** (2004)273-279.

[7] K.C. B Naidu and W.Madhuri, Jour. Bulletin of Materials Science **40** (2017)417-425.

[8] C.B.Naidu Kadiyala &Madhuri Wuppiluri, *Jour.Phase transitions* **90** (2017)847-862.

[9] Yuksel koeoglu, Furkan Alan,Muhammad Tan,Resul Yilgin,Mustafa Ozturk. *Ceramics International* **38** (2012)3625-3634.

[10] U.Naresh, R.Jeevan Kumar, K.Chandrababu naidu, *Jour. Materials Chemistry and Physics* **236** (2019)121807.

[11] A.Muhammad, T.Sato,M.Kriegish, R.Grosinger,F.Kubel, T.konegger. *Journal of Applied Phy.***111** (2009) 013918.

[12] Shyam M. Kodape, Ajit P.Rathod and Shriram S.Sonawane,*IPCBEE* Vol.**46** (2012) IACSIT Press, Singapore. DOI: 10.7763/IPCBEE. 2012. V46. 20.

[13] A.B.Shinder, *International Journal of Innovative Technology and Exploring Engineering* (IJITEE) ISSN: 2278-3075, Volume-**3**, Issue-**4** (2013).

[14] Pawan kumar,M.singh, Jagdish Chand, *Jour. Integrated Ferroelectrics,*134:53-57,2012.

[15] Ragendran Indhrajothi, I. Prakash, M. Venkateswarlu and Nallani Satyanarayana, Lawrence kumar, Manoranjan K. *Ceramics international* **38** (2012), pp.4771-4782.

[16] Elina Manova, Boris Kunev, Daniela Paneva, Ivan Mitov,Lachezar Petrov,Chem. Mater.*Res Exp 16 (*2004**),** pp.5689-5696.

[17] N. Suresh kumar, R.Padma Suvarna, K.Chandrababu Naidu, G.Ranjith kumar, S.Ramesh, *Ceramic International*, Vol.**44** (16), (2018) pp. 19408-19420.

[18] Ragendran Indhrajothi, Ignacimuthu Prakash, Manne Venkateswarlu and Nallani Satyanarayana, *Royal Society of Chemistry*: *New Jour of Chemistry*, **39** (2015), pp.4601-4610.

[19] P Annie Vinosha, L Ansel Mely, G Immaculate Nancy Mary, K Mahalakshmi, S Jerome Das, *Mechanics, Materials Science & Engineering*, Vol **9**, Issue-**1**, April 2017.
[20] Pawan kumar,M.singh, Jagdish Chand, *Jour. Integrated Ferroelectrics*,134 (2012). Pp.53-57.
[21] Verma, O.P. Thakur, C. Prakash, T.C. Goel, R.G. Mendiratta, *Mater. Sci. Eng. B* **116** (2005) pp.1–6.
[22] X. Batlle, A. Labarta, *Jour. Phys. D: Appl. Phys.* **35** (2002) 15–42.
[23] A. Thakur, P. Mathur, M. Singh, *Jour. Phys. Chem. Solids* **68** (2007), pp. 378–381.
[24] E.W Verway. P.W Haajiman, F.C Romeyn, G.W Van Oosterhont, *Res.Rep.***5** (1950) 173.
[25] Z. Beji, L.S. Smiri, N. Yaacoub, J.M. Grenèche, N. Menguy, S. Ammar, F. Fiévet *Chem. Mater.* **22** (2010), pp. 1350–1366.
[26] T. Upadhyay, R.V. Upadhyay, R.V. Mehta, *Phys. Rev. B*. **55** (1997), pp.5585–5588.
[27] K. Raju, G. Venkataiah, D.H. Yoon, *Ceramics. International*. **40** (2014), pp. 9337–9344.
[28] B.D Cullity, C.D Graham, *Introduction to Magnetic Materials*, John Wiley & Sons, 2011.
[29] C.Suvankar, B.Malay, Coercivity of Magnetic Nanoparticles: A Stochastic Model, *Jour. Physics: Condensed Matter* **19** (2007), p 216201.
[30] M.L Kahn, Z. J Zhang, *Appl.Phys. Lett* **78** (2001), p 3651.

Influence of Ni addition on Abrasive wear behaviour of plasma-sprayed Duplex stainless steel coatings

E Edward Anand[1] and K Balasubramanian[2]

[1]Department of Science and Humanities, E.G.S. Pillay Engineering College, Nagapattinam, Tamil Nadu, India
[2]Department of Mechanical Engineering, E.G.S. Pillay Engineering College, Nagapattinam, Tamil Nadu, India

E-mail: dredward@egspec.org

Abstract: Many austenitic and ferritic stainless steels have a higher yield and ultimate tensile strength than duplex steels. The extent to which this is true is determined not only by the alloy's composition but also by how it is treated. This increased strength is usually accomplished without sacrificing the alloy's hardness, as long as the alloy does not contain any of the harmful phasing elements. Thus an attempt is made to study the duplex stainless steel being plasma sprayed on a substrate and study its microstructure, hardness, and abrasive wear resistance. The primary objective is to study the abrasive wear resistance of duplex stainless steel coatings on 316 stainless steel. Ni percentage in coatings is varied at the coating stage in weight percentages of 7.0, 7.5, and 8 % and to study its effect on the wear rate of the coatings. The composition of the coatings is confirmed using XRD. A two-body abrasive wear test is carried out to determine the wear rate of the coatings with varying percentages of Ni and varying loads. Besides, the coatings are characterized by optical microscopy. Weight loss measurements of the tested samples are carried out and the data obtained is analyzed through weight loss plots. Also, weight loss data is used to further calculate the wear rate which is analyzed using wear rate plots. Finally, the microhardness of the coatings is determined.
Keywords: Duplex stainless steels, Abrasive wear, Optical microscopy

1. Introduction

In the chemical and petrochemical industries, duplex stainless steel (DSS) grades are widely used. The key advantages of duplex stainless steels are their excellent resistance to oxidation, corrosion, and stress corrosion, as well as their superior mechanical properties [1-4]. These alloys have general corrosion rates that are similar to, if not lower than, austenitic stainless steel. In dilute sulfuric acid, these alloys have the same or lower general corrosion rates as austenitic stainless steels. This is true for dilute hydrochloric acid as well as caustic solutions used in the pulp and paper industry. Any of these alloys can also be used in organic acid environments. As compared to austenitic stainless steels, many of these alloys have better pitting and crevice corrosion resistance [5-7].

Duplex stainless steels (DSS) exhibit excellent resistance to transgranular and intergranular stress corrosion cracking. Depending on the environment, the composition of the alloy and its microstructure determine the resistance towards corrosion [2]. Instead of austenitic or ferritic stainless steel sometimes Duplex stainless steels which possess excellent mechanical properties are utilized. The hardness of ferrite decreases with decreasing temperature, but since it is only a part of a two-phase alloy, the ductile-to-brittle transformation is not as sharp as in ferritic stainless steels.

Since ferrite has a lower coefficient of expansion and higher thermal conductivity than austenite, the more ferrite present, the lower the coefficient of expansion and the higher the thermal conductivity of the duplex stainless steel. Duplex stainless steels are usually very weldable; in particular, duplex weld metal is used in austenitic stainless steels [8-12]. However, there is no research reported on the microstructure, hardness, and abrasive wear resistance of plasma-sprayed DSS coatings. In light of the above, an attempt was made in this study to investigate the microstructure, toughness, and abrasive wear resistance of plasma-sprayed DSS coatings with varying Ni content. and its influence on microstructure, hardness, and abrasive wear resistance.

2. Materials and Methods

Stainless steel (316 grade) substrate was taken in the plate form of dimensions 76 × 25 × 4 mm. The substrate 316 SS was pretreated and sandblasting was done. DSS grade S31500 (3RE60) in form of wire was coated on the substrate by wire arc spray process at Center of Excellence in Corrosion & Surface Engineering (CECASE) labs, NIT, Trichy. The schematic representation of the wire arc spray process is shown in figure 1. The wire spray process is well explained in the literature elsewhere. After the spray process, the samples are cleaned in acetone and weighed, using a digital balance. To confirm the duplex nature of the coatings XRD was performed on the coatings and the sample coated with 7.0 wt% Ni is shown in figure 2. X-ray Diffraction studies were performed using a Rigaku D/Tex Ultra diffractometer with a Cu K_α radiation of 1.5405 Å. The scanning angle (2θ) ranged from 30° to 120° with a step size of 0.02° and counting time of 2 s/step. The diffraction patterns obtained were indexed and compared with the ICDD-PDF standard data.

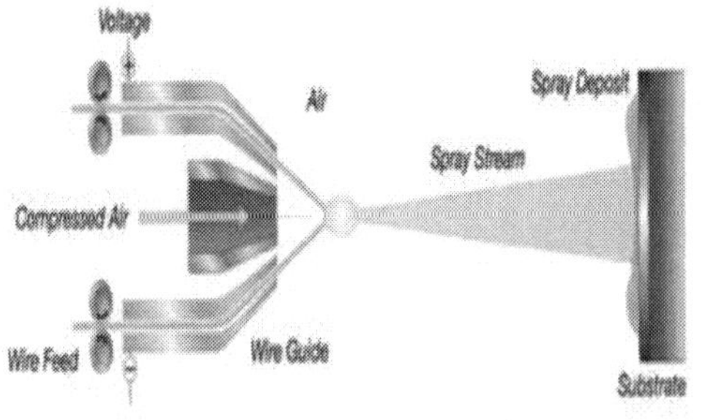

Figure 1. Wire spray process

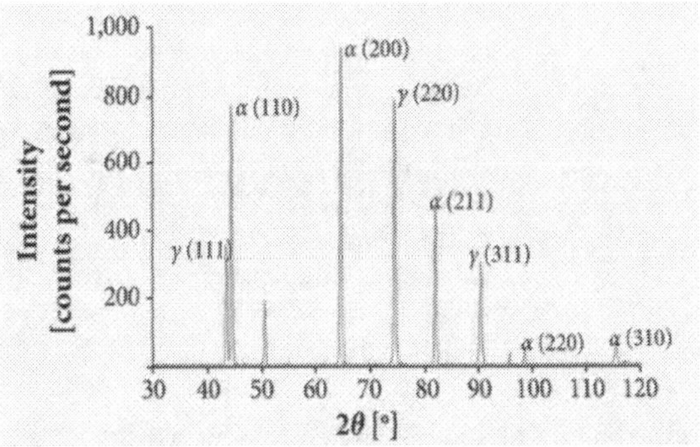

Figure 2. XRD of 7wt% Ni added DSS coating

Vickers microhardness equipment (Matsuzawa MMT-X7 B type, Japan) was used to measure microhardness by performing indentations at a loading force of 100 gf and holding time of 15 s. The samples were polished to remove slight undulations before taking the microhardness tests. At least 5 measurements were conducted and the average microhardness values for the samples are shown in Table 1. Figure 3 shows the optical microscopic image of the indentation on the coated sample.

Figure 3. Optical image of indentation on one of the test sample

2.1 Two-body abrasion test

Before the wear examination, all of the samples were thoroughly washed with acetone and weighed. The tests were conducted on a two- and three-body abrasion testing device provided by DUCOM Instruments, Bangalore, India. Figure 4 depicts the abrasive wear test setup. A spinning wheel with a chlorobutyl rubber beading on its outer surface makes up the apparatus.

The sample holder is constructed so that the sample to be examined is pressed against the spinning rubber wheel and that the applied load can be changed. The rubber wheel is set to turn at a steady speed of 200 revolutions per minute. The abrasive medium is 60-micron silica sand, which is fed between the contact surfaces of the rubber wheel and the sample as shown in figure 4.

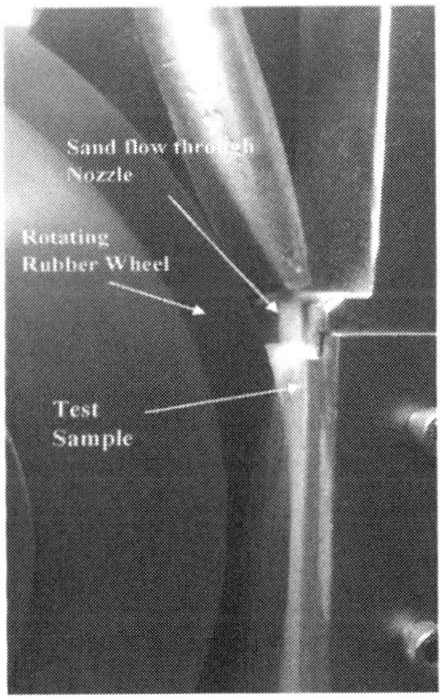

Figure 4. Abrasion test setup

The wear tests were conducted with a constant load of 5 N and 8 N for 3 minutes and 30 seconds. The diameter of the spinning rubber disc and its rotation speed (RPM) was used to measure the sliding time and velocity.

3. Results and Discussion

3.1 XRD analysis

The samples subjected to XRD are analyzed for the composition of the coatings. Figure 2 reveals the XRD pattern obtained for the 7% Ni added coating. Almost all the samples with varying Ni content showed a similar pattern and confirmed the dual-phase normally present in the duplex stainless steels. The presence of ferrite and austenite phases confirmed that the coatings are similar to duplex stainless steels.

3.2 Microhardness test

The microhardness values obtained for the uncoated and coated samples are listed in table 2. The values indicate a slight increase in the H_v value for an increase in Ni percentage. This may be attributed to the fact that the addition of Ni increases the strength and toughness of the materials. Figure 5 graphically represents the variation of hardness to Ni percentage.

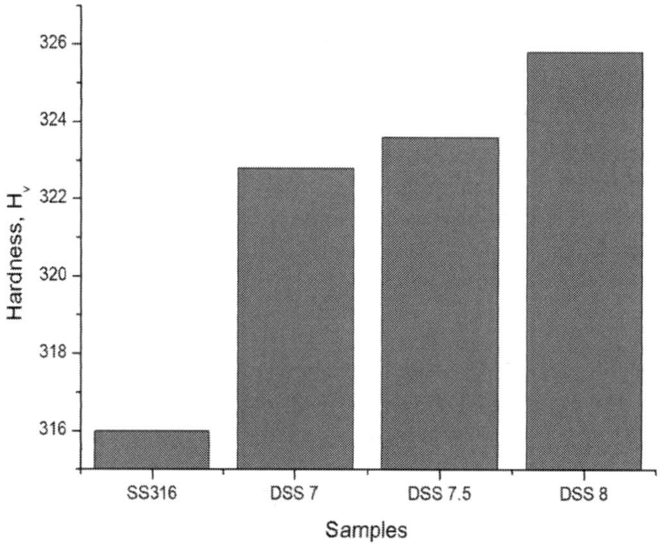

Figure 5. Variation of H_v Vs DSS coatings

Table 1. Microhardness values for uncoated and coated samples

Sample	Hardness value (HV)	Average value	Standard deviation
Uncoated	313	316	4
	321		
	311		
	317		
	318		
Coated with 7wt% Ni	320	322.80	2.588
	321		
	325		
	322		
	326		

Coated with 7.5wt% Ni	322	323.6	1.673
	324		
	322		
	324		
	326		
Coated with 8wt% Ni	324	325.8	1.303
	326		
	325		
	327		
	327		

3.3 Two-body abrasion test

The weight difference between the samples before and after the experiments was determined. The wear rate of various samples was calculated by converting the weight loss in grams into the basic wear rate (mm3/mm). The samples are examined for the effect of Ni addition on the wear rate of the coatings. The values of the test sample under 5N and 8N loads are described in Table 2. The wear mass loss of DSS coatings as a function of Ni content is shown in table 2. The wear mass losses of the coatings initially decrease as the Ni content increases. As compared to the uncoated sample, the wear mass loss of the DSS coating with 8wt% Ni in the coating has the lowest weight loss (0.054 gm), which is 40% less. The values indicate that the wear rate increases marginally as the applied loads' increase. However, the coating with the highest Ni content showed the lowest wear rate once again.

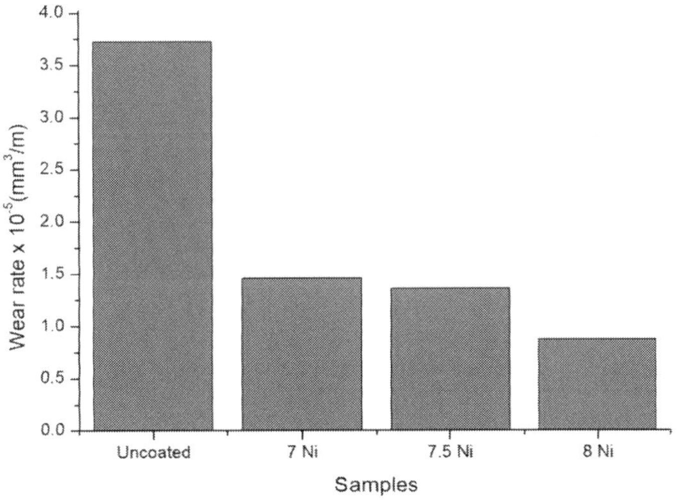

Figure 6. Variation of Wear rate

Figure 6 depicts the difference in wear rate as a feature of Ni percentage in various samples. The wear scars on the samples show that the wear is more abrasive at first. The wear mechanism then shows mixed results as the Ni percent increases due to the ductility imparted by Ni. It begins off as an abrasive and then turns into an adhesive in samples.

Table 2. Wear Test Results

Sample	**Initial weight (gm)**	**Final weight (gm)**	**Weight loss (gm)**	**Volume loss (mm3)**	**Wear rate (mm3/mm) ×10⁻⁵**
5N load					
Uncoated	58.954	58.724	0.23	26.744	3.723
Coated with 7wt% Ni	62.517	62.427	0.09	10.465	1.457
Coated with 7.5wt% Ni	62.528	62.444	0.084	9.767	1.360
Coated with 8wt% Ni	60.591	61.537	0.054	6.279	0.8743
8N Load					
Uncoated	60.869	60.393	0.476	55.348	7.706
Coated with 7wt% Ni	61.527	61.362	0.165	19.186	2.671
Coated with 7.5wt% Ni	62.421	62.281	0.14	16.279	2.266
Coated with 8wt% Ni	61.943	61.835	0.108	12.558	1.748

4. Conclusions

Wire arc spray technique was used to successfully prepare duplex stainless steel coatings of grade S31500 on an SS316 steel substrate. The following are the findings reached as a result of the research:
1) An increase in Ni percent causes dual phases such as ferrite and austenite to emerge in the coating during the wire arc spray process, according to XRD research.
2) As the wt% of Ni in the DSS coatings increased, the microhardness of the coatings increased.
3) The tribological properties of the DSS coating were superior. The coating with a higher Ni percentage demonstrated the best wear resistance in terms of wear mass loss. With a rise in Ni wt percent, the wear mechanisms changed from abrasive to adhesive wear with grooving and ploughing.

REFERENCES

[1] Robert N G 1997 *Duplex Stainless steels-Microstructure, Properties & Applications* 175
[2] Solomon H D 2001 *Encyclopedia of Materials: Science and Technology* 8802

[3] Carsí M et al 2001 *Simulation of the hot forming behavior of super duplex steel by means of torsion testing", Proceedings of the III Seminar on Metallurgical Technologies* 126
[4] Wu M Y and Sherby O D 1984 *Acta Metallurgica* **32** 1561
[5] Rieiro I et al 1998 *J. Mater. Process. Technol.* **78** 177
[6] Adamczyk J et al 1995 *Steel Research* **66** 279
[7] McQueen H J and Ryan N D 2002 *Mater. Sci. Eng. A* **322** 43
[8] Tamura I et al 1988 *Thermomechanical processing of high strength low alloy steels* **48** 17
[9] Sherby O D and Wadsworth J 1989 *Prog. Mater. Sci.* **33** 169
[10] Jiménez J A et al. 1999 *Acta Mater.* **47** 3655
[11] Prasad YV R K et al 1984 *Metall. Mater. Trans. A* **15** 1883
[12] Malas J C and Seetharaman V 1992 *J. Min. Met. Mater. Soc* **44** 8

On the Corrosion behavior of 4A and 5A cast duplex stainless steel under different heat treatment conditions

E Edward Anand[1], C Gopi[2], N Moorthy[3], J Prince Richard[4]

[1, 2, 3 & 4]Department of Science & Humanities, E.G.S. Pillay Engineering College, Nagapattinam, India

E-mail: *alphsedward@gmail.com*

Abstract. In this analysis, corrosion efficiency and mechanical properties of duplex stainless steel castings (DSS-4A and DSS-5A) were compared to their wrought counterparts. Cast duplex stainless steel DSS-4A and DSS-5A are characterized in two separate solution treated settings, while their wrought counterparts are characterized in both solutions treated and forged form. Optical microscopy and X-ray diffractometry were employed to perform the metallurgical characterization. Electrochemical techniques such as Tafel extrapolation and electrochemical impedance spectroscopy were used to investigate corrosion activity. An electrochemical workstation was used to perform the corrosion studies. To compare the stability and changes in corrosion properties of both grades, an ASTM standard B117 salt spray test was performed in a 3.5 percent NaCl medium. Without any intermetallic phases, the XRD showed austenite and ferrite phases. The elements were partitioned further into the phases that they promoted, according to the elemental study of the phases. The ultimate tensile strength and hardness values of DSS-5A are higher than those of DSS-4A. DSS-5A had a higher corrosion resistance than DSS-4A. When comparing the two media used (H2SO4 and NaCl), DSS-5A demonstrated superior corrosion resistance.

Keywords: *Aluminium – DSS, EIS, XRD, Corrosion, EDS*

1. INTRODUCTION

Owing to their improved pitting and stress corrosion cracking resistance, duplex stainless steels (DSS) are being specified for chloride-containing environments. They outperform austenitic stainless steels in terms of corrosion resistance. Duplex stainless steels have better strength properties and come in a variety of wrought and cast forms.

Duplex stainless steels (DSS) have two phases in almost equal proportions: ferrite and austenite. The arrangement of the two phases is highly dependent on their composition and thermal background. Because of its excellent mechanical and corrosion resistance, duplex stainless steel is used in the pharmaceutical, petrochemical, nuclear, marine, and paper industries[1,2]. It has a unique combination of properties, including high corrosion resistance in chloride-ion-rich settings, mechanical strength and ductility, abrasion resistance, and weldability[3-5]. Around 1937, the commercial development of duplex stain-free steels began[6-8].

These alloys outperform austenitic stainless steel in terms of weldability and mechanical properties. Duplex stainless steel has twice the strength of normal austenitic or ferritic stainless steel. They are, however, more vulnerable to phase precipitation and embrittlement than austenitic steels. They have higher durability and ductility than ferritic grades, but not to the same extent as austenitic grades[6]. Duplex stainless steels solidify from a liquid phase to a fully ferritic structure after melting. Around half of the ferritic grains turn to austenitic grains as the materials cool. Decomposition of the ferritic phase in the solution treated condition (at around 1000 C)[7-12] can result in higher hardness, yield stress, and ultimate tensile strength, but at the cost of ductility and toughness.

Even though duplex stainless steels have become increasingly common in recent years, their availability in cast form has lagged behind that of the wrought form. Duplex stainless steel castings are widely used in pumps and valves in several applications, and they are critical components in systems where unexpected service failures can cause major operational problems and costs.

The quality of the performance is crucial. Variability and insufficient performance characteristics of duplex stainless steels in all types, which can be linked to severe in-service issues, are therefore a source of concern. It is also important to have readily available, appropriate methods and procedures for determining performance characteristics of duplex stainless steel cast materials before the operation.

In this analysis, corrosion performance and mechanical properties of selected grades of duplex stainless steel castings (4A and 5A) were assessed. Since there is little literature on the topic, this article aims to compare the performance of cast versus forged DSS. The effect of composition and microstructure on the corrosion behaviour of the samples in 3.5% NaCl and 1.0 N sulphuric acid is discussed in this article.

2. EXPERIMENTAL PROCEDURE

The research was done on cast duplex stainless steel alloys of Grade 4A and 5A. For comparison, forged DSS 4A and 5A samples were used. Tables 1 and 2 show the chemical compositions and pitting resistance equivalent (PRE) numbers of the cast 4A and 5A samples.

Table 1: Chemical composition of Cast 4A (obtained from Spectro analysis)

Elements	C%	Mn%	Si%	P%	S%	Cu%	Ni%	Cr%	Mo%	V%
Weight %	0.026	1.02	1.12	0.022	0.006	0.03	6.33	23.12	2.92	0.064
Elements	W%	Nb%	N%	Al%	Ti%	Sn%	Pb%	B%	Sb%	As%
Weight %	0.016	0.004	0.190	0.007	0.020	0.004	0.002	0.003	0.004	0.014
Elements	Bi%	Ta%	Ca%	Se%	Co%	Fe%	PRE*			
Weight %	0.001	0.010	0.001	0.124	0.035	64.90	35.796			

Table 2: Chemical composition of Cast 5A (obtained from Spectro analysis)

Elements	C%	Mn%	Si%	P%	S%	Cu%	Ni%	Cr%	Mo%	V%
Weight %	0.028	0.77	0.860	0.023	0.002	0.13	7.25	25.30	4.08	0.047
Elements	W%	Nb%	N%	Al%	Ti%	Sn%	Pb%	B%	Sb%	As%
Weight %	0.028	0.004	0.145	0.006	0.014	0.005	0.002	0.003	0.002	0.020
Elements	Bi%	Ta%	Ca%	Se%	Co%	Fe%	PRE*			
Weight %	0.001	0.010	0.001	0.104	0.073	61.10	41.084			

3. MATERIALS AND METHODS

3.1 Melting and Casting

An induction melting furnace was used to melt and cast Duplex stainless steel. The constituent elements were charged in the crucible after being washed with acetone. When all of the elements have melted completely, the melt is poured into the die.

3.2 Hot forging
Cast samples were heated to 1100°C in an oil-fired furnace with a 100 kg capacity and two HP blowers. Three thermocouples are lined with fire brick and fire clay on the three sidewalls. In the pneumatic press, the heated samples were taken one by one and subjected to upset forging. Solution treatment was carried out after hot welding to a reduction of 32 percent.

3.3 Heat treatment
Two different types of heat treatments are applied to the samples. 1) Solution treatment at 1010°C for 1 hour, followed by water quenching; 2) Solution treatment at 1130°C for 1 hour, followed by water quenching.

3.4 Microstructural Analysis
The microstructure of samples is examined using a high-resolution LEICA DM2500M DIC microscope. Optical metallography was performed on samples that had been manually cut out. The specimen was polished using traditional metallographic techniques, which included belt grinding and polishing with emery papers of grades 600, 800, 1000, 1200, 1500, and 2000. Finally, with a legated alumina abrasive on a polishing wheel to achieve a mirror finish. Before testing, the polished specimen was cleaned with double distilled water, degreased with acetone, and dried. Keller's reagent was used to etch the specimen to expose the microstructure.

3.5 X-ray Diffraction Analysis
Rigaku Ultima III XRD equipment was used for the X-ray diffraction study. Cu K radiation with a wavelength of 1.54 Åwas used in the study, with a step size of 0.05 between the values of 20° and 100°.

3.6 Corrosion Test
Corrosion tests are carried out in two types of medium. 1) In 3.5 % NaCl and 2) 1.0 N H_2SO_4.

3.6.1 Electrochemical Measurements. Gill AC ACM instrument is used to perform electrochemical measurements. With a platinum counter electrode and a saturated calomel electrode (SCE) as a reference, a three-electrode compartment with a pyrex glass cell was used. The SCE is referred to for all possible values mentioned. The working electrode was a specimen of Duplex Steel alloy.
Polarization experiments are carried out on the same exposed electrode surface as the electrochemical impedance spectroscopy (EIS) experiments, with no additional surface treatment.

3.6.2 Potentiodynamic Polarization Studies. The corrosion media were exposed to a well-polished Duplex Stainless Steel specimen coupon, which was allowed to create a steady-state open circuit potential (OCP). At a scan rate of 1 mVs–1, the Tafel plots were recorded by polarising the specimen to 250 mV cathodically and +2000 mV anodically relative to the OCP.

3.6.3 Electrochemical Impedance Spectroscopy (EIS) Studies. The application of a periodic small amplitude (10 mV) ac voltage signal with a broad frequency range ranging from 100 kHz to 0.01 Hz was used to calculate impedance at the open circuit potential (OCP). Nyquist plots were used to analyze the impedance data.

3.7 Salt Spray Test
In a chamber temperature of 35°C, test specimens are put in an enclosed chamber and exposed to a continuous indirect spray of salt water (3.5 % NaCl) solution (also known as fog or mist) that falls-out on the specimen at a rate of 1.0 to 2.0 ml/80cm^2/hour. The chamber environment is held at a constant steady-state temperature. The duration of the test is variable, and it is carried out in compliance with the ASTM B117 standard.

4. RESULTS

4.1 Microstructural analysis

Figures 1a - 1d display the microstructures of the as-cast 4A and 5A after solutionizing treatments. Delta ferrite, austenite, and sigma are the three major phases shown by the microstructures. Secondary austenite is more visible when solutionizing at 1010C. In DSS-4A, solutionizing at 1130C shows less secondary austenite and the sigma phase is absent. However, austenite is found within the ferrite in the case of DSS-5A, and the sigma phase is also present.

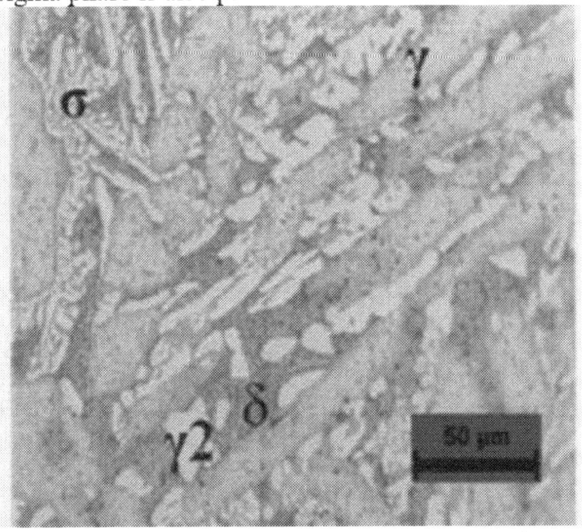

Figure 1a: Microstructure of DSS- 4A after Solutionizing at 1010°C

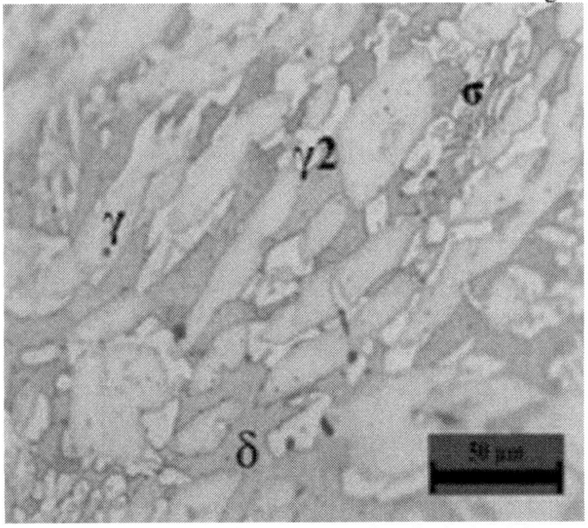

Figure 1b: Microstructure of DSS- 5A after Solutionizing at 1010°C

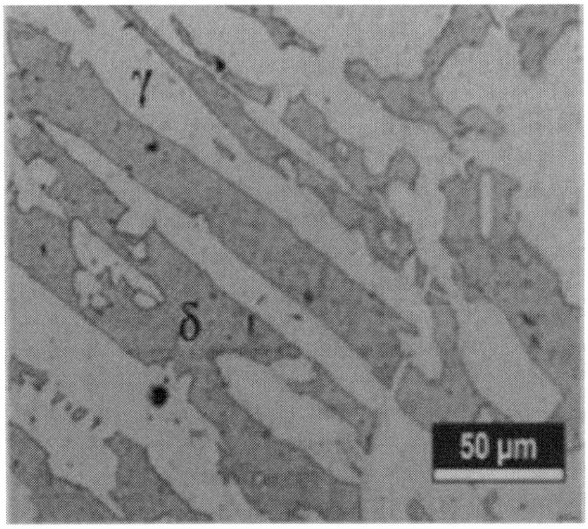

Figure 1c: Microstructure of DSS- 4A after Solutionizing at 1030°C

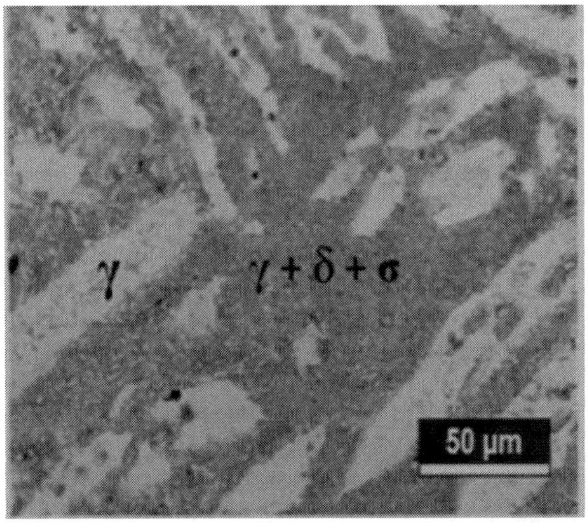

Figure 1d: Microstructure of DSS- 5A after Solutionizing at 1030°C

4.2 X-ray analysis

The XRD patterns of as-cast DSS - 4A and DSS - 5A solutions treated in 1010°C and 1030°C are shown in Figure 2(a-d). The peaks corresponding to austenite and ferrite can be seen in the X-ray diffraction patterns of duplex stainless steel samples. Ferrite and austenite make up the sequence in the DSS-4A and DSS-5A samples. The peak corresponding to ferrite has the highest peak strength in both 4A and 5A. In the case of a solutionizing temperature of 1130°C, the peak density of ferrite increases more than that of austenite.

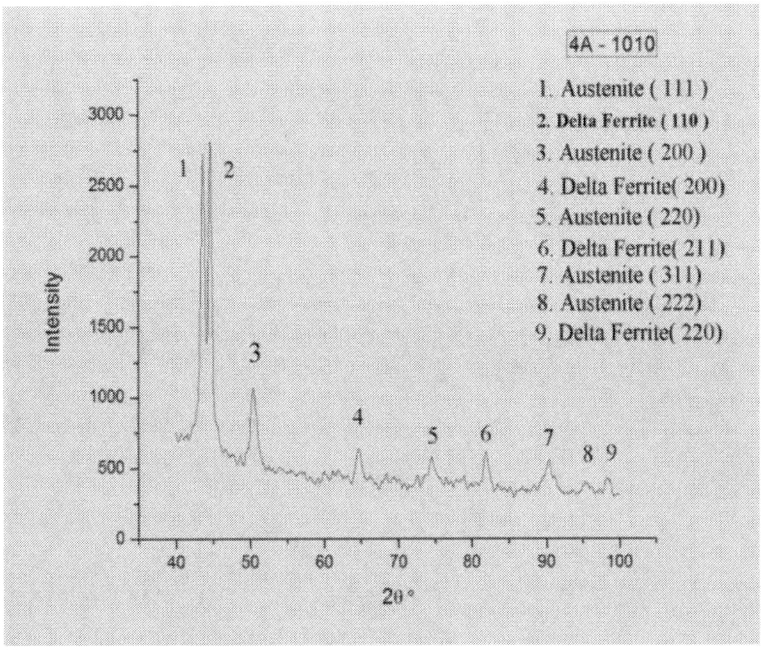

Figure 2a: XRD pattern of as-cast DSS - 4A solutionized at 1010ºC

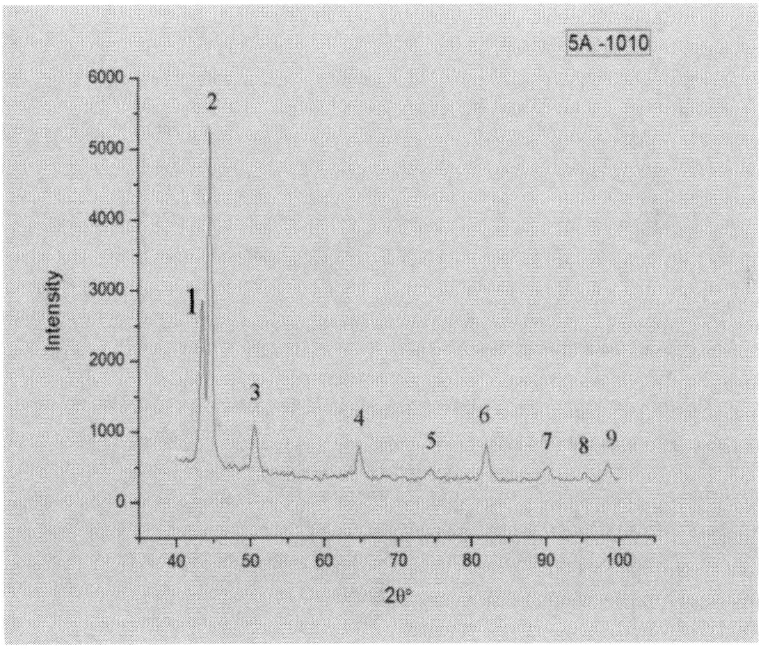

Figure 2b: XRD pattern of as-cast DSS - 5A solutionized at 1010ºC

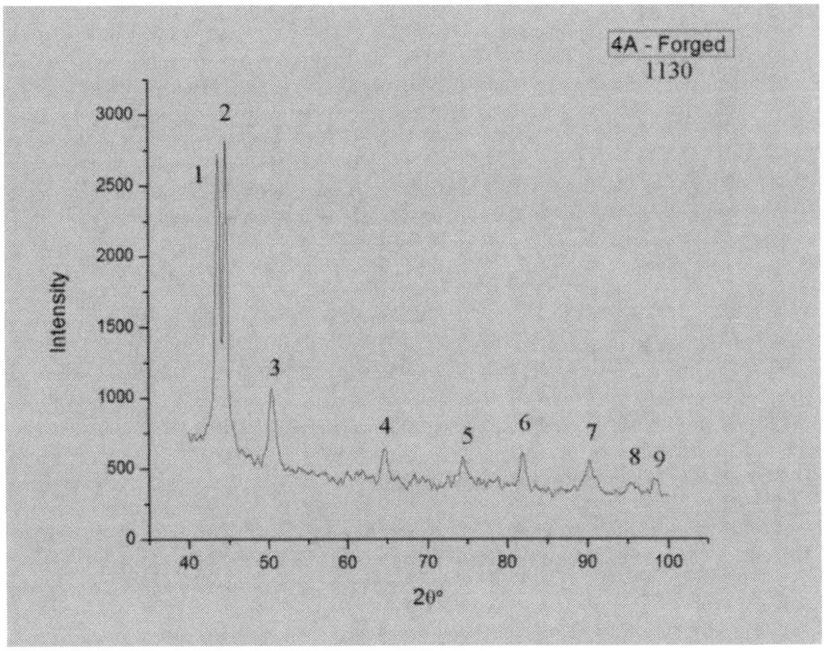

Figure 2c: XRD pattern of forged DSS - 4A solutionized at 1030ºC

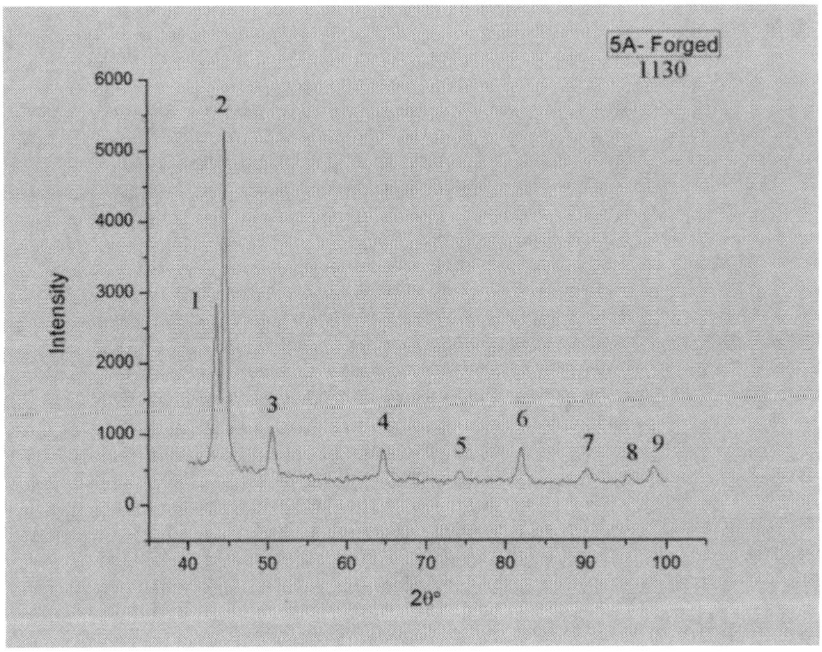

Figure 2d: XRD pattern of forged DSS - 5A solutionized at 1030ºC

4.3 Potentiodynamic Polarization Measurements

Figures 3a-b display potentiodynamic polarization curves for 4A and 5A in 3.5% NaCl solution and 1.0N H2SO4. The polarisation curves for the two specimens are identical, implying similar corrosion mechanisms. Only at cathodic branches of polarisation curves did the linear Tafel actions appear, which are thought to reflect cathodic hydrogen evolution through water reduction.

Since the anodic polarisation curves did not show Tafel activity, they are thought to reflect anodic oxidation. The chemical composition of samples has a major impact on corrosion properties, as shown by a comparison of polarization curves. DSS-4A and DSS-5A have identical potentiodynamic curves for as-cast duplex stainless steels. There were active dissolutions found in the samples. DSS-5A had a higher corrosion potential at first. However, after 120 hours of accelerated corrosion, the corrosion potential values of both specimens are virtually identical. The polarisation curves for both samples initially showed strong passivation, but after 120 hours, no passivation is visible. Both samples have pitting potentials greater than 0.8 V vs SCE. Pseudo-passivity can be seen in the graphs.
Because the corrosion potential is a mixed potential in which the absolute values of the anodic and cathodic currents are identical, and because the cathodic phase corresponds to hydrogen evolution, the corrosion potential is expected to reflect the samples' corrosive propensity.
The amount of chromium and molybdenum in the material determined how resistant it was to pitting attacks. The corrosion resistance was unchanged by the volume fraction. All of the steels used in the study were duplex, with austenite to ferrite ratio of 70:30, which is appropriate.

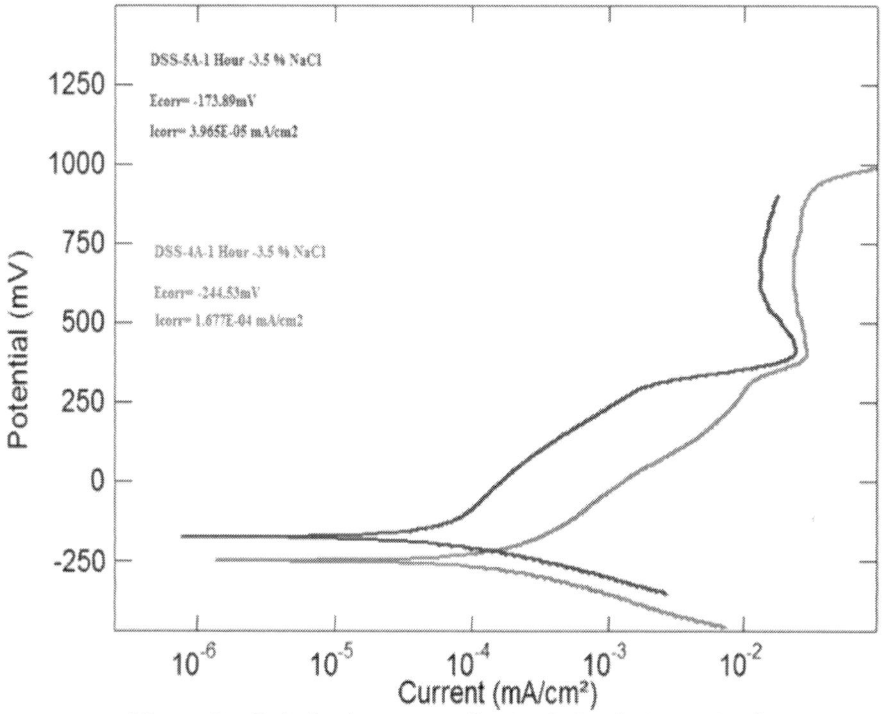

Figure 3a: Polarization curves for 4A & 5A in 3.5% NaCl

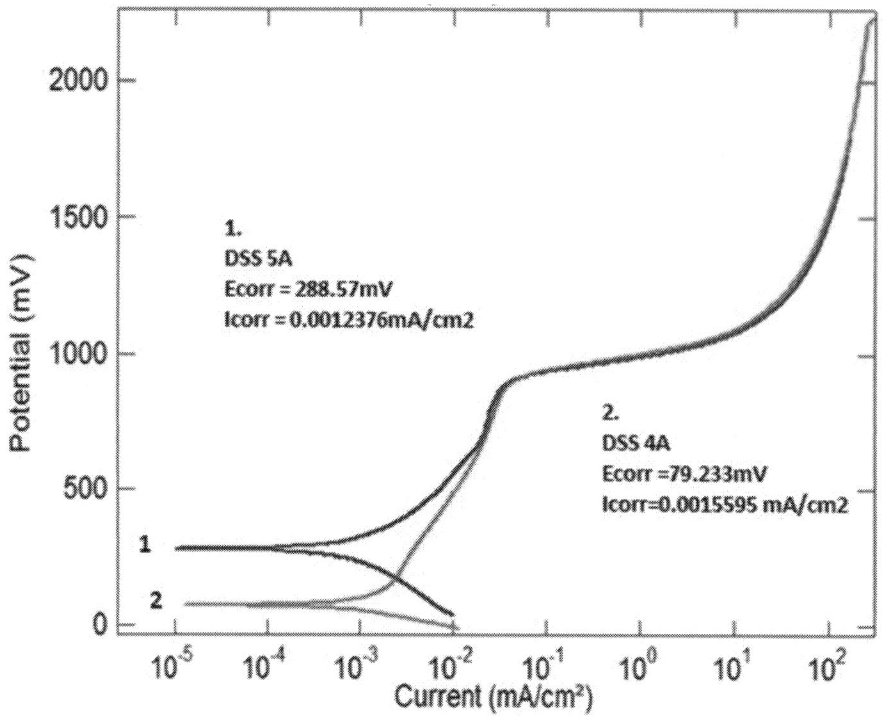

Figure 3b: Polarization curves for 4A and 5A in 1.0N H_2SO_4

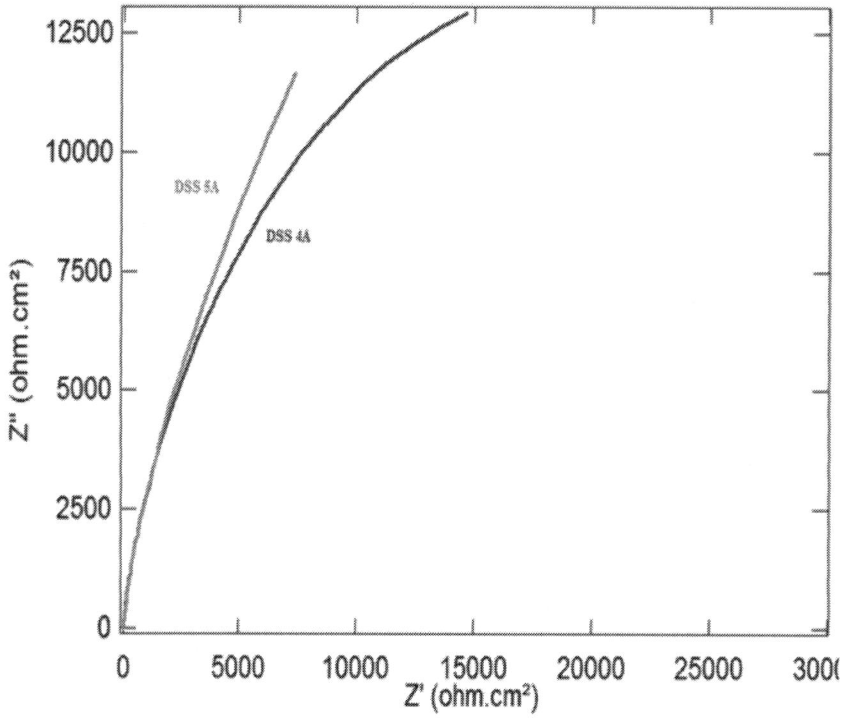

Figure 3c: Nyquist plots for 4A & 5A in 1.0N H_2SO_4

Table 3. EIS parameters obtained from Nyquist plots for 4A & 5A

Parameter	4A	5A
R_{sol} (ohm.cm^2)	1.839	1.839
R_{ct} (ohm.cm^2)	3.721×10^4	5.983×10^4
C_{dl} (F)	5.172×10^{-5}	9.332×10^{-5}

4.4 Electrochemical Impedance measurements

In 1.0N H_2SO_4 solution, electrochemical impedance spectra for 4A and 5A were calculated at the open circuit potential. Figure 3c shows the obtained impedance response as a Nyquist plot, and table 3 shows the parameters extracted from it. Both 4A and 5A have three impedance characteristics in common: a single semi-circle with a real axis intercept that is: at high frequencies, the solution resistance; at low frequencies, the charge transfer resistance.

4.5 Salt Spray Test

According to ASTM specifications, both samples 4A and 5A passed the salt spray inspection. Figures 4a and 4b represent the samples before and after the test. After 120 hours of exposure to 3.5 percent NaCl at a chamber temperature of 35.6°C and an air saturator temperature of 45°C, no weight loss was reported.

The sample is checked for PDP after the salt spray test on the same surface to see if any surface changes have occurred, and the polarization curve is shown in figure 4c. The corrosion resistance of DSS-4A is found to be lower after a 120-hour salt spray test. The corrosion potentials are nearly similar, and both samples have shown pseudo passivation. Passivation is slightly more in the case of DSS – 5A

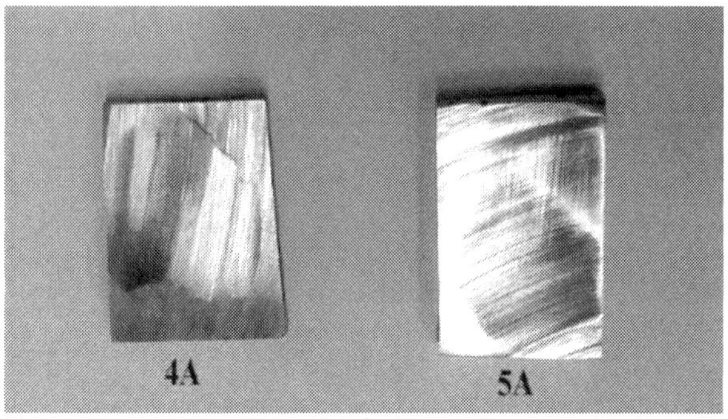

Figure 4a: Samples before salt spray test

4A 5A

Figure 4b: Samples after salt spray test

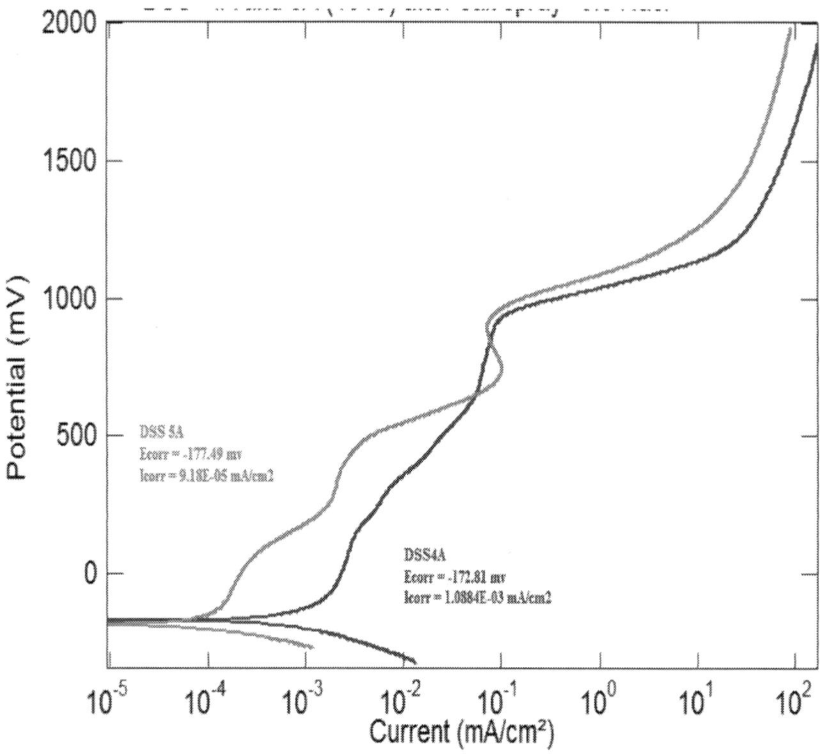

Figure 4c: Tafel plot of DSS 4A and 5A after salt spray test

5. CONCLUSIONS

1. When solutionizing at 10100C, cast DSS 5A is more corrosion resistant.
2. Corrosion activity in 4A and 5A is similar, and pitting potential is virtually identical in both samples.

3. DSS-5A has low corrosion resistance when solutionized at 1130°C, indicating more sigma phase.

4. DSS-5A's supremacy is confirmed by salt spray samples. Both the 4A and 5A steels exhibit equal corrosion resistance in a 3.5 percent NaCl solution when hot forged.

6. REFERENCES

[1] Chandramohan P et al 2007 *J. Mater. Sci. Technol.* **23** 117
[2] Oluwatoyin A O et al. 2014 *Adv. Chem. Eng. Sci.* **4** 89
[3] Solomon H D and Devine T. M. 1983 *Proc. Conf. Duplex Stainless Steels* 693
[4] Khoshnaw F H and Gardi R. H 2006 http://www.stainlesssteelworld.net/duplex.asp
[5] Chance J et al 1983 *Ed., Proc. Duplex Stainless Steel Conf* 371
[6] Desestret A and Charles J 1993 *Eds. Les Editions de Physique* 2006 *2nd Edition*, 2009
[7] Lardon J M et al 1988 *Proc. Conf. High Nitrogen Steels* 280
[8] Davison R M R 1990 *Practical Guide to Using Duplex Stainless Steels. Materials Performance* **29(1)** 57
[9] Atamert S K 1933 *J. Mater. Sci. Lett.* **12(14)** 1144
[10] Nilsson J et al 1996 *Metall. Mater. Trans. A.* **27(8)** 2196

[11] Ogawa K et al 1996 *Study of Weldability of High-Tungsten Duplex Stainless Steel (1st Report). Welding International* **10(6)** 466
[12] Kotecki D J S 1992 *Constitution Diagram for Stainless Steel Weld Metals: A Modification of the WRC-1988 Diagram. Welding Journal* **71(5)** 171

Optimal CRONE controller using meta-heuristic optimization algorithm for shell and tube Heat Exchanger

B.Girirajan and D.Rathikarani

Department of Electronics and Instrumentation Engineering, Annamalai University, Annamalai Nagar, Tamil Nadu, India

giripdy12@gmail.com

Abstract. A meta-heuristic optimization algorithm based third generation CRONE controller is proposed for controlling the thermal interfaces in heat exchanger device. The primary aim of this work is to identify suitable optimization technique among the whale optimization Algorithm (WOA) and grey wolf optimization (GWO) algorithms that would improve the robustness of third order crone controller against the parameter variations. A mathematical model of HE is developed to compute non-integer parameters so as to realize the optimal performance of the controller. The objective function is to minimize cost of fluid flow rate and to maximize the heat transfer rate. The optimization of heat exchanger is complicated by number of process plants, area, and method of connection, pressure drop of fluid, and transfer rate. A competitive result is obtained with multi-objective non-linear WOA for wide range of the process variable variations. The simulation is performed in MATLAB software to verify proposed WOA optimized third order CRONE controller and the performance is compared with the GWO optimized CRONE controller.

Keywords: Shell and tube heat exchanger, CRONE controller, Whale Optimization Algorithm, Gray Wolf Optimization.

1. Introduction

Heat exchangers are commonly used for heat transfer applications in industries. Shell and tube heat exchanger (STHE) provides large space for heat transfer between two fluids compared to other heat exchanger types. The STHE are mostly used in the application of liquid to liquid heat transfer with higher density operating fluids. The working process of nano-fluid shell and tube heat exchanger is employed. The nano-fluid effect on heat exchanger improves the rate of heat transfer is experimentally evaluated [1].Differential Evolution (DE) scheme of STHE optimization design is presented. An enhanced type of Genetic Algorithm (GA) namely Differential Evolution (DE) is employed for design optimization of STHE. Differential Evolution method shows improved performance than Genetic Algorithm [2].For designing heat exchangers an artificial bee colony algorithm (ABC) is introduced. This algorithm is effectively manipulated for optimal design of STHE which gives more accurate and fast method as comparing to conventional trial-and-error method and a conventional GA technique [3]. Particle Swarm

Optimization (PSO) is manipulated from economic perspective to design STHE. The PSO based optimization for STHE is simple in concept, convergence is quick and implementation is simple for different thermal systems [4]. A design approach based on imperialist competitive algorithm (ICA) is presented. This method reduces the cost of capital investment and operating cost that permit quick, good quality solutions which gives designer more freedom with respect to GA approach in the final choice [5].

In heat exchanger, the temperature of outlet liquid is controlled by conventional PID controller [6].With traditional PID controller, a wide range of precise temperature control cannot be reached. So the combination of fuzzy and PID controller is proposed [7]. Genetic Algorithm based proportional integral derivative controller is presented to regulate heat exchanger's output temperature. This controller produces maximum overshoot and higher settling time with high oscillation [8]. To overcome these drawbacks, a new hybrid model reference adaptive supervisory fuzzy controller is developed to regulate output temperature of co-current STHE. This method reduces overshoot problem and settling time is shorter with less oscillation when comparing to fuzzy and PID controller [9].In order to control temperature in heat exchanger a FLC is presented. By using this controller, the mixed fluid maintains constant temperature [10].

The conventional PID controller to CRONE first generation controller is employed on anti-roll scheme to enhance traveler riding consolation in the sense of universal electric vehicle chassis control. For indeterminate perturbed plants, first generation CRONE controller delivers robust fractional order controller [11]. To enhance the behavior of conventional PID controller, the conventional PID controller is replaced by FOPID controller. By reducing the cost function with GA and PSO, the parameters of PID and FOPID are tuned by optimized value. With regard to optimized tuning parameters, PSO-FOPID has minimal integral error compared to GA-FOPID [12]. In STHE, the PID controller is implemented by Z-N tuning technique. Based on FCM algorithm, the fuzzy forward model of STHE is developed. The output model is then compared with real output of STHE [13]. Generally, for linear process PID controller with conventional tuning is employed, but for non-linear process the conventional PID controller provides robustness, steady state error, settling time is long and also not provides better control action. To overcome this problem PID controller is optimized with particle swarm optimization [14]. For integral and unstable systems, a set of tuning rules for Integer order Proportional-Integral and Derivative (IOPID) as well as FOPID are introduced. To reduce the integrated absolute error, the tuning rules are developed based on a simple model of operation. When compared with IOPID, FOPID provides enhanced performance [15].

A meta-heuristic optimization algorithm based third generation CRONE controller is proposed in this work for controlling the temperature of outlet fluid in HE system. To show the performance of third generation CRONE controller, the parameters are tuned by GWO and WOA. The tuned parameters of K_p, K_i, K_d, λ and μ, of WOA are compared with that of GWO, which provides better performance than GWO.

2. Proposed System

The standard chemical heating interaction method contains chemical reactor and STHE. The vapor comes from boiler moves into the pipes while, input liquid flows via the shells of STHE. The input of STHE is input liquid which is stored in storage tank. From boiler, it delivers vapor that heat up the heat exchanger input fluid to a wanted set value. With pump as well as non-returning valve (NRV), the storage tank delivers input fluid into a heat exchanger.

If the rate of inlet flow and outlet flow liquid are equal, the level of fluid is retained constant in STHE, also an insulation wall is neglected due to heat storage capacity. Two kinds of disturbances occur,

IOP Conf. Series: Materials Science and Engineering 1166 (2021) 012058 doi:10.1088/1757-899X/1166/1/012058

first disturbance is input liquid flow variation and second disturbance is input liquid temperature variation. The proposed system's block representation is depicted in Fig 1.

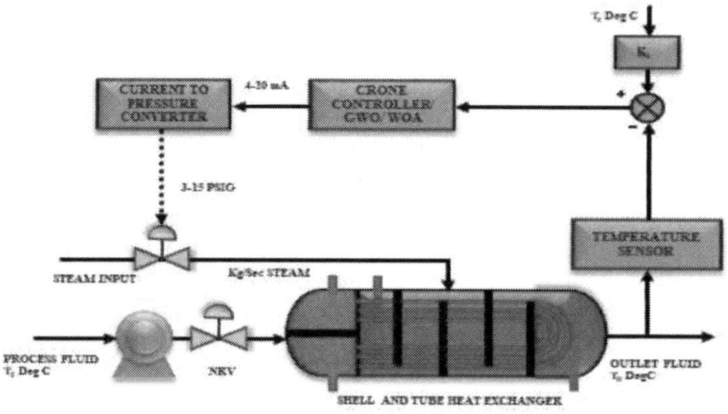

Figure 1. Shell and tube heat exchanger's proposed control scheme

The sensing unit and thermocouple are instigated in the feedback path of control architecture. The thermocouple calculates the temperature of output liquid and sends output of thermocouple into the transmitter device that ultimately transforms output of thermocouple to standardized $(4-20)mA$ signal. The transmitter device outcome is fed to third generation CRONE (TGC) controller. To improve performance of TGC controller, it is optimized by grey wolf optimization and whale optimization algorithm. The output of crone controller is compared with set value and then provides essential command through the actuator element to the last control unit. The actuator element is nothing but it is a current to pressure converter and an air open tap is the last control unit. The TGC controller's outcome is in the range of$(4-20)mA$. This range is given to the input of current to pressure converter and it transforms into a standard pressure signal in the $(3-15)psig$ range. According to control decisions of the controller, the valve actuates.

3. Modeling of Proposed Scheme

3.1 Heat Exchangers (HE)

Heat Exchanger system transfers heat from one medium to other medium. By changing the speed of hot liquid pump, the rate of flow of hot liquid is controlled. By adjusting the percent opening of flow control valve, the rate of flow of cold liquid can be varied. At tube side, the hot liquid flows whereas at shell side, the cold liquid flows. The flow pattern is co-current, indicating that the hot liquid as well as cold liquid flows in similar direction. The main aim of HE is energy transfer.

Therefore, rate of heat flow (Q) is formulated as,

$$Q = \alpha \, . A \, . \Delta T \, . \qquad (1)$$

Where, α denotes coefficient of heat transfer, A denotes area of heat transfer, ΔT denotes difference in temperature.

Instead of local heat flow, the overall heat flow rate for whole heat exchanger is formulated as,

$$Q = U.A.F.LMTD\ . \tag{2}$$

$$\frac{1}{U} = \frac{D_o}{D_i\alpha_t} + \frac{D_o ln\left(\frac{D_o}{D_i}\right)}{2\lambda_w} + \frac{1}{\alpha_s}\ . \tag{3}$$

$$LMTD = \frac{\Delta T_b - \Delta T_a}{ln\left(\frac{\Delta T_b}{\Delta T_a}\right)}\ . \tag{4}$$

Where, U denotes overall heat transfer coefficient, $LMTD$ means Long-Mean Temperature Difference correction factor that depends on flow arrangement, $\alpha_s\ and\ \alpha_t$ denotes heat transfer coefficients on the side of shell as well as tube, $\Delta T_a and\ \Delta T_b$ denotes differences in temperature between two fluids on various sides of heat exchanger.

Usually, the fluids don't flow via heat exchanger, they have to be pumped. The pressure drop (Δp) defines how much energy is required by heat exchanger to pump the fluids.

3.2 Shell and Tube Heat Exchangers (STHE)

STHE contains two kinds of fluids such as "tubeside" fluid and "shellside" fluid. The "tube side" liquid flows inside a set of parallel tubes called" tube bundle". These tube bundles are surrounded inside a shell metal. The "shell side" liquid flows within the shell but around the tubes. The shell metal and pipes are pressurized. During the planned life of equipment, they have to withstand specified pressure design. The STHE schematic diagram is depicted in Fig 2.

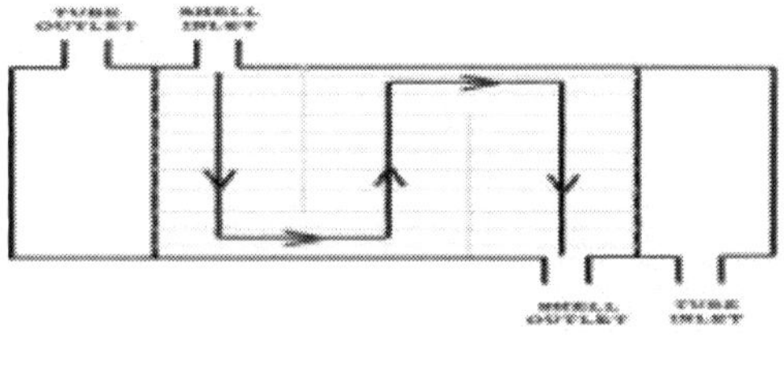

Figure 2. Schematic diagram of shell and tube heat exchanger

To model STHE various methods are used. Generally, TEMA i.e. Tubular Exchanger Manufactures Association method is used. Normally, for classification of shell design TEMA standards are mainly used.

3.3 *Shell Side*

Due to complex geometry as well as flow patterns on side of shell, dissimilar correlations are established for various arrangements. There are two separate shells considered, i.e. TEMA E i.e. one-phase as well as TEMA G i.e. condensation. The first method developed is Kern's technique that predicts the drop in pressure. It is expressed as,

$$\Delta p_s = 2\frac{f_s G_s^{\ 2} D_s(n_b+1)}{D_e \rho_s\left(\frac{\eta_s}{\eta_{sw}}\right)^{0.14}}\ . \tag{5}$$

$$f_s = 0.4475Re^{-0.19} . \quad (6)$$

Heat transfer coefficient for shell side is represented as,

$$\alpha_s = \frac{0.36\lambda}{D_e}\left(\frac{D_e G_s}{\eta_s}\right)^{0.55} {Pr_s}^{1/3}\left(\frac{\eta_s}{\eta_{sw}}\right)^{0.14} . \quad (7)$$

Where, number of baffles is denoted as n_b, number of cross flow segments in HE is denoted as$(n_b + 1)$. The drop in pressure is directly proportional to number of cross-flow segments, which is considered by kern's technique.

The configuration consideration is TEMA G condenser in addition to TEMA E. The geometry is much simpler than that of TEMA E, but the shell side two-phase flow makes modeling more complex.The combined heat transfer coefficient for gravitational as well as heat transfer coefficient for shear is known as heat transfer coefficient.

$$\alpha = \left({\alpha^2}_{shear} + {\alpha^2}_{grav}\right)^{1/2} . \quad (8)$$

3.4 *Tube Side*

Normally, the tube has a simple geometry that leads to easier and improved correlations for heat transfer coefficients as well as drop in pressure. A few examples of heat transfer correlations are Dittus-Boelter & Gnielinski that are given in the expressions (9) & (10).

$$Nu_D = 0.023{Re_D}^{0.8}Pr^n . \quad (9)$$

From equation (9) if tube wall temperature is greater than liquid temperature, the value is mentioned as $n = 0.4$ and if tube wall temperature is lesser than liquid temperature, the value is mentioned as $n = 0.3$. By using friction factor the above equation is written as,

$$Nu_D = \frac{\left({}^f/_8\right)(Re_D - 1000)Pr}{1 + 12.7\left({}^f/_8\left(Pr^{2/3} - 1\right)\right)^{\frac{1}{2}}} . \quad (10)$$

$$f = (0.79ln(Re_D) - 1.64)^{-2} . \quad (11)$$

Where, hydraulic diameter is denoted as D.

By using friction factor the pressure drop Δp is represented as,

$$\Delta p = f\frac{L}{D}\frac{\rho u^2}{2} \quad (12)$$

A mathematical model of HE device is developed on the side of shell and tube in STHE.

IOP Conf. Series: Materials Science and Engineering 1166 (2021) 012058 doi:10.1088/1757-899X/1166/1/012058

4. Design of Third Generation CRONE (TGC) Controller

The structure of third generation CRONE controller is depicted in Fig 3. The design of TGC controller expands CRONE's previous generations by allowing more handling general uncertainties.

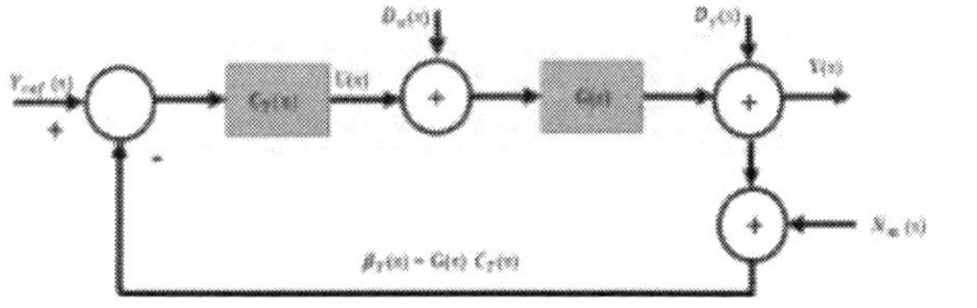

Figure 3. Closed loop structure of third generation CRONE controller

From the diagram, $Y_{ref}(s)$ denotes input reference signal, $C_T(s)$ denotes CRONE controller, $G(s)$ denotes process transfer function, $Y(s)$ denotes plant output, $U(s)$ denotes controller output, $D_u(s)$ denotes input disturbance, $D_y(s)$ denotes output disturbance, $N_m(s)$ denotes sensor disturbance, transfer function of open loop fractional order third generation CRONE controller is denoted as $\beta_T(s)$

The third generation CRONE controller scheme design consists of 3 levels, they are generalized template, optimal template as well as optimization of open loop behaviour. The control scheme design is based on generalized template description to be depicted in Nichols plane by each straight line section in each direction across the crossover frequency $\omega_{cg}$of open loop gain. Generalized template, i.e. Nichols plane is depicted in Fig 4.

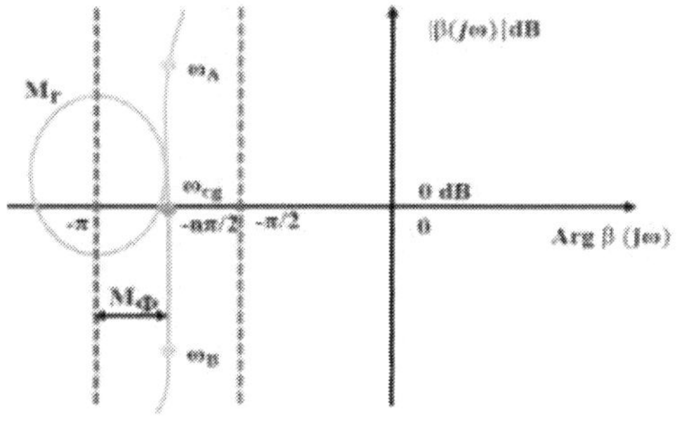

Figure 4. Generalized template (Nichols plane)

From Nichols plane, frequency template phase position specifies the value of real order, i.e. a, and angle to vertical specifies the value of imaginary order, i.e.b. The transfer function of complex fractional order integral is expressed as,

$$\beta_T(s) = C_T(s)G(s) = \left(cosh\left(b\frac{\pi}{2}\right)\right)^{sign(b)}\left(\frac{\omega_{cg}}{s}\right)^a\left(cos\left(bln\left(\frac{s}{\omega_{cg}}\right)\right)\right)^{-sign(b)} . \quad (13)$$

The angle of generalized template to vertical is formulated from the derivatives of the $\beta(j\omega)$i.e. magnitude and phase at frequency i.e. ω_{cg}, as a function of a and b.

$$\frac{d(|\beta_T(j\omega)|_{db})}{d(phase\beta_T(j\omega))} = \frac{-20asign(b)}{ln(10)b\ tanh\left(b\frac{\pi}{2}\right)}. \quad (14)$$

Generalized template transfer function is band-limited and is changed by a common equation.

$$\beta_T(s) = C^{sign(b)}\left(\frac{\omega_l}{s}+1\right)^{n_l}\left(\alpha_0\frac{1+^{s}/_{\omega_h}}{1+^{s}/_{\omega_l}}\right)^{a} \times \left(^{Re}/_{i}\left(\alpha_0\frac{1+^{s}/_{\omega_h}}{1+^{s}/_{\omega_l}}\right)^{ib}\right)^{-qsign(b)}\left(\frac{1}{\left(1+\frac{s}{\omega_h}\right)^{n_h}}\right). \quad (15)$$

So, third generation fractional order CRONE controller is expressed as,

$$C_T(s) = \frac{\beta_T(s)}{G(s)}. \quad (16)$$

The complex fractional order integral transfer function for open-loop $\beta_T(s)$, has eight high level variables they are $n_l, n_h, a, b, \omega_l, \omega_h, \omega_r$ and C. The parameter n_l and n_h are set by control system designer. The parameter ω_r and C are condition of tangency, are mentioned. For the four independent variables, a nonlinear optimization algorithm based on optimal template reduces price function J based on variations of resonant peak and achieves group of shaping constraints.

$$J = M_{rmax} - M_r. \quad (17)$$

Synthesizing such a template that leads the parameterization of 3 dependent variables a, b and c by optimizing three independent parameters ω_l, ω_h and ω_r depicted in Fig 5.

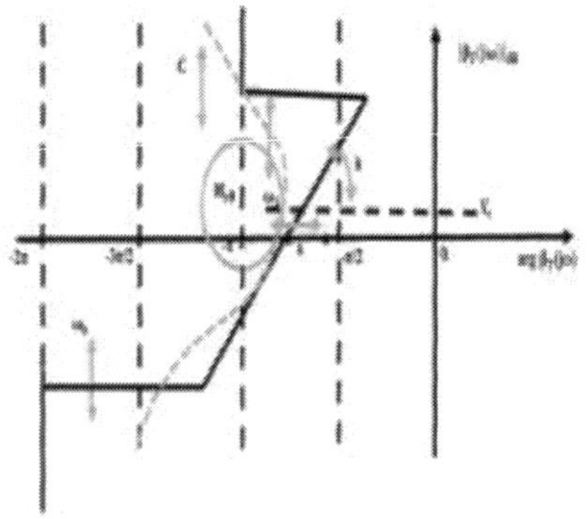

Figure 5. Asymptotic Nichols locus - effect of parameters

The common form of TGC controller is expressed in equation (16). This equation is not in an instigated form, because this equation has terms of fractional order integro-differentiation. Therefore, using recursive distribution method it is transformed into a rational form. Using the technique of recursive distribution, the attainable rational order CRONE controller $C_R(s)$is expressed as,

$$C_R(s) = \frac{\beta_R(s)}{G(s)}. \quad (18)$$

To improve the performance of third generation CRONE controller, the parameters $K_p, K_i \: and \: K_d$ are tuned by means of grey wolf optimization and whale optimization algorithm.

5. Design of Grey Wolf Optimization

To tune the parameters like $K_p, K_i \: and \: K_d$ in CRONE controller, Grey Wolf Optimization technique is used. Mirjalili developed the leadership hierarchy as well as hunting technique that imitate grey wolves in nature. Grey Wolf Optimization consists of four kinds of grey wolves to hunt leadership hierarchy, namely $\alpha, \beta, \delta, \omega$. To achieve improved optimization, the hunting mechanisms are mathematically modelled. Hunting is led through α to search best solution for population and rest of 3 pursues α, β aid α to determine decision making and also act as a bridge from α to less high in position. The $\delta \: and \: \omega$ follows the above two. Grey Wolf Optimization's flow chart is depicted in Fig 6.

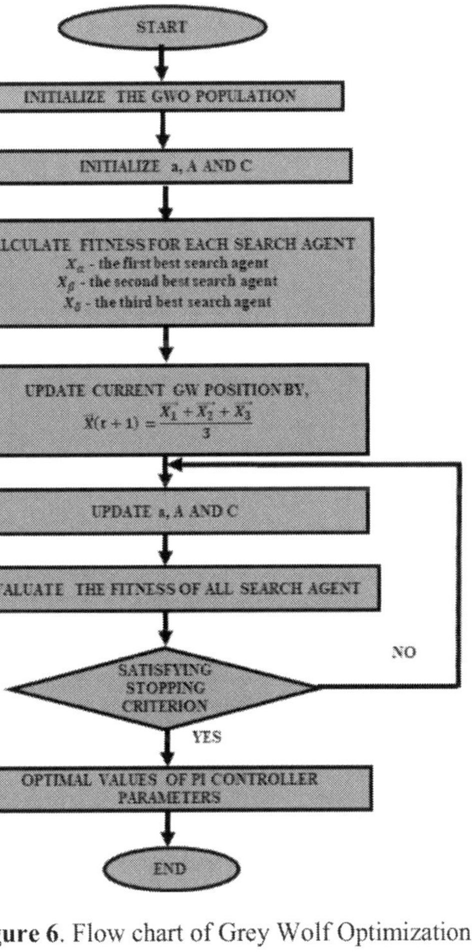

Figure 6. Flow chart of Grey Wolf Optimization

Steps involved in GWO are as follows,

- First initialize $\alpha, \beta, and\ \delta$.
- Search the number of search agents.
- Number of iterations found is maximum.
- Number of positions are chosen for searching the neighborhood and stopping criteria.

The mathematical model of wolf performance is expressed as,

$$\vec{D} = |\vec{C}.\vec{X}_p(t) - \vec{X}(t)|. \quad (19)$$

$$\vec{X}(t+1) = \vec{X}_p(t) + \vec{A}.\vec{D}. \quad (20)$$

Where, current iteration is denoted as t, coefficient vectors denoted as $\vec{A}\ and\ \vec{C}$, prey position vector denoted as $\vec{X}_p$ grey wolf position vector denoted as $\vec{X}$. The expression for $\overrightarrow{A}\ and\ \vec{C}$ are given below,

$$\vec{A} = 2\vec{a}.\vec{r}_1 - \vec{a}. \quad (21)$$

$$\vec{C} = 2.\vec{r}_2. \quad (22)$$

Where, component of $\vec{a}$ linearly decreases from $2\ to\ 0$ over course of iterations and $\vec{r}_1\ and\ \vec{r}_2$ are random vectors in$[0, 1]$.

Update current gray wolf position is mathematically expressed as,

$$\vec{X}(t+1) = \frac{\overrightarrow{X_1}+\overrightarrow{X_2}+\overrightarrow{X_3}}{3}. \quad (23)$$

The third generation CRONE controller parameters are tuned by means of grey wolf optimization algorithm. The results evaluated in this algorithm are compared with whale optimization algorithm.

6. Design of Whale Optimization Algorithm

Meta-heuristic algorithm is nothing but whale optimization algorithm (WOA) that optimizes the third generation CRONE controller parameters like $K_p, K_i\ and\ K_d$ and provides better performance than grey wolf optimization. WOA is inspired from bubble net hunting activities of hump back whales. Generally, the humpback whales are chosen to kill small fishes that are near to water surface. The whales are intelligent and makes 9-shaped path at the period of encircling the prey. The humpback whales identify the location of prey that is near to the sea surface. Whale optimization algorithm's flow chart is depicted in Fig 7.

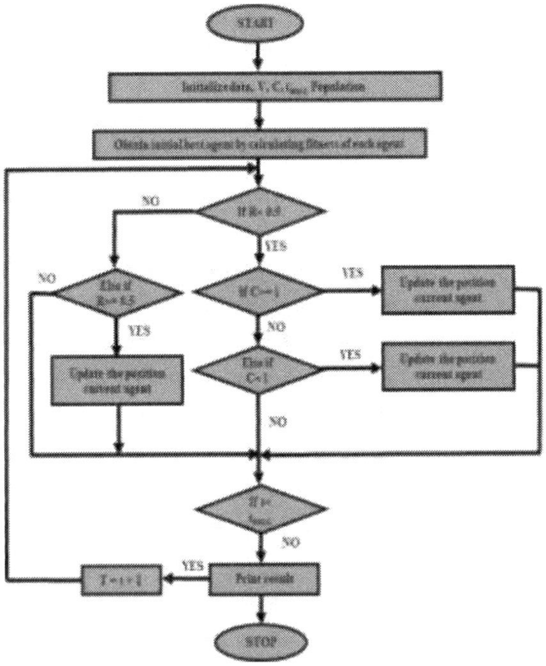

Figure 7. Flow chart of Whale Optimization Algorithm

Following steps include in WOA are,

Step 1: Initialization

- First initialize CRONE controller parameters such as K_p, K_i *and* K_d

. There are some random values allocated that are used to initiate the optimization process.

Step 2: Calculate the fitness function

The aim of whale optimization algorithm is to reduce the Mean square error that is calculated from equation (24).

$$MSE = \sum_{i=1}^{n} \frac{(x_i - y_i)^2}{n} \quad . \tag{24}$$

Where, $x_i$denotes input data, y_i denotes output data, n denotes number of iterations.

Step 3: Update the location

The following equations represents modernized process that are shown below,

$$\vec{U} = |\vec{\gamma} . \vec{p}_{best}(t) - \vec{p}_{best}(t)| \, . \tag{25}$$

$$\vec{p}(t+1) = \vec{p}_{best}(t) - \vec{\omega} . \vec{U} . \tag{26}$$

IOP Conf. Series: Materials Science and Engineering 1166 (2021) 012058 doi:10.1088/1757-899X/1166/1/012058

Where, t denotes current iteration, $\vec{\gamma}$ denotes coefficient vector, $\vec{p}_{best}$ denotes position vector for best description and $\vec{p}$ denotes existing position vector.

The vectors $\overrightarrow{\omega}$ and $\vec{\gamma}$ are expressed as,

$$\overrightarrow{\omega} = 2\vec{\mu}.\vec{r}_i - \vec{\mu}. \tag{27}$$

$$\vec{\gamma} = 2.\vec{r}\,. \tag{28}$$

Here, $\vec{\mu}$ is decreased from 2 to 0 and $\vec{r}_i$ denotes unsystematic vector in $(0,1)$.

$$\vec{p}(t+1) = D.e^{\eta\varphi}.\cos(2\pi\varphi) + \vec{p}_{best}(t) \quad . \tag{29}$$

$$\vec{p}(t+1) = \begin{cases} \vec{p}_{best}(i) - \overrightarrow{\omega}.\vec{U} & if\ R < 0.5 \\ D.e^{\eta\varphi}.\ \cos(2\pi k) + \vec{p}_{best}(t) & if\ R \geq 0.5 \end{cases} \quad . \tag{30}$$

$$\vec{U} = |\vec{\gamma}.\vec{p}_{rand} - \vec{p}|. \tag{31}$$

$$\vec{p}(t+1) = \vec{p}_{rand} - \overrightarrow{A}.\vec{U}. \tag{32}$$

Where, $\vec{p}_{rand}$ denotes random position vector. The search mediators are typically implemented to modernize their position via an ideal search mediator or best description is Recursive method.

The third generation CRONE controller parameters are tuned by whale optimization algorithm and it provides improved performance. The result evaluated in this algorithm is compared with grey wolf optimization. From the comparison it is observe that whale optimization algorithm provides better performance.

7. Result and Discussion

A meta-heuristic optimization algorithm based third generation CRONE controller is proposed for controlling the temperature of outlet fluid, i.e. for hot water in STHE. The third generation CRONE controller for HE system with GWO and WOA optimization technique is performed in this segment. The controller tuning parameters for third generation CRONE controller is attained and the parameters are optimized through GWO and WOA technique and the performance is evaluated. The simulation is performed in MATLAB software to verify proposed WOA optimized third order CRONE controller and the performance is compared with the GWO optimized CRONE controller.

The open loop response is obtained for Step inputs which are depicted in Fig 8. The response of hot water outlet temperature is set at 55^0C, after certain period a sudden step change occurs in both negative and positive directions.

IOP Conf. Series: Materials Science and Engineering 1166 (2021) 012058 doi:10.1088/1757-899X/1166/1/012058

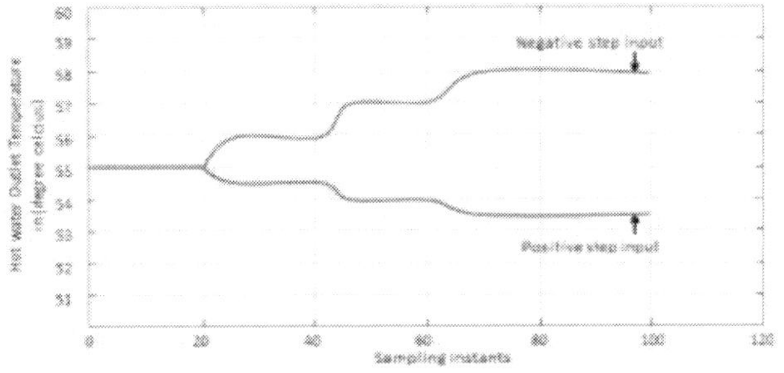

Figure 8 Open loop response for Step input

Step input is depicted in Fig 9. The cold water inflow rate is set at 0.0014 *LPS*, after certain period a sudden step change occurs in both negative and positive directions.

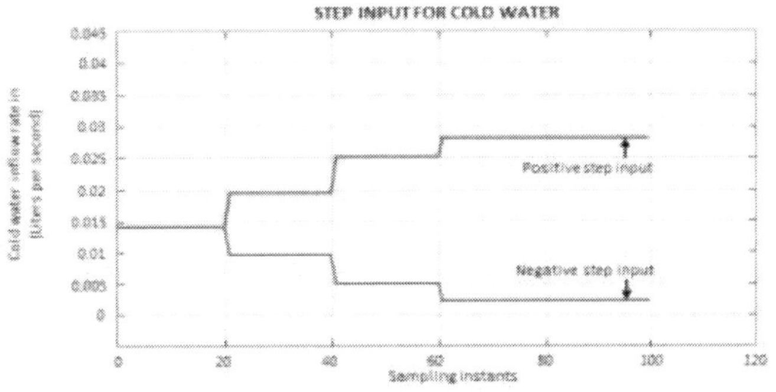

Figure 9. Step input

The comparison waveform of WOA, GWO, and CRONE is depicted in Fig 10. The parameters of third generation CRONE controller are tuned and it is optimized by GWO and WOA techniques. From this waveform it is observed that WOA provides improved performance than GWO with faster settling time and minimize peak overshoot problems.

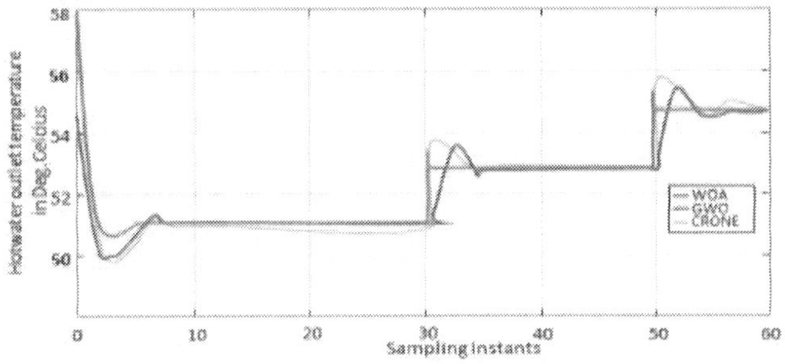

Figure 10. Comparison waveform of WOA-CRONE, GWO-CRONE and CRONE

Table 1. Results for error indices of CRONE controller, WOA- CRONE controller and GWO -CRONE controller for servo response.

REGION/ RANGE [^{o}C]		I (52-54)	II (54-56)	III (56-58)
SETPOINT [^{o}C]		53	55	57
SAMPLING INSTANTS		0-29	30-49	50-65
CRONE	ISE	0.002	0.024	0.038
	IAE	0.21	0.03	0.05
	ITAE	5.32	3.25	3.86
	Tr	2.568	0.01	0.002
	%Mp	0.035	0.601	0.73
GWO – CRONE	ISE	0.02	0.022	0.034
	IAE	0.18	0.03	0.05
	ITAE	5.30	3.22	3.84
	Tr	2.565	0.01	0.002
	%Mp	0.032	0.598	0.68
WOA- CRONE	ISE	0.001	0.016	0.025
	IAE	0.14	0.02	0.03
	ITAE	4.53	2.87	2.34
	Tr	2.452	0.01	0.001
	%Mp	0.031	0.582	0.62

Table 1 lists the result for error indices of CRONE, GWO CRONE and WOA-CRONE controller for servo response from this table, the performance criteria and time responses are evaluated. Through this evaluation, it is clear that crone-controller provides better response than other controllers.

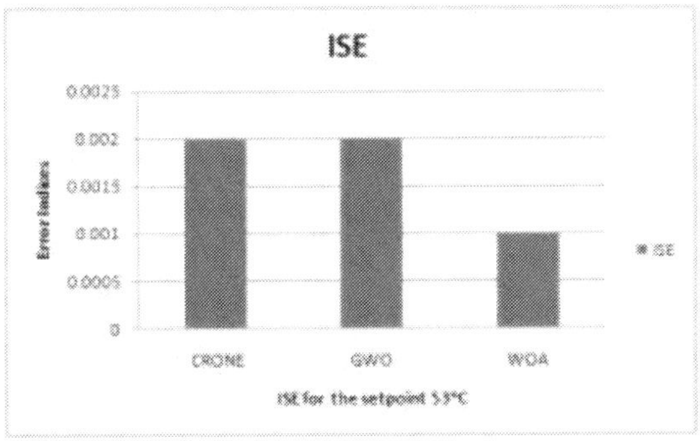

(c)

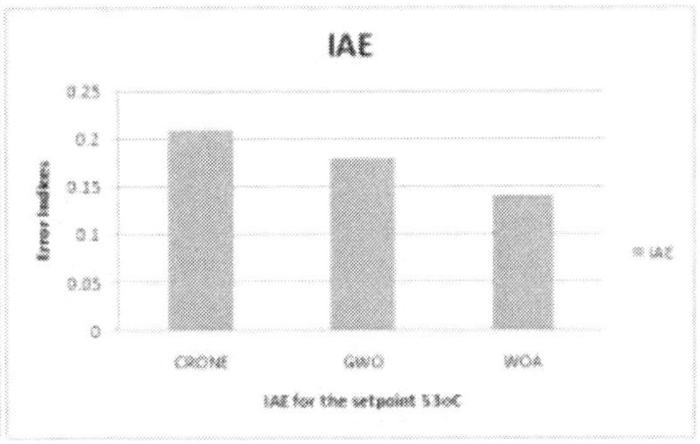

(b)

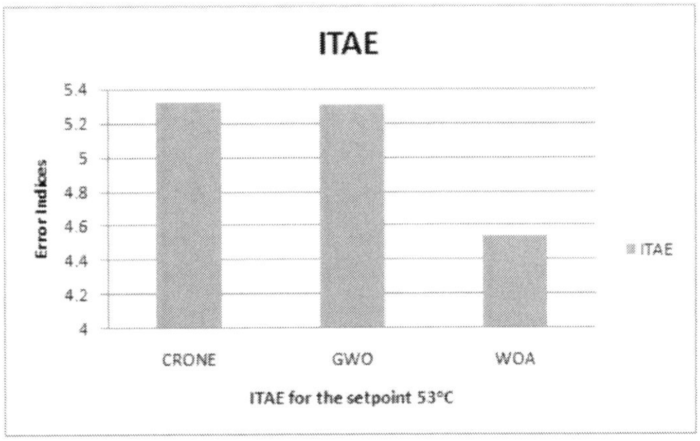

(c)

Fig 11 (a),(b) and (c) Comparison of error indices for the setpoint 53°C.

8. Conclusion

A meta-heuristic optimization algorithm based third generation CRONE controller is proposed for controlling the temperature of outlet fluid in HE device. The primary aim of this work is to identify suitable optimization technique among whale optimization Algorithm (WOA) and grey wolf optimization (GWO) algorithms that would improve the robustness of third order crone controller against the parameter variations. The objective function is to minimize cost of fluid flow rate and to increase the heat transfer rate. The controller tuning parameter of $K_p, K_i,\ K_{d\ ,} \lambda\ and\ \mu$ for third generation CRONE controller is attained through GWO and WOA technique and the performance is evaluated. Drawbacks of GWO are comparatively poor robustness, settling time period is long, peak overshoot problem. To overcome these limitations WOA is employed that provides enhanced performance than GWO. The output is executed through MATLAB software.

References

[1] Kishan H. Maheshwari, Kartik Trivedi, "*A Review on Experimental Investigation of Shell and Tube Heat Exchanger using Nano-fluids*", Journal of Emerging Technologies and Innovative Research (JETIR), **Vol. 2, no. 12**, pp. 145 – 148, Year 2015

[2] Ankit R. Patel, Year: 2013, "*Design and optimization of Shell and Tube Heat Exchanger*", Indian journal of applied research, **Vol. 3, no. 8**, pp. 264 – 266.
[3] Arzu Sencan Sahin, Bayram Kılıc, Ulas Kılıc, Year: 2011, "*Design and economic optimization of shell and tube heat exchangers using Artificial Bee Colony (ABC) algorithm*", Elsevier Ltd, **Vol. 52, pp. 3356 – 3362**.
[4] V.K. Patel, R.V. Rao, "*Design optimization of shell-and-tube heat exchanger using particle swarm optimization technique*", Elsevier Ltd, **Vol. 30, pp. 1417 – 1425**, Year: 2010.
[5] Amin Hadidi, Mojtaba Hadidi, Ali Nazari, "*A new design approach for shell-and-tube heat exchangers using imperialist competitive algorithm (ICA) from economic point of view*", Elsevier Ltd, **Vol. 67, pp. 66-74**, Year: 2012.
[6] Yuvraj Bhushan Khare, Yaduvir Singh, "*PID Control of Heat Exchanger System*", International Journal of Computer Applications, **Vol. 8, no. 6, pp. 22 – 27**, Year: 2010.
[7] P.Sivakumar, D.Prabhakaran, PhD, T.Kannadasan, PhD, "*Temperature control of Shell and Tube Heat Exchanger by using Intelligent Controllers*", International Journal of Computer Applications, **Vol. 59, no. 9,** pp. 5 – 11, Year: 2012
[8] C Somasundar Reddy, K Balaji, "*A Genetic Algorithm (GA)-PID Controller for Temperature Control in Shell and Tube Heat Exchanger*", IOP Conference Series: Materials Science and Engineering, **pp. 1 – 7,** Year: 2020.
[9] M.A. Ahmad, A.A. Ishak, N.K. Ismail, Fakulti Kejuruteraan Kimia , "*New Hybrid Model Reference Adaptive Supervisory Fuzzy Logic Controller for Shell-and-Tube Heat Exchanger Temperature System*", IEEE Control and System Graduate Research Colloquium(ICSGRC), **pp. 49 – 54**, Year: 2012.
[10] Indah Soesanti, Ramadoni Syahputra, "*A Fuzzy Logic Controller Approach for Controlling Heat Exchanger Temperature*", Journal of Electrical Technology UMY (JET-UMY), **Vol. 3, no. 4**, pp. 117 – 124, Year: 2019.
[11] Hussein Termous, Xavier Moreau, Clovis Francis, Hassan Shraim, Year: 2018, "*From the standard PID to the CRONE first generation controller: Application to an anti-roll system for Electric Vehicles*", Elsevier Ltd, **Vol. 51, no. 4**, pp. 733 – 738, Year: 2019.
[12] Nandhini Priyadharshini M, Rakesh Kumar S, Valarmathi R, "*Evolutionary Algorithm based controller for Heat Exchanger*", ARPN Journal of Engineering and Applied Sciences, **Vol. 13, no. 7**, pp. 2369 – 2373, Year: 2018.
[13] B.Dinesh, E.Sivaraman, "*Fuzzy C-means Modeling for Shell and Tube Heat Exchanger*", International Journal of Computer Applications, **pp. 30 – 35**, Year: 2014.
[14] Zwe-Lee Gaing, "*A Particle Swarm Optimization Approach for Optimum Design of PID Controller in AVR System*", IEEE Transactions on Energy Conversion, **Vol. 19**, no. 2, pp. 384 – 391, Year: 2004.
[15] F. Padula A. Visioli, "*Optimal tuning rules for proportional-integral derivative and fractional-order proportional-integral derivative controllers for integral and unstable processes*" IET Control Theory and Applications, **Vol. 6, no. 6**, pp. 776 – 786, Year: 2011.

IOP Conf. Series: Materials Science and Engineering 1166 (2021) 012059 doi:10.1088/1757-899X/1166/1/012059

Identifying non-contact defects in fault bearing

Dr.M.Akila [1], **Mouleswaran.T.S** [2], **Pradeep.P** [2], **Prasanna.R** [2]

KPR Institute of Engineering and Technology, Coimbatore,
Tamil Nadu, India

*Corresponding Author:jayapradeep100@gmail.com

Abstract. Early diagnosis of bearing failures can save time, effort, and money on rotating machine maintenance. On this test, a non-touch type vibration pickup was designed and refined to capture vibration data for bearing fitness tracking under tight load and pace variations, avoiding the bodily connection of vibration pickup to the system tool. The signal was denoised and fault analysis was performed using a Hilbert rework. Principal Component Analysis (PCA) was used to reduce the dimensionality of the extracted capabilities, and then the chosen capabilities for lowering the amount of enter capabilities and discovering the maximum premier function set, the Sequential Floating Forward Selection (SFFS) method was used to rank them in order of significance. Finally, Support Vector Machines (SVM) and Artificial Neural Networks (ANN) were used to determine and classify the numerous faults in bearings. A comparison of SVM and ANN efficacy was carried out. The results reveal that vibration signatures from advanced non-touch sensors (NCS) correspond well with accelerometer data collected under the same conditions. The classification accuracy obtained by combining the advanced NCS with various sensors mentioned in the literature is comparable to that obtained by using the advanced NCS alone. The proposed method could be utilised to detect automated popularity system faults and issue early warnings, preventing unwelcome and unplanned device shutdowns due to bearing failure.

***Keywords*:** Bearing, Principal component of analysis, Support Vector Machine, Artificial Neural Network.

1. Introduction

An orientation is a device aspect that obliges relative motion to simply the suitable motion, and reduces touch among shifting components. The plan of the bearing may, for instance, accommodate unfastened immediate improvement of the shifting component or with the expectancy of complimentary flip round a hard and fast pivot; or, it'd stop a motion through controlling the vectors of common powers that undergo at the shifting components. Most paths inspire the suitable motion through proscribing erosion. Heading are grouped comprehensively as in step with the type of activity, the actions permitted, or to the bearings of the heaps (powers) carried out to the components. Rotational heading keeps pivoting components like shafts or axles interior mechanical frameworks, and flow hubs and outspread burdens from the wellspring of the heap to the development helping it. The simplest sort of bearing, the obvious bearing, consists of a shaft pivoting in an opening. Grease is applied to decrease rubbing. In the metallic ball and curler bearing, to lower sliding erosion, shifting components, for example, rollers or balls with a roundabout cross- location are located among the races or diaries of the bearing get together. An extensive collection of bearing plans exists to allow the requests of the software to be as it should be met for finest effectiveness, dependability, solidness and execution. The expression "bearing" is obtained from the movement word "to undergo"; a bearing being a device aspect that lets in one phase to undergo (i.e., to help) another. The simplest paths are bearing surfaces, reduced or framed right into a phase, with differing tiers of authority over the structure, size, unpleasantness and location of the surface. Different courses are remote devices brought right into a device or device component.

2. Literature Review

Well, it is created based on the Markov chain. Since the genuine issue is more mind boggling than that depicted by the Markov chain model, the noticed occasion isn't balanced compared to the state, yet is connected by a bunch of likelihood disseminations, a particular model is called HMM.

The HMM is a double arbitrary cycle where Markov anchors are utilized to depict changes among states, and general irregular cycles portray measurable connections among conditions and noticed factors. Focusing on the issue that the boundary learning calculation of covered up Markov model will in general meet to nearby ideal arrangements, the hereditary A molecule swarm advancement calculation is proposed to upgrade the underlying boundaries of the covered-up Markov model, with a versatile boundary change strategy is embraced to improve its streamlining execution.

The highlights removed by various strategies are very unique yet higher dimensionality doesn't positively mean higher precision. The search interaction of particles in PSO is influenced by three factors: the impact from the first speed, memory and populace. Input the quantized highlights to be analysed into the prepared model, and the log- probability likelihood estimations of each model are determined.

3. Proposed Methodology

Experiments were carried out on a test rig with a variety of bearing settings in order to give vibration-related statistics for preparation and testing. The vibration indicators had been received utilizing each touch and non-touch kind estimating contraptions in each degree and vertical tomahawks at a trying out recurrence of 12. Eight kHz and 30k examples for exclusive cases. Each trial is rehashed a couple of instances for obtaining the suggested estimation of genuine boundaries.

Using an Electric Discharge Machine, a variety of bearing defects such as Inward Race (IR), Outer Race (OR), and Ball Defect (BD) were induced (EDM). For distinctive types of flaws, marks were achieved for three-rotor velocity within the range of 1600-2000 rpm within the development of rpm and three stacking situations such as no-hoard, four kg, and eight 129 kg. The vibration signals were given with a Healthy (H) bearing, which was taken into account because of the gauge statistics, and was used to investigate why the signals were sent in the first place.

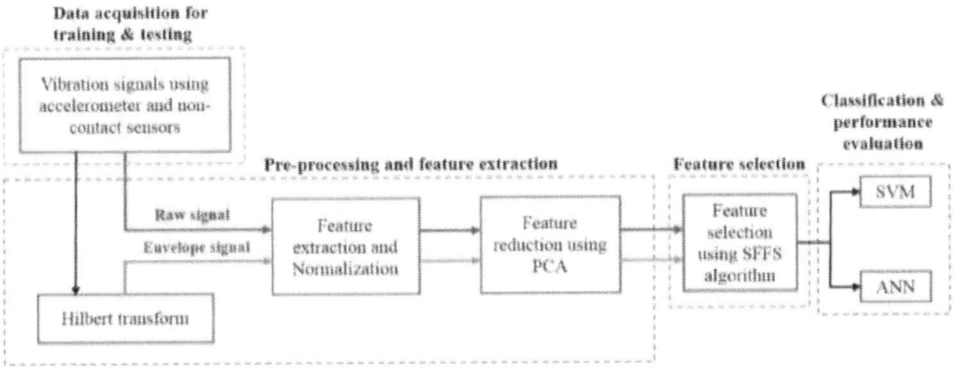

Figure 1. Block Diagram

Data was acquired for the purpose of preparing and testing so that a sufficient number of informational collections could be accessed to precisely analyse the inadequacies. The signals were then pre-processed with Transform, and the highlight extraction approach was shown. PCA was also provided the eliminated list of capabilities for dimensionality reduction. The streamlined component subset was created via feature selection based on the SFFS technique. The use of several classifiers for classification and execution evaluation has been investigated.

4. Methodology

4.1. Testing and Training

SVM and ANN were used to produce and test vibration highlights obtained from raw and pre-processed signals using both contact and non-contact sensors for bearing issue resolution. The examination used absolute events and highlights (11 for crude 295 signal and 13 for envelope signal), as well as factual boundaries for each of the bearing conditions, rotor speed, and the number of burdens used.

4.2. Dimensionality Reduction and Feature Extraction

Highlight extraction is defined as a method for evaluating a few estimations that provide the data included in the sign. The demonstration undertaking in machine wellness monitoring is an issue of example portrayal and example recognition, with include extraction being the most basic progression. Measurements were used to extract a wide range of 13 factual features from both crude and envelope signals, including mean, Standard Deviation (SD), Energy (E), Entropy (En), Skewness (Skew), Smoothness (S), Crest Factor (CF), Kurtosis (Kur), Margin Factor (MF), Impulse Factor (IF), RMS, Peak-Peak abundance (pk-pk), and Shape Factor are some of the terms used in statistics (SF).

4.3. Feature detection using SVM classification with SFFS

A crossover technique for channel and covering highlight choice that exploits an adjusted strategy for consecutive forward skimming search (SFFS) calculation. The separating approach assesses the highlights for foreseeing the yield and supplementing different highlights. The competitor subset produced by the sifting approach is utilized by cross approval of help vector machine (SVM) with client characterized order. SVM classifiers with various part capacities, such as Straight, Quadratic, Cubic, and Gaussian, were used to group the bearing faults using the deleted and chosen highlights obtained from both crude and envelope signals. The piece capabilities were evaluated to see if they were appropriate for the specific problem. The SVM classifier was prepared using the 'one-versus-one' technique, with a there are four classes in all. Correctness was calculated for each mix of highlights arranged by pertinence. A 5-overlay cross-approval conspire was received for the evaluation of the SVM classifiers.

4.4. Feature detection using ANN

This project used a covered up layer with four calculating hubs (for example, 5, 10, 15, and 20). The borders of the applied BPNN are noted for planning purposes. If any of the requirements listed were met, the preparation would come to a halt. The programme introduced the organization's loads and inclinations at random. To evaluate the presentation of neural organisation classifier, the important component framework derived from 316 raw data was divided into three classifications: 70% preparation information, 10% approval information, and 20% testing information. These sets were chosen at random, and the mean estimation of the yield network was calculated after five iterations.

5. Experimental Setup and Result

This section shows the results obtained using SVM and ANN for various bearing deficiency states. The result on a test set is occasionally displayed as a 2D disarray lattice (or possibility table) for multi-class forecasting, with a section and a line for each class.. Every component of 336 the grid uncovers the quantity of preliminary examples for which the anticipated grade is the segment with the genuine grade is the line. The results associate to tremendous numbers downward the primary inclining with 0, little, off-askew components give an exact expectation. The choice of flaw as a trademark for class starts the order interaction with the sequencer yield involves itemized exactness by grade, disarray framework, with evaluation about the ideal arithmetical forecast. The Fine-Gaussian SVM (FG-SVM) classifier's disordered network is reported for recognising the unique bearing inadequacies for crude and envelope signals using ACC and NCS. The disarray network displays a mixture of all vibration highlights for crude sign, and SFFS selected six highlights using ACC that have the highest order exactness of 100 percent and 99 percent, respectively, for moving component deformities. The rate of misdiagnosis was highest for H, as well as deformities.

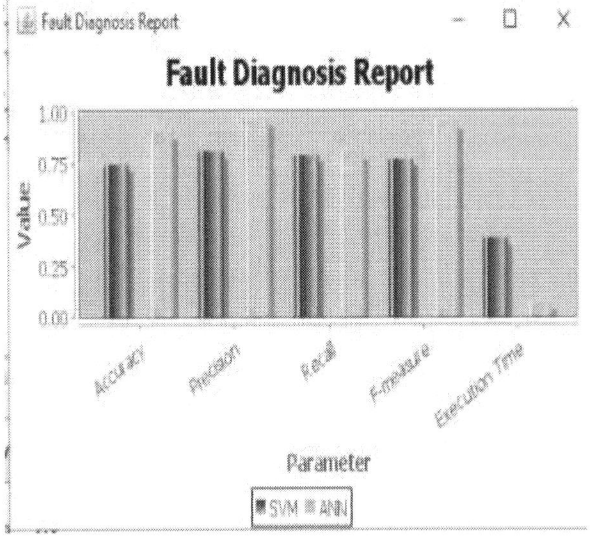

Figure 2. Fault Diagnosis Report

6. Conclusion

For evaluating machine vibrations, a laser shaft based non-contact vibration pickup was proposed and built in this study. A Hilbert change was used to pre-process the obtained signals. Excess highlights were discarded, and highlights were chosen in order of significance using PCA and SFFS. Finally, the selected highlights were characterised and evaluated using SVM and ANN.

The significant discoveries acquired are as per the following:

- The vibration marks obtained from the noncontact sensor contrast well with the accelerometer data obtained in comparable circumstances.
- ANN outperformed SVM for crude vibration signals acquired using NCS and ACC separately by 93.3 percent and 94.2 percent, respectively, and 97.2 percent and 98.3 percent for envelope signals obtained using NCS and ACC.
- The exactness reached for both crude and pre-processed signals produced using ACC and NCS was most extreme for a mix of highlights obtained using the SFFS calculation sorted by significance as studied when all PCA picked highlights were examined at the same time.
- For NCS data, the precision reached utilising envelope signals was 97.2 percent and 96.1 percent, respectively, whereas for ACC data acquired using ANN and SVM independently, it was 98.3 percent and 97.2 percent.

References

[1] Glowacz, "Flaw finding of single-stage acceptance engine dependent “Mechanical Systems and Signal Processing” *on acoustic signs*, vol. **117**, 469 pp. 65–80, 2019.

[2] D. Goyal, B. Pabla, S. Dhami et al., "Non-touch sensor role approach for situation watching of turning machine-components," *Engineering Science and Technology, an International Journal*, vol. **22**, no. **2**, 519 pp. 489–501, 2019.

[3] T. Pilgrimage Mohamad, M. Samadani, and C. Nataraj, "Moving component bearing diagnostics utilizing broadened stage space geography," *Journal of Vibration and Acoustics*, vol. **140**, no. **6**, 2018.

[4] Akshaya S; Kandasamy Sellamuthu; J Anush; S Arunkumar. A Machine Learning Approach to Analyze and Predict the Severity of Road Accidents. *Annals of the Romanian Society for Cell Biology*

[5] Namitha N, Sindhu A S, Neethushree M K, Pavithra P, & Mahima Bhaskar. (2020). *Intelligent Transportation System for Drivers Using IoT*.

[6] S RanjithKumar; Kandasamy Sellamuthu; R S Rajkumar; V Krishnakumar. Certain Investigation on Load Balancing using Cloudlet Assignment and min-max Algorithm. *Annals of the Romanian Society for Cell Biolog*

[7] Vidhya, K., Shanmugalakshmi, R. Modified adaptive neuro-fuzzy inference system (M-ANFIS) based multi-disease analysis of healthcare Big Data. *J Supercomput* (2020)

Churn Prediction in Telecom sector using Deep Neural Network with Flask Application

J.K.Kiruthika[1], B.S.Neshamoney[2], L.Madhan Kumar[3], V.Mugunthan[4]

[1]*Assistant Professor, KPR Institute of Engineering and Technology, Coimbatore, India*
[2,3,4] *Department of computer Science, KPR Institute of Engineering and Technology, Coimbatore, India*

Abstract. Agitate Analysis is one of the overall utilized examination on Subscription Oriented Industries to break down client practices to anticipate the clients which are going to leave the help understanding from an organization. It depends on Deep Learning techniques and calculations and become so significant for organizations in the present business conditions as acquiring another client's expense is more than holding the current ones. The paper audits the significant examinations on Customer Churn Analysis on Telecommunication Industry in writing to introduce general data to perusers about the regularly utilized information mining techniques utilized, results and execution of the strategies and revealing an insight to additional studies. To stay up with the latest, contemplates distributed in most recent five years and basically most recent two years have been incorporated.

Keywords: *Churn; Agitate; Subscription Oriented Industry; Deep learning; client, Mining;*

1. Introduction

Present world Telecom area having gigantic data. In broad, the expectation pace of the typical client in the telecom area is assessed at 3%, which is the complete yearly loss of around few billions. Anticipating agitate clients is multiple times less expensive than drawing in different customers & the expenditure welcoming different customers is few multiple times in excess of continuing to exist clients. Diminishing the beat rate by 5% builds the profit from 25% to 85%.

Anyway the customary parametric model or a solitary man-made consciousness based technique can't accomplish moderately high-exactness forecast, so the foundation of a joined expectation model to improve the expectation precision is an unavoidable pattern to tackle the issue of client beat. In light of this, this paper plans a consolidated forecast model dependent on two models of CNN and RNN calculation.

There are numerous explanations behind beating.

- Unlike post-paid clients, Prepaid clients are not bound to a specialist organization and may beat whenever.
- Churning additionally impacts the general standing of an organization which brings about its image misfortune.

2. Overview

Significant learning is truly a subset of AI. It in truth is AI and capacities essentially however it has different capacities. The essential differentiate among significant and AI is, AI models ended up well consistently be that as it may the show really needs a few heading. Accepting an AI demonstrate returns an off base desire, the designer has to settle that issue explicitly however on account of significant learning, the show does it by him. Programmed vehicle driving system may be a veritable outline of significant learning. Profound learning is a piece of AI with a calculation motivated by the design and capacity of the mind, which is called a fake neural organization.

2.1 Limitation of Deep Learning

2.1.1 Data Labeling
Most current AI models are prepared through "managed learning." It implies that people should name and order the hidden information, which can be a sizable and blunder inclined errand. For instance, organizations creating self-driving-vehicle advances are recruiting many individuals to physically comment on long periods of video takes care of from model vehicles to help train these frameworks.

2.1.2 Obtain Huge Training DataSets
It has been shown that straightforward profound learning methods like CNN can, at times, emulate the information on specialists in medication and different fields. The current flood of AI, nonetheless, requires preparing informational collections that are named as well as adequately expansive and general.
Profound learning techniques required huge number of perception for models to turn out to be moderately acceptable at characterization assignments and, at times, millions for them to perform at the degree of people. Without shock, profound learning is popular in monster tech organizations; they are utilizing enormous information to gather petabytes of information. It permits them to make an amazing and exceptionally precise profound learning model.

3. Deep Learning Process

A significant neural organization gives best in course exactness in various errands, from protest recognizable proof to talk affirmation. They can adjust subsequently, without predefined data explicitly coded by the designers.

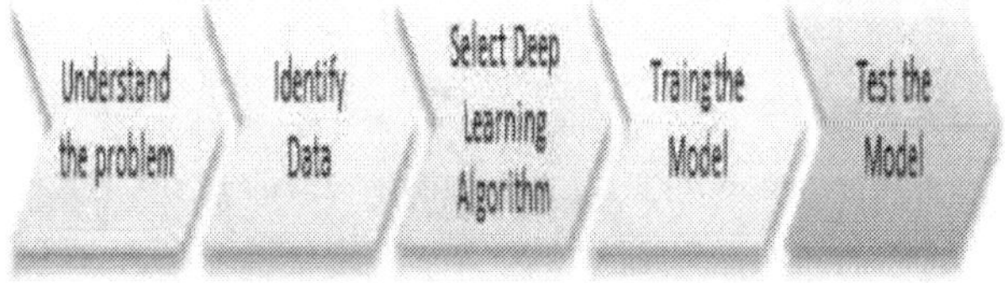

Figure 1. Deep Learning Processing Steps

To urge a handle on the plausibility of significant learning, imagine a family, with a infant child and gatekeepers. The infant centers objects with his small finger and reliably gives the flag 'feline.' As its people are stressed approximately his tutoring, they continue to exhort him 'Indeed, that's a feline' or 'No, that isn't a feline.' The infant child proceeds indicating objects be that as it may turns out to be more correct with 'felines.' The youthful child, where it tallies, doesn't have the foggiest idea why he can say it could be a cat or not. He has as of late figured out how to dynamic frameworks complex highlights concocting a cat by taking a gander at the pet in common and keep on focusing in on nuances just like the tails or the nose some time recently to choose.

A neural organization works an incredible same. Each layer addresses a more profound degree of information, i.e., the order of information. A neural organization with four layers will learn more unpredictable component than with that with two layers.

The learning happens in two stages.

•The first stage comprises of applying a nonlinear change of the information and makes a factual model as yield.

•The second stage targets improving the model with a numerical strategy known as subsidiary.

The neural organization rehashes these two stages hundreds to thousands of time until it has arrived at an okay degree of precision. The rehash of this two-stage is called emphasis.

4. Related Work

Adnan Amin, Babar Shah, Asad Masood Khattak, Thar Baker, Hamood ur Rahman Durani, Sajid Anwar Telecom organizations are dealing with a major issue of client beat because of dramatic development in the utilization of telecom based administrations and the furious rivalry on the lookout. Client stirs are the clients who choose to stop or switch utilization of the assistance or even organization and join another contender. This issue can influence the incomes and notoriety of the telecom organization in the business market. Subsequently, numerous Customer Churn Prediction (CCP) models have been grown; anyway these models, generally concentrate with regards to inside organization CCP. In this way, these models are not appropriate for a circumstance where the organization is recently settled or have as of late embraced the utilization of cutting edge innovation or have lost the verifiable information identifying with the clients. In such situations, Just-In-Time (JIT) approach can be a more common sense option for CCP way to deal with address this issue in cross-organization rather than inside organization beat forecast. This paper has proposed a JIT approach for CCP. Notwithstanding, JIT approach additionally needs some verifiable information to prepare the classifier. To cover this hole in this examination, we constructed JIT-CCP model utilizing Cross-organization idea (i.e., when one organization (source) information is utilized as preparing set and another organization (target) information is considered for testing reason). [1]

A.D.Caigny K.Coussement K.W.D.Bock The inside and out comprehension of clients' conduct is a pivotal methods in CRM for distinguishing its main impetus and creating powerful and customized administration exercises. Moderately little examination sees that it is key significant in the serious market to construct understandable client beat forecast models which can give ventures express client standards of conduct. In this paper, a various bit support vector machines (MK-SVMs) based client agitate forecast model is proposed to exemplify three information revelation errands, which are highlight determination, class expectation and choice standard extraction, into an entire structure A two-stage emphasis of two raised improvement issues is intended for all the while include choice and class expectation. In view of the chose highlights, support vectors are utilized to extricate choice guidelines. An open CRM dataset is utilized to assess the exhibition of this methodology. Trial results show that MK-SVMs accomplish promising execution on the intensely slanted dataset by methods for a re-adjusted system, and the separated standards accomplishes high inclusion and low bogus caution with modest number of preconditions.[2]

Amin,B F. Al-Obeidat B. Shah,Adnan,J. Loo S. Anwar With expanding number of versatile administrators and serious market, beat forecast has become a significant test for telecom organizations. The telecom area is profoundly affected because of adversary rivalry and organizations face loss of income when client change to contender. For holding the client, a precise or exact client beat expectation model is needed to assess the quantity of client who will produce of current specialist organization which will be alluded as churners. With the critical blast in AI research lately and having both force and high dimensional information, AI strategies can be utilized to foresee those churners. This paper gives data on difficulties and method of building the agitate forecast model. This likewise gives a guide to the analysts to handle regulated learning procedures and applying those to settle this present reality problems.[3]

Bhanuprakash P., Nagaraja G.S. With expanding number of versatile administrators and serious market, agitate forecast has become a significant test for telecom organizations. The telecom area is exceptionally affected because of opponent rivalry and organizations face loss of income when client change to contender. For holding the client, a careful or exact client beat expectation model is needed to assess the quantity of client who will produce of current specialist co-op which will be alluded as churners. With the huge blast in AI research lately and having both force and high dimensional information, AI strategies can be utilized to foresee those churners. This paper gives data on difficulties and method of building the agitate expectation model. This additionally gives a guide to the specialists to handle regulated learning methods and applying those to tackle this present reality problems.[4]

5. Proposed Methodology

Client stir is a conspicuous issue confronting organizations. Forestalling client stir, attempting to hold and hold clients has become a significant issue for business activities and development. The joined forecast model is utilized to foresee client beat. By contrasting the expectation exactness of the three models, the legitimacy of the consolidated forecast model is checked.

Utilizing Deep Learning idea for anticipating stir clients Our Deep Learning Algorithm can do the component choice and extraction measure Using Tensor stream and Keras Churn Analysis is one of the overall utilized examination on Subscription Oriented Industries to investigate client practices to foresee the clients which are going to leave the assistance arrangement from an organization. It depends on Deep Learning strategies and calculations and become so significant for organizations in the present business conditions as acquiring another client's expense is more than holding the current ones. The paper audits the important investigations on Customer Churn Analysis on Telecommunication Industry in writing to introduce general data to perusers about the oftentimes utilized information mining strategies utilized, results and execution of the techniques and revealing an insight to additional examinations. To stay up with the latest, examines distributed in most recent five years and fundamentally most recent two years have been incorporated.

6. CNN-Convolution Neural Network

Convolution Neural Network, or else called CNN or ConvNet is a period of neural organizations that has practical experience in preparing material that has a matrix like geography, like a depiction. A computerized depiction is a double portrayal of graphic information by utilizing a CNN. CNN is a neural organization which contains different layers of which some of them are convolution layer pooling layer actuation layer.

6.1 Convolution layer

Convolution is the essential layer to remove highlights from an data picture. Convolution spares the association between pixels by learning picture highlights utilizing small squares of data. It could be a numerical action that takes two information sources, for case, picture organize and a channel or bit.

6.2 Hidden Layer

6.2.1 Strides(Fig 2). Step is the amount of pixels shifts over the data system. At the point when the step is 1 at that point we move the channels to 1 pixel at a time. At the point when the step is 2 at that point we move the channels to 2 pixels all at once, etc.

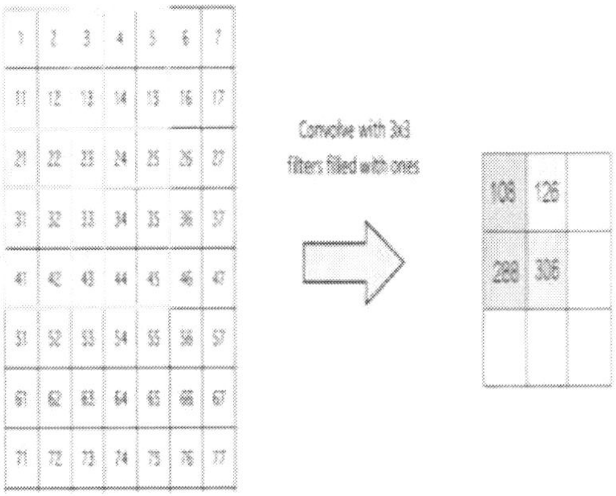

Figure 2. Strides

6.2.2 Padding

Here and there channel doesn't fit totally fit the information picture.
We have 2 options:

- Swab the image with zeros (zero-cushioning) so it fits
- Drop the piece of depiction everyplace the channel didn't fit. Subsequently authentic padding which keeps fair significant piece of the data.

6.2.3 Non Linearity

Rectified Linear Unit for a non-straight activity. The yield is

$$f(Y) = \text{Max}(0,Y).$$

Rectified Linear Unit inspiration is to show non-linearity in Convolution network. Meanwhile, this display reality data would require our Convolution network to memorize will be non-negative coordinate qualities.

6.3 Pooling Layer Zone

Pooling layers zone would diminish the amount of boundaries when the pictures are too much gigantic. Spatial pooling also called sub assessing or down analyzing which reduces the dimensionality of each direct in any case holds noteworthy information. Spatial pooling can be of different sorts:

- Maximum Pool
- Average Pool
- Sumation Pool

Maximum pooling takes the greatest component from the changed highlight outline. Taking the greatest component may moreover take the typical pooling. Sum of all components within the component outline call as total pooling

6.4 Completely Connected Layer

The FC layer, we leveled our lattice into vector and feed it into a completely associated layer like a neural organization (figure 3).

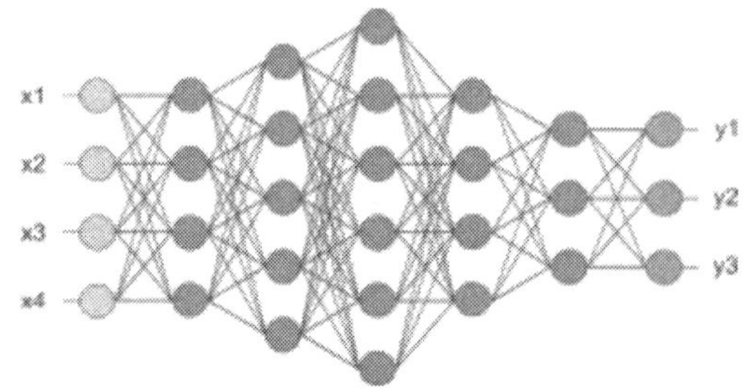

Figure 3. Neural Network

The above chart, the element map grid will be changed over as vector (x1, x2, x3, …). With the completely associated layers, we joined these highlights together to make a model. At last, we have an actuation capacity like delicate max or sigmoid to group the yields as feline, canine, vehicle, truck and so forth,

This Flattened vector is then associated with a couple of completely associated layers which are same as Artificial Neural Networks and play out similar numerical activities!

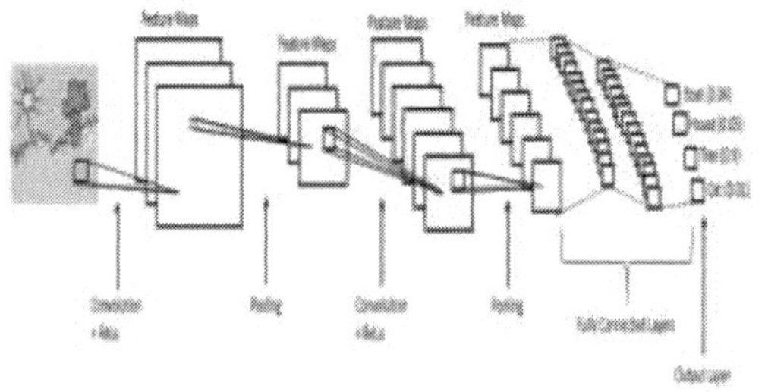

Figure 4. Understanding of Convolutional Neural Network

7. Train Test Split

As we said already, the data we utilize is ordinarily portion into planning data and test data. The planning set contains a realized abdicate and the show learns on this data to be summed up to other data afterward on. We have the test dataset (or subset) to test our model's figure on this subset. We ought to see how to do this in Python. We'll do this utilizing the Scikit-Learn library and expressly the train_test_split procedure. We'll start with bringing within the crucial libraries:

• Pandas — to stack the data report as a Pandas data diagram and investigate the information. Within the occasion simply have to be scrutinize more on Pandas, do not delay to see at my post.

• From Sklearn, I've imported the datasets module, so I can stack an illustration dataset, and the linear_model,
so I can run a coordinate backslide

• From Sklearn, sub-library model_selection, I've imported the train_test_split so I can, all things considered,
part to planning and test sets

Convolution neural organization (CNN) (Figure 4), a class of counterfeit neural organizations that has gotten predominant in different PC vision errands, is drawing in interest across an assortment of areas, including radiology. CNN is intended to subsequently and adaptively learn spatial orders of highlights through backpropagation by utilizing various structure blocks, for example, convolution layers, pooling layers, and completely associated layers. This audit article offers a viewpoint on the fundamental ideas of CNN and its application to different radiological assignments, and talks about its difficulties and future bearings in the field of radiology. Two difficulties in applying CNN to radiological undertakings, little dataset and overfitting, will likewise be canvassed in this article, just as procedures to limit them. Being acquainted with the ideas and benefits, just as limits, of CNN is vital for influence its potential in demonstrative radiology, with the objective of enlarging the exhibition of radiologists and improving patient consideration.

Intermittent neural organizations or RNNs are a group of neural organizations for handling consecutive information. Much as a convolution organization is a neural organization that is specific for handling a framework of qualities Xsuch as a picture, an intermittent neural organization is a neural organization that is specific for preparing a succession of qualities Just as convolutional networks can promptly scale to pictures with huge width and stature, and some convolutional organizations can handle pictures of variable size, intermittent organizations can scale to a lot longer arrangements than would be functional for networks without grouping based specialization. Most intermittent organizations can likewise handle arrangements of variable length.

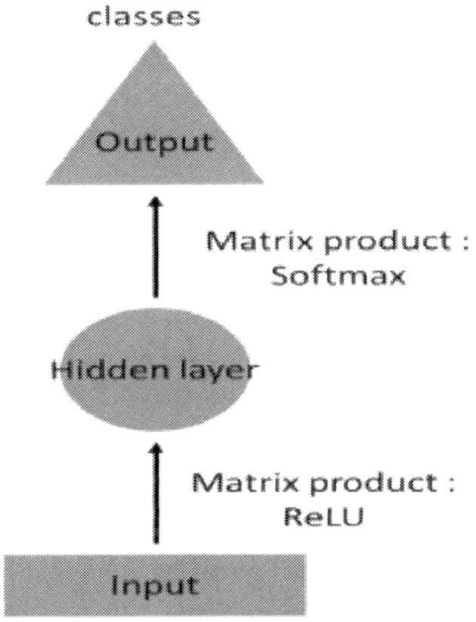

Figure 5. Layer Structure

8. Prediction

What Is Classification Deep Classification includes foreseeing which class a thing has a place with. A few classifiers are parallel, bringing about a yes/no choice. Others are multi-class, ready to arrange a thing into one of a few classifications. Here we are utilizing illness foreseeing

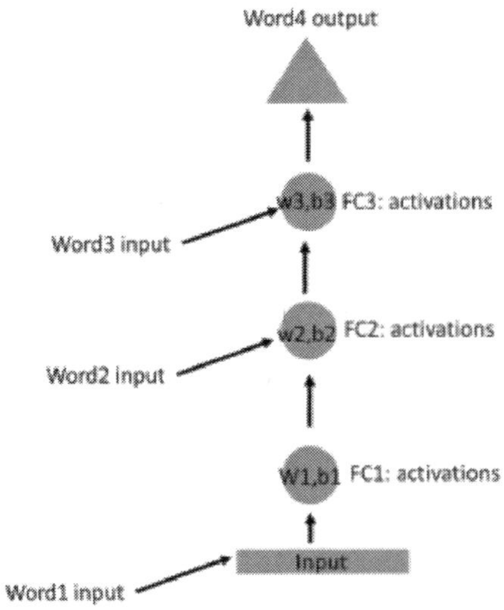

Figure 6. Expected Results structure

9. Experimental Results

9.1 Preprocessing

For this Data Pre-processing content, I will utilize and python explicitly to compose the accompanying code. On the off chance that python isn't as of now introduced when you open up python interestingly, at that point you can undoubtedly introduce it utilizing the UI.

Well the main thought is to eliminate the lines in the perceptions where there is some missing information. Yet, that can be very perilous in light of the fact that envision this informational index contains pivotal data. It would be very risky to eliminate a perception. So we need to sort out a superior plan to deal with this issue. What's more, another thought that is really the most widely recognized plan to deal with missing information is to take the mean of the segments.

In the event that you saw in our dataset, we have two qualities missing, Missing qualities ought to be taken care of during the information examination.

9.2 Feature Extraction

We can utilize text information to separate various highlights regardless of whether we don't have adequate information on Natural Language Processing. So we should examine some of them in this segment. Prior to beginning, how about we immediately read the preparation record from the dataset to perform various assignments on it. In the whole article, we will utilize the twitter estimation dataset from the information hack stage.

9.3 Classification Data

In AI and bits of knowledge, course of action may be a overseen learning approach in which the PC program picks up from the data input given to it and a while later utilizes this figuring out how to gather novel perception. This enlightening collection may fair be bi-class (like recognizing whether the person is male or female or that the mail is spam or non-spam) or it may well be multi-class as well. Many occasions of arrange issues are: talk affirmation, penmanship affirmation, bio metric recognizing verification, record characterization and so on.

10. Conclusion

Within the show genuine advertise of telecom area, beat desire could be a significant issue of the CRM to hold noteworthy clients by recognizing a comparable get-togethers of clients and giving genuine offers/administrations to the person get-togethers. Thusly, in this region, the researchers have been taking a gander at the basic components of beat to hold clients and handle the issues choice tree maker of an organization. In this examination, a client beat show is suited data examination and endorsed through standard appraisal estimations. The procured results appear that our proposed disturb show performed superior by utilizing significant learning strategies. CNN conveyed way better exactness crucial disturb variables from the dataset and performed of the telecom organizations. In future, we are going also look at enthusiastic slanting and listless learning approaches for way better blend desire. The examination can be moreover come to out to examine the changing personal conduct standards of stir clients by applying Artificial Intelligence procedures for expectations and pattern examination.

REFERENCES

[1] V. Lazarov and M. Capota, "Churn forecast," Bus. Butt-centric. Course, TUM Comput. Sci, Technische UnivMünchen, Tech. Rep., 2007. [Online]. Accessible: http://citeseerx.ist.psu.edu/viewdoc/download? doi=10.1.1.462.7201&rep=rep1&type=pdf

[2] R. Vadakattu, B. Panda, S. Narayan, and H. Godhia, "Enterprise membership agitate forecast," in Proc. IEEE Int. Conf. Enormous Data, Nov. 2015, pp. 1317_1321.

[3] V. Umayaparvathi and K. Iyakutti, "Applications of information mining procedures in telecom agitate forecast," Int. J. Comput. Appl., vol. 42, no. 20, pp. 5_9, Mar. 2012.

[4] T. Jahromi, M. Moeini, I. Akbari, and A. Akbarzadeh, "A double step multi-calculation approach for agitate expectation in prepaid broadcast communications specialist co-ops," J. Innov. Sustainab., vol. 1, no. 2, pp. 2179_3565, 2010.

[5] V. Yeshwanth, V. V. Raj, and M. Saravanan, "Evolutionary agitate expectation in versatile organizations utilizing half and half learning," in Proc. 25th Int. Styles Conf., Mar. 2011, pp. 471_476.

[6] G. Nie, W. Rowe, L. Zhang, Y. Tian, and Y. Shi, "Credit card agitate guaging by calculated relapse and choice tree," Expert Syst. Appl., vol. 38, no. 12, pp. 15273_15285, Nov./Dec. 2011.

[7] S. V. Nath and R. S. Behara, "Customer agitate examination in the remote industry: An information mining approach," in Proc. Annu. Meeting Decis. Sci. Inst., vol. 561, Nov. 2003, pp. 505_510.

[8] Y. Zhang, J. Qi, H. Shu, and J. Cao, "A half breed KNN-LR classi_er and its application in client agitate expectation," in Proc. IEEE Int. Conf. Syst., Man Cybern., Oct. 2007, pp. 3265_3269.

[9] H. Yu et al., "Feature designing and classi_er outfit for KDD cup 2010," Dept. Comput. Sci. Inf. Eng., National Taiwan Univ., Taipei, Taiwan, Tech. Rep., 2010, vol. 1, pp. 1_16.

[10] L. Zhao, Q. Gao, X. Dong, A. Dong, and X. Dong, "K-neighborhood most extreme edge highlight extraction calculation for beat expectation in telecom," Group Comput., vol. 20, no. 2, pp. 1401_1409, Jun. 2017.

AUTHOR INDEX